Die Design Handbook

Third Edition

David A. Smith
Editor

Ramon Bakerjian
Staff Editor

Edited from
Die Design Handbook

Second Edition

Frank W. Wilson
Editor-in-Chief

Philip D. Harvey
Editor

Charles B. Gump
Assistant Editor

Die Design Handbook

ISBN No. 0-87263-375-6

Library of Congress Catalog No. 89-063763

Society of Manufacturing Engineers (SME)

Copyright © 1990 by Society of Manufacturing Engineers,
One SME Drive, P.O. Box 930, Dearborn, Michigan 48121

First edition published in 1955 by McGraw-Hill Book Co. in cooperation with SME under earlier Society name, American Society of Tool Engineers (ASTE), and under title *Die Design Handbook*. Second edition published in 1965 again by McGraw-Hill Book Co. in cooperation with SME under earlier Society name, American Society of Tool and Manufacturing Engineers (ASTME), and under title *Die Design Handbook*.

Printed in the United States of America.

PREFACE

The first edition of *Die Design Handbook* appeared in 1955, with the second edition being published in 1965. Both editions were edited by Frank W. Wilson, who imparted a unique practical touch to the work.

The wealth of practical design and troubleshooting information that is responsible for the work's popularity has been retained in this third edition. Nearly all sections have been updated. Major changes and additions include:

- New information on the control of snap-thru energy.
- A new section on product development for deep drawing.
- A new section on progressive die design.
- A new section on fineblanking.
- A new section on multislide tooling.
- Examples of maintenance planning systems.
- Completely revised information on die setting.
- Root causes of stamping process variability.
- Completely revised information on metalworking lubricants.
- Up-to-date information on die sensors.
- Correct installation and applications of tonnage meters.
- Revised welding, heat treating, and surface treatment information.

The individuals and firms who contributed to this edition recognize that there are many benefits beyond any financial recompense to be gained from cooperation and from the mutual assistance. It is recognized, however, that by donating time, talent, and design examples, the contributors and reviewers have indirectly provided incalculable amounts of financial assistance to the development of this project.

This third edition is truly a group effort.

David Alkire Smith
Editor

ABOUT THE EDITOR

David A. Smith is the president and founder of Smith and Associates, a stamping consulting firm in Monroe, Michigan. Mr. Smith has over 30 years of hands-on experience, most notably his 23 years as a diemaker and die tryout group leader for Ford Motor Company. Mr. Smith was also employed with Western Electric.

Mr. Smith is an active member in SME as well as several other societies, including the American Society for Metals International and the Society for Experimental Mechanics. His published SME technical papers include "How to Improve Hit-To-Hit Time With a Tonnage Monitor," "How to Solve Die Impact and Noise Problems With Automotive Pull Rod Shock Absorbers," "Why Press Slide Out of Parallel Problems Affect Part Quality and Available Tonnage," "Reducing Die Maintenance Costs Through Product Design," and "Adjusting Dies to a Common Shut Height."

A frequent speaker at SME clinics, Mr. Smith has taught courses for *Understanding Presses and Press Operations*, *Die and Pressworking Tooling*, *Quick Die Change*, *Sheet Metal Fabrication*, *Press Maintenance*, *Transfer Press and Die Technology*, and *Diesetter Training for Consistent Setups*.

LIST OF CONTRIBUTORS
TO THE THIRD EDITION

James J. Albrecht, Dayton Progress Corporation, Dayton, Ohio

Clyde Bierma, Apprentice Coordinator (retired), Western Electric Co., Columbus, Ohio

Rollin Bondar, President, MPD Welding Inc., Troy, Michigan

John A. Borns, President, Benchmark Technologies Corporation, Toledo, Ohio

Larry Crainich, President, Design Standards Corporation, Bridgeport, Connecticut

Rod Denton, President, Sun Steel Treating Inc., South Lyon, Michigan

Lawrence R. Evans, General Manager, Forward Industries, Dearborn, Michigan

Phillip A. Gibson, District Manager, Atlas Automation Division, Automated Manufacturing Systems, Inc. Atlanta, Georgia

M. Tod Gonzales, General Manager, Product Application Center, National Steel Corporation, Livonia, Michigan

Jeffrey Gordish, Manager, Maintenance Management Consulting, Management Technologies Incorporated, Troy, Michigan

Paul Griglio, Stamping Special Programs Manager, Chrysler Corporation, Detroit, Michigan

Ray Hedding, Stamping Manager, MAZDA Motor Manufacturing (USA) Corporation, Flat Rock, Michigan

Joseph Ivaska, Vice President of Engineering, Tower Oil and Technology Company, Chicago, Illinois

Stuart P. Keeler, Manager, Metallurgy and Sheet Metal Technology, The Budd Company Technical Center, Auburn Hills, Michigan

Karl A. Keyes, President, Feinblänking Ltd, Fairfield, Ohio

George Keremedjiev, Consultant, Teknow Educational Service, Bozeman, Montana

Roman J. Krygier, Stamping Operations Manager, Body and Assembly Division, Ford Motor Company, Dearborn, Michigan

Dan Leighton, Atlas Automation Division, Automated Manufacturing Systems, Inc., Fenton, Michigan

Cecil Lewis, Director of Manufacturing and Engineering, Midway Products Corporation, Monroe, Michigan

Albert A. Manduzzi, Supervisor, Die Design and Standards, Ford Motor Company, Dearborn, Michigan

Michael R. Martin, Application Specialist, Darnell and Diebolt Company, Detroit, Michigan

John McCurdy, President, W. C. McCurdy Company, Troy, Michigan

Carl Meyer, President, Progressive Tool Company, Waterloo, Iowa

Arnold Miedema, President, Capitol Engineering Company, Wyoming, Michigan

Eugene J. Narbut, Specialist, Stamping Manufacturing, MAZDA Motor Manufacturing (USA) Corporation, Flat Rock, Michigan

Angleo Piccinini, Die Design and Die Standards Supervisor, Chrysler Corporation, Detroit, Michigan

Robert W. Prucka, Stamping Manager, Wayne Assembly Plant, Ford Motor Company, Wayne, Michigan

Anthony Rante, Manager, Mechanical Engineering, Danly Machine, Chicago, Illinois

Jerry Rush, U. S. Amada Ltd., Buena Park, California

Edwin Shoemaker, Vice President of Engineering, LA-Z-BOY Chair Company, Monroe, Michigan

Basil Senio, Manufacturing Process and Design Engineering Manager, Ford Motor Company, Dearborn, Michigan

Jeffrey L. Smolinski, Engineering and Manufacturing Technologies Manager, LA-Z-BOY Chair Company, Monroe, Michigan

Edwin A. Stouten, Vice President (retired), Capitol Engineering Company, Wyoming, Michigan

Mark R. Tharrett, General Motors Corp., Warren, Michigan

Maurice Wayne, Director, Corporate Quality Statistical Systems, Opdyke Stamping, Oxford, Michigan

Bernard J. Wallis, Chairman of the Board, Livernois Engineering Corporation, Dearborn, Michigan

Donald Wilhelm, President, Helm Instrument Company, Maumee, Ohio

Lawrence L. Wilhelm, Production Engineering Manager, Advanced Manufacturing Operations, Chrysler Corporation, Detroit, Michigan

Mike Young, President, Vibro/Dynamics, Broadview, Illinois

Timothy Zemaitis, Metallurgical Engineer, Sun Steel Treating Inc., South Lyon, Michigan

LIST OF REVIEWERS
FOR THE THIRD EDITION

Michael Barrilli, Metallurgical Laboratory Supervisor, The Budd Company, Philadelphia, Pennsylvania

Roscoe Brumback, Quality Control Manager, W. C. McCurdy Company, Troy, Michigan

Daniel B. Dallas, Editor (1968-1982), *Manufacturing Engineering*

C. R. Fait, Plant Manager, W. C. McCurdy Company, Troy, Michigan

Leo Geenens, Corporate Quality Control Manager, Midway Products Corporation, Monroe, Michigan

Cass Gizinski, Die Design Standards, Advanced Manufacturing Operations, Chrysler Motors Corporation, Detroit, Michigan

Joseph Hladik, Tool Designer, Western Electric Co., Columbus, Ohio

A. L. Hall, Manager, Stamping Engineering, Ford Motor Company, Dearborn, Michigan

David M. Holley, Chief Engineer, Dadco, Inc., Detroit, Michigan

Norbert Izworski, Product Development Engineer, Body and Assembly Division, Ford Motor Company, Dearborn, Michigan

James Larsen, Die Engineering Standards, Chrysler Corporation, Detroit, Michigan

Gary R. Maddock, Product Metallurgist, High-speed Steels, Crucible Specialty Metals, Syracuse, New York

Gil Novak, Manager, Capacity and Facility Planning, Body and Assembly Division, Dearborn, Michigan

Gerald A. Pool, Senior Manufacturing Engineer, Cadillac Motor Car Div., General Motors Corp., Troy, Michigan

Aniese Seed, President (retired), Toledo Transducers Inc., Toledo, Ohio

Stephen Singleton, Quality Engineer, Teledyne CAE, Toledo, Ohio

Jack L. Thompson, Managing Director, Walker Australia Pty, Ltd., Division of Tenneco Automotive, Australia

Norman L. Vesprini, Applications Engineer, Peerless Steel Supply, Troy, Michigan

Robert L. Wagner, Regional Manager, Helm Instrument Company, Maumee, Ohio

Richard Wilhelm, General Manager, Helm Instrument Company, Maumee, Ohio

LIST OF CONTRIBUTORS
TO THE SECOND EDITION

Howard S. Achler, Vice President, Kaufmann Tool & Engineering Corp., Chicago, Illinois

D. H. W. Allan, Coordinator, Data Processing, American Iron and Steel Institute, New York, New York

Bernard Anscher, President, Mercury Engineering Co., Flushing, New York

W. N. Bachman, President, Bachman Machine Co., St. Louis, Missouri

John A. Barth, Executive Vice President, The Barth Corp., Cleveland, Ohio

E. L. H. Bastian, Senior Engineer, Manufacturing, Shell Oil Co., Chicago, Illinois

K. Beaty, Manager, Tool and Process Engineering, The Vendo Co., Kansas City, Missouri

Roland E. Bechtel, Staff Industrial Engineer, Rockwell Manufacturing Co., Pittsburgh, Pennsylvania

Kenneth C. Butterfield, Manager, Tool Engineering, Chrysler Corp., Detroit, Michigan

Albert Clements, Vice President, Hamilton Div., Clearing Machine Corp., Chicago, Illinois

M. M. Clemons, Vice President, Engineering, Press Automation Systems, Inc., Detroit, Michigan

Ralph Dubey, Metal Stamping Engineering Department, Ford Motor Co., Dearborn, Michigan

Nicholas Dudas, Western Electric Co., Inc., Kearny, New Jersey

Donald F. Eary, Senior Specialist, Mechanical Working, General Motors Institute, Flint, Michigan

E. Warren Feddersen, Director of Manufacturing Engineering, General Dynamics Corp., San Diego, California

Philip Finkelstein, Proprietor, Mercury Tool and Die Co., New York, New York

Charles Girtz, Ryan Aerospace, San Diego, California

Leroy P. Gordon, Chief Tool Engineer, Ekco Containers, Inc., Wheeling, Illinois

Clarence Hall, Die Design Consultant, St. Louis, Missouri

John Hall, Manager of Process Engineering, Chrysler Corp., Detroit, Michigan

Dr. J. C. Hamaker, Vice President, Technology, Vanadium-Alloys Steel Co., Latrobe, Pennsylvania

A. I. Heim, Technical Consultant, Copper Development Association, Inc., New York, New York

E. D. Hinkel, Jr., Assistant Metallurgist, The Carpenter Steel Co., Reading, Pennsylvania

Ernest W. Horvick, Director of Technical Services, American Zinc Institute, Inc., New York, New York

Noble Ida, Explosive Forming Section, Martin Co., Denver, Colorado

Charles R. Isleib, Ductile Iron Section, The International Nickel Co., Inc., New York, New York

A. R. Johnson, Manager, Research Department, Vanadium-Alloys Steel Co., Latrobe, Pennsylvania

J. S. Kirkpatrick, Executive Vice President and Technical Director, The Magnesium Association, Detroit, Michigan

Peter Lecki-Ewing, Manager of Research, Latrobe Steel Co., Latrobe, Pennsylvania

E. E. Lockwood, Senior Research Engineer, North American Aviation, Inc., Canoga Park, California

P. R. Marsilius, Vice President, The Producto Machine Co., Bridgeport, Connecticut

C. R. Maxon, Market Development Div., The New Jersey Zinc Co., New York, New York

Anton F. Mohrnheim, Associate Professor of Metallurgy, University of Rhode Island, Kingston, Rhode Island

Joseph J. Naegelen, Maintenance Engineer, Pittsburgh Railways, Pittsburgh, Pennsylvania

Henry Nida, Manager, Manufacturing Feasibility and Analysis, Metal Stamping Engineering, Ford Motor Co., Dearborn, Michigan

Robert C. Nutting, Chief Engineer, Harig Manufacturing Corp., Chicago, Illinois

F. L. Orrell, Westinghouse Electric Corp., Blairsville, Pennsylvania

Joseph Palsulich, Research Engineer, Western Gear Corp., Lynwood, California

J. R. Paquin, Supervisor, Tool and Die Design, Cleveland Engineering Institute, Cleveland, Ohio

John Pearson, Head, Detonation Physics Group, U.S. Naval Ordnance Test Station, China Lake, California

David D. Pettigrew, Assistant Manager, Power Tool Engineering, Rockwell Manufacturing Co., Pittsburgh, Pennsylvania

W. J. Potthoff, Supervisor of Tooling, The Emerson Electric Manufacturing Co., St. Louis, Missouri

Edward A. Reed, General Supervisor, Drafting, General Motors Corp., Flint, Michigan

Lawrence M. Rheingold, President, Templet Industries, Inc., Brooklyn, New York

Merrill Ridgway, The Minster Machine Co., Minster, Ohio

J. Y. Riedel, Bethlehem Steel Corp., Bethlehem, Pennsylvania

Thomas Riodan, Master Mechanic, Bendix Corp., Utica, New York

E. C. Roark, Precision Forge Co., Santa Monica, California

Ernest J. Ross, Superintendent Master Mechanic, Chevrolet Motor Div., General Motors Corp., Flint, Michigan

David Ryffel, Jr., Sales Promotion Manager, The Producto Machine Co., Bridgeport, Connecticut

Alvin M. Sabroff, Assistant Chief, Metalworking Div., Battelle Memorial Institute, Columbus, Ohio

Robert Sergeson, Chief Metallurgical Engineer, Jones & Laughlin Corp., Warren, Michigan

Floyd E. Smith, President, Automation Devices, Inc., Erie, Pennsylvania

Ray B. Smith, Director, Engineering Standards, Reynolds Metals Co., Richmond, Virginia

Charles H. Stephens, Chief Engineer, Advance Stamping Co., Brighton, Michigan

H. J. Towell, Manufacturing Engineer, Texas Instruments, Inc., Dallas, Texas

F. G. Von Brecht, President, Quick Parts, Inc., Crestwood, Missouri

J. K. Wingard, Chief Engineer, Presses, E. W. Bliss Co., Salem, Ohio

Dr. Louis Zernow, President, Shock Hydrodynamics, Inc., Santa Monica, California

A NOTE ON METRICATION

In some cases (particularly Figures and Tables) in this book the numerical values listed are only in the English system. If you wish to use an appropriate metric value in its place (or convert from metric to English), the following conversion factors are listed below.

Multiply	by	To get
inches	25.4	millimeters
millimeters	0.039	inches
in.-lbf	0.113	newton-meter (N·m)
newton-meter (N·m)	8.851	in.-lbf
lbf	4.448	newton (N)
newton (N)	0.225	lbf
lbf	0.454	kilogram-force (kgf)
kilogram-force (kgf)	2.203	lbf
ton	8.896	kilonewton (kN)
kilonewton (kN)	0.112×10^{-6}	ton
psi	6.895×10^{3}	pascal (Pa)
pascal (Pa)	1.450×10^{-4}	psi
ft-lb	1.356	joule (J)
joule (J)	0.738	ft-lb
ft-lb	1.285×10^{-3}	Btu
Btu	778.26	ft-lb

CONTENTS

SECTION 1
PRESSWORKING TERMINOLOGY

Air draw: A draw operation performed in a single-action press with the blankholder pressure supplied by an air cushion.

Annealing: A process involving the heating and cooling of a metal, commonly used to induce softening. The term refers to treatments intended to alter mechanical or physical properties or to produce a definite microstructure.

Bead: A narrow ridge in a sheet-metal workpiece or part, commonly formed for reinforcement.

Bead, draw: (a) A bead used for controlling metal flow; (b) rib-like projections on draw-ring or hold-down surfaces for controlling metal flow.

Bend allowance: The developed arc length along the neutral axis of bent metal.

Bending: The straining of material, usually flat sheet or strip metal, by moving it around a straight axis which lies in the neutral plane. Metal flow takes place within the plastic range of the metal, so that the bent part retains a permanent set after removal of the applied stress. The cross section of the bend inward from the neutral plane is in compression; the rest of the bend is in tension.

Bend radius: (a) The inside radius at the bend in the work; (b) the corresponding radius on the punch or on the die.

Blank: A precut metal shape, ready for a subsequent press operation.

Blank development: (a) The technique of determining the size and shape of a blank; (b) the resultant flat pattern.

Blankholder: The part of a drawing or forming die which holds the workpiece against the draw ring to control metal flow.

Blankholder, multi-slide: A mechanical device, either spring- or cam-actuated, for firmly clamping a part in place prior to severing from the carrying strip and retaining in place for forming.

Blanking: The operation of cutting or shearing a piece out of stock to a predetermined contour.

Bulging: The process of expanding the walls of a cup, shell, or tube with an internally expanding segmental punch or a punch composed of air, liquids, or semi-liquids such as waxes or tallow, or of rubber and other elastomers.

Burnishing: The process of smoothing or plastically smearing a metal surface to improve its finish.

Bushing, guide-post: A replaceable insert usually fitted in the upper shoe to provide better alignment.

Cam action: A motion at an angle to the direction of an applied force, achieved by a wedge or cam.

Carburizing: A process that introduces carbon into a solid ferrous alloy by heating the metal in contact with a carbonaceous material—solid, liquid, or gas—to a temperature above the transformation range and holding it at that temperature.

Casehardening: Any process of hardening a ferrous alloy so that the case or surface is substantially harder than the interior or core.

Clearance, die: The space, per side, between the punch and die.

Coining: A closed-die squeezing operation in which all surfaces of the work are confined or restrained.

Cold heading: The process of upsetting the ends of bar, wire, or tube stock while cold.

Cold working: Working of a metal, such as by bending or drawing, to plastically deform it and produce strain hardening.

Complimentary output sensor: A solid state sensor having both a normally open output and a normally closed output.

Crimping: A forming operation used to set down or close in a seam.

Cryogenic treatment: A low-temperature steel treatment process used to improve the toughness of hardened tool steel by continuing the transformation of retained austenite into the more desirable martensite, at temperatures ranging from $-120°$ to $-300°$ F ($-84°$ to $-184°$ C).

Cup: Any shallow cylindrical part or shell closed at one end.

Cupping: An operation that produces a cup-shaped part.

Curling: Forming an edge of circular cross section along a sheet or at the end of a shell or tube.

Cylinder, nitrogen: A cylinder containing high-pressure nitrogen gas used as a die spring in pressure pad, draw ring, and cam return applications. These are available in individual piped, manifold, and self-contained styles.

Deep drawing: The drawing of deeply recessed parts from sheet material through plastic flow of the material, when the depth of the recess equals or exceeds the minimum part width.

Deflection: The deviation of a body from a straight line or plane when a force is applied to it.

Dial feed: (a) A press feed which conveys the work to the dies by a circular motion; (b) a mechanism which moves dies under punches by a circular motion and into definite indexed positions.

Die, assembling: A die which assembles and fastens parts together by riveting, press fitting, folding, staking, curling, hemming, crimping, seaming, or wiring.

Die, bending: A die which permanently deforms sheet or strip metal along a straight axis.

Die, blanking: A die for cutting blanks by shearing.

Die, burnishing: A die which improves surface or size by plastically smearing the metal surface of the part.

Die, cam: A die in which the direction of moving elements is at an angle to the direction of forces supplied by a press.

Die, combination: A die in which a cutting operation and a noncutting operation on a part are accomplished in one stroke of the press. The most common type of combination die blanks and draws a part.

Die, compound: A die in which two cutting operations are accomplished in one press stroke. The most common type of compound die blanks and pierces a part.

Die, compound-combination: A die in which a part is blanked, drawn, and pierced in one stroke of the press.

Die, curling: A forming die in which the edge of the work is bent into a loop or circle along a straight or curved axis.

Die, dimpling: A forming die which produces a conical flange (stretch flange) encircling a hole in one or more sheets of metal.

Die, dinking: A die which consists of a press or hand-operated hollow punch with knife-edges for cutting blanks from soft sheet metals and nonmetallic materials.

Die, double-action: A die in which pressure is first applied to a blank through the blankholder and is then applied to the punch.

Die, embossing: A die set which is relatively heavy and rigid for producing shallow or raised indentations with little or no change in metal thickness.

Die, expanding: A die in which a part is stretched, bulged, or expanded by water, oil, rubber, tallow, or an expanding metal punch.

Die, extrusion: A die in which a punch forces metal to plastically flow through a die orifice so that the metal assumes the contour and cross-sectional area of the orifice.

Die, floating (or punch): A die (or punch) so designed that its mounting provides for a slight amount of motion, usually laterally.

Die, forming: A die in which the shape of the punch and die is directly reproduced in the metal with little or no metal flow.

Die, heading: (a) A die used in a forging machine or press for upsetting the heads of bolts, rivets, and similar parts; (b) a die used in a horizontal heading machine for upsetting the flanged heads on cartridges and similar shells.

Die, hemming: A die which folds the edge of the part back over on itself; the edge may or may not be completely flattened to form a closed hem.

Die, inverted: A die in which the conventional positions of the male and female members are reversed.

Die, joggle: A die which forms an offset in a flanged section.

Die, lancing: A die which cuts along a line in the workpiece without producing a separation in the workpiece and without yielding a slug.

Die, multiple: A die used for producing two or more identical parts in one press stroke.

Die, perforating: A die in which a number of holes are pierced or punched simultaneously or progressively in a single stroke of the press.

Die, piercing: A die which cuts out a slug (which is usually scrap) in sheet or plate material.

Die, progressive: A die in which two or more sequential operations are performed at two or more stations upon the work, which is moved from station to station.

Die, recovery: A die that produces a stamping or blank from scrap produced by another operation.

Die, riveting: A die that assembles two or more parts together by riveting.

Die, sectional (segmental): A die punch, or form block, which is made up of pieces, sections, laminations, segments, or sectors.

Die, shaving: A die usually having square cutting edges, negligible punch and die clearance, and no shear on either the punch or the die.

Die, shimmy (Brehm trimming die): A cam-driven die which cuts laterally through the walls of shells in directions determined by the position of cams.

Die, single-action: A drawing die that has no blankholder action, since it is used with a single-action press without the use of a draw cushion.

Die, swaging: A die in which part of the metal under compression plastically flows into contours of the die; the remaining metal is unconfined and flows generally at an angle to the direction of applied pressure.

Die, trimming: A die that cuts or shears surplus material from stock or workpieces.

Die, triple-action: A die in which a third force is applied to a lower punch in addition to forces applied to the blankholder and the punch fastened to the inner slide.

Die, two-step: A drawing or reducing die in which the reduction is made in two stages or levels, one above the other, in a single stroke of the press.

Die, waffle: A type of flattening die that sets a waffle or crisscross design in the blank or workpiece without deforming it.

Die block: (a) A block or plate from which the die itself is cut; (b) the block or plate to which sections or parts of the die proper are secured.

Die cushion: A press accessory located beneath or within a bolster or die block, to provide an additional motion or pressure for stamping operations; actuated by air, oil, rubber, or springs, or a combination thereof.

Die height (shut height): The distance from the finished top face of the upper shoe to the finished bottom face of the lower shoe, immediately after the die operation and with the work in the die.

Die holder: A plate or block upon which the die block is mounted.

Die pad: A movable plate or pad in a female die, usually for part ejection by mechanical means, springs, or fluid cushions.

Die set: A standardized unit consisting of a lower shoe, an upper shoe, and guide pins or posts.

Die shoe: A plate or block, upon which a die holder and in which guide posts are mounted.

Die space: The maximum space within a press bounded by the top of the bed (bolster), the bottom of the slide, and any other press parts.

Dimpling: Localized indent forming of sheet metal, so as to permit the head of a rivet or a bolt to fasten down flush with the surface of the sheet.

Dishing: Forming a large-radiused concave surface in a part.

Double seaming: The process of joining metal edges, each edge being flanged, curled, and crimped.

Downloading: The process of transferring digitalized data from a host computer or CAD system to another digitally controlled device such as a CNC machine tool.

Draft: The taper given to a die so as to allow the part to fall through the die or be removed.

Drawability: (a) A measure of the feasible deformation of a blank during a drawing process; (b) percentage of reduction in diameter of a blank when it is drawn to a shell of maximum practical depth.

Drawing: A process in which a punch causes flat metal to flow into a die cavity to assume the shape of a seamless hollow vessel.

Draw line: A surface imperfection in a drawn part caused by the initial punch contact of a character line on the blank that is subsequently drawn or stretched to a plane surface where it becomes a visible defect.

Draw radius: The radius at the edge of a die or punch over which the work is drawn.

Draw ring: A ring-shaped die part (either the die ring itself or a separate ring), over the inner edge of which the metal is drawn by the punch.

Ductility: The property of a material that permits it to sustain permanent deformation in tension without rupture.

Dwell: The time interval in a press cycle during which there is no movement of a press member.

Eccentric: A machine element that converts rotary motion to straight-line motion.

Elastic limit: The maximum stress to which a material can be subjected, and yet return to its original shape and dimensions on removal of the stress.

Embossing: A process that produces relatively shallow indentations or raised designs with theoretically no change in metal thickness.

Extrusion: The plastic flow of a metal through a die orifice.

Eyelet machine: A multiple-slide press, usually employing a cut-and-carry or a transfer feed for sequential operations in successive stations.

False wiring: Curling the edge of a sheet, shell, or tube without inserting a wire or rod inside the curl.

Feed: A device that moves or delivers stock or workpieces to a die.

Feed, grip (slide, hitch): A type of feed mechanism employing a set of jaws to grip strip stock and feed it to the die.

Feed, indexing: A feed that rotates blanks and parts for various operations, usually visually indicating the position of the blank or part.

Feed, magazine: The combination of a magazine and a mechanism for holding of parts and feeding one unit of the work at a time.

Feed, roll: A die feed operated from the press slide or crankshaft, in which the stock is moved by gripping rollers.

Flaring: (a) The process of forming an outward flange on a tubular part; (b) forming a flange on a head.

Fluting: (a) The forming of longitudinal recesses in a cylindrical part; (b) a surface imperfection in formed or drawn parts; see Luders' lines.

Formability: The ability of a material to undergo plastic deformation without fracture. A quantitative expression of formability, applicable to explosive forming, it is the true strain at fracture.

Form block: A punch or die used in the rubber-pad process to form materials.

Form block, mechanical: A special die used in rubber-pad forming to perform operations which cannot be made with the simpler, regular form blocks.

Forming: Making any change in the shape of a metal piece which does not intentionally reduce the metal thickness.

Forming, high-velocity: Shaping of a workpiece at forming velocities on the order of hundreds of feet per second. This usually requires the use of electrical, magnetic, high-speed-mechanical, or explosive energy sources.

Forming, magnetic: Shaping of a workpiece with the forces generated by the sudden discharge of a current pulse through a coil and the resultant interactions with the eddy currents induced in the conductive workpiece.

Forming, pneumatic-mechanical: Shaping of a workpiece with the force generated by the impact of a ram accelerated by the release of compressed gas.

Four slide machine: A type of multi-slide forming machine having four synchronous rotating shafts on a horizontal plane. The shafts carry cams that power four or more forming slides plus auxiliary motions for feeding materials, piercing and trimming strip stock, severing the blank, and transferring or ejecting the finished part.

Front slide machines: The forming slides on a four slide or vertical slide machine and their respective positions.

Fulcrum slide: A self-contained slide with a fulcrum lever used in a multi-slide forming machine for added power at the sacrifice of stroke length, or conversely, added stroke length at the sacrifice of power.

Gag: A metal spacer to be inserted so as to render a floating tool or punch inoperative.

Gage: A device used to position work accurately in a die.

Gage, finger: A manually operated device to limit the linear travel of material.

Galling: The friction-induced roughness of two metal surfaces in direct sliding contact.

Gibs, adjustable: Guides or shoes designed to ensure the proper sliding fit between two machine parts.

Guerin process: A forming method in which a pliable rubber pad attached to the press slide is forced by pressure to become a mating die for a punch or punches which have been placed on the press bed.

Guide posts (guide pins, leader pins): Pins or posts usually fixed in the lower shoe and accurately fitted to bushings in the upper shoe to ensure precise alignment of the two members of a die set.

Guide, stock: A device used to direct strip or sheet material to the die.

Heel block: A block or plate usually mounted on or attached to a lower die, and serving to prevent or minimize deflection of punches or cams.

Hitch feed: See Feed, grip.

Hole flanging: Turning up or drawing out a flange around a hole; also called "extruding."

Hydropress: A single-action hydraulic press, equipped with a rubber pad.

Hysteresis: In sensor terminology, it is the difference in percentage of effective sensor range or distance between the on and off point when the target is moving away from the sensor face.

Inching: A control process in which the motion of the working members is precisely controlled in short increments.

Incremental deflection factor: The amount that a press deflects due to a given unit of loading. It is usually expressed as the amount of deflection that occurs as a result of one ton per corner (four tons total) increase in tonnage in a straight side press.

Indexing: Rotating a part angularly and sequentially performing a press operation.

Ironing: An operation in which the thickness of the shell wall is reduced and its surface smoothed.

JIC: Joint Industry Conference. Their standards are maintained by NMTBA.

Joggle: An offset surface consisting of two adjacent, continuous or nearly continuous short-radius bends of opposite curvature.

Kingpost: Four slide machine terminology meaning the central forming mandrel or arbor. The kingpost is usually stationary but can be made to cycle up and down if desired.

Kingpost bracket: A bracket mounted to the four slide machine platen for holding the kingpost or forming mandrel along with the stripper slide and related oscillating drive lever mechanisms.

Kirksite: Trade name of a zinc-based alloy used principally in low-production dies.

Knockout: A mechanism for ejecting blanks or other work from a die. Commonly located on the slide, but may be located under the bolster.

Lancing: Cutting along a line in the workpiece without producing a detached slug from the workpiece.

LED: A light emitting diode. LEDs are widely used as indicators of sensor status.

Lightening hole: A hole punched in a part for the purpose of saving weight.

Limit switch: A type of electric switch used to control the operations of a machine automatically.

Luders' lines: Depressed elongated markings parallel to the direction of maximum shear stress, or elevated elongated marking appearing on the surface of some materials, particularly on iron and low-carbon steel, when deformation is beyond the yield point in tension or compression, respectively.

Magazine: A chutelike bin in which parts are uniformly positioned for feeding.

Manifold: (a) A system of piping for connecting several air, fluid, or nitrogen operated devices to a common source. (b) A cross-drilled steel plate fitted with nitrogen cylinders.

Mainframe: A large, often centrally located computer capable of doing a number of tasks at once.

Motion diagram: A graph showing the relative motions of the moving members of a machine.

Necking (necking in): Reducing the diameter of a portion of the length of a cylindrical shell.

Normalizing: A process in which a ferrous alloy is heated to a suitable temperature above the critical range and then cooled in air at room temperature.

Nosing: Forming a curved portion, with reduced diameters, at the end of a tubular part.

Notching: The cutting out of various shapes from the edge of a strip, blank, or part.

Offal: Scrap metal trimmed from stamping operations that is used to produce small stampings in recovery dies.

Oil canning (canning): The distortion of a flat or nearly flat surface by finger pressure, and its reversion to normal.

Olsen ductility test: A test for indicating the ductility of sheet metals by forcing a hemispherical-shaped punch or hardened ball into the metal and measuring the depth at which fracture occurs.

Overcrown: Added crowning of a curved surface in a drawing die for large irregular shapes to compensate for springback.

Overbending: Bending metal to an angle greater than called for in the finished piece, so as to compensate for springback.

Pad: The general term used for that part of a die which delivers holding pressure to the metal being worked.

Parting: An operation usually performed to produce two or more parts from one common stamping.

Perforating: The piercing or punching of many holes, usually identical and arranged in a regular pattern.

Pilot: A pin or projection provided for locating work in subsequent operations from a previously punched or drilled hole.

Plastic flow: The phenomenon which takes place when a substance is deformed permanently without rupture.

Plasticity: The property of a substance that permits it to undergo a permanent change in shape without rupture.

Plastic working: The processing of a substance by causing a permanent change in its shape without rupture.

Platen: The sliding member or ram of a power press.

Powdered metallurgy process: The production of tool steel from finely divided steel powder through fusion at welding temperature in an evacuated steel canister. A much more uniform product is obtained than can be produced from conventional cast ingots.

Press, C-frame: A press having uprights or housing resembling the form of the letter C.

Press, double-action mechanical: A press having one slide within the other, the outer slide usually being toggle or cam-operated, resulting in independent parallel slide movements.

Press, geared: A press whose main crank or eccentric shaft is connected to the drive shaft or flywheel shaft by one or more sets of gears.

Press, hydraulic: A press actuated by a hydraulic cylinder and piston.

Press, inclinable: A press whose main frame may be tilted backward, usually up to 45°, to facilitate ejection of parts by gravity through an open back.

Press, knuckle-joint: A heavy, powerful short-stroke press in which the slide is actuated by a toggle (knuckle) joint.

Press, multiple-slide: (a) A press having individual slides built into the main slide, or (b) a press of more than one slide in which each slide has its own connections to the main shaft.

Press, open-back inclinable: An inclinable press in which the opening at the back between the uprights is usually slightly more than the left-to-right dimension of the slide flange. (See also Press, inclinable.)

Press, punch: (a) Most commonly, an end-wheel gap press of the fixed-bed type; (b) a name loosely used to designate any mechanical press.

Press, rack-and-pinion reducing: A long-stroke reducing press actuated by a rack and pinion.

Press, reducing: A long-stroke, single-crank press used for redrawing (reducing) operations.

Press, rubber-pad: Any single-action hydraulic press with its slide equipped with a rubber pad for rubber pad forming.

Press, single-action: Any press with a single slide; usually considered to be without any other motion or pressure device.

Press, straight-side: An upright press open at front and back with the columns (uprights) at the ends of the bed.

Press, tapering: A press designed to permit placing a blank in a die without the need for a slide plate, and to deliver an exceptionally long stroke.

Press, toggle: (a) Any mechanical press in which a slide, or slides, are actuated by one or more toggle joints; (b) a term applied to double-action and triple-action presses.

Press, toggle drawing: A press in which the outer or blankholder slide is actuated by a series of toggle joints and the inner slide by the crankshaft or eccentrics.

Press, trimming: A special-purpose mechanical press in which shearing and trimming operations are usually done on forgings.

Press, triple-action: A press having three slides with three motions synchronized for such operations as drawing, redrawing, and forming, where the third action is opposite in direction to the first two.

Press, tryout (spotting): A press used in the final finishing of dies to locate inaccuracies of mating parts.

Press, twin-drive: A press having two main gears on the crankshaft meshing with two main pinions on the first intermediate shaft.

Press, two-point: A mechanical press in which the slide is operated by two connections to the crankshaft.

Press, underdrive: A press in which the driving mechanism is located within or under the bed or below the floor line.

Press-brake (bending brake): An open-frame press for bending, cutting, and forming; usually handling relatively long work in strips.

Puckering: A wavy condition in the walls of a deep drawn part.

Punch: (a) The male die part, usually the upper member and mounted on the slide; (b) to die-cut a hole in sheet or plate material; (c) a general term for the press operation of producing holes of various sizes in sheets, plates, or rolled shapes.

Punch holder: The plate or part of the die which holds the punch.

Quill-type punch: A frail or small-sized punch mounted in a shouldered sleeve or quill.

Rack-and-pinion drive: A drive incorporating a rack and pinion and commonly used to actuate roll feeds.

Redrawing: Second and following drawing operations in which cuplike shells are deepened and reduced in cross-sectional dimensions.

Reducing: Any operation that decreases the cross-sectional dimensions of a shell or tubular part; includes drawing, ironing, necking, tapering, and redrawing.

Residual stress: Stresses left within a metal as the result of nonuniform, plastic deformation or by drastic gradients of temperature from quenching or welding.

Restriking: A sizing operation in which compressive strains are introduced in the stamping to counteract tensile strains set up by previous operations.

Riser block: A plate inserted between the top of the bed and the bolster to decrease the height of the die space.

Roll straightener: A mechanism equipped with rolls to straighten sheet or strip stock, usually used with a feed mechanism for pressworking.

Rotary slide machines: A vertical multi-slide forming machine whose cam driven slides are powered by a large sun gear rather than shafts. There can be as many slides as space permits and they can be positioned virtually anywhere around the sun gear and at any angular orientation, affording virtually limitless flexibility.

Rubber pad (blanket): A flat piece of rubber used as an auxiliary tool for rubber forming.

Rubber pad forming: See Guerin process.

Scoring: (a) The scratching of a part as it slides over a die; (b) reducing the thickness of a material along a line to weaken it purposely along that line.

Scrap cutter: A shear or cutter operated by the press or built into a die for cutting scrap into sizes, either for convenient removal from the die or for disposal.

Seam: (a) The fold or ridge formed at the juncture of two pieces of sheet material; (b) on the surface of a metal, a crack that has been closed but not welded; usually produced by some defect either in casting or in working, such as blowholes that have become oxidized, or folds and laps that have formed during working.

Seaming: The process of joining two edges of sheet metal by multiple bending.

Seizing: Welding of metal from the workpiece to a die member under the combined action of pressure and sliding friction.

Sensing range: The distance away from the sensing face at which a nominal target will be detected.

Sensor, analog: A sensor that provides a voltage output that is proportional to the distance from the target.

Sensor, capacitive: An electronic sensor that is triggered when a change in the surrounding dielectric affects the electrostatic field emanating from the sensor.

Sensor, digital output: A sensor that provides an output that is either on or off with no intermediate states.

Sensor, electronic: A device used in die and press automation to sense position by means of a change in an electromagnetic (inductive) or electrostatic (capacitive) field.

Sensor, optical: A device used in die and press automation to sense position by means of reflection or interruption of a light beam.

Setback: The distance from the intersection of two corresponding mold lines to the bend line.

Shank, punch-holder: The stem or projection from the upper shoe which enters the slide flange recess and is clamped to the slide.

Shaving: A secondary shearing or cutting operation in which the surface of a previously cut edge is finished or smoothed.

Shear: (a) A tool for cutting metal and other material by the closing motion of two sharp, closely adjoining edges; (b) to cut by shearing dies or blades; (c) an inclination between two cutting edges.

Shedder: A pin, rod, ring, or plate, operated by mechanical means, air, or a rubber cushion, that either ejects blanks, parts, or adhering scrap from a die, or releases them from punch, die, or pad surfaces.

Shoe: (a) A metal block used in bending processes to form or support the part being processed; (b) the upper or lower component of a die set.

Shut height of a press: The distance from the top of the bed to the bottom of the slide with the stroke down and adjustment up. In general, the shut height of a press is the maximum die height that can be accommodated for normal operation, taking the bolster into consideration.

Sizing: A secondary pressworking operation to obtain dimensional accuracy by metal flow. The final forming in a die of a workpiece which has previously been formed (for example, by free forming) to a shape approximately equal to the shape of the die.

Slide: The main reciprocating press member; also called the ram, the plunger, or the platen.

Slitting: Cutting or shearing along single lines; used either to cut strips from a sheet or to cut along lines of a given length or contour in a sheet or part.

Spotting: The fitting of one part of a die to another, by applying an oil color to the surface of the finished part and bringing it against the surface of the intended mating part, the high spots being marked by the transferred color.

Springback: The extent to which metal tends to return to its original shape or position after undergoing a forming operation.

Staking: The process of permanently fastening two parts together by recessing one part within the other and then causing plastic flow of the material at the joint.

Steel rule die: A metal-cutting die employing a thin strip of steel (printer's rule) formed to the outline of a part and a thin steel punch mounted to a suitable die set. A flat metal plate or block of wood is substituted for the punch when cutting nonmetallic materials and soft metals.

Stock oiler: A device, generally consisting of felt-wick wipers or rolls, which spreads oil over the faces of sheet or strip stock.

Stop pin: A device for positioning stock or parts in a die.

Straight slide: A slide used in multi-slide forming machines whose cam produces direct motion equaling the throw of the cam.

Straightener rolls: See Roll straightener.

Strain: The deformation, or change in size or shape of a body, produced by stress in that body. Unit strain is the amount of deformation per unit length. Deformation is either elastic or plastic. In the literature on forming, the term "strain" is frequently used for the plastic deformation, specifically the quantitative expression for a component of the plastic strain. A strain component may be a normal strain, associated with change of length, either elongation or compression, or it may be a shear (shearing) strain, associated with change of angle between intersecting planes.

Stress: The internal force or forces set up within a body by outside applied forces or loads. Unit stress is the amount of load per unit area.

Stress relief (relieving): A heat treatment which is done primarily for reducing residual stresses.

Stretch (stretcher) forming: The shaping or forming of a sheet by stretching it over a formed shape.

Stripper: A device for removing the workpiece or part from the punch.

Stripper plate: A plate (solid or movable) used to strip the workpiece or part from the punch; it may also guide the stock.

Stripping: The operation of removing the workpiece or part from the punch.

Surge tank: A pressure vessel attached to a die and used in conjunction with a pneumatic or high-pressure nitrogen die pressure system to prevent an excessive pressure buildup during press cycling.

Swaging: A squeezing operation in which part of the metal under compression plastically flows into contours of the die; the remaining metal is unconfined and flows generally at an angle to the direction of applied pressure.

Tapering: A swaging or reducing operation, in which the metal is elongated in compression, for producing conical surfaces on tubular parts.

Tempering (drawing): A heat-treating process for removing internal stresses in metal at temperatures above those for stress relieving, but in no case above the lower critical temperature.

Tensile strength: The ultimate strength measured in pounds per square inch (Pascals) of a material in tension on the original cross section tested, which, if exceeded, causes sectional deformation leading to ultimate rupture.

Transition temperature: The temperature at which there is a transition from ductile to brittle behavior in the fracture of material.

Trimming: Trimming is the term applied to the operation of cutting scrap off a partially or fully shaped part to an established trim line.

Trimming, pinch: Trimming the edge of a tubular part by pinching or pushing the flange or lip of the part over the cutting edge of a draw or stationary punch.

True strain and stress: Terms used in stress and strain analysis, indicating that the strain and stress are calculated on the basis of actual and instantaneous values of length and cross-sectional area. Also known as "natural strain and stress." True strain is also known as "logarithmic strain."

Ultimate strength: The maximum stress which a material can withstand before or at rupture.

Upsetting: A squeezing or compressing operation in which a larger cross section is formed on the part by gathering material in such a way as to reduce the length.

Vent: A small hole in a punch or die for admitting air to avoid suction holding, or to relieve pockets of trapped air which would prevent proper die closure or action.

Yield point: The stress at which a pronounced increase in strain is shown without an increase in stress.

SECTION 2

STAMPINGS DESIGN

No stamping-design work can be considered optimum until, in the judgment of the pressworking department or custom stamper, it holds out the strong probability of achieving the following:

1. A die, or set of dies, that combines maximum production and least maintenance with lowest feasible life cost.
2. Maximum utilization of the least expensive stamping material that will serve satisfactorily.
3. Most efficient pressworking practices.
4. A stamped product that consistently meets sales and service requirements of shape, dimensions, strength, finish, style, and utility.

Design alteration, except to meet changed product or press requirements, or to utilize existing dies with reasonable modifications, is to be avoided as being expensive and time-wasting. Therefore, even preliminary die sketches should be examined in the light of the following data.

GENERAL DESIGN PROCEDURE

The design planning of a new stamped part or assembly should take into account the following steps:

1. Determine exactly what the product is to accomplish in its application.
2. Convert or reduce any vague or generalized specifications to specific descriptions of materials, mechanical units, and dimensions.
3. See that all dimensions have the broadest permissible tolerances, and that initial allowances are made for overtravel, temperature and pressure variations, and other physical factors encountered in product service.
4. Set overall size limits with references to mounting upon or attachment to a machine, a control, or other cooperating parts. These limits may indicate the need of slotted holes, spacers, or similar devices.
5. Are weight limits imposed by the conditions of service? If so, consider reduction of the weight by such means as punching out unneeded material, or by resorting to lighter or thinner-gage material strengthened by stiffening ribs or bosses.
6. Check all critical points, such as bearing points, where high mechanical stresses or excessive wear may occur. Are parts as designed able to resist deformation during ejection from dies, tote-box handling, tumbling, or heat treatment?
7. If very high and fast stamping production is in view, see to what extent the physical proportions, the appearance, or the finish can be altered or compromised to achieve such production.
8. Find out whether there are any unavoidable limitations in available press equipment, fabrication and assembly facilities, or other production factors, and alter stampings design as far as possible to meet the limitations.

REDESIGN OF EXISTING PRODUCTS

When a stamped part or assembly has been in production for some time, it may become feasible or necessary to redesign the product, because of market demands of appearance, functioning, or cost factors. It is especially desirable to consider product redesign if the worn press tools or stamping operations have proven unsatisfactory. The following are good examples of integrated product and die redesign.

Television-tube Electron Gun. The redesign of four electron-gun components shown in Fig. 2-1 resulted in 60% to 75% reduction in manufacturing costs. Top views show original parts design; lower views show them redesigned.

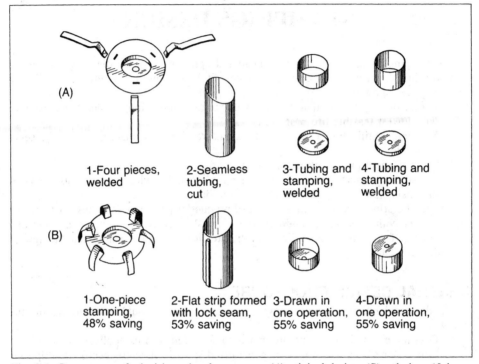

Fig. 2-1. Components of television-tube electron gun: (A) original design; (B) redesign. (*John Volkert Metal Stampings, Inc.*)

The anode (No. 2) was previously made from stainless-steel tubing cut to shape; in redesign it is formed from flat coil stock and lock-seamed in a Multislide machine. The grids (Nos. 3 and 4) were also made from tubing, to which separately blanked and formed lens caps were welded; in redesign they are drawn in one piece from flat stock. The flange spacer or spider (No. 1) was originally fabricated from four pieces welded together; in redesign it is now stamped at the rate of 75 per minute in a six-station progressive die (Fig. 2-2) from automatically fed coil stock. The spacer was improved in redesign by doubling the number of arms. The steps used in producing the flange spacer from stainless-steel strip are (1) pretrimming to permit flow of material in subsequent operations; (2) cupping for the center hub; (3) finish drawing the center hub; (4) finish trimming the six arms and piercing for the center hole; (5) finish sizing and squaring at the hub; (6) cutoff and finish forming.

Elimination of welding on the grids and flange spacer avoided possible loss of accuracy.

Perforated Bulkhead. The original design of a bulkhead (Fig. 2-3A) used conventional geometric layout of holes to provide air flow. Holes had to be kept small in size to minimize distortion is subsequent forming as shown in view C. In this design, the

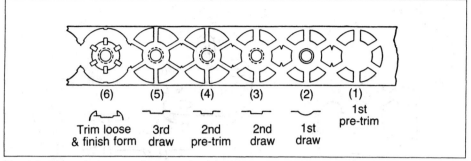

Fig. 2-2. Six-station progressive die strip development for producing electron-gun flange spacer from 1/4 hard fine-grain stainless-steel strip (0.010 × 2.375 in., 0.254 × 60.32 mm). (*John Volkert Metal Stampings, Inc.*)

shearing load of the many small perforating punches was heavy, and the limited capacity of available presses required the perforating and the circumferential blanking to be done in separate operations.

When the die had to be rebuilt, the number of holes was reduced to approximately one-half and made larger to provide equivalent flow area (Fig. 2-3*B*), and the hole layout was so arranged as to avoid the ribs and highly stressed areas involved in subsequent drawing.

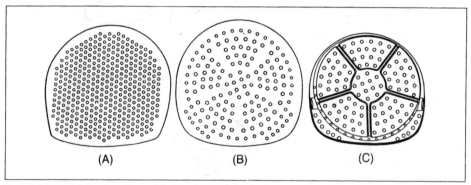

Fig. 2-3. Redesign of perforated bulkhead: (*A*) old design (*B*) new design: (*C*) subsequent forming.[1]

The larger hole size reduced the total perimeter of holes by 33%. The reduced shear load enabled piercing and blanking to be combined in a single two-station die, with the piercing and blanking punches so stepped that the cutting loads at the two stations were not in full action at the same instant.

In the new design, die life was increased 400%. Die cost was reduced approximately 50%.

Where it is known that the production of a stamped part or assembly will be limited, it is highly advisable to consider the possible redesign of the part, so as to achieve minimum tool cost.

In Fig. 2-4, part *A* (of unimportant overall detail) is a hinged key. The hinge is of conventional curved design, and the hinge would require two or more operations. In redesigned part *B*, the only significant change was to move the hinge pin-hole as far as possible from the bent section at *a*. Part *B* was produced in two operations at about half the tooling cost for part *A*.

Also shown in Fig. 2-4 is part *C*, a simple clevis in five design variations. Part C_1 is a casting, heavy and rather high in machining cost. C_2 is milled from bar stock and welded to a stamping; design is better, but the drilling and welding operations increase the cost.

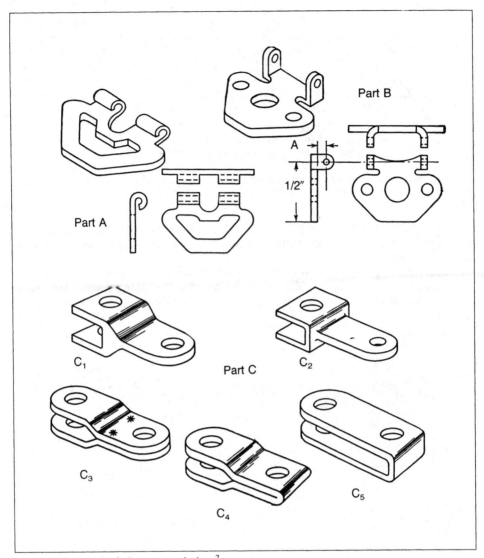

Fig. 2-4. Examples of piece part redesign.[2]

C_3 is a good design of two symmetrical stampings welded together. C_4 is a one-piece stamping involving simple operations and assuring uniform hole alignment. C_5 is probably the ultimate in simplicity of design and economy in tooling and production.

DESIGN FOR EFFICIENT STOCK UTILIZATION

Use of Standard Mill Stock. The size, shape, appearance, and intended use of a stamped product will generally dictate the stampings-materials specifications. However, applied knowledge of the grades, qualities, gages, sizes, physical forms, and finishes available in standard mill stamping stock will frequently assist in securing the best die design and stamping efficiency.

Maximum Volume for Given Area of Material. The calcular principle of maxima and minima can be used to determine the least amount of sheet metal required to form an open-top square-bottomed box of given volumetric capacity. For the box shown in Fig. 2-5, formed from a blank of width a and length a, the volume would be $(x)(a - 2x)^2$, and would be a maximum when

$$x = \frac{a}{6} \qquad (1A)$$

The four corners of x^2 area each, whether cut out or folded in, would be minimum waste under such conditions.

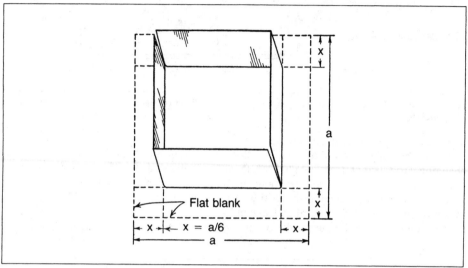

Fig. 2-5. Relation of flat blank to formed box of maximum volumetric capacity.

By the same principles, the minimum amount of metal in a plain cylindrical can of given volumetric capacity, closed at one end, and making allowance for waste in blanking the circular end, but none for crimping the end to the shell, is determined by the formula

$$\text{Height } H = \frac{\text{Diam.}}{2} \qquad (1B)$$

Utilizing Grain Structure. In designing parts to have formed sections, such as lugs or ears, plan if at all possible to form such sections at right angles to the direction of grain in the metal (Fig. 2-6C); otherwise they may crack in forming. Generally, it is preferred practice to design formed sections parallel to each other, as at A and C, or at an angle to each other not exceeding 45°. Where sections must be formed at right angles to one another, blanking them diagonally in the strip, as in view B, will provide some continuous grain running through such sections.

Where the face of a right-angled lug or ear must receive a large thrust load, the part should be so designed that the thrust will be in the direction that induces shear (lug tends to bend in same direction as formed, upper lug, Fig. 2-6C), rather than in the direction that would bend the lug back into the original flat (lower lug, Fig. 2-6C), with danger of cracking. The shear strength of a formed lug is greatest along its narrow edge (view A).

Product Design for Minimum Scrap. Stock layout is one of the most important phases of die engineering, and one of the first steps to be taken.

However, cases frequently occur where a slight to moderate change in product design, while in no way impeding the functioning of the part, will save appreciable materials and tooling expense.

The simple hook-ended flat blank in Fig. 2-7A in conventional layout utilized only 62% of the strip; no other nesting was more efficient. Consultation between the product designer and die designers revealed that the only mandatory dimensions in the piece were A, B, and C. When the piece was redesigned, a new strip layout was possible which

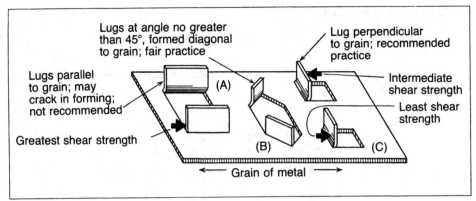

Fig. 2-6. Shear strength of lugs with reference to grain direction; heavy arrows indicate direction of applied force.

utilized nearly 100% of the material.

In the case of the part shown in Fig. 2-7B, a 14% material saving was made just in changing strip layout from single in-line to an alternating double-row arrangement. Further investigation showed that about one-third of each piece could be cut from the piece ahead without changing the utility of the part, and resulting in a 32% saving in material, compared with the original layout.

It is good practice to examine the product design for any corners or flanges projecting

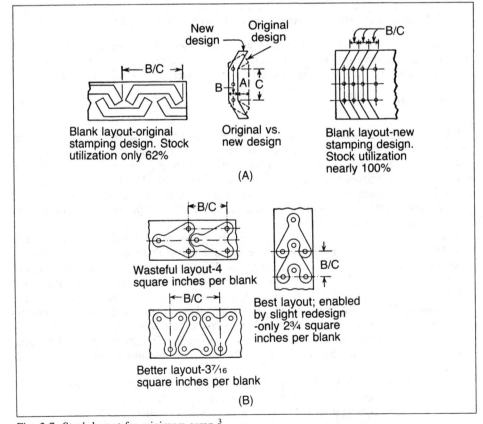

Fig. 2-7. Stock layout for minimum scrap.[3]

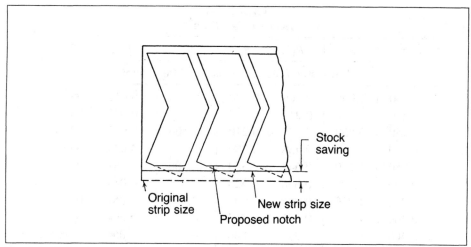

Fig. 2-8. Unnecessary corners cut off to permit use of narrower strip stock.[4]

at the top, bottom, or sides which can be cut off without detriment to strength or necessary welding surface. This practice, as shown in Fig. 2-8, permitted a considerably narrower strip to be used.

Study of some product designs may show that the part can be made in two pieces to reduce scrap.

Aside from the width and length savings in materials, referred to above, savings through use of a thinner stock can commonly be realized through judicious use of beads, ribs, embosses, and flanges.

DESIGN TO INCREASE STRENGTH OF STAMPINGS

Beads, ribs, and flanges may often be used to impart rigidity to stampings which might otherwise be too flexible and weak. Their judicious use may reduce required material thickness by as much as one-half.

Beads and Ribs. The ribs specified in Fig. 2-9[5] and Table 2-1 have been so designed that the angle of the sides will normally allow their use not only on flat surfaces but also on curved and angular surfaces without producing back draft.

In a right-angle bent sheet, the bend is usually a weak point. By the use of stiffening ribs (Fig. 2-10[5]), overall rigidity is increased by 100 to 200%.

If a hole in the flange of a part is functional (not just a lightening hole), a rib as shown in Fig. 2-11 may be formed between the hole circumference and the edge of the part. Since the blank contour is made first, the holes second, and the ribbing last, the ribbing should be as far away from the hole as possible, to minimize distortion of the finished part and to avoid trimming.

A limitation to bead depths may arise from inability to form stretched recesses, such as beads, to the desired depth, since the maximum obtainable depth is a function of the minimum bottom depth R.

$$R = \frac{2TS}{P} \text{ for circular recesses} \qquad (2A)$$

$$R = \frac{TS}{P} \text{ for elongated recesses} \qquad (2B)$$

where R = bottom radius, in. (mm)
T = stock thickness, in. (mm)
S = tensile strength, psi (Pa)
P = forming pressure, psi (Pa)

TABLE 2-1
Bead-Design Data[5,6]

a	A	R_1	R_2	R_3	$T*$	$T\dagger$
Low-carbon Steel, Aluminum, Magnesium						
0.050	0.250	0.109	0.070	0.500	0.016	0.032
0.100	0.500	0.219	0.140	1	0.025	0.040
0.150	0.750	0.328	0.210	1.500	0.032	0.064
0.200	1	0.438	0.280	2	0.040	0.064
0.250	1.250	0.547	0.350	2.500	0.051	0.064
0.300	1.500	0.656	0.420	3	0.064	0.064
0.350	1.750	0.766	0.490	3.500	0.064	0.064
0.400	2	0.875	0.560	4	0.064	0.064
0.600	3	1.312	0.840	6	0.064	0.064
0.800	4	1.750	1.120	8	0.064	0.064
1.000	5	2.188	1.400	10	0.064	0.064
1.200	6	2.625	1.680	12	0.064	0.064
Titanium						
0.38‡	1.33	0.46	0.250	1.12	. . . . .	3.88¶
0.38	1.90	0.94	0.250	1.12		
0.50‡	1.75	0.72	0.250	1.50	. . . . .	2.50¶
0.50	2.50	1.56	0.250	1.50		
0.70	3.50	2.32	0.250	2.00		
0.88‡	3.08	1.46	0.250	2.52	. . . . .	1.76¶
0.88	4.40	3.00	0.250	3.52		
1.12‡	3.92	2.00	0.250	3.00	. . . . .	1.26¶
1.12	5.60	3.82	0.250	3.00		

* Maximum thickness for rubber-pad-formed beads on hydraulic press.

† Maximum thickness for beads formed by punch and die on mechanical press.

‡ Use when edge of bead to edge of sheet does not exceed dimension shown.

¶ Maximum distance between edge of bead and edge of sheet.

See Fig. 2–9.

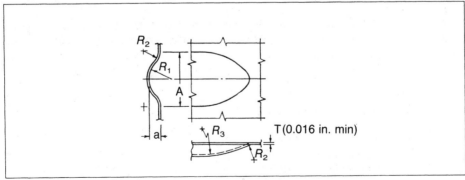

Fig. 2-9. Bead design.[5]

Design Data

Size L	Type	R_1	R_2	R_3	H	M (ref)	Spacing between beads
0.500	1	0.250	0.359	0.187	0.125	0.719	2.500
0.750	1	0.313	0.641	0.266	0.203	1.156	3.000
1.250	2	0.343	0.859	0.328	0.266	1.500	3.500

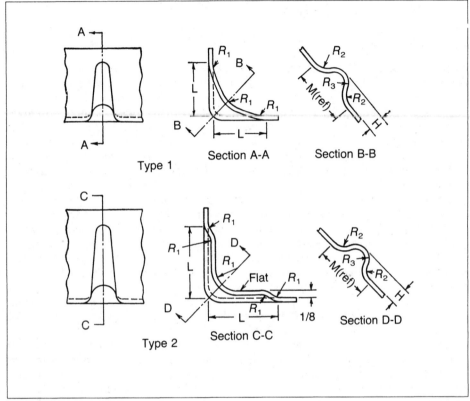

Fig. 2-10. Corner bead designs.[5]

The proportions of locked-in beads (those terminating short of the edge of the sheet) are shown in Table 2-1. If these proportions are maintained, all aluminum alloys (with the heat-treatable alloys in soft condition), all magnesium alloys (condition "O"), low-carbon steels, and annealed corrosion-resistant steels can be formed by the processes listed in the footnotes. The minimum stock thickness is 0.016 in. (0.41 mm), with the maximum as listed in the table.

The beads in the proportions shown may be formed in titanium alloy RE-T-41 and titanium-pure AMS 4901, hot or cold, and in RE-T-32, cold only. The radius R_2 should be 0.312 in. (7.92 mm) for 0.063-in. (1.60-mm) thick RE-T-32 alloy.

Bead ends are blended into the flat sheet surfaces by generous radii to prevent distortion and warping. Figure 2-12 shows sundry other general design suggestions for the use of strengthening ribs and beads.

Internal Beads. This type of bead is produced when the metal is formed into a recess in a form block. In rubber forming of internal beads, the platen contacts the whole web simultaneously, and as pressure builds up, it tends to clamp the sheet in place. As pressure

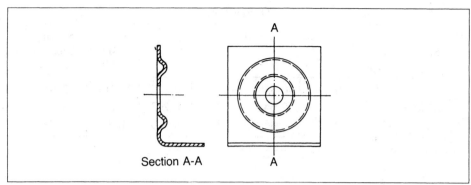

Fig. 2-11. Reinforcing ribs around holes in stampings.

continues to increase, the metal stretches into the bead depression, while the metal around the depression is more tightly locked. Most of the deformation is therefore confined within the bead itself, and distortion of adjacent metal is minimized.

When a mechanical press is used, matched male and female dies are used to form the bead. The male die first contacts the blank and begins to form the bead before a holding contact is established. This so stretches and distorts adjacent material that it is objectionable for some parts. An improvement results from using a double-action die and a spring-loaded pressure pad to keep the metal from wrinkling as the bead is formed in the sheet.

The maximum possible internal bead depth a depends primarily on the bead width A, standard beads commonly having a ratio A/a between 4 and 6. Beads are spaced as closely together as possible to give a maximum strength and to avoid wide flat areas, which are inherently weak. Between parallel beads the minimum spacing is about $8a$ to allow full bead formation without fracturing the metal. Between a bead and flange at right angles to the bead, allow $2a$; between beads at right angles to each other, allow $3a$; and between a bead and a flange parallel to the bead, allow $5a$.

External Beads. In rubber-formed external beads, the highest point of stamping pressure is on the crown of the forming strip. Metal is locked at this point, with increasing pressure, and the area between bead strips is stretched until it bottoms on the form block. Deformation being thus spread progressively over a large area, an external bead can be formed considerably deeper than an internal bead of the same curvature. Somewhat offsetting this advantage, a rather large edge radius is required. Because of the sharper resulting contours, external beads are more efficient stiffeners than internal beads.

Die-formed external beads are formed by metal movement very similar to that for rubber forming of internal beads. However, a very deep external bead designed for rubber forming cannot be produced in a single die operation.

Figure 2-13A gives the minimum recommended center-line distances of external beads parallel to the edge of the sheet and the minimum recommended distance for the end of the bead to the edge of the sheet, using an A/a range between four and six. Dimensions A and C refer to parts with flanges, while B and D refer to parts without flanges.

Figure 2-13B shows the ratio of width A to depth a for beads which can be formed using the usual rubber pressure of between 1,100 and 1,400 psi (7.6 and 9.6 MPa).

Buttons or Bosses. These are flat-bottomed circular depressions or elevations in this sheet (Fig. 2-13 C), used to increase buckling resistance of unstiffened area, and where drawing of the material must be held to a minimum. Distance from center of button to a flanged edge should exceed 60% to 80% of the button diameter, and 70% to 100% to a free edge.

Flanges. A flange as formed on a sheet-metal part is the result of a simple bending operation, or combinations of bending and compression (shrink flanges), or combinations of bending and elongation (stretch flanges). In Fig. 2-14, these are respectively designated as types A, B, and C. Type D flange is a combination of types A, B, and C.

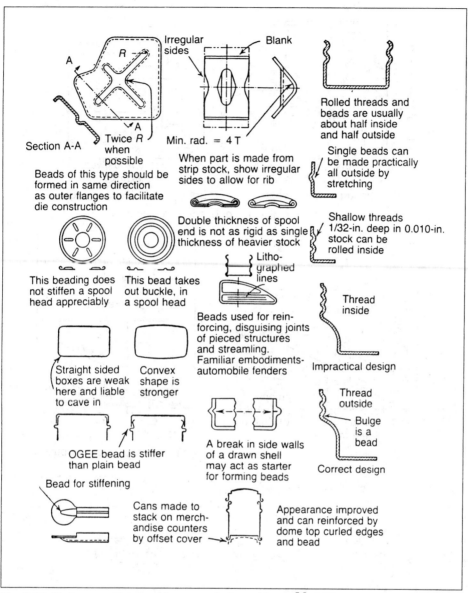

Fig. 2-12. Various considerations for stamped ribs and beads.[7,8] *(Mills)*

Good general flange-design considerations include use of standard bend radii relative to the gage and condition of metal, use of 90° or open-bevel flange wherever possible, and use of standard bend relief for intersecting flanges (Fig. 2-15A).

Curved Flanges. The forming limits of deformation of free edges of flanges depend upon several variables: (1) allowable deformation limits (stretch or shrink) of the metal; (2) radius of the mold line, subtended angle of flange, and restraint at the extremities (tangents) of the curved sections of the flanges; (3) cross-sectional flange dimensions (width and angle); and (4) method of forming.

The limits of stretch depend upon type, gage, and condition of the metal, and the method of forming. The mold-line radius, and the flange width and/or the angle, must be adjusted to meet design requirements and still be within deformation limits of the metal. For limits of allowable deformation, see Table 2-2.

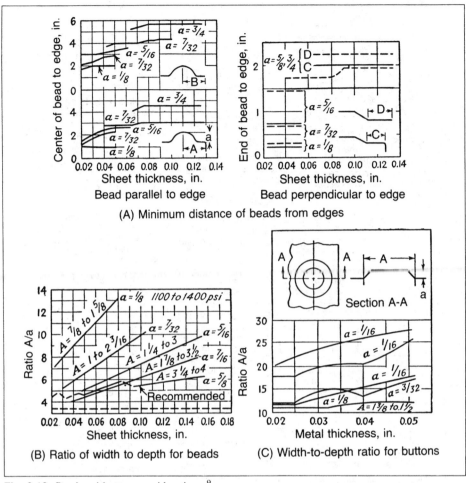

Fig. 2-13. Bead and button considerations.[9]

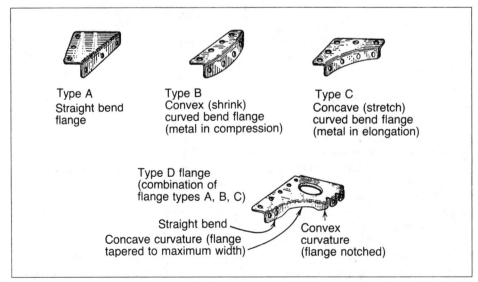

Fig. 2-14. Types of stamped flanges.

TABLE 2-2
Permissible Forming Limits for Curved Flanges

Metal or Alloy	% Elongation (Stretch)		% Compression (Shrink)	
	Rubber Forming	Die Forming	Rubber Forming	Die Forming
Aluminum:				
1100-O	25	30	6	40
1100-T	5	8	3	12
3003-O	23	30	6	40
3003-T	5	8	3	12
5052-O	20	25	6	35
5052-T	5	8	3	12
6061-O	21	22	8	35
6061-T	12	20	8	20
6061-T	5	10	2	10
2024-O and alclad 2024-O	14	20	6	30
2024-T and alclad 2024-T	6	18	0.5	9
2024-T and alclad 2024-T	0	0	0	0
2014-O	14	20	4	30
2014-T	5	6	0	0
7075-O	13	18	3	30
7075-T	0	0	0	0
Steel:				
SAE 1010	. .	38	. . .	10
SAE 1020	. .	22	. . .	10
SAE 8630	. .	17	. . .	8
S.S.302 annealed	. .	40	. . .	10
S.S.302-½H	. .	15	. . .	10
S.S.321	. .	40	. . .	10
S.S.347	. .	40	. . .	10

Data supplied by Curtiss-Wright Corp.

Above allowable deformations are satisfactory for flanges formed in stock, 0.040 in. (1.02 mm) or thicker. Values can be slightly reduced for the lighter gages, particularly for shrink flanges.

Values for compression (shrink) rubber forming may be exceeded by use of notches, controlled wrinkling, special form blocks to compensate for bow effect, etc. Elongation values may be increased for some alloys by polishing edge of the blank.

The approximate deformation of the free edge of the flange is given by

$$D = 100 \left(\frac{R_2}{R_1} - 1 \right) \tag{3}$$

where D = deformation, percent
R_1 = edge radius, in. (mm), before forming (flat pattern radius)
R_2 = edge radius after forming, in. (mm)

Positive values of D indicate elongation (stretch); negative values indicate compression (shrink).

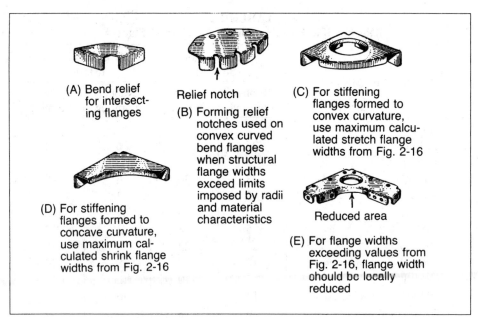

(A) Bend relief for intersecting flanges

Relief notch

(B) Forming relief notches used on convex curved bend flanges when structural flange widths exceed limits imposed by radii and material characteristics

(C) For stiffening flanges formed to convex curvature, use maximum calculated stretch flange widths from Fig. 2-16

(D) For stiffening flanges formed to concave curvature, use maximum calculated shrink flange widths from Fig. 2-16

Reduced area

(E) For flange widths exceeding values from Fig. 2-16, flange width ohould bc locally reduced

Fig. 2-15. Flange widths and reliefs.

Flange Calculations. The relationships of the various dimensional elements for stretch and shrink flanges are shown in Fig. 2-16.

The amount of setback for all flanges can be determined from Fig. 2-17 by connecting the radius scale at the value of R to the thickness scale at the value of T with a straight line.

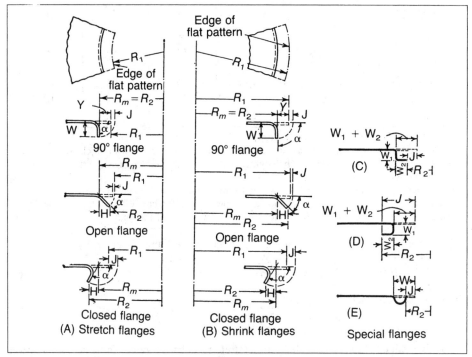

Fig. 2-16. Dimensioning (A) for stretch flanges; (B) for shrink flanges; (C), (D), and (E) for special flanges. For 90° flanges use Fig. 2-18; for other flanges use Fig. 2-19.

The setback value *J* is read at the point where this line intersects the horizontal line representing the bevel of the bead.

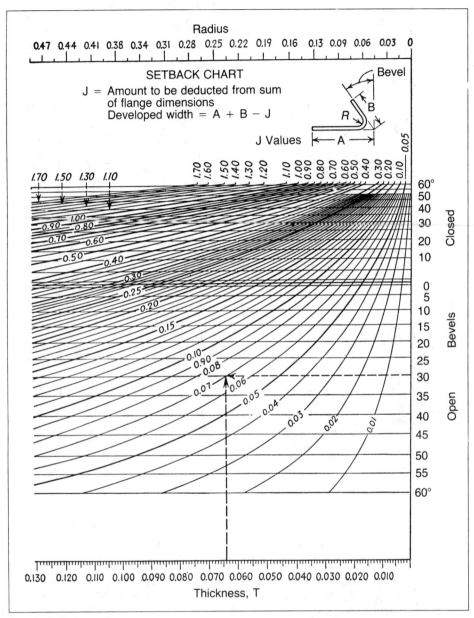

Fig. 2-17. Setback chart.

Dimensions for 90° flanges can be determined from Fig. 2-18, as can be the percentage of elongation (stretch) or compression (shrink) in the metal of a given flange. The use of the chart for determining percentage of elongation is illustrated in Example 1; determining the dimensions of a flange is shown in Example 2. Dimensions for open or closed flanges can be determined from Fig. 2-19. The flange width *W* or the projected flange width *H* can be determined from the lower scale. The approximate deformation of the free edge of curved flanges, in percent, is determined on the upper scale. The use of the chart is shown in Example 3.

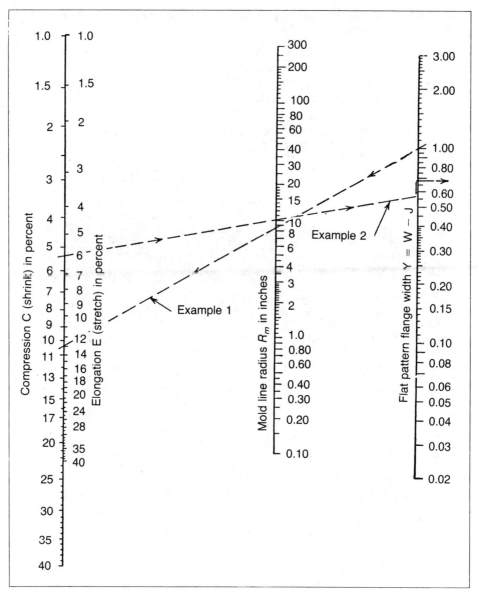

Fig. 2-18. Chart for calculating 90° flange width and percentage of deformation.

EXAMPLE 1: Given: mold-line radius R_m = 8.50 in. (220 mm), flange width W = 1.16 in. (29.5 mm), 90° stretch flange thickness T = 0.064 in. (1.62 mm). Required: percent deformation and permissible material.

Assuming a safe value of $2T$, bend radius R = 0.12 in. (3.0 mm) and, from Fig. 2-17, setback

$$J = 0.13 \text{ in. } (3.3 \text{ mm})$$

Flat pattern flange width $Y = W - J$ = 1.03 in. (26 mm). From Fig. 2-18,
elongation = 13%

select materials from Table 2-2 which have 13% or greater elongation.

EXAMPLE 2: Given: mold-line radius R_m = 10.00 in. (254 mm); 0.040-in. (1.02-mm) thick 2024-T3 aluminum, 90° stretch flange, to be rubber-formed. Required:

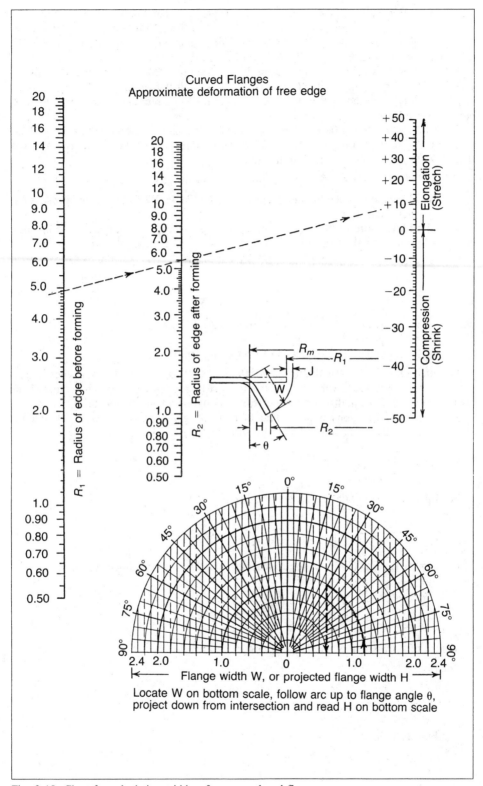

Fig. 2-19. Chart for calculating widths of open or closed flanges.

maximum flange width W. Using value of $3T$ (Table 2-8), bend radius $R = 0.12$ in. (3.0 mm). From Fig. 2-17, setback $J = 0.11$ in. (2.8 mm). From Table 2-2, elongation $E = 6\%$. From Fig. 2-18, $Y = 0.57$ (14.5 mm). Then, maximum flange width $W = 0.57$ (14.5 mm) $+ 0.11$ (2.8 mm) $= 0.68$ in. (17.3 mm).

EXAMPLE 3: Given: mold-line radius $R_m = 6.00$ in. (152 mm), flange width $W = 1.20$ in. (30.5 mm), flange angle $\theta = 30°$ (open), stretch flange thickness $T = 0.064$ in. (1.62 mm), and bend radius $R = 0.19$ in. (4.83 mm). Required: percent deformation and permissible materials.

From Fig. 2-17, $J = 0.06$ in. (1.5 mm), and R_1 (Fig. 2-19) $= 6.00$ (152 mm) $- 1.20$ (30.5 mm) $+ 0.06$ (1.5 mm) $= 4.86$ (123 mm). From Fig. 2-19 (lower graph) projected flange width $H = 0.60$ in. (15.2 mm), and

$$R_2 = 6.00 \ (152 \ \text{mm}) - 0.60 \ (15.2 \ \text{mm}) = 5.40 \ \text{in. (137 mm)}$$

percent elongation $= 11\%$ (upper graph). Consult Table 2-2 for materials possessing 11% or greater elongation.

Permissible Strain in Stretch Flanges. These values, which depend upon edge condition of the metal, flange width (Fig. 2-16), and method of forming strain for right-angle flanges, may be approximated by the formula

$$e = \frac{W}{R_2} \tag{4}$$

where e = elongation (strain) factor at free edge of flange
$\quad\quad W$ = flange width, in. (mm)
$\quad\quad R_2$ = contour (bent-up) radius of flange, in. (mm)

For 2024-O, -T3, and -T4 aluminum 90° flanges, 0.10 is a safe value for e where edges are smooth; 0.06 is a safe value for shear edges. A larger degree of stretch occurs where contour radius R_2 is small or where the stretch flange is adjacent to a shrink flange.

Air Force specifications indicate that there is danger of cracking when elongation exceeds 12% in 2 in. (51 mm). Therefore, for safety, $e = 0.12$, and

$$\frac{R_2 - W}{R_2} = 0.88 \tag{5}$$

For open flanges (angles less than 90°; see Fig. 2-16),

$$e = \frac{W \ (1 - \cos \alpha)}{R_2} \tag{6}$$

For $e = 0.12$, Eq. (6) gives the following values for various open-flange angles:

$\quad\quad \alpha = 30°; \ W = 0.472R_2$
$\quad\quad \alpha = 45°; \ W = 0.290R_2$
$\quad\quad \alpha = 60°; \ W = 0.194R_2$
$\quad\quad \alpha = 75°; \ W = 0.139R_2$

Values of e for some other shaped flanges are as follows:

For Fig. 2-16C,

$$e = \frac{W_1}{R_2} \tag{7}$$

For Fig. 2-16D,

$$e = \frac{W_1 + 2W_2}{R_2} \tag{8}$$

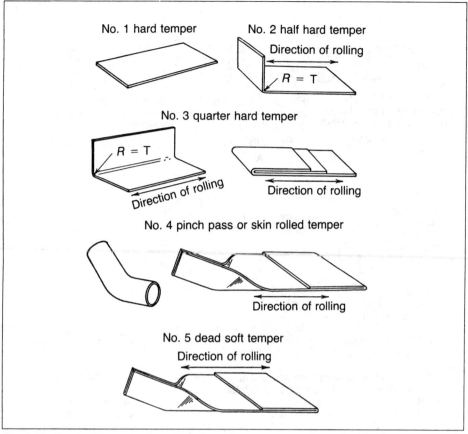

Fig. 2-20. Types of formation for which the various temper numbers of cold-rolled carbon-steel strip are suited.[10]

For Fig. 2-16*E*,

$$e = \frac{J}{R_2} \tag{9}$$

Permissible Strains in Shrink Flanges. Equation (4) for 90° stretch flanges also applies to 90° shrink flanges. Here, however, the metal is in compression, and the sheet must be supported against buckling or wrinkling. With rubber forming, there is practically no support against buckling, and only slight shrinking can be accomplished, so that rubber forming is limited to very large flange radii or very narrow widths. For 2024-O aluminum, without subsequent rework, shrink is limited to not over 2 or 3%; shrink is limited to 0.5% for 2024-T3 and -T4.

Permissible Degree of Cold Forming. Cold rolling changes the mechanical properties of carbon-steel strip and produces certain useful combinations of hardness, strength, stiffness, ductility, and other characteristics. Temper numbers indicate degrees of strength, hardness, and ductility produced in cold-rolled carbon-steel strip. The temper numbers are associated with the ability of each temper to withstand certain degrees of cold forming. Typical illustrations of the forming characteristics of each temper are shown in Fig. 2-20.[10] The No. 1 temper is not suited to cold forming; temper Nos. 4 and 5 are used in producing parts involving difficult forming or drawing operations.

Figure 2-21 illustrates various elements of flange design.

2-19

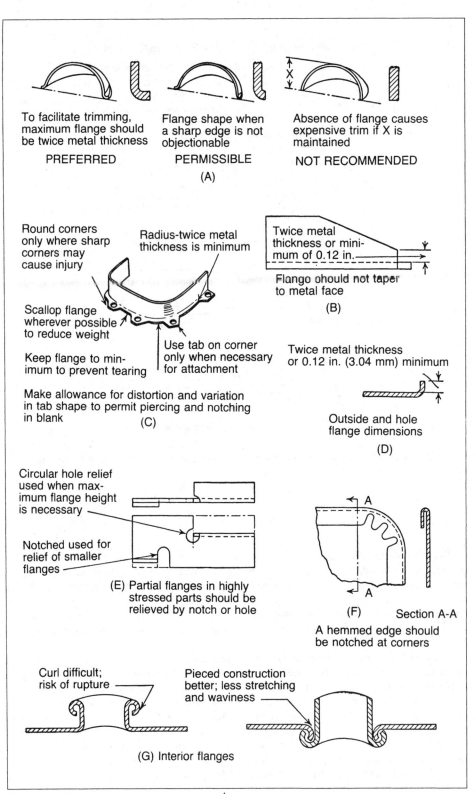

To facilitate trimming, maximum flange should be twice metal thickness

PREFERRED

Flange shape when a sharp edge is not objectionable

PERMISSIBLE

Absence of flange causes expensive trim if X is maintained

NOT RECOMMENDED

(A)

Round corners only where sharp corners may cause injury

Radius-twice metal thickness is minimum

Scallop flange wherever possible to reduce weight

Keep flange to minimum to prevent tearing

Use tab on corner only when necessary for attachment

Make allowance for distortion and variation in tab shape to permit piercing and notching in blank

(C)

Twice metal thickness or minimum of 0.12 in.

Flange should not taper to metal face

(B)

Twice metal thickness or 0.12 in. (3.04 mm) minimum

Outside and hole flange dimensions

(D)

Circular hole relief used when maximum flange height is necessary

Notched used for relief of smaller flanges

(E) Partial flanges in highly stressed parts should be relieved by notch or hole

A

A

(F)   Section A-A

A hemmed edge should be notched at corners

Curl difficult; risk of rupture

Pieced construction better; less stretching and waviness

(G) Interior flanges

Fig. 2-21. Details of stamped flanged design.[4]

TOLERANCES ON STAMPINGS

Dimensional Tolerances. Tolerances are restrictions in specifications and should be made only as close as may be essential or critical in the functioning of the stamped part. Unnecessarily close tolerances will increase tool and production costs and lower the die life, and may make up to 100% inspection mandatory.

Blanking and Piercing Tolerances. Table 2-3 gives blanking and piercing tolerances which can ordinarily be maintained on steel, brass, copper, or aluminum parts. Greater tolerances are usually required on fiber, rubber, and softer materials. Compound dies and fine blanking generally will maintain closer tolerances than listed in Table 2-3 for the life of the tool.

TABLE 2-3
Blanking and Piercing Tolerances, inches (mm)*[11]

Material Thickness in. (mm)	Size of Blanked or Pierced Opening		
	Up To 3 in. (76 mm) Wide	Over 3, Up To 8 in (200 mm) Wide	Over 8, Up To 24 in. (610 mm) Wide
0.025 (0.64)	0.003 (0.08)	0.005 (0.13)	0.008 (0.20)
0.030 (0.76)	0.003 (0.08)	0.006 (0.15)	0.010 (0.25)
0.060 (1.52)	0.004 (0.10)	0.008 (0.20)	0.012 (0.30)
0.084 (2.13)	0.005 (0.13)	0.009 (0.23)	0.014 (0.36)
0.125 (3.18)	0.006 (0.15)	0.010 (0.25)	0.016 (0.41)
0.187 (4.75)	0.010 (0.25)	0.016 (0.41)	0.025 (0.64)
0.250 (6.35)	0.015 (0.38)	0.020 (0.51)	0.035 (0.89)

* All tolerances are plus for blanking, and minus for piercing.

It should be noted that blanking tolerances are indicated as plus; this is because the die opening determines the blank size and, as the die is resharpened, the size of the opening increases. Conversely, piercing punches decrease in diameter with wear, and the pierced holes become smaller (minus tolerance).

Hole tolerance for solid rivets should be plus 0.015 in. (0.38 mm), minus 0.000; for the draw rivets they should be plus or minus 0.005 in. (0.13 mm).[12]

Bending and Form Contour Tolerances. Figure 2-22 suggests allowable tolerances on bent parts.

It may sometimes be inadvisable to specify tolerance on each dimension of a contoured shape, because of a possible objectionable accumulation of these tolerances. In such cases, notation should be clearly made on the drawings, as suggested in Fig. 2-23.

Flatness Tolerances. Parts should be so designed (Fig. 2-24) that straight edges can be maintained on the flat blanks of formed parts wherever possible. This results in economy and ease of production, since the blank can be sheared from flat blanks with relatively inexpensive dies.

Tolerances of 0.005 in. (0.13 mm) per inch (2.54 mm) of length are usual for flatness of punched parts. Where the design will permit use of stippling die, flatness tolerance of 0.001 in. (0.02 mm) can be maintained. For bronze parts thinner than 0.040 in. (1.02 mm), tolerance of 0.010 in. (0.25 mm) is allowable.

Squareness and Eccentricity Tolerances. For squareness of sheared parts, a tolerance of ±0.003 in. (0.76 mm) is allowable for parts dimensions given in decimals, and ±0.010 in. (0.25 mm) on fractional dimensions for each inch of the surface length. This tolerance is for squareness only and is not to be added to the linear tolerance.[11]

The tolerance for eccentricity between hole and OD of washers punched with stock dies is 0.004 in. (0.10 mm) on 1 in. (25.4 mm) OD or less, and 0.008 in. (0.20 mm) on OD greater than 1 in. (25.4 mm).

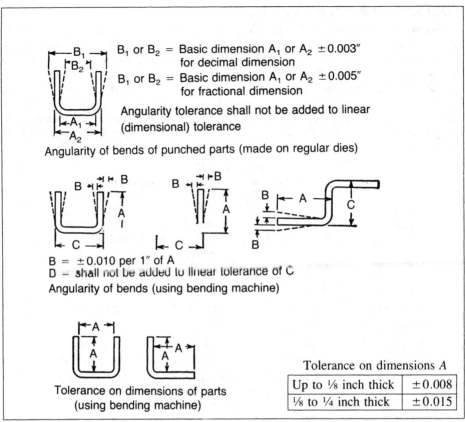

B₁ or B₂ = Basic dimension A₁ or A₂ ±0.003″
            for decimal dimension

B₁ or B₂ = Basic dimension A₁ or A₂ ±0.005″
            for fractional dimension

Angularity tolerance shall not be added to linear
(dimensional) tolerance

Angularity of bends of punched parts (made on regular dies)

B = ±0.010 per 1″ of A
D = shall not be added to linear tolerance of C
Angularity of bends (using bending machine)

Tolerance on dimensions of parts
(using bending machine)

Tolerance on dimensions A

| Up to ⅛ inch thick | ±0.008 |
| ⅛ to ¼ inch thick | ±0.015 |

Fig. 2-22. Tolerance for bent parts.[11]

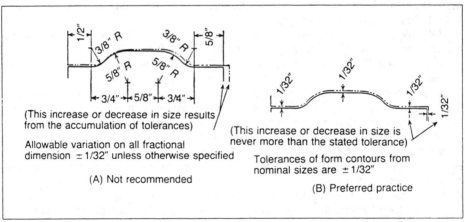

(This increase or decrease in size results
from the accumulation of tolerances)

Allowable variation on all fractional
dimension ±1/32″ unless otherwise specified

(A) Not recommended

(This increase or decrease in size is
never more than the stated tolerance)

Tolerances of form contours from
nominal sizes are ±1/32″

(B) Preferred practice

Fig. 2-23. Indication of form contour tolerances.[12]

Tolerances on Stock Thickness. For standard commercial mill tolerances on stamping strip, sheet, and coil stock, see AISI or suppliers' bulletins.

If high accuracy is necessary in the bending or forming of sheet metal, the overall tolerance on stock thickness must be considered. Cold-rolled strip steel has about 50% closer thickness tolerance than cold-rolled sheet steel. Should the sheet metal vary 0.001 in. (0.02 mm) in thickness, dimension A (Fig. 2-25) would be both a small cumulative error in the bending operation, and an overall variation in the stock thickness.

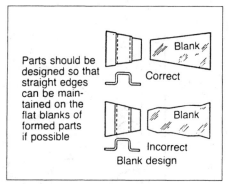

Parts should be designed so that straight edges can be maintained on the flat blanks of formed parts if possible

Correct

Incorrect

Blank design

Fig. 2-24. Design for use of flat stock.

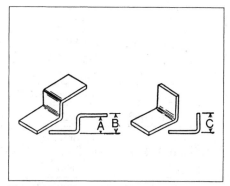

Fig. 2-25. Effect of variation in stock thickness on overall tolerances.[2]

On the other hand, if dimension *B* is taken from the outside of the offset bend, the variation would be twice that of *A*, plus the required bending or forming tolerance.

PRODUCT DESIGN FOR BLANKING AND PIERCING HOLES

In designing stampings with punched holes, it should be remembered that only about one-third the thickness of the metal is sheared cleanly to the size of the punch. The remainder fails in shear by the pressure on the sheared slug, producing a rough hole tapering outward from the diameter of the punch to the diameter of the die (Fig. 2-26). This fact is especially important when the periphery of the hole is intended to act as a bearing surface.

Holes are produced in stamped parts by the use of conventional, extruding, or piercing punches (Fig. 2-27).

Punched Holes. Figure 2-28 presents general guides for punched-hole diameters, center-to-center distances, and distance from edge of blanked part. Tolerances on punched holes are given in Table 2-4.

When holes are to be punched in stampings, for later use in assembling with bolts or screws, minimum edge distances should be so calculated as to utilize the full shear or bearing strength of the bolts or screws (Table 2-5).

Extruded Holes. An extruded hole is usually formed in a single operation by a punch which clean punches a smaller hole, and then follows through to flange the sides. Sometimes a hole flange is specified which is wider than can be formed in one operation.

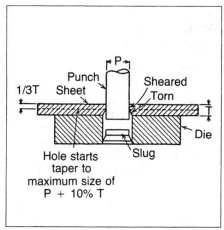

Fig. 2-26. Schematic action of punch in producing a hole.

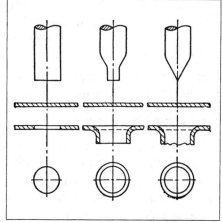

Fig. 2-27. Basic hole forms produced in stampings.

	Punch Press		
R_1	0.16 preferred or 2T		
R_2	2T min		
	T	Nonferrous	Ferrous
D_1 or D_2	Thru 0.062	0.12	0.12
	0.063–0.38	0.12 or 1.5T whichever is greater	2T
D_3	1.0T or 0.098 min except 0.120 min for alloy steels, etc.		
	T	Width	
D_4 or D_5	Thru 0.032	0.06	
	0.033–0.125	2T	
	0.126 0.38	2.5T	

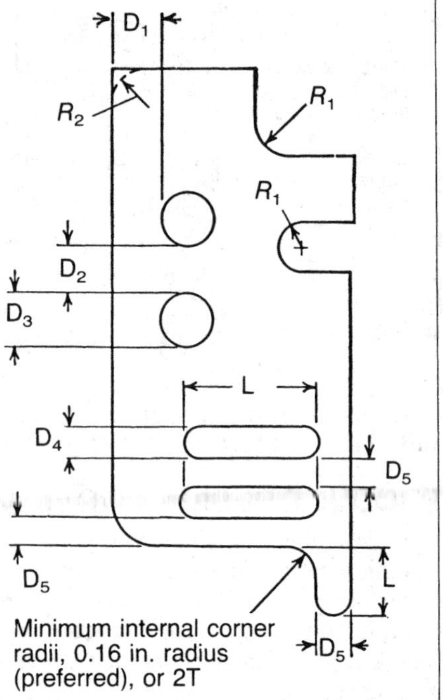

Minimum internal corner radii, 0.16 in. radius (preferred), or 2T

Fig. 2-28. Minimum practical punching and blanking dimension. (*A. H. Petersen*)

This necessitates drawing metal in from outside the hole by first drawing an embossment much larger than the hole, then by successive steps forming the flange and finally punching out the hole. It is a more expensive method, and its use can be limited by keeping the specified flange width to an absolute minimum.

If the extruded hole is to be tapped, the maximum height H (Fig. 2-29) of the extrusion should not exceed one-fifth of the body size of the tap. In view A, with an 8-32 thread, only two threads could be produced. In view B, 3½ to 4 threads would be possible. If permissible, a 10-32 thread in place of the 8-32 would increase the hole size and therefore the height H.

TABLE 2-4
Maximum Punched-Hole Tolerances for
Aluminum and Steel Sheet

Decimal Nominal Diam., in. incl.	Sheet Thickness, in.				
	0.025 through 0.042	0.050 through 0.072	0.078 through 0.093	0.102 through 0.156	0.187 through 0.250
0.125–0.141	+0.002	↑	↑	↑	
0.144–0.228	+0.003	+0.006;	+0.008;	+0.011;	
0.234–0.413	+0.004	− 0.001	− 0.002	− 0.003	↑
0.422–0.688	+0.006	↓	↓		+0.021;
0.703–0.984	+0.009	+0.009		↓	− 0.003
1.000 and up	+0.010	+0.010	+0.010		↓

Data courtesy of A. H. Petersen.

2-24

TABLE 2-5
Recommended Distances, Center of Screw or
Bolthole to Nearest Edge of Part

Connection	Material in which the Bolt or Screw Bears	
	Aluminum Alloys, 8630, Corrosion-resistant Steel	Mg Alloy, Die-cast Materials
Bolts and nonflush screws	1.7 diam. +0.03 in. (+ 0.8 mm)	2.0 diam. + 0.03 in. (+ 0.8 mm)
Flush screws	2.0 diam. + 0.03 in. (+ 0.8 mm)	2.4 diam. + 0.03 in. (+ 0.8 mm)

Data courtesy of A. W. Petersen.
Above values allow a manufacturing tolerance of 0.03 in. (0.8 mm)

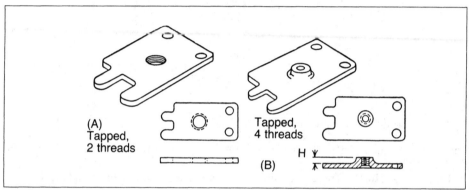

FIg. 2-29. Extruded holes for tapping.[2]

In addition to extruding so as to provide a boss for tapping, tapped holes deeper than the sheet thickness can be built up by means of clinch-on nuts, or by use of a nut spot-welded in line with the punched hole.

Pierced Holes. A pierced hole is made by a sharp-pointed punch which follows through to flange the sides, forming a hole with torn or irregular edges.

Relation of Holes to Right-angle Bends. Because of variations in stock thickness and temper, the relation of right-angle bends to given hole locations is somewhat difficult to maintain.

The most desirable location of a hole adjacent to a right-angle bend is shown in view A 2, Fig. 2-30, the edge of the hole being not closer to the bend than dimension $X = 1\frac{1}{2} T + R$. Where a hole must be located as close as possible to the bend (but not actually in the bend), as in view A 1, the addition of a crescent slot allows the part to be die-cut, blanked, and pierced, with enough flat material around the hole to avoid distortion.

When a hole must be located actually in a right-angle bend, unless piercing is done after bending, hole distortion would be considerable if design were as shown in $B1$, particularly if stock thickness exceeded 0.015 in. (0.38 mm). Redesign, as in $B2$, permits normal sequence of stamping operations.

Holes in a right-angle-bent stamping must sometimes hold a close relationship in assembly with a separate base or part. Thus, the gear-train plate in $C1$ must assemble with another plate on a common base X, with dimensions C and D closely held so that the bearing holes will be in alignment in final assembly. With the design as in $C1$, slight variations in stock thickness or temper could throw the alignment out.

Using the design in $C2$, the ears Y were so formed as to leave a slight clearance F. In

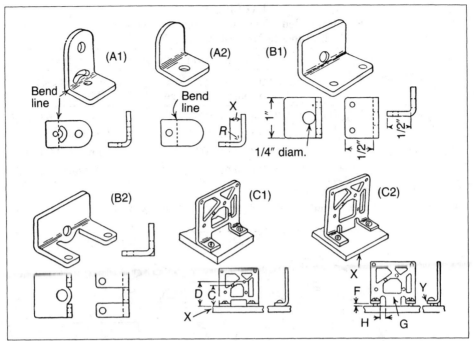

Fig. 2-30. Punched holes relative to right-angle bends.[2]

lining up two such plates, the added section G comes to rest on base X for positive dimensional control.

Lightening Holes. These are inherently associated with their peripheral stretch flanges, to combine saving in weight with increased stiffness (see Product Design for Forming).

SLOTS

Slots of regular shape are employed in design (1) to compensate for inaccuracies of manufacture, or (2) to provide dimensional adjustment. Their basic dimensional considerations are indicated in Fig. 2-28.

NOTCHES; PERFORATIONS

There are two general groups of notches: (1) notches that are part of the product design, and provided for clearance, locating, or attaching; (2) notches added to flanges to facilitate forming of the part.

Notches in highly stressed parts should never be specified with a sharp vee at the vertex because it would provide the starting point of a tear. Instead, it should if at all permissible be rounded (Fig. 2-31A), the minimum radius preferably being $2T$. This type of notch is usually added in the blank; allowance should therefore be made for distortion in any subsequent forming operation.

A sharp vertex in notches is allowable on parts bearing lower stresses, and may aid in lowering diemaking and maintenance costs. If sharp notches are used, allowance for distortion should be made so that the notch can be added in the blank.

If design includes well-defined notches or side-wall perforations (Fig. 2-31B) in a drawn part, they must be made after drawing; otherwise the openings would tend to close up under compressive force.

Relief Notches for Right-angle Bends. If the exterior surface of an unrelieved bend is parallel with the profile of the blank, some tearing or fracturing will occur because of

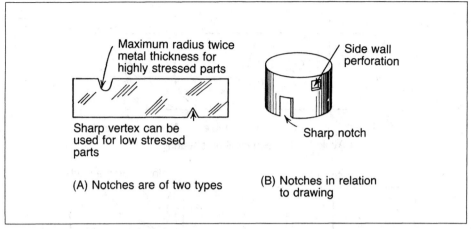

Fig. 2-31. Basic notch considerations.

bending (*A*1, Fig. 2-32); the amount of tearing decreases with increase in peripheral radius of the finished formed part. To avoid such tearing or fracture, relief notches can be made in the flat blank, as at *A*2, and should if possible be at least twice stock thickness in both width and depth. This design will bend better and have a more rigid section.

Such designs as at *B*1 should be avoided for small-lot pilot runs, since the lancing tool not only has to lance the three ears loose, but must also form and set them. With relief notches used as at *B*2, the forming tools merely form and set the ears at almost any desired angle. The radius at which the ears meet the major blank can be readily varied.

When one surface of right-angle-bent stamping meets an adjacent surface at a taper or slant (*C*1, Fig. 2-32), some buckling may result. It is better practice to notch as at *C*2 in order to have work contours meet at 90°.

Figure 2-33 shows some basic considerations in shearing edges of strip stock.

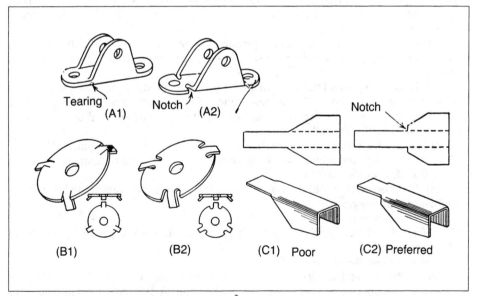

Fig. 2-32. Relief notches for right-angle bends.[2]

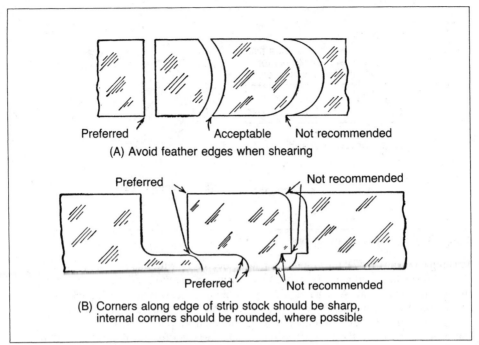

Fig. 2-33. Edge-trimming considerations.

PRODUCT DESIGN FOR BENDING

The term "bending" is here restricted to include only single-curvature or straight-line bends.

Bend Radii. Extremely sharp bends generally cannot be made, because a punch or die having a sufficiently sharp edge to make such a bend would cut the metal. Generally, bend radii should not be less than 0.0625 to 0.03125 in. (1.588 to 0.7938 mm) or metal thickness, whichever value is larger.

For a given metal thickness, circumferential unit stresses in the metal will increase as the bend angle α increases and/or the length of bend L decreases. (see Section 6, Bending of Metals)

When the bend angle and bend length are large, the minimum bend radius is generally limited by the metal thickness T. Minimum bend radii for certain materials and alloys are given in Tables 2-6 to 2-10. Approximate bend radii may be determined by multiplying the stock thickness by the following:

1. 0.3 to 0.7 for soft metals, stock thickness 0.02 in. (0.5 mm) minimum.
2. 0.6 to 1.2 for hard metals, stock thickness 0.06 in. (1.5 mm) minimum.
3. 2.0 to 3.0 for very hard metals.

If cracking or tearing occurs on bends made with these values, the following modifications are suggested:

1. Increase the bend radius.
2. Deburr the blanks, by hand or tumbling, to remove rough edges before bending.
3. Anneal the blanks.
4. Select a more ductile material.

Titanium RC-70 and RC-130A can be bent to a minimum radius of 3T. Annealed and hot-rolled zirconium can be bent to 180° on a 3T radius. A 10% cold working increases the radius to 8T; 20% cold work increases the radius to 12 to 16T.

The circumferential strain, which limits the minimum bend radius, is nearly proportional to the bend angle in the small-angle range, reaching maximum value near 90°. Therefore, bend radius can be smaller if the bend angle is much smaller than 90°.

TABLE 2-6
**Minimum Bend Radii, in. (mm), For Low-Carbon
and Low-Alloy Steels**

Material Thickness, in. (mm)	SAE 1020–1025 in. (mm)	SAE 4130 and 8630 (Annealed) in. (mm)	SAE 1070 and 1095 in. (mm)
0.016 (0.41)	0.03 (0.8)	0.03 (0.8)	0.06 (1.5)
0.020 (0.51)	0.03 (0.8)	0.03 (0.8)	0.06 (1.5)
0.025 (0.64)	0.03 (0.8)	0.03 (0.8)	0.06 (1.5)
0.030 (0.76)	0.03 (0.8)	0.06 (1.5)	0.09 (2.3)
0.035 (0.89)	0.06 (1.5)	0.06 (1.5)	0.09 (2.3)
0.042 (1.07)	0.06 (1.5)	0.06 (1.5)	0.13 (3.3)
0.050 (1.27)	0.06 (1.5)	0.09 (2.3)	0.13 (3.3)
0.062 (1.57)	0.06 (1.5)	0.09 (2.3)	0.16 (4.1)
0.078 (1.98)	0.09 (2.3)	0.13 (3.3)	0.19 (4.8)
0.093 (2.36)	0.09 (2.3)	0.16 (4.1)	0.25 (6.4)
0.109 (2.77)	0.13 (3.3)	0.16 (4.1)	0.31 (7.9)
0.125 (3.18)	0.13 (3.3)	0.19 (4.8)	0.31 (7.9)
0.156 (3.96)	0.16 (4.1)	0.25 (6.4)	0.38 (9.7)
0.187 (4.75)	0.19 (4.8)	0.31 (7.9)	0.50 (12.7)

A fairly sharp contour can be formed on the outside of a flange by first pinching to reduce the section, and then bending (Fig. 2-34). The designer should take into account, however, that this method makes the part weaker, work-hardens the metal excessively, and makes any subsequent forming operations difficult.

Significance of Bend Length. Because of failure in cracking at the convex surface, bending limits are determined by the amount of stretch a metal can undergo under the particular stress conditions.

The bend length (axial line of the bend) directly affects these stresses. For a part of very narrow bend length, the tension S_1 on the convex surface is only circumferential. In addition, where the bend length is finite, there is also present a transverse tension S_2 along the axis of the bend. The ratio S_2/S_1 varies almost linearly with the ratio L/T until, for bend lengths greater than about eight times the thickness ($L/T = 8$), the stress ratio remains practically constant.

Bend Relief Cutouts. Where two or more bend lines intersect (X, Fig. 2-35), or where the bend intersects an edge of the piece at an angle less than 60°, the use of bend relief notches or cutouts is indicated, to prevent distortion due to the proximity of the flanges. The cutout should be so shaped that all portions of the metal are removed which would otherwise have curvature in more than one direction.

The cutout should be extended from 0.3 to 0.6 in. (7.6 to 15.2 mm) past the critical bend tangent line or intersection of bend tangent lines. Internal radii should be a minimum of 0.16 in. (4.1 mm).

W in Fig. 2-35 represents minimum flange width for the composition, heat-treat condition, and thickness of the metal. For rubber forming without auxiliary devices, some practical minimum flange widths are:[13]

2024-O, 7075-O aluminum . . . . . . 0.0625 in. (1.588 mm) + $2.5T$
2024-T3 and -T4 aluminum . . . . . . 0.125 in. (3.18 mm) + $4.0T$
Annealed stainless steels . . . . 0.1875 in. (4.762 mm) + $4.5T$
1/4 H stainless steels . . . . 0.625 in. (15.88 mm)

TABLE 2-7
Minimum Bend Radii for Austenitic Stainless Steels[13, 14]

Type	Temper*	Thickness Range, in.	Bend Radius	Bend Angle, Deg.
301, 302, 304, 316, 321, 347	Annealed (A)	All	½T	180
301, 302, 304	¼H (B)	Up to 0.050	½T	180
		0.051 and over	1T	90
301, 302, 304	½H (C)	Up to 0.050	1T	180
		0.051 and over	1T	90
301, 302, 304	¾H (D)	Up to 0.030	1½T	180
		0.031 to 0.050	1¼T	90
301, 302, 304	Hard (E)	Up to 0.030	2T	180
		0.031 to 0.050	1½T	90
316	¼H (B)	Up to 0.050	1T	180
		0.051 and over	1T	90
316	½H (C)	Up to 0.030	2T	180
		0.031 to 0.050	3T	90
		0.051 and over	2T	90

* Definition of tempers:

	A	B	C	D	E
Tensile strength (min), psi (MPa)	75,000 (517)	125,000 (862)	150,000 (1,034)	175,000 (1,206)	185,000 (1,275)
Yield strength (2% min offset), psi (MPa)	30,000 (207)	75,000 (517)	110,000 (758)	135,000 (931)	140,000 (965)

Nonferrous parts having narrow projecting tabs, small cutouts to reduce air leakage, or slots narrower than 0.75 in. (19.0 mm) may have relief radii smaller that 0.16 in. (4.1 mm). For parts 0.040 in. (1.02 mm) thick or less, minimum radius may be 0.06 in. (1.5 mm); for parts thicker than 0.040 in. (1.01 mm); minimum radius must be 1½T.

Ferrous parts must have a minimum radius of 1½T on both inside and outside corners of the blank.

Joggles. A joggle is an offset bent or formed on a part to provide clearance for adjoining parts. It is generally used in sheet-metal assemblies when a member must be fitted to two parallel surfaces not in the same plane. The decision to joggle will depend on relative stiffness of parts, flushness desired, and the cost compared with alternative provisions for assembly.

Joggles may be provided for surface differences as small as 0.010 in. (0.25 mm). Some aircraft designers set arbitrary height minima at 0.020 to 0.025 in. (0.51 to 0.64 mm) for smooth exteriors, and 0.030 to 0.040 in. (0.76 to 1.02 mm) for interior structures.

Figure 2-36 illustrates some practical joggle design considerations.

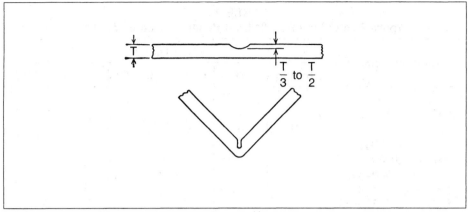

Fig. 2-34. Pinch method for producing sharp bends.[14]

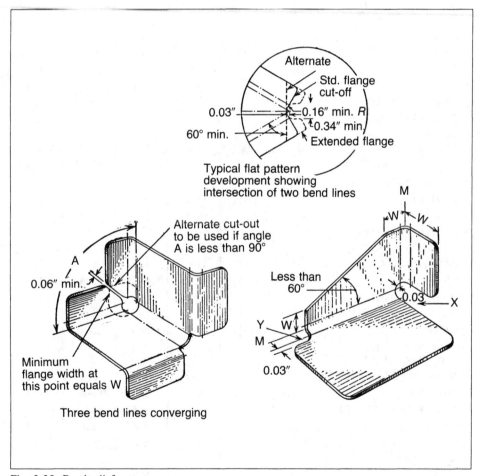

Fig. 2-35. Bend relief cutouts.

TABLE 2-8
Approximate Radii for 90° Cold Bends in Aluminum Sheet*

Alloy and Temper	Thickness, in.					
	0.0156	0.0312	0.0625	0.1250	0.1875	0.250
1100-O, 3003-O, 5005-O, 5050-O, 5457-O	0	0	0	0	0	0
1100-H12, 1100-H14, 2014-O, 2024-O, 3003-H12, 3004-O, 5005-H32, 5052-O 5457-H32, 5457-H25, 6061-O	0	0	0	0	0–1	0–1
3003-H14, 3004-H32, 5005-H34, 5052-H32, 5086-O, 5154-O, 5457-H34	0	0	0	0–1	0–1	½–1½
5050-H32	0	0	0	0	0–1	½–1½
5050-H34	0	0	0	0–1	½–1½	1–2
5083-O	. . . .	. . . .	0–½	0–1	0–1	½–1½
1100-H16, 3004-H34, 5052-H34, 5154-H32, 7075-O, 7079-O, 7178-O	0	0	0–1	½–1½	1–2	1½–3
5083-H32, 5456-H323	. . . .	. . . .	1–2	1½–3	1½–3½	2–4
5083-H34, 5456-H343	. . . .	. . . .	1–2	1½–3	2–4	2½–4½
5086-H112	0–½	0–½	½–1	½–1½	1–1½	1–2
5086-H32	0–½	0–1	½–1½	1–2	1½–2	1½–2½
3003-H16, 5005-H36, 5050-H36, 5154-H34, 5457-H36, 6061-T4	0–1	0–1	½–1½	1–2	1½–3	2–4
5086-H34	0–1	½–1½	1–1½	1½–2½	2–3	2–3
1100-H18, 3004-H36, 5052-H36, 5154-H36	0–1	½–1½	1–2	1½–3	2–4	2–4
6061-T6	0–1	½–1½	1–2	1½–3	2–4	3–4
3004-H38, 3003-H18, 5005-H38, 5050-H38, 5056-H38, 5457-H38	½–1½	1–2	1½–3	2–4	3–5	4–6
5086-H36	. . . .	. . . .	. . . .	2–3½	2½–4	3–4½
5154-H38, 2014-T3, -T4	1–2	1½–3	2–4	3–5	4–6	4–6
2024-T3, -T4	1½–3	2–4	3–5	4–6	4–6	5–7
2014-T6	2–4	3–5	3–5	4–6	5–7	6–10
7075-T6, 7079-T6, 7178-T6	2–4	3–5	4–6	5–7	5–7	6–10
2024-T36, 2024-T81	3½–5	4½–6	5–7	6½–8	7–9	8–10
2024-T86	4½–5	5–7	6–8	7–10	8–11	9–11

* The radii are expressed in multiples of sheet thickness.

2-32

TABLE 2-9
Minimum Bend Radii for Copper Alloys[15]

Alloy	Temper	Nominal Thickness, in.	Minimum Suitable Radius of Punch, in.		
			A*	B†	C‡
Copper	Half hard	0.020	0.031	0.031	0.047
	Extra hard	0.020	0.016	0.031	0.031
Red brass, 85%	Drawing anneal	0.005–0.064	Sharp	Sharp	Sharp
	Half hard	0.020–0.050	Sharp	Sharp	Sharp
	Hard	0.040	0.016	0.031	0.094
	Extra hard	0.040	0.062	0.094	0.219
	Spring	0.040	0.062	0.188	0.500
Low brass, 80%	Hard	0.020	0.031	0.047	0.062
	Spring	0.020	0.031	0.125	0.188
Cartridge brass, 70%	Half hard	0.005–0.050	Sharp	Sharp	Sharp
	Hard	0.040	0.016	0.031	0.047
	Extra hard	0.040	0.031	0.156	0.219
	Spring	0.040	0.062	<0.250	<0.250
	Extra spring	0.040	0.125	<0.250	<0.250
Yellow brass	Half hard	0.005–0.090	Sharp	Sharp	Sharp
	Hard	0.040	Sharp	Sharp	0.031
	Extra hard	0.040	0.047	0.125	0.188
	Spring	0.040	0.047	0.219	<0.250
Medium leaded brass	Half hard	0.040	Sharp	Sharp	Sharp
Phosphor bronze, 5%	Half hard	0.020–0.070	Sharp	Sharp	Sharp
	Hard	0.040	0.062	0.062	0.125
	Extra hard	0.040	0.062	0.125	
	Spring	0.040	0.094		
Phosphor bronze, 8%	Half hard	0.005–0.064	Sharp	Sharp	Sharp
	Hard	0.040	0.031	0.125	
	Extra hard	0.040	0.047	0.156	<0.250
	Spring	0.040	0.094	0.250	0.500
	Extra spring	0.064	0.156	<0.250	<0.250
High silicon bronze	Hard	0.020	0.031	0.031	0.062
	Spring	0.020	0.047	0.094	0.188
Nickel silver 65–18	Half hard	0.040	. . .	0.016	0.031
	Hard	0.040	0.062	0.062	0.062
	Extra hard	0.040	0.125	0.125	0.188
	Spring	0.040	0.156	0.188	0.219
	Extra spring	0.040	0.156	0.219	0.250
Nickel silver 55–18	Half hard	0.040	0.062	0.062	0.062
	Hard	0.040	0.062	0.062	0.094
	Extra hard	0.040	0.125	0.156	0.188

* Bends perpendicular to direction of rolling.
† Bends 45° to direction of rolling.
‡ Bends parallel to direction of rolling.

TABLE 2-10A
Bend Radii of Magnesium Sheet at Various Temperatures

Temperature °F (°C)	Bend Radius in Terms of Sheet Thickness, T					
	AZ31B-0 (Spec. Bend)	AZ31B-0	AZ31B-H24	HK31A-0	HK31A-H24	HM21A-T8
70 (21)	3	5.5	8	6	13	9
200 (93)	3	5.5	8	6	13	9
300 (149)	3	4	6	6	13	9
400 (204)	3	3	3	5	9	9
500 (260)	2	2	2	4	8	9
600 (316)	..	...	..	3	5	8
700 (371)	..	...	..	2	3	6
800 (427)	..	...	..	1	..	4

TABLE 2-10B
Miminum Bend Radii of Round Magnesium Tube for Various D/T Ratios

Alloy	Bend Condition	Minimum Bend Radius* (At Room Temperature)		
		$D/T = 17$	$D/T = 6$	$D/T = 3$
AZ31B	Unfilled	$6D$	$4D$	$3D$
AZ31B	Cerrobend filled	$3D$	$2D$	$2D$
AZ61A	Unfilled	$5D$	$2\frac{1}{2}D$	$2D$
AZ61A	Cerrobend filled	$2\frac{1}{2}D$	$2\frac{1}{2}D$	$2\frac{1}{2}D$

* Bend radius to the tube axis expressed in terms of tube OD.
D = Tube OD, T = wall thickness.

TABLE 2-10C
Suggested Limits for Production
Bending of Magnesium Tube

Alloy	Forming Temperature	Bend Radius*
AZ31B	Room	$4D$
AZ31B	200°F (93°C)	$3D$
AZ61A	Room	$4D$
AZ61A	200°F (93°C)	$3D$
ZK60A-F	Room	$5D$

* Minimum bend radius to axis of tube.
D = Tube OD.

TABLE 2-10D
Suggested Magnesium Extrusion Formability Limits

Alloy*	Typical Bend Radius, Room Temperature	Suggested Limits	
		Time and Temperature	Typical Bend Radius
AZ31B-F	2.4*T*	1 hr at 550°F (288°C)	1.5*T*
AZ61A-F	1.9*T*	1 hr at 550°F (288°C)	1*T*
AZ80A-F	2.4*T*	½ hr at 550°F (288°C)	0.7*T*
AZ80A-T5	8.3*T*	1 hr at 380°F (193°C)	1.7*T*
ZK60A-F	12*T*	½ hr at 550°F (288°C)	2*T*
ZK60A-T5	12*T*	½ hr at 400°F (204°C)	6.6*T*

* 0.090 × 0.875 in. (2.29 × 22.22 mm) extruded flat strip.

TABLE 2-10E
Angular Springback of 90° Bends in Magnesium Sheet*

Temperature, °F (°C)	Die Radius in Sheet Thicknesses	Springback Allowance, deg	
		AZ31B-0	AZ31B-H24
70 (21)	4	8	10
	5	11	13
	10	17	21
	15	25	29
200 (93)	3	4	5
	5	5	7
	10	8	12
	15	13	17
300 (149)	2	1	2
	5	3	4
	10	5	7
	15	8	11
450 (232)	2	0	0
	5	1	1
	10	2	2
	15	4	4
550 (288)	Up to 15	None	None

* Based on sheet in the thickness range of 0.016 to 0.064 in.
(0.41 to 1.62 mm)

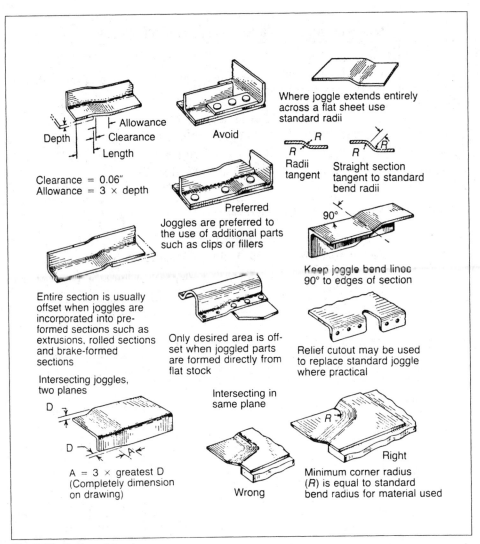

Depth ⊢ Allowance
⊢ Clearance
⊢ Length

Clearance = 0.06"
Allowance = 3 × depth

Avoid

Preferred

Joggles are preferred to
the use of additional parts
such as clips or fillers

Where joggle extends entirely
across a flat sheet use
standard radii

Radii
tangent

Straight section
tangent to standard
bend radii

90°

Keep joggle bend lines
90° to edges of section

Entire section is usually
offset when joggles are
incorporated into pre-
formed sections such as
extrusions, rolled sections
and brake-formed
sections

Only desired area is off-
set when joggled parts
are formed directly from
flat stock

Relief cutout may be used
to replace standard joggle
where practical

Intersecting joggles,
two planes

D

D

A

A = 3 × greatest D
(Completely dimension
on drawing)

Intersecting in
same plane

Wrong

R

Right

Minimum corner radius
(R) is equal to standard
bend radius for material used

Fig. 2-36. Design of joggles.

PRODUCT DESIGN FOR FORMING

The term ''forming'' can generally embrace all noncutting stamping operations. It is here limited to metalworking operations in which characteristically the shape of the punch and die is directly reproduced in the metal, with little or no plastic flow of the metal.

Such limitation cannot be rigidly followed, since the stress characteristics of metal under deep recessing or cupping approach the stress conditions for pure drawing, where the side-wall metal is subjected to combined circumferential compressive stresses and radial tensions. In the following analysis of formed-products design, drawn types are therefore also discussed (see also Product Design for Drawing).

FORMED PARTS CLASSIFICATION BY SHAPE

TYPE 1 — SINGLY-CURVED PARTS

1-A Straight sections	*Long narrow shapes, with readily evident mold lines*	
1-B Straight, flanged	*Flat web and flanges bent in same or opposite directions*	
1-C Smoothly contoured	*Flowing straight bends; no definite mold lines or breaks in the major contour*	

TYPE 2 — CURVED CHANNELS

2-A Curved, symmetrical	*Curved in one or two planes; definite mold lines*	
2-B Curved, non-symmetrical	*Curved in one or two planes; definite mold lines*	
2-C Nonuniform section	*Varied cross-section; uniform or nonuniform*	

TYPE 3 — CONTOURED FLANGED PARTS OF NONUNIFORM SECTION

3-A Stretch flanged	*Flanges on concave profile; may be combined with straight flanges*	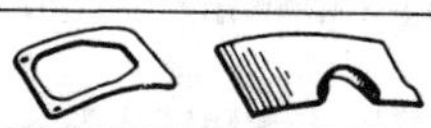
3-B Shrink flanged	*Flanges on convex profile; may be combined with straight flanges*	
3-C Combined stretch and shrink	*Flanges on both convex and concave profiles*	

TYPE 4 — DOUBLE-CURVATURE SMOOTHLY CONTOURED PARTS

4-A Large radius one direction		4-C Opposite contours in two directions: "Saddleback"; corrugated	
4-B Large radius two directions		4-D Small radius at one edge of parts 4-A, 4-B, 4-C	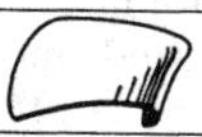

TYPE 5 — DEEP RECESSED PARTS

5-A Cylindrical cup *Vertical walls with or without flange wall cutouts or bottom offsets*	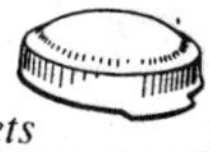	5-E Open parts *Convex curvature; continuous wall absent at one or several places*	
5-B Oval cup *With or without flanges, cutouts or sloping and curved bottoms*		5-F Reentrant contours *Continuous or partial walls*	
5-C Box type *Small corner radii and vertical walls; all curvature convex*		5-G Saddleback bottom *Having reverse contours perpendicular to each other*	
5-D Sloping walls *Curvature convex; sides straight or curved*		5-H Undercut walls *Having recess larger at bottom than at parting plane*	

TYPE 6 — SHALLOW RECESSED PARTS

6-A Dish type *Only one recess; closed or open, and regular or irregular in shape*	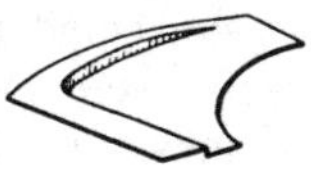	6-B Beaded and embossed *Shapes flat or with single or double curvature; surface broken by several regular or irregular shaped recesses*	

Fig. 2-37. Formed shapes classified in order of increasing severity of operations: class 1, easiest; class 6, most difficult.

ANALYSIS OF FORMED-PARTS DESIGN

Figure 2-37 shows 23 classes of formed parts, classified as to shape and also, from top to bottom, according to the relative severity of tooling and process requirements.

This classification generally applies to all readily formed metals, though especially devised to cover aluminum alloys.

Part types (i.e., shrink flange), whose stamping limitations are chiefly set by compression-type failures, can be produced in many metals to approximately the same limits. However, where a product design makes it inherently liable to tension-type failures (i.e., height of a stretch flange), the processing limits will depend closely upon the work material.

The basis of this classification is the contour of the stamped part, this contour being defined by a number of sections through the part.

Many particular shapes will seem to fall almost equally well into two or more classifications. Sometimes different part types will appear to be similar. In such cases, it is frequently possible by slight design changes to shift the part type from "difficult to form" to "relatively easy to produce."

Again, redesign may permit assembling a single complex product from several simple, easily produced parts.

The following design criteria are specific to the parts types classified in Fig. 2-37.

TYPE 1, SINGLY CURVED PARTS

Class 1-A. Straight Sections. Design considerations are essentially the same as for class 1-B.

Class 1-B. Straight Flanged Sections. See Bend Radii above. Minimum flange widths depend on the forming method. Shorter flanges can be produced by trimming or, in rubber forming, by a wiper-type form block. Flange end cuts should be designed at least 60° to the mold line to avoid pronounced bulging at that point.

Class 1-C. Smoothly Contoured Sections. Because of the large radii of curvature common to these parts, springback is greater than normal, and minimum bends will depend chiefly on the type of forming equipment that can or must be used. Practical radii would be 1.500 to 2 in. (38.10 to 51 mm) on rolls, 1 in. (25.4 mm) using stretch forming, and 0.500 in. (12.70 mm) with a caged rubber die in a power brake.

TYPE 2, CURVED CHANNELS

Class 2-A. Symmetrical Sections. The preferred section, wherever possible to use. Elongation in outer fibers must not exceed the material's limits. One operation contour bending of preformed sections can be done up to about 20% stretch for 2024-O aluminum; for rubber press forming, it is limited to about 12%. Tensile bending strains can be reduced by designing the section so that the neutral axis is close to the outer surface of curvature.

If long, thin-gage parts are required, they should be made of hard or non-heat-treatable material, to resist distortion in heat treatment. Cross sections should have open angles for access of forming tools. For minimum bend radii, see Tables 2-6 to 2-10.

Class 2-B. Nonsymmetrical Sections. Elongation and radii considerations are the same as for class 2-A. These sections commonly distort when removed from the forming mechanism, because the plane of bending does not coincide with the part's principal minor axis. This difficulty may be partly or largely overcome by the product designer's cooperation in changing the cross section. Sometimes, a part with nonsymmetrical cross section can be made symmetrical during forming, either (1) by combining two parts (identical, or right- and left-handed) into a single blank to be trimmed apart after forming (Fig. 2-38A and F), or (2) by adding excess metal and trimming it off after forming (Fig. 2-38C).

Class 2-C. Nonuniform Sections. Elongation and radii considerations same as for

2-38

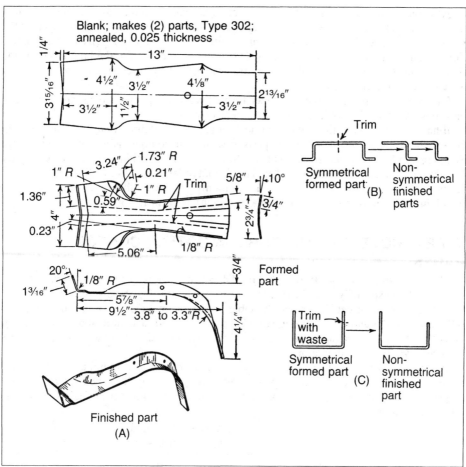

Fig. 2-38. Nonsymmetrical parts produced from symmetrical formed parts.[14]

class 2-A. The cross section should be symmetrical wherever possible. For satisfactory stretch forming, change in cross-sectional shape must be gradual.

TYPE 3, CONTOURED FLANGED PARTS OF NONUNIFORM SECTION

Class 3-A. Stretch-flanged Sections. See Table 2-2 on permissible forming limits; also, accompanying text on dimensional calculations for, and permissible strain in, both stretch and shrink flanges. For bend radii, see Tables 2-6 to 2-10. The designer should remember that stretch increases with increasing flange width, increasing flange-edge movement (increasing bend angle), and decreasing contour radius. Any cutouts made to bring the stretch within rubber-forming limits must extend over the full length of the stretch area.

Class 3-B. Shrink-flanged Sections. Flange widths, bend radii, and forming limits are same as for class 3-A. For rubber forming, high shrink (compressive) strains must be relieved by providing short flange segments with free edges. Hard stock and thin gages do not shrink readily.

Class 3-C. Combined Stretch- and Shrink-flanged Sections. Flange widths, bend radii, and permissible strains are generally the same as for classes 3-A and 3-B, except that strains in adjacent stretch and shrink portions are complex and cannot yet be properly evaluated.

A flange having adjacent stretch and shrink areas can be formed to a severe contour if the contour radii and/or the segment angles are small (under 4 in. (102 mm) and 60°, respectively) since the stretch-and-shrink strains may largely compensate each other. Total strains consist of two components: (1) strains resulting from the contouring of the flanges, and (2) strains created by curving the web. The latter strains are caused by deviations from a flat web.

Shrink areas on reverse-contour rubber-formed parts are always critical, and thin gages or strong, hard stock should be avoided where possible.

It has been observed[9] that stretch portions of reverse-contoured flanges of deep-drawn boxes can be made very high in comparison with their contour radii, without cracking. Such cracking was encountered, however, in a severe-stretch flange adjacent to a shrink flange, which later developed wrinkles. The mutual relief of strains in adjacent shrink and severe-stretch flanges of formed boxes is commonly provided either by closing the box or by adding short portions of convex contour which are trimmed off after forming.

TYPE 4, DOUBLE-CURVATURE SMOOTHLY CONTOURED PARTS

Class 4-A Parts. Large Radius in One Direction. The majority of the parts covered by class 4 are especially suited for stretch forming, permitting the use of hard stock and thus avoiding heat-treat distortion. Suitable materials include 2024-T3 and -T4, and 6061-T4 and -T6 aluminum, and quarter-hard or annealed austenitic stainless steel, for parts formed in one operation; and as-quenched 5052-O and 2024-O aluminum for parts requiring two or more operations.

A practical limit for *longitudinal* stretching may be set at a minimum of twice the maximum transverse extension; broader parts are preferably stretched longitudinally. Approximate limits of elongation (percent in 42 in., 1,067 mm) for certain 0.064-gage aluminum alloys are:

	Sheared Edges, %	Polished Edges, %
2024-O	8.4	10.8
2024-T3, -T4	5.0	7.3
6061-T4, -T6	5.8	12.5
5052-O	10.5	11.6

Minimum bend radii should approximate 0.500 in. (12.7 mm) for double-action die forming, and 1 in. (25.4 mm) for stretch forming. Parts that taper longitudinally are suited for stretching, but greater flash will result because more metal is needed to make the blank symmetrical for forming.

Class 4-B Parts. Large Radii in Two Directions. Data on type of material, bend radii, and maximum stretch are same as for class 4-A. Parts deviating only slightly from flat sheet are difficult to form accurately because of high friction around the edge of the die; also because buckles tend to form along the center of the blank parallel to the direction of stretching. Proper die development can usually correct this buckling tendency.

Class 4-C Parts. Opposite Contours in Two Directions. Suitable materials and forming limits are same as for class 4-A. However, if the maximum contour is at the edge of the part, edge effect will reduce the formability.

The "saddleback" characteristic of this class has contours concave on opposite sides of the sheet. It can be stretch-formed over a block if the one contour is not reversed (curves throughout in one direction); the other contour may be with or without a number of reversals. The direction of stretching must be along the unreversed contour.

To prevent wrinkling, the saddle should be shallow, not exceeding a 0.09 maximum depth-chord ratio. If the part is corrugated (several reversals of contour) the depth-chord ratio of such reversals should not exceed 0.22.

Class 4-D Parts (A, B, C Parts with Small Radii). This class should be designed only for ductile materials. The sharp radius (generally formed in a second operation) should be so proportioned that shrink does not exceed 4% to 6%. If the sharp radius produces a stretch flange, maximum width can be one-twentieth of the profile radius of the part periphery, where the flange is turned through 180°.

Any sharp contour which does not coincide either with the direction of stretching or perpendicular to it requires a separate operation.

TYPE 5, DEEP RECESSED PARTS; TYPE 6, SHALLOW RECESSED PARTS

These two types are here included for classification purposes only (see Product Design for Drawing).

OTHER DESIGN CONSIDERATIONS FOR FORMED PARTS

Lightening Holes in Flanges. Flange design proper has been covered under Design to Increase Strength of Stampings. Lightening holes can be flanged in annealed aluminum alloys up to 0.125 (3.18 mm) thick, and in heat-treated 2024 up to 0.064 in. (1.62 mm) thick. Austenitic stainless steel up to 0.060 in. (1.52 mm) can be formed with external hole flanges, and up to 0.050 in. (1.27 mm) thick, with internal flanges. The quarter-hard stainless steels up to 0.04 in. (1.0 mm) thick can be externally flanged, and internally flanged up to 0.030 in. (0.76 mm) thick.

Rubber-sheared lightening-hole minimum diameters, in relation to aluminum alloy thicknesses, are shown in Table 2-11.

Sheet-metal Box Design. Box constructions shown in Fig. 2-39 represent successful designs in widespread applications.

Lock-seam corners (view A) are excellent for metal 0.018 in. (0.46 mm) or thinner. Manufacturing cost is moderate where tooling cost is justified by the estimated quantity. Lithographing can be done in the flat sheet before forming. Height of box sides is usually less than 3 in. (76 mm), because of press-stroke limitations.

The can-type lock-seam construction (view B) is economical to manufacture but is limited to lightweight stock and round corners. The bottom not being flush with the seamed edge, the construction is not suited for heavy contents but, with double seam, holds powdered products well. This design would require sealing compound for leakproofing.

When design calls for small boxes of fairly stiff metal, a blanked dovetail-fastening construction (view G) will permit the shape to be retained without seaming or welding.

Where simple leakproof design is wanted, a box with ends folded on the outside (view H) may be considered. Ends folded on the inside (view J) make for improved appearance but are not leakproof.

A construction using split corners with tabs (view M) is low in cost, and in widespread use where appearance is not important.

The construction shown in view S presents very attractive appearance, with all corners round and no joint showing on the side where the bottom is connected.

Various welded-box constructions are shown in views U, V, W, and X. Unlike folded designs, spot welding requires spraying or retouching after welding to prevent rusting.

Control of Surplus Stock in Formed Parts. In many parts formed from plain blanks, wrinkling or folding occurs. Sometimes this is not objectionable; more often it is. A common provision is to cut away the surplus stock (Fig. 2-40A). Another good practice, where feasible, is to form the surplus stock into an intentional and controlled shape, either functionally or for styling (Fig. 2-40B).

TABLE 2-11
Lightening Holes for 55° Flange

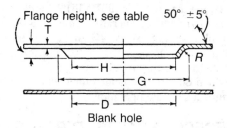

D, in.	H, in.	G, in.	T, in.	R, in.
		0.125-in. Flange Height		
0.445	0.812	1.223	0.020–0.040	0.188
0.400		1.212	0.051–0.072	0.250
0.570	0.938	1.348	0.020–0.040	0.188
0.525		1.337	0.051–0.072	0.250
0.695	1.062	1.473	0.020–0.040	0.188
0.650		1.462	0.051–0.072	0.250
0.820	1.188	1.598	0.020–0.040	0.188
0.775		1.587	0.051–0.072	0.250
		0.156-in. Flange Height		
0.900	1.312	1.800	0.020–0.040	0.188
0.852		1.791	0.051–0.072	0.250
1.150	1.562	2.050	0.020–0.040	0.188
1.082		2.041	0.051–0.072	0.250
1.276	1.688	2.175	0.020–0.040	0.188
1.208		2.166	0.051–0.072	0.250
1.400	1.812	2.300	0.020–0.040	0.188
1.332		2.294	0.051–0.072	0.250
		0.188-in. Flange Height		
1.606	2.062	2.625	0.020–0.040	0.188
1.543		2.617	0.051–0.072	0.250
1.490		2.611	0.081–0.102	0.312
1.856		2.875	0.020–0.040	0.188
1.793	2.312	2.867	0.051–0.072	0.250
1.740		2.861	0.081–0.102	0.312
1.982		3.000	0.020–0.040	0.188
1.919	2.438	2.992	0.051–0.072	0.250
1.866		2.987	0.081–0.102	0.312
2.106		3.125	0.020–0.040	0.188
2.043	2.562	3.117	0.051–0.072	0.250
1.990		3.111	0.081–0.102	0.312
2.356		3.375	0.020–0.040	0.188
2.293	2.812	3.367	0.051–0.072	0.250
2.240		3.361	0.081–0.102	0.312
2.606		3.625	0.020–0.040	0.188
2.543	3.062	3.617	0.051–0.072	0.250
2.490		3.611	0.081–0.102	0.312
2.856		3.875	0.020–0.040	0.188

(Continued)

TABLE 2-11 *(Continued)*
Lightening Holes for 55° Flange

D, in.	H, in.	G, in.	T, in.	R, in.
0.188-in. Flange Height				
2.793	3.312	3.867	0.051–0.072	0.250
2.740		3.861	0.081–0.102	0.312
3.106		4.125	0.020–0.040	0.188
3.043	3.562	4.117	0.051–0.072	0.250
2.990		4.111	0.081–0.102	0.312
3.356		4.375	0.020–0.040	0.188
3.293	3.812	4.367	0.051–0.072	0.250
3.240		4.361	0.081–0.102	0.312
3.606		4.625	0.020–0.040	0.188
3.542	4.062	4.617	0.051–0.072	0.250
3.490		4.611	0.081–0.102	0.312
3.856		4.875	0.020–0.040	0.188
3.793	4.312	4.867	0.051–0.072	0.250
3.740		4.861	0.081–0.102	0.312

These lightening holes may be formed in 5052-H32, 6061-T, 2024-T, 2014-T, R-301-T, and 7075-T aluminum alloy and AMC5250 and FS-Ia magnesium alloy or any softer condition of these alloys.

Figure 2-41 shows several formed-parts designs, and appropriate stock control. At *A*, surplus stock would gather at the smaller of the two diameters, under roll or die forming. By reducing the smaller diameter, in several operations, a slight reduction can be made. Die forming a shell edge, as at *B*, would cause the flange to wrinkle, unless the flange is rolled in from the outside in three stages: 30°, 60°, and 90°. Where design and function permit, curl forming of the shell edge (as at *C*) is a practical method.

Views *D*, *E*, and *F* suggest good product designs wherein forming near an edge, from which the metal can draw freely, tends to cause less tearing of stock than may result from forming in the interior surface.

Covering of Raw Edges. The design of a part to be formed can often, without requiring extra operations, utilize functional hemming, curling, or beading to eliminate raw edges (Fig. 2-42).

Container Top and Bottom Attachment. Figure 2-43 illustrates some common methods of attaching tops and bottoms on sheet-metal containers. A majority of the applications involve the use of interlocking seams; in some cases, snap or press fits are quite satisfactory. Although the designs shown are generally produced in special automatic machines, a number of the designs can be stamped in adapted dies of certain types described in the sections Forming Dies and Assembling Dies.

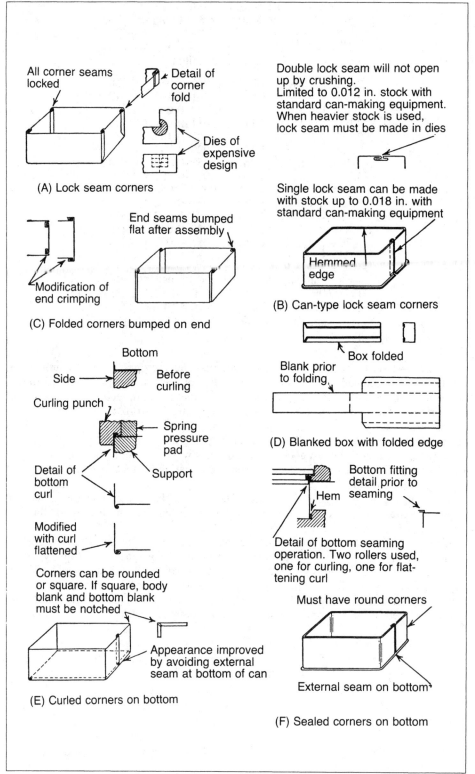

All corner seams locked

Detail of corner fold

Dies of expensive design

(A) Lock seam corners

Modification of end crimping

End seams bumped flat after assembly

(C) Folded corners bumped on end

Bottom

Side — Before curling

Curling punch

Spring pressure pad

Detail of bottom curl

Support

Modified with curl flattened

Corners can be rounded or square. If square, body blank and bottom blank must be notched

Appearance improved by avoiding external seam at bottom of can

(E) Curled corners on bottom

Double lock seam will not open up by crushing.
Limited to 0.012 in. stock with standard can-making equipment. When heavier stock is used, lock seam must be made in dies

Single lock seam can be made with stock up to 0.018 in. with standard can-making equipment

Hemmed edge

(B) Can-type lock seam corners

Box folded

Blank prior to folding

(D) Blanked box with folded edge

Bottom fitting detail prior to seaming

Hem

Detail of bottom seaming operation. Two rollers used, one for curling, one for flattening curl

Must have round corners

External seam on bottom

(F) Sealed corners on bottom

Fig. 2-39. Design of sheet-metal boxes.[16]

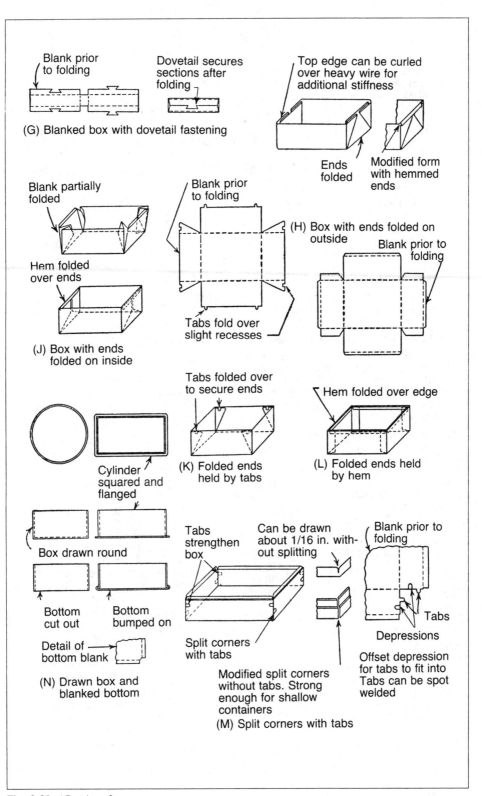

Blank prior to folding

Dovetail secures sections after folding

(G) Blanked box with dovetail fastening

Top edge can be curled over heavy wire for additional stiffness

Ends folded

Modified form with hemmed ends

Blank partially folded

Hem folded over ends

(J) Box with ends folded on inside

Blank prior to folding

Tabs fold over slight recesses

(H) Box with ends folded on outside

Blank prior to folding

Cylinder squared and flanged

Tabs folded over to secure ends

(K) Folded ends held by tabs

Hem folded over edge

(L) Folded ends held by hem

Box drawn round

Bottom cut out

Bottom bumped on

Detail of bottom blank

(N) Drawn box and blanked bottom

Tabs strengthen box

Can be drawn about 1/16 in. without splitting

Blank prior to folding

Split corners with tabs

Modified split corners without tabs. Strong enough for shallow containers

(M) Split corners with tabs

Tabs

Depressions

Offset depression for tabs to fit into Tabs can be spot welded

Fig. 2-39. (*Continued*)

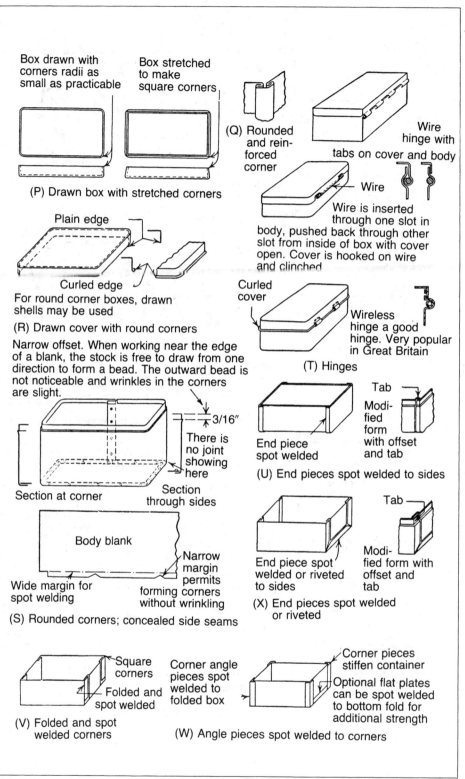

Box drawn with corners radii as small as practicable

Box stretched to make square corners

(P) Drawn box with stretched corners

Plain edge

Curled edge

For round corner boxes, drawn shells may be used

(R) Drawn cover with round corners

Narrow offset. When working near the edge of a blank, the stock is free to draw from one direction to form a bead. The outward bead is not noticeable and wrinkles in the corners are slight.

3/16″

There is no joint showing here

Section at corner

Section through sides

Body blank

Wide margin for spot welding

Narrow margin permits forming corners without wrinkling

(S) Rounded corners; concealed side seams

(Q) Rounded and reinforced corner

Wire hinge with tabs on cover and body

Wire

Wire is inserted through one slot in body, pushed back through other slot from inside of box with cover open. Cover is hooked on wire and clinched

Curled cover

Wireless hinge a good hinge. Very popular in Great Britain

(T) Hinges

Tab

Modified form with offset and tab

End piece spot welded

(U) End pieces spot welded to sides

Tab

End piece spot welded or riveted to sides

Modified form with offset and tab

(X) End pieces spot welded or riveted

Square corners

Folded and spot welded

(V) Folded and spot welded corners

Corner angle pieces spot welded to folded box

Corner pieces stiffen container

Optional flat plates can be spot welded to bottom fold for additional strength

(W) Angle pieces spot welded to corners

Fig. 2-39. (*Continued*)

2-46

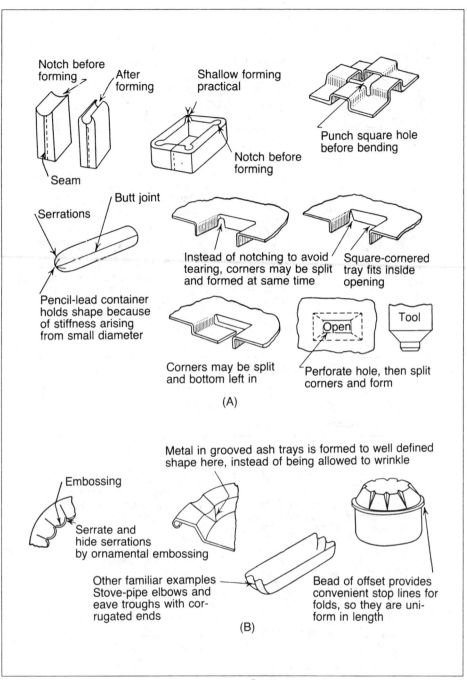

Notch before forming

After forming

Seam

Serrations

Butt joint

Pencil-lead container holds shape because of stiffness arising from small diameter

Shallow forming practical

Notch before forming

Instead of notching to avoid tearing, corners may be split and formed at same time

Corners may be split and bottom left in

Punch square hole before bending

Square-cornered tray fits inside opening

Open

Tool

Perforate hole, then split corners and form

(A)

Embossing

Serrate and hide serrations by ornamental embossing

Metal in grooved ash trays is formed to well defined shape here, instead of being allowed to wrinkle

Other familiar examples — Stove-pipe elbows and eave troughs with cor- rugated ends

Bead of offset provides convenient stop lines for folds, so they are uni- form in length

(B)

Fig. 2-40. Forming design for surplus stock control.[7] *(Mills)*

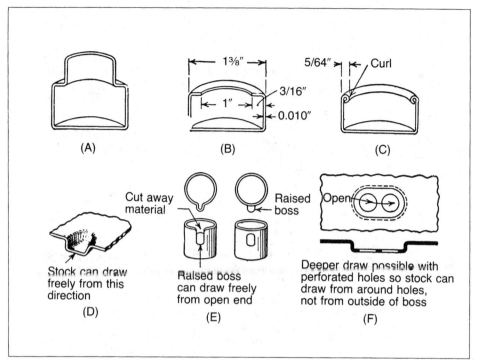

Fig. 2-41. Controlling surplus stock in roll and die forming.[7] *(Mills)*

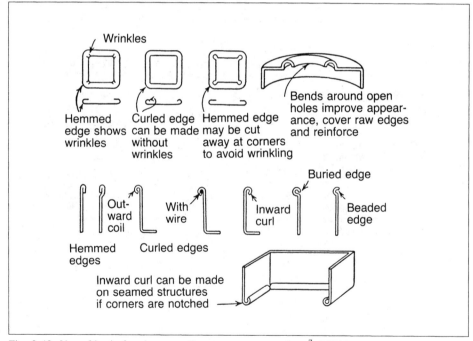

Fig. 2-42. Use of basic forming operations to cover raw edges.[7] *(Mills)*

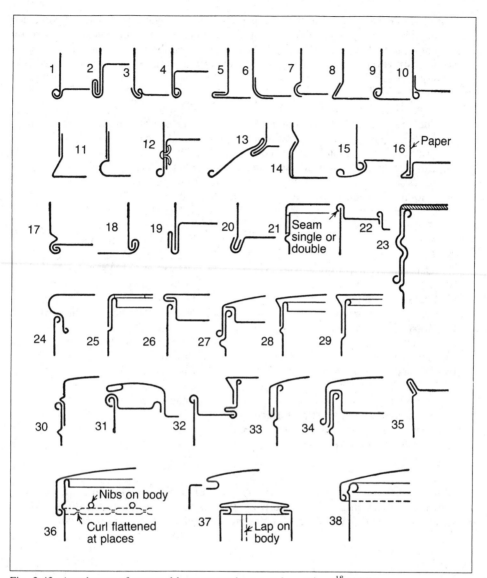

Fig. 2-43. Attachment of tops and bottoms to sheet-metal container.[18] (*Mills*)

Container-lid Nibs and Catches. Figure 2-44 shows a variety of detent designs for removable lids on sheet-metal boxes, cans, and cabinets. Catches are commonly made by forming nibs or small projections in both the container and the lid. A listing of the various types of construction for attachments is as follows (item numbers correspond to those in Fig. 2-43):

1. Single-seam bottom (rapidly produced; no inside support required)
2. Double-seam bottom (for liquid or heavy contents; rapidly produced)
3. Bumped-on bottom (inside support required)
4. Curled-on bottom
5. Flanged bottom
6. Floating bottom
7. Snapped-on bottom
8. Clinched bottom
9. Outside-curled bottom
10. Curled floating bottom
11. Pressed-in bottom
12. Removable bottom
13. Tobacco-can bottom
14. Rolled clinch bottom
15. Snap-fit bottom
16. Seamed-metal bottom
17. Flour-sifter bottom
18. Flat-seam bottom (for teapots, wash boilers, etc; must be formed over a solid plug; slower production than for single or double seam)
19. False-seam bottom
20. Squeezed and soldered seam for rectangular cans
21. Slip cover
22. Friction plug
23. Screw cover
24. Slip plug
25. Offset slip cover
26. Two-piece plug
27. Slip cover and friction plug
28. Beaded slip cover
29. Offset beaded slip cover
30. Inside slip cover
31. Friction plug
32. Necked with beaded slip cover
33. Hemmed slip cover
34. Two-piece inside-fit cover
35. Rolled seam
36. Lock cover
37. Tea-can cover
38. Snap-on cover

Product Design for Drawing. Under Analysis of Formed-parts Design, in a preceding section, a classification for all formed parts was described and illustrated (Fig. 2-37). It was there stated that classes 5 and 6 more properly relate to drawing operations.

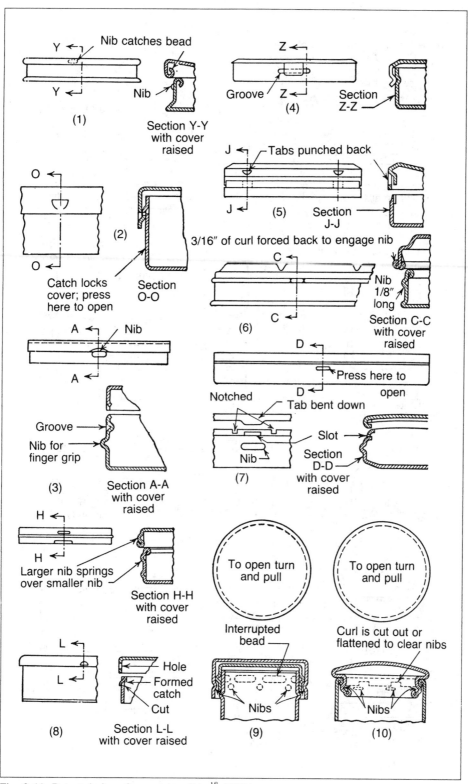

Fig. 2-44. Detent designs for container lids.[18] (*Mills*)

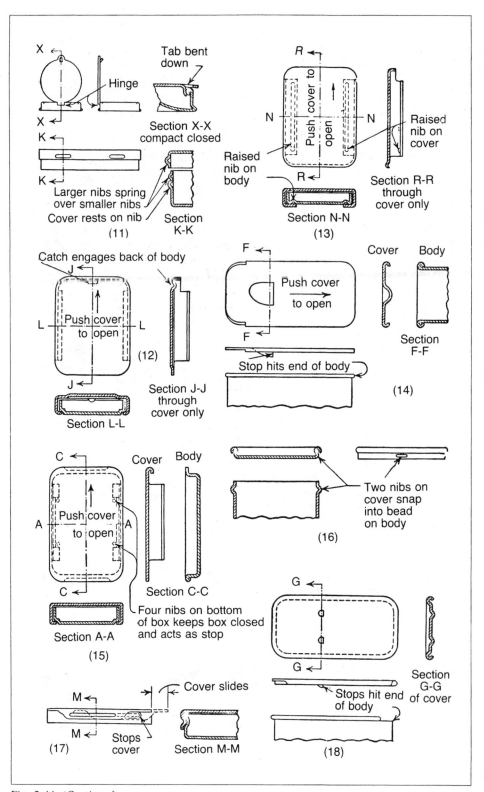

Fig. 2-44. (*Continued*)

TYPE 5, DEEP RECESSED PARTS

Type 5-A. Cylindrical Cups. These are characterized by having vertical walls, with or without cutouts in walls of flanges or offsets at the bottom. For deep drawing with double-action dies, the material strength should not exceed 30,000 psi (207 MPa). For higher-strength materials of the same gage, the maximum single depth of draw must be reduced. The diameter and height of a part should be such that the ratio of punch diameter to blank diameter will be at least 0.5 for deep-draw 1010 steel and 2024-O or 5052-O aluminum alloys, and 0.3 for 6061-T4 and -T6 alloy. Draw radii should be within the range of $5T$ to $10T$, with $8T$ optimum. The maximum depth of draw is equal to the sum of the part depth and the required flange widths, if any.

Type 5-B. Oval Cups. With or without flanges, cutouts, or curved or sloping bottoms. The same general design provisions prevail as for class 5-A.

Type 5-C. Rectangular Box Types. Includes parts with vertical walls and small corner radii; all curvature convex; regular, irregular, or sloping bottom. Draw depth is limited to seven times any corner radius from 0.500 to 1 in. (12.70 to 25.4 mm), or 12 times corner radii of 0.250 in. (6.35 mm) or less. Both the draw and the vertical corner radii must be $5T$ minimum; bottom radii can range from $2\frac{1}{2}T$ to $8T$. For 3003-O, 2024-O, 5052-O, and deep-drawing 1010 steel, the box height-width ratio is about 0.6 for corner radii up to 0.375 in. (9.52 mm), and 0.75 for radii exceeding 0.500 in. (12.7 mm). Included corner angles of less than 60° reduce the feasible depth of draw.

Type 5-D. Sloped-wall Parts. Have convex curvature, straight or curved sides, and flat or irregular bottoms. Draw and punch radii requirements similar to those for classes 5-A and 5-B. The punch-blank area ratio is optimum at about 0.33. 3003-O, 2024-O, 5052-O, and other highly ductile materials draw best in this class.

Type 5-E. Open Parts. Characterized by the absence of a continuous wall at one or several locations, and convex curvature. For single draws the draw radius should be as large as $10T$ to $12T$, especially where there is an abrupt change in the parting plane. The part should be designed with fairly uniform depth, or else gradual change of depth, to avoid wrinkling. The punch area-blank area ratio should be a minimum of 0.25 for a single draw.

Type 5-F. Reentrant Contour Parts. Whether walls are partial or continuous, the radius of the reentrant area should be not less than the depth of the part, to avoid excessive tool wear and breakage. The punch area-blank area ratio must be not less than 0.30 for 2024-O, 5052-O, and other ductile materials. Draw radius on the reentrant contour should be 0.50 in. (12.7 mm), or greater. Where design permits, sloping walls permit easier forming.

Type 5-G. Saddleback-bottom Parts. Generally, saddlebacks should be avoided because of marked wrinkling tendencies in the saddle center. Such wrinkling is minimized if moderate contours can be held. Draw radii, punch area-blank area ratios, and maximum contours are same as for class 5-E. Saddleback design is best formed in low-yield-strength materials such as 3003-O and 2024-O.

Type 5-H. Parts with Undercut Walls. Characterized by having a recess that is larger at the bottom than at the parting plane. The part design should permit the undercut to be produced in one operation by tilting the part to permit the access of conventional tooling. Punch area-blank area ratio should be at least 0.30. Draw depth and radius, contour, and materials are the same as for classes 5-A and 5-B.

TYPE 6, SHALLOW RECESSED PARTS

These parts are characterized by having one or more recesses with depth shallow in relation to the major dimensions of the recess. The depth-width ratio must be less than 0.15. Careful parts design, with optimum distribution of recesses, is required to avoid the tendency to buckle in the flat area between recesses. To overcome buckling, beads or other shallow recesses may have to be added to those required by the design. Parts of this design can be formed from as-quenched 2024; in some cases, 6061-O can be satisfactorily formed within eight hours after quenching.

Class 6-A. Dish-type Parts. Typically, have only one recess; shape may be open or closed, and regular or irregular. Since most of these parts are formed by pure stretch, all radii should be generous for greatest distribution of strain. Walls of the recess should slope as much as possible.

When the cross section is a circular arc, the following are good values of permissible stretch for the indicated recess depth-width ratios:

Ratio	0.10	0.15	0.20	0.25	0.30
Stretch, %	3	6	10	15	22

Class 6-B. Beaded and Embossed Parts. This class includes a wide variety of shapes, flat or with single or double curvature, having the surface broken by several regular or irregular-shaped recesses. It is essential, when forming 2024-O or 5052-O, that required stretch limits do not exceed 10% for rubber forming, or 15% for die forming, unless the recess can be located near a free edge where the material can feed into the recess. Hard stock, such as 2024-T3 and -T4, can be rubber-formed where most of the part area is covered by closely spaced parallel beads.

Recess mold line radii should be about 1 in. (25.4 mm) minimum for properly distributing strains over a large area. Where the part periphery has a surrounding wall, outside corner radius should be at least half the blank radius. Whenever forming is done by pure stretch, the slope at a recess corner should be less than the slope along the sides.

OTHER DESIGN CONSIDERATIONS FOR DRAWN PARTS

Deep-drawn Thin-walled Cylindrical Cups. Drawability in thin-walled cups (blank diameter of $20T$ to $30T$ or larger) is influenced chiefly by the punch and the die radii; the strain properties depend rather little upon the metal thickness. Figure 2-45, valid for brasses and many other ductile materials, shows that an average initial-draw reduction of somewhat over 50%, from blank diameter to drawn-shell diameter, can be realized if the die radius is between $5T$ and $10T$, and the punch radius is at least $5T$.

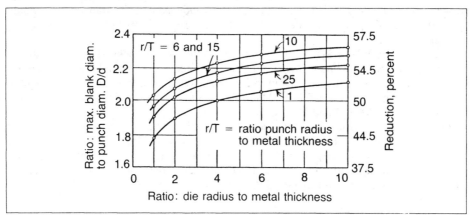

Fig. 2-45. Drawability factors for 63/37 brass sheet, 0.040 in. (1.02 mm) thick, 1.2 in. (30.5 mm) punch diameter.

Deep-drawn Thick-walled Cylindrical Cups. The deep drawing of very thick blanks involves interrelated complex limiting factors. The limitations are much more severe for steel than for such ductile metals and alloys as brass. The ironing phase in cup drawing chiefly determines the maximum feasible reduction. A maximum reduction of 65% in cross-sectional area is theoretically possible. Experimentally, wall reductions of 60% of blank thickness, or more, are obtainable.

Deep-drawn Boxes. Where boxes have no flanges, the depth and corner radii are the chief limiting factors. Tests from which Table 2-12 was compiled showed that the corner radius has little effect on maximum depth for the more ductile alloys 1100-O, 3003-O, and 5052-O. For other alloys such as 2024-O and 6061-T6, permissible depth would be reduced 50% or more if the corner radius is reduced to about 10% of box width.

The limits in cupping box-shaped parts of austenitic stainless steels are indicated in Figs. 2-46 and 2-47. From Fig. 2-47, it is seen that, for a R/d range from 0.03d to 0.5d (cylindrical cup), the maximum box height for stainless (also carbon) steels is greater than 0.8D. Figure 2-46 shows that, in the drawing of stainless steel, the depth and/or the corner radius are also limited by the tendency to stress cracking.

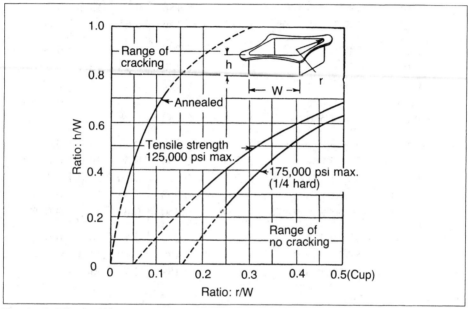

Fig. 2-46. Effects of depth and corner radius on srress-cracking tendency for austenitic stainless steels of various tempers.[14]

The above considerations do not apply to boxes having reentrant corners, since in such designs the metal is stretched in the corners, so only metals having high elongation are suitable.

Drawn Shallow Recessed Parts. These parts (class 6), being produced almost entirely by stretching, the average forming strain (stretch) e involved is given by the equation

$$e = \frac{L_1 - L_0}{L_0} \tag{10}$$

where L_1 = recessed contour length
L_0 = its developed length in the blank

For the most common, circular type of contour recess, average stretch is a function of the ratio between the depth h and the length L_0 of the recess (Fig. 2-48).

TABLE 2-12
Dimensions of Trimmed Rectangular
Aluminum Alloy Boxes[9]
Drawn From Circular Blanks in a Single Operation

Thickness, in. T	Size, in. $W \times L$	Trimmed Depth, in. h	Corner Radius, in. r	Ratios		
				h/w	h/r	r/w
0.032	2 × 3	1.750	0.156	0.88	11.2	0.078
0.040	4.750 × 6	2	0.250	0.42	8.0	0.053
0.051	5.750 × 9	2.125	0.156	0.37	13.7	0.027
0.051	5.875 × 6.500	2.500	0.188	0.51	13.3	0.037
0.051	3.688 × 5	2.375	0.156	0.65	15.2	0.042
0.051	5.875 × 7.500	4	0.250	0.68	16.0	0.043
0.051	3.688 × 5	2.875	0.188	0.78	15.3	0.051
0.051	4.625 × 4.875	3	0.250	0.65	12.0	0.054
0.051	7 × 9.875	2.250	0.250	0.36	10.0	0.036
0.064	2 × 4.500	1.750	0.250	0.88	7.0	0.125
0.064	3 × 3	2	0.250	0.67	8.0	0.083
0.064	5 × 5	2.500	0.250	0.50	10.0	0.050
0.064	5 × 11.312	2.500	0.500	0.50	5.0	0.100
0.064	5.500 × 6	3	0.250	0.55	12.0	0.045
0.064	5.500 × 9.500	3.500	0.250	0.64	14.0	0.045
0.064	4 × 7.125	3.625	0.250	0.91	14.5	0.063
0.064	5.750 × 9	2.125	0.188	0.39	11.3	0.034
0.064	5.250 × 5.625	4	0.188	0.76	21.0	0.037

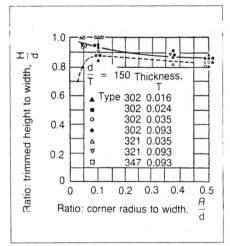

Fig. 2-47. Effects of corner radius and sheet thickness on drawability of stainless-steel boxes.[13] (*Lockheed Aircraft Corp*)

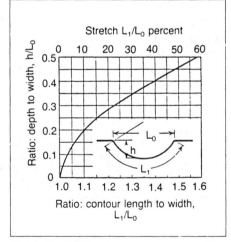

Fig. 2-48. Relation between average stretch and dimensions of a recess of circular contour.

PIECED VS. UNIT CONSTRUCTION

Figure 2-49 illustrates a number of good product-design principles involving resort to pieced construction.

Restriction of Metal Flow. In a double-cup part of integral construction (view 1), the metal must draw simultaneously over two forming punches and there is strong likelihood of stretching and tearing; pieced construction is preferable in such a case. In a circular shell (view 2), it is well to design the part so that the stock can draw freely from all directions and, preferably, to have the draw near the periphery. Where design permits a center hole (view 3), difficult draws can be made more easily; however, possible tearing limits the amount of draw from inside.

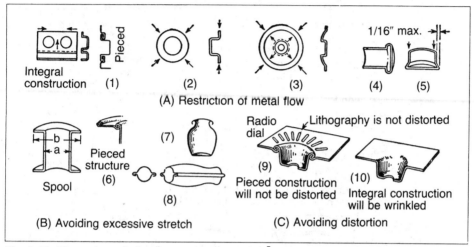

Fig. 2-49. Comparison of pieced vs. unit construction.[7] (*Mills*)

Avoidance of Excessive Stretch. The spool and the jar designs shown in views 6 and 7 are not practical because of risk of tearing. Stretching of the spool from diameter *a* to flange diameter *b* risks rupture in thin stock; pieced structure avoids this. The jar in view seven can be formed in thin metal only to a limited extent by bulging with rubber or hydraulic dies; but pieced construction as in view 8 is practical for die forming.

DESIGN FOR STRONG, SIMPLE TOOLING

A little thought and discretion on the part of the product designer frequently can greatly facilitate optimum tool design and operation. In Fig. 2-50A, views 1 and 2 emphasize the need to avoid designs that require tools which are either too fragile or else must be ground to a contour for sharpening.

A product design that does not permit interior access for a forming tool is impractical. View 5 shows a condition where a fiber strip (or other soft material) is to be assembled in a drawn shell, where it is to be held in by an inwardly formed bead. Here, product design must permit entry of a hard tool in order to form the sharp groove.

Where a sleeve or tube is designed to be filled with soft material, as in view 6, interior tool access is prevented. The top can be flanged over but, without an interior hard tool to bump against, there will be wrinkling.

Too small an opening, as in view 7, will also prevent insertion of a forming tool, whereas the product design in view 8 does provide interior tool access.

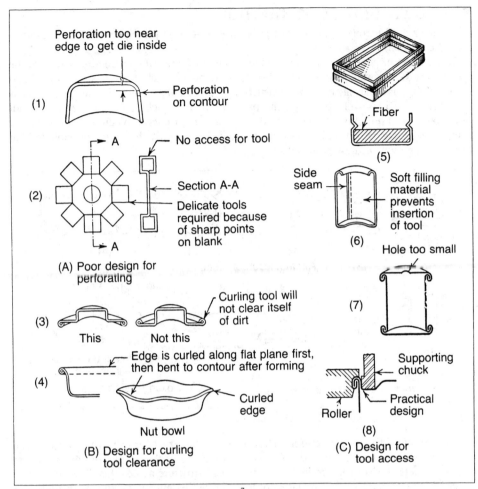

Fig. 2-50. Design consideraions for tool access.[7] *(Mills)*

DESIGN OF METAL PARTS FOR PORCELAIN ENAMELING

Gage of Metal. Use of the proper gage of metal is all-important for avoiding such processing defects as chipping, hairlining, warpage, and sagging. It is much better to use somewhat too heavy a stock than one that is too light. Table 2-13 is a safe guide for parts of moderate size, intended to be enameled in white or light colors, and required to have only moderate rigidity and flatness.

The required flatness for large architectural porcelain enamel panels may call for heavier gages than Table 2-13 indicates, going perhaps to 18 or 16 gage (1.2 or 1.5 mm). For formed metal plumbing ware, necessary rigidity calls for stock not lighter than 14 gage (1.9 mm).

TABLE 2-13
Metal Gages for Average Porcelain Enameling Requirements[19]

Gage, in. (mm)	Width, in. (mm)	Total Area, ft²(m²)
24 (0.024, 0.61)	6 (152)	0.5 (0.05)
24	12 (305)	3.0* (0.3)
24	18 (457)	5.0* (0.5)
22 (0.030, 0.76)	6 (152)	1.0 (0.09)
22	12 (305)	1.5 (0.14)
22	18 (457)	6.0* (0.56)
22	24 (610)	8.0* (0.74)
20 (0.036, 0.91)	6 (152)	1.5 (0.14)
20	12 (305)	5.0 (0.5)
20	18 (457)	8.0* (0.74)
20	24 (610)	10.0–15.0* (0.93 – 1.4)

* Should be embossed, flanged, or otherwise suitably reinforced.

Materials. Ordinary stamping grades are not suited as basis metals, because of impurities that may cause blistering or chipping of the coat. Commonly used materials include:

1. "Enameling iron" steel sheets
2. Good grades of drawing steel of 20% carbon or less
3. Non-heat-treatable aluminum alloy sheet
4. Castings: gray iron, malleable iron, aluminum, etc.

Specific design considerations are indicated in Fig. 2-51.

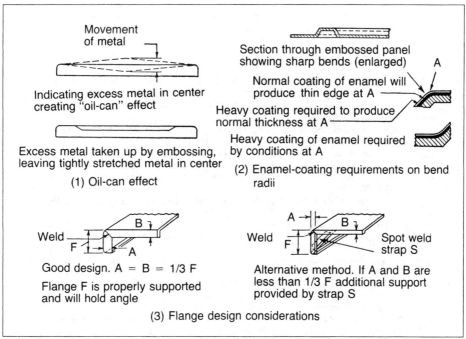

Fig. 2-51. Product-design suggestions for parts to be porcelain-enameled.[20]

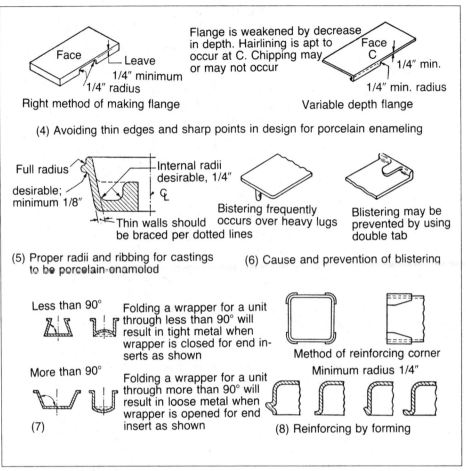

Fig. 2-51. *Continued.*

References

1. A.F. Murray, "Redesign to Simplify Tooling," *Machine Design* (September 1951).
2. D.A. Rodgers, "Basic Designs for Small Part Castings," *Product Eng.* (January 1952).
3. C.W. Hinman, *Pressworking of Metals*, 2d. ed. (New York: McGraw-Hill Book Company, 1950).
4. General Motors Drafting Standards.
5. Bernhard Rogge, "Beading Techniques for Strengthening Sheet Metal Parts," *Product Eng.* (July 1955).
6. "Forming Data for Part Details," *Am. Machinist* (Dec. 16, 1957).
7. W.C. Mills, "Practical Design of Thin-metal Stampings," *Am. Machinist* (Oct. 23, 1947).
8. W.C. Mills, "Sheet-metal Stamping—Principles of Design," *Product Eng.* (November 1945).
9. "Classification and Analysis of the Forming of Various Parts," issued by National Defense Research Comm. of O.S.R.D., 1943.
10. *Steel Products Manual—Cold Rolled Strip*, (New York: American Iron and Steel Institute, 1960).
11. L.M. Neilsen, "Shop-run Tolerances," *Product Eng.* (May 1948).
12. *Facts about Stampings*, Pressed Metal Institute.
13. G. Sachs, *Principles and Methods of Sheet-metal Fabrication*, (New York: Reinhold Publishing Corporation, 1951).
14. *Forming of Austenitic Chromium-nickel Stainless Steels*, The International Nickel Co., Inc., 1948.
15. John L. Everhart, "Designing Metal Stampings," *Mater. Design Eng.* (September 1958).
16. W.C. Mills, "Designs for Sheet Metal Boxes," *Product Eng.* (February- April 1947).
17. R.B. Schulze, "How to Rubber-form Light Metals," *Am. Machinist* (Mar. 10, 1949).
18. W.C. Mills, "Design Work Sheets," Eighth Series, *Product Eng.*
19. *Design and Fabrications of Metal Parts for Porcelain Enameling*, Porcelain Enamel Institute, 1944.
20. Society of Manufacturing Engineers, *Tool Engineers Handbook*, 2d. ed. (New York: McGraw-Hill Book Company, 1959).

SECTION 3

DIE ENGINEERING—PLANNING AND DESIGN

FACTORS IN PRESSWORKING STAMPING MATERIALS

Early in the product design process, the decision must be reached whether the contemplated component, as-designed or with permissible redesign, will be stamped or produced by some other process. The decision calls for the concerted attention of the design, materials, methods, tooling, manufacturing, and any other functions that have a responsibility for producing the product. Certain practical guiding criteria follow.

Product design factors (see Sections 2 and 11):

1. Shapes are limited to those which may be produced by the cutting, bending, forming, compressing, compacting, or drawing of materials.

2. With the exception of high-velocity explosive forming operations, which are limited only by the types and sizes of available materials, maximum sizes are limited chiefly by press capacity. There are few limitations upon minimum size; sections as thin as 0.003 in. (0.08 mm) are possible, with parts so small that 10,000 may be held in one hand.

3. Tolerances are very good. For small stampings made with progressive or compound dies, plus or minus 0.002 in. (0.05 mm) is common, and closer limits are possible on small and thin parts. Precision fineblanked parts typically are stamped to much closer tolerances.

4. The weight factor is highly advantageous. Parts formed from sheet metal and stampable plastics have favorable weight-to-strength ratios.

5. Surface smoothness is excellent, since surface condition usually is not affected by the forming operation. Pre-painted steels are in common use as stamping materials. The surface finish of sheet steels can be optimized by special texturizing of the finish mill rolls for maximum paint luster on the finished stamping.

6. A wide choice of materials is available, including any in sheet form and not so brittle as to break.

7. Design changes are usually costly, if required after the original tooling has been completed. Careful planning for ease of manufacturability early in the design stage is essential.

Production factors:

1. Except for temporary or low-production tooling, the lead time required for tooling is long compared with some other production methods. Die design, tryout, and development may take months.

2. Output is very high; over 600 large automotive body stampings per hour are commonly produced on automated transfer presses. Small stampings produced on high speed equipment often have production rates well in excess of 10,000 pieces per hour.

Economic factors:

1. Stamping-materials costs are comparatively low. A favorable cost factor is the minimum scrap loss achieved through careful selection of stock and skilled strip layout.

2. Tool and die costs are high, often higher than tooling for comparable parts that are to be die-cast. Costs are most favorable where large production is planned.

3. Direct labor costs depend upon the part size and shape and extent of automation. Usually these costs are characteristically very low.

4. Presses, except for the small manual punch presses, are typically more costly than standard machining equipment such as lathes and grinders, and require a higher machine-hour rate.

5. Finishing costs are low. Often no other finishing is required than normal painting or plating.

6. Inventory costs can be quite low. Quick die change equipment and techniques permit a different part to be produced with a change over time of under ten minutes.

BASIC PROCEDURE FOR PRESSWORKING PROCESS PLANNING

Once a decision has been reached to produce a part by stamping, the process planner may have three major areas of responsibility:

1. Planning the sequence of operations, the specifying of the metalworking equipment, and the gaging necessary to produce high-quality parts economically at the specified production rate.

2. Coordination of the allied processes such as heat treatment, metal finishing, and plating.

3. Integrating the required material handling and operator movement paths.

The second and third responsibilities are often performed by experts in plant facilities and layout, and are not considered further here.

The following steps illustrate a typical procedure for the planning of a pressed-metal manufacturing process.

1. Analyze the Part Print. To aid this step it may be desirable to have enlarged layouts, additional views, experimental samples, models, and limit layouts. It will be helpful to chart assigned responsibility for carrying out the specifications.

a. What Is Wanted? The product designer must establish explicit detailed specification for size, shape, material type, material condition, and allied processes. The process planner must be left in no doubt as to all the specifications and how they relate to each other.

b. List Manufacturing Operations and Allied Processes. A typical listing would be:

1. Pierce hole, 0.501 in. (12.72 mm), + 0.002, − 0.000.
2. Flange 90° ±2°.
3. Buff external surfaces.
4. Blank.

At this point, there is no need to list the operations in their proper sequence, or to combine the operations, but only to make a preliminary survey of basic operational requirements. Each listed item should be checked off on the part print drawing.

c. Determine Manufacturing Feasibility. Consider the possible die operations that could produce the part with the specified surface relationships. Computer-aided formability analysis should be employed at this point. Close co-operation with the product designer is required at this point. A hole close to a flange, an unreasonably small radius, a draw requiring annealing, an area of a drawn part having an excessive percentage of elongation, a blank that cannot be economically nested—these and other conditions can frequently be improved by the product designer without affecting the functional requirements (see Sections 2 and 11).

d. Write Recommendations to Product Engineering. Upon completion of part print analysis, recommendations should be written to the product designer. All accepted recommendations call for necessary engineering changes in the part print.

2. Determine the Most Economic Processing. For the stamping of the same part, there are usually several alternative production methods. The method selected should be the one which will result in the lowest overall cost of producing a finished part of the required quality. The cost includes material, tooling, direct and indirect labor, and overhead burden.

Determination of the most economical processing can be accomplished by comparing two or more feasible processes for producing the given pressed-metal part. The comparison of unit costs for each such process, for equal production quantities, will give a break-even point which is a guide to selecting the most economical tooling. Productive labor costs and burden rates are estimated from past performance, and by the use of standard time data.

A graphical presentation of the break-even point is useful where the spread between processes is small. Where the spread is small, but increased production requirements in the future are likely, it may be preferable to use the higher-cost process.

Unless new pressworking equipment can be amortized over a rather short period, or have future value as standard equipment, it may prove more economical to use existing available equipment even though production costs would be higher.

Likewise, simple dies may be favored over the high-production dies which seem indicated by the anticipated requirements, because of lack of the special skills required to design, construct, and maintain high-production dies. Also, the simpler single-operation dies may permit interchangeability of tooling for different parts which have several common shape and/or size specifications.

3. Plan the Sequence of Operations. Operations planning done only on the basis of past experience can prove costly if seemingly minor details are overlooked.

a. Determine Critical Specifications. Dimensional specifications which have comparatively close tolerances or the limitations placed upon the specifications for allied processes are known as "critical" specifications. Study of the comparative effects of specifications upon surface relationships, with the aid of a limit layout, will reveal the critical specifications from a manufacturing standpoint.

b. Select Critical Areas. Most critical specifications pertain to measurement of surface relations within specified close tolerances. Critical areas are those areas or surfaces from which the measurements for all specifications can be taken so as to determine the geometry of the part. Limit layouts serve also to determine the critical areas.

In ideal planning, critical areas should be established first, provided that they are "qualified" as surfaces of registry, in a location system, for subsequent operations and allied processes (see 3-2-1 System for Locating, and Tests for Qualified Areas in subsequent text).

c. Determine Critical Manufacturing Operations. An operation is designated as "critical" when it is required in order to establish a critical area from which subsequent operations or allied processes can be gaged or located. The required degree of control over such stock variations such as width, thickness, camber, and mechanical properties, should be specified within obtainable commercial limits. Stock variability results in product variation, especially as regards the amount of springback. These are basic factors in defining a critical manufacturing operation. The ideal critical manufacturing operation would establish the critical areas in a single operation from the sheet, strip, or coiled stock.

d. Accomplish Critical Manufacturing Operations. This is the major responsibility of the die designer, working to the process plan. However, the process planner must know the basic types of dies (see Sections 5, 7, 9, and 12) and their general applications. Close cooperation with the die designer and production facility is very desirable in order to get needed input regarding the ability to accomplish critical operations.

e. Determine Secondary Manufacturing Operations. These operations are intermediate between the critical manufacturing operations and the finished part. Limitations in these operations are those imposed by the workpiece specification, and by the amount of metal flow and/or movement (see Sections 4, 6, 8, and 10). Additional secondary operations, such as restrike, may sometimes be necessary to coordinate with an allied process or to reestablish a critical area for subsequent operations.

f. Accomplish Secondary Manufacturing Operations. This follows the same procedural pattern and involves the same responsibilities for the planner as for critical manufacturing operations.

g. Determine Allied Processes. These processes are determined during part print analysis, except as they arise through emergency or necessity. Thus, an annealing operation may become necessary when secondary manufacturing operations have been determined. In some cases, the process planner may avoid annealing by recommending a change in material specifications. All possible elimination of annealing, plating, cleaning, and other allied processes will appreciably reduce total manufacturing costs.

h. Accomplish Allied Processes. These are often the function of specialists, but the pressed-metal process planner must cooperate by delivering a workpiece in suitable condition for the allied process. The specialist should advise the processor of the effect the allied process will have on the workpiece.

4. Specify the Necessary Gaging. Gaging here includes (1) the gaging of material as-received, and (2) gaging of the workpiece in process. Gaging of material will include width, thickness, and camber for specified tolerances, and mechanical properties. The planning of in-process gaging for use during manufacture follows the same procedure as used to select critical areas, or areas from which measurements can be taken to define the geometry of the workpiece. The workpiece should not be allowed to continue through the sequence of operations if it is defective from a previous operation.

5. Specify the Necessary Press Equipment. A press should be specified according to the size and actuation requirements of the die, the type of pressworking operation to be performed, properties of the workpiece material, required tonnage, and the required production accuracy (see Section 27, Press Data).

6. Prepare Routing of Process. The operational machine routings vary in form throughout industry but must meet two common essential requirements: (1) description of the operation must be accurate and complete; (2) the nomenclature used should be according to accepted practice. A pictorial sketch of the part, shown on the operational machine routing sheet, aids considerably.

3-2-1 SYSTEM FOR LOCATING AND TESTS FOR QUALIFIED AREAS

Qualified areas are those areas which fulfill the requirements of arithmetical, mechanical, and geometrical tests in order to serve as surface of registry.

1. Arithmetical Tests. The selected surface of registry must not cause a limit stack. If surfaces of registry cannot be selected, they must be qualified, i.e., must be produced to a tolerance closer than those required and specified by the product engineer.

2. Mechanical Test. The size, shape, and finish of the selected surfaces of registry must permit a seat of registry design so each will withstand the operating forces exerted, and also the necessary holding forces.

3. Geometrical Tests. This test pertains to the distribution of the surfaces of registry so that the workpiece will be positionally stable. If surfaces of registry are not thus qualified, the process planner must consider suitable redesign with the product engineer. In the 3-2-1 locating system (Fig. 3-1), six points are the minimum number of points required to fix a square or rectangular shape in space. Three points establish a plane, two points define a straight line, and one point for a point in space, give a combined total of six points. A small pyramid symbol is used to designate a locating point. In Fig. 3-2, this symbol is used to illustrate a locating system for a rectangular solid. Variations of the illustrated system can be used to fix location of a cylinder, cone, disk, or other geometric shape. The surface of the device used by the die designer to establish the locating point specified by the process planner is known as a "seat of registry." The corresponding area

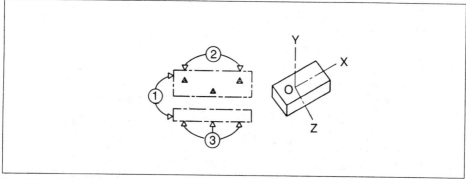

Fig. 3-1. 3-2-1 locating system.

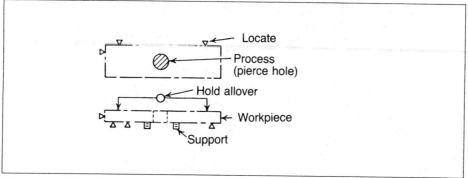

Fig. 3-2. Application of process symbols.

on the workpiece is known as a "surface of registry." Process-planning symbols can be used to avoid lengthy writing in the preliminary stages of planning to utilize critical areas for critical and secondary manufacturing operations. Figure 3-2 illustrates the use of such symbols.

APPLICATION OF PROCESS PLANNING TO SPECIFIC PRESSED PARTS

Sound general principles are always of value. An example of how good process planning is applied to a pressed steel part that is being successfully produced illustrates the method.

Cross-shaft Bracket. Figure 3-3 is the part print of a cross-shaft bracket, an approved component of an assembly with a forecast production of approximately 100,000 per year. In some plants the process planner might be expected at the outset to furnish a preliminary routing so that unit costs may be estimated. Such routing might indicate: (1) "blank and pierce," (2) "first flange," (3) "finish flange," and (4) "inspect," together with other normally required information and specifications. Although such preliminary routing may satisfy estimating needs, it does not always ensure quality parts, economically produced. For such goals, it is essential to apply an accepted planning procedure.

Analyze Part Print. Each specification, dimensional or noted, must be studied to understand exactly what is specified by the product engineer. One method used by the experienced planner is to check off each specification on the print, once it has been interpreted to his or her satisfaction. For the less experienced planner, a tabular analysis such as that in Fig. 3-4 may prove more effective, because it is a graphical record of information on all specifications shown or implied on the part print.

Terms used as column heads, "Material," "Die," and "Processing" in Fig. 3-4 are

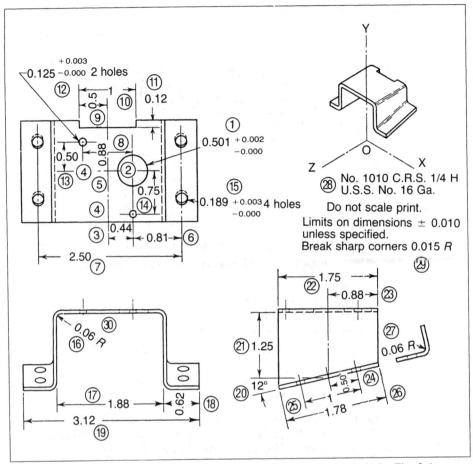

Fig. 3-3. Cross shaft bracket. Circled entries refer to entries on tabular analysis, Fig. 3-4.

listed in the following annotation.

"Material" refers to the material of the part to be pressworked, and it should be considered when a surface relationship depends upon a variable of the material. For example, variation in thickness would affect a forming operation.

"Die" refers to the tool which is anticipated for obtaining a required surface relationship. For example, the size of a pierced hole and the obtainable limits depend upon the skill of the diemaker in producing the punch to proper size.

"Processing" refers to the work of the process planner as it affects surface relationships. For example, if two holes were pierced simultaneously, the accuracy of surface relationship would depend upon the die; but if one hole were pierced and the second hole is located from the first, then the accuracy of surface relationship would depend in part, on the skill of the process planner.

This technique in process planning classifies all specifications concerned with surface relationships and reveals whether the relationships depend upon the material, the die, the processing, or some combinations of these factors.

The process planner is primarily concerned with the "Die" and "Processing" columns but, in order to plan an acceptable sequence of operations, the items in the "Remark" column must be clarified in consultation with the product engineer. The product engineer must also be consulted on any changes in materials specifications which might avoid manufacturing difficulties without affecting the part's functioning requirements. In extreme cases, a study of the materials considerations might show the wisdom or even the necessity of changing to some process other than pressed metalworking. In short, discussions between the process planner and the product engineers should firmly resolve the difference between "What Is Specified" and "What Is Wanted".

Operational Requirements. The basic pressed-metalworking operations such as cutting, forming, or drawing, which are used to obtain the surface relationships as indicated in the tabular analysis (Figure 3-4), must be resolved after all engineering decisions have been made on specifications which were noted in the "Remarks" column.

Operational requirements for the cross-shaft bracket, without regard to final determined sequence, are indicated by stars in the tabular analysis. The basic operations are:

1. Cut: (1 notch) . . . . . . 0.12 × 1 in. (3.0 × 25.4 mm)
2. Cut: (1 blank) . . . . . . 3.12 × 1.75 × 1.78 in. (79.2 × 44.4 × 45.2 mm)
3. Cut: (pierce 1 hole) . . . 0.501 +0.001, −0.000 in. (12.72 +0.02, −0.00 mm)
4. Cut: (pierce 2 holes) . . . 0.125 +0.003, −0.000 in. (3.18 +0.08, −0.00 mm)
5. Cut: (pierce 4 holes) . . . 0.189 +0.003, −0.001 in. (4.80 +0.08, −0.02 mm)
6. Form: (2 bends) . . . . . .0.06 in. (1.5 mm) radius
7. Form: (2 bends) . . . . . .0.06 in. (1.5 mm) radius

A further analysis must now be made of each listed operation, in the light of "Feasibility for Manufacturing" and "Economics of Tooling," previously discussed. The operational requirements (excluding 1 and 2) will now be examined briefly, in the above listed order.

3. (Pierce one hole, 0.501 +0.002, −0.00 in.): No problems are apparent in producing this hole. Tolerance is close, but not impossible with properly maintained tools. The natural break from piercing will provide adequate surface in the hole to meet functional requirements.

4. (Pierce two holes, 0.125 +0.003, −0.000 in.): No problems are apparent in producing the holes. The natural break from piercing will provide adequate surface in the hole to meet functional requirements.

5. (Pierce four holes, 0.189 +0.003, −0.001 in.): No problems are apparent in producing the holes. The product engineer objected to the suggestion of keeping all hole axes in the same plane, which might have simplified tooling and processing.

6 and 7. (Form four bends, 0.06-in. radius): These forming operations are similar in some respects and, combined, would constitute the complete forming of the part. Therefore, they evidently merit being analyzed together.

The forming operations consist of working a flat blank into a shape which will meet part print specifications (Fig. 3-3), plus any decisions reached between the process planner and the product engineer.

Controllability of the blank, the metal flow and movement, and quality of finish are factors in the "Feasibility for Manufacturing" and the "Economics of Tooling" for the bracket-forming operations.

Fig. 3-5 shows three possible methods of forming the flat blank to obtain the final shape of the cross-shaft bracket.

In the single-operation method shown at A, the developed blank would be placed in the die having a suitable locating system. In a single stroke, the metal would be worked over almost the entire surface in forming the blank to shape. In such a method, the surface on the die radius would be subject to excessive wear and, under production conditions, the part might have a distorted surface. Also, it would be difficult to control springback and the part symmetry because of variables of the material, even though the pad would hold the blank securely against the punch face. The 12° flange on the part (Fig. 3-3) would cause a localized flow of material, upon initial closing of the die, which would also distort the part.

In the two-die method shown at B of Fig. 3-5, the developed blank would be flanged on the ends in one die; then the sides would be formed by a single-pad flanging die. Methods A and B both provide fair control of the part, but springback control is difficult in both methods because the flanges are parallel.

TABULAR ANALYSIS

PART NAME—*Cross shaft bracket*
PART NUMBER—*P-460 529*

DATE—*Feb. 1, 1964*
PLANNER—*E.A. Reed*

No.	Specifications	Between Surfaces — Depend On			Remarks	Operational Requirements
		Material[1]	Die[2]	Processing[3]		
1	$0.501\ ^{+0.002}_{-0.000}$	✓	✓		Squareness implied 90°	★
2	℄ Part to ℄ 0.501 hole			✓		
3	0.44 to 0.501 & 0.125 holes			✓	Possible limit stack with spec No. 6	
4	℄ to ℄		✓		Implied coincident	
5	℄ ⊥ ℄		✓		Implied 90°	
6	0.81	✓	✓	✓		
7	2.50	✓	✓	✓		
8	0.88		✓	✓		
9	0.50		✓	✓		
10	1	✓	✓		Squareness implied 90°	★
11	0.12	✓	✓	✓		
12	$0.125\ ^{+0.003}_{-0.000}$ (2) holes	✓	✓			★
13	0.50 to 0.125 hole		✓	✓		
14	0.75 to 0.125 hole		✓	✓		
15	$0.189\ ^{+0.003}_{-0.001}$ (4) holes	✓	✓		Squareness implied 90°	★
16	0.06 Radius (2)	✓	✓		Implied 90° bend	
17	1.88	✓	✓			★
18	0.62	✓	✓		Possible limit stack with spec. No. 17	

Fig. 3-4. Tabular analysis of cross-shaft bracket specifications.

In the method shown at *C*, using the same blank, a single solid-type die could both form the end flanges and establish the break lines for the sides. In the final forming stage, the sides are flanged in a single-pad die. Blank control would be good, and springback in the end flanges could be compensated for by overbending. Since both the break lines are established in the same operation, with minimum metal flow and movement in the die, better dimensional control should be secured in the part.

It therefore appears to the process planner that the method shown in *C* of Fig. 3-5

No.	Specifications	Between Surfaces — Depend On			Remarks	Operational Requirements
		Material[1]	Die[2]	Processing[3]		
19	3.12	✓	✓		Implied symmetrical	◄┐
20	12°		✓	✓	No tolerance on L	
21	1.12		✓	✓	No tolerance on flatness or parallelism	
22	1.75	✓	✓	✓		◄— ★
23	₵ to 0.88			✓		
24	₵ to 0.50			✓		
25	1		✓	✓		
26	1.78	✓	✓	✓		◄┘
27	0.06 Radius (2)	✓	✓		Implied 90° bend	★
28	1010 C.R.S. ¼ H U.S.S. No. 16 Ga.	✓			Tolerance on thickness	
29	Break sharp corners 0.015 R	✓		✓	Question need	Not needed

Fig. 3-4. *Continued.*

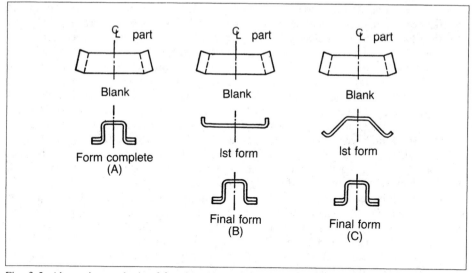

Fig. 3-5. Alternative methods of forming the cross-shaft bracket.

would be the most satisfactory and would permit combining the blank and pierce operations if hole location tolerances are sufficiently large.

1 and 2 (Blank and notch): The general dimensions of the outline of the part were listed in the tabular analysis (Fig. 3-4). In studying a part print, the process planner will determine if an initial blanking operation to prepare a flat piece of material for the subsequent operations (i.e., a blank) must be developed. The blank for the cross-shaft bracket is shown in Fig. 3-6. When the blank has been developed, the process planner must immediately consider utilization of the material, grain direction as related to blank nesting, and any other pertinent factors.

The shape of the ends of the blank requires a skeleton of material between the blanks. However, if the product engineer can be induced to decide that the "optional" blank end, shown in Fig. 3-6, will not affect the part's functional or structural requirements, then both the material and the die costs can be lowered.

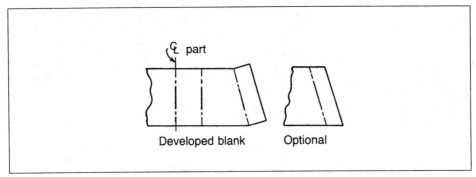

Fig. 3-6. Developed blank for the cross shaft bracket.

Grain direction will not be a problem in this plan, because the specified quarter-hard SAE 1010 cold-rolled steel will form satisfactorily either across or parallel to the grain direction.

The 0.125 × 1-in. (3.18 × 25.4-mm) notch is, in effect, a portion of the outline of the blank. There being no metal flow or movement in the notch area, the decision will be to include the notch with the blanking operation.

Planning the Operations Sequence. The major decisions on the basic operations for the cross-shaft bracket have now been made. The next step is to determine the critical specifications, and so establish the critical areas which can be used as surfaces of registry after they have been accomplished by the proper critical manufacturing operations.

Since close tolerances are an indicator of critical specifications, the process planner will first consider the 0.501 in. +0.002, −0.000 (12.72 mm +0.05, −0.00) hole and the two 0.125 in. +0.003, −0.000 (3.18 mm +0.08, −0.00) holes. The four 0.189 in. +0.003, −0.001 (4.80 mm +0.08, −0.02) holes might also be considered on the basis of close tolerance, except for the facts that there is metal flow between the planes of the two groups of holes and that the planes of the 0.501 in. +0.002, −0.000 (12.72 mm +0.05, −0.00) hole were established as common to all flanging operations in the form of the part. Hence, the surfaces of the 0.501-in. and 0.125-in. (12.72-mm and 3.18-mm) holes are decided to be the critical areas from which may be selected the surface of registry which will constitute the locating system required.

Since a minimum of six points or surfaces of registry are required to locate the workpiece, the process planner will consider the inside surface of the top side of the cross-shaft bracket (provides three points), the 0.501-in. (12.72-mm) hole (provides two points), and the 0.125-in. (3.18-mm) hole (provides one point).

This system should be checked to determine whether the selected surfaces of registry are qualified (1) arithmetically, (2) mechanically, and (3) geometrically:

1. *Arithmetically*, the selected areas are qualified because of their close tolerances. The holes have a maximum 0.004-in. (0.10-mm) tolerance, as compared with location dimensions having plus or minus 0.010-in. or 0.020-in. (0.25-mm or 0.51-mm) tolerance.

2. *Mechanically*, the surface of the 0.501-in. (12.72-mm) hole in the *YOZ* and *XOY* planes (Fig. 3-3) qualifies satisfactorily, since a 0.50 in. (12.7-mm) diameter pin is known to be structurally adequate for the two seats of registry, one in each plane. The surface of the hole will be only 30% of the thickness of the stock, because of the metal action in piercing the hole, but this will not disqualify the surface, since only one point in each plane is needed.

The 0.125-in. (3.18-mm) hole qualifies for the other surface of registry in the *YOZ* plane, except that so small a hole does not permit a structurally adequate locating pin for a seat of registry. Therefore, it is necessary to select some other surface of registry in the YOZ plane, such as the edge of the blank. This edge is qualified because it is to be produced in the same die as will produce the holes.

Although the 0.125-in. (3.18-mm) hole should theoretically be used, it was necessary to move the symbols of the surfaces of registry to the edges. Since a workpiece of this type would be nested, the equivalent of a nest is indicated in the operation diagrams (Fig. 3-7C, D, and E) by a dashed "Locate" symbol.

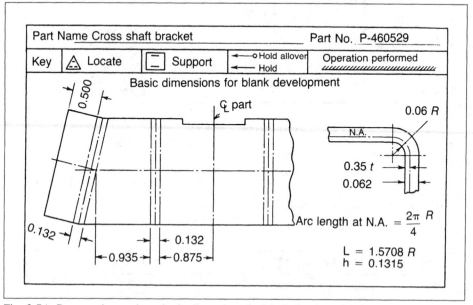

Fig. 3-7A. Process picture sheet; basic dimensions for blank development.

3. *Geometrically*, the 0.501-in. (12.72-mm) hole would be qualified to provide the necessary surfaces of registry to locate the workpiece in the YOZ and XOY planes, since the inside metal surface in the ZOX plane provides the required three points.

The 0.125-in. (3.18-mm) hole is geometrically qualified in relation to the 0.501-in. (12.72-mm) hole, because it provides an adequate distance between surfaces of registry for locating purposes. However, it was previously disqualified mechanically because of small size.

Critical manufacturing operations are those required so as to obtain critical areas for secondary operations. The holes could be pierced first, and then used to locate for a blanking operation. However, it is obviously practical to blank and pierce in the same operation.

Since the same locating system can be used both for the forming operations and for piercing the 0.189-in. (4.80-mm) holes, this information is now passed along to the die designer in the form of process picture sheets (Fig. 3-7A to E, inclusive), which supplement the usual machine and tool routing sheet.

Process picture sheets show the workpiece for each operation. Only those views are shown which are necessary to specify the surfaces of registry and any dimensions needed other than those given on the part print. On each sheet showing an operation, the required press specifications are given.

The machining and tool routing sheet will indicate the gages which must be either designed or selected from commercial standard sizes.

Allied processes, such as heat treating or tumbling, were considered and determined as not needed in producing the cross-shaft bracket.

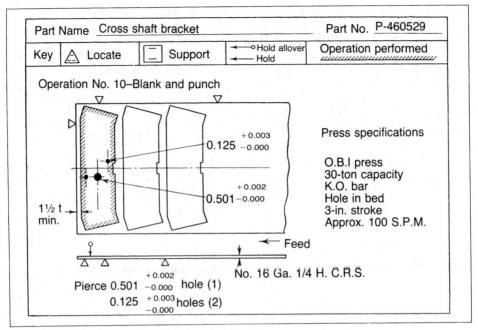

Fig. 3-7*B*. Process picture sheet; blank and pierce

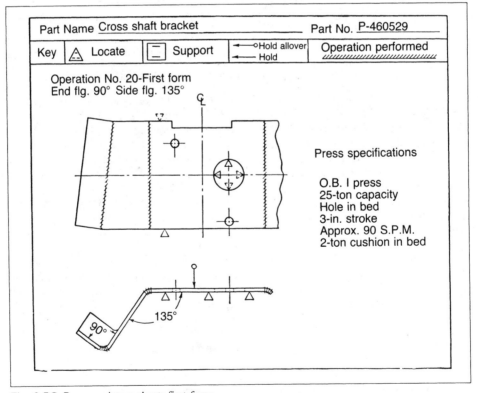

Fig. 3-7*C*. Process picture sheet; first form.

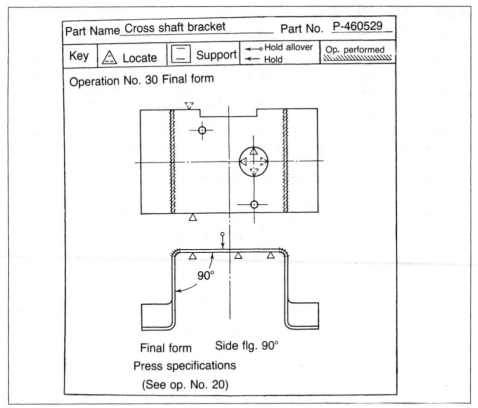

Fig. 3-7D. Process picture sheet; final form.

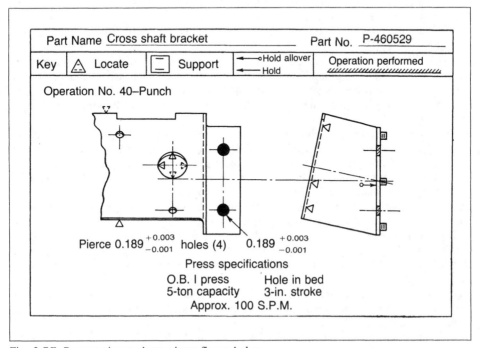

Fig. 3-7E. Process picture sheet; pierce flange holes.

PLANNING FOR PANEL PROCESSING

The basic procedures and principles of process planning for sheet metal parts are in general the same for all classifications of parts. However, the techniques used for the preliminary steps of processing a panel are considerably different from those applied to the processing of a symmetrical geometric part such as a drawn cup. In both cases the part material is subjected to the same types of tensile and compressive stresses, and the material strain characteristics vary with the severity of the drawing operation.

However, the source of information and specifications for planning production of an irregularly shaped panel is basically different from that for planning production of a simple drawn shape. Although the large outer panels of a automobile have structural value, one of the important design considerations is styling for sales appeal. In contrast, some functional components of the same product are formed from sheet metal for reasons of economy.

The modern automobile offers the best example of the application of both types of sheet metal components. The front fenders of a passenger car become an integral part of the stylist's design, which is worked out as an artist's rendering, in chalkboard drawings, in clay models, and finally in styling drawings.

Simultaneous Engineering. The process engineer, in order to meet his responsibilities within the allotted lead time for the product, must get preliminary information on the size and shape of the fender directly from the clay model. The clay model at this stage is also an aid to the process engineer and others in the evaluation of product manufacturability. Many minor changes are agreed to by the stylist, the design engineers, and the process engineer to keep the product within the known range of material and sheet metal working techniques. When the clay model is approved, templates and critical measurements are made to develop styling drawings. The styling drawings are used to develop the master surface plates as well as advance information for dies and experimental evaluation of drawn shapes. The master surface plate, which is the source of size and shape specifications for the die plate or aluminum draft, the master model, checking fixtures, and prototype tooling, is actually the starting point for manufacturing's responsibility for part specifications. Up to this point, styling has the responsibility, as can be see in the flow chart shown in Fig. 3-8.

Actually, to conserve lead time, the master model is built while the styling drawings are being developed.

Critical Path Analysis. The determination of the time required from design concept until an automobile is in production is based upon the number of tasks that depend upon the completion of a previous task to permit their commencement. The series of such sequentially dependent tasks that requires the greatest amount of total time is called the critical path.

Simultaneous engineering should be applied to the tasks in the critical path. Once the effect of simultaneous engineering on reducing the time required for the sequentially dependent tasks has been estimated, a new critical path can be determined.

This procedure can be re-applied as often as desired to shorten the total time required to complete the total project. Several popular computer programs are available to do critical path analysis as well as estimate the cost of the project. The cost of accelerating any aspect of the total project can also be estimated.

Digitalizing the Contoured Surface. It is important to recognize that models made of easily worked, yet temperature- and humidity-stable, materials provide a means of transferring specifications on contour and profile which would be difficult to define by coordinate points.

Electronic and optical measuring and recording systems have alleviated this problem. In some cases, digitalized data has taken the place of costly models which are subject to slight dimensional instabilities.

It is important for the die engineer to have advance information from the clay model

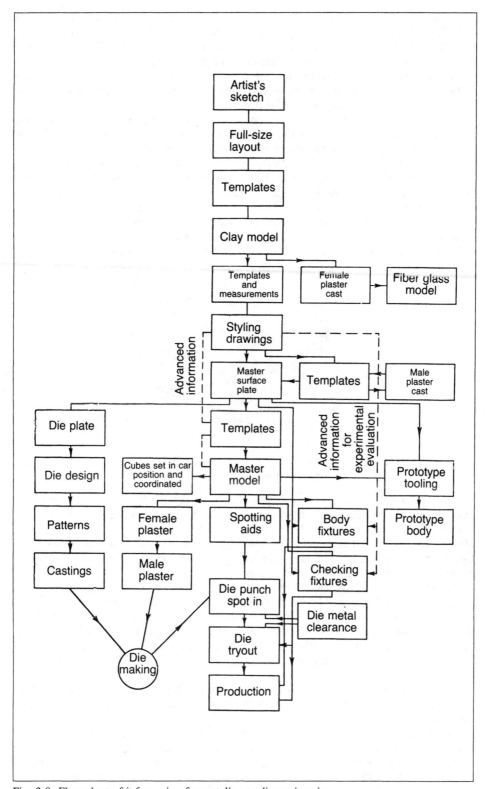

Fig. 3-8. Flow chart of information from styling to die engineering.

in the form of a female plaster and design lines so that work can be started on die positions and patterns for the major die castings well in advance of completion of the master model. When the master model is completed with all significant engineering changes (including those recommended by the process engineer when viewing the clay model), female plasters for making male models for machining and spotting aids for finishing such castings as the draw-die punch are made. Other body panels would follow the same procedure for dies, checking fixtures, and prototype work.

The trend in the industry is to rely less on tooling aids and plaster models for machining and fitting critical die surfaces and to develop precision cutter paths for profile milling directly from the digitalized data taken from the clay model. Allowances for springback and metal thickness can be made with computer-aided design (CAD) equipment.

The procedure used in the development of a draw die for an irregularly shaped panel, such as a fender, requires several steps to establish die position, draw-die wrap, and the necessary allowances for subsequent trim- and flange-die conditions.

As mentioned in the testing procedure, a female plaster is made from the original clay model. A male model as shown in Fig. 3-9 is made from this female plaster. The accuracy of these models is adequate for preliminary transfer of size and shape specifications.

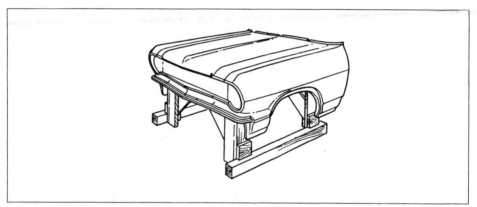

Fig. 3-9. Male model of automobile hood and fenders.

The plaster models take from this preliminary model of the hood and fenders, for example, are used to develop the draw-die position and wrap and the flange and trim lines. The draw setup shown in Fig. 3-10 shows the male plaster removed from the female plaster. The method of developing the flange and trim lines is shown in section. The flange-line development is important to panel appearance, and the trim-line development in the draw is important to secondary operations.

A female plaster as shown in Fig. 3-11 is taken from the male. The draw-ring and punch-entry angle is planned to give a controlled flow of metal. The flow of metal in this configuration of die is not readily computed as in the cup shape and therefore must be studied and determined as outlined. Computer-aided design equipment is an aid in developing a blankholder configuration that will minimize kinks in the blank upon blankholder closure. These kinks often result in a serious surface defect in the drawn panel known as a buckle.

In severe cases experimental dies are built, preferably to full size, to develop the optimum metal flow conditions and in some cases to establish the sheet-stock requirements for the panel.

Concurrent with this development, steps are taken to finalize the master surface plate and the master model as previously listed. The wood and plastic model shown in Fig. 3-12 is the master for a fender "cubed" in car positions. This final model is used to prepare the final spotting aids and machining models used to build the dies. In many cases, to remain within the lead time, the patterns for the dies, particularly the draw die, are built

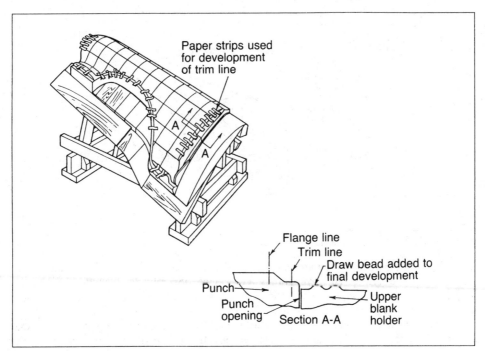

Fig. 3-10. Preliminary draw plaster.

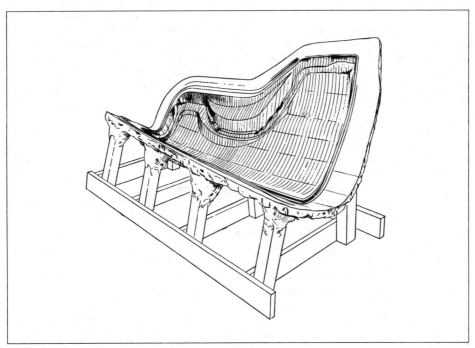

Fig. 3-11. Female plaster with preliminary binder development.

from preliminary specifications and then finished to the final specifications of the master model.

Critical Points for Analysis. The process planning engineer is confronted with manufacturability problems which result from product design or from die capabilities and

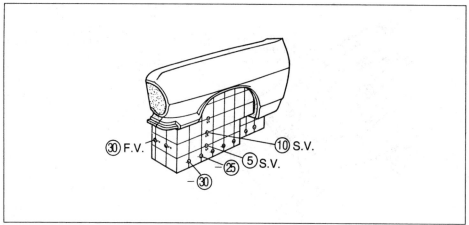

Fig. 3-12. Final wood and plastic model.

conditions; in most cases they are interrelated, but a good die condition does not always permit attractive styling. However, out of experience, several rules have been developed which should be considered when the conditions are recognized in the analysis of a proposed panel or in the design of the dies:[1]

1. Avoid localized contact of punch with metal. The part should be positioned so that the punch contact area is distributed for the best utilization of the strain range of the material.

2. Draw part to final shape in a minimum number of operations. Figure 3-13*A* illustrates the tipping of a part to completely form the right-hand side while the left-hand side requires additional operations. Had the designer tipped the part so as to partially form the left-hand side, additional operations would have been required on the right-hand side, resulting in possible marking of the appearance surfaces.

3. Favor critical characteristics. Figure 3-13*B* illustrates how the more critical line is favored in the positioning of a part in a draw die. The more rounded surface of the protrusion on the left-hand side will cause less noticeable draw lines than the lower and sharper surface.

4. Overcrown to compensate for springback. Large areas having plan surfaces with nonsupporting characteristics often require special treatment. Springback and a weak supporting structure may be compensated for by additional crown to the punch. Development of this crown depends on experience with the type of panel and on whether the surface configuration offers any supporting strength.

5. Avoid an excess or deficiency of metal from blankholder action. The wrap and the draw die-blankholder relation must not cause metal excess or deficiency in appearance surfaces because such strains cannot be removed by drawing action.

6. Work the metal to establish shallow shapes. The length of the draw-ring or binder line should be less than the punch-part length to assure that stretch will occur, thus elongating the metal to the permanent fabricated shape. A perfect wrap should be avoided on the face of the punch before drawing starts. Generally 2 to 3% minimum elongation is required to remove the normal surface waviness caused by mill rolls.

7. Minimize the effect of the length of adjacent contours. Drastic differences in lengths of lines in adjacent contours might require stretching the metal excessively through line *X* of Fig. 3-14 to prevent wrinkles along line *Y*. The problem is improved by more generous radii to provide metal in the deficient areas.

8. Avoid sharp radii for flow of appearance surfaces. The initial contact of the punch at *Y* (Fig. 3-15) locally stresses the metal, and as the punch carries the metal into the die, the metal is further elongated by the restraining action of the blankholder and continues to move over point *Y*. The effect known as a draw line is usually noticeable on the

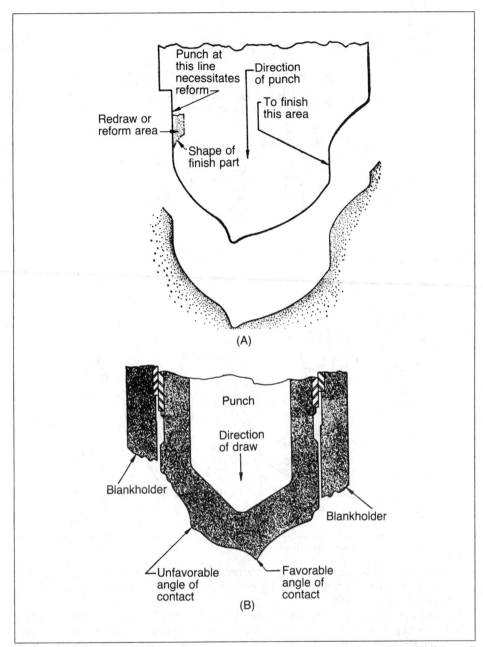

Fig. 3-13. Punch tipping (*A*) to equalize punch control area, and (*B*) to favor character lines and angle of forming.

appearance side of the panel.

9. Allow for subsequent flanging and trimming operations. In Fig. 3-16, note that material has been provided for the flange which has been partially formed. The trim line is accessible for a conventional trim-die operation instead of by cam or indirect trimming.

10. Control location of contact marks. Figure 3-17 shows the contact marks on a panel made in an incorrect die design carried into the exposed flange. With the proper die design shown, the mark is outside the specified 0.62 in. (15.7 mm) width.

11. Control metal flow close to critical shapes. When difficult shapes, such as the

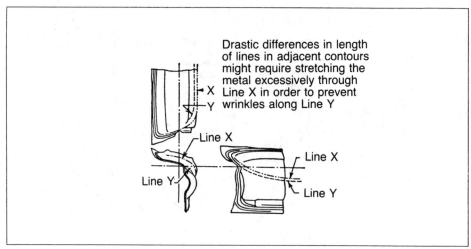

Drastic differences in length of lines in adjacent contours might require stretching the metal excessively through Line X in order to prevent wrinkles along Line Y

Line X

Line Y

Line X

Line Y

Fig. 3-14. Adjacent contour effect.

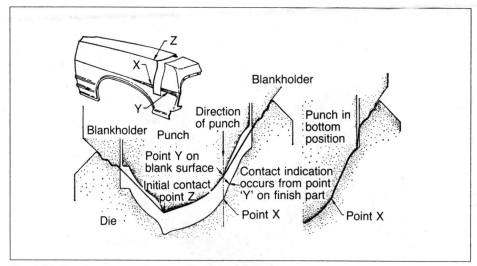

Fig. 3-15. Punch contact and wear points.

pyramidal shape shown in Fig. 3-18, are formed, keep the draw ring as close as possible to the shape, and then control the metal flow by balancing the restraining action by beads and blankholding pressure.

Special Procedure for Transfer Press Dies. The large transfer presses designed specifically to produce automotive body panels have fixed limits as to the number of dies that can be accommodated. A six die limit is fairly common in the double slide six column presses used for large closure panels. As a consequence, the design of the panel must not require more than six dies. The practice is to combine operations into a single die through the use of combination die designs. Such dies make extensive use of large complicated aerial cams for trimming, piercing, and flanging. This can result in additional tooling economy by reducing the number of dies that would otherwise be required. In such cases an idle station or stations can be used in the transfer press.

Panel Design for Economical Production. Product design for ease of manufacturability is very important in the design of automotive body panels. Avoiding deep drawing conditions in such parts as fenders reduces costs in several ways.

1. Steel costs are reduced because less strain is involved.

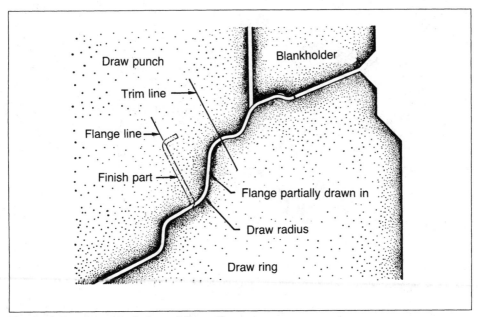

Fig. 3-16. Development of trim and flange lines.

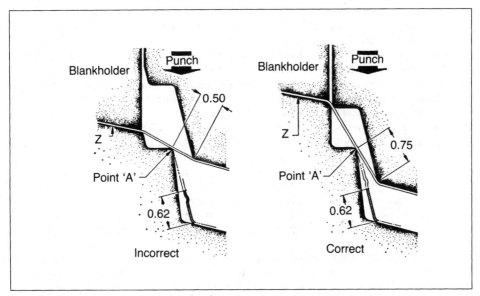

Fig. 3-17. Trim-ledge and draw-ring development.

2. The use of an inverted stretch draw may be feasible.

3. If a stretch draw can be used, a substantial metal savings may be realized.

4. A stretch draw operation in an inverted draw die will eliminate the need to turn the part over before placing it into the trim die. The turnover in a tandem line is often the limiting factor of the attainable production rate.

5. An inverted stretch draw die used in conjunction with a press cushion or nitrogen springs may not require a double action press. If this is the case, all operations may be done in a transfer press.

As experience is gained from model year to model year, it may become apparent that two body panels can be drawn as a single panel. A good example of this is the trend to

3-21

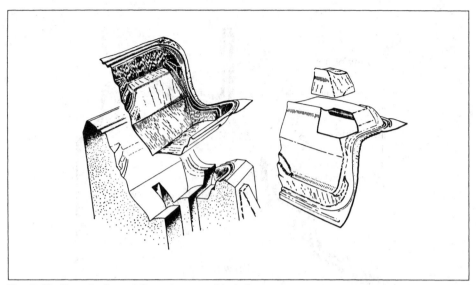

Fig. 3-18. Control of metal flow close to critical shapes.

combine the quarter lock panel with the quarter panel. The economy is partly offset because the quarter panel then is a more severe deep draw than would otherwise be the case.

Computer-aided drawing feasibility studies are a good means to identify operations that are candidates for combining into single panels.

Optimizing Casting Weight for Die Economy. Most of the die shoe sizes and casting weights shown in this handbook reflect conservative design practices. In order to reduce tooling costs, lighter castings are often specified. To achieve lighter construction, rib and shoe thicknesses are reduced, often by a factor of 50% or more.

While some designers feel that acceptable die deflection is the controlling factor in how thin a cast section can be made, other factors should be considered.

In spite of all precautions, dies are subjected to some abuse. Automation arms and transfer fingers can and do get caught in dies in spite of elaborate die safety interlock systems. Misfeeds still do occur. Press cushion pins do get placed in the wrong holes. Extra casting weight can equal the difference between a die with deformed surfaces and one with broken castings in the event of an accident.

Another advantage of a safety margin in specifying casting weight is that heavier die shoes are easier to repair with draw rods, dutchmen, and boiler plate inserts than shoes that are excessively thin.

Press Alignment Considerations. One reason for the practice of overdesigning automotive dies has been the unstated fact that the die is expected to align the press slide upon heel block entry. Such heel blocks are very massive and provided with generous leads. Heavy die shoes also permit the use of oversized equalizer blocks that can compensate for large misalignments in the press.

For lightweight die casting construction to function well in older presses, the gibbing clearances and general alignment of the presses must receive careful attention.

PRESSWORKING COST COMPARISONS

The previously described methods of presswork process planning, consistently followed, will in most cases narrow down the final decisions to a very few of many seeming possibilities. But several alternatives may remain for combining operations, making second passes through a die, selecting from available presses of different types and capacities, and other factors.

Under these conditions, comparisons of costs for different feasible dies and processing methods may quickly reveal the combination that will result in lowest total cost per pressworked part.

Let N_T = total number of parts to be produced in a single run

N_B = number of parts for which the unit costs will be equal for each of two compared methods Y and Z ("break-even point")

D_Y = total die cost for method Y

D_Z = total die cost for method Z

P_Y = unit pressworking cost for method Y

P_Z = unit pressworking cost for method Z

C_Y, C_Z = total unit cost for methods Y and Z, respectively

Then

$$N_B = \frac{D_Y - D_Z}{P_Z - P_Y} \tag{1}$$

$$C_Y = \frac{P_Y N_T + D_Y}{N_T} \tag{2A}$$

$$C_Z = \frac{P_Z N_T + D_Z}{N_T} \tag{2B}$$

EXAMPLE 1: While the dollar amounts shown in the following studies are out of date because of inflation, they are excellent examples of how to apply the formulas.

The aircraft flap nose rib shown in Fig. 3-19, is to be made of 0.02-in. (0.5-mm) 2024-T3 Alclad. The producer had the option to form this part by Hydropress, drop hammer, Marforming, drawing in a steel draw die, and hand forming. For such reasons as die life, equipment available, and hand work required, the choice was narrowed down to Hydropress vs. steel draw die. With Hydropress, the flanges had to be fluted, and for piece quantities over 50, a more expensive steel die had to be used.

TABLE 3-1
Cost Comparison of Methods For
Producing Aircraft Flap Nose Rib

N_T*	5	25	50	100	500	760†
P_Y	\$ 3.00	\$ 1.18	\$ 1.11	\$ 1.05	\$ 1.05	\$ 1.05
P_Z	4.40	2.05	1.96	1.85	1.85	1.85
D_Y	810.00	810.00	810.00	810.00	810.00	810.00
D_Z	103.00	103.00	103.00	103.00	202.00	202.00
C_Y	165.00	33.60	17.30	9.15	2.67	2.12
C_Z	25.00	6.17	4.02	2.88	2.25	2.12

* Symbols in this column are the same as in Eqs. (2A) and (2B).
† Extrapolated.

Actual die and process costs for both methods are listed in Table 3-1. P_Y and P_Z are processing costs, and D_Y and D_Z are die costs, for the steel draw die and Hydropress methods, respectively.

Figures in the last column of Table 3-1 were not stated in the original report but can properly be extrapolated on the basis of apparent stability of P_Y and P_Z at $N_T = 500$, and

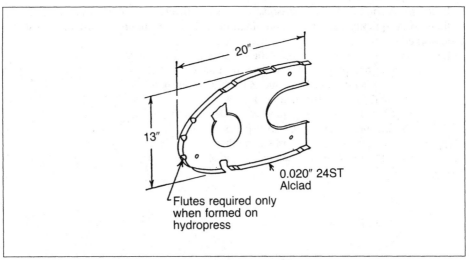

Fig. 3-19. General specifications for aircraft flap nose rib.[6]

assuming their stability at higher production.

On the basis of listed figures at $N_T = 500$, and from Eq. (1), the production at which total unit costs C_Y and C_Z will be the same for both methods is

$$N_B = \frac{810 - 202}{1.85 - 1.05} = 760 \text{ pieces}$$

Combined Operations. Under certain conditions, operations can be advantageously combined. The total cost of tooling may be reduced, or production costs, or both. A further advantage may be gained in combining a fast operation with another operation which, separately performed, might be slower. A precaution, however, is that setup and maintenance costs may increase.

Table 3-2 gives cost comparisons for a three-setup method vs. a single-setup method for blanking a 1-in. by 2-in. by 0.062-in. (25.4-mm by 51-mm by 1.59-mm) rectangular cold-rolled-steel part, and perforating holes in it. Lots are sufficiently small to require no estimates of maintenance cost.

Method II is clearly the lowest in total cost for 500 parts. At 5,000 parts, costs are virtually the same, and the choice of methods may be chiefly determined by the likelihood of future required reruns or other factors. At 10,000 parts, the advantage is heavily with method I and will increase as production goes higher.

TABLE 3-2
Comparison of Single-Setup vs. Three Setup
Pressworking Methods[3]

Lot Quantity	Method I			Method II		
	500	5,000	10,000	500	5,000	10,000
Tool cost	$30.00	$30.00	$30.00	$15.00	$15.00	$15.00
Setup cost	1.00	1.00	1.00	3.00	3.00	3.00
Processing cost	0.75	2.50	5.00	1.60	15.00	30.00
Total cost	$31.75	$33.50	$36.00	$19.60	$33.00	$48.00

Method I (single setup): Blank and perforate from roll stock.
Method II (three setups): Blank and perforate from sheet stock, sheared into 2-in. (51-mm) -wide strips, and further sheared into 1-in. (25.4-mm) -wide strips.

Table 3-3 reflects a case where the cost of combined tools was less than the total cost of the separate tools. The combined operation was done at the speed of the blanking operation.

TABLE 3-3
Costs of Combined vs. Separate Operations[3]

Costs	Blanking Operation Alone	Forming Operation Alone	Total Blank and Form	Combined Operation
Tool cost	$40.00	$30.00	$ 70.00	$50.00
Setup cost	2.00	2.00	4.00	3.00
Maintenance cost	2.00	. . . .	2.00	2.00
Processing cost	4.00	30.00	34.00	4.00
Total cost	$48.00	$62.00	$110.00	$59.00

MATHEMATICAL ANALYSIS OF DIE COMPONENTS*

There are often several approaches to the analysis of a single die component. The analysis of a complex component can usually be approximated by using the nearest simple shape. Since all calculations require additional design time and increase design costs, they should be simplified whenever possible.

The calculations are based on accepted engineering formulas.

Estimating Side Thrust in Cutting Dies. An approximation of the side thrust generated when performing straight shearing is given by

$$\frac{C}{T - P} = \frac{F_H}{F_V} \tag{3}$$

where P = penetration, usually 0.5 times thickness T
$\quad\ \ C$ = clearance
$\quad\ \ F_V$ = cutting force
$\quad\ \ F_H$ = side thrust

When applying this formula, adjustment may need to be made for the type of material and actual die conditions. Figure 3-20 shows that shearing actually involves some bending. Excessive clearances and dull die steels can generate very high side thrusts (Fig. 3-21). The side forces can result in large side deflections unless the die alignment system is strong enough to limit the resultant deflection to acceptable values.

In extreme cases, the deflection will result in clearances so large that the die may shatter due to excessive pressures and interference of steels. This can result in injury to press-room personnel. The alignment system must be rigid enough to prevent excessive deflections from occurring.

Other Sources of Side Thrust. Side thrusts other than the one illustrated for cutting exist in dies. Similar thrusts occur during right-angle bending and cup-drawing operations. Large side thrusts are present when cams are used. Neglecting friction, with 45° cams the side thrust equals the vertical force of the cam driver. Large side thrusts also occur when nonsymmetrical shaping is performed in forming and drawing dies. Examples

* Assistance with the engineering concepts, application of engineering equations, and text was provided by Mr. Anthony Rante, Manager, Mechanical Engineering, Danly Machine, Chicago, Illinois.

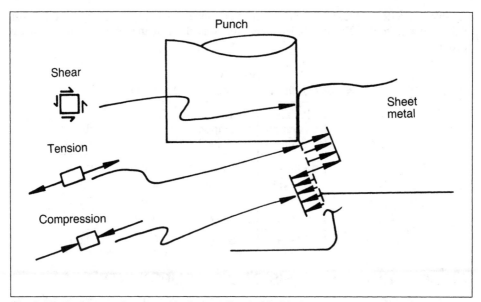

Fig. 3-20. Cutting metal creates side forces. (*Danly Machine*).

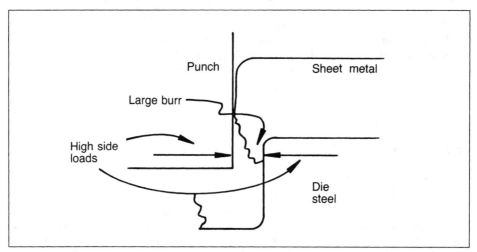

Fig. 3-21. Clearance and edge conditions affect the amount of side thrust. (*Danly Machine*)

of thrust are shown in Fig. 3-22.

A press ram which is not parallel to the bed or lower die shoe is another cause of side-thrust conditions in dies. Causes of side thrusts in dies can be itemized as follows:

1. Side thrusts due to the clearance used in cutting, forming, and drawing dies.

2. Side thrusts necessary to align the press ram.

3. Side thrusts due to poor alignment of components during die construction.

4. Side thrusts due to angular contacts between surfaces such as cams and die and punch steels.

5. Side thrusts created by nonsymmetrical forms or draws where punch and die are loaded off-center at initial contact.

6. Side thrusts resulting from the use of unbalanced shear or angular cutting faces to reduce force requirements.

7. Side thrusts occurring in cutoff, trim, bending, and flange dies where forces act on only one edge of the steels.

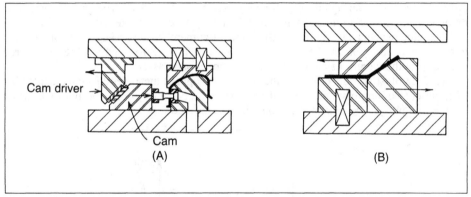

Fig. 3-22. Horizontal thrusts in dies: (A) cam pierce die; (B) bending die.

Alignment-system Design. It would be difficult to mathematically design each die component individually without any reference to to other components since most die components are interrelated as far as function is concerned. Therefore, the better approach is first to consider the die as a whole.

Clearance Requirements. An important requirement of any die operation is the maintenance of alignment between components. Alignment is necessary to maintain proper clearance between the die components shaping the metal. Clearance is commonly measured at one side rather than overall. The purpose of the clearance varies with each operation.

Proper cutting clearance is required to obtain a clean fractured edge. Excessive clearance allows pull-in of metal and large burr formation. Insufficient clearance causes improper fracture and a resultant jagged edge with excessive burnishing.

Extrusion clearance is the die gap which creates and determines the wall thickness of the shell produced.

In many dies, the thrusts due to unsymmetrical work load clearance are the only ones present. In a symmetrical, well-constructed blanking die with balanced shear and a well-aligned press ram, side thrusts will be at a minimum. The die alignment system is required to keep side thrust from causing unacceptable clearance variations.

The amount of clearance in a die is determined by the size of the steels and the accuracy of their location during assembly to the shoes. Good component alignment is necessary to limit the variation of clearance. Clearance control within acceptable limits is the most realistic way to analyze an alignment problem.

Guide-pin and Heel Design. Guide pins and die sets were originally used to aid in die construction, die maintenance, and die setting. Small guide pins can provide very limited alignment when the die is exerting forces to shape metal. The pins must be quite large if they are expected to guide the press slide. Even today, in some stamping plants, die sets are considered a luxury. The methods of providing alignment vary in advantages. General rules which should be followed are:

1. Guide pins should be as short as possible.

2. When die forces occur, have the guide pin entered into the bushing as far as possible.

3. Heels offer more rigidity than pins, particularly when U-shaped heels are used.

4. Care must be exercised to assure that all heels make contact under side thrust conditions.

5. The press must be in good alignment.

6. Any side thrust generated within the press should be controlled by the press gibbing.

7. The pressure occurring between the components of a die alignment system must not be so high as to cause rapid pin, bushing, and wear plate wear.

Pin Sizes Needed for Static Alignment. In this case, the guide pins are used only to provide alignment during die construction, setting, and repair. The press frame and gibbing maintain die alignment when metal shaping forces are exerted.

The guide pins must be rigid enough to hold alignment during assembly of components to the shoes, setting of the die in the press, and bench work for repairs. The main force occurring is the weight of components to the upper shoe assembly.

Fig. 3-23 illustrates the effects of the outboard weight of the upper die on the guide pins of a dieset. To simplify analysis and provide a safety factor, only one pin of a dieset having two pins at the rear of the shoe will be analyzed.

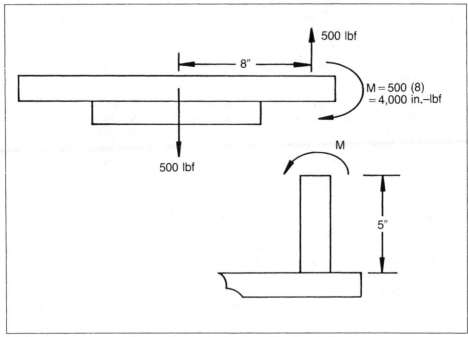

Fig. 3-23. Static outboard load on guide pins in a trim die. (*Danly Machine*)

A moment study of the pin provides the following:

$$\delta = \frac{ML^2}{2EI} \tag{4}$$

where δ = deflection, inches (mm)
  M = Moment, in.-lbf (N•m)
  L = Length, inches (mm)
  E = Modulus of elasticity, psi (Pa)
  I = $\dfrac{\pi d^4}{64}$ = Moment of inertia, in.^{4} (mm^4)

  d = Diameter of the guide pin, in. (mm)

For 0.0001-in (0.002-mm) deflection:

$$d^4 = \left(\frac{4{,}000 \text{ in-lbf} \times (5 \text{ in.})^2 \times 64}{2(30 \times 10^6 \text{ lbf/in.}^2) \, \pi \, (0.0001 \text{ in.})} \right) = 339.5$$

$$d = 4.293 \text{ in.}$$

The above equation is correct for a cantilever beam. For short beams, shear deflection should be included.

$$\delta_{\text{Total}} = \frac{FL^3}{3EI} + \frac{2FL \, (Q/D)}{EI} \tag{5}$$

$$\text{Bending} \qquad \text{Shear}$$

where F = Force, lbf (N)
Q = First area moment, in.^{3}(mm^3)
D = Thickness of the cross section at the neutral axis, in. (mm)

For circular cross sections:

$$\frac{Q}{D} = \frac{d^2}{12}$$

$$\delta_{\text{Total}} = \frac{800 \, (5)^3}{3 \times 30 \times 10^6 \left[\dfrac{\pi}{64} (3.9)^4 \right]} + \frac{2(800)5 \left(\dfrac{(3.9)^2}{12} \right)}{30 \times 10^6 \left[\dfrac{\pi}{64} (3.9)^4 \right]}$$

$$\delta_{\text{Total}} = 0.0001 + 0.00003 = 0.00013$$

As a practical consideration, shear deflection allowance is not required for the pin in our example.

If four guide pins are used and the left two pins are considered nonexistent, then $d^4 = 170$ or $d = 3.6$ in. (92 mm).

Notice that pin diameter can be reduced only slightly when increasing the number of pins. When the upper shoe is clamped to the ram, pin deflection caused by weight will be removed, and the clearance increase allotted to the pins can then be used to provide a partial correction for minor ram misalignment. If a safety allowance is not desired, smaller pins of about 2 or 2.5 in. (51 or 64 mm) diameter could be used.

Pin Sizes Required for a Known Side Thrust. In this example, the guide pins are expected to both provide alignment to the ram and resist internal die forces. A disadvantage to this design method is that more money is spent on each die placed in the press.

The die used in this example is illustrated in Fig. 3-24. The side thrust is calculated with Eq. (3).

Compute the vertical trimming force:

Shearing strength = 42,000 psi (290 MPa)
$$P = SLt$$
$$P = 42,000 \times 10 \times 0.035 = 14,700 \text{ lbf (65 kN)}$$

Include a safety factor of 300%:

$$P = 14,700 \times 3 \quad = \quad 44,100 \text{ lbf (196 kN)}$$

Compute the side thrust due to trimming:

Clearance = 10%
Penetration = 40%
Clearance = $0.1 \times 0.035 = 0.0035$ in. (0.089 mm)
Penetration = $0.4 \times 0.035 = 0.0140$ in. (0.356 mm)

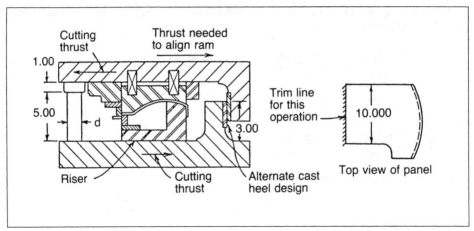

Fig. 3-24. Alignment of a trim die.

From Eq. (3), the side thrust = 7,350 lbf (32.7 kN) for trimming +12,650 lbf (56.3 kN) for shifting the ram = 20,000 lbf (90.0 kN). The side thrust per pin is 20,000/2 = 10,000 lbf (44.5 kN).

From Eq. (4), the pin diameter d = 5.4 in. (137 mm). If four pins are used, then d^4 = 424 or d = 4.5 in. (115 mm).

Since the guide-pin sizes are large, heels should be considered for alignment purposes. Assume that two sets of 5-in. (127-mm) square heels are used.

Using Eq. (4) and $I = bh^3/12$, the deflection of the heels is 0.00008 in. (0.0020 mm), which is well within the prescribed limit of 0.0001 in. (0.002 mm).

To obtain the full effect shown by the calculations, both lower heels must be to the left of the upper heels; however, standard die design practices often place the lower heels on opposite sides of the upper heels. As deflection occurs, one set of heels does not contact and is therefore inactive. An alternative construction using pins would be to lower the bushing on the upper shoe or to make the bushing longer. The pin can then be shorter and more rigid. Also, smaller-diameter pins could be used. Often, die designers combine guide pins and heels in one die. The heels shift the ram first, and then the pins resist internal die forces.

Punch and Die-steel Design. The punch and die steels for trimming will now be designed by allowing each component the specified portion of clearance increase. Each steel will be allowed a 0.0001-in. (0.002-mm) deflection. Since the designs of both steels are similar for the die under study, only the upper steel will be analyzed. Clearance increase can be caused by the trim steel in any or all of three ways: (1) by shifting, (2) by tipping, and (3) by bending.

Prevention of Shifting. Shifting of the punch steel is prevented through the use of static friction as shown in Fig. 3-25. The cutting force and screws act together as the normal force to create enough friction to prevent any shifting. By this means, the proper size and quantity of screws in the steel can be calculated from the screw manufacturer's data.

There is a misconception that dowel pins and keys prevent shifting. Since dowels and keys are elastic steel components, they will give under any load and allow some shift. Therefore these details will only restrict or limit shifting to a minimum value provided the loads do not result in plastic deformation. The original use of dowels was to align components in the no-load situation to aid in die assembly and maintenance.

The coefficient of friction between die components varies, depending on the surface finish as well as the presence of oils, burrs, and foreign matter. To provide ease of calculation and an additional safety factor, the vertical forces acting on the trim steel are not considered.

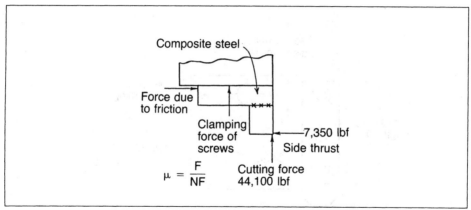

Fig. 3-25. Forces acting to prevent shifting of die steels.

$$\mu = \frac{\text{Friction}}{\text{Normal force}} \tag{6}$$

where μ = coefficient of friction

A general rule is that the clamping force of the screws must be 6.7 times the side thrust based upon a coefficient of friction of 0.15.

Size and Quantity of Screws. The die designer should follow the torque-vs-tension graphs for the screws used by his or her plant. Many screw manufacturers provide tables and slide charts for this purpose. Because of the wide variety of threaded fasteners that are used in diemaking and die setting, the individual manufacturer's data should be followed as regards strength, fatigue strength, and tightening requirements.

Proper torque tightening of screws prevents premature loosening from die cycling. Screws which are tightened randomly may be stretched into permanent yield. Screws have broken from fatigue in die operations because of overstressing. For critical applications, torque wrenches should be used to obtain the full value of each screw. When randomly tightened, only the tighter screws are carrying the load.

Limitation of Shifting. Calculations can be made to determine the dowel and key sizes needed to restrict shifting to a given value. Such an approach may be necessary when extreme side thrusts occur in form or draw dies. The prevention of shifting may not be economical in such cases, because the required screw sizes would be very large.

Considering the die steel shown in Fig. 3-26, extreme side thrusts may occur on the angular surface because of metal-shaping forces, particularly when the die bottoms. Slight adjustments in slide position will change the force significantly. For this die, a side thrust of 55,000 lbf (245 kN) occurs when the stop blocks are in contact. Less force will occur if a shim is used on the stop block during die setting. In a die of this type, the wedge action creates a side thrust greater than the vertical force.

The modulus of elasticity under a shear stress is used to solve this problem. This is sometimes call the "shear modulus of rigidity". For either plain carbon or alloy steel in the soft or hardened state, this modulus is about 12,000,000 psi (83 GPa).

A major factor in key design is the size of chamfer on the back corner of the die steel. Because the chamfer sets the distance between the shear forces it should be kept small.

An example of a key calculation is as follows. The formula is valid for SI units as well. The key is 6 in. (152 mm) long. The chamfer on the die steel is 0.060 in. (1.5 mm) × 45°. The side thrust, P, is 55,000 lbf (245 kN), the shear modulus, G, is 12,000,000 psi (83 GPa), the allowed shift, s, is 0.0001 in. (0.002 mm), and the effective length, l, of the

$$\text{key} = f - (e \times \text{no. of screws})/2 = 6.00 - (0.375 \times 3)/2 = 5.44 \text{ in.} \tag{7}$$

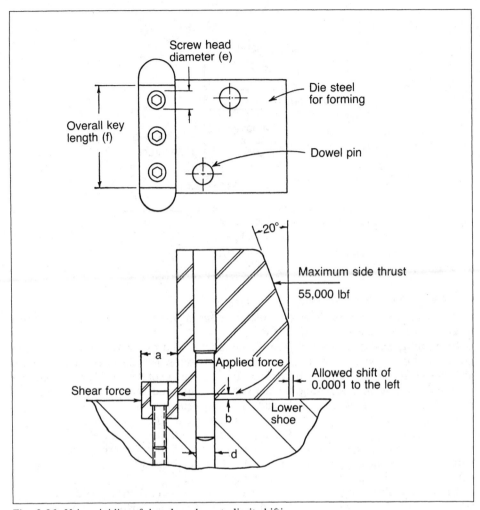

Fig. 3-26. Using rigidity of dowels or keys to limit shifting.

$$G = \frac{Pb}{As} = \frac{Pb}{als} \quad \text{or} \quad a = \frac{Pb}{Gls}$$

$$a = \frac{55,000 \times 0.060}{12 \times 10^6 \times 5.44 \times 0.0001}$$

$$= 0.506 \text{ in.}$$

A check should be made to assure that the elastic limit of the key material has not been exceeded, which would result in permanent deformation.

Yield strength, shear = 0.577 yield strength, tension   (8)

Thus, for SAE 1018 cold-rolled steel,

Yield strength, shear = 0.577 × 80,000 = 46,000 psi (317 MPa)

The stress on the key is

$$S_s = \frac{P}{A} = \frac{55,000}{5.44 \times 0.5} = 20,220 \text{ psi} \tag{9}$$

The key is therefore stressed elastically since the stress on the key is less than the yield strength of shear. If the elastic limit is exceeded, either the width must be increased or the allowed shift decreased. If two dowel pins only were used to limit shifting, their diameter would be found as follows:

$$G = \frac{Pb}{2As} = \frac{Pb}{2\frac{\pi}{4}d^2s} \tag{7A}$$

$$d^2 = \frac{4Pb}{2\pi Gs} = \frac{4 \times 55,000 \times 0.060}{2\pi \times 12 \times 10^6 \times 0.0001} = 1.75$$

$$d = 1.32 \text{ in.}$$

The distance b should be equal to zero if the die block and lower shoe are final reamed in assembly and the dowel has a tight press fit. The shear stress in the dowels is 20,000 psi (138 MPa) and should be reduced.

The dowels and key could be used together, thus reducing the size of both. The shear strength of the screws cannot be used in such an analysis since the screws are placed in oversize holes in the die steel, and shear loads would not occur on the screws until the die steel shifted to take up the gap. A combination of key shear strength, dowel shear strength, and static friction by screws would allow the smallest size of all fastening components. In this case, the side thrust could be divided among the three components, and calculations made as per the examples. An alternative solution is to place the die steel in a pocket or against a shoulder milled in the shoe. The large cross section of the shoe offers high shear strength and rigidity.

Prevention of Tipping. Properly placed screws of correct size and tightness will prevent tipping or revolving of a die steel about its back lower edge. The forces acting on the form steel and their distances from point x are illustrated in Fig. 3-27. There is room for three screws in the die steel. The size of the screws needed to prevent tipping may be found as follows: Setting the moments about x equal to zero,

$$2.5T_v + 1F + 2F \times 2 - 4T_H = 0$$

where $T_v = T_H \tan 20° = 55,000 \tan 20° = 20,018 \text{ lbf}$

$$20,018 \times 2.5 + 1F + 2F \times 2 - 55,000 \times 4 = 0$$
$$F = 33,991 \text{ lbf}$$

Two very important rules may be derived from the analysis shown. First, screws should be placed as near the working edge of a steel as is practical. Those screws have a greater moment arm and therefore are better for restricting tipping. The greater number of screws in a trimming die section, for example, are placed near the cutting edge. Secondly, tipping is not as likely to be a problem if the height-to-width ratio of the steel is less than, or equal to, 1. The vertical forces will then tend to counteract side thrusts without the aid of screws. Tipping is more of a problem in form dies, where the side thrusts exceed the vertical force. The reverse is true in cutting dies.

Limitation of Bending. The side thrusts on a die steel cause it to bend or deflect in a manner similar to a cantilever beam. Deflection is limited by controlling the height-to-width ratio of the component. Short or stubby components are more rigid. The composite trim steel from Fig. 3-24 is used for this study. The analysis of deflection will use Fig. 3-28 for reference. Because of shut-height requirements, the tool-steel blade must be about 1 in. (25.4 mm) high. The thickness of the blade h can be found to restrict deflection to the 0.0001 in. (0.002 mm) previously determined. Deflection of the soft-steel base is negligible because of its very stubby and rigid shape.

$$\delta = \frac{F_v L^3}{3EI} = \frac{F_v L^3}{3Ebh^3/12} \qquad (4A)$$

$$h^3 = \frac{F_v L^3 12}{3Eb\delta} = \frac{7,350 \times 1^3 \times 12}{3 \times 30 \times 10^6 \times 10 \times 0.0001} = 0.98$$

$$h \cong 1.00 \text{ in.}$$

Some of the analytical procedures shown can be applied to other components under similar loads. These would include such components as spacers, gages, keepers, and cam drivers.

Base-plate Design. Most die steels are mounted on base plates, which are commonly called "shoes". The lower shoe is fastened to a bolster plate, which is usually attached to the press bed. Most beds have large openings to permit the installation of air cushions or the drop-through of scrap, blanks, or parts. The bolster and/or lower shoe must be rigid and not deflect excessively into the bed opening. Otherwise, the clearance or die alignment may be impaired. The thickness requirement of the lower shoe can be computed with this deflection approach. The upper shoe is usually made equal to, or thinner than, the lower shoe.

Plate Thickness for a Stated Loading and Deflection. This procedure is useful to determine the correct thickness of die shoes and other similarly loaded die members such as some pressure pads.

The mathematical analysis is greatly simplified if the lower shoe is considered to be on parallels. The parallels would be spaced about equal to the bed hole size. The shoe thickness is made large enough to restrict deflection between the parallels. Dies are sometimes placed on parallels to allow removal of parts or scrap which cannot drop through the press bed.

This approach of using parallels for the analysis provides a considerable safety factor in case parallels are not used beneath the lower shoe.

For this example, assume a lower shoe to be a simply supported beam with a concentrated load of 40,000 lbf (178 kN) at the center. The distance between the parallels is 10 in. (254 mm), the width of the lower shoe is 20 in. (508 mm), and there is a maximum deflection of 0.001 in. (0.02 mm), which would cause the die clearances to decrease.

$$\delta = \frac{FL^3}{48EI} = \frac{FL^3}{48Ebh^3/12} \qquad (4B)$$

For a steel shoe, $E = 30 \times 10^6$ psi

$$h^3 = \frac{FL^3}{48Eb\,\delta/12} = \frac{40,000 \times 10^3 \times 12}{48 \times 30 \times 10^6 \times 20 \times 0.001} = 16.67$$

$$h = 2.55 \text{ in.}$$

For a cast-iron shoe, $E = 20 \times 10^6$

$$h^3 = 25$$
$$h = 2.92 \text{ in.}$$

The lower rigidity of the cast iron necessitates the use of a thicker shoe. These calculations can be duplicated if the die forces are known and are on the center of the shoe. The bed opening is used as the distance between the parallels. The allowed deflection can be determined by using similar triangles and the height of the steel. If the die shoes are to be fastened directly to the press bolster and slide, thinner shoes are permissible. For a given plant, a table could be made covering the tonnage range of the dies and the press bed openings. A listing of lower-shoe thicknesses is found in Table 3-4. If the lower shoe is not placed on parallels, the values of the table may be used for the combined thickness of lower shoe and bolster plate. SI units may be used with equation (4B).

If a distributed load is assumed, the correct formula to use is

$$\delta_{\mathcal{C}} = \frac{5FL^3}{384EI} = \frac{FL^3}{76.8EI} \qquad (4C)$$

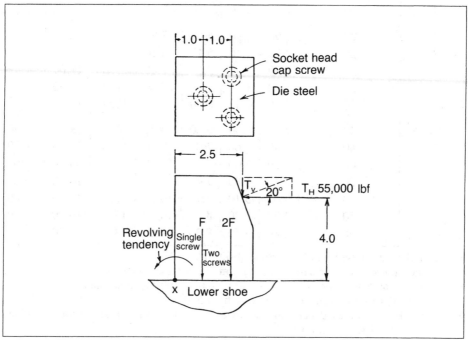

Fig. 3-27. The use of screws to prevent tipping of a die steel.

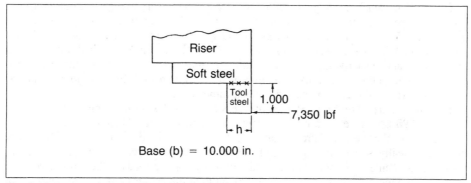

Fig. 3-28 Design of die steels to limit deflection.

3-35

TABLE 3-4
**Thicknesses of Steel Lower Die Shoes Having A
Centrally Applied Load**

Load, tons	Shoe Width, in.	Distance Between Parallels, in.						
		5	10	15	20	25	30	40
10	10	1.30	2.60	3.80				
20	10	1.60	3.20	4.80				
30	10	1.80	3.70	5.50				
40	10	2.00	4.10	6.10				
50	20	1.70	3.50	5.20	6.90			
60	20	1.80	3.70	5.50	7.40			
70	20	1.90	3.90	5.80	7.80			
80	20	2.00	4.10	6.10	8.10			
90	30	1.80	3.70	5.50	7.40	9.20	11.00	
100	30	1.90	3.80	5.70	7.60	9.50	11.50	
150	30	2.20	4.40	6.50	8.70	10.90	13.10	
200	40	2.20	4.40	6.50	8.70	10.90	13.10	17.50
250	40	2.40	4.60	7.00	9.40	11.80	14.10	18.80
300	40	2.50	5.00	7.50	10.00	12.50	15.00	20.00
350	50	2.40	4.80	7.40	9.80	12.20	14.70	19.60
400	50	2.60	5.10	7.70	10.20	12.80	15.30	20.40
450	50	2.70	5.30	8.00	10.60	13.30	16.00	21.30
500	50	2.80	5.50	8.30	11.00	13.80	16.50	22.00

- Calculations are based on a deflection of 0.001 in.
- To obtain thicknesses for cast-iron shoes, multiply table values by 1.15.
- For an allowable deflection of 0.002 in. (0.05 mm), multiply table values by 0.785. For 0.005-in. (0.13 mm) deflection, multiply by 0.580.
- If parallels are not used beneath the lower shoe, the value may be the combined thickness of the shoe and bolster.

DIE DIMENSIONS

Die-block dimensions are governed by the strength necessary to resist the cutting forces and will depend on the type and thickness of the material being cut. On very thin materials, 0.5-in. (12.7-mm) thickness should be sufficient, but except for temporary tools, finished thickness is seldom less than 0.875 in. (22.22 mm), which allows for blind screw holes and resharpening and also builds up the tool to a common shut height for quick die change considerations. The empirical data given in Tables 3-5 to 3-7 may be used to determine the approximate die-block dimensions.

Design of Large Die Sets. Studies of stress distributions in dies weighing more than 100 tons (890 kN) have been made to determine their optimum thicknesses and width.[5]

Nearly all the vertical stress components are contained in the stress boundary diagrammatically shown in Fig. 3-29A. The stress studies were made for both edge-to-edge and circular loading. For assumed uniformly applied loads at the top of a die, the vertical component P_y at the bottom is a maximum at the center ($x = 0$) and a minimum where $x = w/2$, as shown in Fig. 3-29B.

TABLE 3-5
Die Thickness Per Ton of Pressure[4]

Stock Thickness, in.	Die Thickness, in.*
0.1	0.03
0.2	0.06
0.3	0.085
0.4	0.11
0.5	0.13
0.6	0.15
0.7	0.165
0.8	0.18
0.9	0.19
1.0	0.20

* For each ton per square inch of shear strength.

TABLE 3-6
Factors for Cutting Edges Exceeding 2 Inches[4]

Cutting Perimeter, in.	Expansion Factor
2–3	1.25
3–6	1.5
6–12	1.75
12–20	2.0

TABLE 3-7
Minimum Critical Area vs. Impact Pressure[4]

Impact Pressure, tons	Cross-sectional Area of a Die Between the Cutting Edge and Outside Border of a Die, in.^{2}
20	0.5
50	1.0
75	1.5
100	2.0

Optimum dimensions of a large die can be determined with the use of Fig. 3-30.
Examples involving edge-to-edge and circular loading are as follows:

EXAMPLE 2: (circular loading) Given the circular loading on an area of 40 in. (1 m) diameter, $c = 20$ in. (0.5 m), load $P_0 = 50,000$ psi (345 MPa), find the size of the die, so that the maximum stress transmitted to the press bed shall not exceed 30,000 psi (206 MPa).

Solution:

1. Compute ratio $P_Y/P_0 = 30,000/50,000 = 0.600$ and locate on ordinate in Fig. 3-30.

2. Follow along horizontal line from point $P_Y P_0 = 0.600$ to intersect with curve P_Y (max).

3. Extend vertical line down to intersect with curve P_Y (min). The point of intersection

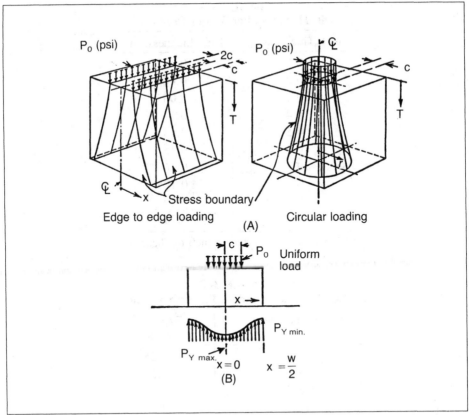

Fig. 3-29. Stress considerations in dies: (A) stress distributions for circular and edge-to-edge loading; (B) maximum and minimum vertical stress components.[6]

corresponds to a value of $r/c = 1.51$ or $r = 1.51c, r = 151 \times 20 = 30.2$ in. (767 mm), and

$d = 2r = 60.4$ in. (1.53 m)

This is the required diameter or width of the die.

4. To determine the thickness of the die set, extend the vertical line farther to intersect abscissa T/c. This point of intersection corresponds to $T/c = 1.78$ or

$T = 1.78c = 1.78 \times 20 = 35.6$ in. (904 mm)

This is the required thickness of the die.

5. From intersection with curve P_Y(min) read horizontally the ordinate value

$P_Y/P_0 = 0.324$

or $P_Y = P_Y$ (min) $= 0.324 \times P_0 = 16,200$ psi (112 MPa). This is equal to the minimum stress at the edge of the die being transmitted to the press bed.

EXAMPLE 3: (edge-to-edge loading) Given the loading on rectangular area of $2c \times L$ as shown in Fig. 3-29, intensity of load $P_0 = 50,000$ psi (345 MPa). Assume the thickness of die is limited to a value of $T/c = 0.800$ and $c = 31$ in. (787 mm). Required to find the width of the die, w, and the maximum stress P_Y transmitted to the platen below the die for these conditions.

Solution:

1. Ratio $T/c = 0.800$. Locate this point on abscissa of Fig. 3-30.

2. Extend vertical line from this point upward to intersect curve P_Y (min). This point corresponds to a value of $x/c = 1.700$. Hence $x = 1.700c$ or $w = 2xc = 3.4c = 105.4$

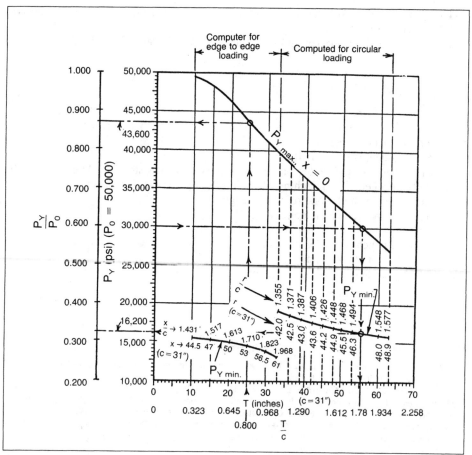

Fig. 3-30 Graph for determining optimum die proportions.[6]

in. (2.68 m). This is the required width of the die.
3. Extend vertical line $T/c = 0.800$ farther to intersect curve P_Y (max).
4. Read horizontally on ordinate P_Y/P_0 = value equal to 0.872, or

$$P_Y = P_Y \text{ (max)} = 0.872 \times P_0 = 43,600 \text{ psi (301 MPa)}$$

This is the maximum stress transmitted to the platen for the specified loading condition. The maximum stress is at the center of the die at the plane of contact with the press platen.

PUNCH DIMENSIONING

The determination of punch dimensions has been generally based on practical experience.

When the diameter of a pierced round hole equals stock thickness, the unit compressive stress on the punch is four times the unit shear stress on the cut area of the stock, from the formula

$$\frac{4S_s}{S_c} \frac{t}{d} = 1 \tag{10}$$

where S_c = unit compressive stress on the punch, psi (Pa)
S_s = unit shear stress on the stock, psi (Pa)
t = stock thickness, in. (mm)
d = diameter of punched hole, in. (mm)

The diameters of most holes are greater than stock thickness; a value for the ratio d/t of 1.1 is recommended.[1]

The maximum allowable length of a punch can be calculated from the formula

$$L = \frac{\pi d}{8} \sqrt{\frac{E}{S_s} \frac{d}{t}} \tag{11}$$

where d/t = 1.1 or higher

E = modulus of elasticity

This is not to say that holes having diameters less than stock thickness cannot be successfully punched. The punching of such holes can be facilitated by:

1. Punch steels of high compressive strengths.
2. Greater than average clearances.
3. Optimum punch alignment, finish, and rigidity.
4. Shear on punches or dies or both.
5. Prevention of stock slippage.
6. Guiding the punch with the stripper.

References

1. E.J. Ross, "Drawability as Influenced by Product Engineering and Processing," SP63-101, Society of Manufacturing Engineers, Detroit, 1963.
2. J. Van Hamersveld, "Cost Control Engineering," *Machine Design* (March 1951).
3. G.M. Foster, "Low CostTooling. . . Estimating and Economics," *The Tool Engineer* (November 1949).
4. F. Strasser, "Should Die Thickness Be Calculated?", *American Machinist* (February 19, 1951).
5. American Society of Tool Engineers, "Tool Engineers Handbook," *A*. Product Development (Sec. 2), *B*. Tool Engineering Economics (Sec. 3), *C*. Production Analysis and Cost Estimating (Sec. 4) (NY: McGraw-Hill Book Co., 1949).
6. J.L. Lebach, "A Method for the Determination of Die-set Dimensions for Closed-die Forging Processes," Hydropress, Inc., and Loewy Construction Co., Inc., New York, 1951.

SECTION 4
SHEAR ACTION IN METAL CUTTING

The cutting of metal between die components is a shearing process in which the metal is stressed in shear between two cutting edges to the point of fracture, or beyond its ultimate strength.

The metal is subjected to both tensile and compressive stresses (Fig. 4-1); stretching beyond the elastic limit occurs, then plastic deformation, and reduction in area. Finally, fracturing starts through cleavage planes in the reduced area and becomes complete.

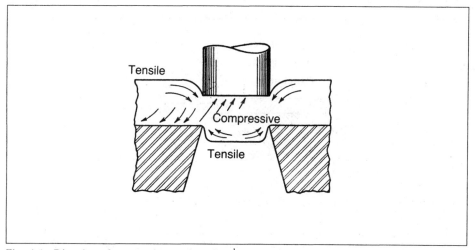

Fig. 4-1. Direction of stresses in metal cutting.[1]

The fundamental steps in shearing or cutting are shown in Fig. 4-2. The pressure applied by the punch on the metal deforms it into the die opening. When the elastic limit is exceeded by further loading, a portion of the metal is forced into the die opening in the form of an embossed pad on the lower face of the material and a corresponding depression on the upper face, as indicated at A. As the load is further increased, the punch will penetrate the metal to a certain depth and force an equal portion of metal thickness into the die, as indicated at B. This penetration occurs before fracturing starts and reduces the cross-sectional area of metal through which the cut is made. Fractures will start in the reduced area at both upper and lower cutting edges, as indicated at C. If the clearance is suitable for the material being cut, these fractures will spread toward each other and eventually meet, causing complete separation. Further travel of the punch will carry the cut portion through the stock and into the die opening.

Clearance. Clearance is the measured space between the mating members of a die set. Proper clearance between cutting edges enables the fractures to meet and the fractured portion of the sheared edge has a clean appearance. For optimum finish of a cut edge,

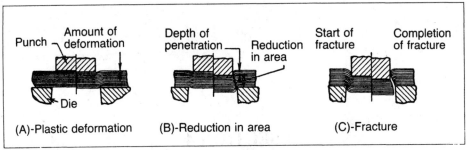

Fig. 4-2. Steps in shearing metal.[2]

proper clearance is necessary and is a function of the kind, thickness, and temper of the work material. Clearance, penetration, and fracture are shown schematically in Fig. 4-3.

In Fig. 4-4, characteristics of the cut edge on stock and blank, with normal clearance, are schematically shown. The upper corner of the cut edge of the stock (indicated by *A*) and the lower corner of the blank (indicated by *A-1*) will have a radius where the punch and die edges, respectively, make contact with the material. This is due to the plastic deformation taking place. This edge radius will be more pronounced when cutting soft metals. Excessive clearance also will cause a large radius at these corners, as well as a burr on opposite corners.

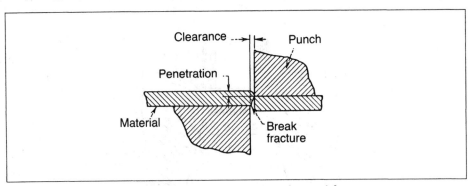

Fig. 4-3. Schematic drawing illustrating clearance, penetration, and fracture.

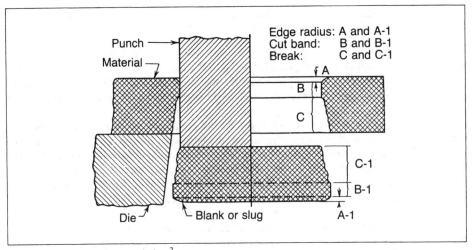

Fig. 4-4. Cut-edge characteristics.[2]

In ideal cutting operations, the punch penetrates the material to a depth equal to about one-third of its thickness before fracture occurs, and forces an equal portion of the material into the die opening. That portion of the thickness so penetrated will be highly burnished, appearing on the cut edge as a bright band around the entire contour of the cut adjacent to the edge radius—indicated at *B* and *B-1* in Fig. 4-4. When the cutting clearance is not sufficient, additional bands of metal must be cut before complete separation is accomplished, as shown at *B* in Fig. 4-5. When correct cutting clearance is used, the material below the cut will be rough on both the stock and the slug. With correct clearance, the angle of fracture will permit a clean break below the cut band because the upper and lower fractures extend toward one another. Excessive clearance will result in a tapered cut edge since, for any cutting operation, the opposite side of the material which the punch enters will, after cutting, be the same size as the die opening.

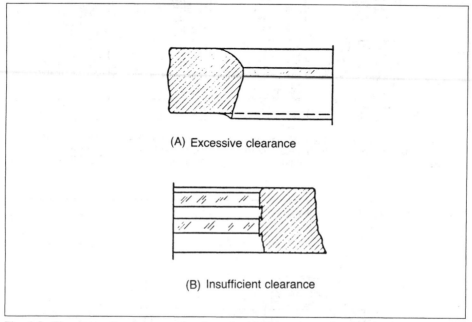

(A) Excessive clearance

(B) Insufficient clearance

Fig. 4-5. The greater the clearance, the closer the condition approaches forming instead of cutting.[2]

The width of the cut band is an indication of the hardness of the material, provided that the die clearance and material thickness are constant; the wider the cut band, the softer the material. The harder metals require larger clearances and permit less penetration by the punch than ductile metals; dull tools create the effect of too small a clearance as well as a burr on the die side of the stock. The effects of various amounts of clearance are shown in Figs. 4-5 to 4-7. Defective or nonhomogeneous material cut with the proper amount of clearance will produce nonuniform edges.

The edge conditions *C* and the hypothetical load curves *B* (Fig. 4-6, Cases 1, 2, 3, and 4) are shown, as well as the amount of deformation and extent of punch penetration.

Location of the proper clearance (Fig. 4-7) determines either hole or blank size. Punch size controls hole size; die size controls blank size.

At *A*, which shows clearance *C* for blanks of a given size, make die to size and punch smaller by the amount of total clearance 2*C*. At *B*, which shows clearance for holes of a given size, make punch to size and die larger by the amount of the total clearance 2*C*.

The application of clearances for holes of irregular shape is diagramed in Fig. 4-8; at *B*, the hole will be a punch size, while at *A*, the blank will be of the same dimensions as the die.

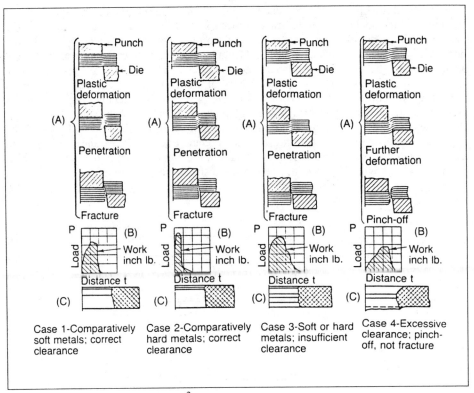

Fig. 4-6. Effect of certain clearances.[2]

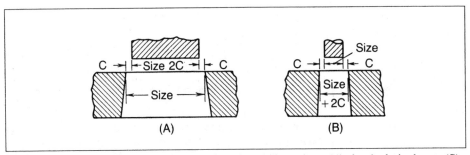

Fig. 4-7. Clearance location related to part, punch and dimensions: (A) slug is desired part; (B) slug is scrap.

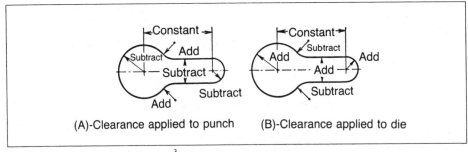

Fig. 4-8. How to apply clearances.[3]

One manufacturer charts clearances per side for groups of materials up to and including thicknesses of 0.125 in. (3.18 mm).

The die-clearance chart (Fig. 4-9) may be used to find the recommended die clearance to be allowed, and to be provided for, in designing a die for service as determined by the materials groups which follow, and for the preestablished percentage of material thickness of the original part which the die is designed to produce.

Group 1. 1100 and 5052 aluminum alloys, all tempers. An average clearance of 4.5% of material thickness is recommended for normal punching and blanking.

Group 2. 2024 and 6061 aluminum alloys; brass, all tempers; cold rolled steel, dead soft; stainless steel soft. An average clearance of 6% of material thickness is recommended for normal punching and blanking.

Group 3. Cold-rolled steel, half hard; stainless steel, half hard and full hard. An average clearance of 7½% is recommended for normal punching and blanking.

An example is found in Fig. 4-9. Here, it is seen that, for a nominal stock thickness of 0.080 in. (2.03 mm), the die clearance for any Group 1 materials would be 0.0036 in. (0.091 mm); for any Group 2 materials, 0.0048 in. (0.122 mm); for any Group 3 materials, 0.006 in. (0.15 mm).

Interchangeable Use of Same Die. It is often good economy to use a die originally designed for a given type and thickness of material, for punching or blanking material of a different type and/or thickness. A convenient chart for such a purpose is provided in Fig. 4-10.

Unlike Fig. 4-9, which is based on single average clearance values, this chart is based

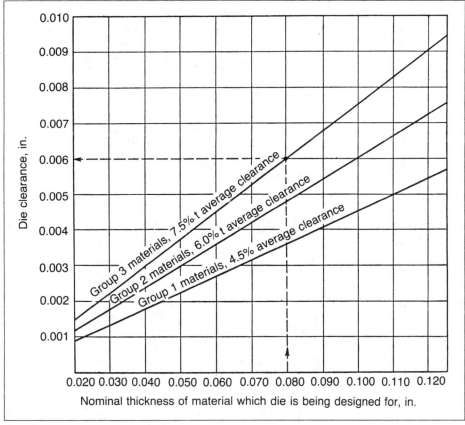

Fig. 4-9. Die-clearance chart by groups of materials, using the recommended percentage of metal thickness (indicated clearances are per side).[4]

on clearance ranges of 3.4% to 6.8%; 4.5% to 9.0%, and 5.6% to 11.2% for material Groups 1, 2, and 3, respectively. Materials groups are the same as for Fig. 4-9.

EXAMPLE 1: Die made for 5052 aluminum alloy 0.030 in. (0.76 mm) thick. Can it be used for the same material, 0.064 in. (1.62 mm) thick? This is a Group 1 material. From Fig. 4-9, the clearance is 0.0012 in. (0.030 mm). On the 0.0012-in. (0.030-mm) abscissa of Fig. 4-10, find minimum and maximum points for Group 1 materials. Horizontal extensions from these points show a permissible material thickness range of 0.020 to 0.040 in. (0.51 to 1.02 mm); 0.064-in. (1.62 mm) thickness would therefore exceed the permissible maximum, and the die should not be used, since clearance on the original die would be below the range minimum for group 1 materials with 0.064-in. (1.62-mm) nominal stock thickness.

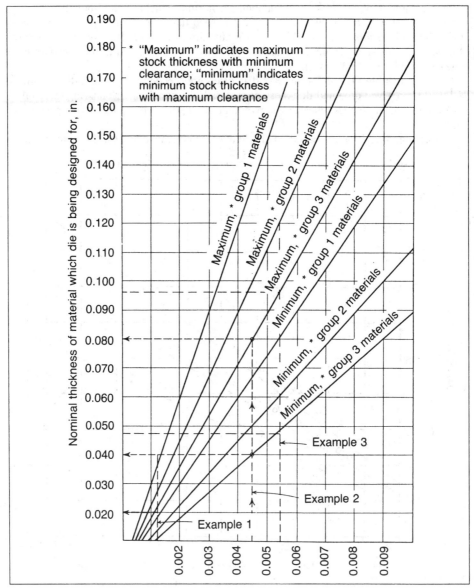

Fig. 4-10. Interchangeable material thickness chart, based on clearance for which a die was originally built (indicated clearances are per side).[4]

EXAMPLE 2: Die made for 2024 aluminum alloy, 0.090 in. (2.29 mm) thick. Can it be used for cold-rolled steel, half hard, 0.042 in. (1.07 mm) thick? The die was originally made for a Group 2 material with a clearance of 0.0054 in. (0.137 mm) (from Fig. 4-9). The intended material is Group 3. On the 0.0054-in. (0.137-mm) abscissa of Fig. 4-10 find maximum and minimum points for Group 3 materials. Horizontal extensions from these points show a permissible thickness range of 0.048 to 0.096 in. (1.22 to 2.44 mm) for the cold-rolled steel, half hard. Since the intended thickness is 0.042 in. (1.07 mm), or less than the range minimum, the die should not be used, since clearance on the original die would be below the range minimum for Group 3 materials of 0.042-in. (1.07-mm) nominal stock thickness.

EXAMPLE 3: Die made for cold-rolled steel, half hard, 0.060 in. (1.52 mm) thick. Can it be used for 6061 aluminum alloy, 0.064 in. (1.62 mm) thick? The die was originally designed for a Group 3 material and a clearance of 0.0045 in. (0.114 mm) (Fig. 4-9). The intended material is Group 2. On the 0.0045-in. (0.114-mm) abscissa of Fig. 4-10 find maximum and minimum points for Group 2 materials. Horizontal extensions show a permissible thickness range of 0.05 to 0.10 in. (1.27 to 2.5 mm) for the alloy. Since the intended thickness of 0.064 in. (1.62 mm) lies within this range, the die may be used.

Working only from the known clearance for which a die was originally built, Fig. 4-10 may be used to find the thicknesses of a material of any group, for which the die may be used.

EXAMPLE 4: Given a clearance of 0.0045 in. (0.114 mm), can a die with this clearance be used for stainless steel, full hard, 0.062 in. (1.57 mm) thick? This is a Group 3 material. Along the 0.0045-in. (0.114 mm) abscissa of Fig. 4-13, find maximum and minimum points for Group 3 materials. Horizontal extensions from these points show a permissible thickness range for the stainless steel full hard, of 0.040 to 0.080 in. (1.02 to 2.03 mm); since the intended thickness lies within the range, the die may be used.

Clearance ranges to produce uniformly good cuts for aluminum are listed in Table 4-1.

TABLE 4-1
Total Clearance for Blanking Aluminum

Alloy	Clearance Range, % of Total Stock Thickness
1100-O	7–16
1100-T	8–17½
2017-T	10–18

Clearances for punching electrical steel laminations are listed in Table 4-2 arranged in the order of decreasing silicon content. The data indicate that, the greater the silicon content, the greater is the required die clearance. A softer stock will require smaller die clearance, but greater angular clearance to prevent scoring of die walls. Angular clearances, per side in 1.5-in. (38.1 mm) length, ground after hardening, are 0.001 to 0.002 in. (0.02 to 0.05 mm) for hard stock, and 0.002 to 0.003 in. (0.05 to 0.08 mm) for soft stock.

TABLE 4-2
Per-side Clearance, Inches, for Lamination Dies

Grades of Steel	29 gage (0.0155 in.)	26 gage (0.0186 in.)	24 gage (0.0249 in.)
Transformer Grades	0.0007	0.00085	0.001
Dynamo Special	0.0007	0.00085	0.001
Dynamo	0.0006	0.00075	0.0009
Electrical	0.0006	0.00075	0.0009
Armature	0.0005	0.00065	0.0008
Export Armature	0.0005	0.00065	0.0008

CLEARANCES FOR NONMETALLIC MATERIALS

For nonmetallic materials other than cellulose acetate, cloth, and paper, the total clearance between punch and die should be 2.5% of stock thickness. For mica (data supplied by New England Mica Co.), the die set should be built to a shear fit, that is, tissue paper will be cut cleanly; the weight alone of the lower half of a new die is not sufficient to open the die set. Normal die wear ordinarily will allow the die set to be separated by shaking, without tapping with a hammer.

Clearances for fully cured C-stage thermosetting laminated plastics should be approximately 0.0005 in. (0.013 mm) for sheet thicknesses of 0.015 to 0.032 in. (0.38 to 0.81 mm) and 0.0015 in. (0.038 mm) for sheets 0.040 to 0.093 in. (1.02 to 2.36 mm) thick.

CLEARANCES AND ALLOWANCES FOR SHAVE DIES

Where only one shave is required, a leading manufacturer's standard clearance is 0.001 in. (0.02 mm) per side; in some cases, clearance can be 1.5% of stock thickness. This same manufacturer uses a shaving allowance per side of 10% of stock thickness plus 0.002 (0.05 mm) ± 0.001 in. (0.02 mm) tolerance; minimum allowance is 0.005 in. (0.13 mm).

Small gears should have an allowance of 0.0035 in. (0.089 mm) for each 0.031-in. (0.79-mm) thickness of the blank; two-step shaving operations should remove two-thirds of such allowance in the first shave, and the balance in the second shave.[5]

Allowances for shaving the softer metals are larger than for the harder metals; compare Table 4-3 and Table 4-4, which were taken from another manufacturer's standard data sheets.

Die-cut holes of less than 1 in. (25 mm) diameter tend to close in; blanks under 1 in. (25.4 mm) diameter tend to swell. Allowances should be added to the punch diameter, or subtracted from the die diameter (Table 4-5). One-half the given figure should be added or subtracted all around a punch or die of irregular contour.

TABLE 4-3
Shaving Allowances Per Side, Inches,
for Steel, Brass, and German Silver Stock[6]

Thickness of Blank, in.	Steel			Brass and German Silver
	Hardness, 50–66 R_B	Hardness, 75–90 R_B	Hardness, 90–105 R_B	
Where One Shave Is Necessary				
0.0468	0.0025	0.003	0.004	0.005
0.0625	0.003	0.004	0.005	0.006
0.078	0.0035	0.005	0.006–0.007	0.007
0.0938	0.004	0.006	0.007–0.008	0.008
0.1094	0.005	0.007	0.009–0.011	0.010
0.125	0.007	0.009	0.012–0.014	0.014
Where a Second Shaving Operation Is Necessary				
0.0468	0.00125	0.0015	0.002	0.0025
0.0625	0.0015	0.002	0.0025	0.003
0.078	0.00175	0.0025	0.003–0.0035	0.0035
0.0938	0.002	0.003	0.0035–0.004	0.004
0.1094	0.0025	0.0035	0.0045–0.0055	0.005
0.125	0.0035	0.0045	0.006–0.007	0.007

TABLE 4-4
Shaving Allowances for Aluminum[2]

Thickness, in.	First Shave Allowance, in.	Final Shave Allowance, in.
0.03	...	0.004
0.05	...	0.006
0.06	...	0.007
0.08	0.007	0.003
0.100	0.008	0.004
0.125	0.010	0.005
0.175	0.013	0.007
0.250	0.020	0.010

TABLE 4-5
Compensation Allowance
for Part-size Change In Round Holes[3]

Stock Thickness, Gage	Allowance,* in.
Less than 22	0.001
22–16	0.0015
16–10	0.002

* Add to punch diameter or subtract from die diameter.

ANGULAR CLEARANCE

Angular clearance is defined as that clearance below the straight portion of a die surface introduced for the purpose of enabling the blank or the slug (punching operation) to clear the die (Fig. 4-11). Angular clearance is usually ground from 1/4° to 3/4° per side but occasionally as high as 2°, depending mainly on stock thickness and the frequency of sharpening.

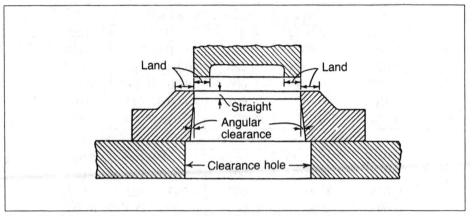

Fig. 4-11. Schematic drawing showing straight, land, and angular clearances.

Other definitions pertinent to die clearances are:

Land: A flat surface contiguous to the cutting edge of a die, its purpose being to reduce the area to be ground and reground in maintaining a sharp cutting edge (Fig. 4-11).

Straight: That portion of the die surface between the cutting edge and the angular clearance in a blanking or punching die (Fig. 4-11). It has been found to be good practice to maintain a minimum straight-wall height of 0.125 in. (3.18 mm) on all materials less than 0.125 in. (3.17 mm) thick. Straight-wall height for thicker materials, equal to material thickness, has proved to be good practice. These rules hold generally good for all classes of dies.

Draft: Draft is the amount of taper placed on a die to enable the severed slug or blank to drop through without binding. In Fig. 4-13, it is shown as A, the projection of the angular clearance. When a die is sharpened below the straight, the die opening is increased. This increase per side can be calculated as

$$A = B \tan \alpha$$

where α = clearance or draft angle, deg.

Table 4-6 lists the increase per side in die openings, ground from the die face below the straight for small angles. This table is useful for ensuring that regrinding in the tapered area does not result in excessive clearance.

EXAMPLE 5: Calculate how much a die, initially ground with minimum hole clearance, can be reground below the straight, and still keep within maximum clearance limits, for the following three materials, all 0.0312 in. (0.794 mm) thick.

Stock A: 1100-O aluminum alloy, with 3.4% to 6.8% T-die-clearance range.
Stock B: Stainless steel, soft with 4.5% to 9% T-die-clearance range.
Stock C: Cold-rolled steels, half hard, with 5.6% to 11.2% T-die-clearance range.

Since minimum clearance was ground on the die, the remaining available clearances for stocks A, B, and C are 0.00106 in. (0.0269 mm), 0.00141 in. (0.0358 mm), and 0.00175 in. (0.0444 mm), respectively. In Table 4-6, selecting the exact or next lower listed values, the maximum permissible depths of regrind are found as shown in Table 4-7.

TABLE 4-6
Die-Opening Increases (Draft), Inches Per Side, Due To Resharpening

Amount Ground Below Die Straight, in.	Draft or Clearance Angle											
	0° 15'	0° 30'	1°	1° 30'	2°	2° 30'	3°	4°	5°	6°	7°	8°
0.010	0.00004	0.00009	0.00018	0.00026	0.00034	0.00043	0.00052	0.00070	0.00087	0.00105	0.00123	0.00140
0.015	0.00006	0.00013	0.00026	0.00039	0.00052	0.00066	0.00079	0.00104	0.00131	0.00158	0.00184	0.00211
0.025	0.00011	0.00022	0.00044	0.00065	0.00087	0.00109	0.00131	0.00175	0.00219	0.00263	0.00307	0.00351
0.035	0.00015	0.00030	0.00061	0.00092	0.00122	0.00153	0.00183	0.00245	0.00306	0.00369	0.00430	0.00492
0.040	0.00017	0.00038	0.00070	0.00105	0.00139	0.00174	0.00210	0.00280	0.00350	0.00420	0.00491	0.00562
0.045	0.00020	0.00039	0.00078	0.00118	0.00157	0.00196	0.00236	0.00315	0.00394	0.00473	0.00552	0.00632
0.050	0.00022	0.00044	0.00088	0.00131	0.00174	0.00218	0.00262	0.00350	0.00437	0.00525	0.00614	0.00703
0.060	0.00026	0.00052	0.00105	0.00157	0.00209	0.00262	0.00314	0.00419	0.00525	0.00631	0.00737	0.00843
0.070	0.00030	0.00061	0.00122	0.00183	0.00244	0.00301	0.00367	0.00489	0.00612	0.00736	0.00859	0.00984
0.080	0.00034	0.00070	0.00140	0.00210	0.00309	0.00349	0.00419	0.00559	0.00700	0.00841	0.00982	0.01124
0.090	0.00039	0.00078	0.00158	0.00236	0.00314	0.00392	0.00472	0.00629	0.00787	0.00946	0.01105	0.01265
0.100	0.00044	0.00087	0.00175	0.00262	0.00349	0.00437	0.00524	0.00699	0.00875	0.01051	0.01228	0.01405
0.125	0.00055	0.00109	0.00218	0.00327	0.00436	0.00546	0.00655	0.00874	0.01094	0.01314	0.01535	0.01757
0.150	0.00065	0.00131	0.00262	0.00393	0.00523	0.00654	0.00786	0.01049	0.01312	0.01576	0.01842	0.02108
0.175	0.00076	0.00153	0.00305	0.00458	0.00611	0.00764	0.00917	0.01224	0.01531	0.01839	0.02149	0.02459
0.200	0.00087	0.00175	0.00350	0.00524	0.00698	0.00873	0.01048	0.01399	0.01750	0.02102	0.02456	0.02811
0.225	0.00098	0.00196	0.00393	0.00589	0.00786	0.00982	0.01179	0.01573	0.01968	0.02365	0.02762	0.03162
0.250	0.00110	0.00218	0.00438	0.00655	0.00872	0.01091	0.01310	0.01748	0.02187	0.02628	0.03069	0.03513
0.275	0.00120	0.00240	0.00480	0.00720	0.00960	0.01201	0.01441	0.01923	0.02406	0.02890	0.03376	0.03865
0.300	0.00131	0.00262	0.00523	0.00785	0.01048	0.01310	0.01572	0.02098	0.02625	0.03153	0.03683	0.04216
0.325	0.00142	0.00284	0.00567	0.00850	0.01135	0.01419	0.01703	0.02273	0.02843	0.03416	0.03990	0.04567
0.350	0.00153	0.00305	0.00611	0.00891	0.01222	0.01528	0.01834	0.02447	0.03062	0.03678	0.04297	0.04919
0.375	0.00164	0.00327	0.00654	0.00916	0.01309	0.01638	0.01965	0.02622	0.02381	0.03942	0.04604	0.05270
0.400	0.00174	0.00349	0.00698	0.01047	0.01397	0.01746	0.02096	0.02797	0.03500	0.04204	0.04911	0.05622
0.425	0.00185	0.00361	0.00742	0.01113	0.01484	0.01855	0.02227	0.02972	0.03718	0.04467	0.05218	0.05973
0.450	0.00196	0.00393	0.00785	0.01178	0.01571	0.01965	0.02358	0.03147	0.03937	0.04729	0.05525	0.06324
0.475	0.00207	0.00415	0.00829	0.01243	0.01659	0.02074	0.02489	0.03322	0.04156	0.04992	0.05892	0.06676
0.500	0.00218	0.00436	0.00872	0.01309	0.01746	0.02183	0.02620	0.03496	0.04374	0.05255	0.06139	0.07027

TABLE 4-7
Maximum Regrind Depths, Inches, for Clearance Angles and Stocks Listed

Clearance Angle	Stock A*	Stock B†	Stock C‡
15′	0.225	0.300	0.400
30′	0.100	0.150	0.200
1°	0.060	0.080	0.100
1°30′	0.040	0.050	0.060
2°	0.025	0.040	0.050
2°30′	0.015	0.025	0.040
3°	0.015	0.025	0.025
4°	0.015	0.015	0.025
5°	0.010	0.015	0.015
6°	0.010	0.010	0.015
7°	No regrind	0.010	0.010
8°	No regrind	0.010	0.010

* 1100 aluminum alloy, 0.03125 in. (0.7938 mm) thick.
† Stainless steel, soft, 0.03125 in. (0.7938 mm) thick.
‡ Cold-rolled steel, half hard, 0.03125 in. (0.7938 mm) thick.

SHEAR

Shear is the amount of relief ground off the face of a die or punch, primarily: (1) for reducing the required shearing force, (2) to reduce stress on the tool, (3) to enable thicker or more resistant stock to be punched on the same press, or (4) to permit use of lower-rated presses.

Relation of Forces to Amount of Shear. Forces, but not work done, vary with various amounts of shear (Fig. 4-12).

View A of Fig. 4-12 shows a cutting operation in which the cutting edges are parallel, i.e., shear is zero. Stock thickness is indicated by t. Since the cut takes place on the entire periphery at once, it is obvious that the operation will entail maximum load. The load diagram at the right shows the rapid rise at maximum pressure, then the sudden load release, sometimes severe on both press and dies, as the cut is completed.

View B of the same figure shows a punch ground with shear. If the punch were to be ground so that the height of the angle equals one-third of the metal thickness, i.e., shear equals 1/3 t, its leading edge would start cutting before the rest of the punch made contact. With this condition, only part of the punch would be cutting at any one instant. While cutting load would be decreased, the punch would have to progress further through the stock to complete the cut. With the punch in position shown, cutting pressure would be at maximum for this amount of shear, and because the cut is more or less complete the total cutting load would be less than that required for the condition shown in view A. Pressure is slightly less than when shear is zero, but as the work is the same, the distance through which pressure is applied is greater. Load release also is somewhat less sudden.

View C of the same figure shows that, if shear is ground on the punch so that the height of the angle equals the full thickness of the metal cut, the shear would be equal to 1t. When the leading edge has progressed entirely through distance t, the trailing edge is just making contact with the metal. Maximum pressure would be at this position of the punch, and since most of the cut is complete the total load would be about half that required when shear is zero. The distance through which the punch load functions is greater still than that indicated in view B.

Concave shear has been cylindrically ground on the long sides of the die of Fig. 4-13 A, which produces a rectangular blank. The punch first cuts along the short sides of the

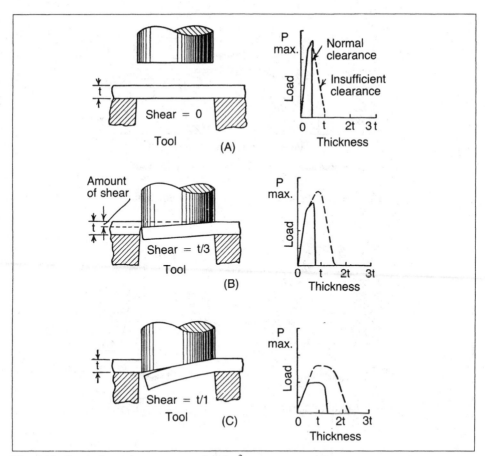

Fig. 4-12. Loads for various amounts of shear.[2]

blank, and at these locations the punch is supported as it cuts the long sides toward their centers. The stock is held more securely than by first cutting the long sides, with the shear ground on the short sides of the die. If the blank is round, a series of scalloped or wave-like shear areas should be ground on the die. The amount of shear varies from less than $1t$ for heavy ferrous material to $2t$ for thin stock. Flat blanks also will be produced by a die with convex shear (Fig. 4-13B), but the upper part of the die section will be weak. The punched-out metal (Fig. 4-13C, D, and E) will be distorted, but the stock will be flat.

A disadvantage of adding unbalanced shear to a punch is that a lateral force is created that will cause the punch to deflect. In some cases, this deflection can result in the stock being displaced. Also, punch deflection will result in improper clearances and die damage. A preferred method of providing shear on a punch is indicated in Fig. 4-14.

In the die shown in Fig. 4-15, flat areas around the corners at D provide a level support for the stock and prevent its slippage at the time of punch engagement. With the amount of shear B and angle of shear A, cutting progresses from the outside to the center, producing a flat blank.

Shear location is determined by confining distortion to unwanted metal: grind shear on the die if a flat blank is required; grind shear on the punch if the punched-out metal is to be scrap.

Shear is sometimes applied to a die for the purpose of forming a portion of the blank to the required shape. The die illustrated in Fig. 4-16 produces the clip shown at the right. The angular portion of the punch strikes the strip first, shearing the tongue of the piece part and simultaneously curling it. When the flat part of the punch contacts the strip, the

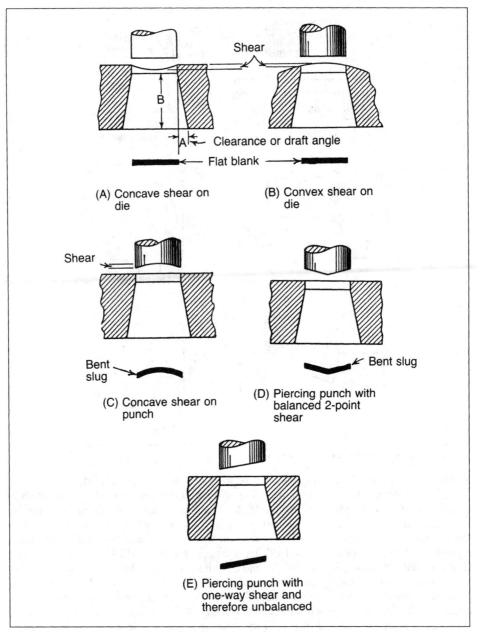

Fig. 4-13. Use of shear on dies and punches (exaggerated views).

tab of the part is blanked out. The angle of the punch must be determined by trial and error. After it has been established, a gage is made for use in future punch-sharpening operations.

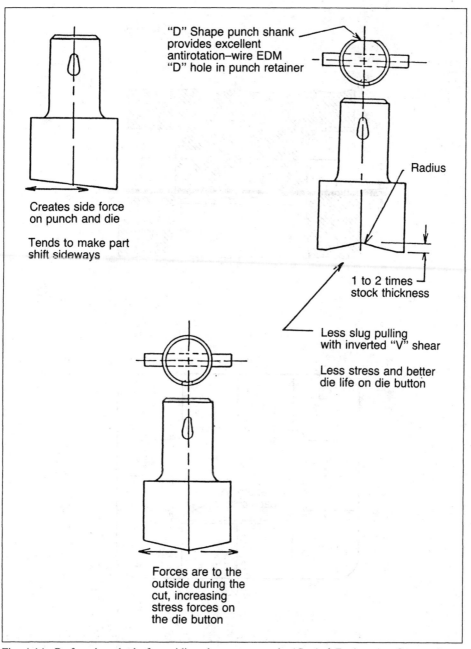

"D" Shape punch shank provides excellent antirotation–wire EDM "D" hole in punch retainer

Creates side force on punch and die

Tends to make part shift sideways

Radius

1 to 2 times stock thickness

Less slug pulling with inverted "V" shear

Less stress and better die life on die button

Forces are to the outside during the cut, increasing stress forces on the die button

Fig. 4-14. Preferred method of providing shear on a punch. (*Capital Engineering Company*).

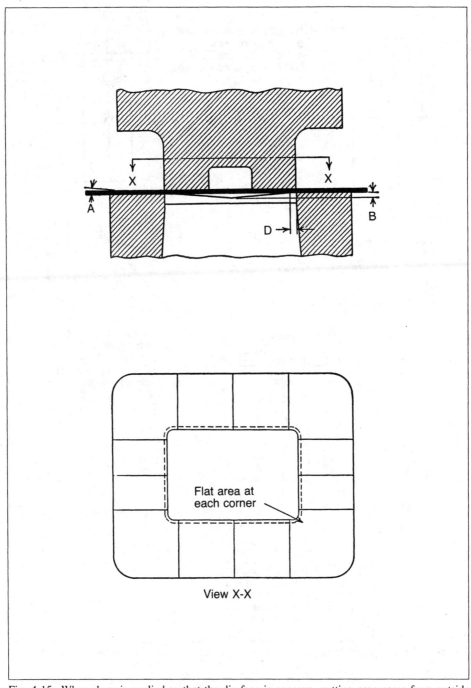

View X-X

Fig. 4-15. When shear is applied so that the die face is concave, cutting progresses from outside to center.

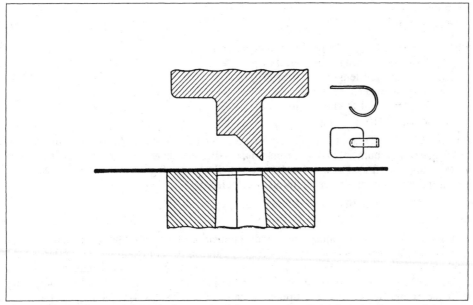

Fig. 4-16. Angular part of punch shears tab of the part and curls it; then the flat part of the punch blanks the completed piece.[7]

PUNCHING AND BLANKING PRESSURES

Theory of Cutting. When sheet metal is being cut by blanking and punching, the force applied to the metal is basically a shear force. The total force is comprised of equal and opposite forces spaced a small distance (clearance) apart on the metal. The ability of a material to resist this force is called shear strength. The shear strength of a material is directly proportional to its tensile strength and hardness. The shear strength increases as the tensile strength and hardness of a material increase.

Two types of shear strength that must be considered are true shear strength and die shear strength. The strength of a metal being cut on a squaring shear is referred to as the true shear strength. True shear strength is generally estimated to be equal to a metal's tensile strength.

Cutting with a punch and die is more efficient, in terms of energy, than cutting with a shear. For this reason, a punch and die needs less shear force to make a given cut than squaring shears. This lower amount of shear strength required is referred to as the die shear strength and can be estimated to be about 0.75 times the true shear strength for most carbon steels and aluminum alloys. For example, for a 1020 steel with a tensile strength of 60,000 psi (414 MPa), the true shear strength is 60,000 psi (414 MPa) and the die shear strength is 0.75 × 60,000, or 45,000 psi (310 MPa).

There are occasions when it is necessary to calculate cutting force. Some of these occasions may be: (1) when specifications are being prepared for ordering a new press: (2) when equipment for a stamping operation is being selected from existing equipment, or (3) when it is necessary to determine whether a particular thickness or type of metal can be cut on a given machine.

Blanking or Punching Force. The cutting force formula is derived from the basic relationship that force equals area times shearing stress. Blanking or punching force for various metals and gages can be calculated with the aid of Table 4-8 and Table 4-9.

Equations. Assuming no shear on the punch or die, the formula for calculating the force required to blank or punch a given material for any shape is:

$$F = SLT \qquad (1)$$

for round holes:

$$F = S \pi DT \tag{2}$$

where F = blanking or punching force, lbf (N)
S = shear strength of material, psi (Pa)
L = sheared length, in. (mm)
D = diameter, in. (mm)
T = material thickness, in. (mm)
π = 3.1416

EXAMPLE 6: Calculate the force required to punch a 0.50 in. (12.7 mm) diameter hole in 20 gage aluminum 2024-T4. From Table 4-10, the shear strength of the material is 41,000 psi (283 MPa). From Table 4-10, the thickness of 20 gage is 0.038 in. (0.96 mm). Using these values, the force required is calculated:

$$F = 41,000 \times 3.1416 \times 0.50 \times 0.038$$
$$= 2,446 \text{ lbf} = 1.2 \text{ tons} = 10.9 \text{ kN}$$

For convenience, the tonnage required to punch or blank round holes in various metal thicknesses has been calculated and charted in Table 4-10. For example, a 0.050 in. (1.27 mm) diam hole punched through 0.25 in. (6.35 mm) thick mild steel requires 9.8 tons (87.2 kN). The chart also can be used for materials other than mild steel by multiplying the chart figure by a "chart multiplier" obtained from Tables 4-11, 4-12 and 4-13. For example, the chart multiplier for 6061-T6 aluminum is 0.58 which means that its shear strength is 58% of the value for mild steel. Therefore, a 0.050 in. (1.27 mm) diam hole punched through 0.25 in. thick 6061-T6 aluminum would require 9.8 × 0.58 or 5.68 tons (50.5 kN). Force requirements for holes above 1 in. (25.4 mm) diam can be obtained by adding the forces for two diameters that total the desired diameter.[8]

TABLE 4-8
Shear and Tensile Strengths[8]

Material	Shear Strength		Tensile Strength	
	psi	(MPa)	psi	(MPa)
Aluminum 1100-H14	11,000	(75.8)	18,000	(124.1)
Aluminum 20224-T4	41,000	(282.7)	68,000	(468.9)
SAE 3240	150,000	(1,034.2)	105,000	(723.9)
SAE 4130	55,000	(379.2)	90,000	(620.5)
SAE 4130	65,000	(448.2)	100,000	(689.5)
SAE 4130	75,000	(517.1)	125,000	(861.8)
SAE 4130	90,000	(620.5)	150,000	(1,034.2)
SAE 4130	105,000	(723.9)	180,000	(1,241.1)
Stainless (18-8)	70,000	(482.6)	95,000	(655.0)
Steel (0.1 C)	45,000	(310.3)	60,000	(413.7)
Steel (0.25 C)	50,000	(344.7)	70,000	(482.6)
Copper-hard	35,000	(241.3)	50,000	(344.8)
Brass-hard	50,000	(344.7)	78,000	(537.8)
Tin	6,000	(41.4)	5,000	(34.5)
Zinc	20,000	(137.9)	24,000	(165.5)

When shear strength is unknown, but tensile strength is given, the shear strength can be estimated by taking a percentage of the tensile strength. The percentage varies with the material type and thickness. Figure 4-17 gives the percentages obtained from sampling three types of material.

TABLE 4-9
English and Metric Equivalents of U.S. Gage Numbers[8]

U.S. Gage No.	Inch	(mm)	U.S. Gage No.	Inch	(mm)
0000000	0.5	(12.7)	16	0.063	(1.59)
000000	0.4689	(11.91)	17	0.0564	(1.43)
00000	0.4378	(11.11)	18	0.05	(1.27)
0000	0.406	(10.32)	19	0.04	(1.11)
000	0.375	(9.53)	20	0.038	(0.95)
00	0.344	(8.73)			
0	0.313	(7.94)	21	0.0344	(0.87)
			22	0.031	(0.79)
1	0.281	(7.14)	23	0.028	(0.71)
2	0.266	(6.75)	24	0.025	(0.64)
3	0.25	(6.35)	25	0.022	(0.56)
4	0.234	(5.95)			
5	0.219	(5.56)	26	0.019	(0.48)
			27	0.017	(0.44)
6	0.203	(5.16)	28	0.016	(0.40)
7	0.188	(4.76)	29	0.014	(0.38)
8	0.172	(4.37)	30	0.013	(0.32)
9	0.156	(3.97)			
10	0.141	(3.57)	31	0.011	(0.28)
			32	0.010	(0.26)
11	0.125	(3.18)	33	0.009	(0.24)
12	0.109	(2.78)	34	0.009	(0.22)
13	0.094	(2.38)	35	0.008	(0.20)
14	0.078	(1.98)			
15	0.070	(1.79)	36	0.007	(0.18)
			37	0.007	(0.17)
			38	0.006	(0.16)

Nomograph. Punching force for various materials can also be determined with customary inch-pound units from the nomograph in Fig. 4-18. For example, if a 3-in. diameter hole is to be punched in a 0.051 in. 2024-T4 aluminum part, the punching force can be found using the following information and guidelines:

1. Using 9.42 in. (πD) as the length of sheared edge, find 0.051 in. on Scale 1 (at Point 1) and 9.42 in. on Scale 4 (at Point 2).
2. Connect Points 1 and 2.
3. Note that the line connecting Points 1 and 2 intersects Scale 2 at 0.480 in. (12.19 mm), or Point 3.
4. From Table 4-3, determine the shear strength for the material. In this case, the shear strength for 2024-T4 aluminum is 41,000 psi.
5. From Point 4 on Scale 5 (representing 41,000 psi), draw a line to Point 3 on Scale 2.
6. Note that the line connecting Points 3 and 4 intersects Scale 3 at point 5, giving a reading of 9.8 tons.

TABLE 4-10
Force Required to Punch Mild Steel[8]

Metal Gage	Metal Thickness, in. (mm)	Hole Diameter, in. (mm) Force Required, tons (kN)														
		1/8 (3.2)	3/16 (4.8)	1/4 (6.4)	5/16 (7.9)	3/8 (9.5)	7/16 (11.1)	1/2 (12.7)	9/16 (14.3)	5/8 (15.9)	11/16 (17.5)	3/4 (19.1)	13/16 (20.6)	7/8 (22.2)	15/16 (23.8)	1 (25.4)
20	0.036 (0.91)	0.35 (3.1)	0.53 (4.7)	0.71 (6.3)	0.88 (7.8)	1.1 (9.8)	1.2 (10.7)	1.4 (12.5)	1.6 (14.2)	1.8 (16)	1.9 (17)	2.1 (18.7)	2.3 (20.5)	2.5 (22.2)	2.7 (24)	2.8 (25)
18	0.048 (1.22)	0.47 (4.2)	0.71 (6.3)	0.94 (8.4)	1.2 (10.7)	1.4 (12.5)	1.7 (15)	1.9 (17)	2.1 (18.7)	2.4 (21.3)	2.6 (23)	2.8 (25)	3.1 (27.6)	3.3 (29.4)	3.5 (31)	3.8 (33.8)
1/16 or 16	0.060 (1.52)	0.59 (5.2)	0.89 (7.9)	1.2 (10.7)	1.5 (13.3)	1.8 (16)	2.1 (18.7)	2.4 (21.3)	2.7 (24)	2.9 (25.8)	3.2 (28.5)	3.5 (31)	3.8 (33.8)	4.1 (36.5)	4.4 (39)	4.7 (41.8)
14	0.075 (1.90)	0.74 (6.6)	1.1 (9.8)	1.5 (13.3)	1.9 (17)	2.2 (19.6)	2.6 (23)	2.9 (25.8)	3.3 (29.4)	3.7 (33)	4.1 (36.5)	4.4 (39)	4.8 (42.7)	5.2 (46.3)	5.5 (49)	5.9 (52.5)
12	0.105 (2.67)	1.0 (8.9)	1.6 (14.2)	2.1 (18.7)	2.6 (23)	3.1 (27.6)	3.6 (32)	4.1 (36.5)	4.7 (41.8)	5.2 (46.3)	5.7 (50.7)	6.2 (55.2)	6.7 (59.6)	7.2 (64)	7.7 (68.5)	9.3 (82.7)
1/8 or 11	0.120 (3.05)	1.2 (10.7)	1.8 (16)	2.4 (21.3)	3.0 (26.7)	3.5 (31)	4.1 (36.5)	4.7 (41.8)	5.3 (47)	5.9 (52.5)	6.5 (57.8)	7.1 (63.2)	7.7 (68.5)	8.3 (73.8)	8.8 (78.3)	9.4 (83.6)
10	0.135 (3.43)		2.0 (17.8)	2.7 (24)	3.3 (29.4)	4.0 (35.6)	4.6 (41)	5.3 (47)	6.0 (53.4)	6.6 (58.7)	7.3 (65)	8.0 (71)	8.6 (76.5)	9.3 (82.7)	10.0 (89)	10.6 (94.3)
3/16	0.187 (4.75)		2.8 (25)	3.7 (33)	4.6 (41)	5.5 (49)	6.5 (57.8)	7.4 (65.8)	8.3 (73.8)	9.2 (81.8)	10.2 (90.7)	11.1 (98.7)	12.0 (106.8)	12.9 (114.8)	13.8 (122.8)	14.8 (131.7)
1/4	0.250 (6.35)			4.9 (43.6)	6.2 (55.2)	7.4 (65.8)	8.6 (76.5)	9.8 (87.2)	11.0 (98)	12.3 (109.4)	13.5 (120)	14.8 (131.7)	16.0 (142.3)	17.2 (153)	18.5 (164.6)	19.7 (175.3)
5/16	0.312 (7.92)				7.8 (69.4)	9.2 (81.8)	10.8 (96)	12.3 (109.4)	13.8 (122.8)	15.4 (137)	16.9 (150.3)	18.4 (163.7)	20.0 (178)	21.5 (191.3)	23.0 (204.6)	24.6 (218.8)
3/8	0.375 (9.52)					11.1 (98.7)	13.0 (115.6)	14.8 (131.7)	16.6 (147.7)	18.5 (164.6)	20.3 (180.6)	22.1 (196.6)	24.0 (213.5)	25.8 (229.5)	27.7 (246.4)	29.5 (262.4)
1/2	0.500 (12.70)						17.2 (153)	19.7 (175.3)	22.1 (196.6)	24.6 (218.8)	27.1 (241)	29.5 (262.4)	32.0 (284.7)	34.4 (306)	36.9 (328.3)	39.4 (350.5)
5/8	0.625 (15.87)									30.8 (274)	33.8 (300.7)	36.9 (328.3)	40.0 (355.8)	43.0 (382.5)	46.1 (410)	49.2 (437.7)
3/4	0.750 (19.05)										40.6 (361.2)	44.3 (394)	48.0 (427)	51.9 (461.7)	55.4 (492.8)	59.0 (525)
7/8	0.875 (22.22)											51.6 (459)	56.0 (498)	60.2 (535.5)	64.6 (574.7)	69.0 (613.8)
1	1.00 (25.40)												64.0 (569.3)	68.8 (612)	73.8 (656.4)	78.8 (701)

Note: Force requirements are based on mild steel having a shear strength of 50,000 psi (345 MPa). For punching materials having a different shear strength, use a multiplier (see Table 4-11). For example, to punch a 15/16" (23.8 mm) diam. hole through 1" (25.4 mm) thick ASTM A36 steel having a shear strength of 60,000 psi (414 MPa), the value from this table, 73.8 tons (656.6 kN), would be multiplied by 1.2 (from Table 4-11) to obtain 88.6 tons (788 kN) as the force required.[8]

TABLE 4-11
Multiplying Factor for Shear Strength
of Various Metals Compared to Mild Steel[8]

Material	Chart Multiplier	Material	Chart Multiplier
Aluminum		Steel	
1100-0	0.19	Mild A-7 Structural	1.00
1100-H14	0.22	Boiler Plate	1.10
3003-H14	0.28	Structural A-36	1.20
2024-T4	0.82	Structural Cor-Ten	
5005-H18	0.32	(ASTM-A242)	1.28
6063-T5	0.36	Cold-rolled C-1018	1.20
6061-T4	0.48	Hot-rolled C-1050	1.40
6061-T6	0.58	Hot-rolled C-1095	2.20
7075-T6	0.98	Hot-rolled C-1095, Annealed	1.64
Brass, Rolled Sheet		Stainless 302, Annealed	1.40
Soft	0.64	Stainless 304, Cold-rolled	1.40
1/2 Hard	0.88	Stainless 316, Cold-rolled	1.40
Hard	1.00		
Copper			
1/4 Hard	0.50		
Hard	0.70		

TABLE 4-12
Shear Strength of Various Nonmetallic Materials

Material	Shearing Strength, psi (MPa)	Material	Shearing Strength, psi (MPa)
Asbestos Board	5,000 (34)	Leather, Rawhide	13,000 (90)
Cellulose Acetate	10,000 (69)	Mica	10,000 (69)
Cloth	8,000 (55)	Paper*	6,400 (44)
Fiber, Hard	18,000 (124)	Bristol Board	4,800 (33)
Hard Rubber	20,000 (138)	Pressboard	3,500 (24)
Leather, Tanned	7,000 (48)	Phenol Fiber†	26,000 (179)

* For hollow die, use one-half value shown for shearing strength.
† Blank and perforate hot.

Metric Nomograph. If, in the preceding example, all quantities are expressed in SI metric terms, the metric version of the nomograph (Fig. 4-19) can be used as follows to determine punching force:
1. Find the total length of the cut (3.14×76 mm $= 0.24$ m) on Scale 2, or Point 1.
2. Find the metal thickness (1.29 mm) on Scale 4, or Point 2.
3. Connect these two points with a straightedge to find the area in shear on Scale 3, or Point 3.
4. From this point on Scale 3, draw a line to the metal shear strength (283 MPa) on Scale 1, or Point 4.
5. Extending this line, read the force requirement in kilonewtons (87) on Scale 5, or Point 5.
6. To check the answer for agreement with the result obtained when using the original nomograph in Fig. 4-19, divide the result (87 kN) by the conversion factor (8.9). The result should be 9.8 tons.

TABLE 4-13
Scrap Allowance for Blanking

Material	When length of skeleton segment between blanks or along edge is equal to or less than 2T			When length of skeleton segment between blanks or along edge is greater than 2T		
	Thickness of stock (T), in.	Edge of stock to blank, in.	Between blanks, same row, in.	Thickness of stock (T)	Edge of stock to blank, in.	Between blanks, same row, in.
Cloth, paper (bond, manila, red rope, etc.)	All	3/32–5/32	3/32–5/32	All	3/32–5/32	3/32–5/32
Felt, leather, soft rubber	Under 0.062 Above 0.062	1/16 T	1/16 T	Under 0.062 Above 0.062	1/16 T	1/16 T
Hard rubber, celluloid, Protectoid	All	0.04T* or 0.040 min.	0.04T* or 0.040 min.	All	0.04T* or 0.040 min.	0.4T* or 0.040 min.
Metals, general: Standard strip stock	Under 0.021 0.022–0.055 Above 0.055	0.050 0.040 0.7T	0.050 0.040 0.7T	Under 0.044 Above 0.044	0.050 0.9T	0.050 0.9T
Extra-wide stock and weak scrap skeleton	Under 0.042 Above 0.042	0.060 1.4T	0.050 1.2T	Under 0.033 Above 0.033	0.060 1.8T	0.050 1.6T
Stock run through twice	Under 0.042 0.043–0.055 Above 0.055	0.060 1.4T 1.4T	0.050† 0.040 0.07T	Under 0.033 0.034–0.044 Above 0.044	0.060 1.8T 1.8T	0.050† 0.040 0.9T
Stock run through twice and first and second rows of blanks interlock	Under 0.042 Above 0.042	0.060 1.4T	0.050* 1.4T	Under 0.033 Above 0.033	0.060 1.8T	0.050* 1.8T†
Mica, Micanite, phenol fabrics, and phenol fiber	All	0.6T* or 0.060 min.	0.6T* or 0.060 min.	All	0.6T* or 0.060 min.	0.6T* or 0.060 min.
Permalloy	All	0.060	0.060	All	T‡	T‡
Pressboard, asbestos board	Under 0.031 Above 0.031	1/16 2T	1/16 2T	Under 0.031 Above 0.031	1/16 2T	1/16 2T
Steel (silicon, spring, stainless)	Under 0.042 Above 0.042	0.060 min. 1.4T	0.060 min. 1.4T	Under 0.033 Above 0.033	0.060 min. 1.8T	0.060 min. 1.8T
Vulcanized fiber	All	0.8T* or 0.080 min.	0.8T* or 0.080 min.	All	0.8T* or 0.080 min.	0.8T* or 0.080 min.

* Allow 0.060 between blanks at first and second rows.
† Allow 0.060 between blanks in same row and also between blanks of first and second rows.
‡ When the blank edge parallels the edge of stock or between blanks more than four times the thickness, the allowance of 1.8 thickness of stock shall apply.

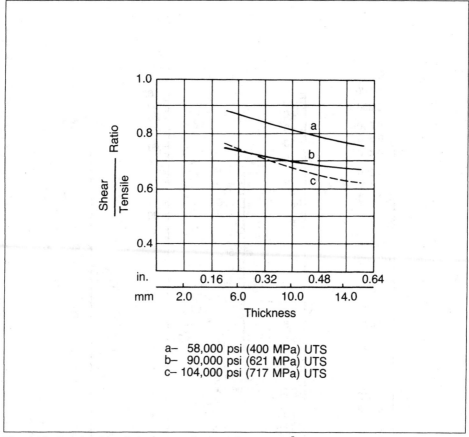

Fig. 4-17. Relationship of shear strength to tensile strength.[8]

STRIPPING FORCE

The punching process requires two actions, punching and stripping (extracting the punch). A stripping force develops due to the resiliency or springback of the punched material that grips the punch. Additional friction is created by cold welding and galling that occurs on the punch surface if tooling is not maintained properly.

Stripping force is generally expressed as a percentage of the force required to punch the hole. This percentage greatly changes with the type of material being punched and the amount of clearance between the cutting edges. Figure 4-20 is a graph showing the effect of these two conditions.

The surface finish of the punch changes with the number of holes being punched; and if a lubricant is not used, the effect can be very severe. A good grade of die cutting lubricant substantially reduces the stripping force.

Stripping Force Equation. The force required to strip a punch is difficult to determine, since it is influenced by the type of metal pierced, the area of metal in contact with the punch, the punch/die clearance, punch sharpness, spring positioning with respect to the punch, and other factors. The following empirical equation is used for rough approximations:

$$F = L \times T \times 1.5 \tag{3}$$

where F = stripping force, ton
L = length of cut, in.
T = material thickness, in.

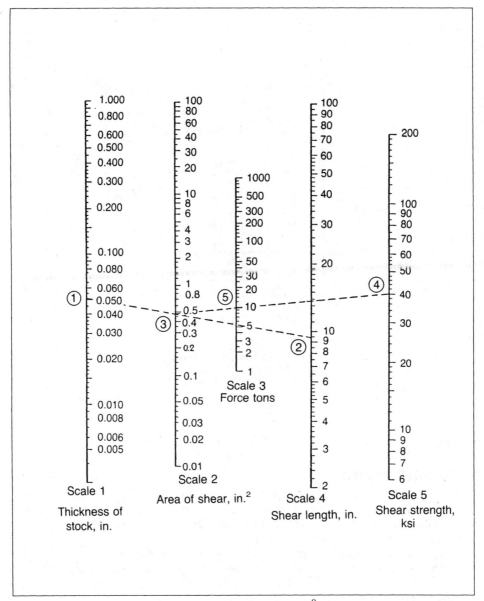

Fig. 4-18. Nomograph for determining punching force in tons.[9]

When using metric units, the equation is:

$$F = L \times T \times 20{,}600 \tag{4}$$

where F = stripping force, kN
L = length of cut, m
T = material thickness, m

Stripping Methods. As shown in Fig. 4-21, stripping can be accomplished with a spring (urethane or die rubber) loaded plate that holds the material firmly against the die, or a positive stripper plate set a short distance above the material. With a positive stripper, the material is lifted with the ascending punch until the material contacts the stripper and is freed of the punch.

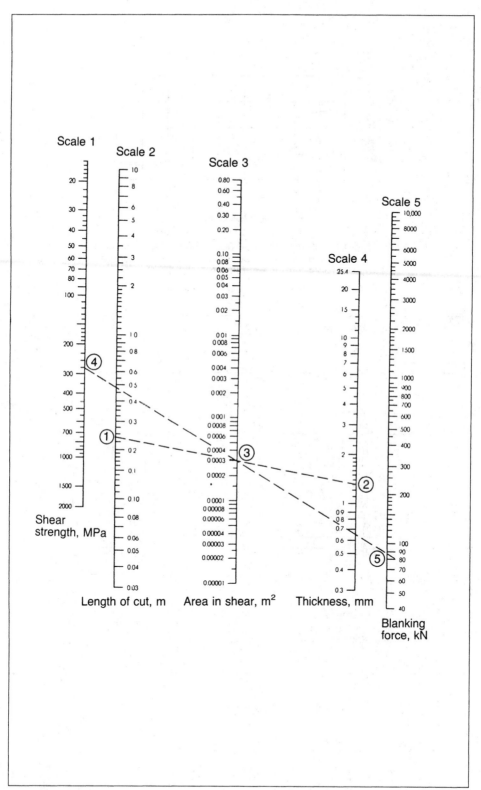

Fig. 4-19. Nomograph for determining punching force in kilonewtons.[10]

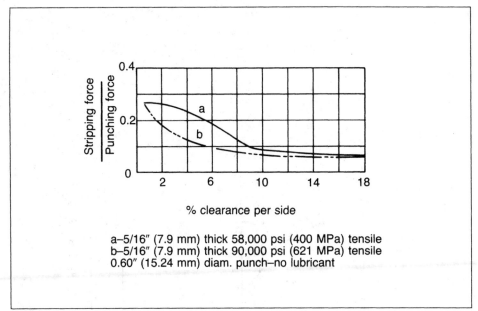

Fig. 4-20. Stripping force varies with clearance.[11]

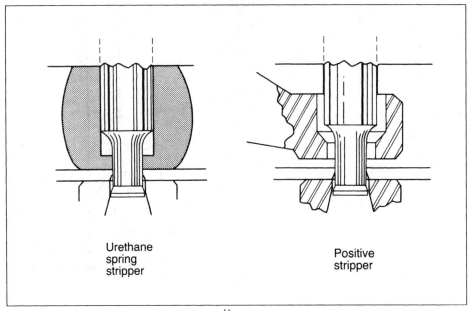

Fig. 4-21. Spring and positive type stripping.[11]

SHRINKAGE ALLOWANCE FOR HEATED PHENOL FABRICS AND PHENOL FIBER

In general, the expansion coeffecient of phenol fabrics and phenol fiber is 16.7 parts per million (ppm) per degree Fahrenheit change in temperature or 30 ppm per degree centigrade change in temperature. When the material is blanked or perforated hot, it will (in most cases) shrink by the above amount.

SHOCK PROBLEMS IN PUNCHES

Figure 4-22 shows how the compressive loading on a punch is released suddenly when the punch breaks through the stock. The stress which results concentrates at the point where the punch head joins the punch body. In severe cases, this stress concentration can result in punch head breakage. Figure 4-23 shows measures that can be taken to solve head breakage problems.

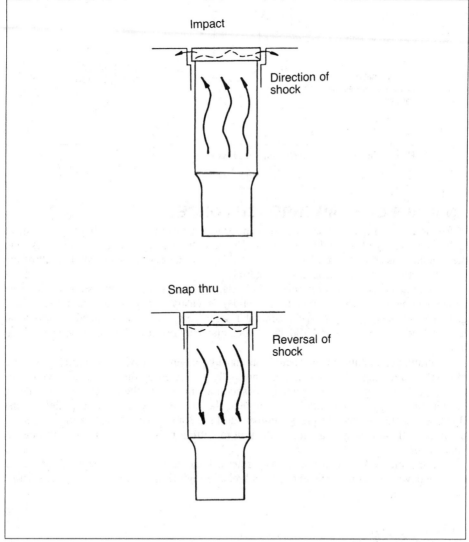

Fig. 4-22. Head breakage may be caused by excessive load or pumping.[12] (*Dayton Progress Corporation*)

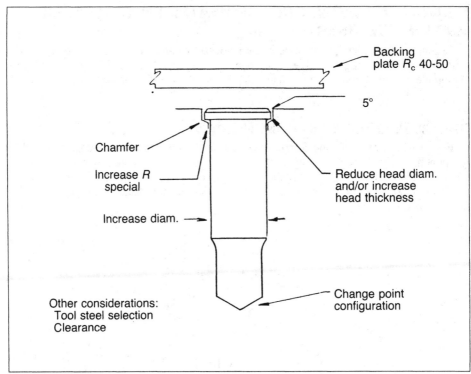

Fig. 4-23. Head/point breakage.[12] (*Dayton Progress Corporation*)

ANALYSIS OF SNAP THROUGH FORCES

The loud boom that characterizes a snap-through problem when cutting through metal is the direct result of the sudden release of the potential energy that is stored in press and die members as strains or deflection (Fig. 4-24). The deflection is a normal result of the pressure required to cut through the material.

In extreme cases, the energy released can damage the press. The press connection (the attachment point of the pitman to the slide) is easily damaged by the reverse load generated by snap-through. As a general rule, presses are not designed to withstand reverse loading of more than 10%. This shock also may result in die components working loose.

In timing punch entry or die shear, care must be given to provide for a gradual release of developed tonnage. The shock is not normally generated by the impact of the punches on the stock. In fact, when the punches first contact the stock, the initial work is done by the kinetic energy of the slide. To complete the cut, energy must be supplied by the flywheel. As this occurs, the press members deflect. An analysis of the quantity of energy involved will show why a gradual reduction in cutting pressure prior to snap-through is very important.

The magnitude of the actual energy released increases as the square of the actual tonnage developed at the moment of final breakthrough (Figs. 4-25 and 4-26). The actual energy is given by:

$$E = \frac{F \times D}{2} \tag{5}$$

4-28

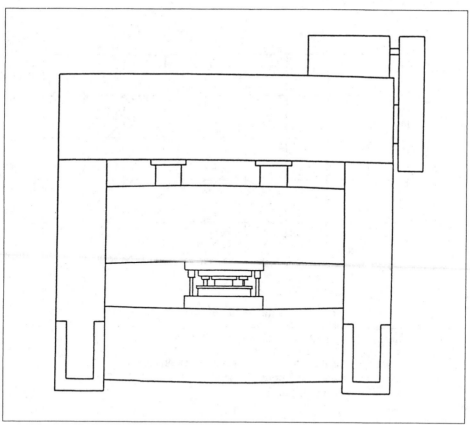

Fig. 4-24. A press deflects under load (exaggerated view). (*Smith & Associates*)

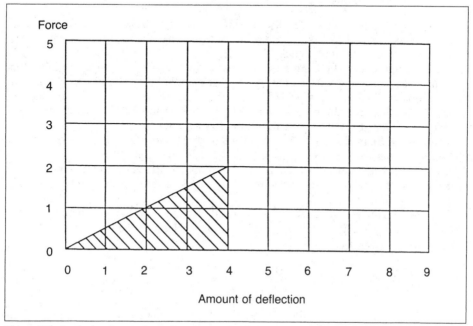

Fig. 4-25. Analysis of snap-through energy. (*Smith & Associates*)

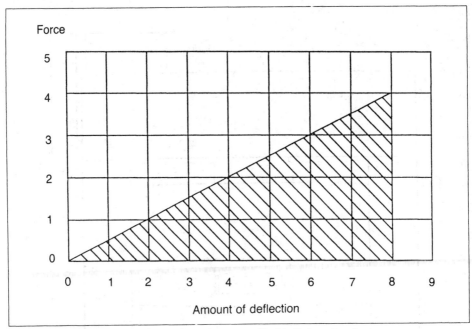

Fig. 4-26. Further analysis of snap through energy. (*Smith & Associates*)

where F = Pressure at moment of breakthrough in standard tons.
D = Amount of total deflection in inches.
$E \times 166.7$ = Energy in foot pounds

or

F = Pressure at moment of breakthrough in Metric tons (kgf $\times$ 1000).
D = Amount of total deflection in millimeters.
$E \times 9.807$ = Energy in Joules or Watt Seconds

For example, if 400 tons (3,560 kN) which resulted in 0.080 in. (2.03 mm) total deflection was required to cut through a thick steel blank, the energy released at snap through would be 2,667 foot pounds (3,616 J).

If careful timing of the cutting sequence results in a gradual reduction in the amount of tonnage at the moment of snap-through so that only 200 tons (1,780 kN) is released, the reduction in shock and noise would be dramatic. Half the tonnage would produce half as much deflection or 0.040 in. (1.02 mm). The resultant snap-through energy would only be 667 foot pounds (904 J), or one fourth the former value.

Minimizing Snap-through Energy. Timing shear and punch entry sequences to provide a gradual release of tonnage prior to snap-through is the most effective way to reduce the shock and noise associated with this problem. Blanking presses should be of robust construction to minimize deflection. The energy released has a direct relationship to the amount of deflection.

Figure 4-27 shows the large reverse load resulting from incorrect punch entry timing. The result of proper entry timing to provide a gradual release of the tonnage prior to final breakthrough is shown in Fig. 4-28.

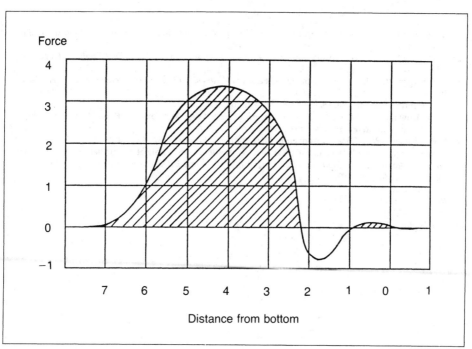

Fig. 4-27. No timing of entry. (*Smith & Associates*)

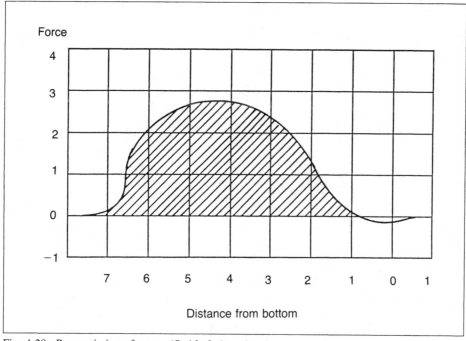

Fig. 4-28. Proper timing of entry. (*Smith & Associates*)

References

1. E. A. Reed, "Progress Planning for Pressed Metal Manufacturing," *General Motors Institute Report* (August 1949).
2. J. W. Lengbridge, "The Theory and Practice of Pressing Aluminum," *The Tool Engineer* (June 1948 to July 1949).
3. J. R. Paquin, "It's Easy to Calculate Die Clearances," *American Machinist Reference Book Sheet* (June 16, 1949).
4. "Engineering Standard Sheets," Sperry Corporation.
5. A. A. Dowd and F. W. Curtis, *Tool Engineering* (New York: McGraw-Hill Book Company, 1925).
6. *Tool Engineering Reference Sheets* (IBM).
7. J. R. Paquin, "Why Shear is Applied to Dies," *American Machinist* (August 25, 1949).
8. Charles Wick, John W. Benedict and Raymond F. Veilleux, eds., "Sheet Metal Blanking and Forming," *Tool and Manufacturing Engineers Handbook*, 4th ed., Volume 2 (Dearborn, MI: Society of Manufacturing Engineers, 1984), p. 4-19.
9. *Ibid.*, p. 4-20.
10. *Ibid,* p. 4-21.
11. *Ibid*, p. 4-22.
12. *The Tooling,* Dayton Progress Corporation.

SECTION 5
CUTTING DIES

PIERCING AND PERFORATING DIES

Single-station Piercing and Perforating Dies. Dies under this classification punch regular or irregular shapes from metal stock. The punched stock is typically scrap.

Typical Single-station Piercing Die. One manufacturer classifies the die shown in Fig. 5-1 as a typical piercing die. General specifications included in this manufacturer's

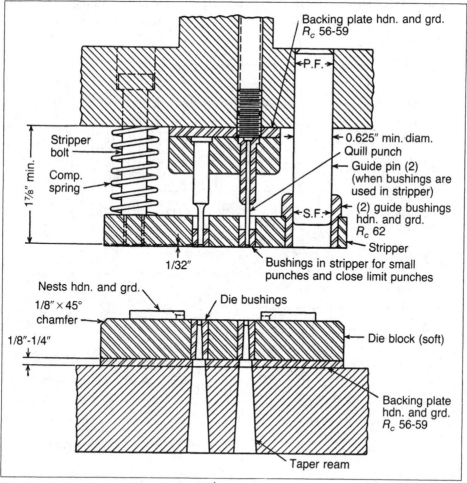

Fig. 5-1. Typical single-station piercing die.[1]

manual of tool engineering are noted on the drawing.

Magnetic Locating Piercing Die. The die shown in Fig. 5-2 pierces one hole in a cold-rolled steel keyboard part which has been blanked, slotted, and formed. Four permanent magnets (*D2*) in bronze bushings (*D1*) hold the piece part after it has been located on the gage pins (*D3*) (*D* indicates detail number on the drawings). These gage pins are set in removable blocks (*D4* and *D6*) so they can be relocated after the die has been sharpened. As the ram descends, a positive locator or wiper contacts the cam block (*D5*) and holds the work against the side of the die as the punch pierces the hole. Nominal punch diameter is 0.09455 in. (2.4016 mm), allowing 0.00085 in. (0.0216 mm) for punch wear.

Die for Piercing and Slotting a Key. Two end slots, two holes, and a central slot are separately cut by punches (*D1*, *D2*, and *D3*) at one press stroke in the die shown in Fig. 5-3. The workpiece is held in nests (*D4* and *D5*). The two punches for notching the end slots are heeled (Section A-A). A cam stripper (*D6*) is provided which is bushed for the guide pins (*D7* and *D8*).

Piercing Die for Press Brake. This die (Fig. 5-4) is one of several set up in a press brake for a number of operations on steel strip used in blank books. A specially designed feeder (not shown) facilitates strip movement between the series of dies which make up a progressive arrangement of independent single-station dies, any one of which can be removed from the brake without disturbing the remaining dies. Strip stock is positioned by guides (*D5*) and removed from the punches (*D1*, *D2*, and *D3*) by a stripper (*D6*). The

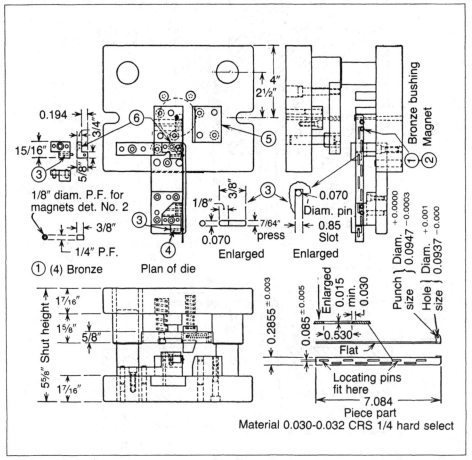

Fig. 5-2. Magnetic locating piercing die. (*National Cash Register Co.*)

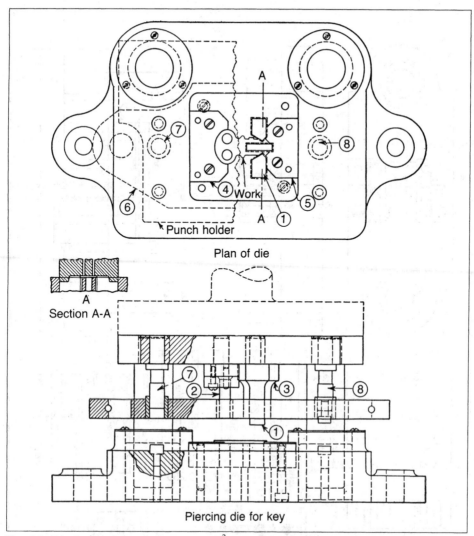

Fig. 5-3. Die for piercing and slotting a key.[2]

die sections (D4) are of split design for easier grinding and construction.

Tube-piercing Die. This die (Fig. 5-5) pierces tubing without inside support, such as a mandrel or horn. Spring-loaded upper and lower pads enclose and clamp the outside diameter of the work, before the opposed punches pierce the holes from the outside. This design results in slight depressions around the punched holes; countersunk or embossed (flanged) holes may be produced with suitable punches.

Sectional Slot and Pierce Die. A light-gage aluminum cup is radially slotted and pierced to close tolerances in the die shown in Fig. 5-6. Seven closely spaced punches (D3, D4, D5, and D6), are well guided by the combination stripper plate and pressure pad. Pressure is applied to the plate by a spring through shedder pins. The piercing and slotting punches are backed up with a hardened steel plate.

Slotting punches and dies (D1 and D2) are of sectional design to facilitate construction and sharpening.

Inverted Pierce Die. A stainless-steel (0.031-in. (0.78-mm) deep-drawing quality) drip pan is held on a spring-loaded horn (Fig. 5-7, D3) by flat springs. A spring-loaded shedder pin (D2) pushes the oval slug off the positive knockout pin (D4), which ejects the

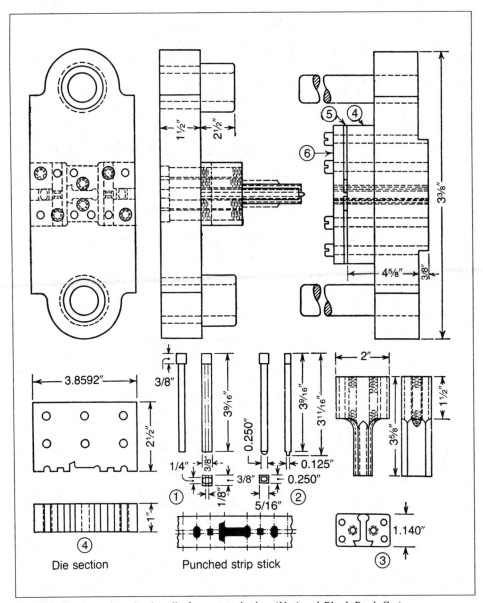

Fig. 5-4. Single-station piercing die for a press brake. (*National Blank Book Co.*)

slug from the inverted die (*D*1). There is some extrusion (flanging) around the oval hole.

Piercing Holes of Less Than Stock Thickness. Eight holes, closely spaced, whose diameters are less than stock thickness are punched in stainless steel in the die shown in Fig. 5-8. Intermeshing sleeved punches (*D*1, *D*2, *D*3, and *D*4) ensure maximum rigidity and support for the entire length of each punch. A manual ejector (*D*5) is used.

Side-piercing Cam-actuated Die. Horizontally operating cams are frequently used when the shut height of a press is limited or when it is necessary to pierce two or more holes whose center lines are not parallel. In the die shown in Fig. 5-9, three side ("dog-leg") cams (*D*1) transmit pressure to the side-piercing punches (*D*3). A single cam (*D*2) operates two opposed slides (*D*4 and *D*5). These slides drive punches (*D*6) through pivoted arms (*D*7) shown in Section B-B. The sliding punch-holder blocks actuated by the arms are returned to the open position by springs as shown in Section A-A. The work is removed from the locating post by three stripping hooks (D8) attached to the upper shoe.

Combined Vertical and Side-piercing Die for Heavy Stock. Sliding punch blocks (Fig. 5-10, Section A-A, D5), to which eight horizontal punches (D13) are mounted, are actuated by vertical cams (D2), backed up by heel blocks (D3). Hardened wear plates are provided for all wear surfaces of moving parts of the cam assemblies. The punches enter die buttons (Fig. 5-10, Section A-A, D1). They cut four 0.406-in. (10.31 mm) diameter holes in the flanges at each end of the part, an automotive-transmission support (Fig. 5-10A). Spring-loaded strippers (Fig. 5-10, Section A-A, D4) strip the ends of the part from these punches. Three vertical punches (D9 and D10) shown in Section C-C, Fig. 5-10, cut oval holes in the middle of the part. A double pry-bar assembly (Fig. 5-10, D7 and D8) shown as Section D-D, Fig. 5-10, raises the center part of the part out of the die as the operator pushes a handle (Fig. 5-10, D6) downward.

Cam Die for Perforating 1,080 Holes. Six holes of 0.59375-in. diameter and 1,074 holes of 0.1875-in. diameter are pierced outward in a cylindrical tub by the die shown in Fig. 5-11A. An indexing mechanism rotates and locates the tub so a total of 180 holes are perforated every 60°. Two cam blocks (Fig. 5-11B, D12) carrying punch holders (D4), each equipped with 90 punches (D11), and strippers (D2), are located inside the tub. Two sliding die blocks (D5), actuated by right-hand and left-hand cams (D1), each carrying 90 die buttons (D6), slide inward. Six holes (19/32-in. diameter) are embossed as well as pierced per complete revolution of the ball-bearing indexing and supporting table (D8) for the tub. A spring-loaded pad (Fig. 5-11B, D3) clamps the tub. Spring pins (Fig. 5-11B, D10) locate the tub to the index plate or table. Felt washers (D9) retain lubricant for the revolving indexing table. A heel block (Fig. 5-11B, D7) takes the side thrust of the right and left vertical cams. One automatic cycle of six press strokes per minute in a 300-ton (2,670 kN) Verson mechanical press pierces all 1,080 holes. Other views of the die blocks and punch plates are shown in Fig. 5-11C, Section A-A. The part is the inner tub of an automatic washing machine.

Piercing Small Closely Spaced Slots. The component produced by this tool is the slow wave structure for a very high frequency electron tube. The structure, known as the "ring and bar," is shown in Fig. 5-12A and consists of a series of rings joined on alternate sides by stubs which also serve to support the structure. The section was built from two half-plates either brazed or pressure-welded together as shown in view B. To form these half-plates, it was necessary to cut 81 slots 0.004 in. (0.1 mm) wide and four slots 0.035 in. (0.889 mm) wide in a plate 0.004 in. (0.1 mm) thick. In the cylindrical section, the slots are separated by as little as 0.004 in. (0.1 mm).

The two halves of the plate are identical, and the slots line up accurately when the halves are assembled back to back. One half is turned over about an axis parallel to the slots as shown in C. To function satisfactorily, the slot width must be within 0.0001 in. (0.003 mm), and the random error in the position of any one slot must be within 0.0005 in. (0.013 mm).

The tool consists of a punch, complete with a stripper plate, and a die. Punch blades for piercing the slots are 0.0039 in. (0.099 mm) thick and are separated and supported by 0.0041-in. (0.104-mm) spacers. The stripper plate is carried on the punches, with 0.0041-in. (0.104-mm) spacers provided to assure the required separation and support. The die is built in a similar manner from 0.004-in. (0.1-mm) die segments separated by 0.004-in. (0.1-mm) spacers and free knockout blades. These parts are shown in Fig. 5-12 E. The punches and stripper plate are ground after assembly, leaving the semicylindrical form standing; similarly, the female form is ground in the die.

The slots are extended 0.005 in. (0.127 mm) beyond the outer radius of the rings, as shown in Fig. 15-12A. This extension was provided to permit strengthening the die knockouts at this point. Had the slots not been extended, conditions would have been as shown in view D.

Years ago, it was found that the material used for razor blades was nominally 0.004 in. (0.1 mm) thick ±0.0001 in. (0.0025 mm), approximately. By careful selection from a

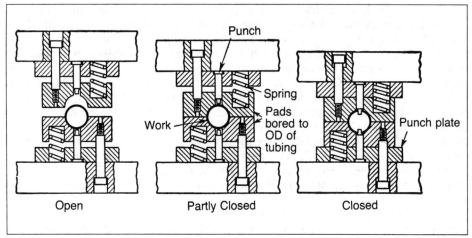

Fig. 5-5. Tube-piercing die. (*O.H. Mathisen, SME.*)

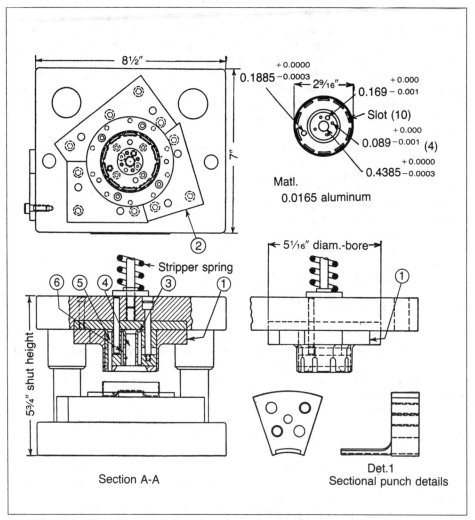

Fig. 5-6. Sectional slot-and-pierce die. (*National Cash Register Co.*)

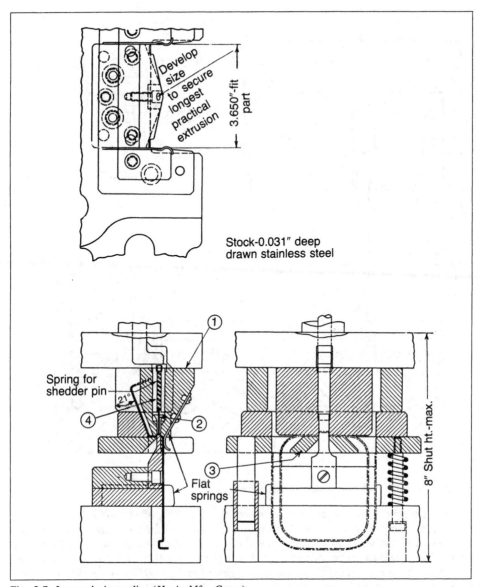

Fig. 5-7. Inverted pierce die. (*Harig Mfg. Corp.*)

large number of razor blades, the blades and spacers for the punch, die, and stripper were made without further sizing on thickness or any heat treatment.

A sectional view of the complete tool is shown in Fig. 5-13. The punch holder (*D*1) is located accurately in relation to the die nest (*D*2) by four posts (*D*3); the stripper plate (*D*4) also is guided by these posts. The punch blades and spacers (*D*5) and the die knockouts (*D*6) and segments and spacers (*D*7) are retained by end clamps. Two straps (*D*8) located in a recess ground half in the die segments and half in the die nest hold the spacers and segments rigidly in position. A spring pressure pad (*D*9) actuates the knockouts for scrap removal. The stripper-plate spacers (*D*10) like the die spacers, are held in place by straps (*D*11). For stripping the pierced blank from the punches, the stripper plate is halted by the link arms (*D*12); a stop (not shown) ensures that the punches are not raised clear of the stripper.

Die for Piercing Round Cold-rolled-steel Rods. According to the designers of the die

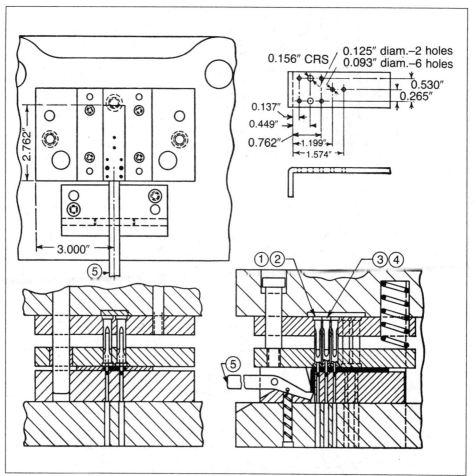

Fig. 5-8. Piercing holes of less than stock thickness.[3]

illustrated in Fig. 5-14, holes with diameters that are 40% of the rod diameter can be punched with some slight bulging of the rod, if the center of the hole is not less than stock thickness (rod diameter) away from the end of the rod. Both the die button (D1) and the punch sleeve (D2) fit the outside of the rod. Two limit pins (D3) check punch overtravel, and heavy stripper springs (D4) minimize rod distortion. A 0.09375 in. (2.3812 mm) hole is pierced 0.188 in. (4.76 mm) from the end of a 0.188 in. (4.76 mm) rod.

Die for Piercing Small Closely Spaced Holes. Conventional design of small individual punches for cutting twin slots in the part (Fig. 5-15) would result in frequent breakage. Both sets of piercing (D2) and slotting punches (D1) are inserted in upper and lower intermeshing sleeves. The twin slotting punches are contained in upper and lower halves made in three pieces and assembled with a ring (D3) to keep them together.

Cutting Slots Narrower than Stock Thickness. Buckling of the thin blade of the sleeved punch (Fig. 5-16, D1), equipped with a heel, is prevented because it enters the die button before the slot is cut. The gage plate (D2), thinner than the workpiece, allows the stripper plate to clamp the work firmly, to prevent distortion of the part on upstroke and downstroke. Punch travel is controlled by a limit stop pin (D3).

Die for Cutting Holes in a Bent Part. A die like the one shown in Fig. 5-17 incorporates accurately guided punches to minimize punch breakage on the downstroke. The part closely fits the die (D5) so with the help of stripper pressure, there will be no spring-back of the part or punch breakage on the upstroke. The stripping action is severe

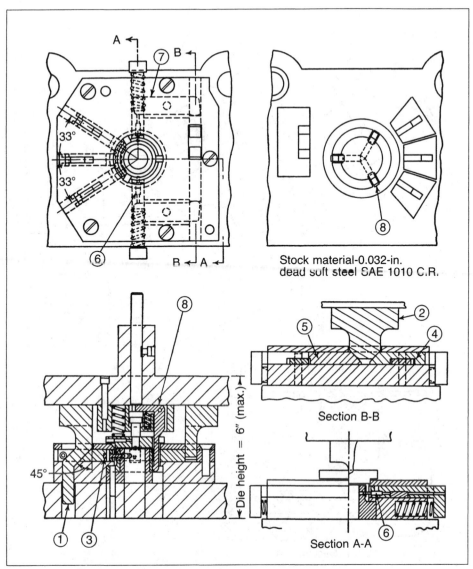

Fig. 5-9. Side-piercing cam-actuated die. (*Harig Mfg. Corp.*)

and additional cushioning springs (D4) are provided to prevent breaking of the stripper bolts (D3). Heavy springs (D2) enable the stripper (D1) to clamp the part to the die, but its pressure would be insufficient to stop spring-back if the part and die were not well mated.

Cam Die Design for Accurate Punching. Precise alignment of holes in the workpiece (for shaft bearings) is ensured by using guide pins (Fig. 5-18, D7) and sleeved punches (D6). A spring-loaded pad (D1) clamps the work before the cams (D3) actuate the slides to which punch pads (D5) are fastened. Spring-loaded strippers (D2) strip the punches. The heel block (D4) minimizes side thrust of the cam.

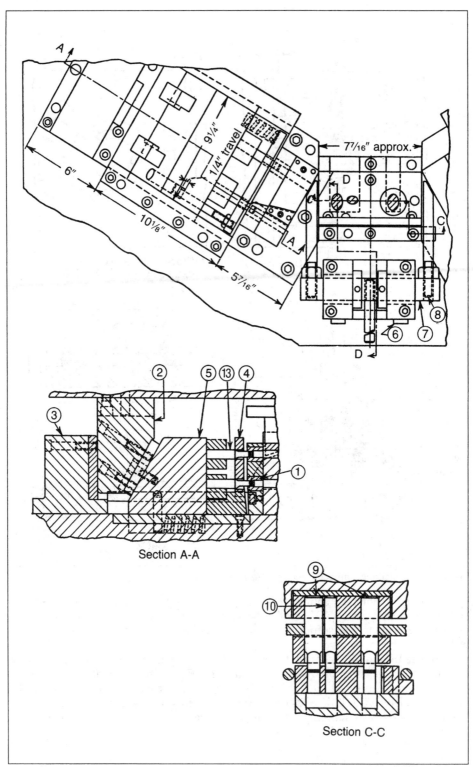

Fig. 5-10. Combined vertical and side-piercing die for heavy stock. (*Buick Division, General Motors Corp.*)

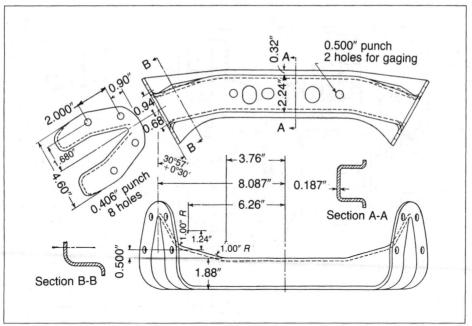

Fig. 5-10*A*. Automobile-transmission support part drawing for Fig. 5-10.

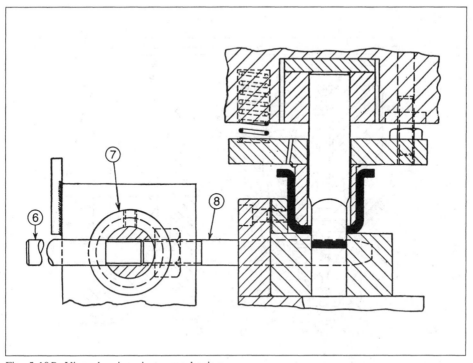

Fig. 5-10*B*. View showing ejector mechanism.

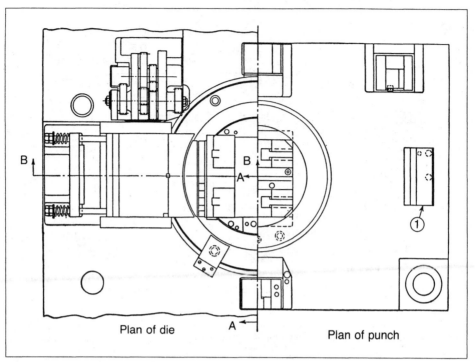

Fig. 5-11A. Plan view of cam die for perforating 1,080 holes. (*Maytag Co.*)

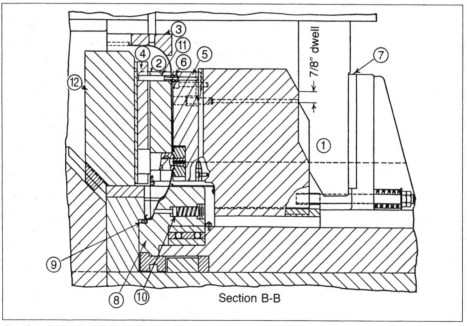

Fig. 5-11B. Section B-B of Fig. 5-11A.

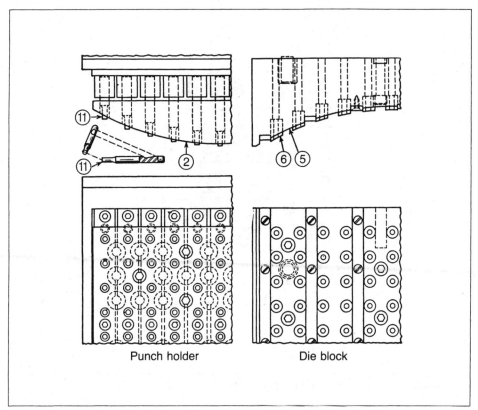

Punch holder   Die block

Fig.5-11C. Punch and die blocks of cam die shown in Fig. 5-11A.

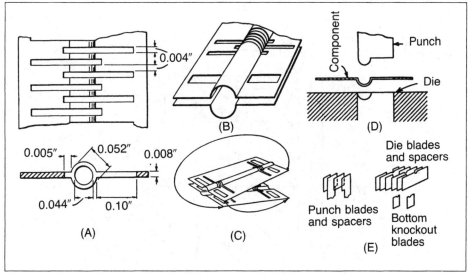

Fig. 5-12. Ring and bar structure for a high-frequency traveling-wave tube, and some punch and die details. [4]

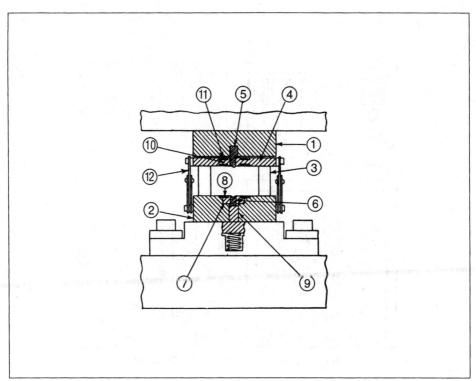

Fig. 5-13. Sectional view of die for the part shown in Fig. 5-12. [4]

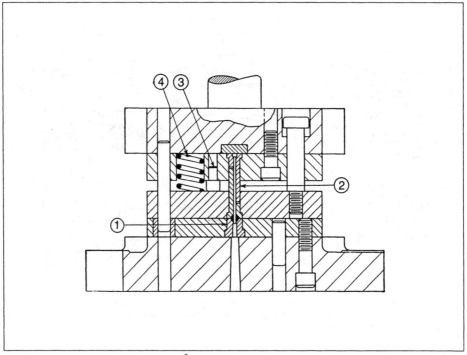

Fig. 5-14. Die for piercing round rod. [3]

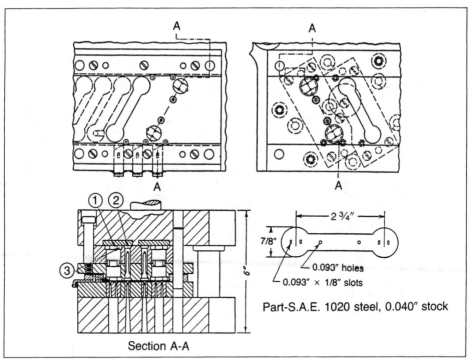

Fig. 5-15. Die for piercing small holes close together.[3]

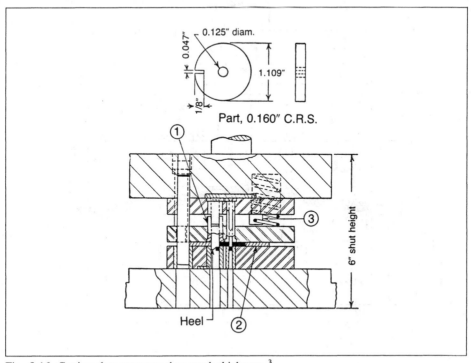

Fig. 5-16. Cutting slots narrower than stock thickness.[3]

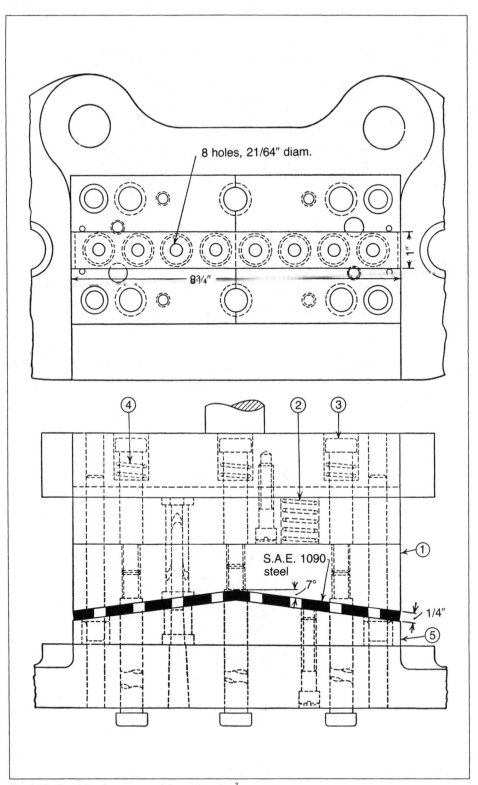

Fig.5-17. Punching heavy material at an angle. [3]

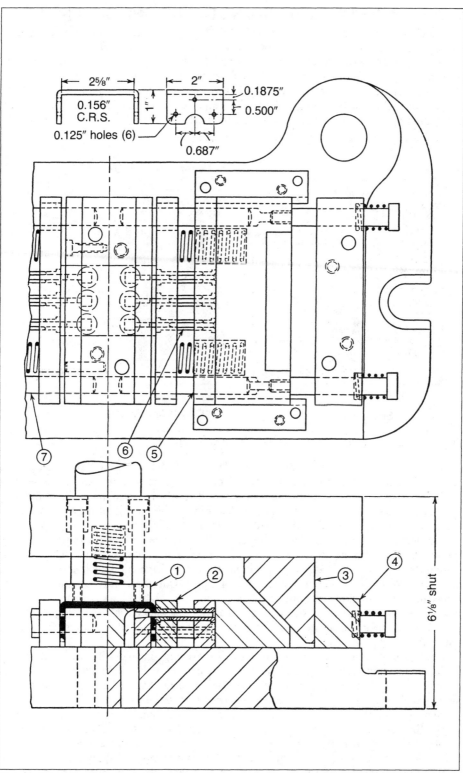

Fig. 5-18. Cam design with guided slides and sleeved punches.[3]

5-17

MISCELLANEOUS SLOTTING, NOTCHING AND CUTTING DIES

Knockout Dies. Typical single and double dies for punching knockouts in conduit boxes, enclosing boxes, and similar products are shown in Fig. 5-19.

Bumper blocks between the punch holder and die shoe (Fig. 5-19, single knockout die) limit punch penetration to a predetermined depth. To ensure a clean cut around the knockout periphery without straining the holding tabs, penetration is limited to the depth at which daylight can be seen around the edge of the cut. This usually requires a minimum penetration of about two-thirds the material thickness or more.

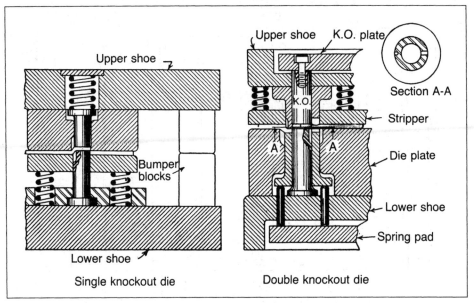

Fig. 5-19. Single and double knockout dies. (*General Electric Co.*)

Knockout slugs should be readily removable from the outside of the box so care should be exercised to see the material is pierced from the proper side in view of subsequent operations. On single knockouts, the slug should be pushed toward the inside of the box. For double knockouts, the outer ring is pushed toward the outside of the box, while the center slug and the balance of the blank remain on the same plane.

Die clearance in knockout dies is generally reduced to one-half the customary clearance for pierce dies, making it about 3% of the material thickness per side when cutting medium rolled steel. A flattening operation is used to force the knockout slug or ring partially back into the material. This can be combined with subsequent forming operations or performed as an individual operation. In general, the slug or ring is forced back until it protrudes a maximum of one-half the material thickness for material 0.03125 in. (0.7938 mm) thick and under, and a maximum of 0.03125 in. (0.7938 mm) for heavier materials.

Holding tabs bridge the knockout slugs to the parent metal. Tabs are produced by providing matching grooves in the punch and die periphery which do not cut the material at these points. Fig. 5-19A, upper view, shows two tab shapes commonly used. The straight tab is preferred, because it does not work the material so severely and can be made smaller. The width of the straight tab is made to a minimum of the material thickness.

Holding tabs for double knockouts must be located so the inner slug can be removed without loosening the outer ring. In addition, more tabs are used in the larger sizes. Figure 5-19A, lower view, shows recommended location and number of straight tabs in double knockouts of various sizes. In the case of single knockouts, it is customary to use one holding tab on diameters up to 1.5 in. (38.1 mm) and two tabs on larger diameters.

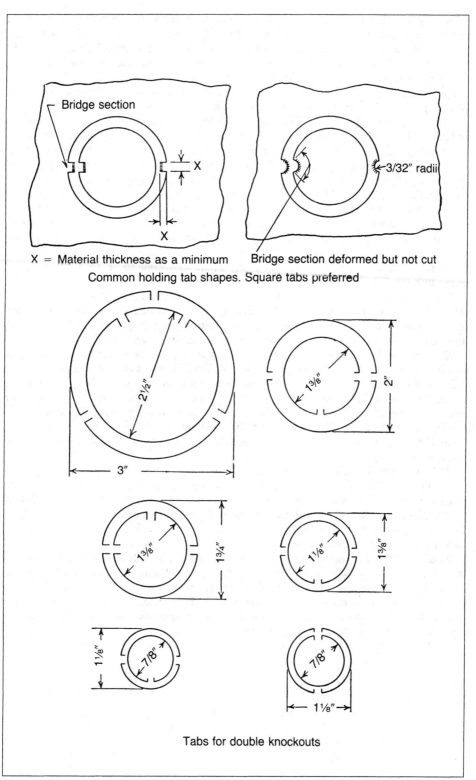

Fig. 5-19A. Tab shapes and sizes of knockouts commonly used. (*General Electric Co.*)

For small-lot production, a knockout die may cut different thicknesses of material, in which case specific rules are followed to establish die clearance and width of straight tabs. For example, a die made to cut materials from 0.062 to 0.13 in. (1.6 to 3.2 mm) thick will use the mean thickness of material cut to determine die clearance. For medium rolled steel, the clearance per side would be 3% of the mean thickness, 0.09375 in. (2.3812 mm), or 0.0028 in. (0.07 mm). The width of a straight-type tab would be equal to the greatest material thickness to be cut, or 0.125 in. (3.18 mm).

Louvering Die. The die of Fig. 5-20 should be classified with dies in which cutting and noncutting operations are combined. It is included here to illustrate punch design for the shearing function in fabricating louvers. Four inverted punches ($D1$), secured with screws and a low-melting-point alloy, are provided with a cutting or slitting edge as well as a forming surface. A nest plate ("foolproof" plate, $D2$) and a locating pin ($D3$) position the workpiece. This die was made to form the same louver pattern in four different parts. One of the parts is an opposite hand to the one shown and the other two are a right- and left-hand pair with a blank of a different contour. The foolproof nesting plate is turned over and the locating pins, which are held in place by a setscrew, are moved to the opposite position to make the part of opposite hand. For producing the other two parts, a similar foolproof plate was made to match the contour of the part and is used in the manner described above.

Notching Die. In the die illustrated in Fig. 5-21, the workpiece is shifted by the handle ($D1$), allowing 10 slots to be punched at each position per stroke. Twenty slots each 0.018 in. (0.46 mm) wide could not be punched at one stroke, since 20 punches, 0.250 in. (6.35 mm) apart, center to center, could not be mounted in the punch block.

Slotting Die. Bullet-nosed pins are located in bushings mounted on the die slide of the die shown in Fig. 5-22. Holes previously punched in the fan-blade stock, of eight different-sized blades, are engaged by the pins, so slots of eight different lengths of constant width are cut by the elongated nibbling punch. Setup time and tooling costs for eight conventional slotting dies are reduced by this die design. There are four bushings in the forward end of the slide plate, and five bushings in the rear end, to receive the removable locating pins. Two pins are placed in the proper set of holes to position the blank for notching.

Tube-slotting Die. Two slots are cut in the die shown in Fig. 5-23, to a slot alignment of 0.005 in. (0.127 mm) in SAE 1137 steel tubing. Positive-return cams actuate slide blocks, which are provided with clamp inserts to hold the workpiece. The work is revolved after the first slot is cut until the locator pin enters the slot, aligning the tubing for punching the second slot by the hollow-ground punch. Turning the key depresses the locating pin and allows part removal.

Notching Die. Two dissimilar notches are cut out of the ends of a box-shaped part in the die shown in Fig. 5-24. Cams ($D4$) actuate internal sliding punches ($D1$, $D2$) on the opposite ends of the die, which leave the burrs on the outside of the part and also drive the slugs outwardly to openings in the die block and die shoe. Heavy die-section design ensures long life. Parts are clamped in the die by a spring-loaded pad ($D3$). Production is 500 parts per hour in a No. 4 Niagara press.

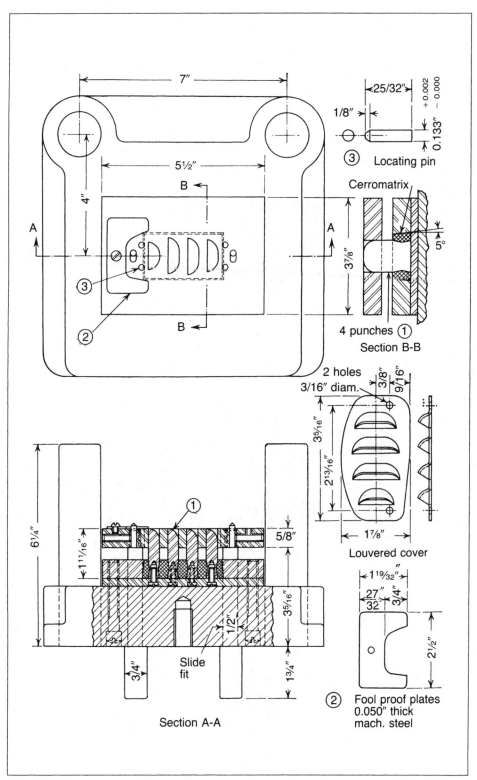

Fig. 5-20. Louvering die. (*Century Electric Co.*)

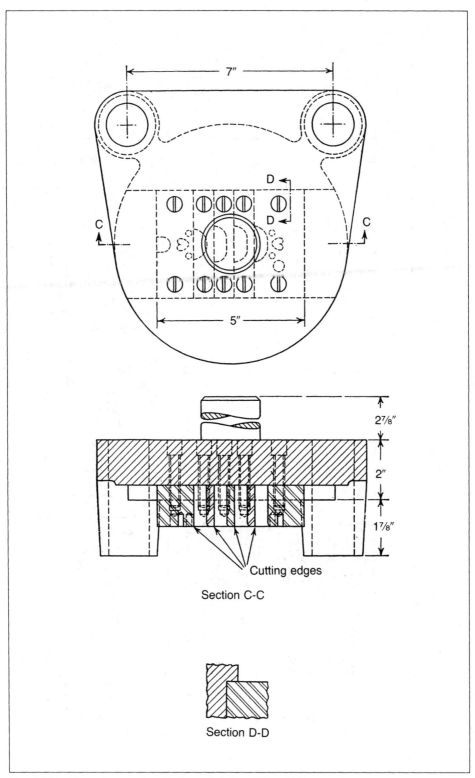

7"

5"

D

C

C

D

2⅞"

2"

1⅞"

Cutting edges

Section C-C

Section D-D

Fig. 5-20A. View of upper shoe for louvering die of Fig. 5-20.

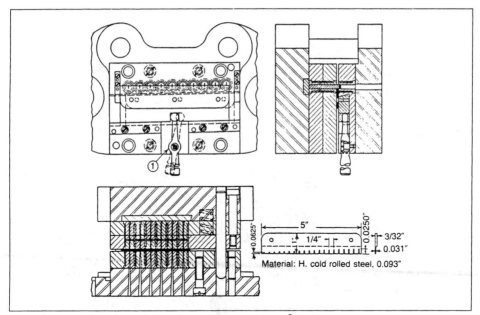

Fig. 5-21. Notching die with lever for shifting workpiece. [3]

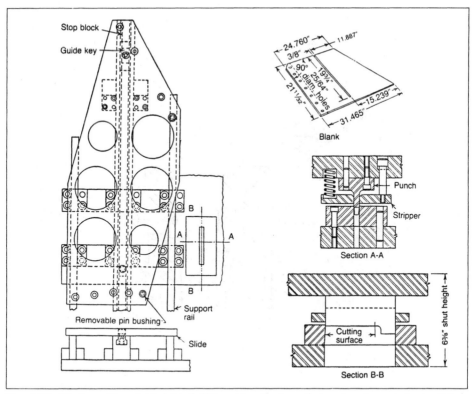

Fig. 5-22. Die for cutting eight slots of different lengths in a single setup. Large holes in slide reduce weight. [2]

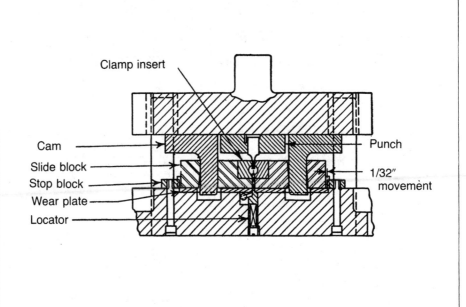

Clamp insert

Cam

Slide block

Stop block

Wear plate

Locator

Punch

1/32″ movement

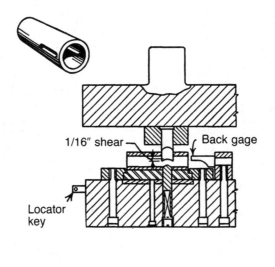

1/16″ shear

Back gage

Locator key

Fig. 5-23. Tube-slotting die.[5]

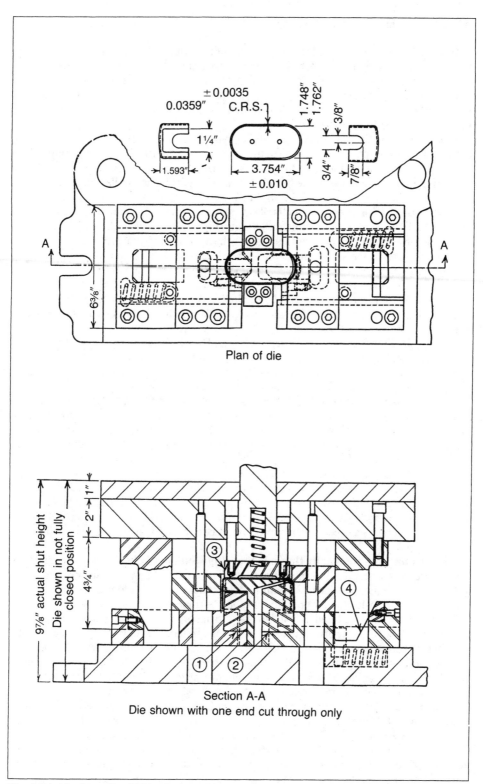

Fig. 5-24. Notching die. (*White-Rodgers Electric Co.*)

HORN-TYPE CUTTING DIES

Horn Die for Slotting Tubes. A horn (mandrel) may support a tube or shell (Fig. 5-25A) during the die-cutting operation. Two notches are cut in shell or tube ends with this die. The part is slipped over the locating plug, the upper edge entering a slot cut into the punch. On the ram descent, both notches are cut. The lower slug drops through the die; the upper slug is raised flush with the locating plug and is blown off. The free end of the work is supported by a rest pad.

Indexing Horn Die for Slotting. Another type of horn die (Fig. 5-25B) notches one slot at a time. The part is placed over the horn and located by a pilot entering a hole previously pierced in the flange of the part. A hardened block is set into the horn and does the cutting in conjunction with the heeled punch. The part is then pulled back and rotated so that the pilot enters the next locating hole.

Horn-type Die for Piercing Shells. Two opposed holes are accurately located by the use of a sliding locating pin in this die design (Fig. 5-25C). With the sliding plate (on which the pin is mounted) pushed to the right, the shell is placed over the horn. After the first hole is pierced, the shell is rotated and the slide is moved left to a stop pin. With the locating pin engaged in the first pierced hole, the press is again tripped to pierce the second hole. This is inexpensive construction for dies of this type.

Horn-type Piercing Die for Shallow Shells. A die for this purpose (Fig. 5-25D) incorporates a horn secured to an angle block. A flat on the bottom of the horn provides slug relief.

Horn-type Piercing Die for Large Shells. The die design shown in Fig. 5-25E permits piercing of shells larger in diameter than the maximum opening between ram and bolster plate by incorporating a horn which overhangs the bolster plate. The eccentric punch is well guided for close alignment with the die button in the horn.

Die for Punching Opposite Holes in a Shell or Tube in One Stroke. This operation (Fig. 5-26A) uses the slug from the upper hole as the punch for the lower hole. Fracture will be fairly clean, but there will be a burr on the lower side of the part. A spring stripper contacts the work first, causing the spring-supported horn to bottom on the die block. Further ram descent results in piercing of the upper and lower holes. A spring raises the horn to the loading position as the ram ascends.

Die for Piercing Irregular Shells. A horn on the die shown in Fig. 5-26B is mounted on an angle plate or bracket so punch travel is at a right angle to the surface to be punched. Cast iron can be used for the bracket if light piercing is involved; cast steel is used for shells of thick cross section. A weldment is not recommended, since it may not withstand repeated shock.

Die for Punching Internally Burred Opposing Holes in Shells. A spring-loaded horn is provided with a stop to limit horn travel upward on its guide pin in the design shown in Fig. 5-26C. Punches located in both punch holder and die holder are stripped, by a conventional spring-loaded stripper and a rubber stripper.

Horizontal Cam-actuated Horn Die. A die incorporating horizontal cam-actuated punches (Fig. 5-26D) is useful when shut height is limited. The horn is bored out for slug drop-out. When the ram ascends, punch slides are retracted by rods under spring tension.

Horn-die Design for Heavy Pressures. In this design (Fig. 5-26E) a spring-loaded backup block is used to overcome any misalignment due to the overhanging horn and severe pressures. On the ram ascent, the backup block returns to the idle position for removal of the work. The punch is guided in a solid stripper to further ensure alignment.

Horn Die for Long Parts. A swinging horn is engaged by a short guide pin after the long workpiece is slid on the horn in the die shown in Fig. 5-26F. The part is loaded by raising the horn until it clears the short guide pin, then swinging it to one side. In the piercing position the horn is supported by solid pads. The long guide pin could be equipped with a spring and a limit stop at its upper end for ease of part insertion and removal.

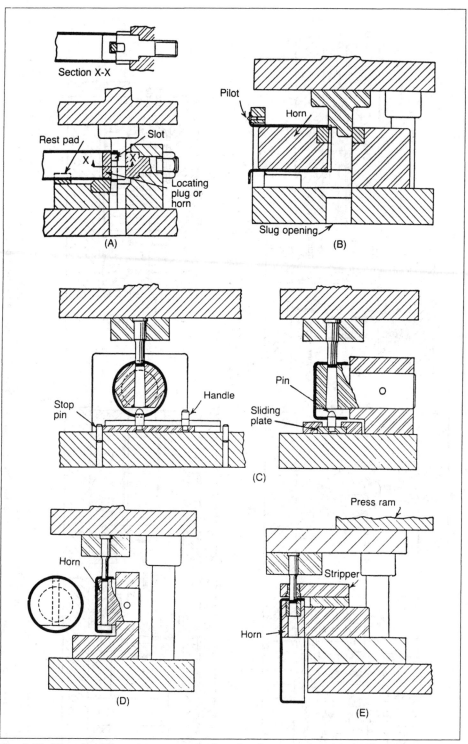

Fig. 5-25. Horn dies (*A*) cutting two opposed notches in one stroke with only one punch; (*B*) notching in relation to hole in flange; (*C*) with movable gage pin to locate second hole.[6,7] Horn dies (*D*) for small-diameter shells; (*E*) part overhangs bolster plate to accommodate large shells.[6,7]

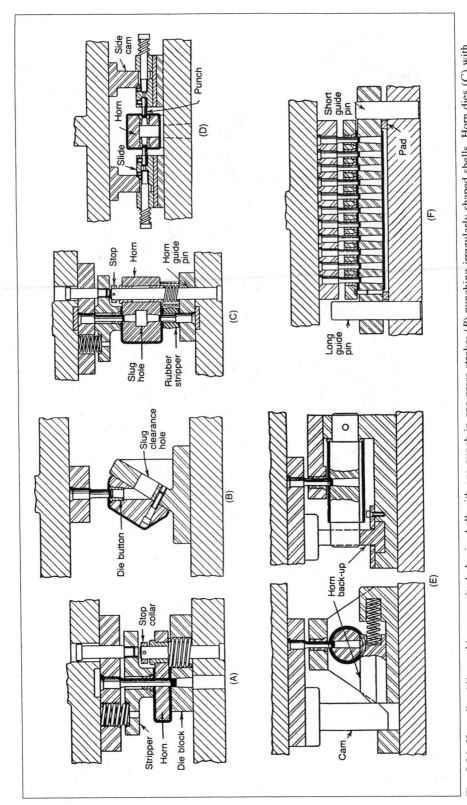

Fig. 5-26. Horn dies (A) punching opposite holes in shell with one punch in one press stroke; (B) punching irregularly shaped shells. Horn dies (C) with floating horn and (D) cam-actuated die. Horn dies with backup block to support burr (E). Horn die for long parts (F).

5-28

Lever-actuated Punching Die. This die (Fig. 5-27) is similar to a conventional cam die because the vertical movement of the punch holder (D2) actuates slides (D3), carrying horizontal punches (D4) by levers (D1) and slots in camlike members (D5), which are backed up by heel blocks (D6). The ejector mechanism's operating lever (D8) is provided with an idle time slot, so that ejection is not accomplished until all punches are retracted.

Die Design for Incorporating Entire Cam Mechanism in the Punch Holder. A spring-loaded cam plate, in a die of this type (Fig. 5-28), descends ahead of the punch holder to clamp the workpiece first. Punching action begins when the cam blocks strike the heads on the punches. Each punch is spring-retracted for stripping.

Index Slotting Die. The die shown in Fig. 5-29 pierces nine slots in a tubular part. The upper shoe is spring-loaded, and its upward motion is restricted by four shoulder screws (D1). This prevents the overhanging punch assembly (D2) from damaging the workpiece on the upstroke. The die block (D3) serves as a locator and has two pins (D4) for lateral location of the workpiece. The index plate (D5) is threaded into the workpiece and is located radially by the spring-loaded lever and pin assembly (D6). The slots in the index plate are beveled to assure lateral location. The punch (D7) is held in the holder by a cap screw; its radial location is maintained by a plate bearing on a flat. A fixed stripper (D8) holds the tube as the punch is retracted.

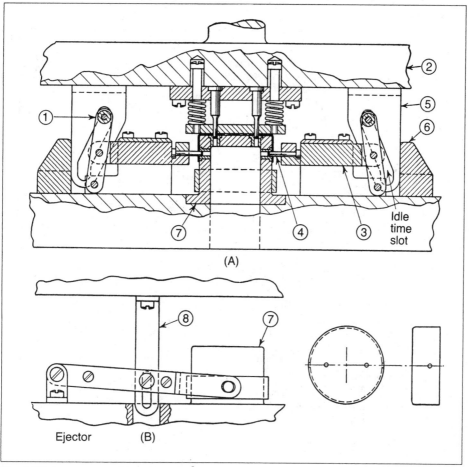

Fig. 5-27. Lever-actuated punching die.[8]

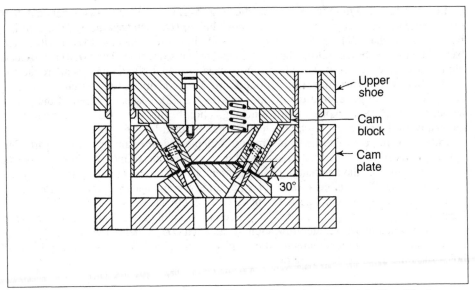

Fig. 5-28. Cam-punching die design for movable cams integral with punch holder. (*Design Service Co.*)

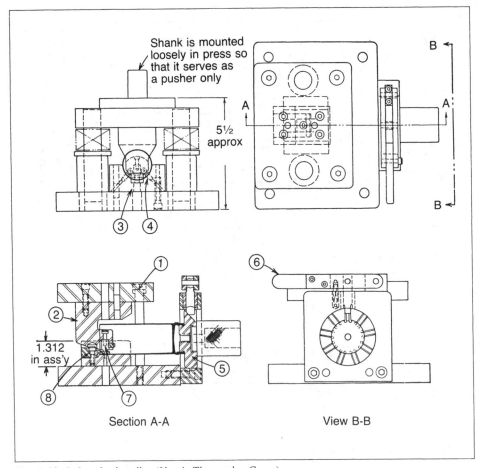

Fig. 5-29. Index-slotting die. (*Norris-Thermador Corp.*)

BLANKING DIES

Single-station Blanking Dies. Blanking heavy (0.75 in. or 19.05 mm) cold-rolled steel stock 1.375 in. (34.92 mm) long around part of its perimeter reduces scrap but necessitates the use of a heel block (Fig. 5-30, *D*1). A spring-loaded plunger (*D*2) clamps the stock against the heel block. The punch holder (*D*3) is machined to fit the punch blade (*D*4) to which it is welded. Blanking pressure is reduced by grinding shear equal to one-third stock thickness on the die only. Since the finished blanks fall through the die, this die design can be classified as a drop-through die.

Inverted Blanking Die. The blank for the part shown in Fig. 5-10 *C* is produced from 0.187-in. (4.75 mm) SAE 1008 steel stock by the die of Fig. 5-31. All holes are cut after the part is drawn or formed with the exception of two piloting holes cut by two heavy punches, one of which is shown in Section A-A; the resulting slight elongation of these holes is permissible. The stripper on the upper shoe strips the part from the perforating punch and the inverted sectional blanking die. A stripper plate on the lower shoe removes the blank from the sectional blanking punch. The stock is fed into the front of the die and positioned from left to right by guide strips and front to back by the spring-loaded stop pins shown in Section B-B. The scrap skeleton is chopped by two scrap cutters whose construction is shown in Section C-C.

Steel Rule Dies. Many types of steel rule dies are regularly used over a wide range of industrial applications. All make use of low-cost materials, and they are usually constructed in a fraction of the time needed to manufacture a conventional die for the same part. Two distinctly different operating principles separate the various types into the classifications of single-element cutting dies and two-element metalworking dies.

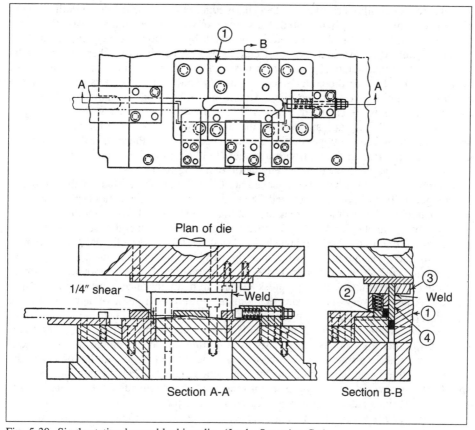

Plan of die

Section A-A   Section B-B

Fig. 5-30. Single-station heavy blanking die. (*Leake Stamping Co.*)

5-31

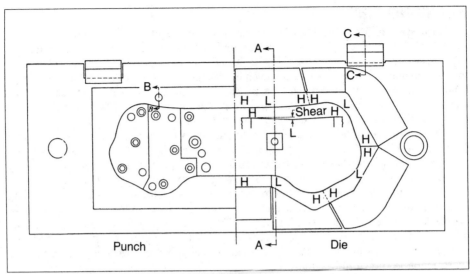

Fig. 5-31. Inverted blanking die. (*Buick Division, General Motors Corp.*)

Single-element Cutters. All the older, better-known forms of rule-die tooling fall within this category, which includes dinkers, clickers, one-piece forms, block dies, and cutting and creasing dies. For the most part, these are single-element tools consisting of a die section only and without opposing punchlike cooperating elements. They employ printer's rule or similar strip steel, with a sharp edge, bent or formed to the outline shape of the part desired. Suitably mounted, the shaped knives are operated against a flat metal platen or a hard wood block and produce the blank by a cutting or cleaving action.

Dies of this type are widely used for cutting papers, cardboards, fibers, rubber, felt, leather, and similar relatively soft materials and occasionally for blanking thin, soft metals. The printer's or die-cutter's rule is suitably held within a kerf sawed from a die board consisting of five to seven plies of gum or maple wood, as shown in Fig. 5-32. Cutting-rule heights are standard. The kerf is often cut intermittently with gaps tying in the framework. When gaps are used in the die board, in one-piece forms, the rule is notched or bridged so the upper cutting surface is continuous. Pieces of rule may also be used and are inserted in the kerf with a mallet. In clicker dies, the ends of the rules are brazed to preserve continuity and strengthen the cutting form.

The rules should be inserted or attached at 90° ± 1/4° to the surface of the die board and are generally hardened to Rockwell C52 to 55, although this may not be necessary for some materials. These dinking rules are purchased in soft to hard condition. They are readily worked in the Rockwell C40 to 50 range and are generally cut to size in all prehardened conditions. Stripper material is usually 0.03125 to 0.375 in. (7.938 mm to 9.53 mm) thick, extending somewhat over the height of the cutting rules, and is usually of Neoprene, cork sheet, or rubber. These single-element tools may be mounted in almost any form of die-cutting or stamping press and are, in many instances, operated in printing presses. Although very useful and economical within their area of application, the operating principle limits their application in the metalworking industries.

Two-element Metalworking Rule Dies. Unlike earlier forms of rule dies, the metal-working rule die does not produce parts by a cutting action but employs rule to exert pressure to produce a fracture in shear to the outline desired. Although not precisely the same, the action is similar to the conventional punch and die. Construction methods and specifications vary for these two-element dies, and developments have resulted in the granting of various patents on dies and associated equipment.

Steel rule and wood, or a similar substance, may be used for the upper die section, but somewhat more attention is given to construction detail, and better grades of materials are

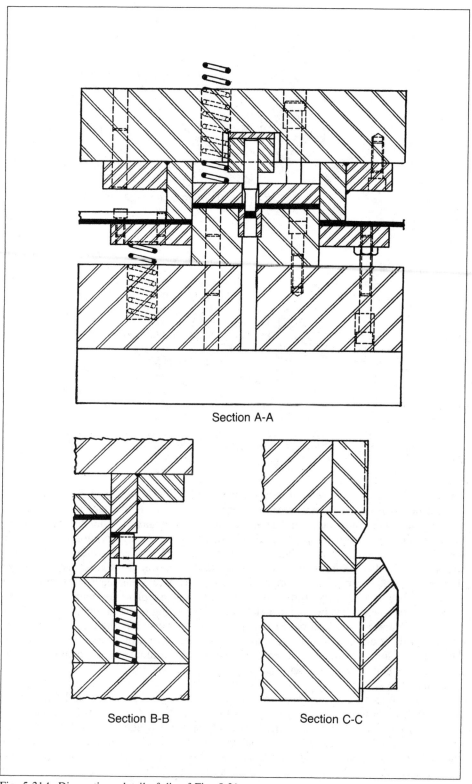

Section A-A

Section B-B

Section C-C

Fig. 5-31A. Die-sections detail of die of Fig. 5-31.

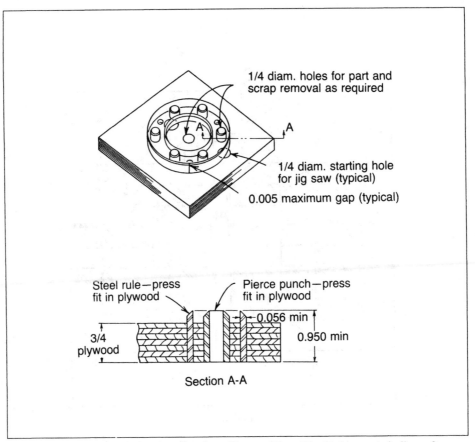

1/4 diam. holes for part and scrap removal as required

A

1/4 diam. starting hole for jig saw (typical)

0.005 maximum gap (typical)

Steel rule—press fit in plywood

Pierce punch—press fit in plywood

0.056 min

3/4 plywood

0.950 min

Section A-A

Fig. 5-32. Steel rule die for cutting of papers, fibers, rubber, felt, leather, and similar soft materials.

employed. The punchlike section is made from relatively thin steel plate, usually of a hardenable steel, which is mounted on a subplate. Because of the stresses involved in shearing steel sheet and plate, considerable know-how is required to construct and operate this type of tool properly. Neoprene, springs, or some suitable ejectors are used to remove the blank and strip the waste. Punches may be installed to pierce while blanking the workpiece. Metalworking rule dies are best operated in specially designed master die sets (Fig. 5-33), which permit interchangeability and corrective adjustment for stability and alignment. While commonly utilized as compound blank and pierce or notching tools, the metalworking rule die has been extended to areas permitting combined operations in dies which blank, form, trim, and pierce in a single hit, such as illustrated in Fig. 5-34.

Various other designs of metalworking rule tooling, such as progressive and trimming dies, have been used with some success. In general, this tooling is not recommended for small parts. However, on larger parts, tool economies show an increased saving proportional to size.

One steel rule die user recommends the steel rule thicknesses listed in Table 5-1. A land is ground on the steel rule as shown in Fig. 5-35, with a minimum bevel length of 0.015 in. (0.38 mm). Standard methods of applying shear to the die may be used if the press tonnage is exceeded. Conventional punch and die clearances are used for the material being blanked. The shear plate shown in Fig. 5-34 may be made of high-carbon steel such as SAE 1080 or 1095 or alloy steel such as SAE 4130 or 8630 and heat-treated to Rockwell C39 to 44. Large sections may be flame-hardened.

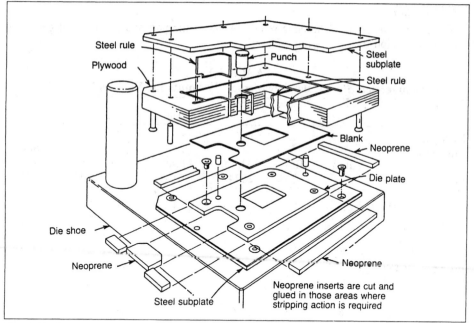

Fig. 5-33. Exploded assembly view of a steel rule die.

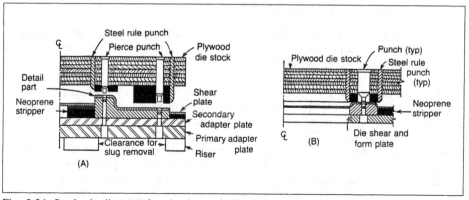

Fig. 5-34. Steel rule dies; (A) for piercing and trimming a preformed workpiece; (B) for forming, piercing, and blanking a workpiece. *(Template Industries, Inc.)*

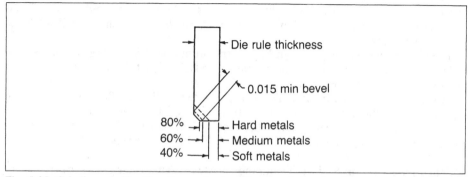

Fig. 5-35. Grinding specifications for steel rule. *(Template Industries, Inc.)*

TRIM DIES

Trim Dies for Shells. The edge requirements of the part and its size will generally indicate what kind of a die is most feasible.

TABLE 5-1
Recommended Steel Rule Thicknesses By Points for Various Materials*

Material Thickness		Material						
in.	(mm)	Alum., soft	Brass ½ H	Alum. 2024-T	Alum. 7075-T	Steel, mild	Steel 4130	Stainless 302
0.010	(0.25)	3	4	4	4	4	4	4
0.031	(0.79)	4	4	4	4	4	4	4
0.062	(1.57)	4	4	4	4	4	4	6
0.078	(1.98)	4	4	4	4	4	6	6
0.093	(2.36)	4	4	6	6	6	6	6
0.125	(3.18)	6	6	6	6	6	8	8
0.150	(3.81)	6	6	8	8	8	8	

* One point equals approximately 0.014 in. (0.36 mm)

Pinch Trimming. A pinch trim can be incorporated with a standard drawing or reducing operation, eliminating a separate operation. Clearances between the punch and die must be held to a minimum so pinch-off is even and the formation of sharp or rough edges on the part is prevented. Figure 5-36 shows a die in which drawing and pinch trimming are combined. Pinch trimming may also be done in a push-through die (Fig. 5-37). In this type of die the sharp cutting edge of the punch trims the shell flange against the radius on the die opening, where there is some ironing action. The shell is pushed entirely through the die.

Wipe-down Trimming. This type of trimming, commonly known as a trim and wipe-off operation, allows a previously drawn part to be produced with a trimmed edge normal (at 90°) to the center line of the part. This operation leaves a wipe-off line or shallow depression on the outside of the part. The finished edge will be identical with the original edge of the flange. A wipe-down and trim die is shown in Fig. 5-38. The shell is pushed through the die and is stripped by the edges of the lower side of the die.

Flush Trimming. This type of trimming leaves a slight projection on the outside rim of the shell (due to clearance), a condition often desirable. Flush-trim dies for long shells and shells of average length are shown in Fig. 5-39. Shells to be flush-trimmed are drawn with a small flange.

Burr-location Considerations in Flange-trimming-die Design. The desired location of burrs on a flanged shell may alone determine the type of die design, since burr location and direction are important in many assemblies.

Inverted Dies for Trimming Flanges. The inverted die (Fig. 5-40A) trims the inverted shell and leaves the burr on the top of the flange. Scrap cutters at the base of the punch split the scrap ring. Another inverted die, shown at B, receives the shell in the upright position, but the burr is left on the underside of the flange. Scrap cutters (not shown) are provided.

Horn-type Die for Square Trimming of Shells. This design (Fig. 5-41) can be classified as a notching die, since only a segment of the shell is cut at one press stroke.

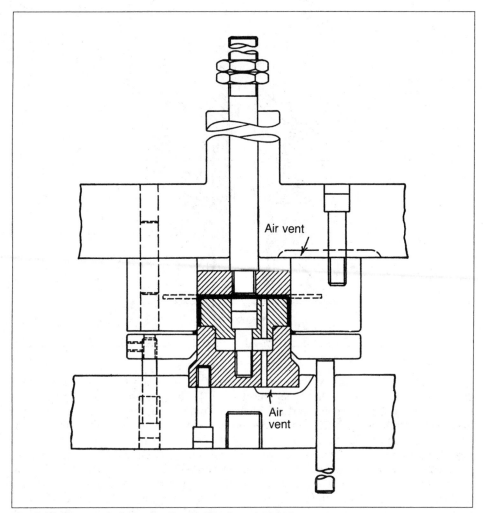

Fig. 5-36. Draw-and-pinch-trim die. (*Acklin Stamping Co.*)

The shell is rotated on the horn after the punch trims a peripheral section. This die design is suitable for low-production requirements.

Cam Trimming a Continuous Flange. A die design for this operation (Fig. 5-42) includes six cams (D1 and D3) provided with trim punches (D5). A hooklike extension (D2) allows overlapping of adjoining trim punches. The ring or collar of scrap cutoff is split into six segments by six scrap cutters (D7 and D4). There is considerable distortion of the scrap, but not of the part, which is clamped securely by a spring-loaded stripper (D9). The part is cut between the two sharp edges of the cutting or trim punches (D5 and D6). The part is lifted by a spider-shaped lifter harness (D8), operated by an air cylinder properly synchronized with the upstroke of the ram. The part is manually or mechanically pulled or pushed off the lifter before the end of the upstroke. The mechanism for such part removal is not shown. The vertical cam is fitted with a wear plate (D13). The six scrap segments slide down six radiating cored passages (D10), whose walls are indicated by the lines D11 and D12.

Shimmy Die. A cup drawn from 0.025-in. (0.64-mm) thick stock is set in a holder (Fig. 5-43, D9 and D10) clamped by spring-loaded plungers (D6 and D8) and trimmed along its top from the inside by a horizontally moving punch (D7). During a 3-in. (76 mm) press stroke the angular surfaces of the cams (D1, D2, D3, and D4) move downward to

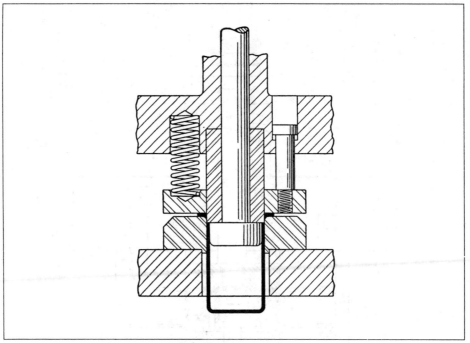

Fig. 5-37. Push-through pinch-trim die.[6]

contact the outside surface of the cam follower ($D5$), to which the punch is attached. Horizontal movements of the punch result only from angular contacts between the follower and the four cams; vertical contacts produce no motion. There are eight horizontal punch movements which complete the trimming: the initial movement, actuated by cam $D1$, starting at the center of the punch-motion diagram (Fig. 5-44) is directly backward; numbered consecutively the remaining movements are: (2) to the right, actuated by cam $D2$; (3) to the right and front by cams $D2$ and $D3$; (4) to the front by cam $D3$; (5) to the front and left, by cams $D3$ and $D4$; (6) to the left, by cam $D4$; (7) to the left and backward, by cams $D4$ and $D1$; (8) to the back by cam $D1$. Numbered spaces (Fig. 5-44) between the lines drawn on the cam profiles are the cam surfaces which generate the correspondingly numbered punch movements.

A variation of the shimmy-trimming die incorporates the cam-actuated portion in the lower die shoe and may have an air-cushion arrangement to control the cam descent instead of springs. Dies using an air cushion do not require an opening in the bolster, but they usually have a greater shut height. In this die construction the workpiece oscillates, and the trimming punch remains stationary. Quality of production is the same in either of the dies.

Combined Vertical and Horizontal Trimming. A partially flanged part is trimmed on three sides (Fig. 5-45, at $D1$, $D2$, and $D3$) by direct (vertical) action, and on the fourth side ($D4$) by the outside cammed steel ($D5$) and the inside steel ($D6$). The latter is backed up by the shoulder in the die shoe ($D7$). The design incorporates three scrap cutters ($D8$, $D9$, and $D10$), two gage plates ($D11$ and $D12$), and a spring-loaded cam stripper ($D13$), which locate the part. The cam casting ($D15$) is guided and held down by hardened steel gibs ($D16$), and its back travel is limited by a stop block ($D14$). The ample initial contact area between the vertical driver cam ($D17$) and the cam casting prevents rapid wear of the wear plate.

Die for Trimming Rings. A cam-actuated trim die (Fig. 5-46) incorporates a spring pad ($D1$) to clamp the workpiece (a ringed plate of a loose-leaf blank book) and trims the overhanging ring between the cam-actuated cutter slides D2 (which are forced outward)

5- 38

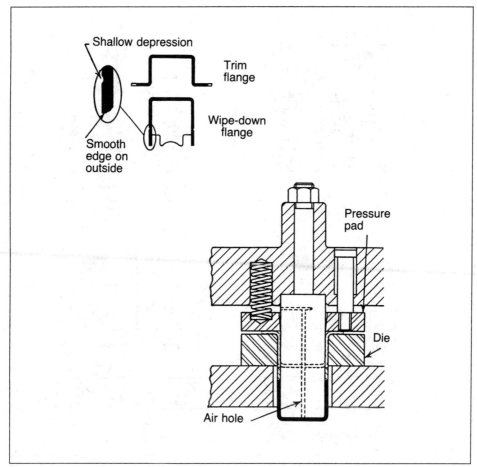

Fig. 5-38. Wipe-down trim die.[6]

and the upper cutters (D4). The cutter slides are moved by cams (D3) on opposite sides of the die. Spring-loaded pins (D5) prevent the part from adhering to the upper die member.

SHAVING DIES

Die for Shaving Slots. Approximately 10,000 parts (0.094 in. (2.39 mm) thick, SAE 1010 steel) per grind are shaved 0.006 in. (0.15 mm) in the die shown in Fig. 5-47. Simultaneous and equal shaving of the slots and the outside of the part opposite, with no distortion, is accomplished through the use of adjustable spring locators. The part remains in the die block until a positive knockout pin ejects it at the top of the stroke, preventing the spring locators from pushing the shavings back into the part. The part and the chips are simultaneously blown off the die. Maximum accuracy and straightness of holes are secured when the burr side of the hole is toward the punch, so that the shaved chip becomes smaller as the punch progresses.

Die for Shaving at an Angle. It is necessary that the part (a flapper valve) have a sharp and true edge at a 15° angle. Guide rings (Fig. 5-48) provide rigidity for shave die and punch by enclosing both before actual shaving begins.

Die for Shaving a Portion of a Blank. Cam-actuated backup blocks (Fig. 5-49, D1) prevent movement of the part in shaving only a portion of its perimeter. Parts uniformly wide are shaved with less movement of the backup block (view B); those which vary in

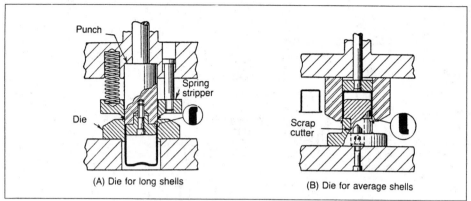

Fig. 5-39. Flush-trim dies for shells.[6]

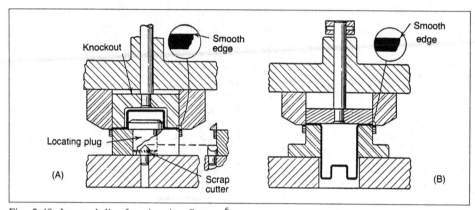

Fig. 5-40. Inverted dies for trimming flanges.[6]

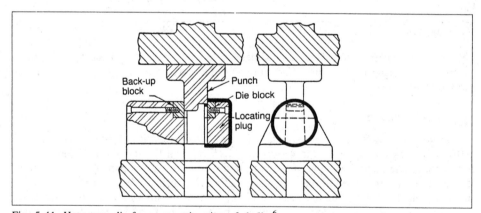

Fig. 5-41. Horn-type die for square trimming of shells.[6]

width are shaved with more movement (view *A*).

Die for Shaving Small Gears. Leveling studs (Fig. 5-50, *D*1) contact the face of the die (*D*2), resulting in uniform and accurate contact with the work, a small gear (*D*3). An enlarged view of the shedder pin (*D*4) is shown. The punch (*D*6) is counterbored for the gear hub. Stripper pins (*D*7) actuate a stripper plate (*D*5).

Die for Shaving Holes at an Angle. The shaving die shown in Fig. 5-51 is designed to shave 12 holes angularly in a steel disk circumferentially. A 10-station dial feed is incorporated so six holes may be shaved at a time in each of two of the 10 stations (the

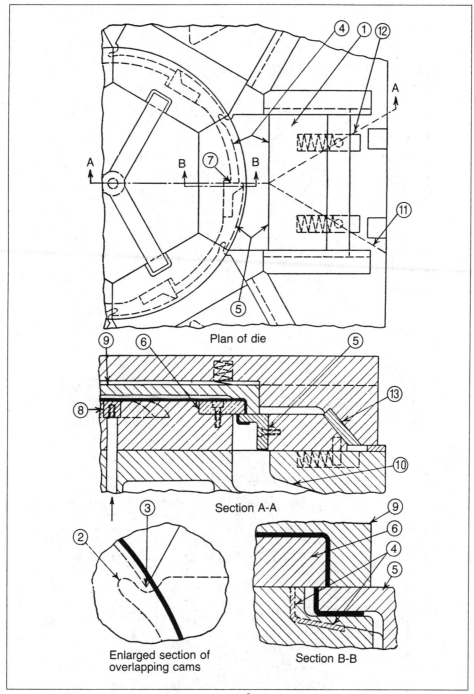

Plan of die

Section A-A

Enlarged section of
overlapping cams

Section B-B

Fig. 5-42. Cam-trimming die for cylindrical parts.[9]

remaining eight stations are idle). The six well-lubricated sliding punches and the oil reservoir are shown in Fig. 5-51A. The spring-loaded pilot and microswitch assembly prevent operation of the press if the disk is not located properly.

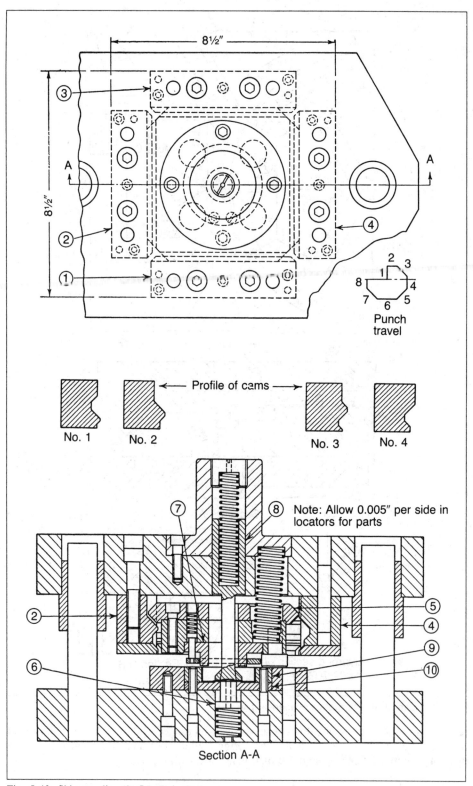

Fig. 5-43. Shimmy die. (*Lufkin Rule Co.*)

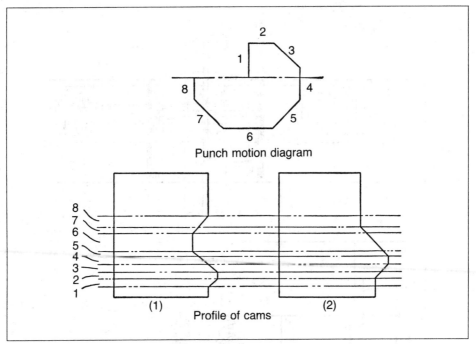

Fig. 5-44. Profile of the cams in the die of Fig. 5-43.

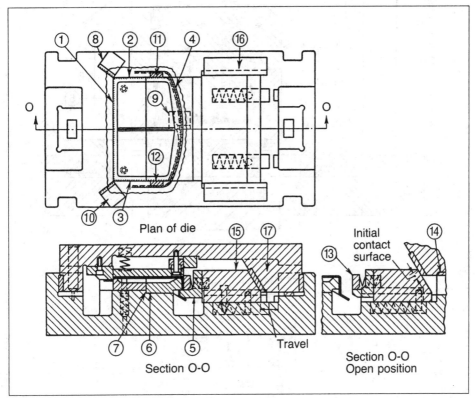

Fig. 5-45. Vertical and side-cam trim die.[9]

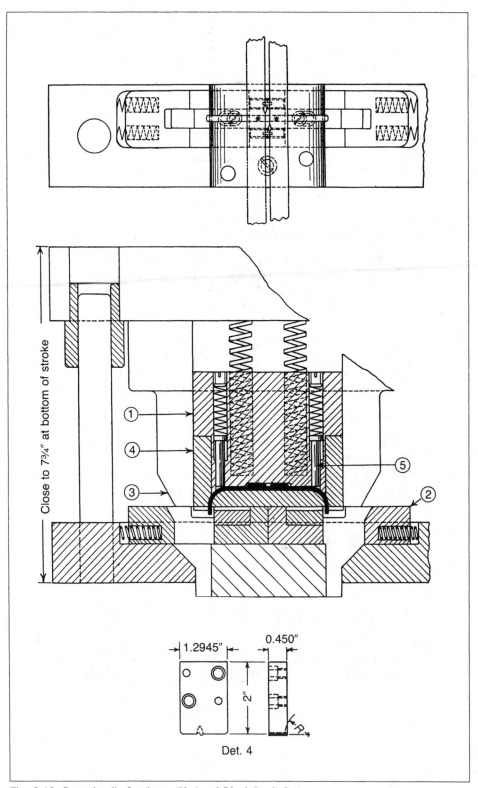

Fig. 5-46. Cam-trim die for rings. (*National Blank Book Co.*)

5- 44

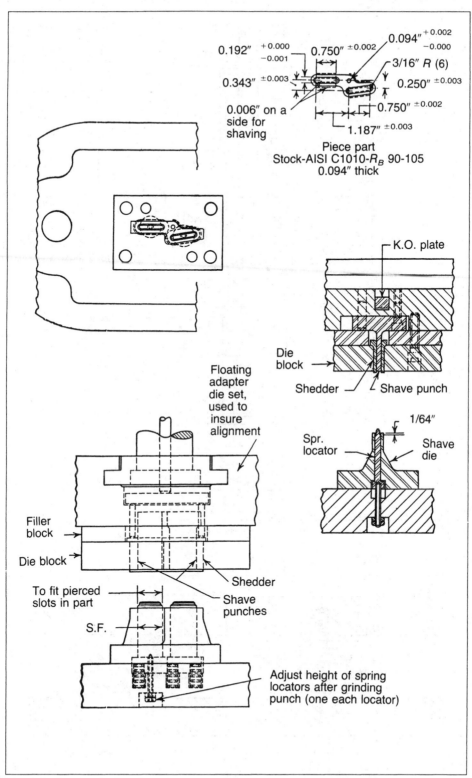

0.192″ $^{+0.000}_{-0.001}$

0.094″ $^{+0.002}_{-0.000}$

0.750″ ±0.002

0.343″ ±0.003

3/16″ R (6)

0.250″ ±0.003

0.006″ on a side for shaving

0.750″ ±0.002

1.187″ ±0.003

Piece part
Stock-AISI C1010-R_B 90-105
0.094″ thick

K.O. plate

Die block

Shedder

Shave punch

Floating adapter die set, used to insure alignment

1/64″

Spr. locator

Shave die

Filler block

Die block

Shedder

To fit pierced slots in part

S.F.

Shave punches

Adjust height of spring locators after grinding punch (one each locator)

Fig. 5-47. Die for shaving slots. (*International Business Machines Corporation*)

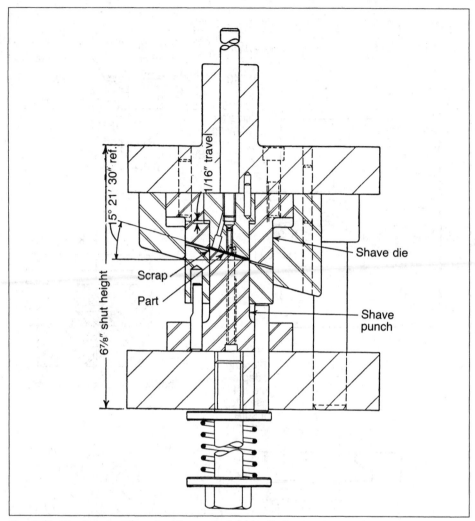

Fig. 5-48. Shaving at a 15° angle. (*Carter Carburetor Co.*)

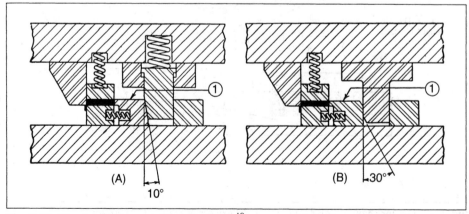

Fig. 5-49. Dies for shaving a portion of a blank.[10]

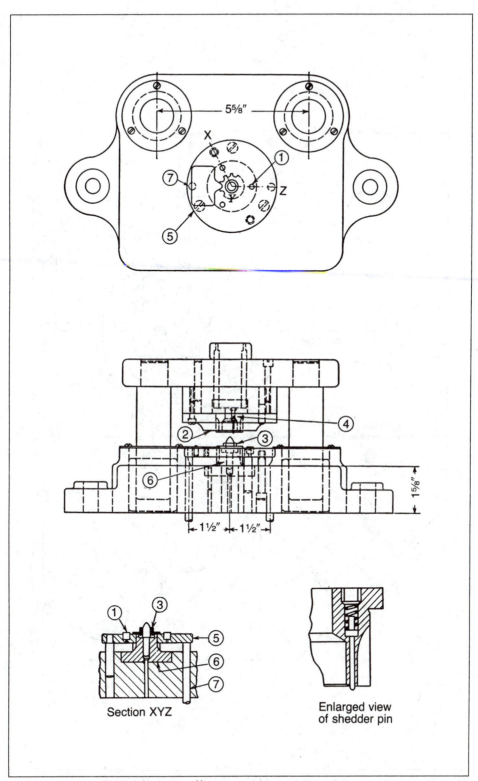

Fig. 5-50. Shaving die for small gears.[11]

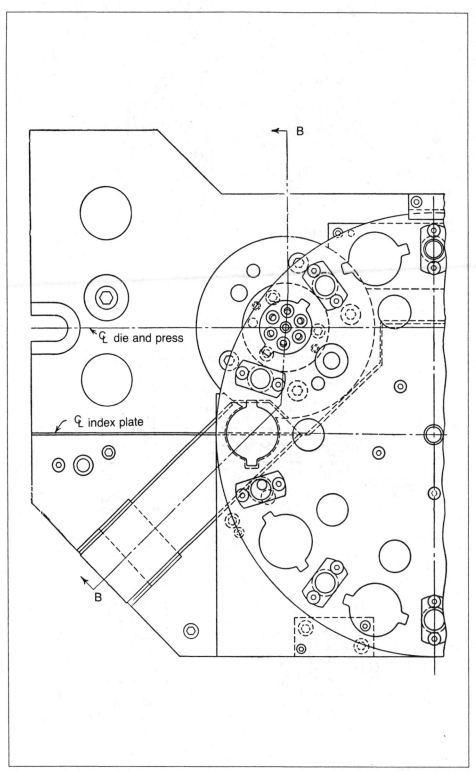

Fig. 5-51. Plan view of die for shaving multiple holes at an angle. (*Oldsmobile Division, General Motors Corp.*)

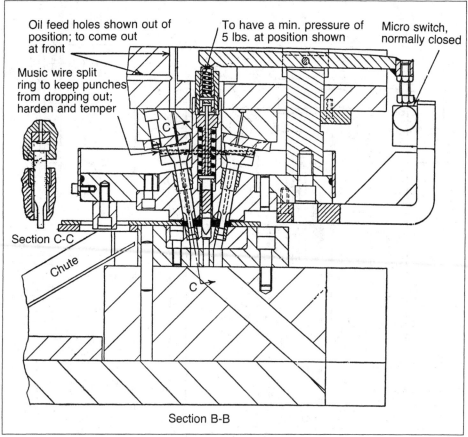

Section C-C

Section B-B

Fig. 5-51A. Section B-B of die shown in Fig. 5-50.

CUTOFF DIES

Cutoff Die for Tubing. A sliding block (Fig. 5-52, *D2*) is actuated by a cam (*D3*) to clamp the work (0.035-in. (0.89 mm) cold-rolled-steel tubing) before the punch blade descends. The shape and thickness of the blade (*D1*) and the contour of the clamping surfaces determine the amount of possible dimpling or distortion of the work.

This die is designed to cut off oval-shaped tubing with the punch first contacting the narrow width. Similarly shaped punches function equally well with round and square tubing, as illustrated in examples which follow. To minimize distortion, the workpiece must be completely confined and securely clamped for a short distance on each side of the cutoff blade.

Triangular-blade Cutoff Die. Spring-loaded pressure pads closely fit the outside of the tubing and clamp it before the blade descends in the die illustrated in Fig. 5-53. The point of the blade leaves a small indentation at the top of the tubing. The ring slug drops through the die, requiring no scrap cutter.

Cutoff Die for Square Tubing. The die design of Fig. 5-54 incorporates a thin punch with the contour shown. Square tubing is confined and clamped by a spring-loaded pad. The pointed end of the lancing punch pierces the top of the tubing and progressively shears the sides and bottom of the tube as the punch descends. The careful fitting of the punch in the slot ensures the success of this die.

Slug and Shearing Dies. A type of cutoff die, commonly known as a parting or slug die (Fig. 5-55, view *A*), completely trims the outline of both sides or ends of the blank, producing the burr on the same side of the part at both edges. In contrast, the no-scrap

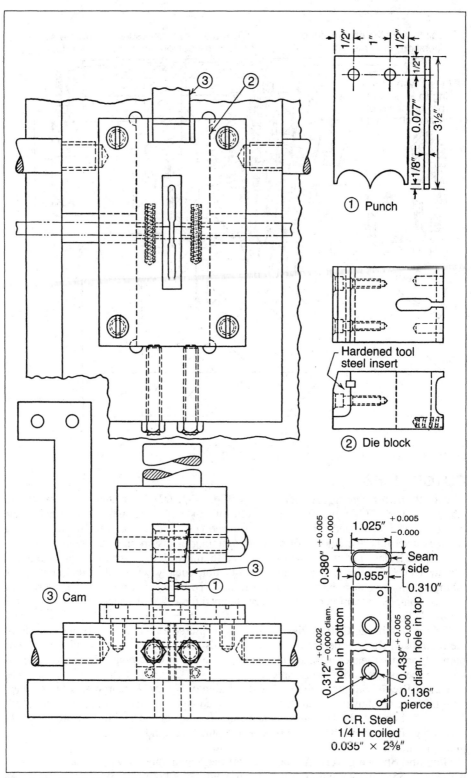

① Punch

Hardened tool
steel insert

② Die block

③ Cam

① Punch dimensions: 1/2", 1", 1/2", 1/2", 3½", 0.077", 1/8"

1.025" +0.005 −0.000
0.380" +0.005 −0.000
0.955"
0.310"
Seam side
0.312" +0.002 −0.000 diam. hole in bottom
0.439" +0.005 −0.000 diam. hole in top
0.136" pierce

C.R. Steel
1/4 H coiled
0.035" × 2⅜"

Fig. 5-52. Cutoff die for tubing.

5-50

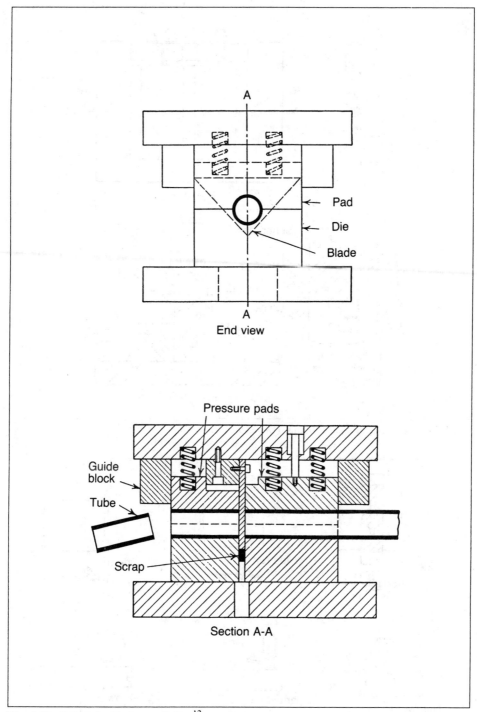

Fig. 5-53. Triangular-blade cutoff die. [12]

shear dies shown at *B* and *C* will produce burr on opposite sides or ends of a given part. Wide-run shearing will reduce handling of material strips. But, where parts are to be bent in a subsequent operation, strips are run the narrow way so that the bends will be across the grain.

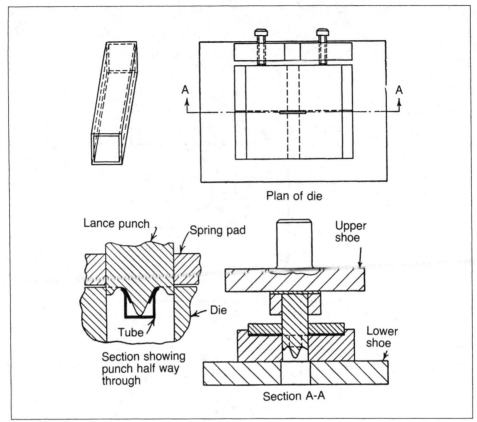

Fig. 5-54. Cutoff die for square tubing. (*General Electric Co.*)

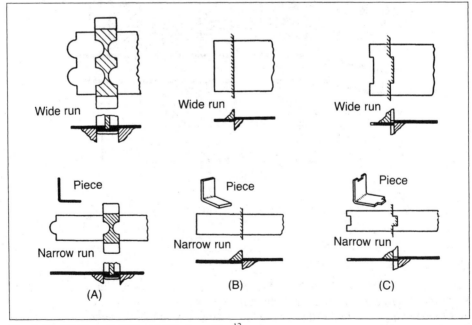

Fig. 5-55. (*A*) Slug die; (*B*) and (*C*) shear dies.[12]

BROACHING DIES

Die for Trimming Square Ends on a Shaft. A square section is broached on the end of a round bar in two operations in the die illustrated in Fig. 5-56. A heeled punch (*D3*) broaches the first pair of flats in the right-hand nest (*D6*). The part is removed, turned 90°, and inserted by hand in the left-hand nest (*D5*) for the second broaching operation. Springs (*D2*) hold the nests open until the upper nest is forced down by the pin (*D4*) to clamp the work in place. The second operation cannot be done until the first is completed,

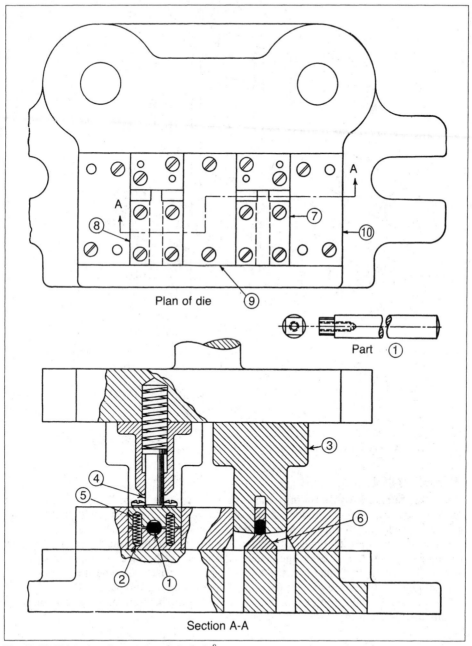

Plan of die

Part ①

Section A-A

Fig. 5-56. Trimming flats on a round shaft.[8]

5- 53

because of the shape of the nests.

Die for Broaching Slots in a Round Shaft. Toothed broaching blades of high-speed steel (Fig. 5-57, D1) cut opposed slots in round rod. The chip is approximately 0.003 in. (0.08 mm) per tooth. A stop pin (D2), a workholder (D3), and broaching-blade heel blocks (D4) are incorporated.

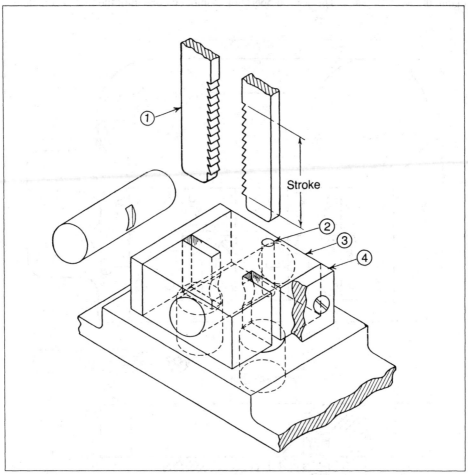

Fig. 5-57. Broaching slots in a round shaft. (*O.H. Mathisen, SME.*)

FINEBLANKING

Smooth Surfaces by Blanking. Blanking processes developed in Switzerland and Germany are capable of producing smooth surfaces without an additional shaving operation. The sudden rupture and ragged surfaces which characterize conventional blanking processes are eliminated by extending the plastic deformation over the entire period of shearing. In Fig. 5-58, the blankholder preloads the sheet without damaging its surface. To accomplish blanking, the blanking punch must overcome the resistance of material plus the resistance of the hydraulically preloaded backup punch. The total effect of this setup is the induction of superimposed hydrostatic pressure on the shear zone that inhibits the initiation of fracture in the sheared edges.

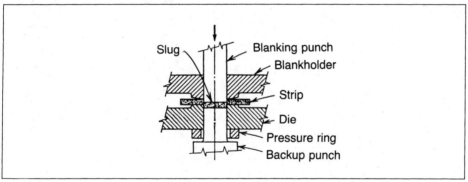

Fig. 5-58. Blanking die to produce smooth edges on both the blank and strip.[14]

References

1. *Tool Engineering Reference Sheets*, International Business Machines Corp.
2. F.E. Truaned, "Nibbling Die Slides away from High Costs," *American Machinist* (August 18, 1952).
3. *Reference Book*, Vol. 4, Durable Punch & Die Co.
4. L. Hodge and R. Sheppard, "A Piercing Tool for Producing a Number of Small Closely-spaced Slots", *Machinery* (London) (December 16, 1959).
5. R.C. Johnson and J. Ingraham, "Hollow-ground Punch Slots Die-supported Tubing," *American Machinist* (March 6, 1950).
6. J.R. Paquin, "How to Choose Trim Dies for Drawn Shells," *American Machinist* (October 30, 1950).
7. J.R. Paquin, "Horn Dies Will Do Many Jobs," *American Machinist* (April 3, 1950).
8. K.L. Blues, "Die-Grams," *Western Machinery & Steel World* (August 1948).
9. C.R. Cory, "Die Design Manual." Part 1 (1949).
10. J.R. Paquin, "What to Watch For in Shaving Design," *American Machinist* (August 11, 1949).
11. A.A. Dowd and F.W. Curtis, *Tool Engineering* (New York: McGraw-Hill Book Co., 1925).
12. K.L. Stoltenberg, "Slitting Punch and Die Cut Off Tubing," *American Machinist* (April 4, 1951).
13. J.R. Paquin, "Choose from Twenty-four Cut-off Dies," *American Machinist* (November 27, 1950).
14. International Manufacturing Research. Part II, "Advances in Forming," *The Tool and Manufacturing Engineer* (June 1963).

SECTION 6
BENDING OF METALS*

Bends are made in sheet metal to gain rigidity and produce a part of desired shape to perform a particular function. Bending in several directions can produce parts that otherwise would require a drawing operation.

The simplest form of bending, air bending, is illustrated by a beam supported at two points, with a load applied at the midpoint, as shown in Fig. 6-1.[1] The load produces compressive stresses in the inner layers of the bend and tensile stresses in the outer layers. If the stress exceeds the material yield strength, the beam takes a permanent set or bend. Not all of the material in the bend zone is not equally stressed. The material in the inner and outer surfaces is stressed the most, and the stress gradually diminishes toward a neutral axis between the two surfaces. At that point, the stress is zero, and there is no length change.

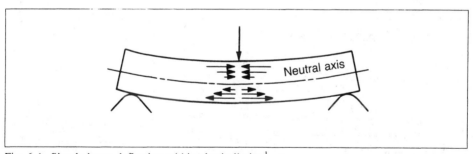

Fig. 6-1. Simple beam deflection within elastic limits.[1]

A pure bending action does not reproduce the exact shape of the punch and die in the metal; such a reproduction is one of forming.

Terms used in bending are defined and illustrated in Fig. 6-2.

Strain Distribution. Circumferential strain on the convex side of a bend is considerably larger and on the concave side almost identical to strains calculated from theory. This explains the decrease in metal thickness with increasing curvature (Fig. 6-3).

Circumferential strains are dependent upon the bend angle as illustrated in Figs. 6-4 and 6-5. Values in these two graphs are expressed in the so-called natural strain ϵ, not the conventional strain e, which is the absolute change in unit length.

The natural strain ϵ is expressed as follows:

$$\epsilon = \log_e (L + e) \qquad (1)$$

where L = length of bend.

The transverse distribution of circumferential strain is shown in Fig. 6-6, which

* Assistance with the text was provided by Mr. Cecil Lewis, Director of Manufacturing and Engineering, Midway Products Corporation, Monroe, Michigan.

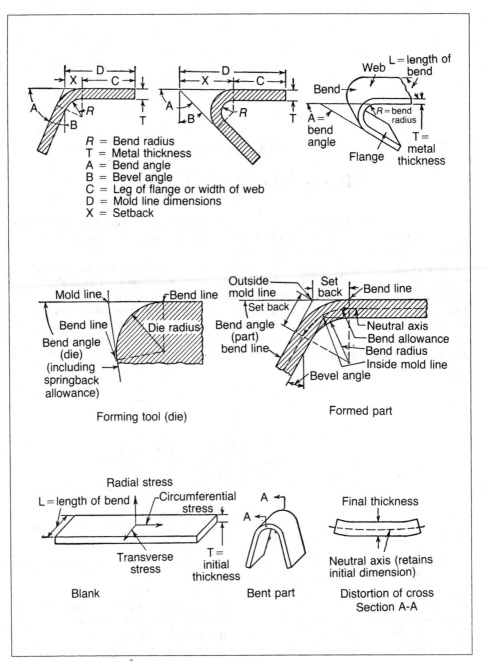

R = Bend radius
T = Metal thickness
A = Bend angle
B = Bevel angle
C = Leg of flange or width of web
D = Mold line dimensions
X = Setback

Fig. 6-2. Bending terms.[2]

explains the familiar trapezoidal section (Fig. 6-2, Section *A-A*) of a bent strip or bar. The transverse strain is zero if the bend length is sufficiently long. The test pieces were various widths of 0.125-in. (3.18-mm) thick 2024-T3 aluminum. The distribution of transverse and circumferential strains (ϵ_1 and ϵ_2) as affected by different bend lengths (width of sample) of the aluminum specimens is shown in Fig. 6-7, where the ratios of bend lengths to the stock thickness (L/T) are, respectively, 1, 2, 4, and 8.

Longer bend lengths of a given metal strip of a given thickness are less ductile than the same metal strip of shorter bend lengths (narrow parts) as shown in Fig. 6-8.

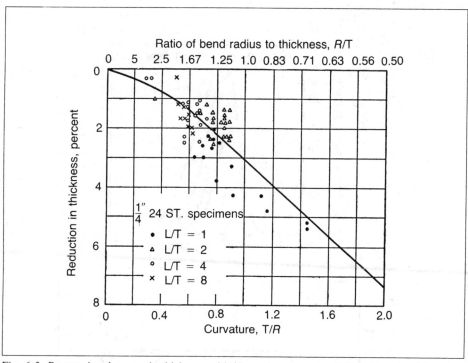

Fig. 6-3. Progressive decrease in thickness with increasing curvature (L/T is the ratio of length to thickness).[2]

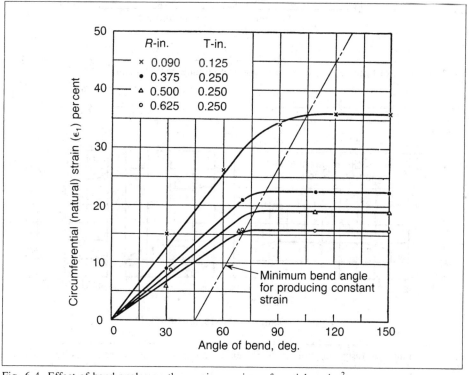

Fig. 6-4. Effect of bend angles on the maximum circumferential strain.[2]

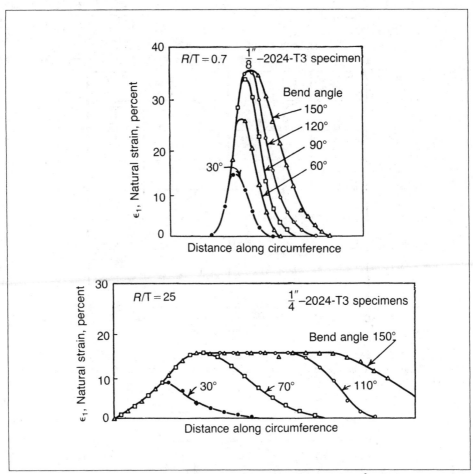

Fig. 6-5. Effect of bend angle on the distribution of circumferential strain.[2]

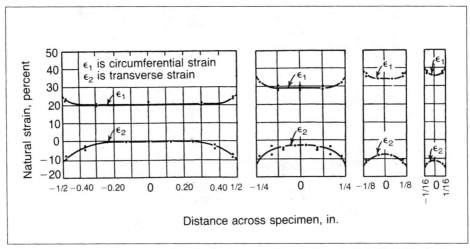

Fig. 6-6. Transverse distribution of strains in bent 0.125-in. (3.18-mm) thick 2024-T3 aluminum sheet.[2]

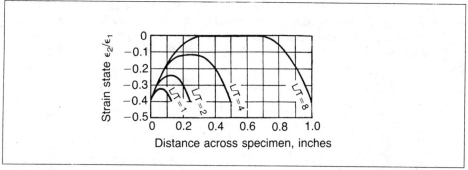

Fig. 6-7. Effect of the bend length on the distribution of strain states in 0.125-in. (3.18-mm) thick 2024-T3 aluminum sheet.[2]

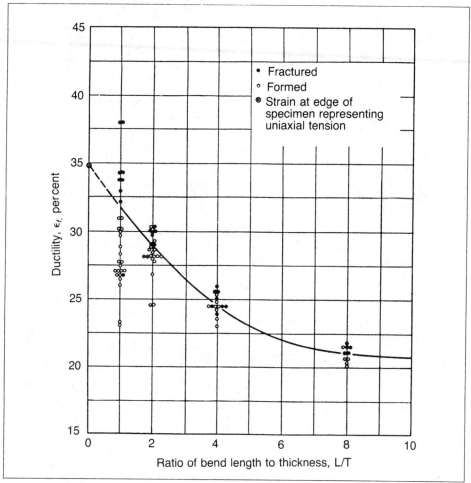

Fig. 6-8. Effect of bend length on ductility.[2]

MINIMUM BEND RADII

The minimum bend radii vary depending on the type of metal. Most annealed metals can be bent to a radius R, equal to the thickness t, and sometimes to $R = t/2$, for a given angle and bend length. R/t is constant for most metals, but this ratio increases for some aluminum alloys (Fig. 6-9).

The die designer will have to consider the change in cross section at the center of the bend when choosing gaging, nesting, and piloting methods for bending narrow parts.

The minimum bend radius is affected by bend length (Fig. 6-10); the minimum bend radius is nearly constant when the bend radius length is eight or more times the metal thickness.

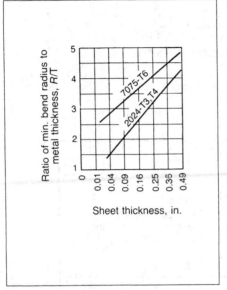

Fig. 6-9. Effect of sheet thickness on minimum bend radii of heat-treated aluminum alloys.[2]

Fig. 6-10. Effect of bend length on minimum bend radii of 0.025-in. (3.18-mm) thick heat-treated aluminum alloys.[2]

The minimum bend radius that can be formed in most metals is dependent on the condition of the edges (Fig. 6-11). The effect of poor edge conditions is of greater concern in the case of short bend lengths relative to stock thickness.

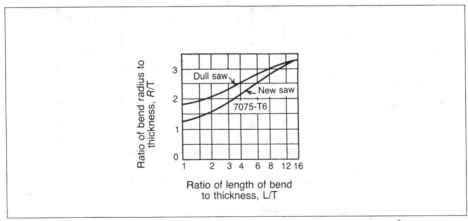

Fig. 6-11. Effect of edge condition on minimum bend radii of an aluminum alloy.[2]

Blanking Angle. Directionality in sheet or strip metal limits the minimum bend radius. Blanking angles (the angle between the bend axis and the direction of rolling) can govern the orientation of all die elements with respect to the direction of feeding; a blanking angle of 90° allows most, but not all, metals to be bent to the smallest possible radii.

BEND ALLOWANCES

For close work the exact length of metal required to make a bend is often determined by trial and error. The assumed neutral axis (Fig. 6-2) varies depending upon the bending method used, the location in the bend, and the type of stock being bent.

Direction of grain in a steel strip relative to the bend also has a slight effect on the length of metal required to make a bend. Bending with the grain allows the metal to stretch more easily than bending against the grain, however this results in a weaker stamping.

Bend allowance depends more upon the physical properties of the material such as tensile strength, yield strength, and ductility than on the metal from which it is made.

Empirical rules. The exact bend allowance is the arc length of the true neutral axis of the bend (above the neutral axis metal is stretched; below it metal is compressed). The problem is that neutral axis can only be approximated. Many manufacturers assume the neutral axis is 1/3 stock thickness from the inside radius of the bend for inside radii of less than twice stock thickness. For inside radii of two times stock thickness or greater, the neutral axis is assumed to lie 1/2 stock thickness from the inside radii. One reason for relatively less metal being required to make a tight bend is that the sharp radius tends to be drawn slightly.

Many experts believe that the location of the true neutral axis from the inside radius varies from 0.2 to 0.5 times stock thickness. An important factor that determines the neutral axis is how the bend is accomplished. Less metal is required for a bend made by a tightly wiped flange than for an air bend on a press brake. Wiping the flange tends to stretch the metal.

Formulas and Data. Tabular data for inside of bend radii from 0.015 to 1.250 in. (0.38 to 31.75 mm), including metric equivalents for various materials and types of bends, are available. The formulas used to develop these tables are based on extensive experimental data. Such tables were very useful before the availability of the electronic calculator.[3]

For 90° bends the radius of the assumed neutral axis is multiplied by 1.57 (the number of radians in 90°) to determine the assumed amount of metal required to make the bend. For bends that are not exactly 90° multiply the number of degrees of bend times 0.0175 (the number of radians in a degree) and substitute the result for the coefficient 1.57.

Trial-and-error Determination. Because of the uncertainty of the exact location of the neutral axis, trial-and-error methods are often employed when close tolerance stampings are being developed. Section 21 describes a method of blank development that uses a CNC laser cutting process.

When time permits, it is best to build the blank die for a developed blank last, after the exact dimensions of the blank have been determined. The CNC digital data used to laser cut the trial blanks can be used to design and build the die. This procedure is also useful when building progressive dies (Section 16).

SPRINGBACK COMPENSATION

Whenever metals are formed, some springback occurs. The cause of this springback is the residual stress that is an inevitable result of cold working metals. For example, in a simple bend, residual compressive stress remains on the inside of the bend, while residual tensile stress is present on the outside radius of the bend.

When the bending pressure is released, the metal springs back until the residual stress forces are balanced by the material's ability to resist further strain. This is mainly a function of the material's modulus of elasticity. This is why materials with a high modulus of elasticity (as compared to tensile strength), such as mild steel, tend to spring back less than materials with a lower modulus but equal tensile strength, like hard aluminum alloys.

Factors That Affect Springback. Some factors that increase springback are:
1. Higher material strength.

2. Thinner material.
3. Lower Young's modulus.
4. Larger die radius.
5. Greater wipe steel clearance.
6. Less irregularity in part outline.
7. Flatter part surface contour.

If a flanged part has sufficient irregularity with respect to either the outline or surface contour, the springback will be very slight. An approximation of the springback of a straight bend for low-carbon, low-strength steel and a zero wiping-steel clearance is shown in Fig. 6-12. The springback for large wiping-steel clearances can be several times the amount shown on the graph.

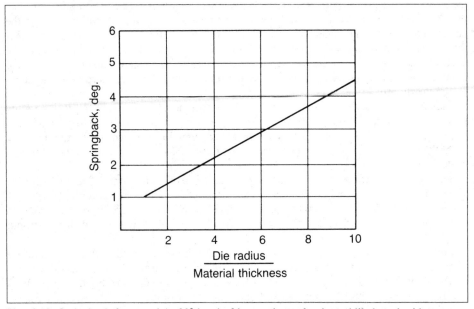

Fig. 6-12. Springback for a straight 90° bend of low-carbon, aluminum-killed steel with zero wiping steel clearance.[1]

Overbending to Correct for Springback. A common method to compensate for springback is to overbend the part enough to allow it to spring back to the desired shape.

A common example of overbending for springback compensation is air-bending, as is commonly done in a press-brake (Fig. 6-13). This method of bending also requires minimum tonnages for the work performed. Exact repeatability of ram travel is required to maintain close duplication of the bend-angle. Changes in stock characteristics, such as yield strength or thickness, will cause a change in the angle-of-bend. Compensation for any change of condition affecting the angle-of-bend is easily provided by adjusting the ram travel.

Coining the Bend to Eliminate Springback. Figure 6-14 illustrates a press-brake die designed to coin the bend in order to obtain a precise angle-of-bend. This coining-action eliminates the root cause of springback, which is the tensile and compressive residual stresses. This is accomplished by bringing the entire thickness of the metal in the bend area up to the yield point of the material.

This method has the advantage of producing sharp bends with less sensitivity to material conditions than air bending. The principal disadvantages are tonnage requirements many times that of air-bending and accelerated die wear.

The Effect of Die Wear on Springback. If a V-bend die for forming 90° bends in mild steel is ground to provide the correct overbend to offset the effect of springback, the bend

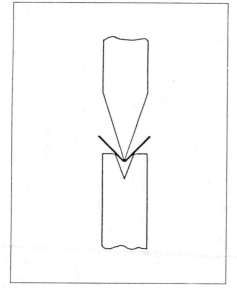

Fig. 6-13. Overbending to obtain a 90°
finished bend.

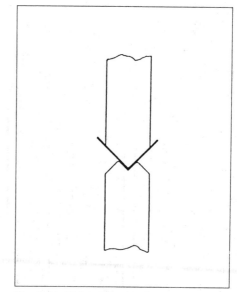

Fig. 6-14. Coining the bend area to
eliminate springback.

can be formed at tonnage values that agree with calculated values. As the forming surfaces become worn, it may be found that 90° bends can still be made if the tonnage is increased to several times the initial value.

This occurs because the additional tonnage will produce pressures on the thickness dimension of the stock exceeding the yield point of the material. In this case, some areas of the part are actually coined. This relieves the residual stresses that cause springback.

This same effect is observed in forming and reforming (restrike) dies. As the dies wear, tonnage must be increased to produce a part of the required form. What happens is the die surfaces providing the overbend are being worn away. The die must then be used as a coining die to achieve the desired form.

Such high pressures may exceed press capacity, and can result in accelerated die wear and possible die damage. The solution is to build the die of materials that will resist rapid wear on critical forming surfaces. The die should be reworked to restore the original form whenever higher than normal tonnages are required to form the part.

The Effect of Stock Thickness on Springback. Producing close-tolerance stampings requires that stock thickness variations be held to a minimum. Figure 6-15 illustrates the effect of thickness variations on the angle-of-bend. The stamping on the right is under-bent due to inadequate wiping and bottoming action, while the one on the left is excessively coined and subject to galling due to insufficient clearances.

The Effect of Out-of-parallel Conditions. Off-angle flanges can result from a press or die out-of-parallel condition. Figure 6-16A illustrates a section through a flange die. The wiper steels bottom on the bend to control springback. The desired shape of the finished part is illustrated in Fig. 6-16B.

Figure 6-16C illustrates die closure with an out-of-parallel condition. The part (Fig. 6-16D) is misformed. The right flange is under-bent due to the wiper steel not bottoming on the bend radius to correct for springback. The left flange is over-bent, and also exhibits sidewall curl. The over-bending is a result of the wiper steel bottoming excessively hard on the bend radius, while sidewall curl is caused by the flange steel wiping too hard due to the lateral forces generated by the out-of-parallel condition.

The solution to this type of quality problem is to maintain all dies in a parallel condition. A regular program of checking and adjusting the alignment of all presses is also required.

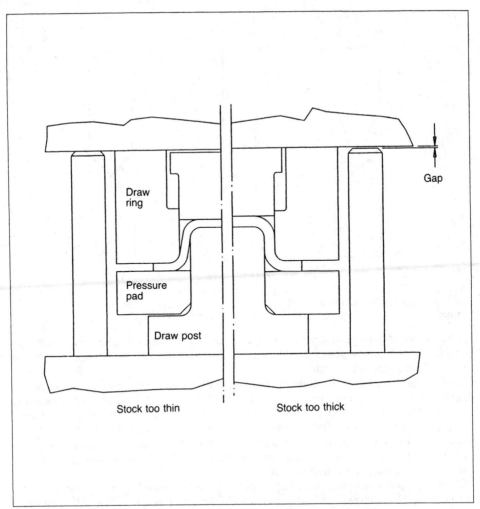

Draw ring

Gap

Pressure pad

Draw post

Stock too thin

Stock too thick

Fig. 6-15. The effect of stock thickness variation on springback. (*Capitol Engineering*).

The Effect of Pad Pressure. Pad pressure can be used to adjust the exact bend-angle in a flanging operation. Figure 6-17A illustrates a flanging die used to form two 90° bends in mild steel stock. The die is operated in an OBI press with an adjustable air cushion which is used to supply pad pressure.

High pad pressures result in less than 90° bends, while very low pressures result in substantial overbending. In this case, a recoil condition as the die closes and the bends are formed (Fig. 6-17B) is flattened out when the pad bottoms, forcing metal into the bends. The metal involved in the recoil does not reach yield-point pressures.

Fine adjustment of pad pressure can vary bend-angles several degrees, permitting compensation for stock variations. This springback control method is used in conjunction with an on-line quality control program and can be considered part of a process control feedback loop system.

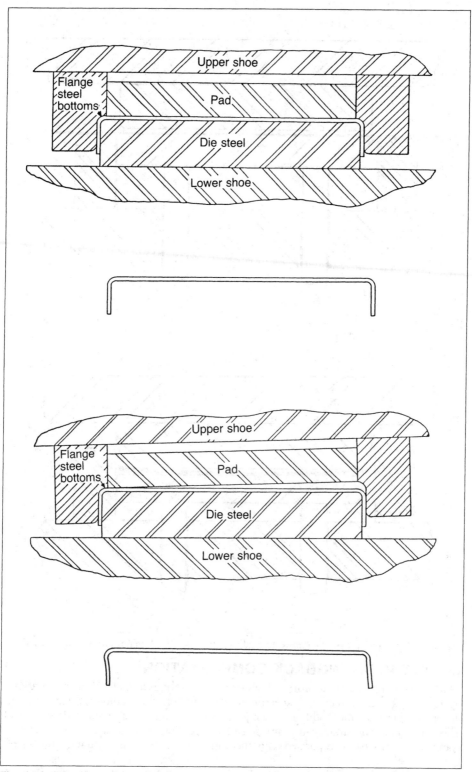

Fig. 6-16. The effect of press and die out-of-parallel conditions on the bend angle of a U-shaped section. (*Midway Products Corp.*)

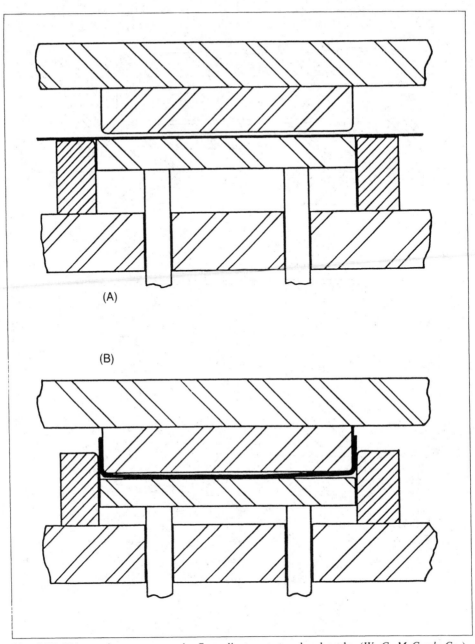

(A)

(B)

Fig. 6-17. Using pad pressure to make fine adjustments to a bend angle. (*W. C. McCurdy Co.*)

EMPIRICAL SPRINGBACK COMPENSATION

Many complex factors determine the amount of springback that will occur in a given operation. As a result, data for a specific material and forming method is often developed under actual production conditions to aid process control and future product development. This sub-section illustrates how one manufacturer developed data for the amount of compensation to be incorporated into the form blocks used in hydraulic rubber-pad presses.

In Fig. 6-18, the amount of springback depends primarily upon the ratio of angles A/A_1 and the temper of the metal.

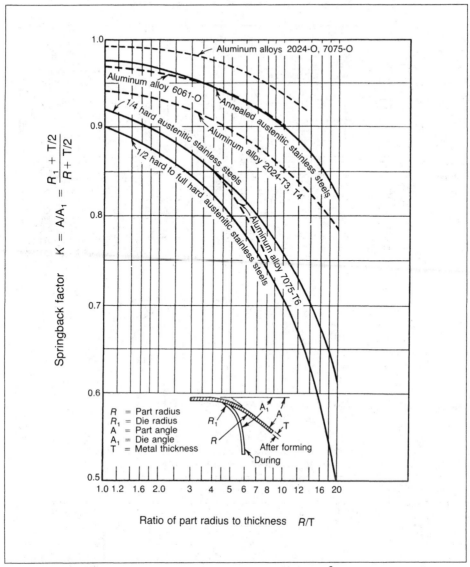

Fig. 6-18. Springback data for aluminum alloys and stainless steels.[2]

Springback factor K is expressed as

$$K = \frac{A}{A_1} = \frac{R_1 + T/2}{R + T/2} \tag{2}$$

where A = bend angle of part, deg.
A_1 = bend angle of part during bending, deg.
R = part radius
R_1 = die radius
T = metal thickness

Higher values than those calculated are found in actual practice (Fig. 6-19), due to variations in workhardening rates, die clearances, and the actual physical properties of a particular metal. Variations from the calculated angle may be as high as $\pm 2°$.

Springback allowances for 90° bends in aluminum are listed in Table 6-1.

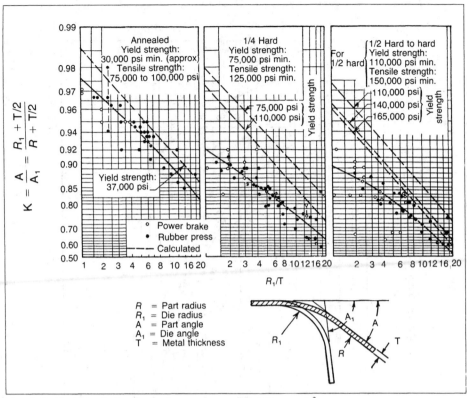

Fig. 6-19. Springback data for various tempers of stainless steel.[3]

TABLE 6-1
Springback Allowance in Degrees for 90° Bends
in 2024-O and 7075-O Aluminum

Sheet Thickness, in. (mm)	Bend Radius, in. (mm)							
	0.094 (2.38)	0.125 (3.18)	0.188 (4.78)	0.250 (6.35)	0.312 (7.94)	0.375 (9.52)	0.438 (11.11)	0.500 (12.7)
0.020 (0.51)	3	4	5.500	7.500	8.500	9	9.500	12
0.025 (0.64)	2.750	3.750	5.500	6.500	8	8.250	8.750	10.750
0.032 (0.81)	2.250	3	4.750	6	6.750	7	7.500	9.500
0.040 (1.02)	2	3	4	5	6	6.250	6.750	8.750
0.051 (1.30)	2	2.500	3.500	4	5	5.250	5.750	7.500
0.064 (1.62)	1.500	2	2.750	3.750	4.500	5	5.500	6.750
0.081 (2.06)	1	1.500	2	2.500	3.250	3.500	4	4.750
0.094 (2.39)			1.750	2.500	3	3.250	3.750	4.500
0.125 (3.18)			1.500	2	2.250	2.750	3	3.750

Courtesy of Emerson Electric Co.

For angles other than 90° the springback allowance for 90° will be multiplied by the factors set forth in Table 6-2. To find springback allowance for an 80° flange, proceed as follows:

TABLE 6-2
Springback Factors in Bending Aluminum
To Other Angles Than 90°

Angle, deg.	2024-O and 7075-O Aluminum	2024-T3 Aluminum
60	0.6	0.7
65	0.6	0.7
70	0.7	0.8
75	0.7	0.8
80	0.8	0.9
85	0.8	0.9
95	1.05	1.02
100	1.05	1.05
105	1.1	1.05
110	1.2	1.1
115	1.2	1.1
120	1.3	1.2

Courtesy of Emerson Electric Co.

From Table 6-1, the springback allowance in 2024-O Dural sheet 0.040 in. (1.02 mm) thick for a 0.125-in. (3.18-mm) bend radius of the flange is 3° for a 90° flange. The 3° allowance is then multiplied by 0.8, the factor opposite 80° in Table 6-2. The product, 2.4°, is the springback allowance for an 80° angle.

Springback allowances for 90° bends in 2024-T3 stock are listed in Table 6-3. To obtain the springback allowance for angles other than 90°, the allowance for 90° should be multiplied by the factors shown in Table 6-2.

TABLE 6-3
Springback Allowance in Degrees for 90° Bends in
2024-T3 Aluminum

Sheet Thickness, in. (mm)	Bend Radius, in. (mm)							
	0.094 (2.38)	0.125 (3.18)	0.188 (4.78)	0.250 (6.35)	0.312 (7.94)	0.375 (9.52)	0.438 (11.11)	0.500 (12.7)
0.020 (0.51)	10	12	15.500	19	22.500	24	27.250	33.500
0.025 (0.64)	8.750	10.500	14	16.750	17.750	21	23	28.500
0.032 (0.81)	7.750	8.750	12	14.500	16.750	17.750	19.250	24
0.040 (1.02)	7.250	8.250	10.750	12.750	14.500	15.250	17	20.500
0.051 (1.30)			9	10.500	12.250	13	14.500	16.750
0.064 (1.62)			8	9.750	11.250	12	12.750	15
0.081 (2.06)					9.500	10.500	11.250	13
0.094 (2.39)					8.750	9.750	10.500	12

Courtesy of Emerson Electric Co.

Tables 6-1 to 6-3, inclusive, are based upon the results of thousands of tests and should be accurate to within 2° for average operating conditions on a hydraulic press with an 8-in.

(200-mm) rubber pad having a Shore durometer hardness of 60 to 70.

The operating pressure of the press was 2,500 tons (22,240 kN) and a 1-in. (25.4-mm) throw pad having a Shore durometer hardness of 60 to 70 was located over the blanks on the form blocks during the forming operations.

Tables 6-1 to 6-3, inclusive, were calculated from tests on straight flanges. In view of the many possible contours of curved flanges it is difficult if not impossible to pre-determine accurately the springback allowances for such a flange. As a practical working basis, the tables for straight flanges can be employed as a guide for determining the springback allowances for curved flanges. However, the exact springback allowance for curved flanges must be developed by trial and error.

An empirical equation[3] for predicting angular deviation in curved flanges is

$$X = 100\ AC + \frac{0.095\ (A-B)}{0.0005}\ D + EG + F \tag{3}$$

where X = total degrees of angular deviation (springback)
  A = punch and die setting (clearance when closed), in. (mm)
  B = minimum stock thickness, in. (mm)
  C = factor for difference between die setting and maximum stock thickness
  D = factor for difference between die setting and minimum stock thickness
  E = factor for hardness variation
  F = constant for difference between die setting and minimum stock thickness
  G = difference between high and low points of Rockwell B scale in the stock specifications

Values of factors C, D, E, and F for certain materials are listed in Table 6-4.

TABLE 6-4
Deviation Factors for Springback in Curved Flanges[4]

Material (half hard only)	Factor C*	Factor D*	Factor E*
Steel	0.08	0.19	0.05
Brass	Zero	0.13	0.04
Aluminum	0.12	0.12	Zero

Material	If minimum stock thickness is	Factor F*
Steel	More than 95% of the punch-and-die setting	0.50
Steel	Less than 95% of the punch-and-die setting	1.40
Brass	More than 95% of the punch-and-die setting	0.80
Brass	Less than 95% of the punch-and-die setting	2.00
Aluminum	Any thickness	0.05

*Factors for use in Eq. (3).

EXAMPLE 1: To form a 90° angle in 0.032-in. (0.82-mm) half-hard steel, with a thickness variation of 0.003 in. (0.08 mm) and minus 0.002 in. (0.05 mm), Rockwell range of B71 to 82, and a die setting of 0.035 in. (0.89 mm), substitute the the known values in the formula and find the angular deviation to be 2.4°.

Form-block Contours. A graphic method of determining the contour of form blocks to compensate for springback is illustrated in Fig. 6-20. The part is divided into a few lengths, L', L'', etc., each possessing an approximately common radius R', R'', R'''. These radii are determined graphically, and the ratios R/t (t = metal thickness) are found.

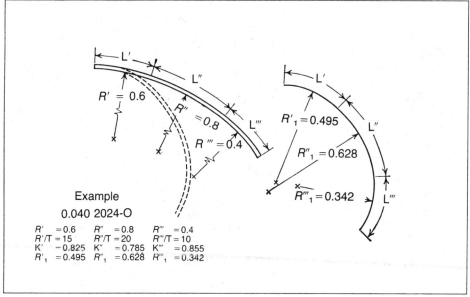

Example
0.040 2024-O

R' = 0.6	R'' = 0.8	R''' = 0.4
R'/T = 15	R''/T = 20	R'''/T = 10
K' = 0.825	K'' = 0.785	K''' = 0.855
R'_1 = 0.495	R''_1 = 0.628	R'''_1 = 0.342

Fig. 6-20. Graphical method for determining form-block contours.[2]

The springback factors K', K'', K''' are taken from Fig. 6-16. Die radii $R_1' = K'R'$, R_1''' = $K'''R'''$ are then calculated.

L' is laid out on the perimeter of a circle of radius R_1', L'' on a circle of radius R_1'' and L''' on a circle of radius R_1'''. These three segments are faired, or smooth blended, in agreement with the desired part contour.

For large radii the data derived from Figs. 6-18 and 6-19 would be inapplicable. Instead, the springback must be determined from special tests on the workpiece material.

FLANGING

Flanging is a forming operation in which a narrow strip at the edge of a sheet is bent along a straight or curved line. Flanges can be open, at 90°, or at an acute angle. A flange is used for appearance, rigidity, or edge strengthening, as well as for an accurately positioned fastening surface.

Types of Flanges. The three major types of flanges are the straight, stretch, and shrink flanges, as illustrated in Fig. 6-21. The joggled flange in view F is a combination of all three major types. The reverse flange, view G, is a combination of the stretch and shrink flanges; and the hole flange, view H, is a special case of the stretch flange.

Straight Flange. The straight flange is a simple bend with no longitudinal stresses imposed on the material except at the bend radius; therefore, within reason, a flange of any desired width can be made.

Stretch Flange. The stretch flange is unique in that the flange has been stretched in the length direction during the flanging operation. The greatest longitudinal stretch occurs at the edge of the flange and diminishes to zero at the bend radius. A stretch flange forms with a concave outline, as shown in Fig. 6-21, view B; a concave surface contour, as shown in view C; or a combination of the two conditions.

Edge splitting can be a problem when stretch flanging. The susceptibility to splitting is a function of the tensile stress at the edge, the material properties, and the edge condition resulting from shearing or trimming. Tensile stress can be lessened by reducing the flange width or by providing notches or scallops as shown in Fig. 6-22. This method has the effect of reducing the flange strength. Deep-drawing quality steel with a high strain ratio (r value) withstands edge splitting better than steels with lower r values. Section 11 covers edge conditions in detail.

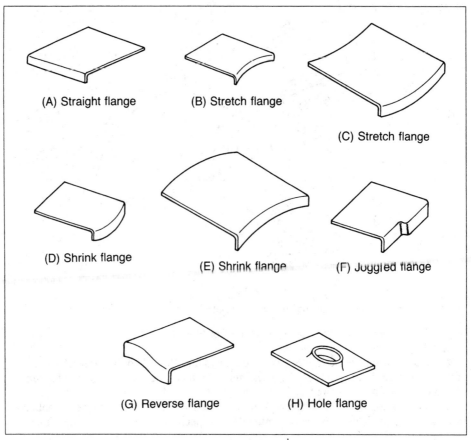

Fig 6-21. Examples of straight, stretch, and shrink flanges.[1]

A simple general equation expresses the strain at the edge of the stretch flange, where most failures begin:

$$e_x = \frac{R_2}{R_1} - 1 \qquad (4)$$

where e_x = strain at flange edge
 R_1 = flange edge radius before forming
 R_2 = flange edge radius after forming

Circle Grid Analysis (CGA) is an excellent method for determining the actual amount of strain on a flanged edge. Materials may also be tested by expanding a drilled, deburred hole with a lubricated conical punch to determine the forming limit.

Figure 6-23 illustrates the effect of burrs on flange formability. The burr height/thickness is the ratio of burr height above the edge to material thickness. The formation of burrs requires cold-working of the metal which also reduces formability.

Tabs and notches at the edge of a flange localize strains and can greatly reduce the forming limit. The larger the tab and smaller the root radius, the worse the reduction.

Shrink Flange. The shrink flange is the opposite of the stretch flange, having shrunken in length during the flanging operation. The greatest compressive stress occurs at the edge of the flange and diminishes to zero at the bend radius. A shrink flange forms with a convex outline, as shown in Fig. 6-21, view *D*; a convex surface contour, as shown in view *E*; or a combination of the two conditions.

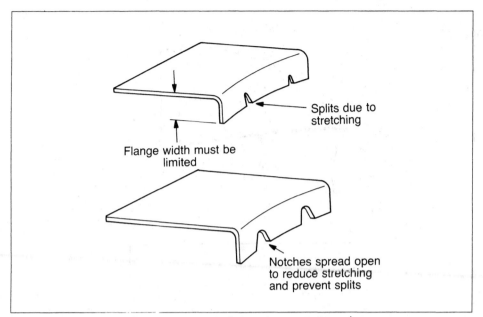

Fig. 6-22. Notches or scallops can prevent splitting of a stretch flange.[1]

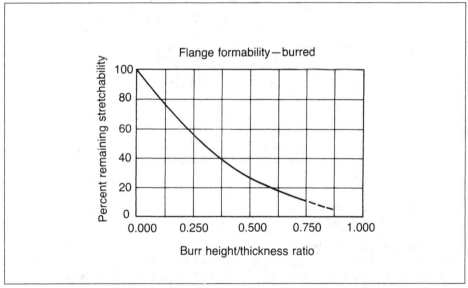

Fig. 6-23. The effect of burr height on formability. (*Chrysler Corporation*)

A common problem of shrink flanges is wrinkling. This is a buckling phenomenon and is a function of the ability of the flange to resist the compressive stresses during the forming operation while the material is not confined. The condition can be improved by limiting flange width, increasing material thickness, providing offsets in the flange to take up excess metal, and reducing clearance between the wiper steel and the male die steel to iron out wrinkles that occur. The ironing solution may result in galling conditions on the wiper steel.

Tooling to Make Flanges. Typical flange tooling is shown in Fig. 6-24. Normally, a part is formed before flanging. After the formed part is placed on the male punch and positioned by appropriate locators, the pad descends to clamp the part. The wiping steel

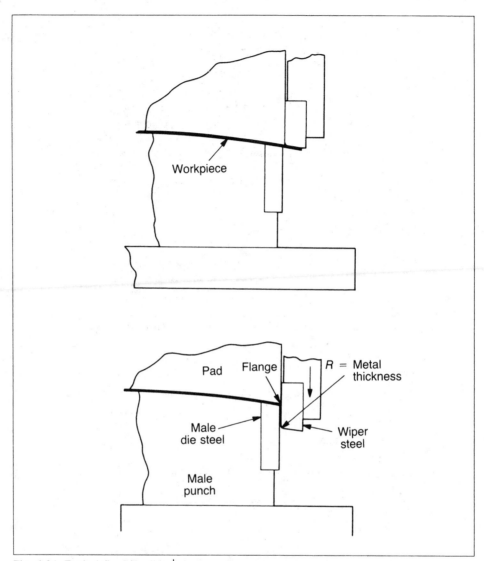

Fig. 6-24. Typical flange tooling.[1]

then descends wiping the edges down over the male steels.

The male steels should be sufficiently hard to maintain the proper flange radius and tough enough to resist corner chipping. The radii of flanges that are to be subsequently hemmed are kept fairly sharp.

Flanging Pads. To avoid recoil, the flanging pad must grip the part firmly to ensure that the part will remain in tight contact with the male die half during flanging. The effect of recoil is shown in Fig. 6-25. This is a particular problem when appearance is of great concern, as in the case of automotive outer panels. A recoil condition of this type can increase the amount of springback since the recoil-area stores residual stress much like a clock spring. Recoil is caused by insufficient pad pressure, a worn pad, or a pad that gives insufficient part coverage.

Figure 6-26A illustrates a hardened tool-steel insert in a pad to resist wear from the normal recoil forces present during flanging. Coining of the top portion of a flange to control springback is shown in Fig. 6-26B. Depending upon the pressure involved, the latter method can actually accomplish several degrees of overbend. If coining in this

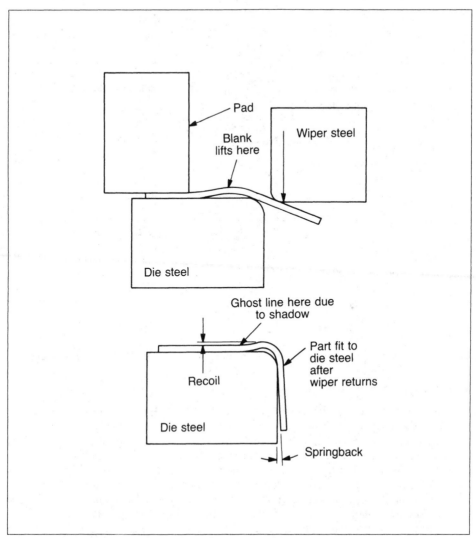

Fig. 6-25. Metal recoil and springback during flanging.[1]

manner is used to control springback, great care should be taken to insure that the press and die are parallel in order to obtain uniform coining pressures.

Causes of Distorted Flanges. Both stretch and shrink flanges are strained nonuniformly during the flanging operation. The residual stress from such a flanging operation can sometimes result in undesired elastic deformation and distortion of a previously formed part. In some cases it is possible to change the contour of the formed part prior to flanging to compensate for this effect. Often an easier solution is to compensate by changing the contour of the male punch and pad of the flanging die.

Determining the Required Pad Pressure. A rule of thumb for determining sufficient pad force is as follows:

$$\text{Pad force} = SLt/3 \tag{5}$$

where F = force, lbf (N)
$\quad S$ = ultimate strength of the material, psi (Pa)
$\quad L$ = flange length, in. (mm)
$\quad t$ = material thickness, in. (mm)

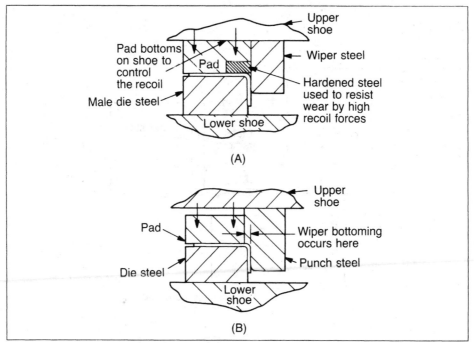

Fig. 6-26. The use of a hardened-steel insert: (A) to reduce pad wear. Coining a flange edge (B) to control springback.[1]

HEMMING OPERATIONS

A hem is a flange that has been bent more than 180°. Hems are primarily used for appearance and for the attachment of one sheet metal part to another. They are not as rigid or accurate as a flange, but they very effectively remove a dangerous sheared edge. They are used extensively in automobiles to join inner and outer closure panels.

Types of Hems. Four different types of hems are shown in Fig. 6-27. The tear drop hem is used for materials that do not have the ductility required to form the flattened hem. The open hem and rope hem are used for attachment to other sheet metal parts. The rope hem is used for materials with insufficient ductility to form the open hem.

Tooling for Hems. Although flanges and hems can be made in one operation, in order to gain repeatability hemming is often performed on a part that has been flanged in a previous operation.

The sequence of a typical hemming operation after bending a 90° flange in the outer panel is to place the inner panel in the flanged outer panel and bend the outer panel an additional 45°. The assembly die to perform this operation is often called a 45° die. The partially assembled panels are next transferred to the hem die where the final assembly takes place by flattening the pre-bent hem.

It is possible to combine the 45° and hem flattening operation into one die. For many applications a combination die is not desirable because of increased die maintenance problems and more difficult die tryout.

Quality of Hem. Some hem quality problems, such as edge splitting, wrinkling, and part distortion, are the same as those for flanging, and the solutions are similar. However, unlike flanging, a major difficulty with hemming is maintaining an accurate part outline. The reason is that the amount of roll-in or roll-out is difficult to predict because it is affected by many factors including all of the following:

1. Material thickness.
2. Material strength.
3. Flange die radius.

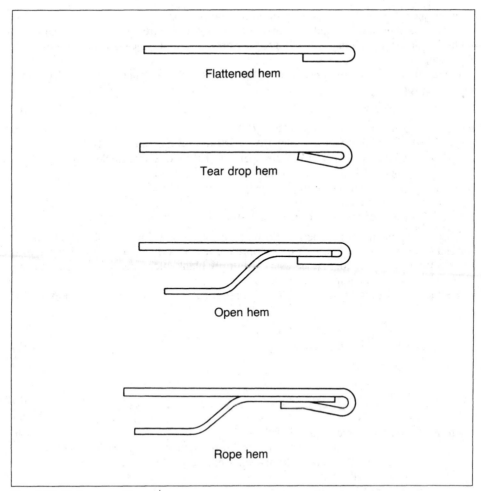

Fig. 6-27. Four types of hems.[1]

    4. Outline curvature—direction and degree.

    5. Surface contour curvature—direction and degree.

    6. Flange length.

Because of the unpredictability of the location of the hemmed outline, the solution to maintaining an accurate part outline is a trial-and-error changing of the position of the flange outline and length. The experience of the die tryout operator is invaluable in reducing the time involved in this tedious job.

A well-timed hemming operation, in which the pre-hem steel hits most of the flanged edge at the same time, is essential to a good part. Mistiming can result in recoil and misforming due to the part shifting in the die.

Hemming Pressures. Hemming pressures, including seaming pressures, will generally amount to seven times the forming pressures required for 90° bends and may be as high as a ratio of 40:1, depending upon stock thickness, tensile strength, size of area to be flattened or hemmed, and tightness of the hem.

BENDING PRESSURES

The amount of pressure required depends upon the thickness of the stock, the length of the bend, the width of the die, whether a lubricant is used, and the amount of wiping, ironing, or coining present. V-dies in which the punch does not bottom (air-bending), commonly used in press brakes, require minimum pressures.

For U-forming and channel bending, the pressures required will be approximately double while edge bending will require pressure as little as one half that required for V-bending.

Formula for V-bends and U-forming. A simple formula used to determine V-bending forces

$$F = KLSt^2/W \tag{6}$$

where F = bending force required, tons (N)
    K = die-opening factor: varies from 1.20 for a die opening of 16 times metal thickness, to 1.33 for a die opening of 8 times metal thickness
    L = length of bent part, in. (mm)
    S = ultimate tensile strength, tons per sq in. (N/mm^2)
    t = metal thickness, in. (mm)
    W = width of the V-, channel, or U-forming lower die, in. (mm)

The use of equation (6) for deriving bending pressures is valid for V-shaped dies only. For channel forming and U-forming, multiply the result by 2. In forming a channel with a flat bottom, a blankholder is necessary. Multiply blankholder area in square inches by 0.15 and add to the bending force derived in equation (6).

Formula for Flanging Forces. The force required to bend or flange a sheet as shown in Fig. 6-28 is:

$$F = 0.167 \, \frac{SLt^2}{A} \quad \text{theoretical} \tag{7}$$

$$F = 0.333 \, \frac{SLt^2}{A} \quad \text{for wiping dies} \tag{8}$$

where F = bending force required, lbf (N)
    K = a constant that varies from 0.167 for large die radii and clearances to 0.333 for sharp die radii and high plastic working stresses
    t = sheet metal thickness, in. (mm)
    L = length of bend, in. (mm)
    r_1 = punch radius, in. (mm)
    r_2 = die radius, in. (mm)
    C = die clearance, in. (mm)
    S = ultimate tensile strength, psi (Pa)
    A = $r_1 + C + r_2$

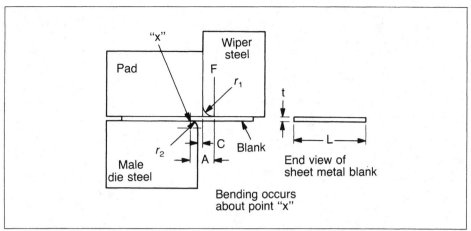

Fig. 6-28. Factors that are entered to determine flanging force requirements.[1]

References

1. Charles Wick, John W. Benedict, and Raymond F. Veilleux, eds., "Sheet Metal Blanking and Forming," *Tool and Manufacturing Engineers Handbook*, 4th ed., Volume 2 (Dearborn, MI: Society of Manufacturing Engineers, 1984).
2. G. Sachs, *Principles and Methods of Sheet-metal Fabricating* (New York: Reinhold Publishing Corporation, 1951).
3. G. Sachs and U. N. Krirobok, *Forming of Austenitic Chromium-nickel Stainless Steels*, The International Nickel Co., Inc., New York, 1948.
4. R. S. Fries and J. A. Thorud, "Shop Tolerances That Can Be Met," *Factory Management and Maintenance* (September 1951).

SECTION 7
BENDING DIES

PRESS-BRAKE DIES

A stamping operation—sometimes classified as either bending or forming—may actually include both, as exemplified in the press-brake dies shown in Fig. 7-1.

1. 90° Forming Dies. Figure 7-1A shows a typical 90° forming die, which is one of the most common dies used in press brakes. Most 90° dies are bottoming dies and, in using them, characteristics of bottoming must be remembered. In general, the radius of the bend should be not less than the thickness of the material, and the V-die opening should be eight times the metal thickness. High-tensile materials require larger radii and wider vees than this, and plates over 0.5 in. (12.7 mm) thick also require V-die openings of more than eight times the metal thickness.

2. Acute-angle or Air-forming Dies. The dies shown in Fig. 7-1B are known as "acute-angle" dies because of the acute angle they can form, and as "air-forming" dies because they usually do not bottom but form in the air. They may be used for 90° bends where accuracy is not too important. They also may form a wide range of both acute and obtuse angles simply by adjusting the ram of the press, which in turn determines how far the punch enters the die.

3. Gooseneck Dies. Figure 7-1C shows a typical gooseneck or return flanging die. These are essentially simple V-bend dies with clearance for return flanges. Care must be taken in using these dies because they are usually out beyond the center line and can easily be bent by overloading.

4. Offset Dies. An offset can be formed by making two bends with a 90° acute-angle die. However, for long runs or sharp offsets, dies of the type shown in Fig. 7-1D are generally used. The pressures required are sometimes dangerously high, being from four to eight times that for a single bend, depending on the nature of the offset. Where each bend is more than 90°, these dies are usually referred to as Z-dies.

5. Hemming Dies. The edges of a sheet are sometimes hemmed or turned over to provide stiffness and a smooth edge. Hemming can be done in two operations, starting with an acute-angle die and finishing with a flattening die such as shown in Fig. 7-1E. However, most hemming is done on regular hemming dies, which are two-stage dies combining an acute-angle die with some sort of flattening arrangement. One type is shown in Fig. 7-1F. Pressures required will vary greatly with thickness of the hem and the degree of flatness.

6. Seaming Dies. Seams in sheets or tubes can be made in a variety of ways. A set of dies for making simple seams is shown in Fig. 7-1G.

7. Radius Dies. These dies are usually employed where the radius exceeds four times the thickness of the material. Such bends can be made with a V-die machined to less than 90° and a full-radius punch. However, better results are usually obtained by means of spring loaded dies such as that shown in Fig. 7-1H. Instead of spring pads, rubber pads may be used. The angle of bend is adjusted by varying the distance that the punch enters the die. Punches with different radii may be used with the same die to get different radii

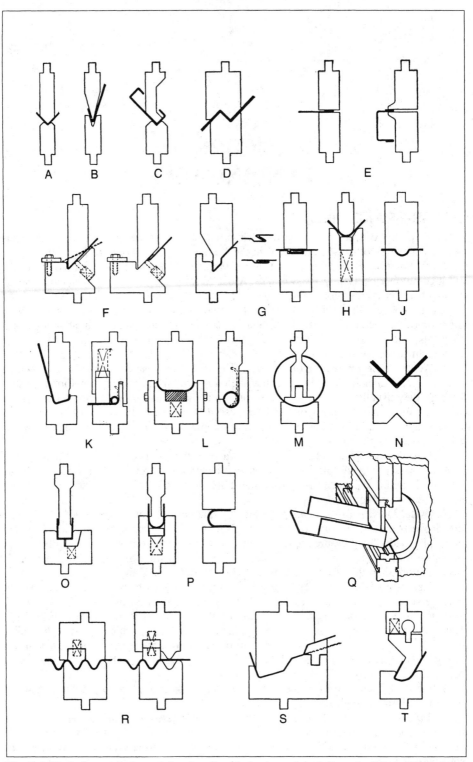

Fig. 7-1. Typical press-brake and forming dies, derived or adapted from the Verson Allsteel Press Co. *(A, F, G, H, J, L, M, O, P, R, S, T)*; Cincinnati Shaper Co. *(B, D, N, Q)*; Dreis and Krump Mfg. Co. *(E, K)*[1].

on the bend. The inside radius on air bends is commonly controlled by the die opening. This inside radius, with normal dies, is very nearly 5/32 of the die opening.

8. Beading Dies. Beads serve as a stiffening means in flat sheets and sometimes permit the use of thinner material than would be possible otherwise. Beads are of two general types: open beads extending from edge to edge of the sheet, and closed or blind beads that fade out in the sheet. Open beads are usually formed by simple dies, such as that shown in Fig. 7-1*J*. Closed beads, on the other hand, require the use of spring pressure pads at the ends, which fade out to minimize wrinkling of the metal.

9. Curling Dies. These dies provide a curl or coiled up end to the piece. Hinge dies make use of a curling operation. The curl may be centered, or it may be tangent to the sheet, as shown in Fig. 7-1*K*.

10. Tube- and Pipe-forming Dies. These are similar to curling dies. The edges of the metal must be bent as a first operation. The piece will then roll up properly. Figure 7-1*L* shows a two-operation die for forming small tubes. For larger tubes, a bumping die such as in Fig. 7-1*M* is necessary. For accurate work such bumped tubes should be sized over a sizing mandrel. Seams can be formed on the edges of these tubes before they are rolled.

11. Four-way Die Blocks. For small production runs or for a job shop, the four-way die block as shown in Fig. 7-1*N* is quite useful and represents a material saving in tool cost.

12. Channel-forming Dies. Channels may be formed in gooseneck dies or in single stroke channel dies as shown in Fig. 7-1*O*. Such dies are commonly made with a spring pressure pad release of some sort to eject the formed part from the die. Strippers are sometimes provided to strip the part from the punch.

13. U-bend Dies. U-bend forming is similar to channel forming, but spring back is usually more pronounced and means must be provided to overcome it. One way of accomplishing this task is shown in Fig. 7-1*P*.

14. Box-forming Dies. While boxforming consists of simple angle bending, there are problems peculiar to the nature of the work. In general, a high punch and a low die are required as seen in Fig. 7-1*Q*. Sometimes the punch is cut into sections so the side of the box can come up between them. Certain shapes of boxes may be formed on horn presses.

15. Corrugating Dies. Corrugating dies can be provided to produce a variety of corrugations. From one to four corrugations are made at a single stroke. Figure 7-1*R* shows such a die.

16. Multiple-bend Dies. Multiple-bend dies offer an infinite variety of possibilities. Commonly used on large production runs, one die can accomplish, in a single stroke, an operation that would require several operations with single-bend dies. Figure 7-1*S* shows such a die. This type of die requires much greater forming pressures than do dies for the individual operations.

17. Rocker-type Dies. Rocker-type dies can form parts that would be impossible with a die acting only vertically. A typical example is shown in Fig. 7-1*T*.

REPRESENTATIVE PRESS-TYPE BENDING DIES

Typical die designs to compensate for spring-back are shown in Fig. 7-2 and Fig. 7-3. Variations in sheet temper and thickness cannot be exactly calculated; hence the desired bend angle in the part may be secured by handwork.

Flexible V-die Design. A practical V-die has four sizes of punches (Fig. 7-4, *D*1, *D*2, *D*3, and *D*4) with sliding gages *D*6, *D*7, and *D*8 to allow flexibility in the positioning and forming of various-sized blanks into V-bends up to approximately 90°. A punch of the desired size is placed in the holder *D*5, and the mating V-die is located in line for performing the bending operation.

Flexible U-die Design. Blanks of various sizes are accommodated by changing the size of the die opening through the insertion of "floaters," Fig. 7-5, (*D*1) into the die block. These are steel blocks 4.375 in. (111.13 mm) long and 0.875 in. (22.23 mm) thick. They vary in width from 0.311 to 2.248 in. (7.9 to 57.1 mm) in increments of 0.062 in.

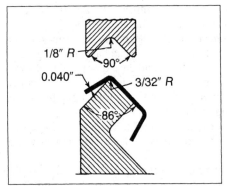

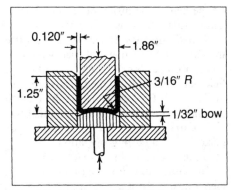

Fig. 7-2. V-die for accurate setting of a bend angle.[2]

Fig. 7-3. Square bending with bowed spring pad to prevent distortion of the web.[2]

(1.57 mm). Punches (D3) of various widths bend the work into channel or U-shapes. Possible combinations of floaters and punches provide a wide range of thicknesses and sizes of workpieces that can be bent. The floaters also function as ejectors actuated through ejector pins (D2).

Rotary Twisting Die. A plunger with a helical groove (Fig. 7-6, D2, D3) revolves and twists the work through 90° through the engagement of a hardened pin in the groove on the downward movement of the ram. On the upstroke, a cam (D4) prevents plunger movement until it is disengaged from the part, when the stop (D5) contacts the cam, allowing the spring (D7) to return the plunger to its original position. The work is held in position by the lever (D6), which, along with a stationary block, has a slot to position and prevent rotation of the part. The finished part is shown as D1.

Rotary Bending Dies. Two cam surfaces on the end of the punch (Fig. 7-7, D3) rotate a ring (D1) counterclockwise by contacting the cam rollers (D2) so that the piece part (a carpenter's brace handle made of 0.375 in. (9.52 mm) cold-rolled steel rod) is bent between two hardened rollers (D4 and D5). A conventional press setup provided too much spring-back to allow free turning of the hand grip sleeve.

Bending and Setting a Straight Flange. There is some wiping action as the flange of the workpiece is bent to a 90° angle (Fig. 7-8). A sliding punch (D1) is cam actuated to "set" the 0.406-in. (10.31 mm) dimension. Stock of uniform thickness must be used; otherwise, the die can be wrecked. The part is positioned by the nest plates (D2 and D3). This die could be classified as either a bending or forming die since both actions are used.

Cam-actuated Double-flanging Die. An H-shaped cutout is made in the blank to provide tabs which, when bent 90°, produce a flange extending above and below the surface of the finished part as seen in Fig. 7-9. The section through the bending die shows the reliefs cut in the pressure pad to allow these tabs to swing downward as the flange is being bent upward.

The blank is positioned in the die by the nesting blocks fastened to the sliding die blocks (D1) and is supported by the pressure pad. The flange is bent upward as the forming punch pushes the blank down between the sliding die blocks. The space between, and squareness of, the flanges is assured by the action of the punch (D3) on the sliding die block through the wedge block (D2).

Tube-bending Die. Steel tubing (0.060-in. walls, 0.625 in. OD) dipped in heavy lubricant is forced through the die shown in Fig. 7-10 and is ejected by the succeeding tubing. Die clearance is 0.010 in. (0.25 mm), punch clearance 0.003 in. (0.08 mm). Production is 60 pieces per hour. The lower block is made in two parts to facilitate machining. Numerous lower blocks can be made to attach to the guide block for forming the same diameter tubing to different radii, making a more universal tool.

Die for Hemming an Auto Fender. An auto fender, with a 90° flange, is clamped against the die section (Fig. 7-11, D5) by a spring loaded hold down (D8) as the punch

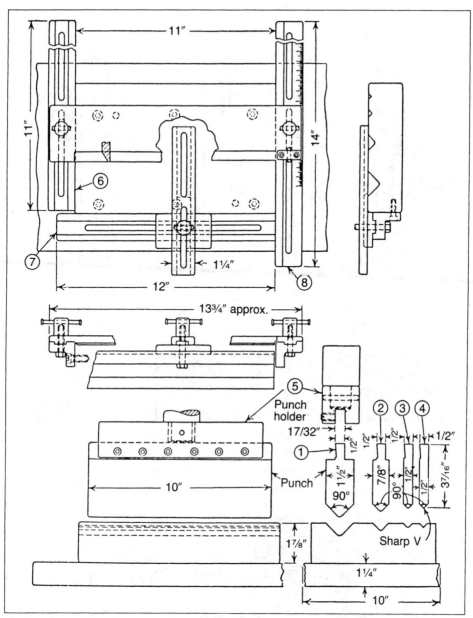

Fig. 7-4. Standard V-die. (*General Electric Co.*)

block (*D2*) descends. A sliding punch (*D7*) is forced to the left by the cammed plunger (*D1*) and the cammed block (*D12*), so that the flange is bent to the left at a 45° angle. A vertical punch (*D9*) descends to flatten the flange against a die block (*D6*) and to flatten it back on itself just after the sliding punch is forced to the right by the cammed plunger. The cammed block (*D11*) is not essential in providing correct timing for the vertical and sliding punch movement. The timing of the right hand or return movement of the sliding punch must be such that the sliding punch does not obstruct the downward movement of the flattening punch. A wear plate (*D4*) is fitted to the plunger, which slides between two die members (*D3* and *D13*).

Flanging and Hemming Die. A flanged part is placed between a spring loaded pad

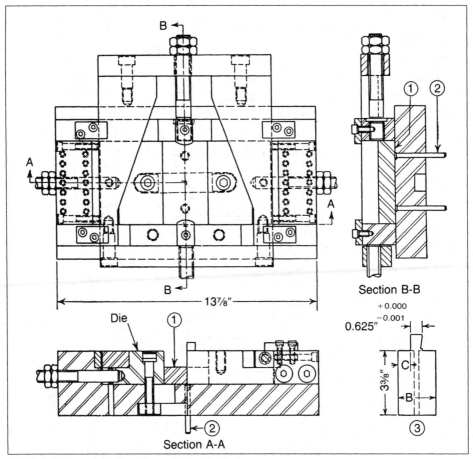

Fig. 7-5. Universal U-die. (*General Electric Co.*)

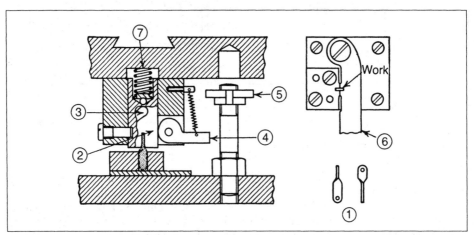

Fig. 7-6. Rotary twisting die.[3]

and an air pad (Fig. 7-12, $D1$, $D2$), and an angular flange ($D5$) is bent by the punch ($D3$). The hem is flattened by the action of the lower die member ($D4$) after the air pad compresses. If the angular flange were nearly horizontal it would buckle instead of being flattened.

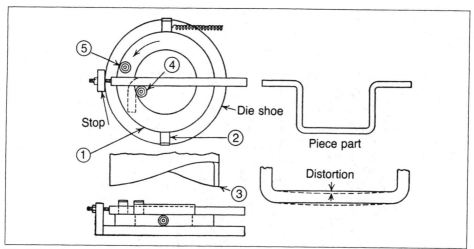

Fig. 7-7. Rotary bending die.[4]

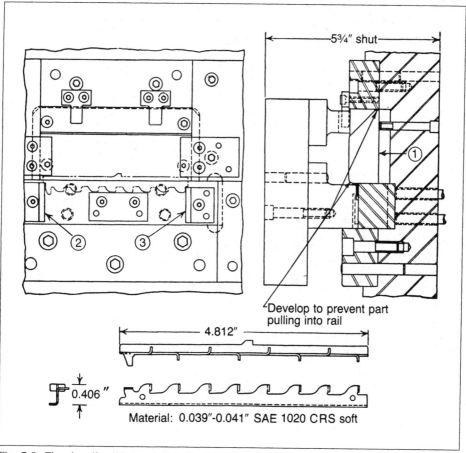

Fig. 7-8. Flanging die. (*National Cash Register Co.*)

7 - 7

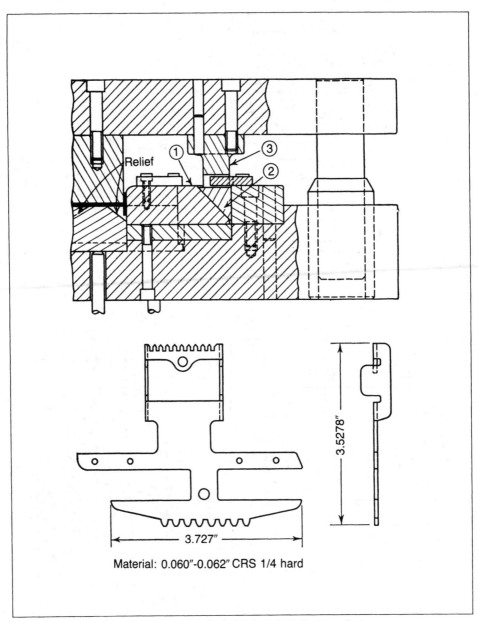

Relief

① ③ ②

ммммммм
○
○ ○ ○ ○
○
ммммммм
← 3.727″ →

Material: 0.060″-0.062″ CRS 1/4 hard

3.5278″

Fig. 7-9. Cam-actuated double-flanging die. (*National Cash Register Co.*)

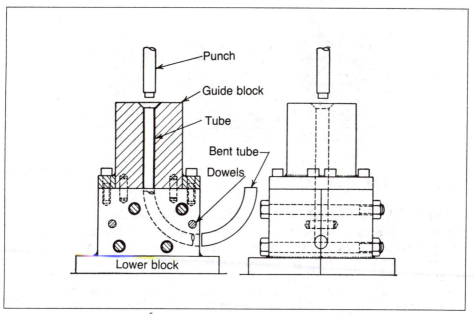

Fig. 7-10. Rotary bending die.[5]

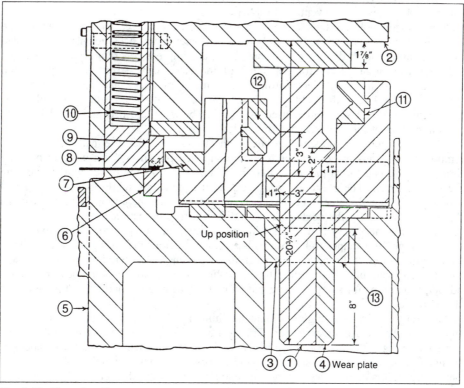

Fig. 7-11. Hemming die.

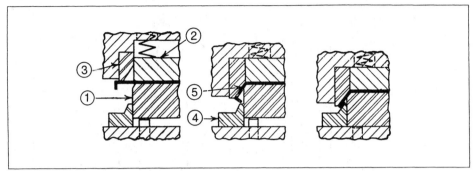

Fig. 7-12. Flanging and hemming die.[6]

URETHANE TOOLING FOR PRESS-BRAKE DIES

Specially formulated urethane materials for metalworking have advanced soft die tooling techniques in many applications where setup time, flexibility, and nonmarring of parts are important factors. In press brakes, a urethane pad functions to a considerable extent as a universal female die; it can bend and form precise shapes with a variety of punches, metals, and gages. Only the punch is machined for each application.

Urethane performs like a solid liquid under confined conditions, such as in a metal die-pad retainer. When the elastomer is properly controlled under compression, more uniform pressures can be developed to form heavier gages, more complex shapes, and sharper radii than with conventional rubber.

Standardized tooling setups for press brakes utilize various sizes and grades of urethane die pads which are used interchangeably in metal die retainers. An air channel is provided in the retainer beneath the pad to distribute internal compressive strain evenly throughout the pad; also, it helps control deflection of the elastomer, permitting deeper punch penetration and more efficient forming. The metal retainer must be designed with sufficient strength to withstand the pressures transmitted by the elastomer under load. A thin urethane wear pad is often used as a buffer between die pad and metal blank to prevent cutting of the urethane pad during long runs.

When the punch depresses the metal sheet into the flat die pad, the urethane temporarily deforms, producing high bottom and side pressures. This uniform pressure throughout the pad forms the metal into the precise shape of the punch without marring the surface finish. The elastomer wraps around the punch so tightly that spring back is virtually eliminated. When the punch is withdrawn, the pad returns to its original dimensions, ready for the next stroke or a different forming application.

Maintaining necessary bottom and side pressures in the die pad for a particular range of metal gages and shapes to be formed is a primary factor when designing a tool setup. The configuration of the punch, its penetration, and the characteristics of both pad and metal are all variables contributing to the problem of determining how best to provide adequate pressures for sharp forming but without exceeding the deflection limitations of the urethane.

Generally, punch penetration should be limited to one-third of the combined pad thickness and air space under the pad. For most uses, hardness should be in the 85 to 95A durometer range. Because of the universality of urethane, there is usually considerable latitude when selecting pad sizes and grades. However, when switching from light to heavy gages, it is often necessary to change to a harder and/or thicker die pad. To extend service life, overdeflection of die pads should be avoided.

The machinability of urethane is an advantage when it is necessary to accommodate various application requirements. Since it is often important that the lengths of the punch and the die pad be approximately equal, the same die retainer can be used for short or long punches by adjusting the length of the urethane pad by sawcutting the pad for shorter punches or butting short sections of urethane together in the retainer for longer punches (they perform like one piece under compression). However, when cutting a longer pad to a shorter punch length, fill in the vacant end, or ends, of die retainer with bar stock. This end packing is necessary to maintain proper forming pressure in the pad.

Figure 7-13*A*, *B*, and *C* shows typical V- and U-forming applications. Under compression of the punch and blank, the elastomer bulges slightly above the retainer walls and also into the stress relief air channel under the pad. The solid hydraulics principle of this method tends to equalize the forces throughout the pad to provide uniform pressure at bottom of stroke for sharp, precise bending or forming.

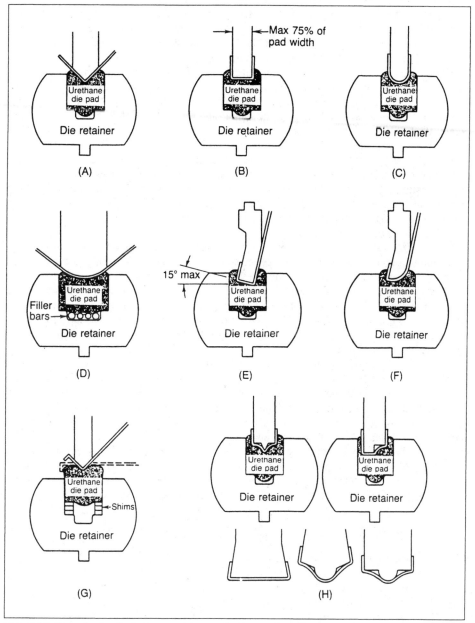

Fig. 7-13. Press-brake dies using urethane as the lower die. (*Kaufmann Tool & Engineering Corp.*)

Since the accuracy of forming to the punch is governed by the amount of final bottoming, greater pressures are required in U-forming than in V-bending. Also, to prevent overdeflection, it is important that the punch width (at bottoming) not exceed a maximum of 75% of die pad width.

The flange width and the inside radius affect the forming of a flange with a urethane die. For mild steel, the minimum flange width is four times stock thickness plus the inside radius, or 0.125 in. (3.18 mm), whichever is greater. For softer metals, shorter flanges can be formed; with harder metals, longer flanges are required.

Radius-forming applications (Fig. 7-13D) require a wider die pad and retainer. This may also require an adjustment of the timing of bottoming pressure. The simplest method is to decrease the area of the stress relief channel beneath the pad by inserting filler bars. This causes the elastomer to deflect upward with greater force, permitting faster bottoming pressure with less penetration. This same procedure is often used when forming heavier gages (with the same tooling) than otherwise recommended.

Figure 7-13E and F shows typical setups for forming with gooseneck punches. Urethane is somewhat limited in these applications since punch design is restricted to the approximate maximum angle shown in the drawing. The angle of tipping is governed by the length of the punch, the length of the panel, and the width of the ram face. An angle greater than 15° requires considerable penetration of the punch into the urethane die, thus reducing its useful life.

Special side pressure forming can be accomplished by using preshaped urethane die pads. For good forming definition, a cast or machined-to-shape die pad is used, roughly approximating the configuration of the punch. This design provides adequate side pressure at the bottom of stroke for these difficult bends. The urethane pad acts as a conventional female steel die until the punch is near bottom; then it deflects and forms the blank entirely around the punch.

When using standard tooling for reverse bend forming as shown in Fig. 7-13G and also for forming bends requiring deeper penetration, a method of delaying the timing of bottoming pressure is necessary. The urethane pad is elevated with shims to provide necessary clearance during the second stroke of a reverse bend. This also creates a greater air gap in the stress relief channel for deflection of the elastomer, increasing side pressure when deeper penetration of the punch is required. However, this design usually limits use of the die to lighter gages since a corresponding loss of compressive pressure occurs at the unretained top of the pad.

Contour forming with concave areas such as shown in Fig. 7-13H presents a special problem because the urethane cannot form into sharp recesses. However, the lighter the metal gage, the sharper the definition possible.

A wipe-down die tooling design using high-durometer (hard) urethane-forming blocks is shown in Fig. 7-14. Here, instead of a universal female die, the specially formulated urethane elastomer functions as a nonmarring forming punch. These machinable forming blocks can be set for minimum clearances since the urethane will deflect slightly to accommodate stock thickness variations and will then resume accurate dimension. Also, to eliminate the use of springs, other standard urethane materials are saw-cut and bonded or bolted into place to function as pressure pads during the wiping action.

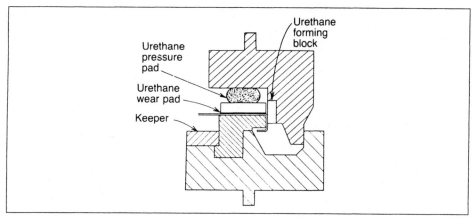

Fig. 7-14. Flanging die with urethane forming block and pressure pad. (*Kaufmann Tool & Engineering Corp.*)

References

1. Society of Manufacturing Engineers, *Tool Engineers Handbook*, 2nd ed. (New York: McGraw-Hill Book Company, 1959).
2. G. Sachs, *Principles and Methods of Sheet-metal Fabricating* (New York: Reinhold Publishing Corporation, 1951).
3. L. L. Lee, "A Simple Twisting Die," *American Machinist* (April 18, 1926).
4. J. ᴾ lnick, "Rotary Punch Bends Brace Handles," *American Machinist* (June 12, 1950).
5. F. E. Riley, "Press Tool Bends Tubing," *American Machinist* (July 28, 1949).
6. C. R. Cory, *Die Design Manual* Part II (Dearborn, MI: Society of Manufacturing Engineers, 1958).

SECTION 8

METAL MOVEMENT IN FORMING

Forming. Forming is a metalworking process in which the shape of the punch and die is directly reproduced in the metal with little or no metal flow. Forming, bending, or drawing actions may be combined in a die; a die is classified according to the predominant action.

Forming Limits. Conventional mechanical tests which accept or reject a given metal do not determine forming limits exactly, although they may be generally indicative as to type and extent of feasible forming operations.

The decision to use a form die, instead of a draw die, will depend considerably upon the part shape. Geometrical criteria are stated in Section 2.

The use of a draw die may be indicated if the use of a form die would (1) cause the metal to tear because of excessive tensile strain, or (2) form objectionable wrinkles because of an excess of metal.

Forming limits depend also on the metalworking method; a double-action die may successfully fabricate a certain part, where a single-action die could not.

Forming Stress. Forming limits are functions of various stress distributions and amounts; these stresses are also treated in Sections 6, 10 and 11. All stresses can be resolved into either tensile or compressive type for any single metalworking operation or combinations thereof. Analysis of combined stresses in a multiple operation can be quite difficult.

FORMING OF STRETCH AND SHRINK FLANGES

Figure 8-1 shows the two basic types of flanges—stretch and shrink. The reverse-type flange is a combination of these two. Any stretch flange can be considered as a segment of a flanged hole.

In stretch flanging, tensile strain increases from zero at the flange "break" line (axis of bending) to a maximum at the flange edge. For that reason, any tearing will start from the flange edge. The amount of tensile strain increases with (1) increasing forming angles and (2) increasing flange heights. Ability to withstand strain without tearing increases with the metal thickness and is affected by the composition and treatment of the material and by its edge smoothness.

In shrink flanging, the tendency to wrinkle increases from zero at the flange break line to a maximum at the flange edge. Generally, wrinkling will be avoided if the flanging operation can be done in the same direction as that of the flange movement, so that any excess flange metal is ironed or "wiped" out to the edge of the flange as added flange height. To get such ironing or wiping effect, the blank may have to be tipped so that the flange is vertical in the die. Wrinkles cannot be forcibly spanked out of thin-gage flanges, but large wrinkles can sometimes be distributed into a greater number of small, unobjectionable wrinkles or may be changed into folds of metal.

Critical Strain. The distribution and the magnitude of strains in the forming of stretch flanges are mainly functions of the part shape.

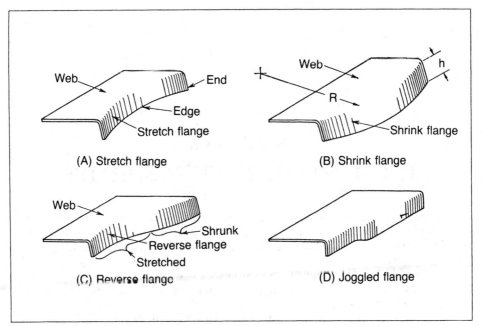

Fig. 8-1. Basic types of flanges.

Failures in Flanged Parts. The major failures in rubber-formed (Guerin process) flanges are elastic sine-wave buckling, splitting, and insufficient pressure. Sine-wave buckling, shown in Fig. 8-2A, occurs when the forming pressure is removed from the part, and is due to the residual stresses remaining in the flange. These buckles can generally be pressed out by subsequent finish forming if the buckling is not too severe.

Flange splitting, shown in Fig. 8-2B, is dependent on the elongation character properties for a given material and is eliminated only by decreasing the h/R value to formable limits.

A hump in the bend area occurs when there is insufficient pressure to force the part to the die (Fig. 8-2D and E) and is common for parts with large h/R and small h/t values. This distortion can be minimized by using higher forming pressures and smaller material thickness.

Springback (Fig. 8-2F) and crown (Fig. 8-2G and H) occur on all formed parts and depend on the material and the shape of the part. They can be eliminated by hot finish forming, by allowing for springback and crown on the form tool, and by hand working.

The usual amount of springback allowance is 2° for steel, 3° for soft aluminum, and 6° for tempered aluminum. However, because of the several factors involved, such as the kind and condition of material and the amount of stretch or shrink, final determination of springback must generally be left to die tryout.

Shear buckling (Fig. 8-2J and K) is an end condition that occurs at high h/R values for materials with a low E/S_t value, where E is the modulus of elasticity, and S_t is the tensile yield strength. Control of this distortion can be made only by allowing excessive material so that the ends can be trimmed.

Flange-edge Thickness Changes. The forming of stretch or shrink flanges is always accompanied by change in metal thickness at the flange edge, thickness being decreased for a stretch flange and increased for a shrink flange. This thickness change is usually insignificant, especially in the case of steel parts. Die tryout is generally necessary for accurate determination, but an approximation is:

$$\text{Thickness change } \Delta t = \frac{te}{2} \tag{1}$$

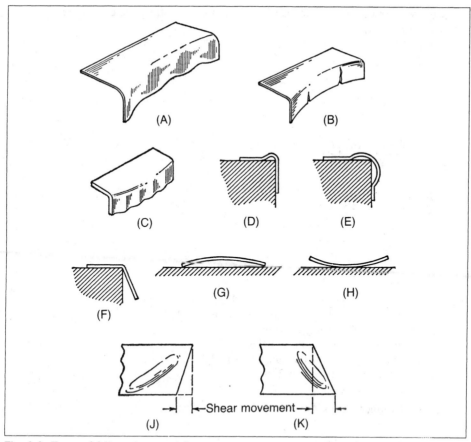

Fig. 8-2. Types of failures in curved flanges.

Formability Limits. Composite graphs plotted on log-log paper composed of empirically determined forming-limit curves are shown in Fig. 8-3 and Fig. 8-4. In addition to splitting and buckling, there is a minimum flange-height limit below which the flange will not form down to the die. This limit is based on a maximum machine limit of 1,925 psi (13.3 MPa).

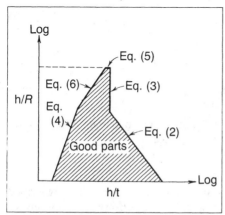

Fig. 8-3. Forming-limit curves for rubber-formed stretch-flanged parts.

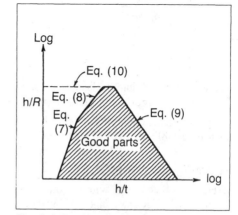

Fig. 8-4. Forming-limit curves for rubber-formed shrink-flanged parts.

For rubber-formed stretch flanges, the equations for plotting the curves are:

$$\frac{h}{R} = \frac{E_t}{S_t} \frac{0.09}{(h/t)^2} \tag{2}$$

$$\frac{h}{t} = \left(\frac{11.4\ E_t}{S_t}\right)^{2/5} \tag{3}$$

$$\frac{h}{R} = \frac{6.99 \times 10^6}{S_t^{4.9}} \left(\frac{h}{t}\right)^{17} \tag{4}$$

$$\frac{h}{R} = 0.421 \ \ln 13.5e \tag{5}$$

$$\frac{h}{R} = \frac{0.19}{(2.72 \times 10^{-8} \ S_t^{4.9})^{0.0816}} \tag{6}$$

where h = height of flange, in. (mm) (see Fig. 8-1)
R = curvature of flange, in. (mm) (see Fig. 8-1)
t = material thickness, in. (mm)
S_t = tensile yield strength, psi (Pa)
E_t = modulus of elasticity, tensile, psi (Pa)
e = elongation, percent in 2 in. (51 mm)
$\ln$ = natural logarithm

For rubber-formed shrink flanges, the equations are:

$$\frac{h}{R} = \frac{6.99 \times 10^6}{S_c^{4.9}} \left(\frac{h}{t}\right)^{17} \tag{7}$$

$$\frac{h}{R} = \frac{0.19}{2.72 \times 10^{-8} \ S_c^{4.9}} \left(\frac{h}{t}\right)^{1.17} \tag{8}$$

$$\frac{h}{R} = \frac{0.0225\ E_c}{S_c(h/t)^2} \tag{9}$$

$$\frac{h}{R} = \frac{1.4}{S_c} \tag{10}$$

where E_c = modulus of elasticity, compression, psi (Pa)
S_c = compression yield strength, psi (Pa)

In solving the equations and plotting the curves, minimum and maximum values for h/t must be established, based on the anticipated working range.

FLANGING OF HOLES

Holes in sheet metal with surrounding flanges of various shapes are frequently termed "extruded," "countersunk," "burred," or "dimpled" holes. However, they are all holes with stretch flanges.

Dimpling. Dimpling is a process of forming a recessed area for a fastener so that the fastener head can be flush with the sheet surface. The process is used where the material thickness will not permit countersinking.

There are two methods of dimpling. One uses a forming punch and die (Fig. 8-5A), and the other uses a coining ram (Fig. 8-6).

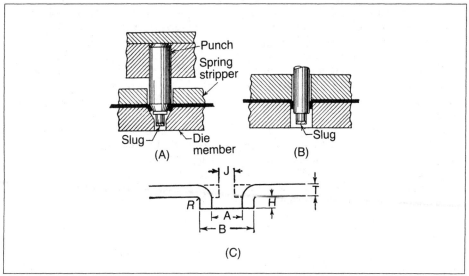

Fig. 8-5. Hole-flanged dies. (A) For dimpling. (B) For 90° flanges. (C) For tapping: J, diameter of original pierced hole; T, thickness of stock; R, bend radius of flange; A, ID of hole flange; B, OD of hole flange; H, width of flange.

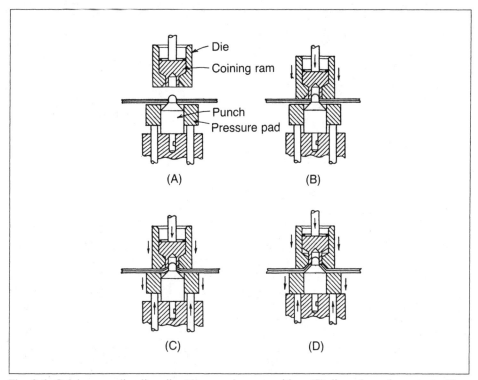

Fig. 8-6. Coining-ram dimpling die: (A) normal open position; (B) die and ram in contact with metal; (C) metal clamped between die and pressure pad; (D) die fully closed.

The coining-ram method controls hole stretch and balances internal strains, eliminating radial and internal shear cracks. The confining action of the pad face, die face, and coining ram forces the material into the exact configuration of the tool geometry.

There are four primary forces at work during the forming of a dimple which tend to make it crack. The first is stretching at the hole in brittle material and in thin gage where there is insufficient material to accommodate the stretch. This causes radial cracks which start at the edge of the dimple and grow outward. The second force, bending over the die cavity, sets up tensile stresses in the upper portion of the dimple, causing circumferential cracks around the dimple. The third force is heavy shear loads below the top of the dimple which cause internal circumferential shear cracks. Both types of circumferential failures can be prevented by the proper application of sufficient temperature and pressure. The fourth force is the coining-ram pressure, which, if excessive, causes compressive hole cracks.

The major limitations in dimpling are machine pressure and die temperature. Many of the high-strength alloys display poor definition at room temperature because of insufficient pressure, and may be dimpled at elevated temperatures.

Production parts made of materials in the heat-treated condition should be dimpled in this condition because subsequent heat treatment will cause a misalignment of holes.

Formability Limits. The predictability equation to construct the formability curve of Fig. 8-7 is

$$\frac{h}{R} = \frac{0.444e^{0.253}}{1 - \cos \alpha} \tag{11}$$

where e is the percentage of elongation in 2 in. (51 mm). The remaining terms are defined in Fig. 8-8.

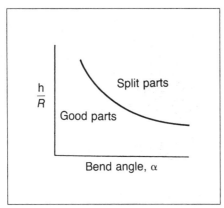

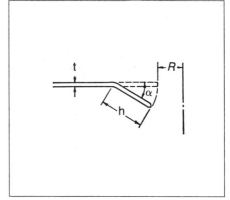

Fig. 8-7. Formability curve for ram-coined dimples.

Fig. 8-8. Cross section of a dimple showing geometric parameters.

Table 8-1 lists some observations made during an experiment on dimpling of various materials, using a bend angle of 40° and an approximate h/R value of 1.2. The data indicate whether the materials can be dimpled at room temperature. Also shown are the experimental results, giving the temperature of the test and whether an acceptable dimple can be formed at the test temperature. The last column indicates the recommended elevated temperature for fabrication of standard dimples, based on published elevated-temperature properties.

90° Hole Flanging. Forming a flange around a hole at a bend angle of 90° is nothing more than the formation of a stretch flange at the angle.

A 90° hole flange can be pierced and flanged in most metals by one press stroke, using a two-step punch as shown in Fig. 8-5*B*.

TABLE 8-1
Experimental Results on Dimpling Various Materials

Material	Condition	Dimpled At Room Temperature	Experimental Tests Temp. °F (°C)	Results	Recommended Elev. Temp. °F (°C)
HM21XA	T8	No	800 (425)	Yes	750 (400)
2024 Aluminum	T3	Yes	RT*	Yes	No
17-7 Ph	TH 1050	No	RT	No	No
PH 15-7 Mo	TH 1050	No	RT	No	1,000 (540)
AM 350	Age-hardened 850° F (450° C)	No	RT	No	1,000 (540)
A-286	Age-hardened 1,325° F (719° C)	Yes	RT	Yes	1,450 (790)
USS-12-MoV	Age-hardened 800° F (425° C)	No	RT	No	No
Titanium (6Al-4V)	Mill-annealed	No	800 (425)	Yes	900 (480)
Titanium (13V-11Cr-3Al)	Age-hardened 900° F (480° C)	No	800 (425)	Yes	900 (480)
Vascojet 1000 (H-11)	Hardened	No	RT	No	1,200 (650)
Beryllium (pure)	Condition ''C''†	No	1200 (650)	No	1,450 (790)‡
René 41	Age-hardened 1,400° F (760° C)	Yes	RT	No§	Any temp.
Inconel X	Age-hardened 1,300° F (705° C)	Yes	RT	Yes	1,800 (980)
Hastelloy X	Solution-treated	Yes	RT	Yes	Any temp.
L-605	Solution-treated	Yes	RT	Yes	800 (425)
J-1570	Age-hardened 1,400° F (760° C)	Yes	RT	Yes	No
Columbium (10Mo-10Ti)	As received	No	800 (425)	No	1,000 (540)
Molybdenum (0.5% Ti)	Hot-rolled, stress relieved	No	800 (425)	No	1,800 (980)
Tungsten (pure)	As received	No	800 (425)	No	1,200 (650)

* Room temperature.
† Material is rolled and stress-annealed for 10 min at 1,400° F (760° C).
‡ Form dimple at 2 in. (51 mm) per min strain rate.
§ Insufficient pressure.

One manufacturer has standardized flange widths (Fig. 8-5C, dimension H) for holes to be tapped in low-carbon-steel stamping stock as follows:

$$B = A + \frac{5T}{4} \quad \text{when } T \text{ is less than 0.045 in. (1.14 mm)}$$

$$B = A + T \quad \text{when } T \text{ is more than 0.045 in. (1.14 mm)}$$

$$H = T \quad \text{when } T \text{ is less than 0.035 in. (0.89 mm)}$$

$$H = \frac{4T}{5} \quad \text{when } T \text{ is 0.035 to 0.050 in. (0.89 to 1.3 mm)}$$

$$H = \frac{3T}{5} \quad \text{when } T \text{ is more than 0.050 in. (1.3 mm)}$$

$$R = T/4 \quad \text{when } T \text{ is less than 0.045 in. (1.14 mm)}$$

$$R = T/3 \quad \text{when } T \text{ is more than 0.045 in. (1.14 mm)}$$

$$J = \frac{TB^2 + 4TA^2 + 4HA^2 - 4HB^2}{9T} \tag{12}$$

FORMING OF BEADS

Beads are rather shallow recesses in thin sheet-metal panels for stiffening purposes. The possible depth and contour of beads that can be successfully formed depend upon the permissible amount of stretching of the metal, the method of forming, and the available pressures. The rubber forming of beads is usually restricted to the forming of low-strength materials such as aluminum and magnesium in their soft condition. This restriction is imposed on the process because of the requirement of high rubber pressure to produce parts of good bend definition.

The criteria for failure in rubber bead forming are splitting and insufficient pressure. Both types of failure are due to the physical properties of the material, the applied rubber pressure, and the geometric variables. For the purpose of this evaluation, parts are considered unacceptable because of insufficient pressure if there is less than 0.50 in. (12.7 mm) flat area between beads. Longitudinal buckling and the free form radius also may be considered limiting factors. Longitudinal buckling generally occurs in the heavy-gage materials where the bead is of insufficient length. The free form radius, the radius between the bead and the flat area, may be considered a limiting factor because a large free form radius reduces the stiffening characteristics of the bead.

The predictability equations for rubber forming of beads used in plotting the curves of Fig. 8-9A are:

$$\frac{R}{L} = 0.065 \left(\frac{\frac{1}{S_t} \times 10^4}{0.065} \right)^{0.216} \left(\frac{R}{t} \right)^{0.29} \tag{13}$$

$$\frac{R}{L} = 6.0 \times 10^{-21} \, (eS_u)^{9.3} \left(\frac{R}{t} \right)^{-12.4} \tag{14}$$

where R = inside radius of bead, in. (mm)
L = center-line spacing of beads, in., (mm)
t = material thickness, in. (mm)
S_t = tensile yield strength, psi (Pa)
S_u = ultimate strength, psi (Pa)
e = elongation, percent in 2 in. (51 mm), expressed as a decimal

There is no developed predictability equation for the dashed portion of the Eq. (14) curve extending from the $R/L = 0.10$ to line of Eq. (13). This changes with the material being formed and must be plotted from empirical data.

The predictability equation for die or drop-hammer forming of beads used in plotting the curve of Fig. 8-9B is

$$\frac{R}{L} = 60.6 \, (e_1)^2 \left(\frac{R}{t}\right)^{-1} \tag{15}$$

where $e_1 = \ln^{-1}\left[(e_L)_{0.5} - \frac{(e_w)^2_{0.5}}{(e_L)_{0.5}}\right] - 1$

$$(e_L)_{0.5} = \frac{\Delta L}{0.5} \qquad (e_L)_{0.5} = \ln\,[1 + (e_L)_{0.5}]$$

$$(e_w)_{0.5} = \frac{\Delta W}{0.5} \qquad (e_w)_{0.5} = \ln\,[1 + (e_L)_{0.5}]$$

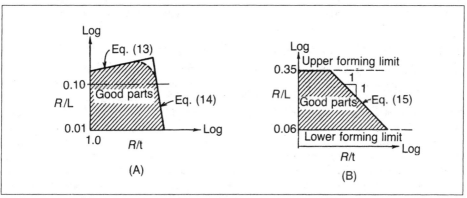

Fig. 8-9. Formability curves for beads: (*A*) rubber-formed; (*B*) drop-hammer-formed.

The value of e_1 is found from a standard 0.50 in. (12.7 mm) wide tensile specimen. Figure 8-10 shows where the various dimensions can be taken by a cathetometer.

For die- or drop-hammer-formed beads, the upper forming limit for the ratio R/L is 0.35 when the punch radius is 0.50 in. (12.7 mm). The lower limit of R/L is 0.060. Beaded panels below this ratio would have the effective stiffness reduced, thus lowering the structural efficiency of the panel.

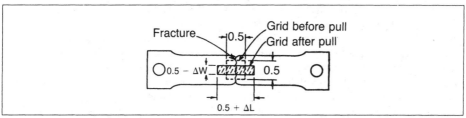

Fig. 8-10. Tensile specimen for obtaining measurements when calculating formability of drop-hammer beads.

HIGH VELOCITY METAL FORMING

High-velocity forming refers to the processes of shaping workpieces at forming velocities that may exceed 300 feet per second (100 m/s). This requires the use of explosive, electrical, or high-speed mechanical energy sources. These processes can be contrasted with conventional or low-velocity forming processes which use press brakes, mechanical presses, or drop hammers. Deformation velocities for the low-velocity forming processes normally range from a few tenths of a foot per second to about 20 fps (6 m/s).

Explosive forming largely is used in the aerospace and aircraft industries and has been successfully employed in the production of automotive-related components. In general, explosive forming has found its greatest potential in limited-production prototype forming and in forming large parts for which conventional tooling costs are high.

In electrohydraulic forming, electrical energy, discharged in a liquid medium, generates a very high pressure and a shock wave similar to the one generated by a high explosive. In electromagnetic forming, electrical energy discharged through a coil generates an intense magnetic field which interacts with the electrical currents induced in the conductive workpiece to apply a force to the workpiece and to the coil.

Pneumatic-mechanical forming is a process that can be applied to the production of parts by forging, impact forming, extruding, and powder compacting. Some applications of this process take advantage of the fact that many metals exhibit essentially hydrodynamic behavior when heated to a plastic condition and formed at high velocity. Energy stored in the form of compressed gases is used to accelerate the forming machine ram.

EXPLOSIVE METALWORKING OPERATIONS

High-explosive Forming. In the standoff method of high-explosive forming, the explosive charge is located at some predetermined distance from the workpiece (Fig. 8-11). Water is generally used as the energy transfer medium to ensure a uniform transmission of energy and to muffle the sound of the blast. The mass of the water also serves to back the female die cavity during impulsive loading. After detonation of the explosive, a pressure pulse of high intensity is produced. A gas bubble also is produced which expands spherically and then collapses until it vents at the surface of the water. When the pressure pulse impinges against the workpiece, the metal is displaced into the die cavity with a velocity of 60 to 400 fps (18 to 122 m/s). To ensure proper die filling, the die cavity must be evacuated. A vacuum of 28 in. (711 mm) of mercury is sufficient, in most cases, to prevent burning or welding from adiabatically heated air.

Measurements indicate that the pressure pulse accounts for 60% of the energy available, the first oscillation of the gas bubble for 25%, and the remaining oscillations for the other 15%. These values, however, vary with charge weight and type of explosive used. For many forming operations, the charge weight and standoff distance combinations are chosen so that the bubble breaks over the workpiece during the first expansion.

When a high explosive is detonated, a violent chemical reaction proceeds through the explosive material at a high velocity which can exceed 26,000 fps (7,900 m/s). This reaction generates very high pressures, typically over 3 million psi (20,700 MPa). These pressures in the explosives in turn generate a very high-pressure pulse in the surrounding medium. The pressure pulse propagates outward from the source initially with a supersonic velocity in the surrounding medium. As the pulse amplitude decays, the velocity decreases until, at low-pressure amplitudes, the velocity of the pressure pulse approaches sonic velocity in the medium. It is primarily the action of this pressure pulse on the workpiece that is responsible for forming the part.

Explosives such as dynamite should be avoided when possible, and only stable explosives should be used to insure good repeatability. Some of the more commonly used high explosives are trinitrotoluene (TNT), cyclotrimethylene trinitramine (RDX), and pentaerythritol tetranitrate (PETN).

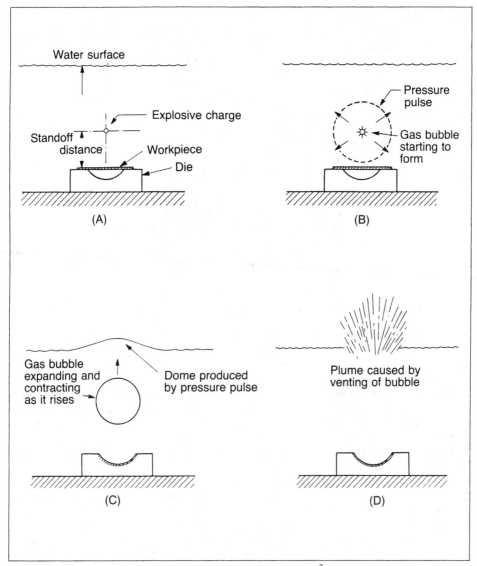

Fig. 8-11. Sequence of underwater explosive forming operations.[2]

Parameters for controlling pressure magnitude at the workpiece include the type and amount of explosive, the charge configuration, the standoff distance, the type of energy-transfer medium, and the effects of pressure focusing or confinement. Working pressures normally range from several thousand to several hundred thousand psi. The time during which the energy is delivered to the workpiece depends upon the size of the explosive charge and its distance from the workpiece. Delivery times in the range of tens to hundreds of microseconds are common. These times should not be confused with the time required for the workpiece to deform to its final shape, which is usually a few milliseconds.

A typical system for forming metal parts with a standoff explosive charge consists of four basic parts: (1) an explosive charge, (2) an energy-transfer medium, (3) a die assembly, and (4) a workpiece. Many operations also may involve auxiliary components such as a forming tank, an air compressor, a vacuum pump, a hydraulic press and heavy-duty equipment for handling the dies and workpiece.

COMBUSTIBLE GAS FORMING

In combustible gas forming, the workpiece is placed in a special closed die (Fig. 8-12). Upon ignition, working pressures from 1,000 to 110,000 psi (7 to 760 MPa) can be obtained by varying the gas mixture and initial gas pressure.

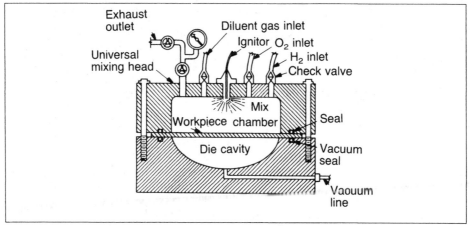

Fig. 8-12. Schematic of typical gas forming arrangement.[2]

High-energy-rate forming using combustible gases differs considerably from high-explosive forming. The major difference is that the strain rate is much slower for combustible gas forming. The longer deformation times in combustible gas forming also can result in the workpiece being heated by the hot gas.

The fuel gases commonly used are: hydrogen, ethane, and methane. The energy release can be reduced by the addition of dilutant gases: nitrogen is a common dilutant, although helium, carbon dioxide, and argon also are used. The oxidizer is commonly oxygen, air, or an air-oxygen mixture.

Appropriate Applications. Explosive forming is particularly applicable in the following instances:

- When the application of uniformly applied strain will result in desirable characteristics in the formed part. An example is the precontouring of a specific thickness pattern in a blank.
- Where the formed parts are exceedingly large, requiring large and/or heavy presses. The primary limitation on size is the size of the blank available.
- When a small number of parts are required and/or wide flexibility is desirable in the design of the tooling to permit economical and rapid design changes to be incorporated.
- For the forming of specific materials which exhibit better formability at ambient temperatures at high strain rates than is exhibited under conventional strain rates.
- For the forming of welded blank material where an additional quality check of the weld is desired.
- Where it is desirable to hold the residual stresses in the formed part to a minimum.

Sizing of Complex Parts. The use of explosive forming process for the truing of parts is of significant value when high precision in contour, configuration, and diameter is desirable. Complex shapes which do not readily permit the use of male rams may be trued by the process. The inherently uniform straining will permit truing of welded preforms without detriment to welds due to the frictional effects which arise when male tooling constrains the material to be formed. The welds must be high quality, or the process will cause their failure. Welding often causes distortion in the welded parts, especially in unsymmetrical locations. Explosive truing often has been successfully applied to correct such welding distortion.

Although *forming* normally is a term applied to relatively thin workpieces, pieces as thick as 4 in. (101.6 mm) or more have been formed by the explosive technique.

Piercing the Workpiece During Forming. The perforation of a workpiece is readily added as an auxiliary operation of the explosive process. Explosive perforation carried out from within a cylindrical form will normally produce a radius on the outside diameter of the perforation, thus eliminating burrs. More than one hole can be perforated at a time, provided that the web thickness between the holes in the die is adequate to maintain the necessary die strength.

The hole diameter and location depend upon the die and are thus less subject to variation than may be the case with secondary operations in conventional tooling.

Explosive perforation of brittle materials such as tungsten is an economical process compared with electrical discharge machining (EDM). The associated hazard, in conventional perforating, of initiating cracks in the areas that are being perforated is reduced by explosive perforation.

The limitations of explosive perforating are:
1. The hole diameter must exceed the material thickness.
2. The die design for perforation must be adequate for repetitive shots.
3. The web thickness between the holes in the die must be adequate to maintain die strength in that area.

Forming Beads. Stiffening beads made by explosive forming differ from those formed in a closed die or by rubber pad forming. There is less time for lateral movement of metal into the bead grove with the consequence that metal in the bead may be thinner than would be the case with conventional tooling. The process is therefore essentially controlled by the condition of rigid clamping modified by the fact that some metal flow may occur in the surrounding area. The amount of metal flow possible is determined by the overall geometry of the bead arrangement and is not necessarily the same around all beads within a group.

A result of the uneven metal flow is that the plate, after forming, contains residual stresses. These may not rise to structurally dangerous levels but can be sufficient to require attention with respect to possible warpage if large portions of the plate are removed in trimming operations after forming.

The typical metal flow pattern for a plate with parallel beads is shown in Fig. 8-13*A*. The length of the arrows indicates the relative amount of metal flow. At the semicircular ends, the flow is approximately the same as that for forming of a circular dome. Along the straight parts of the bead, the metal flow becomes perpendicular to the bead axis. Each two adjacent beads share the material between them very evenly, and the metal flow becomes highly uniform.

In the two outer areas, where there is practically no restraint, the metal flow is much greater than between beads, and increases considerably from the ends toward the middle.

Fracture may be expected to occur in the three critical locations shown in Fig. 8-13*B*. They are: (1) a line along the center of the bead, (2) two lines near the edges of the die cavity, and (3) the center of the semicircular end of the bead. The first two fracture hazards are caused by tensile strain; the last hazard involves bending of the sheet and is particularly serious if the radius on the edge of the die cavity is inadequate.

The approach to successful explosive forming of beads must be established from case to case depending upon the material, depth-to-width ratio, thickness-to-edge radius ratio, and bead spacing.

Explosively formed beads show a peculiar thickness variation pattern, with near-constant thinning almost all the way across the bead and with a slight localized maximum at the center, as shown in Fig. 8-14. The thinning runs from 25% to 50% for the common bead configurations, and this must be taken into consideration when the beaded panel constitutes a structural member.

For estimating purposes, the following rule may be applied: maximum thinning is approximately twice the average thinning. Average thinning is the thinning that would

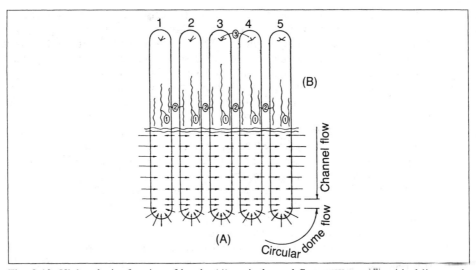

Fig. 8-13. High-velocity forming of beads: (A) typical metal flow patterns; (B) critical lines and locations of fractures.

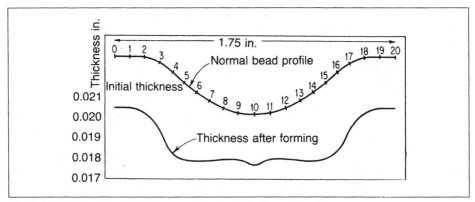

Fig. 8-14. Thickness variation pattern of explosively formed beads.

result if all the material across the bead, including the flat between the beads, were stretched evenly.

ELECTROHYDRAULIC AND ELECTROMAGNETIC OPERATIONS

These forming processes utilize direct application of electrical energy. In electrohydraulic forming, electrical energy, discharged through either a spark or a conductor located in a liquid medium (usually water), generates a very high pressure and a shock wave much like that generated by a high explosive. In electromagnetic forming, electrical energy discharged through a coil generates an intense magnetic field which interacts with the electrical currents induced in the conductive workpiece to apply a force to the workpiece and to the coil. Each of these processes derives its electrical energy from a high-voltage capacitor bank source and has a switching device to release the stored energy. Equipment manufacturers may incorporate additional features such as automatic cycling, time delay, and remote switching devices.

Electrohydraulic Forming. Figure 8-15 shows a typical example of an electrohydraulic forming system. A typical example of tooling is shown in Fig. 8-16, an

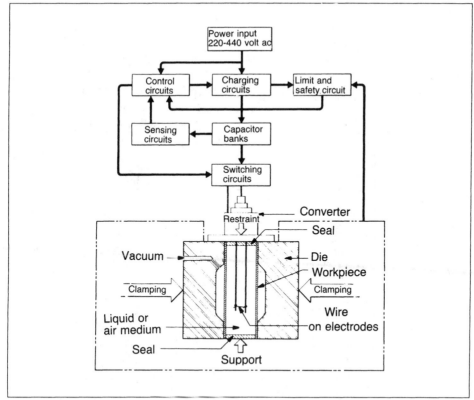

Fig. 8-15. Diagram of the electrohydraulic forming system.[2]

electrohydraulic forming die together with a workpiece.

The shock wave is produced by either a spark gap or an exploding wire. The liquid medium is not limited to water, since the material to be formed may require elevated temperatures. Forming has been done by using hot oils and molten salts.

The spark-gap technique requires more than 10,000 volts to create a shock wave of sufficient magnitude to form relatively large parts. This technique also provides a convenient means of applying repetitive shock waves of equal or varying magnitude without removing the workpiece from the die. By programming the spark gap-voltage relationship, parts of varying size and shape can be formed automatically.

The exploding-wire technique permits use of lower voltages than the spark-gap technique but requires an increase in capacitance in order to maintain the desired energy level. An advantage of this process, although it is not as readily adaptable to automation, is the capability of directing the shock wave by shaping the exploding wire. This makes simultaneous forming, blanking, and piercing possible with one blow.

Electromagnetic Forming. The electromagnetic forming process produces an intense magnetic field, providing pressures typically as high as 50,000 psi (345 MPa), which can be used either to expand or to compress the conductive workpiece.

Forming is accomplished by either permanent or expendable coils. Figure 8-17 shows the types of tooling and coils that are used for electromagnetic forming.

The expendable-type coil is a simple winding of insulated copper wire, which explodes during the forming operation. The coils are wound on mandrels having the same shape and dimensions as the workpiece. This type of coil is generally used for any experimental forming operation or when only a limited quantity of production parts is to be made. Protection must be provided against flying debris when expendable coils are used.

Permanent-type coils consist of a heavy copper helix, insulated with epoxy potting

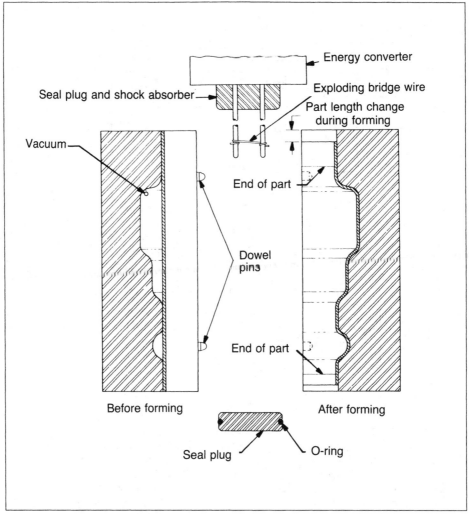

Fig. 8-16. Cross section of a die used for electrohydraulic forming showing workpiece orientation before and after forming.[3]

material. The cost of a permanent coil is high; therefore, the quantity of parts must be sufficient to justify this type of construction.

Advantages. The electrohydraulic and electromagnetic forming processes have a basic capability similar to those employing many types of conventional forming equipment, (e.g., punch presses and swaging machines). The advantage is that a simpler die arrangement can be used because the shock wave functions as the punch of a conventional die.

The electrohydraulic process lends itself readily to bulging, expanding, beading, and flanging, and can perform complicated forming operations that would otherwise require the use of oil, grease, tallow, beeswax, or rubber as a forming medium.

Each of these processes lends itself to automation, including the programming of time and power requirements to perform the operation. The two processes may be used in combination, thereby expanding their capabilities.

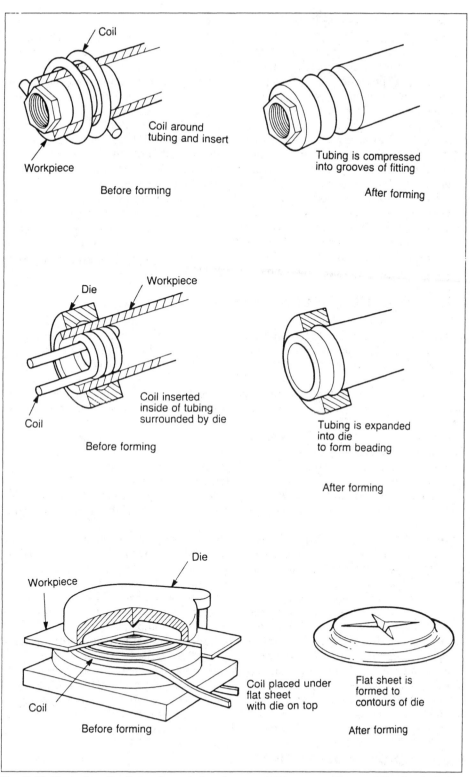

Fig. 8-17. Typical applications of magnetic pulse compression, expansion, and flat forming coils.[2,3]

PNEUMATIC-MECHANICAL OPERATION

This process can be applied to the production of parts by forging, impact forming, extruding, and powder compacting. The machines transmit high-level thrust loads by the controlled release of energy stored initially in the form of compressed gases.

The sudden application of the high pressure to a large area on a piston accelerates the piston (or ram) to high velocities in very short travel distances. The net thrust on the working piston changes form zero to high values in about one millisecond. This high-magnitude thrust is applied so rapidly that the working piston can develop a velocity of around 100 fps (30 m/s) in only a few inches of piston travel. Most operations last only a few milliseconds from the time of contact to the formation of the finished pieces.

Many metals exhibit essentially hydrodynamic behavior when heated to a plastic condition and formed at high velocity. The combination of heat and high pressure is responsible for the highly plastic behavior. Pneumatic-mechanical equipment utilizes this behavior in the production of extrusions, forgings, and molded parts with extremely thin webs and flanges. This method has been found to be particularly useful in the forging of refractory metals and alloys as well as in the fabrication of tough steel alloys. If the workpiece is heated, it is essential that it be ejected from the die as rapidly as possible after the stroke to avoid drawing the hardened steel dies with the latent heat of the part.

ADDITIONAL CONSIDERATIONS

Products formed by high-velocity methods are usually for critical applications. Because of this, material imperfections are not acceptable. Despite this, the explosive forming process has successfully formed workpieces with inclusions, and even holes, whereas deep-drawing presses operating with the same defective material would yield fractured pieces.

In most deep-draw explosive forming operations, the initial (as-received) condition of material is the fully annealed state, with subsequent heat treatment after forming, if required.

Lubrication. If lubrication is required, extreme-pressure pressworking lubricants are normally used for high-velocity forming operations. Due to the material-forming rate, any lubricant must be used sparingly. If an excess is used, the lubricant will be trapped between the material and die, resulting in distortion of the finished part.

The lubricant should be noncorrosive. Extreme-pressure lubricants are typically sulfurized or chlorinated fatty oils, or paraffin waxes in concentrated form or diluted with mineral oil.

Material-handling Problems. The extreme pressures and the high forming velocities involved in explosive forming operations require that reasonable care be exercised in handling the material, blank, or preform to avoid nicks and scratches. Vacuum cups or magnetic devices are commonly used for lifting. Handling the formed part involves other problems which may require the use of slings with rubber-covered hooks.

During the loading operation prior to forming, dirt or other foreign material must not be permitted to enter or be left between the die and the workpiece.

In the application of the various explosive processes, material-handling problems arise in the forming and truing of exceedingly large and heavy parts. By the nature of the process, especially where large and heavy parts are involved, the fabrication will necessarily take place in open areas. This permits the use of cranes and fork-lift tractors without a major limitation on capacity or mobility. The use of steel clamping devices, vacuum lifts, vacuum cups, and specifically designed vacuum handling equipment will contribute to efficient and safe handling of these large parts.

DIES AND DIE DESIGN CRITERIA

Forming Dies. The choice of material used in the construction of dies for high-velocity metal forming is governed by production requirements. Both ferrous and nonferrous, as well as nonmetallic materials, are used for dies. Sufficient strength must be built into the die to prevent permanent plastic deformation during the forming operation, although some elastic deformation is desirable and in fact contributes to the tolerance-holding capabilities of the die. If the die is permanently deformed even slightly during each forming operation, it will soon yield parts outside of tolerance.

The type of material to be formed and its condition and thickness, the amount of blank deformation, the number of parts to be made, and the dimensional tolerances on the workpiece have a great influence on the selection of the high-velocity forming-die material. The thickness of the workpiece material, together with its basic specifications, determines the basic load which will be exerted on the forming die. A safety factor should be designed into the die so that it will withstand the high dynamic load exerted during the forming operation. The effect of the shock wave transmitted to the die and the velocity of impact of the blank material on the die as it is formed in the die are both functions of the size and type of explosive charge. These also are determining factors for the selection of the die material.

Metallic Dies. Ferrous and nonferrous dies may be used for high-velocity forming operations. Ductile iron and high-strength steel castings, high-strength steel forgings, and dies made of laminated steel plates have been used. Low-melting-point alloys, zinc alloys, aluminum castings, and lead also may be used, depending upon the requirements and applications.

It is not possible to formulate a specific selection basis which may be applied to every fabrication problem. Metallic die materials, in descending order of anticipated die life, are: high-strength steel forgings, laminated steel plates, high-strength steel castings, ductile iron, cast iron, aluminum, zinc alloys, low-melting-point alloys, and lead.

Dies made of zinc and low-melting-point alloys are satisfactory for limited-production runs of light-gage material where the forming pressure does not exceed the yield strength of the alloy. If dies become obsolete, the material may be reclaimed and reused. These alloys will give good service provided there are no high-pressure points, the workpiece material is less than 0.125-in. (3.18-mm) thick, and a low production quantity is needed. The inherently greasy nature of these alloys is most desirable since dies containing them require little or no lubrication of the draw radius.

Patterns must be made for low-melting-point alloy castings, and good foundry practice must be exercised to produce a casting without porosity. These alloys machine easily and may be polished to a high luster. Dies of these materials are often made with a minimum wall thickness of 8 in. (203 mm) and must be well-supported to prevent any possible distortion. An outer casing of steel commonly is used.

Where the number of parts required is large, the workpiece material is greater than 0.125-in. (3.18 mm) thick, or the shape of the parts will result in high-pressure points, steel inserts or a wrought-steel, ductile-iron, or cast-steel die should be used. A draw ring, when required, should be made of heat-treated high-strength steel to reduce galling on the draw ring and to prevent draw-ring deformation.

Ferrous Die Materials. Some die designers believe that an important consideration in the selection of die materials is damping capacity, or the ability to absorb energy within the material. Pearlitic cast iron and low-carbon cast steel have a high capacity to damp out internal vibrations. Ductile iron works well from the standpoint of inherent strength and low elongation.

Because of the high strength of cast iron, the die may be shaped to the approximate contour of the finished part, as shown in Fig. 8-18. The low-elongation property of ductile iron makes it a desirable die material where the finished part has a moderate dimensional tolerance. This property, however, also may be a disadvantage since the minimum wall thickness should be at least 8 in. (203 mm). The exact choice of wall thickness depends upon the thickness of the workpiece material and a safety factor of 10 to withstand the dynamic load exerted on the forming die during the forming operation.

Ductile iron also is used for heated high-velocity forming dies because of its low

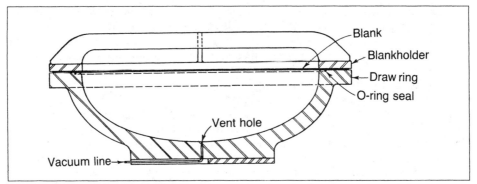

Fig. 8-18. Ductile-iron or cast-steel high-velocity forming die. (*Ryan Aeronautical Co.*)

coefficient of expansion and high resistance to oxidation at elevated temperatures. High-quality pattern making and foundry practice is required to produce ductile iron alloy castings of nodular quality without porosity. The advantages of ductile iron are its dimensional stability, high-impact strength, ease of machining, high corrosion resistance, self-lubricating qualities, and wear resistance. One limitation is the amount of ductile iron that can be poured at one time.

Normalizing of the finished casting prior to machining is recommended because normalizing generally imparts greater impact resistance, as well as higher yield and ultimate strength, than full annealing.

Medium-carbon low-alloy cast steel has excellent weldability, medium strength, and good durability and machinability. Because of the medium elongation factor, cast-steel forming dies require a greater wall thickness than ductile iron dies to maintain the deflection within the workpiece tolerance. This depends upon the workpiece thickness and material specification. A safety factor of 12 to 15 is recommended for the dynamic loading during the forming operation.

Quenching for hardening is not recommended. In medium-carbon steels, only a thin shell or case will be effectively hardened. The die is more likely to crack under a high impact.

High-strength steel dies or components should be heat treated to 270 to 290 Bhn to reduce die dimensional growth during use, and to eliminate excessive brittleness leading to die failure. Because of defects in castings which are sensitive to transmitted shock, forgings are preferred as die components by some users.

Nonmetallic Dies. The choice of nonmetallic materials depends upon production requirements and thickness and strength of the workpiece material. Common materials are thermosetting plastics, phenolic casting resins, and fiber-glass laminates. In most cases, because of the pressures generated during the high-velocity forming operation, these materials must be backed up with a material of sufficient mass and strength to prevent distortion and minimize deflection. Low-melting-point alloys and reinforced concrete often are used. Plastic die faces are light in weight and present but few problems in casting or layup, since shrinkage allowance is unnecessary. The material gives good results in the area of workability and stability of contour, regardless of changes in temperature and humidity. However, most plastic dies tend to chip or break down when subjected to a number of high-energy impacts. Plastic or plastic-reinforced dies should be considered only when the workpiece material is thin or of low yield strength or where the number of parts to be formed is very low. The advantages of this type of die are the low initial cost of fabrication and the light weight.

The surface of the plastic face must be free from porosity and have no internal defects. Since a plaster pattern is required to lay up reinforced plastics or to pour casting plastics, the surface condition and the dimensions of the die will be determined by the respective conditions of the plastic pattern. Close inspection should be maintained on the plaster pattern, the plastic mix specifications, and the method of fabricating the plastic surface. Maintenance of the plastic surface has few requirements. The requirements include: (1) good cover during outside storage and (2) diligent care in preventing the dropping of objects upon the plastic face (which might cause indentations or chipping).

When reinforced concrete backs up a plastic surface, the plaster pattern is inverted, a top pressure plate approximately 3 in. (76.2 mm) thick is positioned around the pattern, and the casting plastic or glass-reinforced plastic is applied. A container of approximately 0.250-in. (6.35-mm) thick steel is used to contain the concrete. Two courses of 0.50 in. (12.7-mm) diameter steel reinforcing rods are positioned, as shown in Fig. 8-19, around the plastic surface and are welded to the top pressure plate. While the partially completed die is still inverted, lightweight concrete is poured around the plastic exterior surface and allowed to cure. Within seven days the die is turned over, and the plaster pattern is removed. The exterior of the die is then poststressed, using prestressed cables. The glass-reinforced or plastic-faced die is ready for use approximately 14 days after the concrete is poured.

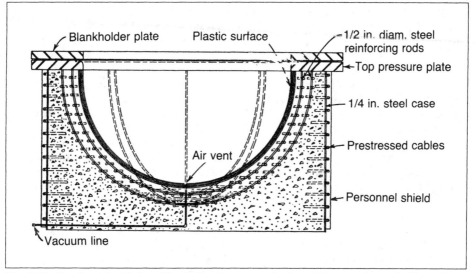

Fig. 8-19. Plastic-surfaced die with reinforced concrete backup. (*Ryan Aeronautical Co.*)

All exposed metal parts, including the prestressed cable, should be painted with rustproof paint to prevent corrosion. A shield should cover the prestressed cable as a safety precaution.

Die Design Considerations. In designing explosive forming dies, the die dimensions are held near the lower tolerance limit of the finished part. This allows for some growth of the die dimensions due to repeated use, while maintaining the finished workpiece within dimensional tolerance.

Allowances for springback are not needed in most high-velocity forming dies. The effects of springback are reduced by using a charge large enough to expand the die beyond the nominal dimensions of the part but within the elastic limits of the die. Elastic recovery of the die as well as recovery of the metal in the workpiece will produce a part to the nominal dimensions.

To reduce casting defects such as internal piping and cavitation to a minimum, it is desirable that a thick uniform cross-section be maintained.

In all forming dies and especially in truing dies, provisions must be made for air evacuation between the blank and the die. A vacuum of 28 in. (711 mm) of mercury is sufficient, in most cases, to prevent burning or welding from adiabatically heated air. Vent holes, approximately 0.250 in. (6.35 mm) in diameter, leading to a vacuum pump are usually drilled at the lowest part of the die or in an area where the blank velocity is at a minimum. The vent holes may be so located that the remaining entrapped air in the die is pushed by the shock front into the vent hole ahead of the material being formed. However, because of the speed of metal movement, simple venting of the die cavity is usually unsatisfactory.

The vacuum line may enter the die in the trim area, or the vacuum-line entrance into the die cavity may be plugged with a porous sintered powder metal insert. Parts can be designed so that forming operations are staged to prevent disfiguring the part at the vent hole. However, vent holes can be designed so that they are small enough not to cause difficulty.

Sealing of the blank to the die to prevent water from leaking behind the blank and to provide for air evacuation is mandatory. The die-cavity seal is a very important factor in high-velocity forming die design. Sealing may be accomplished with O-rings, tape, Permagum, drawing compound, or lubricant placed between the draw ring and the blank. For dish- or hemispherical-shaped dies, an O-ring seal can be made by machining a groove in the flat draw-ring surface as shown in Fig. 8-18. Cylindrical and conical expanding dies may have a more complex seal problem.

Blankholder. Wrought steel is commonly used as blankholder and draw-ring material. The thickness depends upon the workpiece material thickness and yield strength. In some cases, a simple truss arrangement across the blankholder must be employed to prevent distortion of the blankholder during the forming operation (as shown in Fig. 8-18).

Pressure must be applied to the blankholder to prevent wrinkles or puckering in the edge of the formed part and to facilitate sealing the die cavity for air evacuation. Blankholder pressure may be applied by hydraulically actuated lever clamps, hydraulically actuated deep-throated C-clamps, a hydraulic hold-down fixture, or bolts.

In forming thin material, it may be necessary to retard metal flow during the forming operation. Beads, similar to those used in conventional draw dies, may be used intermittently or continuously around the periphery, depending upon where the metal flow rate must be retarded to form the part.

Precautions must be taken to refrain from introducing threads or other stress riser configurations within the die in a direction perpendicular to the direction of shock travel. This precaution includes sharp corners on the exterior surface of the die.

Die-wall finish is critical. Die-wall finish affects the finish of the formed part. Tool marks, scratches, or foreign bodies, if present, will be embossed in the workpiece.

Inspection. Inspection procedures for metallic explosive forming dies must be flexible to conform to the die size requiring inspection. Inspection must be designed to detect internal piping and cavitation in the die material. On smaller dies, where it may be utilized, X-ray inspection is adequate. On larger dies or die components where X-rays will not be effective, ultrasonic inspection will be necessary, particularly in the thicker die sections and in the apex.

Experience has shown that imperfections, cavitations, and piping under 0.078 in. (1.98 mm) wide, with a maximum allowable number of imperfections of four per inch and a minimum allowable distance between imperfections of 0.031 in. (0.79 mm), will not be detrimental to the die life.

Maintenance. Maintenance of explosive forming dies is chiefly concerned with removing burrs on the blankholder and draw rings, replacing die components which become distorted, and maintaining the surface finish of the die wall.

Most fatigue failure in actual service occurs in the structure subject to stress concentration. Any steel or iron which has been allowed to corrode—and is then subjected to repeated stresses—will have a lower endurance limit than if corrosion had not taken place. The exterior surface of the die should be painted with a good exterior paint to prevent rust, oxidation, or corrosion. The interior of a die or forming surface should be protected for temporary outside storage or prolonged inside storage with one or two coats of an oil-based, zinc-chromate corrosion inhibitor. The dies should be covered so that a falling object cannot mar or indent the forming surface and to keep dirt and other foreign objects out of the die.

Die Weights and Associated Problems. Normally, lifting equipment or ramps are provided to handle dies being lowered into, and raised from, water-filled forming tanks. It is not feasible to raise and lower extremely large dies. There are two solutions to the problem.

First, by placing the die in an empty forming tank, assembling the blank to the die in the tank, filling the tank with water from a reservoir, forming the part, pumping the water from the tank to the reservoir, and finally disassembling the formed part.

Second, by placing the die above ground level and building a water forming tank that is either temporary or expendable, such as a plastic swimming pool, around the die. Appropriate site preparation may permit recovery of the water for further use.

FORGING AND EXTRUSION DIES

High-velocity forging dies are mostly of the closed type, although the more conventional flash and gutter impression type also are employed. The dies are very much like those used in conventional impact extrusion or cold forming. The most successful tooling assemblies incorporate shrink-fitted or press-fitted inserts into the dieholders. A fairly common tool set, consisting of a lower shoe, a shrink-type insert holder, an interference-fitted die insert, and an ejector, is shown in Fig. 8-20. The upper die consists of a simple punch having a clearance at the edge to permit the flash to flow around it.

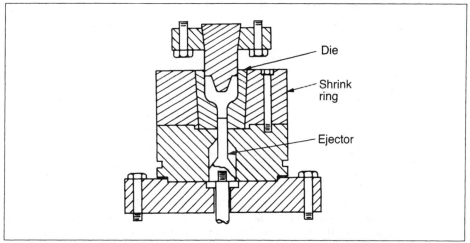

Fig. 8-20. Cross section of simple tooling assembly.[2,3]

Material-selection Criteria. Though not necessarily listed in order of importance, the following conditions should be taken into account in selecting the die material for a high-velocity forging die: mechanical shock, thermal shock, abrasion resistance, forging tolerances, number of forgings required, cost of die steel, die sinking sequence (before or after hardening), and type of material being forged.

Successful high-velocity forging dies have been made from chromium tool steels H11, H12, and H13 in applications where high stresses and long production runs are a consideration. Molybdenum high-speed tool steel M2 is used in warm forming operations. Cold forming operations often use D2 and D3 tool steel hardened to R_C 58-60 for dies and A2 or D2 hardened to R_C 56-58 for the punches. Nonstandard low-alloy tool steels (such as 6F2 and 6F3) are usually selected where parts requiring low forging stresses and short production runs are hot forged.

Type 6F3 is an all-purpose die steel with very good abrasion resistance and good shock resistance at high hardness. Resistance to heat checking is quite good. This material can be purchased in the heat-treated condition. It is commercially practical to sink prehardened die blocks as hard as 477 Bhn.

Type H11 is a medium-alloy, hot-work steel that has good resistance to heat checking when water-quenched during use. Shock resistance and abrasion resistance are fair to good, depending upon the hardness of the die block. This material has very good resistance to cracking and crack propagation.

Type H13 has a relatively high vanadium content, which makes this steel very resistant to abrasion. It has excellent impact strength and can be water-cooled in service. Resistance to heat checking is very good.

Type S1 has excellent toughness, which makes it a good punch material. Machinability is fairly good in the hardened condition. Shock resistance and abrasion resistance of this steel are very good.

Cropping blades are generally made of D2 and A2 tool steels hardened to R_C 54-56.

Fig. 8-21 shows the materials and hardness of a tooling assembly used for forging gear shafts.

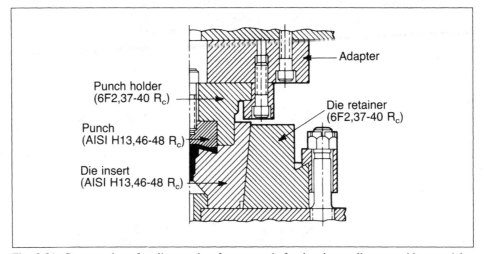

Fig. 8-21. Cross section of tooling used to form gear shafts showing tooling assembly material and hardnesses.[3]

Heat Treatment. Standard commercial practice should be followed when heat treating any die steel for high-velocity forging dies. It is advisable to give the die blocks a triple tempering to ensure that all the austenite has transformed. Stress-induced transformation of retained austenite to brittle martensite is particularly harmful to high-velocity forging dies as they are subjected to high-fatigue loads.

All heat-treating operations should be performed in a protective atmosphere to prevent decarburization and scaling. Where the die impression has been sunk prior to heat treatment, care must be taken to ensure that the furnace atmosphere has not been contaminated with nitrogen. Contamination may cause cracking.

Geometrical Design Considerations. Draft is generally not required on inside diameters. In some instances, one-half to one degree draft is required on the outside. There are several cases where reverse draft (hooking) is required on the outside to prevent the punch from sticking in the forging, as shown in Fig. 8-22. When a die is reverse-drafted, the forging is held by the walls of the bottom die while the punch is withdrawn. As the forging cools, it shrinks, thus allowing easy removal from the die.

In most cases, radii and fillets need not be greater than 0.125 in. (3.18 mm). Where there is danger of severe die wash, 0.25-in. (6.4-mm) fillets and radii should be used in the problem areas.

A completely confined web, surrounded by ribs, will require a blocked die to form the

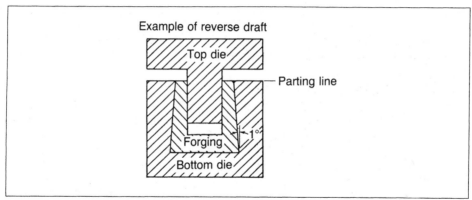

Fig. 8-22. Example of reverse draft in a closed forging die. (*Western Gear Corp.*)

web section. Otherwise, the web will move through the ribs, causing cracking.

The forging of disk configurations requires blocker dies when the height-to-diameter ratio exceeds 3. There is a tendency to form an hour-glass configuration when upsetting if the ends of the billet are not confined. By confining the ends of the billet, the center of the billet must move in an outward direction.

The distance between the die cavity and the outside of the die block should be minimum 2 in. (51 mm). Otherwise, the high compressive loading to which the dies are subjected during forging may cause premature die failure.

On some draftless forgings where the billet is extruded far into the bottom die, the top die may have a very coarse thread machined into it extending up about 3 in. (76 mm). Thus, when the punch is retracted, the forging is pulled from the bottom die and may then be unscrewed from the punch. This technique requires additional billet material and involves the cost of removing the threaded shank and the added die cost.

Finish of Die Cavity. The surface finish of the die cavity is determined by the number of forgings to be made and the finish desired on the forging. Because of the extremely rapid movement of metal during the forging operation, the minute voids in the die wall are not filled as they are in conventional forging operations.

On jobs where the number of forgings is less than 250, a highly polished die surface is not required. However, when the number of forgings exceeds 250, it is most advantageous to polish the die surfaces. A highly polished die surface leads to increased die life. Under no circumstances should there be any tool marks on the die surface.

Acceptance Inspection of Dies. A careful check should be made of a lead or plaster cast of the die impression. The dimensions of the casting should conform to the drawing of the forging to be made, taking into account a shrink factor added to the size of the die impression. This shrink factor will vary from material to material. The average allowance for steel forgings is 1.56%. Shrinking of the lead casting upon cooling or of the plaster while setting must be kept in mind. When it appears that there may be a lack of normal shrinkage allowance in the lead casting, the die impression should be checked very carefully by direct measurement. In general, the same procedures are followed for high-velocity forging dies that are used for drop-hammer dies.

Maintenance. Little maintenance is required for high-velocity forging dies. Where the billet material has welded to the die surface, it is removed by grinding. Occasional polishing may be required on long-run tooling.

Die Life. Experience indicates that die life for high-velocity forging dies is five times that of conventional forging dies. Usually the hot billet is in contact with a die for only a few seconds, which lessens the tendency for the die to become overheated and thus be drawn back or soften. Since most parts are made in one or two blows, the scale buildup on the billet is much less than would be encountered on billets forged under conventional circumstances. As the strain rate increases, so also does the yield strength of steel. Dies utilized on high-velocity forging equipment are subjected to very high loading rates, which tend to make them more resistant to deformation.

Following good die design principles should result in a die life of from three to seven times that of conventional hammer forging dies of a similar configuration.

SAFETY PRECAUTIONS

While explicit definitive information about safe practices and procedures is beyond the scope of this handbook, a few common-sense points should be considered by anyone considering the use of high-velocity forming. Everyone involved in a high-velocity forming process must understand and follow all applicable laws and rules regarding safe practices. All high-velocity metalworking processes depend upon stored energy, often in forms that may be unfamiliar.

Explosive Forming. The experienced explosives worker has been trained in the safe storage and handling of explosives. He or she is provided with the equipment necessary to carry out the operation reliably as well as safely. Proper training is a must before a new worker is permitted to operate the equipment.

One of the most important aspects of explosives safety which needs careful attention is the action to be taken in case of a misfire. Careful, detailed plans for such an occurrence should be developed in advance and followed in detail when required.

Much valuable assistance in both the training of personnel and the provision of the proper safe firing devices can be obtained from the manufacturers of industrial explosives, and their assistance—or that of other experienced personnel—should be sought by anyone planning to undertake such an operation.

Electrical Forming. The operator using the electrohydraulic or electromagnetic process must realize that the electric power generated is much like that contained in high explosives when released. It is recommended that the operator be protected by a heavy shield. There is always the possibility of a die fracturing and shrapnel-like fragments flying in all directions. When using an expendable coil for a swaging operation, it is mandatory that a shield be placed over the workpiece to limit the flying fragments.

Because of the danger of the capacitor being overcharged or the case being weakened by repeated firings, to protect the operator, the cabinet must be constructed of heavy material that will contain the metal fragments in the events of a capacitor blowout.

Manufacturers and users of power supply and switching consoles for electrohydraulic and electromagnetic forming equipment must take the necessary measures to protect the operator from the dangers of stored high-voltage electrical energy.

Pneumatic-mechanical. The general safety measures associated with the pneumatic-mechanical process are similar to those drop-forge, punch-press, or hydraulic-press operation. A proper shield should be mandatory to protect the operator from flash during forging or extruding. All closed-cavity dies must be vented to ensure against ruptures caused by hydraulic pressures developed with semimolten metals. Punch and die clearances must be carefully checked since the punch must not bottom. Serious ruptures of hardened punches have occurred.

SCALING LAWS

General Considerations. Generally, scaling permits the development of the necessary forming parameters on a scale model and the subsequent accurate application of the developed data to a full-scale component. For scaling to be applied effectively, the following requirements must be met.

1. Complete geometrical similitude is required between the scale model and the full-scale components.
2. The ratio between the explosive standoff distance and the die-opening diameter must remain constant.
3. The same explosive must be used on the full-scale component as on the model.

4. The blank yield strength, modulus of elasticity, and ductility must be the same for all sizes.
5. The ratio of the blank thickness to the die-opening diameter must remain constant.
6. The clamping-system stiffness must remain the same.
7. The coefficient of friction must remain constant.
8. The hold-down force increases by the square of the scale factor. A force of 5,000 lbf (22.2 kN) on a 6 inch (152 mm) die would require 500,000 lbf (2,220 kN) on a 60 inch (1,524 mm) die.

The application of scaling to high-explosive forming has permitted the economical development of many parts that would otherwise have required considerable experimental data on full-scale parts. Further steps to refine high-explosive forming by computerizing the major factors affecting metal deformation have resulted in computer programs for predicting metal-deforming parameters.

References

1. W.W. Wood, et al.: "Theoretical Formability," Vol. II, Chance-Vought Corp., Dallas, 1961 ASD TR 61-191 (II).
2. E.J. Bruno, et al., *High-Velocity Forming of Metals* (Dearborn, MI: Society of Manufacturing Engineers, 1968).
3. Charles Wick, John W. Benedict, and Raymond F. Veilleux, eds., *Tool and Manufacturing Engineers Handbook*, 4th ed., Volume 2 (Dearborn, MI: Society of Manufacturing Engineers, 1984).

SECTION 9
FORMING DIES

Simple Flanging Dies. Figure 9-1 shows a die for simultaneously forming and stretch-flanging a right-hand and a left-hand part from 2024 clad aluminum alloy, 0.064 in. (0.163 mm) thick, in a 250-ton (2,200-kN) double-action hydraulic press. The previously developed blank was annealed at 650°F (343°C), with edges polished at A and B to prevent cracks from forming. Deformations were: stretch, 43% at A, 55% at B, 12% at C; decrease in thickness, 22% at A, 28% at B, 11% at C.

The spring-loaded pad attached to the inverted die holds the web of the blank as the

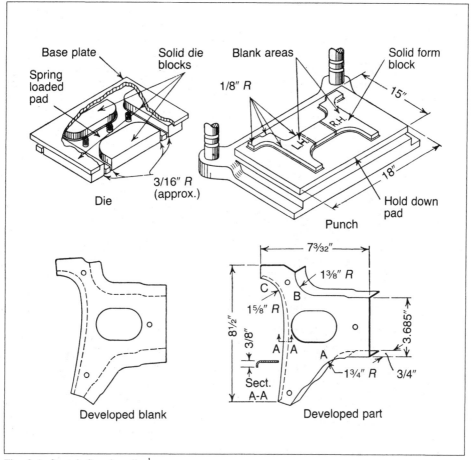

Fig. 9-1. Stretch-flanging die.[1]

solid die blocks form the flanges down. Gages or nests for this elementary die are not shown. It is not capable of high production, and the finished part is removed by hand. A high-production design would incorporate a pressure pad attached to the lower shoe and an automatic stripper (such as stripping hooks) to remove the part from the solid form blocks.

Figure 9-2 shows tooling for shrink-flange forming of a part from 2024 clad aluminum alloy, 0.032 in. (0.081 mm) thick, in a 40-ton (356-kN) crank press. The part is produced from a developed flat blank. The major shrink strain at the closed end of flange was 25%, necessitating use of deep-drawing technique. Pressure is supplied by a spring pad and pins from a cushion.

Drawing and/or stretching of the metal may occur in the die, depending upon the amount and location of the net hold-down pressure exerted.

Die designs for the forming of curved flanges commonly incorporate a spring or pneumatic pad for clamping the web, to prevent slippage or distortion. Sometimes, adequate clamping pressure can be secured by forming right-hand and left-hand parts in the same blank, and subsequently trimming them apart (Fig. 9-3).

Flanging an Automotive Part. A design for the quantity forming of a slightly bowed and tabbed flange is shown in Fig. 9-4. Gage pins (*D6*) enter the holes in the blank, and gages (*D7*) position the blank and prevent its slipping. The flange (*D1*) and its tab (*D2*) are formed upward and outward by the punch (*D4*) against the die (*D3*) as the punch forces the pad (*D5*) down. Pressure is supplied to the pad through air pins (*D8*) by a die cushion. The part falls out of the back of the die used in an inclinable press.

Hole-flanging Die. The formation of a 0.312-in. (7.92-mm) hub or circular flange around a 0.250-in. (6.35-mm) -diameter hole is accomplished in the die shown in Fig. 9-5. The inverted flanging punch (*D1*) enters the hole in the workpiece (a fan of 0.037-in. (0.94-mm) -thick deep-drawing hot-rolled steel).

Slitting and forming of the blades as well as spot welding the fan back are done in subsequent operations. A die cushion actuates the stripper (*D3*) and the part is removed by a knockout (*D2*).

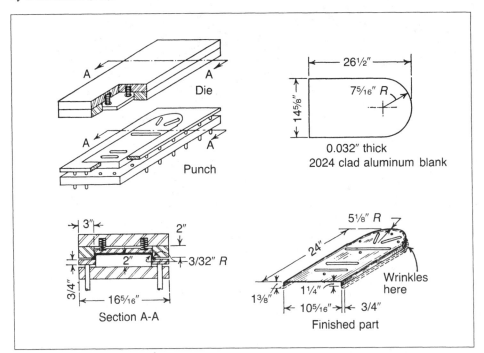

Fig. 9-2. Shrink-flanging die.[1]

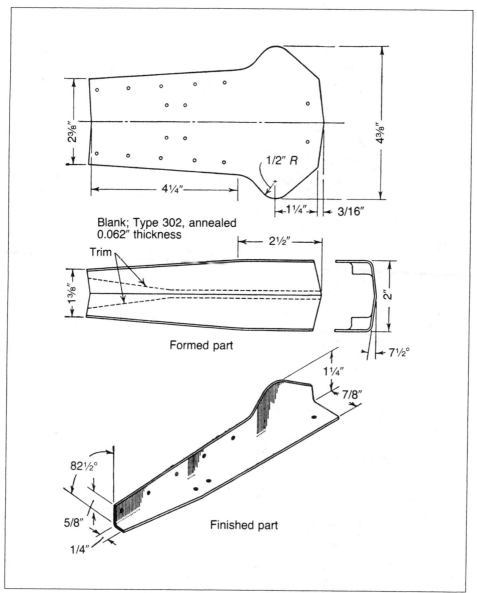

Fig. 9-3. A right-hand and left-hand part laid out for simultaneous forming of stretch flanges.[2]

Flanging Large Flat Parts. The forming of flanges around the periphery of refrigerator doors, home-freezer lids, and similar parts is frequently done in dies equipped with a number of cams. The cam action shown in Fig. 9-6 is not in the same direction as the action of the vertical forming punch. Cam return (not shown) is by spring action or by the positive return stroke of the driver.

Double-cam Flanging Die. The die shown in Fig. 9-7 has, in addition to the out-side cam, an inner cam which collapses when the die is open. The inside cam action is required whenever the part is locked on the punch by interfering flanges, offsets, or shapes.

Forming a Flanged Cylinder. A circular flat blank is formed into a flanged cylinder in the die shown in Fig. 9-8. A double-action press, equipped with a rubber die cushion, performs a combined drawing and flanging operation. The circular blank is loaded in the nest ($D1$) and centered on the pin on the lower knockout ($D3$). The outer ram of the

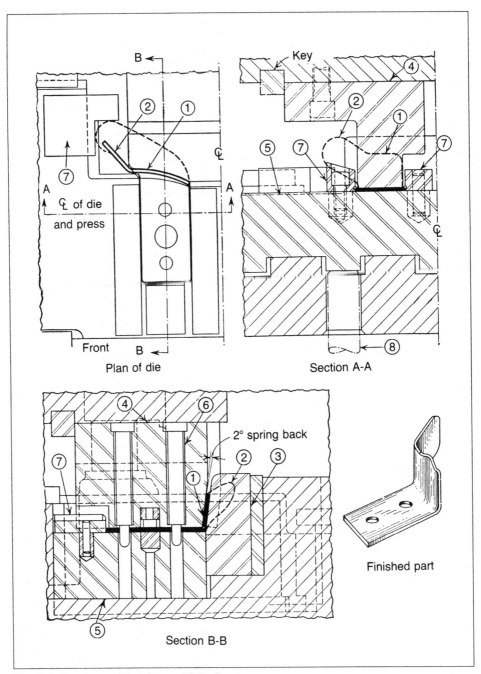

Fig. 9-4. Forming a tabbed flange. *(C.R. Cory)*

double-action press descends to lower the spring pad (*D*4) and draw sleeve (*D*5). The spring pad grips the blank as the draw sleeve draws the part to the 1.4145-in. (35.928-mm) dimension. The part is then gripped between the draw sleeve and the step in the die, and the inner punch (*D*6) descends to draw the 1.1785-in. (29.934-mm) dimension and complete the part.

Flanging Shim Stock. Two-inch strip shim stock 0.003 in. (0.08 mm) thick is blanked and pierced by 0.069-in. (1.75-mm) -diameter nail punches (*D*1, *D*5) which also forms

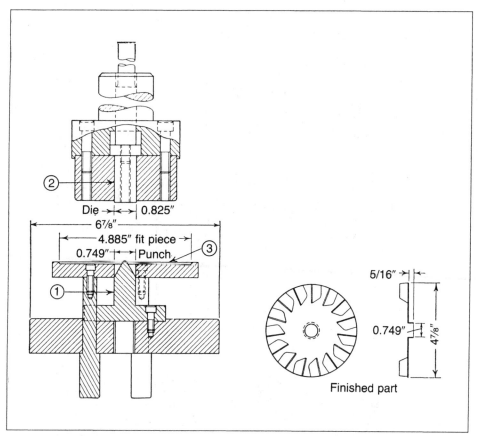

Fig. 9-5. Hole-flanging die. *(Century Electric Co.)*

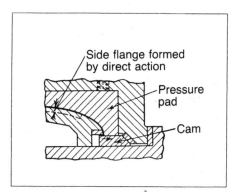

Fig. 9-6. Cam flanging die.[3]

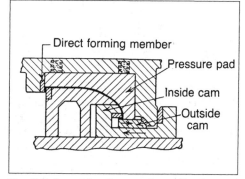

Fig. 9-7. Internal and external cam flanging die.[3]

flanges around the hole to a height of 0.036 in. (0.92 mm) (Fig. 9-9).

The holes are pierced without the cutting of any slugs, by the use of a four-sided tapered point on the punches. The 0.069-in. (1.75-mm) flanged holes shrink to a diameter of 0.067 in. (1.7 mm) after the flanges are formed and the punches withdraw. Spring-loaded ejector pins (*D2*, *D6*) are used in this die design, which achieves piercing and flanging in opposite directions by the use of a lever (*D3*) actuated by a pin (*D4*).

Forming an Inward and a Straight Flange. An inward flange is formed by a cam

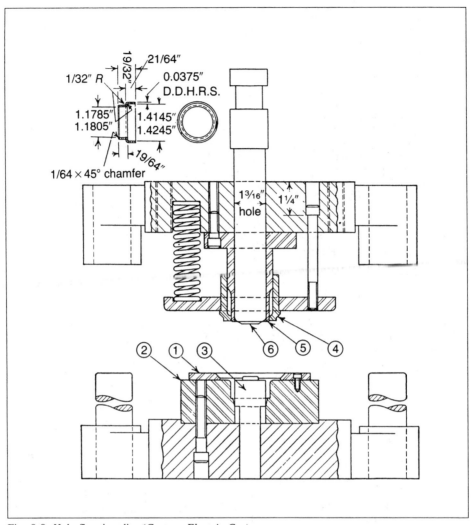

Fig. 9-8. Hole-flanging die. *(Century Electric Co.)*

forming member sliding on a horizontal surface of the die shoe in the die shown in Fig. 9-10. The cam driver, with a driving angle equal to or greater than the angle of the flange to be formed, forces the cam flanging member inward as the air-pressure pad is forced down. The punch is split into two parts: a spring-loaded punch, and a solid punch, which forms the straight flange. This design prevents the part from sticking to the split punch. On the upstroke, the pad travels upward carrying the flanged part above both the solid flanging member and the cam flanging member so the finished part can be removed.

Return Flanging Die. The dies shown in Figs. 9-11 and 9-11A, for flanging a freezer-lid panel, are notable for their low required shut height of 16 in. (406 mm), as compared with 24 in. (610 mm) required by conventional design.

The 0.036-in (0.91–mm) cold-rolled-steel part (previously pierced, drawn, and redrawn) is placed in the die with its flanges down. The die incorporates a spring pad traveling ahead of any action. Six side and end cam drivers (D6) and four corner cam drivers (Fig. 9-11, D7) arrive simultaneously, pulling all the center collapsible sections outward. The cam drivers dwell while the outer cam slide flange sections travel inwardly, forming the four flanges under one press stroke.

On the upstroke, the outer cam slides move outward after the inner collapsible sections

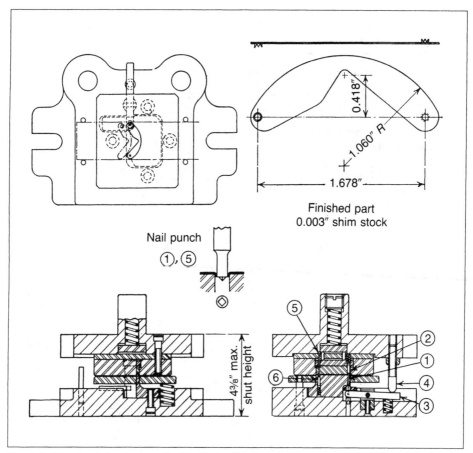

Fig. 9-9. Flanging holes in opposite directions. *(Harig Mfg. Corp.)*

recede under spring pressure toward the center of the die, permitting the part to be unloaded. Figure 9-11A shows the cam drivers (D1) and the corresponding external and internal collapsible or retracting forming members (D2, D3) for the rounded corners of the lid. Similar sets of forming members (D4, D5) flange the straight sides of the lid. This die in a 150-ton (1,334-kN) mechanical press attended by two operators produces 120 pieces per hour.

Roller Die for Flanging Inwardly. The swiveling die members (D1) in the die shown in Fig. 9-12 are rollers which flange 0.031-in. (0.79-mm) copper blanks placed between the location pins (D2). The work is bent to form the completed clip by the revolving rollers, actuated by the punch (D3). The part is slid endwise off the punch.

Typical Standard Cam Forming Die. Cam-actuated forming-die designs are standard designs with many manufacturers; Fig. 9-13 illustrates a typical design. Either spring shedder pins or conventional strippers are incorporated (not shown) if the flange angle is close to 90°. The die can be wrecked if stock thickness is not uniform.

Double-cam Forming and Setting Die. After the vertical punch (Fig. 9-14, D1) has completed its forming function, horizontal punches (D2) actuated by two cams (D3) set the legs of the part. A floating member (D4) with a vertical travel of 0.750 in. (18.95 mm) (A) limits cam movement through pad action to allow for thickness variations in the part. Legs are set to ±0.003 in. (±0.08 mm) from the center line. Nest plates (D7, D8) position the part, and it is removed by stripping brackets (D5, D6).

Die for Press-brake Flanging. Flanges and joggles in aluminum strip (Alclad 7075) are formed in the press-brake die shown in Fig. 9-15. The blanks are heat-treated and

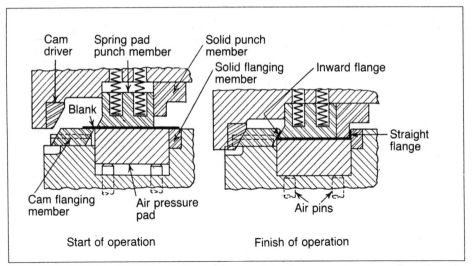

Fig. 9-10. Forming an inward and a straight flange (C.R. Cory)

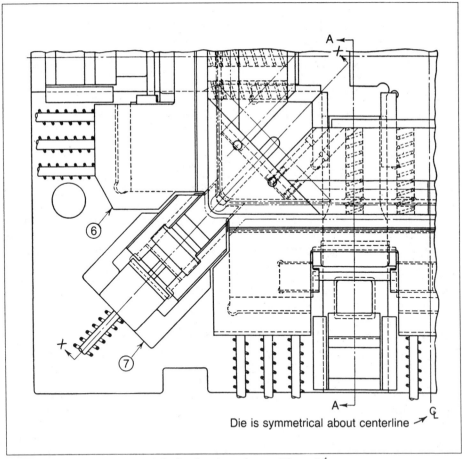

Fig. 9-11. Quarter-plan view of die for flanging a freezer-lid panel.[4]

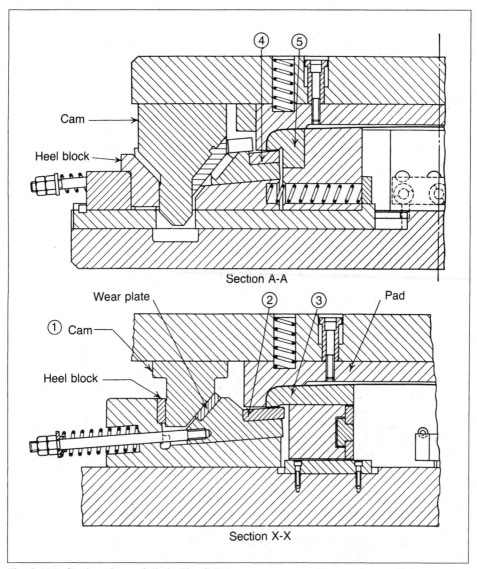

Fig. 9-11*A*. Section views of die in Fig. 9-11.

stored at approximately 0° F (−18° C) so they will not age-harden before forming is completed. An insert (*D*6), in the joggle die (*D*2), in conjunction with a pressure pad (*D*4), the flange die (*D*3), and the punch (*D*5), form the flanges and joggles. The part is positioned at each end by locating pins (*D*1) and removed by the stripper (*D*7).

Embossing and Flanging Die for Channel. The die shown in Fig. 9-16 performs the first (embossing) and third (flanging) operations in producing a channel structural member of 0.064-in (1.62-mm) 3003-H14 sheet aluminum. A second (trimming) operation is performed in a die not shown. The embossing is done with a long shim block (*D*1) inserted between the die shoe and the embossing die. The third (flanging or channel-forming) operation is performed with the shim block (*D*1) removed (section *B-B*), permitting the punch member to enter the lower die member to form the channel sides.

Overform Flanging Die. Cabinet sheet steel is overformed from a flat blank positioned by suitable nests (not shown) in the die illustrated in Fig. 9-17. The pressure pad (*D*2) is actuated by pressure pins (*D*1). The 90° flange is formed in another similar

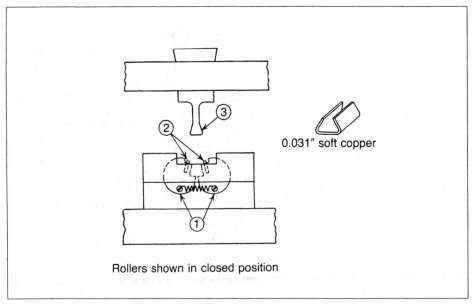

0.031" soft copper

Rollers shown in closed position

Fig. 9-12. Swiveling-roller die for flanging.

die, followed by a restrike operation to set the angle and dimensions of the 90° flange.

Forming Permanent Magnets. Magnet steel 0.320 in. (8.13 mm) sq. by 5.25 in. (133 mm) long is formed between hinged form blocks ($D1$) and the punch ($D2$, Fig. 9-18) into a U-shape, after the blank is placed between the gages ($D3$). A cam-actuated knockout ($D4$) strips the work from the vertical punch and is retracted by springs ($D5$). The part is snapped to the right and comes to rest around the two locating pins ($D6$). The U-shaped blank is sized, on the next press stroke, between the flattening punch and die ($D7$ and $D8$).

Forming Brush-holder Bracket. The four flanges on the brush-holder bracket are formed simultaneously in the die of Fig. 9-19. The precut blank is prelocated by two channel-shaped locators ($D1$). To prevent the mislocating of the blank in the die, a foolproofing pin ($D2$) must enter the slot in the flat blank. As the die closes, locating pins ($D3$) in the punch ($D4$) enter holes in the blank to precisely locate it during the forming operation. A forked stationary stripper ($D5$) assures the removal of the workpiece from the punch.

A spring-loaded pressure pad ($D6$) holds the blank firmly against the punch during the forming operation and ejects the part from the die block ($D7$).

To facilitate machining, the die block is made of four pieces of oil-hardening tool steel and is hardened and ground. The punch, pressure pad, and locators also are made of oil-hardening tool steel and are hardened and ground.

The punch and die blocks are cap-screwed and doweled to a standard two-post commercial die set.

Die Forming of Gold Sheet. A blank made from 0.028-in. (0.71-mm) -thick 14K gold is placed in a nest (Fig. 9-20, $D9$). As the ram descends, the part is U-formed by the spring-loaded punch ($D6$). Near the bottom of the stroke, this punch dwells while the double-faced cam ($D1$) forces two slides ($D4$, $D5$) inward to offset the part. A cam-actuated knockout ($D3$) strips the part from the punch and ejects it toward the rear of the die. The knockout has a spring return ($D7$, $D8$).

Die Design for Tube Flanging. Steel tubing is clamped by a die member ($D1$) and a spring-loaded pad ($D2$) in the die shown in Fig. 9-21. Retreating gages ($D3$) return automatically after the tubing is flanged and expanded on the right and left ends, respectively, by suitable punches ($D4$, $D5$), which are actuated by cams ($D6$, $D7$). Note

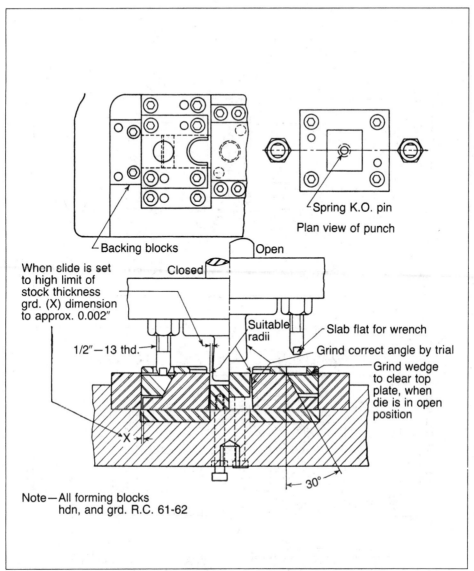

Fig. 9-13. Standard cam forming die.[5]

that the right-hand flanging die is cammed into the part and then dwells for 0.188 in. (4.78 mm) of the press stroke so the left-hand flanging die can completely enter the part, then both dies can continue on in, flanging the tube on both ends.

Forming and Flattening Operations on Tubing. The die shown in Fig. 9-22 is used with a die cushion to form stainless steel preformed tubing and also to flatten a section by swaging operation. Fractures at the corners have resulted from the use of non-homogeneous stock.

Forming Die for Motor Frame Foot. A developed blank (Fig. 9-23A) is engaged by two pilots incorporated in the forming die of Fig. 9-23. The second and final forming of the part is completed in a die of identical design except that punch and die radii are shorter and it is equipped with suitable punches and die inserts (Fig. 9-23) for forming welding projections. The part is trimmed and slotted in two subsequent operations.

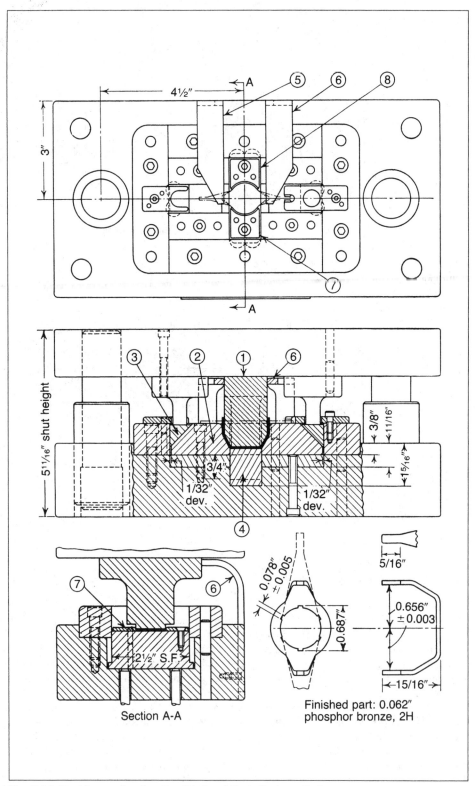

Fig. 9-14. Double-cam forming die. *(National Cash Register Co.)*

9- 12

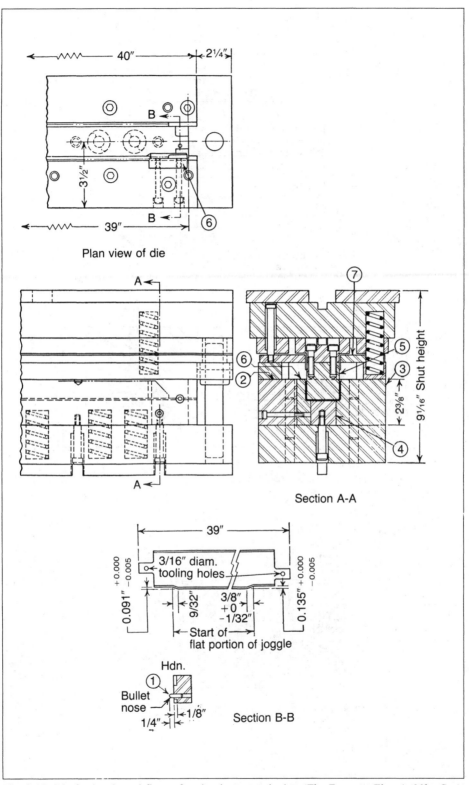

Fig. 9-15. Die for joggle and flange forming in a press brake. (*The Emerson Electric Mfg. Co.*)

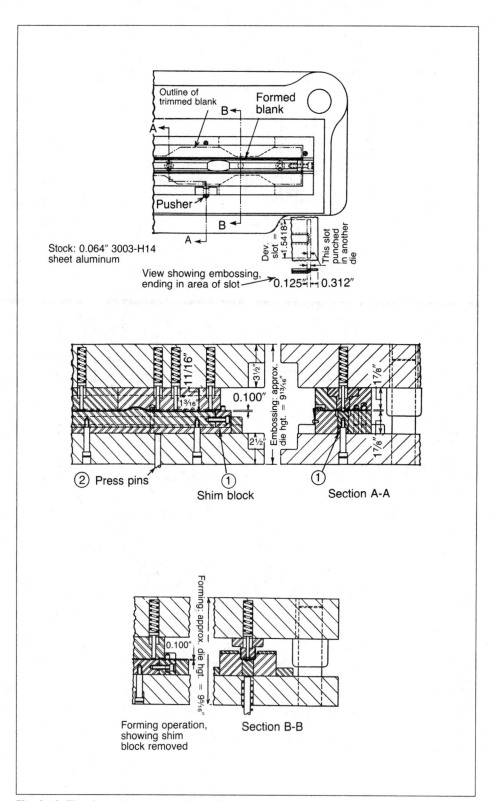

Fig. 9-16. Flanging and embossing die. *(Harig Mfg. Corp.)*

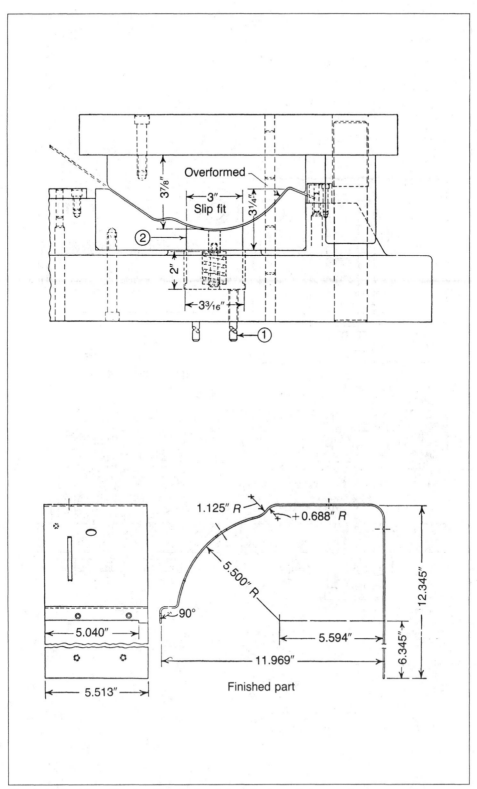

Fig. 9-17. Overform flanging die. *(National Cash Register Co.)*

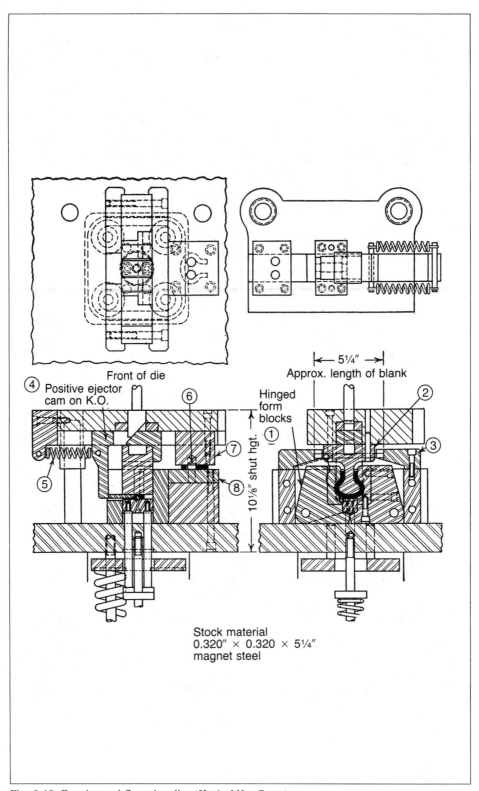

Front of die

④ Positive ejector
cam on K.O.

⑤

⑥

⑦

⑧

|← 5¼" →|
Approx. length of blank

Hinged
form
blocks

①

②

③

10⅛" shut hgt.

Stock material
0.320" × 0.320 × 5¼"
magnet steel

Fig. 9-18. Forming and flattening die. *(Harig Mfg. Corp.)*

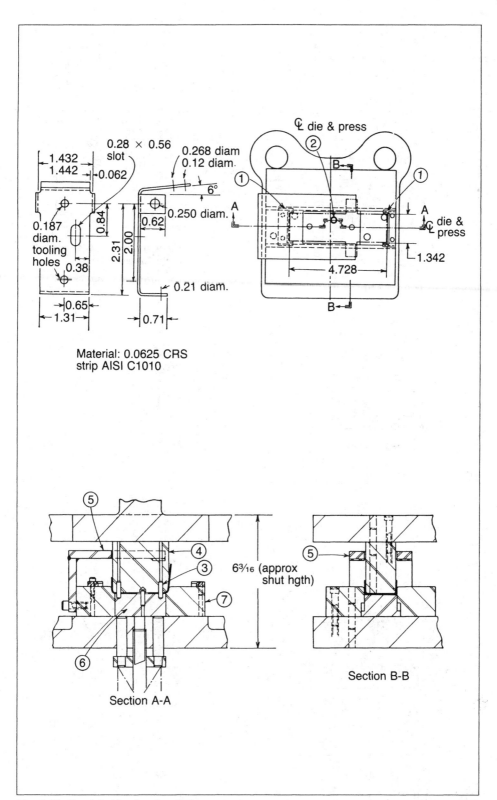

Fig. 9-19. Brush holder forming die.

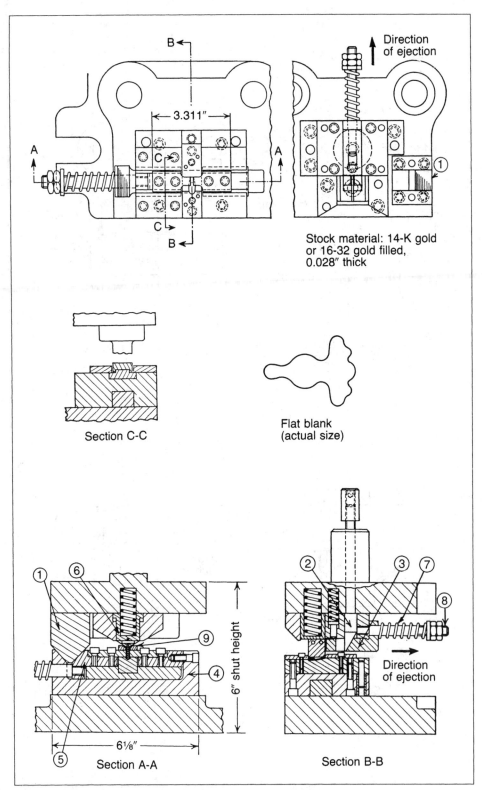

Fig. 9-20. Single-cam double-slide forming die. *(Harig Mfg. Corp.)*

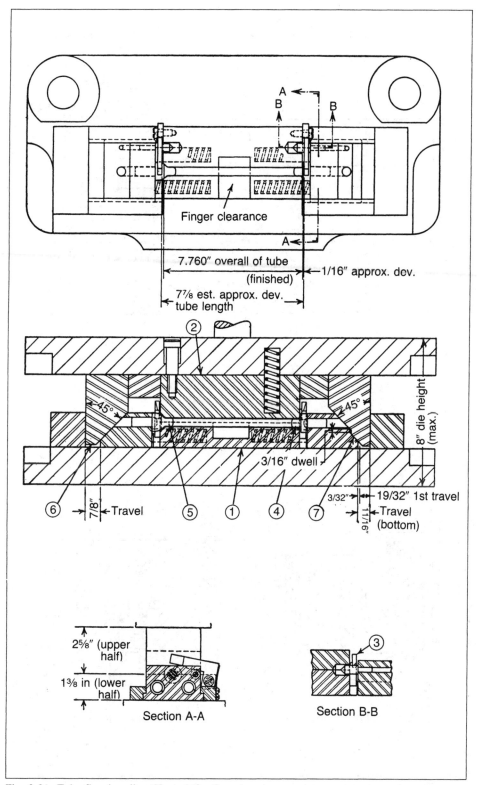

Fig. 9-21. Tube-flanging die. *(Harig Mfg. Corp.)*

9- 19

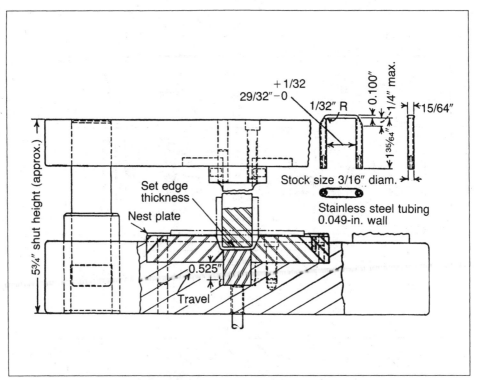

Fig. 9-22. Forming and swaging die. *(National Cash Register Co.)*

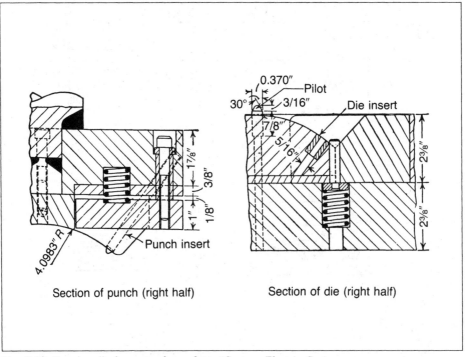

Fig. 9-23. Forming die for motor-frame foot. *(Century Electric Co.)*

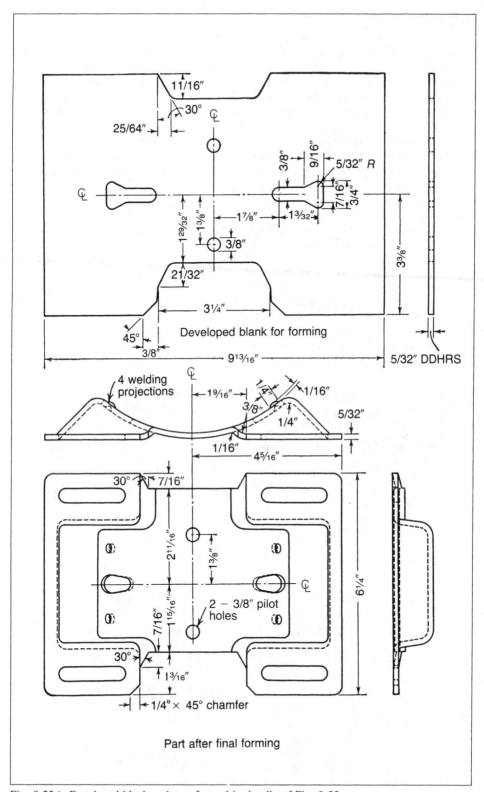

Fig. 9-23A. Developed blank and part formed in the die of Fig. 9-23.

CURLING DIES

Curling the edges of a flat or curved sheet, tube, or shell is essentially a flanging operation. It consists of an expansion of the metal followed by its reduction. The term *reduction* is not to be confused with its use in connection with drawing operations.

Curling-punch Design. The diameter of the groove in a curling punch should equal the curl diameter of the part. The groove diameter should not be less than 0.062 in. (1.57 mm) for a stock thickness of 0.010 in.(0.25 mm); the groove diameter should not be less than 0.125 in. (3.2 mm) for 0.018-in. (0.46 mm) stock thickness, according to Mills. Crane states that the curl diameter should be from 10 to 20 times metal thickness. When the groove is too large, the curl forms to its own diameter (Fig. 9-24A).

Burr and Flare Direction. The optimum condition of a shell or tube for curling occurs when both its burr and flare are in the direction of curling.

Outward curling of a shell periphery is the easiest curling operation, particularly if the shell has been drawn in a single-action die.

Inward curling on an interior flange (Fig. 9-24B) is easier to accomplish, when the flare is inward, than an outward curling.

Die-design principles for the inward curling of shallow and deep shells are shown in Fig. 9-25. If the piece edge has been nipped, or has a flare in the right direction, the use of a tapered plug may be unnecessary.

Inward Curling of a Deep Shell. This operation is achieved in the die of Fig. 9-26, which is equipped with an expander (D1) that slides on the top surface of a supporting plug (D2). A heavy spring (D5) presses against a tapered plug (D3) to provide dwell for holding the expander open before the curling punch engages the work. After the curling

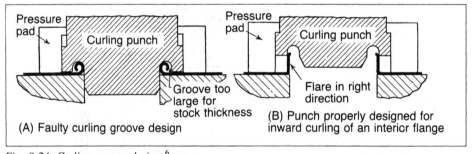

Fig. 9-24. Curling-groove design.[6]

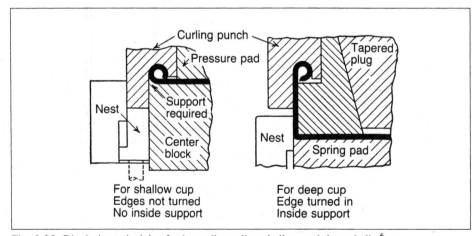

Fig. 9-25. Die-design principles for inwardly curling shallow and deep shells.[6]

9- 22

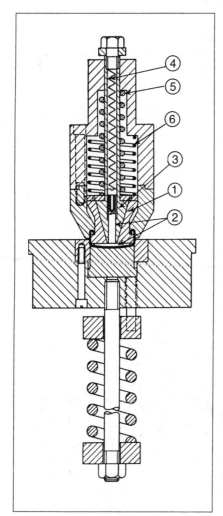

Fig. 9-26. Inward-curling die for deep shells.[6]

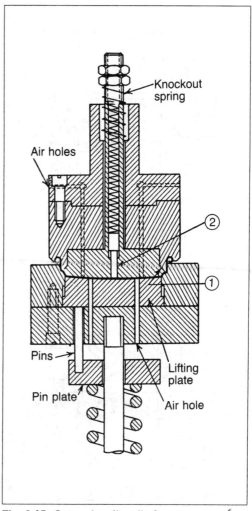

Fig. 9-27. Outward-curling die for a can cover.[6]

punch has cleared the work on the upstroke, a spring ($D6$) causes the expander to dwell until it is collapsed by the upward movement of the curling steel. The light spring ($D4$) merely retains the supporting plug ($D2$).

Curling a Can Cover. A heel on the curling punch (Fig. 9-27, $D1$) supports the inside of the work (a can cover) which is to be curled outwardly. A pocket in the die supports the outside of the work, which is stripped by an adjustable-spring counter-balanced knockout and oil-seal breaker pins ($D2$). During the working stroke, the part is gripped between the lifting plate and the knockout. On the upstroke, the lifting plate ensures positive part movement upward with the punch so that it can be ejected.

Die for Forming Partial Curls. Cam-actuated punches ($D1$, $D2$) form two partial curls in 0.012-in. (0.3-mm) -thick 8H phosphor bronze in the die shown in Fig. 9-28. The part is gaged into location without manual adjustment. The hand ejector mechanism, pushing the part out from the back of the die, functions only when the part is fully formed. Note that the part has been nipped to facilitate producing the curl on the right-hand side.

Curling and Flanging Die with Retractable Holding Jaws. A holding clamp (Fig. 9-29, $D1$) opened by a plunger ($D2$) allows a curl to be formed on one end of the part by a curling punch ($D3$). A flanging punch ($D4$) flanges the sides of the part (a pen clip).

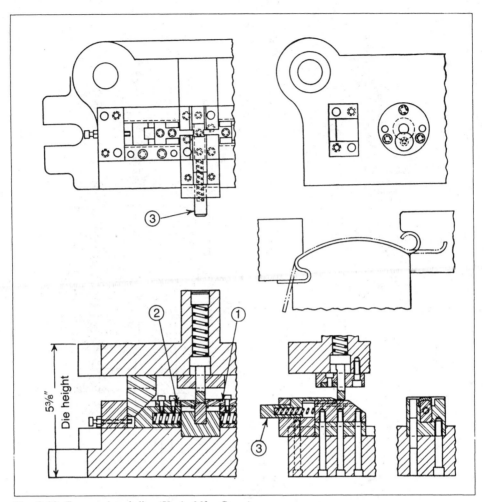

Fig. 9-28. Form-and-curl die. *(Harig Mfg. Corp.)*

Hinge-forming Die. A revolving lower curling steel (Fig. 9-30, *D*1) is loaded while in the horizontal position. The part to be curled is located over two locating pins (*D*4) and backed up by a support plate (*D*3). The anvil block (*D*8) has a slot to provide clearance for the locating pins (*D*4). This anvil block is free to move in a vertical direction and is held against its upper stops by springs (*D*9). On the downstroke, a plunger (*D*2) swings the lower curling steel (*D*1) to a vertical position. The lower curling steel is then forced against the anvil (*D*8) by the heel (*D*5) on the die block (*D*6). The upper curling die then contacts the anvil and drives it downward, curling the part as it goes. The entire force of the curling operation is carried by the support plates (*D*3) which must be securely doweled in place. Since the work is loaded and unloaded in a horizontal position and away from the closing members of the die, this design provides for operator safety.

Curling Channel Edges. Slots in a typewriter part (Fig. 9-31) are gang-milled after three die operations curl the edges. The first operation (not shown) forms the flat blank into a channel shape, with radii equal to one-half of the ID of the final curl (0.1575 in., 4 mm). The part is placed, legs down (view *A*) in a nest consisting of a split die and cam-actuated forming punches. The spring-loaded plunger (*D*1) forces the part and the expanding die steels (*D*2) downward. The die steels are expanded by an adjustable wedge (*D*3) and hold the part while the outer die halves are cammed in to form the start of the inward curl. The curl is completed by a solid die and a conventional curling punch (view *B*).

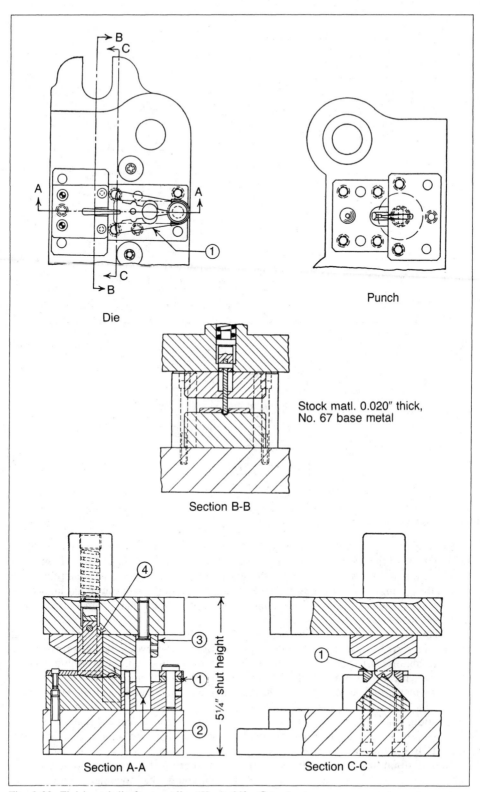

B
C
A · · · A
1
C
B

Die

Punch

Stock matl. 0.020″ thick,
No. 67 base metal

Section B-B

4
3
1
2

5¼″ shut height

Section A-A

1

Section C-C

Fig. 9-29. Finish-curl die for pen clip. *(Harig Mfg. Corp.)*

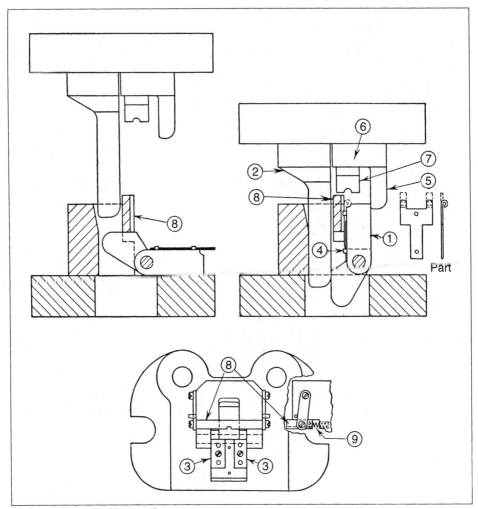

Fig. 9-30. Hinge-forming die.[7]

Inward Curling of a Shallow Part. This operation will distort and bulge the curl (Fig. 9-32A) if any of the following conditions are present: the curling punch is too loose or too tight; there is too much or too little material; the pressure pad does not clamp the work before the curling punch engages the work; or the flange of the workpiece has high spots. Rubber suction cups, assisted by spring-loaded pins, lift the part from the die and a knockout removes the part from the cups.

Wiring Dies. Strickly speaking, a wiring die is a curling (or false-wiring) die in which a wire is inserted in the edge to be curled either by hand or by a suitable spring pad, so the part edge is curled around the wire. Such an operation is simply forming a shrink flange which can be formed without wrinkles provided the flange width is less than 10 times the metal thickness; it is commonly done on annealed metals such as the wiring of pails, pots, and similar utensils.

Seaming Dies. Seaming dies usually join or crimp straight or curved flanges together, although it may be an operation of hemming only, in which case the flange is bent back upon itself and flattened to give the part increased strength.

Longitudinal Seaming of Tubes. Horn-type die designs for seaming round tubes are shown in Fig. 9-33. The edges of the tube are bent and hooked together before it is slid on the horn. A slot is machined in the die and the punch is radiused for inside seaming.

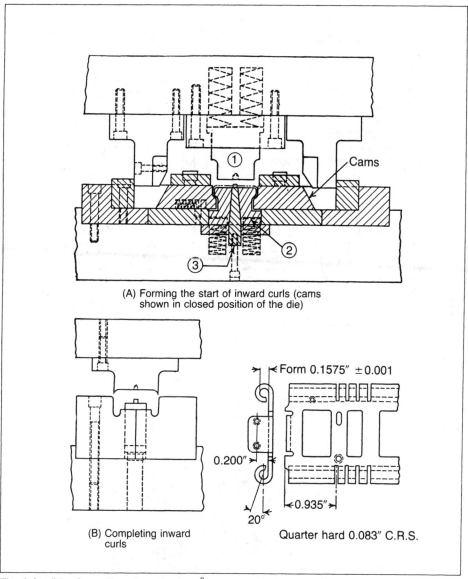

(A) Forming the start of inward curls (cams shown in closed position of the die)

Cams

(B) Completing inward curls

Form 0.1575″ ±0.001

0.200″

0.935″

20°

Quarter hard 0.083″ C.R.S.

Fig. 9-31. Dies for curling channel edges.[8]

The outside-seaming die is provided with a smooth horn and a suitable slot in the punch.

"Zipper" Closing and Seaming Die. A preformed U-shaped blank with bent edges (Fig. 9-34, section A-A) is formed into a cylinder and the edges are lock-seamed in the die shown. The driving shoulder of the arbor-like punch forces the blank inside the guide plates and into the die block. Chilled-iron inserts (sections C-C to G-G) fold and seam the edges. Rollers (sections H-H and I-I) flatten the seam. The finished tube is forced out of the die by the following workpiece. The special vertical hydraulic press used has a 44-in. (1.2-m) (maximum) stroke.

Forming Lightening-hole Flanges. Figure 9-35 shows a die for blanking out and forming a flange or bead around a lightening hole. The part is located by a 0.125-in.(3.18 mm) -diameter pin (D1) which is press-fitted into the tool-steel punch (D2). A spring-loaded pressure pad (D3) holds the sheet against the lower forming die (D4) while the combination blanking die and forming punch (D5) performs its functions. The blanking

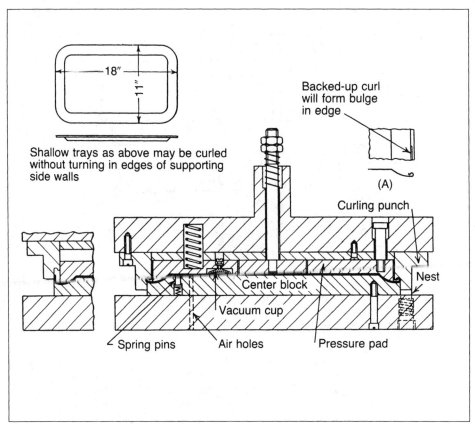

Fig. 9-32. Inward-curling die for shallow parts.[6]

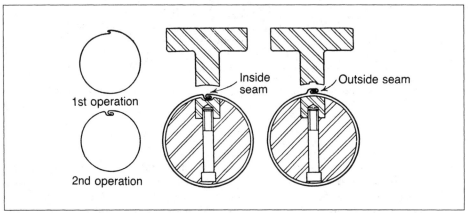

Fig. 9-33. Dies for seaming round tubes.[9]

punch has an air vent to assist in removing the blank from the locating pin. The positive knockout device (D6) ejects the blank from the blanking die. The forming of the flange pulls the metal away from the blanking punch. No stock lifters are needed in that area.

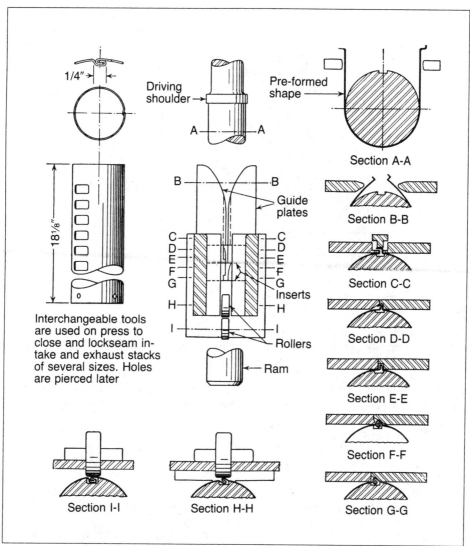

Fig. 9-34. "Zipper" closing and seaming die.[10]

BEADING DIES

Beads are rather shallow recesses in a metal sheet, usually rounded troughs of uniform width which may be either straight or curved. The possible depth and contour of beads that can be successfully formed depends upon the permissible amount of stretching of the metal and the method of forming.

The formation of beads, corrugations, bosses, and similar recesses in parts of certain types is practical by drawing (Section 12), by forming (Sections 8 and 14), or by special methods of combinations thereof (Section 21). Suitable techniques and die designs are included in the sections listed.

Press-brake Beading Dies. A typical die for this operation is shown in Fig. 7-1*J*; a die for corrugating (forming multiple beads) in Fig. 7-1*R*.

"Piano" Corrugating Die. A corrugation (consisting of four bends in the metal) can be formed as a single operation but becomes tedious if many corrugations (parallel beads) are to be formed. If more than one corrugation is to be formed simultaneously, rupture will occur unless the corrugation is shallow. A "piano" die (Fig. 9-36), having one or

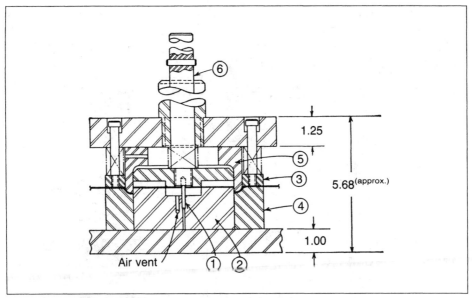

Fig. 9-35. Forming flanges or beads around a lightening hole.

more spring-loaded punches and a rigid punch, progressively forms one corrugation at a time, confining any stretching to the cross section of one corrugation, and allowing metal to be "sucked in" from the unformed portion of the sheet.

Beading a Doubly Curved Part. The part shown in Fig. 9-37, having a pronounced double curvature, is first formed to that curvature, and the beads are formed in a second operation. The beads are formed in a double-action press so there is both a drawing and stretching action in their formation.

Beading a Rectangular Box. In the die shown in Fig. 9-38, the nest and the inner expanding jaws hold the part (a rectangular box) and descend together. A tapered expanding plug drives the expanding jaws outward to force the metal into the groove in the nest, forming an external bead. Stock is pulled upward and downward to form the bead. Coil springs in two grooves encircling the jaws collapse them on the upstroke. The work is stripped from the punch by the expanding plug and pushed off the jaws by a centrally located spring pin and out of the die by the bottom spring stripper.

Beading a Can Cover. A can cover is supported on the inside by the spring-loaded plug in the die shown in Fig. 9-39. Downward pressure is applied by the hardened ring to prevent buckling of the can walls and to roll-form the bead outward at the corner of the cover. On the upstroke, the work is forced from the punch by the plug. A spring-loaded pin pushes the work downward and off the knockout.

Nib-forming Die. The punch closes the interlocking jaws from the outside, and the center plug closes them from the inside, to form a nib as shown in Fig. 9-40. The spring-loaded jaws are released before the punch ascends, since the jaws are housed in a cage which slides up and down in the punch.

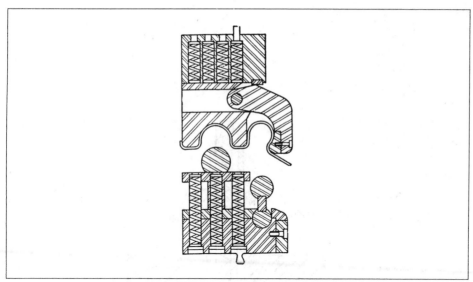

Fig. 9-36. "Piano" corrugating die.[11]

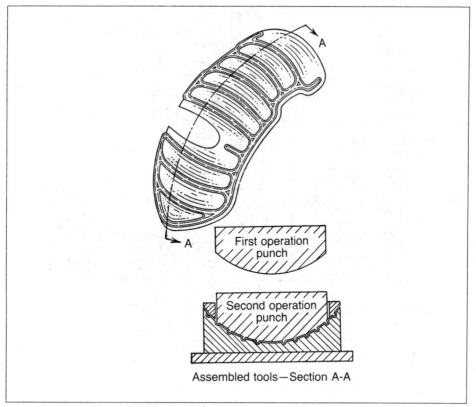

First operation punch

Second operation punch

Assembled tools—Section A-A

Fig. 9-37. Beading a doubly curved part.[1]

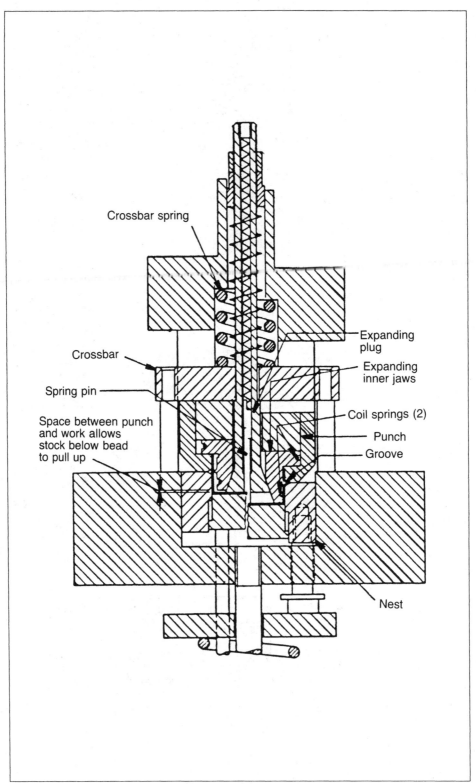

Fig. 9-38. Beading a rectangular box.[7]

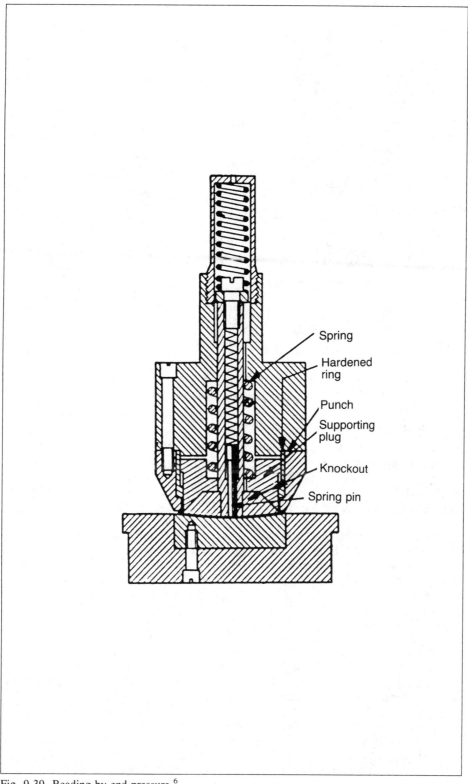

Spring

Hardened
ring

Punch

Supporting
plug

Knockout

Spring pin

Fig. 9-39. Beading by end pressure.[6]

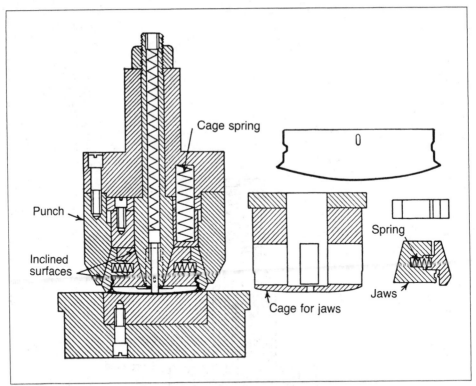

Fig. 9-40. Nib-forming die.[6]

FORMING CYLINDRICAL PARTS

Forming Thin Metal Cylinders. Metal cylinders can be formed in one press stroke by U-forming a blank over a spring-supported horn and driving the horn into a U-shaped die to complete the operation. To perform this operation successfully, the metal must be thin enough to be U-formed on spring pressure, and the proportion of tube diameter to tube length must be such that the horn will have sufficient rigidity to allow the forming of the part.

Forming Tubes with Butt Joints. A typical die for forming small short tubes (Fig. 9-41A) first bends the blank into a U-shape on a spring-supported horn, utilizing a concave punch with sharp edges extending 0.032 in. (0.81 mm) below center to allow for wear. The punch and horn continue down as a unit, and the tube is finish-formed in the lower die. It may be more convenient to feed the blank under the horn; the inverted die design (view B) then can be used.

Tube-forming Die with Retractable Arbor. Figure 9-42 shows a die with arbor for forming a tubular end to an OD of 0.250 in. (6.35 mm) on 0.031-in. (0.79-mm) stainless-steel strip. Forming is done between the inverted punch (D2) and the die (D3). The left end of the arbor (D1) slides in the retaining block (D6) and is held above (D2) by the spring assembly (D7). As the die closes, the spring pad (D5) contacts the arbor, forcing it downward and U-forming the blank. As the die continues to close, this pad holds the stock down and the die completes the curling operation. A hand-operated lever (D4) forces the arbor to the left and out of the formed tube.

Cam-actuated Arbor or Horn-type Forming Die. A preformed rounded channel-shaped blank (0.025-in. (6.35-mm) 1010 cold-rolled steel) is formed into an electric-iron-plug receptacle in the die (Fig. 9-43). The work is formed around an arbor (D1) by cam-actuated forming punches (D2) as it is clamped in position by a spring-loaded

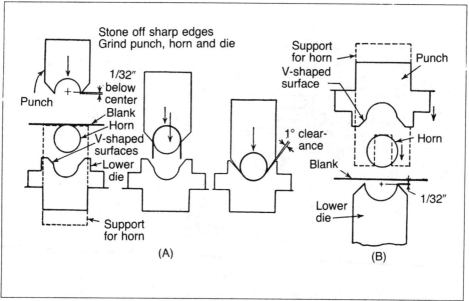

Fig. 9-41. Forming small tubes.[12]

plunger (D3). A vertical spring-loaded punch (D4) snaps up after the two sliding punches are home to set the part. Ejection is manual; these elements are shown: D5, D6, D7, and D8.

Forming a Clamp Ring. Forming a hose clamp ring can be done in two die operations. The blank is first formed (Fig. 9-44A) with the proper allowance for springback on the punch and die radius. The distance X is found by the equation

$$X = (ID + t)1.414 \qquad (1)$$

where   ID = inside diameter of the ring, in. (mm)
          t = stock thickness, in. (mm)

The part is held by the nests (D1) and lower and upper spring-loaded clamping plungers (D3 and D2); the latter extend beyond the punch radius to position the part before forming starts. The first-formed part is then final-formed around a mandrel (view B), being gripped by a clamping plunger (D4) during this operation. The finished part is stripped from the mandrel by hand.

Removable-mandrel-type Forming Die. The blank is dropped in the trough of the die shown in Fig. 9-45. A loose mandrel (D1) is then dropped in so each end rests on a spring pad (D3). The punch (D2) curls the blank around the mandrel and spanks it to a round shape. The mandrel is removed and the part stripped from it by hand.

Forming Lapped-joint Tubes. For forming a tube with a lap joint, the design shown in Fig. 9-41 can be changed so the longer branch of the U-formed blank strikes a raised section of the die, forcing the branch against the horn (Fig. 9-46).

The shorter branch is formed outside to overlap. The raised die section is spring-supported. The bottom of the horn or mandrel is slotted to accommodate the overlapping joint. If desired, one overlapping flap could be designed with raised nibs for projection welding.

Forming a Yoked Tube. A yoked tube of mild steel (0.0625 in. (1.59 mm) thick) is formed around a mandrel (Fig. 9-47, D2) by hinged punches (D1). The blank is positioned by guides (D3) and gage pins (D6). Vertical punch travel is limited by stop blocks (D7) after the squeezing action by contact with the 30° angle of the die (D5) is completed. The latch (D4) supports the mandrel at the back of the die but is opened when

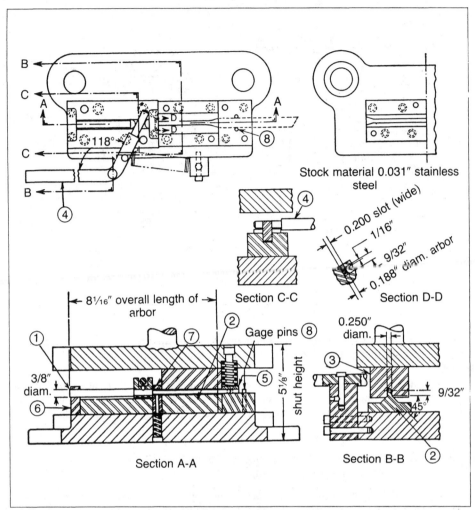

Fig. 9-42. Die with retractable arbor for forming a tubular section on the end of a strip. *(Harig Mfg. Corp.)*

a blank is inserted and returned by a spring (*D8*). Yokeless tubes can be formed by reducing the distance between the guide blocks (*D3*), and designing the hinged punches so their ends will meet under the mandrel at the end of the downstroke.

Forming Dies with Movable Horns. An inexpensive gravity-fed inverted die design for an inclined press is shown in Fig. 9-48. The pivoted horn is supported during the forming action by the spring-loaded C-block. No stripper is provided since the cylindrical parts slide off by gravity. This inverted die is economical to build.

A hinged lift horn or mandrel (Fig. 9-49, *D1*) forms a tube of rectangular cross section from preformed 0.034-in. (0.87-mm) annealed steel. The horn forms two corners only as it slides down into the die against the force exerted by the die cushion through the pins (*D2*). Pins (*D3*) push the horn and pad down while allowing the part to fold around the horn.

Another die design uses a pivoted mandrel with its pivot (Fig. 9-50, *D1*) located at the back of the die. A fountain-pen nib is formed around the horn by two spring-loaded horizontal sliding punches (*D3*) cam-actuated, in conjunction with the vertical punches (*D2*, *D5*). A hand-operated ejector strips the finished nib off the arbor and out the front of the die. As the die closes, the spring-loaded plunger (*D8*) forces the mandrel (*D9*) down

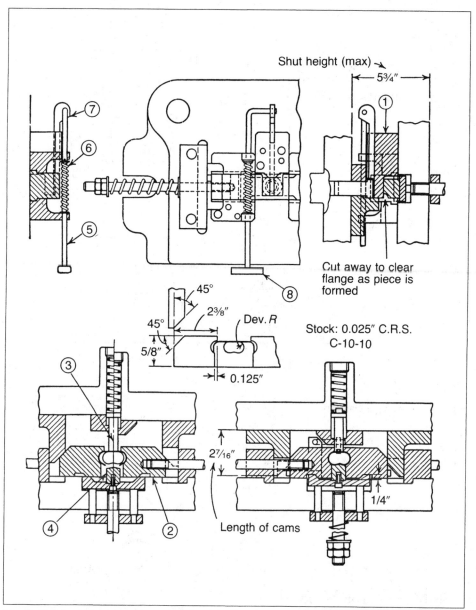

Fig. 9-43. Mandrel-type cam-actuated forming die. *(Harig Mfg. Corp.)*

and U-forms the part against the pad (D2). The sliding punches (D3) are then cammed in to form the part around the mandrel. While the cams dwell, the upper punch (D5) closes the part and at the bottom of the stroke the sliding punches are again cammed in to restrike the part and make it round.

Low-grade steel is formed into a core for rolls of cash-register paper strip (Fig. 9-51 A). The slight indentation made by the blade (D2) is of no importance in this inexpensive but high-production part, but it nullifies spring-back. The die incorporates a typical floating mandrel (D1) in which a flat blank is placed between the nests (D3) and formed around the mandrel by the punch (D4) and die (D5). Stripping is automatic through the movement of the stripping collar (D6), which is actuated by the bell cranks (D7) pivoted at D8. On the upstroke, the studs (D9) contact the upper part of the bracket, forcing the

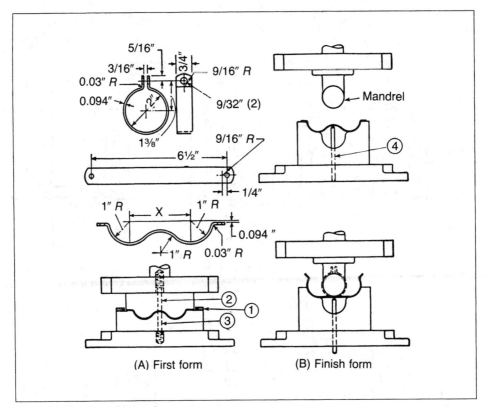

Fig. 9-44. Forming a clamp ring.

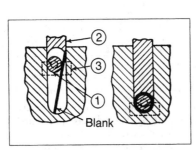

Fig. 9-45. Removable-mandrel-type forming die.[3]

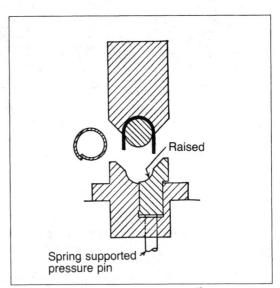

Fig. 9-46. Forming lapped-joint tubes.[12]

stripping collar to the right and allowing the operator to remove the finished core from the mandrel.

Forming Open Tubes. Forming open tubes (Fig. 9-52) allows the use of a webbed horn. In both dies, the blanks are retained by guides and ledges to prevent tipping. At *A* the left-hand guide is spring-loaded; at *B* the blank fits the guides loosely.

Another die design for forming open tubes, using a vertical mandrel, is shown in Fig. 9-53. The flat blank is placed in a slot in the cover of the die. The first form cam forces the first punch toward the front, forming the blank into a U-shape. Further descent of the ram retracts the first form punch; the mandrel (pin) descends, and the two second-form punches are forced inward by the two side-acting second-form cams to form the finished open tube. The vertical shouldered mandrel forces the work through the die to size and iron it. Shoulders on the second-form punches act as strippers.

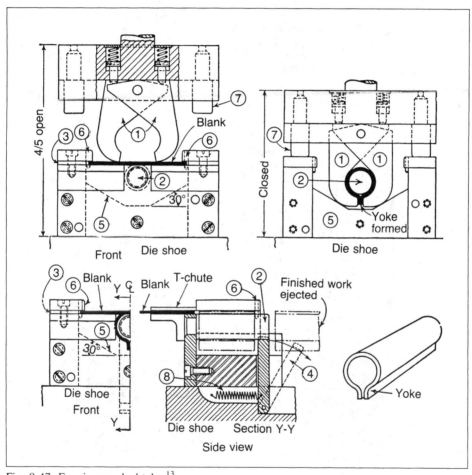

Fig. 9-47. Forming a yoked tube.[13]

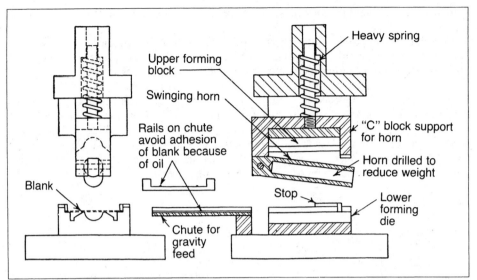

Fig. 9-48. Gravity-fed swinging-horn type forming die.[12]

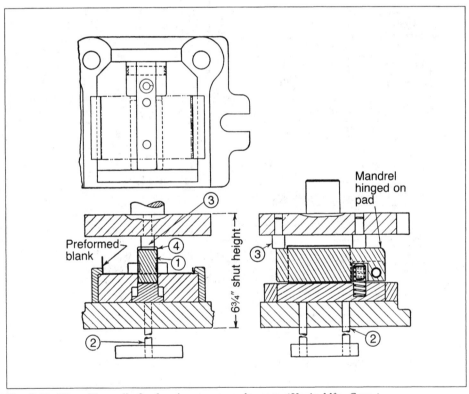

Fig. 9-49. Hinged-horn die for forming a rectangular part. *(Harig Mfg. Corp.)*

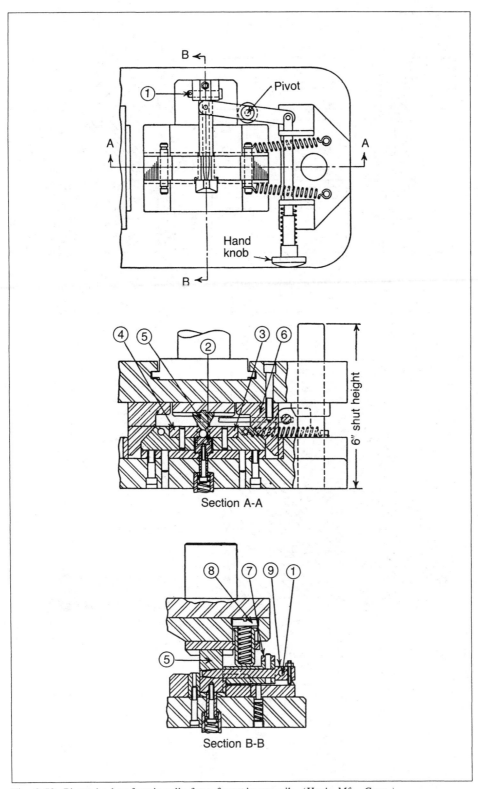

Fig. 9-50. Pivoted-arbor forming die for a fountain-pen nib. *(Harig Mfg. Corp.)*

9- 41

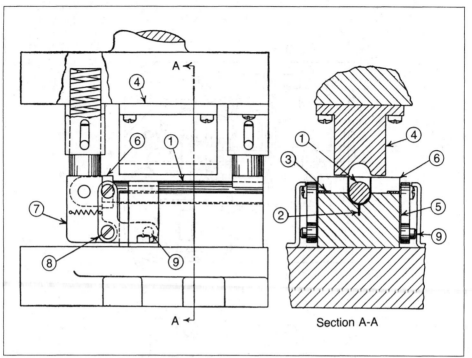

Fig. 9-51. Floating-arbor tube-forming die.[7]

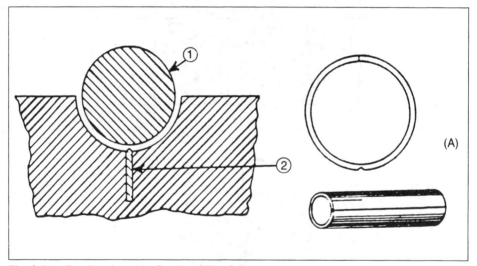

Fig. 9-51A. Detail and product for die of Fig. 9-51.

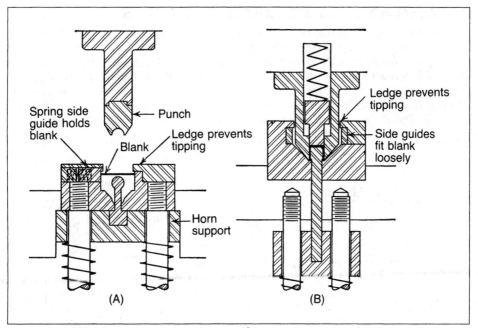

Fig. 9-52. Horn-type die for forming open tubes.[12]

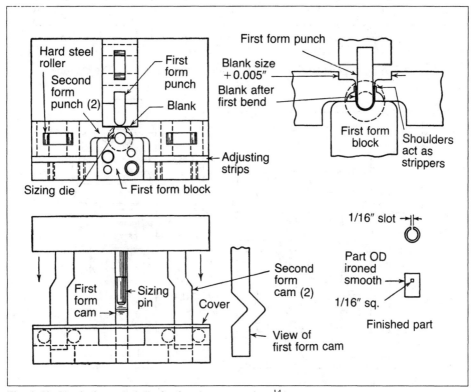

Fig. 9-53. Vertical mandrel die for forming open tubes.[14]

FORMING RINGS, CLIPS, AND SIMILAR PARTS

Ring-Forming Die. A flat blank is placed in the disappearing nests (Fig. 9-54, *D*1) and is formed into a U-shape around the upper half of the mandrel (*D*2). Four pins (*D*6), actuated by a die cushion, hold the blank between the mandrel and the upper punch. The inverted U-formed blank is carried down by the mandrel and the upper forming punch (*D*4) to where the ends of the blank are closed by the lower punch (*D*5) to form the finished ring. Air-actuated stripping pins (*D*3) push the ring from the mandrel on the upstroke. This die is identical in principle to the die shown in Fig. 9-42.

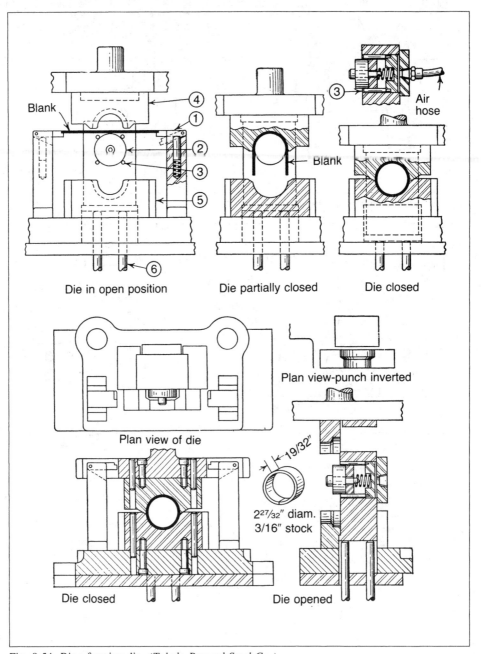

Fig. 9-54. Ring-forming die. *(Toledo Pressed Steel Co.)*

Forming a Flattened Ring. A channel-shaped blank is placed under the forming anvil (Fig. 9-55, *D*1) with the ends of the blank resting on two grooved rolls (*D*2), which are mounted on cranks (*D*5). As the ram descends, the part (with flanges down) is held tightly between the bottom of the anvil (mandrel) and the top of the die. Racks (*D*3) attached to the punch then engage pinions (*D*4) to turn the cranks.

The rolls are forced around the anvil to form the flattened ring around it; one rack is timed ahead of the other to permit one end of the ring (a rope guide for a clothesline pulley) to lap over the other.

Forming a Phosphor Bronze Clip. A relieved forming punch (Fig. 9-56, *D*1) forms the ears on a phosphor bronze blank (0.014 in. (0.36 mm) thick) with clearance of stock thickness only. The bottom of the part is embossed to a depth of 0.031 in. (0.79 mm) and to a diameter of 0.139 in. (3.53 mm) by an embossing punch or pin (*D*3). The spring-loaded pad (*D*4) for the pin travels 0.250 in. (6.35 mm) against the pins (*D*8), equal to the depth of the die (*D*5). The blank is placed in a suitable nest (*D*2). The finished part is ejected by a sliding ejector (*D*6) which strips the clip from the horn-like forming punch; the ejector is actuated by a knockout rod (*D*9). The punch (*D*1) is spring-mounted

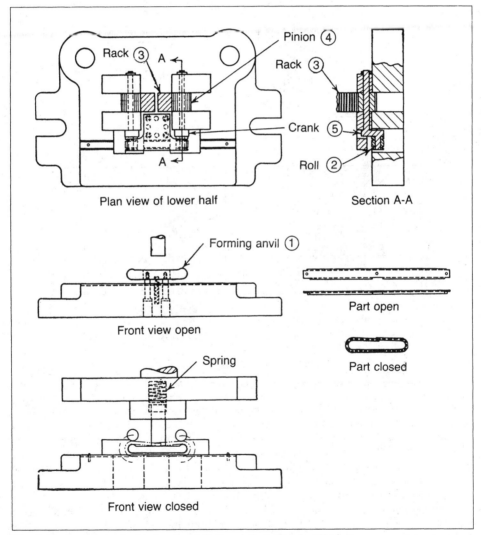

Plan view of lower half

Section A-A

Front view open

Part open

Part closed

Front view closed

Fig. 9-55. Anvil-type die for forming a flattened ring. *(Stanley Works.)*

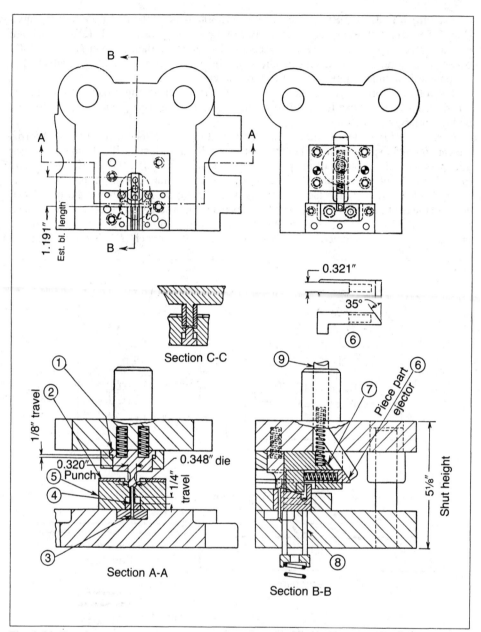

Fig. 9-56. Forming a phosphor bronze clip, using a relieved punch. *(Harig Mfg. Corp.)*

so after it has U-formed and embossed the part, it can dwell while the form punch (D7) continues downward and bends the angle on the tab at the rear of the part.

Toggle-action Forming Die. Another design of a forming die, using toggle joints to actuate horizontal forming punches or plates, is shown in Fig. 9-57. A flat blank is placed in the nest, and small clips and cylindrical parts are formed around the mandrel after the blank becomes U-shaped in the die or lower plate. This design is a variation of the conventional cam or wedge die with sliding horizontal punches.

Forming Die with Air Actuation. Clips, cylinders, and rings are formed in the die of Fig. 9-58, in which air at 80 psi (550 KPa) is used instead of springs to return the horizontal forming punches (D1). The same air pressure actuates the pressure pin (D2).

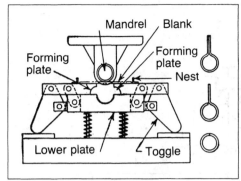

Fig. 9-57. Toggle-action forming die.[16]

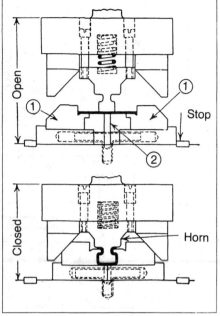

Fig. 9-58. Air-actuated forming die.[15]

BULGING DIES

The design of various products requires the lower portion of some parts be expanded to a size larger than the top; this expansion may be either symmetrical or nonsymmetrical. This type of work cannot be accomplished by a conventional drawing operation and must be done in a separate operation. To facilitate the removal of the parts from the die after bulging, the die elements must be split, sectional, or of a fluid type. The fluid portion of the punch may be water or oil, pumped into the shell under sufficient pressure to expand it. Rubber, heavy grease, or tallow also are used as the bulging medium. Bulging punches of segmental design normally leave slight flats on the finished part which might be objectionable.

A bulging operation is limited by the amount of cold working the workpiece material will withstand before fracturing. Annealing the metal between cold-working operations increases the amount of cold working the material will withstand.

Bulging Die for Pot Cover. The die in Fig. 9-59 forms a bead around a steel shell. The finished part is a cover for a cooking utensil; since the cover fits inside the utensil, the bead acts as a shoulder to hold the cover in place.

A ring of rubber (D1) is the bulging medium in this die. The plate D2, machined to fit the inside contour of the drawn shell, forces the rubber to be displaced outwardly. The size and location of the bulge are determined by the location and contour of the recess formed by the two rings (D3 and D4).

Other examples of dies for expanding cover shapes are shown in Fig. 9-60. View A shows a die for producing a bead similar to the example in Fig. 9-59. The open edge of the shell is forced down to assist the rubber in bulging the metal. View B shows a die for producing a closed flange around the cover. The operation is based on controlling the failure of a band of skirt metal, which is made to collapse outward and is then flattened. In this type of tool, the metal in the upper part of the cover and the lower part of the skirt is more or less confined. To make a uniform flange, the collapse should take place along a plane parallel to the open edge of the shell and about in the center of the band. A small groove is sometimes rolled in the skirt to control the location of the collapsing band.

Die to Bulge a Rectangular Cover. The bulging of a bead in a rectangular shaped

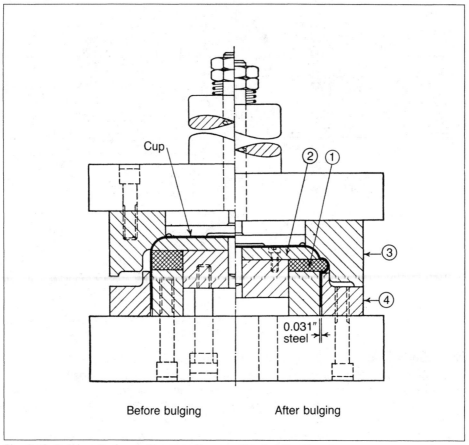

Cup

② ①

③

④

0.031"
steel →|←

Before bulging   After bulging

Fig. 9-59. Bulging die for cooking-utensil cover. (*Harig Mfg. Corp.*)

cover is illustrated in the die shown in Fig. 9-61. In this die, the straight sides are bulged by confining the upper and lower parts, but the corners are formed by sliding wedges (*D1*), which are beveled on one end and formed to the contour of the bulged corner on the other end. A beveled edge on the stationary heel plate (*D2*) forces the sliding wedge outward as the die closes. The sliding wedges are contained in a spring-loaded blankholder, and the stationary heel plate is mounted on the lower shoe.

Die for Bulging a Brass Tube. Short lengths of tubing also can be bulged by using rubber as a bulging medium. The die in Fig. 9-62 uses this principle. The tube is slipped over the location plug (*D1*) and, as the die closes, the rubber (*D2*) bulges the metal into the recess formed by the two rings (*D3* and *D4*). The locating plug can be elevated to eject the part from the lower ring.

Beading an Automotive Distributor Insert. The distributor insert which receives the spark-plug wire, shown in Fig. 9-63, is extruded and beaded in one operation. The slug of copper or aluminum is placed in the die and, as it closes, the punch forces the metal to be extruded as shown at *A*. Further closing of the die starts the bead to be formed as at *B*; view *C* shows the bead finished. Since the part is both extruded and beaded in this die, the cavity that forms the bead must be in the punch holder.

Bulging Die for Double-action Press. The bulging of a shell in a die designed for use in a double-action press is shown in Fig. 9-64. The contour of the cavity is machined in two blocks, the upper half being fastened to the blankholder slide and the lower half to the press bed. The bulging tool is attached to the inner slide. In operation, the drawn shell is placed in the lower half of the die cavity and the blankholder slide is lowered. As the inner

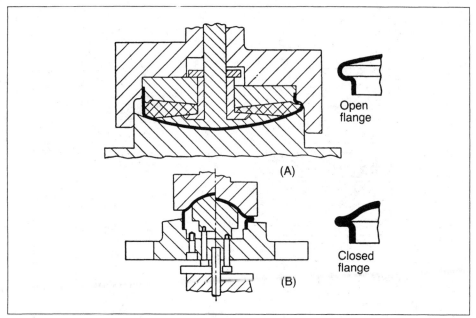

Fig. 9-60. Dies for expanding flanges on cover shapes: *(A)* open-flange tool; *(B)* closed-flange tool.[17]

slide moves the punch down into the shell, its bottom face comes in contact with the bottom of the shell in the cavity. Further downward movement compresses the rubber, which moves outwardly to force the metal into the shape of the cavity. When the punch moves up, the rubber pad resumes its original shape and permits withdrawal from the shell.

Bulging a Drawn Shell. The die illustrated in Fig. 9-65 is for bulging the bottom of a textile-machinery part. The drawn shell for this operation is produced in three conventional drawing operations. Before being placed in this die, the shell is partially filled with a measured amount of melted tallow which is allowed to solidify. The amount of tallow must be carefully measured, because too little would result in an incompletely formed shell, and too much might cause serious damage to the tools.

The tallow-filled shell is placed upside down on the locating plug (D1), and the split-die-cavity block (D2) is placed in the clamping ring (D3). The sleeve (D4) surrounding the punch (D5) is spring-loaded to hold the split-die block in place and, as the punch descends, the part is bulged into the cavity provided. The tallow supports the shell while it is changing shape. The ejector pins extending to the die cushion lift the die-cavity blocks from the clamp ring at the end of the operation.

Straight-sided shells which are to be bulged at the bottom, similar to the one in Fig. 9-65, do not require the split die to permit their removal from the die. Such parts can be formed in a fixed die block, and the die cushion ejects the finished part from the die.

Bulging with a Liquid. A shell may be expanded by placing it within a die cavity and partially filling with fluid. A punch closely fitted to the mouth of the shell is inserted, and its downward travel displaces the incompressible fluid. With the shell resting upon the bottom of the die, the pressure set up by the ram is transmitted to the fluid; in following the path of least resistance, it expands the shell within the die walls. Success of the operation depends upon the thickness and annealed condition of the shell material. When using split-die-cavity blocks not encased in a clamping ring, the hinge and latching device must be quick and easy to operate and hold tight enough that the internal pressure will not force the halves apart and mark the shell. The punch should fit closely within the neck of the shell and enter at least half its diameter before expansion begins.

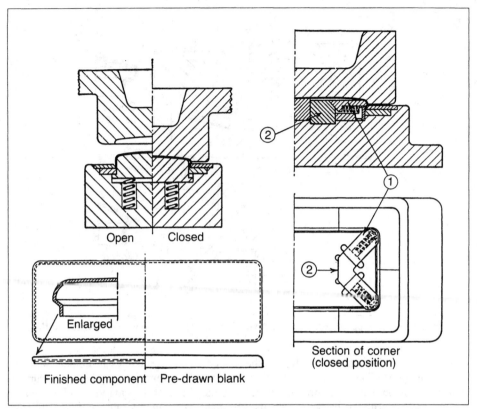

Open | Closed

Enlarged

Finished component | Pre-drawn blank

Section of corner
(closed position)

Fig. 9-61. Die to bulge rectangular-shaped cover. *(Husqvarna Vapenfabriks Aktiebolag, Husqvanena, Sweden.)*

Bulging Die with Interchangeable Cavities. The die shown in Fig. 9-66 has interchangeable parts to accommodate two different shells. The predrawn shell is placed in an inverted position over the locating plug (D1) and the leather cup (D2). As the ram descends, surfaces of the cam (D3) on the upper die engage segments (D4) on the lower die, moving these segments radially in around the work and sealing the mouth of the drawn shell. The segments are moved outward by springs to permit loading and unloading of the die. After the die is closed, the fluid is pumped into the shell to expand it to the shape of the cavity. Air vents are provided in the die cavity to allow the air to escape during the bulging operation. Any air trapped in the die cavity outside the shell may collapse the shell when the internal pressure is released. To position each predrawn shell properly, the locating plug in the lower die is interchangeable as well as the die cavity of the upper die.

This die performs the bulging operation on the coffee maker shown in Figs. 15-18, 15-19, and 15-20.

Bulging by Hydraulically Expanded Rubber Die. The bulging of a shell may be done by hydraulically expanding a rubber bladder. A preformed shell can be placed in a cavity, the rubber bladder being then inserted, locked in place, and expanded by pumping fluid into it. The die cavity and locking arrangement should be carefully designed to withstand pressure.

Molded-rubber Expanding Die. The die shown in Fig. 9-67 is used to form the flutes on light reflectors. A molded-rubber form is contained in the upper die cavity and expands the shell inwardly against the accurately machined punch. The flutes which are machined in the punch are formed in the reflectors as the pressure is applied by closing the die halves together.

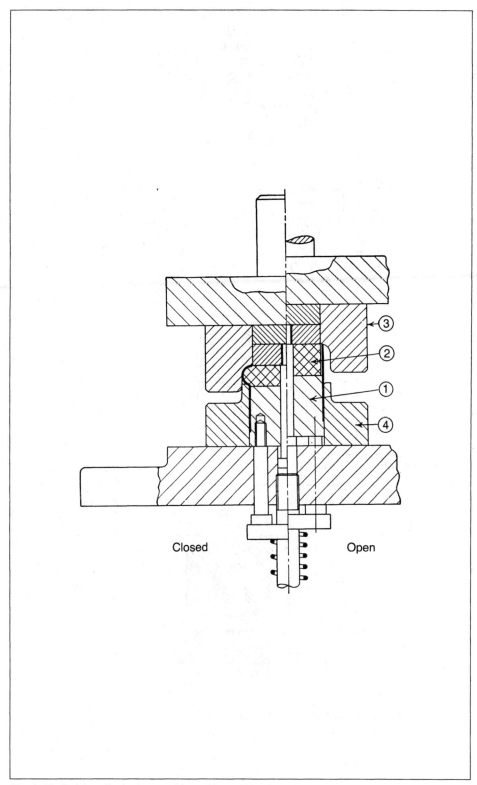

Closed   Open

Fig. 9-62. Bulging die for brass tubing. *(Harig Mfg. Corp.)*

9-51

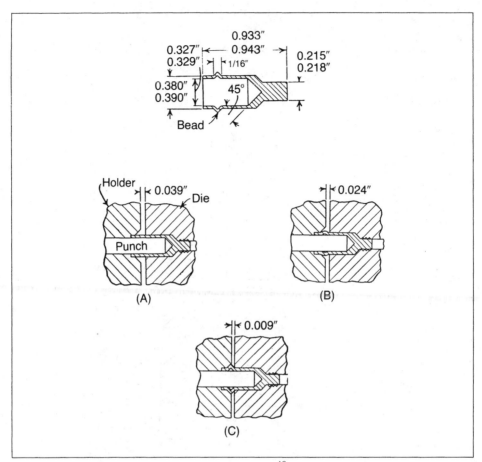

Fig. 9-63. Beading die for automotive distributor insert.[18]

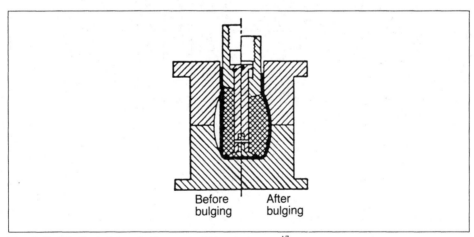

Fig. 9-64. Die for bulging shell on a double-action press.[17]

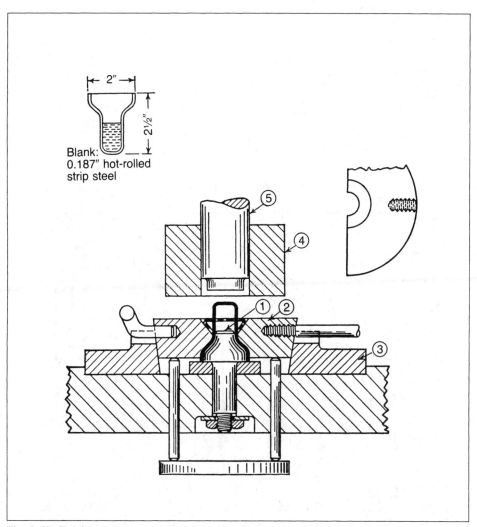

Blank:
0.187″ hot-rolled
strip steel

Fig. 9-65. Bulging a drawn shell in a die using tallow as a bulging medium. *(F.W. Curtis.)*

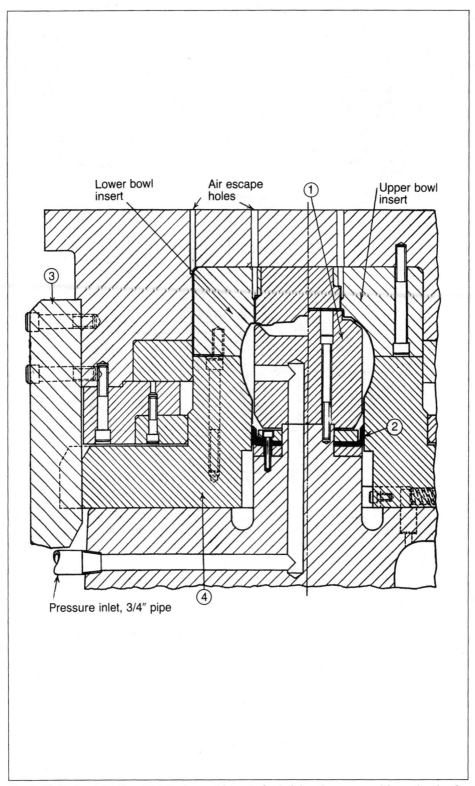

Fig. 9-66. Hydrostatic die with interchangeable parts for bulging the upper and lower bowls of a coffee maker. *(Knapp-Monarch Co.)*

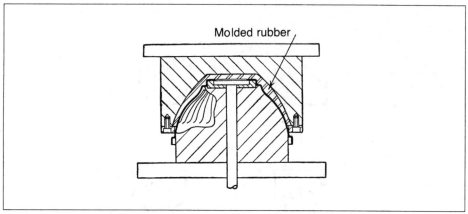

Fig. 9-67. Molded rubber expanding die for lighting reflector. *(General Electric Co.)*

REDUCING AND NECKING DIES

These dies are used for operations subsequent to drawing-die operations. Shells are placed in these dies to reduce the cross-sectional dimensions for part of their length. The reducing operation may be performed on either the open or closed end. *Reducing* is often referred to as work performed on the closed end of the shell, and *necking* as work performed on the open end.

Reducing Die for Doorknobs. The development of the part, and a die for reducing the shank end of a doorknob from a drawn cup, are shown in Fig. 9-68. The dies for the reducing operations 2, 3, 4, and 5 are designed to be placed in a universal die shoe for mounting into a press. The first operation uses a combination blank-and-draw die to produce the cup for the doorknob. Dies progressively reduce the shank to size in three steps. Decorative and functional steps are embossed, in addition to reducing the shank to its final diameter, in the die shown in view *A*. The shank is ironed to a uniform thickness and the steps embossed in the previous die are sharpened by coining by the die shown in view *B*. Operations on this part include piercing the bottom and closing in the top to a semispherical shape.

Die for Necking Steel Tubing. The necking of SAE 4140 steel tubing is illustrated in Fig. 9-69. Before being placed in the die for forming the tubes are cut, centerless ground, and bored. The section of the die shown is one station of a nine-station indexing die. The die has one unloading, two loading, and six working stations.

The tube is placed in the nest (*D*1) at one of the loading stations and the indexing plate moves the part through each successive die. The tube is forced into the die (*D*2), containing a carbide insert for the cavity, by the punch (*D*3) and the face plate (*D*4). Lubricating oil is injected into each die by the tube (*D*5). After the forming operation, the parts are lifted out of the die cavity by the ejector rod (*D*6) operated by a lifter plate which is fastened to the press ram. In the unloading station, the tube is pushed down through the die shoe and slid out the front of the press, passing to a conveyor which takes the parts to a degreaser to prepare them for subsequent operations.

Necking Dies for Cartridge Cases. Dies similar to those shown Fig. 9-70 often taper and reshape a neck on the end of a cartridge case. A plain taper is formed as at *A* by the first die; the neck is completed by the second die as at view *B*.

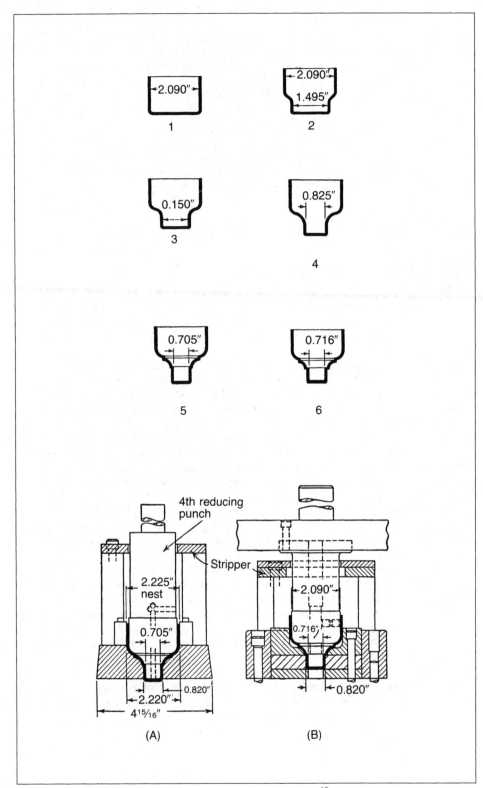

Fig. 9-68. Series of reducing dies for a tubular doorknob shank.[19]

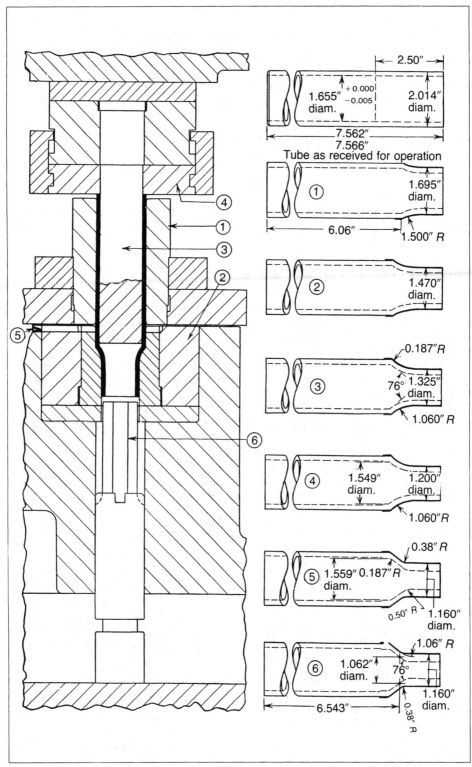

Fig. 9-69. Tube necking die and the progressive shapes to which the tube is formed. *(Oldsmobile Division, General Motors Corp.)*

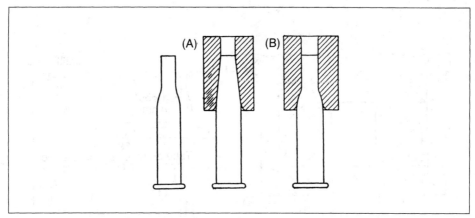

Fig. 9-70. Necking dies for cartridge cases: *(A)* first operation, form taper; *(B)* second operation, complete neck shape.[20]

HIGH-VELOCITY FORMING DIES

Bulkhead Die. The dish-shaped die shown in Fig. 9-71 forms bulkheads 2 in. (51 mm) thick, 43 in. (1.1 m) in diameter, and 8.5 in. (216 mm) deep. Hold-down clamps are not required because the weight of the blank is sufficient to create an airtight seal. The vacuum vent hole is at the center of the die, and the outlet, through pipe fittings, is at the lower outside front. A vacuum seal is made between the blank and the die by an O-ring. The bulkheads are formed in two stages using 3 lb (1.4 kg) of commercial dynamite.

This 7,000-lb (3,175-kg) die is handled by the cradle-and-sling arrangement instead of lifting attachments to the die.

Engine-cooling-shroud Die Inserts. The die inserts shown in Fig. 9-72 have a 5° taper on the outside to mate with a master retaining die. The inserts are made of heat-treated steel to withstand the high pressures necessary to form the small ($2T$) female radii of convolutions. Vacuum seals use O-rings and are actuated by various plates and machined rings designed to compress the O-rings as required. This die requires four seals. The part is expanded in one shot from a cylindrical blank. The shroud has been made of five different metals, including Hastelloy X, Roselyn metal, and type 310 stainless steel.

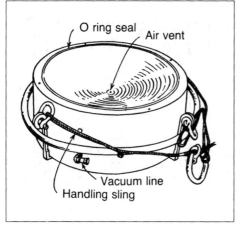

Fig. 9-71. Die for explosive forming of a 43-in.-diameter bulkhead. *(General Dynamics Corp.)*

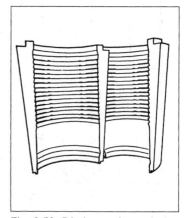

Fig. 9-72. Die inserts for explosive forming engine-cooling shroud. *(General Dynamics Corp.)*

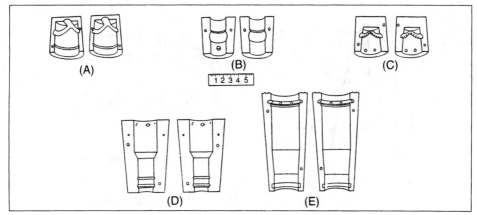

Fig. 9-73. Die inserts for electrohydraulically formed parts. *(General Dynamics Corp.)*

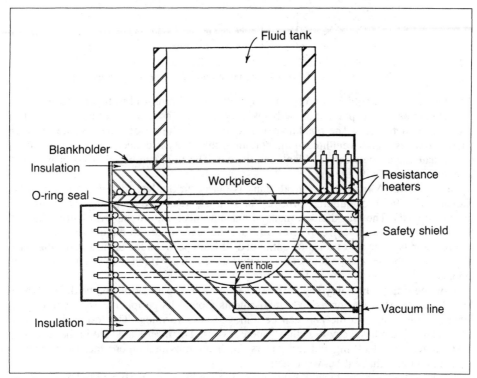

Fig. 9-74. Ductile-iron die for explosive forming parts at elevated temperatures. *(Ryan Aeronautical Co.)*

Dies for Electrohydraulically Formed Parts. Figure 9-73 illustrates five tapered die inserts used on master retaining dies on electrohydraulic forming equipment. Vent holes are on the interior of the die, while dowel pinholes are on the flat surfaces of the die half-blocks. Cutting surfaces and edges are located at the recessed grooves in the upper portion of dies *A, C,* and *E.*

Hot-forming Die. Some metals, such as magnesium, are formed more readily at elevated temperatures than at ambient temperatures. The construction shown in Fig. 9-74 enables the die to be heated during the forming operations. Resistance heaters are embedded around the die cavity and in the blankholder. Insulation is applied around the

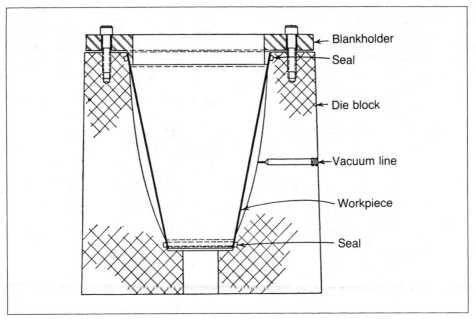

Fig. 9-75. Explosive bulging die of low-melting-point alloys. *(Ryan Aeronautical Co.)*

die block and blankholder to protect the operator from burns and to retard heat loss.

A standpipe is incorporated in the blankholder and serves as a water tank in which the explosive is detonated. The blankholder is made of machine steel, and the die block is ductile-iron casting. A vent hole at the bottom of the cavity is connected to a vacuum line to evacuate the air between the blank and the die. The fluid can be oil or water which is heated by contact with the die.

Bulging Die. A truncated conical blank is placed in the low-melting-point alloy die shown in Fig. 9-75 for bulging. Grooves for seals are machined as the top and bottom of the cavity walls. The blankholder, made of machine steel, has an integral tapered ring for holding the blank in place. Cap screws secure the blankholder to the die block.

Provision is made for evacuating entrapped air from between the blank and the die. The die is open at the top and bottom so the cavity can be quickly filled with water and quickly emptied.

Low-melting point Alloy Die with Steel Inserts. The part formed in the die shown in Fig. 9-76 has reentrant curves and embossments around its periphery. This type of part requires split inserts so the part can be removed after forming. The low-melting-point alloy retaining die has a cavity with 5° tapered walls. The inserts are held in the retaining die by a machine-steel ring and cap screws. Seals are provided on one face of the inserts to ensure an airtight seal between them.

The blankholder has a tapered ring to position the blank properly and a flat surface to compress the upper seal to ensure an airtight compartment between the blank and die. Openings at the top and bottom permit ease of filling with, and emptying of, water.

The tapered inserts are made of low-melting-point alloy metal, while the embossing inserts are of heat-treated tool steel.

Plastic-faced Die. The use of a plastic facing on an explosive forming die is illustrated in Fig. 9-77. The parabolic-shaped part with vertical beads is formed from a conical-shaped blank. The die block is made of zinc alloy cast around the plastic form. A groove is machined at the top of the cavity for a seal. The vent hole is place in one of the beads where the blank velocity is expected to be the lowest value. A machine-steel blankholder, secured by cap screws, positions the blank and assures an airtight seal.

Plastic-faced Forming and Perforating Die. The die shown in Fig. 9-78 is used to

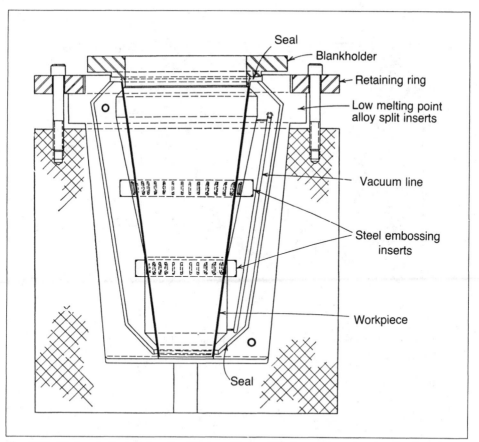

Fig. 9-76. High-velocity bulging die of low-melting-point alloy with steel embossing inserts. *(Ryan Aeronautical Co.)*

bulge a tube slightly, form a bead on each end, and perforate a 0.750-in. (19-mm) -diameter hole. The plastic material is backed up with a 6-in. (152-mm) -diameter by 0.50-in. (12.7-mm) -thick steel casing. Eyebolts close the forming die. Because of the gentle shape of the part, vent holes are not required.

A hardened steel bushing is inserted in the area of the perforation.

Sealing disks are customarily provided at the top and bottom of this type of die; however, since no vacuum is necessary on this die, sealing is required only against leakage of the interior fluid.

HIGH PRESSURE NITROGEN MANIFOLD DIES

Nitrogen manifold systems operate at gas pressures as high as 1500 psi (10.3 MPa). The individual cylinders, which screw into the manifold are available in several sizes with pressure ratings up to 6 tons (53 kN).

High pressure nitrogen systems have several important advantages when compared to springs, air cylinders, and press cushions. Generally, much higher pressures per unit volume are possible than is the case with die springs or air cylinders. Unlike springs, the full rated pressure is available at the start of the cylinder stroke. Greater flexibility of die design is another advantage. The design is not dependent on the location of air cushion pin holes, a factor that often dictates the use of sub-plates.

The nitrogen manifold can be designed to also function as a die shoe, spacer plate, or other die member. Manifold systems are much less subject to leak than are individual cylinders interconnected with high pressure hoses to an external surge tank. Drilled

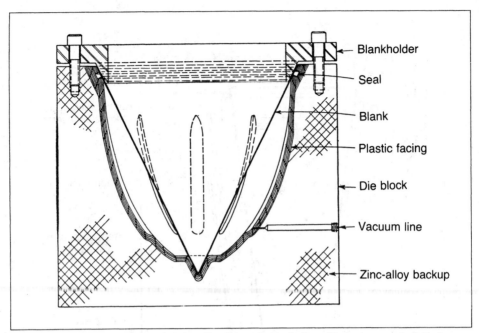

Fig. 9-77. Zinc-alloy explosive forming die with a plastic facing. *(Ryan Aeronautical Co.)*

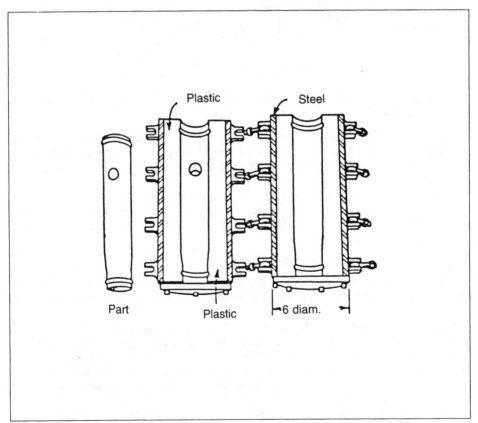

Fig. 9-78. Plastic high-velocity forming die surrounded by a 6-in. (152-mm) -diameter steel casing *(General Dynamics Corp.)*

passages in the manifold serve as a surge tank to prevent excessive pressure increases as the cylinders are stroked. Large port plugs with O-ring seals are used to close the ends of the drilled passages.

An additional advantage of using a nitrogen manifold system is that the pressure is adjustable to allow compensation for stock variations.

Corrugating Step Form Die. Figure 9-79A shows a die for corrugating a large metal panel in one press stroke in the open position. The nitrogen manifold also serves as the upper die shoe.

The step form sequence begins in the middle to allow the material to flow freely in from the ends as the cylinders collapse. This sequence continues until the part is finished, and prevents the stock from stretching or tearing.

Figure 9-79B shows the die in closed position. The same stroke length cylinders are used with pressure pins of different length. This provides for complete interchangeability of cylinders, thus reducing the spare parts inventory. Cylinders with rod extensions are available, but their use should be avoided. A long shaft extension transmits greater lateral force to the rod guide bushing resulting in reduced cylinder life.

The finished part is shown in Fig. 9-79C. The function of this die is similar to the "piano" corrugating die shown in Fig. 9-36.

Double Action Form Die. The die shown in Fig. 9-80 forms the part in two separate actions. As the die closes, the right side flange is formed up by the pressure transmitted throughout the nitrogen cylinders. As the flange is completed, the cylinders stroke in while supplying the clamping force required for the form punch to stretch-form the center section of the part. The nitrogen manifold also serves to support the form punch.

Forming Up and Down. The ability to adjust the nitrogen pressure in a manifold system can permit exact fine tuning of the sequence of operations. Figure 9-81 shows a die with essentially identical upper and lower halves. The upper manifold is charged to 800 psi (5.5 MPa) while the lower manifold is charged to a higher pressure of 1,200 psi (8.3 MPa). As a result, the flange on the right side is formed first when the upper manifold's cylinders collapse. The flange on the left side is not formed until the upper pad bottoms out. As the die opens, the lower pad is first to return to the die open position.

This principle has important applications when an exact sequence of operations must be maintained. For example, the pads could be used to partly form embosses in a flat sheet of metal upon initial contact. Next the flanges could be formed in sequence. Third, the part could be lanced or pierced. Finally, the embosses could be firmly set by bottoming out the pads.

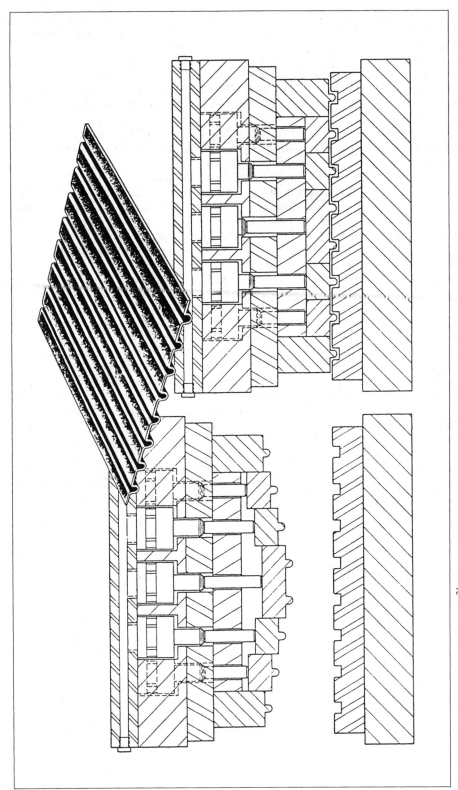

Fig. 9-79. Forming a corrugated panel. [21] (*Forward Industries*)

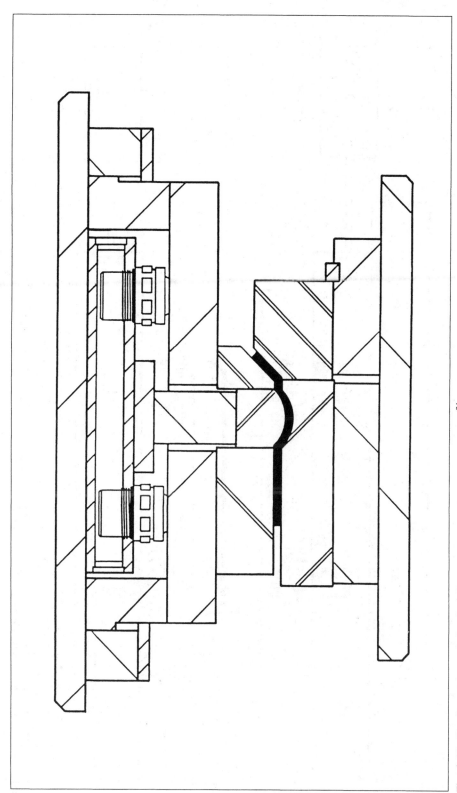

Fig. 9-80. Bending and stretch-forming die using a nitrogen manifold system.[21] *(Forward Industries)*

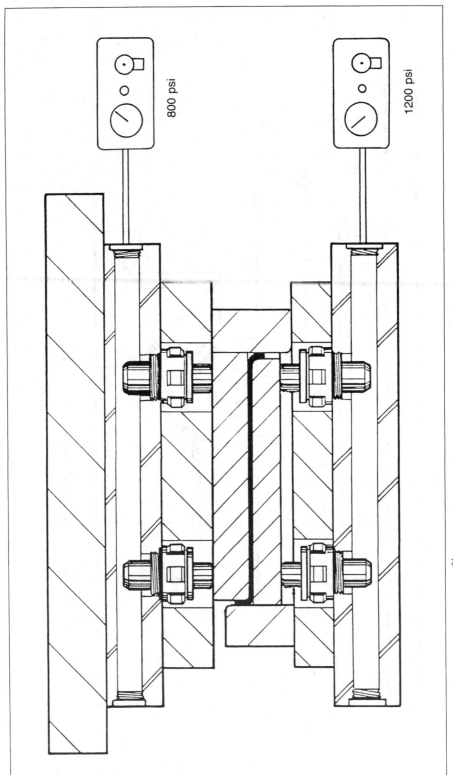

800 psi

1200 psi

Fig. 9-81. Up and down flanging operation.[21] (*Forward Industries*)

References

1. *Correlation of Information Available on the Fabrication of Aluminum Alloys* (National Defense Research Committee, 1943).
2. V.N. Krivobok, and G. Sachs, *Forming of Austenitic Chromium-nickel Stainless Steels* (New York: The International Nickel Co., Inc., 1948).
3. C.R. Cory, *Die Design Manual*, Part II (Dearborn, MI: Society of Manufacturing Engineers, 1958).
4. E.N. Sorenson, "Dies for Home Freezer-lid Panels," *The Tool Engineer* (September 1949).
5. *Tool Engineering Manual*, International Business Machine Corp.
6. W.C. Mills, "Thin-stock Dies for Secondary Operations," *American Machinist* (October 24, 1946).
7. K.L. Bues, "Die-Grams," *Western Machinery & Steel World* (January 1950; November 1949; June 1948).
8. E.E. Bangs and C. V. Seagers, "Secondary Operations on Stampings Show Ingenious Tool Engineering," *American Machinist* (April 3, 1950).
9. J.R. Paquin, "Horn Dies Will Do Many Jobs," *American Machinist* (April 3, 1950).
10. H.A. Waldon, "Zipper Die Closes and Crimps Tube Seam," *American Machinist* (March 20, 1950).
11. G. Sachs, *Principles and Methods of Sheet-metal Fabrication* (New York: Reinhold Publishing Corp., 1951).
12. W.C. Mills, "Forming Thin-metal Cylinders," *American Machinist* (December 4, 1947).
13. C.W. Hinman, "An Analysis of Forming Die Designs," *Modern Machine Shop* (June 1950).
14. C. Morgan, "Cam-controlled Punch Forms Accurate Tube," *American Machinist* (September 29, 1952).
15. C.W. Hinman, "An Analysis of Bending and Forming Dies," *Modern Machine Shop* (May 1950).
16. F.W. Curtis and C. B. Cole, *Tool and Die Design*, (Chicago: American Technical Society, 1932).
17. J.W. Lengbridge, "Types and Functions of Press Tools," *The Tool Engineer* (June 1949).
18. R.L. Kessler, W. A. Fletcher and W.P. Bowman, "How Delco-Remy Cold-forms Metals," *American Machinist* (July 20, 1953).
19. H.E. Nagle, "Precision Dies Draw Brass Lock Parts," *American Machinist* (December 30, 1948).
20. F.A. Stanley, *Punches and Dies*, 4th ed. (New York: McGraw-Hill Book Co., 1949).
21. L.R. Evans, *Nitrogen Die Cylinders—The Modern Method of Pressure Placement* (McLean, VA: National Machine Tool Builders Association, 1989).

SECTION 10
DISPLACEMENT OF METAL IN DRAWING

Drawing is a process of cold forming a flat, precut metal blank into a hollow vessel without excessive wrinkling, thinning, or fracturing. The various forms produced may be cylindrical or box-shaped, with straight or tapered sides or a combination of straight, tapered, and curved sides. The parts may vary from 0.250 in. (6.35 mm) diameter or smaller to aircraft or automotive parts large enough to require the use of mechanical handling equipment.

The term "redrawing" is used for a variety of operations in which a part is reduced in its lateral dimensions by means of single- or double-action dies without reducing the wall thickness. Regular redrawing is done by positioning the part under the punch which forces the cup into the die, reducing the bottom dimensions, and increasing the side-wall height. Reverse or inside-out redrawing is done by positioning the cup over a die ring, and the punch contacts the outside of the bottom, turning the part inside out into the die opening.

METAL FLOW

Deep Drawing of Cylindrical Cup. When the punch of a drawing press forces a portion of a metal blank through the bore of the draw ring, different forces come into action (Fig. 10-1) causing a complicated plastic flow of the material. The volume and the thickness of the metal remain constant, and the final shape of the cup will be similar to the contour of the punch. The progressive stages of cupping are schematically shown in Fig. 10-2. After a small stroke of the punch, cupping stage *A*, the metal volume element (2) of the blank is bent and wrapped around the punch nose. Simultaneously, the outer portions of the blank, depicted by sections 3, 4, and 5, move radially toward the center of the blank, as shown in cupping stages *B* and *C*. The various volume elements decrease in circumferential length (not shown) and correspondingly increase in radial length until they reach the bore of the draw ring. They then bend over, conforming to the edge of the die. After becoming part of the shell wall, the elements are straight. During drawing, area 1, for the specific example illustrated, is unchanged in the bottom of the cup. The areas which become the sidewall of the shell (2, 3, and 4) change from the shape of angular segments to longer parallel-sided shapes as they are drawn over the inner edge of the draw ring; from this point no further metal flow takes place.

In general, the metal flow by cupping may be summarized as follows:

1. Little or no metal deformation takes place in the blank area which forms the bottom of the cup. This is indicated by the unchanged distances between marking lines and the radial boundaries of the annular quadrants remaining radial in the base of the shell.

2. The metal flow taking place during the forming of the cup wall uniformly increases with cup height. This is indicated by the marking lines, which remain concentric but also show increasing distances between each other. Radial boundaries of the blank segments

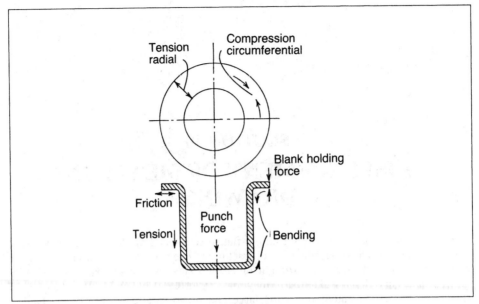

Fig. 10-1. Forces involved in metal flow during cupping.

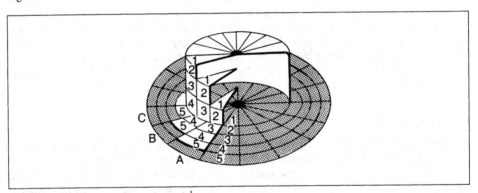

Fig. 10-2. Step-by-step flow of metal.[1]

become parallel when they are drawn over the inner edge of the draw ring, where they assume their final dimensions in the cup wall.

3. The metal flow of the volume elements at the periphery of the blank is extensive and involves an increase in metal thickness caused by severe circumferential compression (Fig. 10-1). This increase in wall thickness at the open end of the cup wall, though significant in practice, is not shown in the figures. The increase is usually slight because it is restricted by the clearance between the punch and the bore wall of the die ring.

Metal Flow in Rectangular Shells. The drawing of a rectangular shell involves varying degrees of flow severity. Some parts of the shell may require severe cold working and others simple bending. In contrast to circular shells in which pressure is uniform on all diameters, some areas of rectangular and irregular shells may require more pressure than others. True drawing occurs at the corners only; at the sides and ends metal movement is more closely allied to bending. The stresses at the corner of the shell are compressive on the metal moving toward the die radius and are tensile on the metal that has already moved over the radius. The metal between the corners is in tension only on the sidewall and flange areas.

The variation in flow in different parts of the rectangular shell divides the blank into two areas. The corners are the drawing area, which includes all the metal in the corners of the blank necessary to make a full corner on the drawn shell. The sides and ends are the forming area, which includes all the metal necessary to make the sides and ends full depth. To illustrate the flow of metal in a rectangular draw, the developed blank in Fig. 10-3B has been divided into unit areas by two different methods. In Fig. 10-3A the corners of the shell drawn from the blank in view B are shown. The upper view is the corner area which was marked with squares, and the lower view is the corner area which was marked with radial lines and concentric circles. The severe flow in the corner areas is clearly shown in the lower view by the radial lines of the blank being moved parallel and close together, and the lines of the concentric circles becoming farther apart the nearer they are to the center of the corner and the edge of the blank. The relatively parallel lines of the sides and ends show little or no flow occurred in these areas. The upward bending of these lines indicates the flow from the corners to the sides and ends to equalize the height where these areas on the blank were blended to eliminate sharp corners.

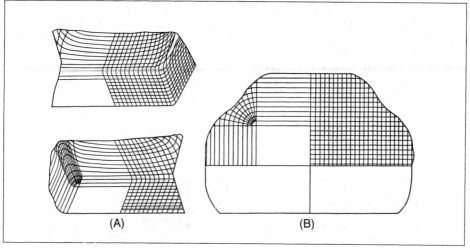

(A) (B)

Fig. 10-3. Displacement of marking lines by metal flow in rectangular draws: (A) the two box corners showing the lines displaced by deep drawing; (B) portion of blank showing evenly spaced lines.[1]

Wrinkling and Puckering. The shaping of a shell necessitates severe cold working and involves plastic flow of the metal. Any condition retarding the necessary flow must be avoided to minimize stresses to which the metal is subjected. The metal may buckle rather than shrink in any location of the blank if it is very thin and if a sufficiently wide area is free to move away from the tools. These buckles are called "wrinkles" if they occur at the edge of the blank and "puckers" if they appear in any other part of the blank. The formation of wrinkles in the flange area is to be expected since the direction of the stresses is circumferential; this wrinkling must be controlled because it may adversely affect the normal metal flow. Because relatively thin metals have a high wrinkling tendency, higher blankholding pressures are necessary. When the thickness-diameter ratio of the blank is low, high blankholding pressure is required; when this ratio is high, little or no blankholding pressure is required. Also, as a general rule, as the thickness-diameter ratio of the blank decreases, the amount of drawing possible decreases correspondingly, and the tools for these draws must be finished with greater care.

The shape of the shell section governs, to some extent, whether wrinkles or puckers will be most prevalent under conditions of poor control. Straight-sided shells are typical shapes in which wrinkles would occur, while puckers are most likely to appear in domed or tapered shells. If the die radius and/or the punch radius is too large, even though the sides are straight, the conditions come close to domed shapes, and both wrinkles and puckers have a tendency to occur. In Fig. 10-4, a series of different shell shapes is

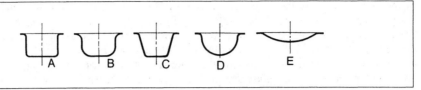

Fig. 10-4. Series of different shell shapes in the order of increasing pressure requirement for the blankholder.[1]

arranged in order of increased blankholder pressures required. *A*, *B*, *C*, and *D* have the same diameters but require different blankholder pressures to control metal flow. Shell *A* will tend to wrinkle without sufficient pressure; shell *B* will wrinkle and possibly pucker because of the large punch and die radii; shells *C* and *D* will wrinkle and pucker. Shell *E* will tend to pucker because there is very little metal flow, and a very high blankholder pressure will be required to pull the material tightly around the punch.

Controlling Metal Flow. The two draw dies shown in Fig. 10-5 illustrate good control of flow (*A*) and poor control of flow (*B*). In view *A*, the tool faces are in close contact with the blank at all points, but insufficient blankholder pressure may encourage wrinkles to occur in the shell. In view *B* there is poor control of the metal flow because only the tip of the punch is in contact with the blank, leaving much of the center area of the blank free to pucker. Depending upon the material, increased blankholder pressure may not produce a good shell.

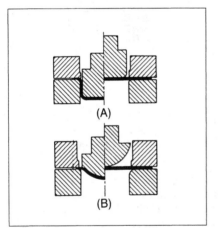

Fig. 10-5. Set of draw dies showing (*A*) good and (*B*) bad control of metal flow in cupping; the contour of the punch determines its initial contact with the blank.[1]

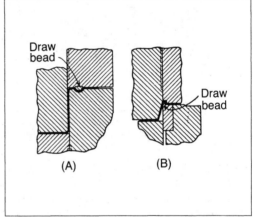

Fig. 10-6. Two different draw beads to control metal flow during deep drawing: (*A*) molding type, offset in blankholder; (*B*) lock type, inserted in the draw ring.[2]

Draw beads are sometimes included in the blankholder faces to provide more resistance to metal flow, aiding control of metal movement into the die bore. The beads need not be continuous around the die, and more than one is sometimes placed in areas requiring great retarding of metal flow. Figure 10-6 shows two common types of draw beads. Draw beads may be mounted in a die in several different ways. Draw beads are further discussed in Section 13.

Speeds for Drawing. The drawing speeds are greatly influenced by uniformity and physical characteristics of stock. It is usually necessary to determine by experimentation the best speed at which a particular stock can be worked. The metal must be given sufficient time to flow, otherwise fractures are likely to occur. The tentative drawing speeds in Table 10-1 may be used for average conditions, and adjusted up or down to suit the specific applications.

TABLE 10-1
Tentative Drawing Speeds

Material	Drawing, fpm		Ironing or Burnishing, fpm
	Single Action	Double Action	
Aluminum	175	100	
Strong aluminum alloys	. .	30–40	
Brass	200	100	70
Copper	150	85	
Steel	55	35–50	25
Steel (in carbide dies)	. .	60	
Steel, stainless	. .	20–30	
Zinc	150	40	

BLANK DEVELOPMENT FOR CYLINDRICAL SHELLS

Development of Blanks. The development of the approximate blank size should be done (1) to determine the size of a blank to produce the shell to the required depth and (2) to determine how many draws will be necessary to produce the shell. This is determined by the ratio of the blank size to the shell size. Various methods have been developed to determine the size of blanks for drawn shells. These methods are based on (1) algebraic calculations; (2) the use of graphical layouts; (3) a combination of graphical layouts and mathematics. The majority of these methods are for use on symmetrical shells.

It is rarely possible to compute any blank size to close accuracy or maintain perfectly uniform height of shells in production, because the thickening and thinning of the wall varies with the completeness of annealing. The height of ironed shells changes with commercial variations in sheet thickness, and the top edge differs from square to irregular, usually with four more or less pronounced high spots resulting from the effect of the direction on the crystalline structure of the metal. Thorough annealing should largely remove the directional effect. For all these reasons, it is usually necessary to figure the blank sufficiently large to permit a trimming operation. The drawing tools should be made first, then the blank size should be determined by trial before the blanking die is made. There are times, however, when the metal required to produce the product is not immediately available from stock and must be ordered at the same time as the tools. This situation makes it necessary to estimate the blank size as closely as possible algebraically or graphically to know what sizes to order.

Algebraic Method. The following equations may be used to calculate the blank size for cylindrical shells of relatively thin metal. The ratio of the shell diameter to the corner radius (d/r) can affect the blank diameter and should be taken into consideration.

When d/r is 20 or more,

$$D = \sqrt{d^2 + 4dh} \tag{1}$$

When d/r is between 15 and 20,

$$D = \sqrt{d^2 + 4dh} - 0.5r \tag{2}$$

When d/r is between 10 and 15,

$$D = \sqrt{d^2 + 4dh} - r \tag{3}$$

When d/r is below 10,

$$D = \sqrt{(d - 2r)^2 + 4d(h - r) + 2\pi r(d - 0.7r)} \qquad (4)$$

where D = blank diameter
     d = shell diameter
     h = shell height
     r = corner radius

The above equations are based on the assumption that the surface area of the blank is equal to the surface area of the finished shell.

In cases where the shell wall is to be ironed thinner than the shell bottom, the volume of metal in the blank must equal the volume of the metal in the finished shell. Where the wall-thickness reduction is considerable, as in brass shell cases, the final blank size is developed by trial. A tentative blank size for an ironed shell can be obtained from the equation

$$D = \sqrt{d^2 + 4dh\,\frac{t}{T}} \qquad (5)$$

where t = wall thickness
     T = bottom or blank thickness

Simple Graphical Method. A simple graphical method (Fig. 10-7) for determining the diameter D of a circular blank, knowing the height h and the diameter d of a cylindrical shell to be drawn, is as follows:

1. From a level reference plane, raise a perpendicular of height h.
2. From top of the perpendicular, draw a hypotenuse of length $h + (d/2)$, to intersect the reference plane.
3. The horizontal component X, between the intersections on the reference plane, equals the radius of the necessary circular blank of diameter D.

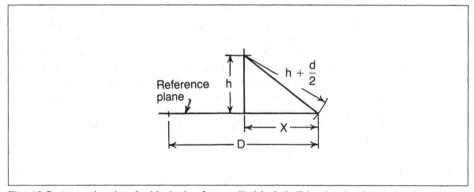

Fig. 10-7. Approximating the blank size for a cylindrical shell by the simple graphical method.

Center-of-gravity Method. The blank size for a symmetrical drawn cup as shown in Fig. 10-8 may be closely determined by the rule of Guldinus, which states that the area is equal to the length of the profile, multiplied by the length of the path of its center of gravity. With the area known, it is simple to calculate the diameter of the blank. The length of the lines C, L_1, L_3, and L_4 (taken along the neutral axis) is known, with the locations of their centers of gravity known in relation to axis A-B. Arc length L_2 and the position of its center of gravity with respect to axis A-B are unknown. Angle α may be found by using the function of the sine of the angle.

The length of the circular arc is

$$L_2 = 0.01745r(2\alpha) \qquad (6)$$

10 - 6

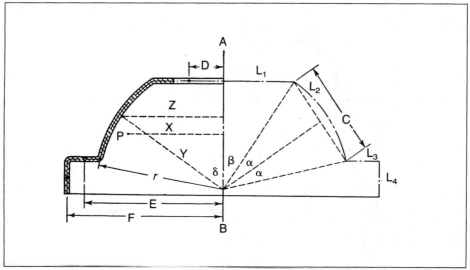

Fig. 10-8. Determining the blank size by the center-of-gravity method.[3]

The center of gravity of arc L_2 is located on the line that bisects the arc, at a distance Y from the center of the circle where C is the chord length:

$$Y = \frac{Cr}{L_2} \tag{7}$$

To find the horizontal distance Z, between the center of gravity of arc L_2 and the axis A-B, angle β may be found by using the function of the sine of the angle. Then,

$$\delta = \beta + \alpha \tag{8}$$

and

$$Z = Y \sin \delta \tag{9}$$

To find the horizontal distance X between the combined center of gravity P and the axis A-B, divide the sum of the moments of each section by the combined lengths of the sections:

$$X = \frac{L_1 D + L_2 Z + L_3 E + L_4 F}{L_1 + L_2 + L_3 + L_4} \tag{10}$$

Applying the rule of Guldinus (also known as Pappas' second theorem),

$$A = (L_1 + L_2 + L_3 + L_4)2\pi X \tag{11}$$

Since the area is unimportant, the desired blank diameter D can be solved directly, instead, by using Eq. (12),

$$D = \sqrt{8X(L_1 + L_2 + L_3 + L_4)} \tag{12}$$

Area-of-element Method. The blank diameter for complex circular shells, as the one in Fig. 10-8, may be divided into simple elements of shape, such as the elements numbered 1, 2, 3, and 4 in Fig. 10-9. Element 1 is a cylinder, 2 is a flat ring, 3 is a radiused ring or a portion of a sphere, and 4 is a disk. The area of each element may be found by using the equations in Fig. 10-10. From the total of these areas, the diameter of the blank may be determined by

$$D = 1.128 \sqrt{A} \tag{13}$$

10 - 7

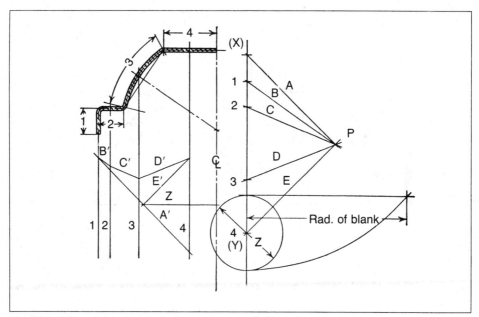

Fig. 10-9. Determining the blank size by the area-of-elements method.

Layout Method. A graphic method of determining the blank diameter of the same shell shown in Fig. 10-8 is illustrated in Fig. 10-9. The procedure to determine the blank is as follows:

1. Make an accurate layout of the part, including a line through the center of the stock.

2. Number each dissimilar section starting at the extreme edge of the part.

3. Draw vertical line *X-Y* and mark off the length of each section accurately starting with section 1 at the top of the line. Number each section to correspond with the same section of the shell.

4. Through the center of gravity of each section, draw a line downward parallel to line *X-Y*. The center of gravity of an arc lies on a line which is perpendicular to and bisects the chord and is two-thirds of the distance from the chord to the arc.

5. From point *X* draw line *A* at 45° to point *P*, which is about midway between *X* and *Y*. Draw line *A'* parallel to line *A* intersecting the lines drawn in step 4.

6. Connect *P* to the ends of the sections on line *X-Y* obtaining lines *B*, *C*, *D*, and *E*. Draw parallel lines *B'*, *C'*, *D'*, and *E'*. Note that *B'* starts where *A'* intersects the first center-of-gravity line and so on until where *E'* starts where *D'* intersects the fourth center-of-gravity line and continues to intersect *A'*.

7. Through the intersection of *A'* and *E'* draw a horizontal line *Z* to the center line of the shell. Construct a circle using *Y* as the center point and *Z* as the diameter. Using point *X* as the center point, scribe an arc tangent to the small circle.

8. Draw a horizontal line tangent to the top of the small circle until it intersects the large arc. The distance from this intersection to line *X* − *Y* is the radius of the required blank.

Figure 10-11 shows a group of equations for blank diameters of differently shaped symmetrically drawn shells.

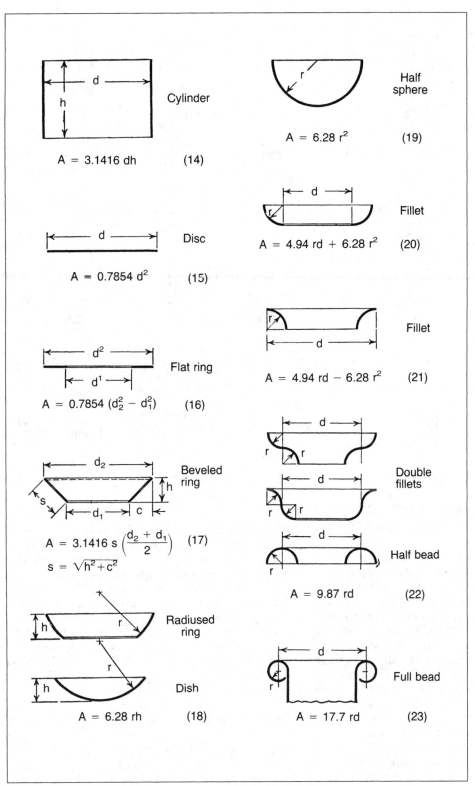

Cylinder

$$A = 3.1416\ dh \qquad (14)$$

Disc

$$A = 0.7854\ d^2 \qquad (15)$$

Flat ring

$$A = 0.7854\ (d_2^2 - d_1^2) \qquad (16)$$

Beveled ring

$$A = 3.1416\ s \left(\frac{d_2 + d_1}{2}\right) \qquad (17)$$
$$s = \sqrt{h^2 + c^2}$$

Radiused ring

Dish

$$A = 6.28\ rh \qquad (18)$$

Half sphere

$$A = 6.28\ r^2 \qquad (19)$$

Fillet

$$A = 4.94\ rd + 6.28\ r^2 \qquad (20)$$

Fillet

$$A = 4.94\ rd - 6.28\ r^2 \qquad (21)$$

Double fillets

Half bead

$$A = 9.87\ rd \qquad (22)$$

Full bead

$$A = 17.7\ rd \qquad (23)$$

Fig. 10-10. Equation for computing the areas of cylindrical shells.

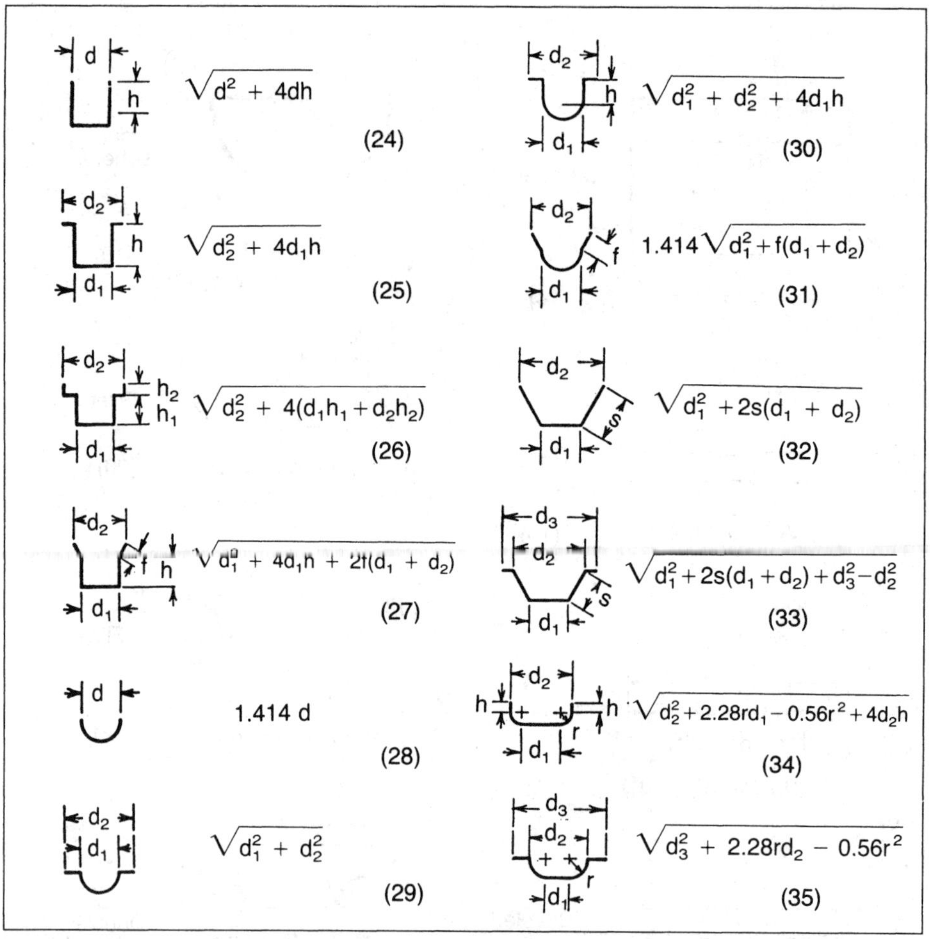

$$\sqrt{d^2 + 4dh} \quad (24)$$

$$\sqrt{d_2^2 + 4d_1h} \quad (25)$$

$$\sqrt{d_2^2 + 4(d_1h_1 + d_2h_2)} \quad (26)$$

$$\sqrt{d_1^2 + 4d_1h + 2f(d_1 + d_2)} \quad (27)$$

$$1.414\,d \quad (28)$$

$$\sqrt{d_1^2 + d_2^2} \quad (29)$$

$$\sqrt{d_1^2 + d_2^2 + 4d_1h} \quad (30)$$

$$1.414\sqrt{d_1^2 + f(d_1 + d_2)} \quad (31)$$

$$\sqrt{d_1^2 + 2s(d_1 + d_2)} \quad (32)$$

$$\sqrt{d_1^2 + 2s(d_1 + d_2) + d_3^2 - d_2^2} \quad (33)$$

$$\sqrt{d_2^2 + 2.28rd_1 - 0.56r^2 + 4d_2h} \quad (34)$$

$$\sqrt{d_3^2 + 2.28rd_2 - 0.56r^2} \quad (35)$$

Fig. 10-11. Expressions for computing the blank diameter.

MEASURES OF DRAWING

After the approximate blank size has been determined, consideration is given to the number of draws that will be required to produce the shell and the best reduction rate per draw. For diameter reduction, the area of metal held between the blankholding faces must be reasonably proportional to the area on which the punch is pressing, since there is a limit to the amount of metal which can be made to flow in one operation. The greater the difference between blank and shell diameters, the greater the area that must be made to flow, and the higher the stress required to make it flow.

After each differential of draw depth, the material has a new group of physical properties resulting from cold working. Elastic limit, hardness, yield point, and, to a lesser extent, ultimate strength are increased, and the plastic range is decreased. The total depth of draw is not limited by the plastic range. As a result, only the depth in one operation is restricted. Annealing may be resorted to after a draw to restore, almost entirely, the original plasticity.

Plastic deformation during these operations must be kept below the strain at the ultimate strength of the material, which is the forming limit. But optimum high strains should be achieved to obtain the best physical and mechanical properties of the product and for forming economy.

Important specific factors associated with metal displacement and strain relations in drawing and cupping operations require means of measurement and an understanding of certain dimensional changes and interrelationships. These may be determined mathematically by using the equations which follow.

Height of Cups or Tubular Parts. During cupping, the outside portion of the circular blank—that is, a ring having the outside diameter D and the inside diameter d—is folded and becomes the wall of the cup. The width of this ring H is theoretically the portion of the height of the cup that is not elongated and may be considered the quasi-initial height of the cup; thus $H = (D - d)/2$.

The depth or height of a cup can be derived from simple geometric considerations if, for the purpose of simplification, the following assumptions are made: (1) the cup has sharp corners, (2) the thicknesses of the cup wall and cup bottom are the same as the blank thickness, and (3) that terms involving the squares and the cube of the thicknesses can be ignored (these are inconsequential in practical computations). Therefore:

$$h = \frac{(D^2 - d^2)}{4d} \tag{36}$$

where h = height (depth) of cup, in. (mm)
D = diameter of circular blank (disk), in. (mm)
d = mean diameter of cup, in. (mm)

The height of a tubular part after n redrawing operations is

$$h_n = \frac{(D^2 - d_n^2)t}{4d_n t_n} \tag{37}$$

The diameter D of a blank of thickness t is approximated from the tubular part of a reduced shell wall t_n after n redrawing operations, by transposing Eq. (37).

Severity of Draw. The severity of draw or cupping ratio is often expressed by the ratio

$$\frac{D}{d} \quad \text{or} \quad \frac{d}{D} \tag{38}$$

Also widely accepted as a measure is the percentage reduction R_c of the diameters:

$$R_c = 100 \frac{D - d}{D} \quad \text{or} \quad 100 \left(1 - \frac{d}{D}\right) \tag{39}$$

The practical maximum reduction per drawing operation is approximately 50% if optimum conditions of tooling and metal are involved.

Strain Factor of Cupping. The traditional measures of draw severity (Eqs. (38) and (39)) are arbitrary and do not express the effective actual strain of the metal due to elongation during deep drawing. The plastic deformation or strain resulting from cupping can best be expressed by the strain factor E_c or $E_{c,n}$ after n redrawing operations:

$$E_c = \frac{h}{H} = \frac{(D^2 - d^2)2}{4d(D - d)} = \frac{D + d}{2d} = \frac{D/d + 1}{2} \tag{40}$$

$$E_{c,n} = \frac{h_n}{H_n} = \frac{(D^2 - d_n^2)2t}{4d_n(D - d_n)t_n} = \frac{(D + d_n)t}{2d_n t_n} \tag{41}$$

$$E_{max} = 1 + \frac{e}{100} \tag{42}$$

where e is maximum elongation at fracture, percent.

For computing the mean diameter d of a cup from the blank diameter D and the strain factor of cupping E_c, the following equation can be derived from Eq. (40):

$$d = \frac{D}{2E_c - 1} \tag{43}$$

10 - 11

Recommended maximum cupping reductions for a number of metals and alloys, under favorable conditions of part and tool design, are assembled in Table 10-2 and converted into the strain factor of cupping E_{max}.

TABLE 10-2
Recommended Maximum Reductions for Cupping[4]

Metal	Reduction in Diam., % (max)*	Cupping Ratio D/d	Strain Factor $E_{c,n}$	Reduction in Area, % (max) †
Aluminum alloys	45	1.80	1.40	28
Aluminum, heat-treatable	40	1.60	1.30	23
Copper, tombac	45	1.80	1.40	28
Brasses, high, 70/30, 63/37	50	2.00	1.50	33
Bronze, tin	50	2.00	1.50	33
Steel, low-carbon	45	1.80	1.40	28
Steel, austenitic stainless	50	2.00	1.50	33
Zinc	40	1.60	1.30	23

* $= 100(1 - d/D)$.
† $= 100(1 - a/A)$.

The relationship between the cupping ratio D/d and the strain factor of cupping E_t is shown in Fig. 10-12. This is a simplified log-log graph showing the logarithmic relationship as a straight diagonal line. For example, when D/d is 2.00, E_C is 1.50.

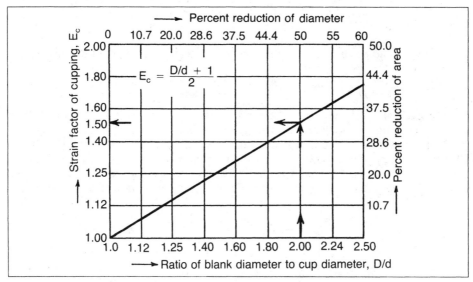

Fig. 10-12. Relationship between cupping ratio D/d and the strain factor of cupping E_c (logarithmic basis).[4]

Multiple Redrawing. When a series of n redraws is performed, the total strain factor E_t is the product of the respective single strain factors:

$$E_t = E_1 \times E_2 \times E_3 \ldots E_n \qquad (44)$$

The maximum value of the total strain factor E_t for n redraws depends mainly on the ductility of the material and its rate of strain hardening during cold working. When the maximum strain factor is reached before the final shape is produced, the metal being drawn requires annealing to prevent fracture during subsequent operations. The maximum of the single strain factors, however, is influenced by the ratios of the punch and die radii and the metal thickness, the lubrication, and other variables. Because of the cumulative work hardening of the material for each succeeding redraw, each strain factor must be smaller than the preceding one.

For many materials the first redrawing strain factor may be computed from the first drawing or cupping strain factor by $E_1 = E_c^m$, the exponent ranging from 0.4 to 0.8. For example, if $E_c = 1.32$ and the exponent is 0.6, then $E_1 = 1.18$. If the exponent is 0.4, $E_1 = 1.12$. A strain factor ranging from 1.12 to 1.18 is recommended for several equal redraws of most materials.

Strain Hardening. An increase in the strain factor of cupping E_c increases the elastic limit, the hardness, the yield point, and, to a lesser extent, the ultimate strength of the material. In the general practice of deep drawing, the plastic limit or ultimate strength, which is determined by the strain-hardening exponent N, is usually not utilized. The reasons for this are the practical considerations of tool design which limit a single drawing operation. A subsequent draw or even several draws after cupping may be possible before annealing becomes necessary to restore the original plasticity of the material. The tensile strength S_c of a drawn cup is given by

$$S_c = S_0 \log_e E_c^{\,N} \qquad (45)$$

According to its definition, the strain factor $E = 1.0$ is associated with the soft (annealed) metal having the original or yield strength S_0. When the metal is work- or strain-hardened to its limit, the ultimate tensile strength S_u occurs, with which the ultimate strain factor E_{max} [Eq. (42)] is associated. In general, the characteristic data of these specific mechanical properties are supplied with the material by the mill, or they can be determined by the standard tensile test.

The relationship between strain factor and tensile strength, for practical purposes and for most metals, is linear if plotted on a log-log graph. This linear relationship is shown in Fig. 10-13 for a material having the original strength $S_0 = 50,000$ psi (345 MPa) at $E = 1.0$ (annealed blank) and the ultimate strength $S_u = 90,000$ psi (620 MPa) at $E_{max} = 1.6$ where the metal fractures. Similar graphs can be drawn for most metals, using their characteristic mechanical data.

If die and drawing conditions permit the strain factor of cupping of $E_c = 1.32$, then the tensile strength of about $S_c = 71,000$ psi (489 MPa) will occur in the upper portion of the cup wall, which is strained most during cupping. This tensile strength is found by following the reading direction indicated by the arrows of the graph.

In most cases, the ultimate strength of the material is not reached by cupping, as shown in Fig. 10-13, and the cup can be further strained by redrawing without anneal. The ultimate value of strain E_{max}, however, cannot be safely utilized since fracture is imminent. A somewhat smaller value, for example, $E_t = 1.5$, may be chosen as the safe total strain factor. This value, as shown by the corresponding arrows of the figure, will produce the tensile strength of about $S_r = 82,000$ psi (565 MPa) in the upper portion of the redrawn cup wall. The required strain factor for the redraw is given by Eq. (44) and is for this example: $E_r = E_t / E_c = 1.5/1.32 = 1.12$.

Strain Limits. Cupping involves various amounts of displacement (flow) of the blank metal. The metal flow is uniform for each volume unit of metal lying at the same distance from the die bore; however, metal flow is different for units of different distances from the die bore. Therefore, the introduced strain factor of cupping represents only the integrated effect of strain of all volume units taking part in the total metal flow. The periphery of the blank is subjected to maximum flow and is most compressed when it moves over the inlet of the die to its bore. This causes a decrease of surface area of the blank and a corresponding increase in thickness. However, the final thickness is limited to the clearance between the punch and the die, not considering small amounts of

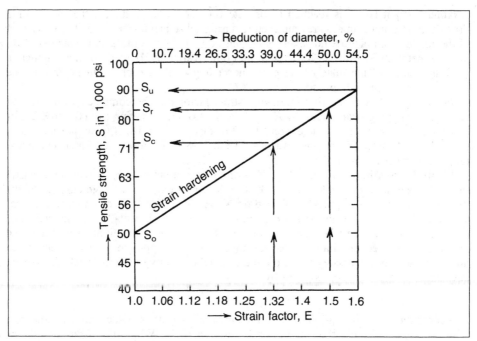

Fig. 10-13. Relationship between strain factor, strain hardening, and tensile strength.

springback. Depending upon this clearance, the wall may be thicker at the open end of the cup than the original blank. Variations in thickness of the shell wall can be removed by subsequent ironing (wall thinning).

In addition to the influence of the cupping ability of the metal (Table 10-2), the ratio of the blank or previous shell diameter to the metal thickness D/t determines the maximum strain factor which can be applied to the cupping, the first redraw, and any subsequent redrawing operations, as well as whether single or double-action deep drawing can be applied. Figure 10-14 shows this relationship schematically. For example, if the ratio of blank diameter to metal thickness (left ordinate) is relatively small, $D/t = 25$, a single-action die and only a relatively small strain factor of about 1.5 will be permissible, even for quite ductile metals. For a larger ratio, for example, $D/t = 50$, a total strain factor of 1.7 to 1.8 is possible, and cupping may be followed by a single redraw operation (E_cE_1) without anneal. If the ratio D/t is larger than 63, a second single redraw $(E_cE_1E_2)$ may follow the first redraw without anneal.

For double-action dies the minimum ratio D/t is about 160, and additional redraws may be possible if D/t is larger than 250. The values shown are for orientation and first approximations only, since the limits depend upon the material, the geometry of the dies, and the drawing conditions employed.

Cupping Force and Work. The draw force required in cupping F_c is a function of the strain factor and the other variables shown in the following equation:

$$F_c = a_w S_y \eta_c \ln E_c \qquad (46)$$

where F_c = cupping force, lbf (N)
a_w = cross-sectional area of cup wall ($a_w = \pi dt$), sq in. (sq mm)
S_y = yield strength of material, psi (Pa)
η_c = deformation efficiency of cupping
$\ln E_c$ = natural logarithm of strain factor E_c

Deformation efficiency η_c is a factor accounting for friction and bending and decreases with increasing strain factor (Fig. 10-15). The expression $D/d - C$, instead of the product

10 - 14

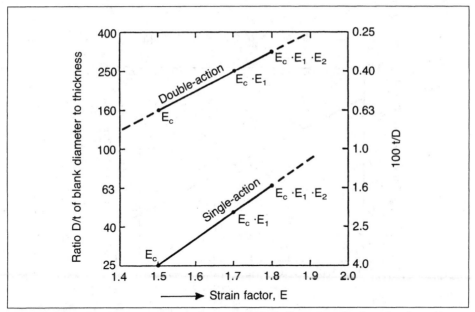

Fig. 10-14. Relationship between strain factors and blank dimensions for deep drawing with single- and double-action dies.

$\eta_c \ln E_c$, can be used in Eq. (45), where C is an empirical constant varying from 0.6 to 0.7. Thus

$$F_c = \pi d t \, S_y (D/d - C) \tag{46A}$$

Equation (46) does not include the force needed to overcome the hold-down force of the blankholder. The total force F_t required for drawing a cup from a blank which is held down by a blankholder is approximated by the sum of the draw force F_c and the hold-down force for the blank F_h:

$$F_t = F_c + F_h \tag{47}$$

F_h is an inverse function of the blank thickness t. By expressing this relationship as a draw-force factor X_t, as shown in Fig. 10-16, the total force F_t can alternatively be expressed for convenience by

$$F_t = F_c X_t \tag{47A}$$

The work to be done must be known for the design of draw presses where spring, pneumatic, or hydraulic actions are employed, and the amount of work is approximated by the product of F_t, the height of the cup h (in feet), and the work factor X_w:

$$\text{Work} = F_t h X_w \text{ in foot-pounds (J)} \tag{48}$$

The work factor is a simple function of the strain factor, and the deformation efficiency X_w, as shown in Fig. 10-15, is smaller than 1.0, since only a fraction of the final cup height is derived from actual straining. This fraction is shown as a percentage of the total cup height on top of the graph. The complement, that is, the difference between 100% and the percentage shown, derives from the width of the folded outer ring of the original blank, as previously mentioned.

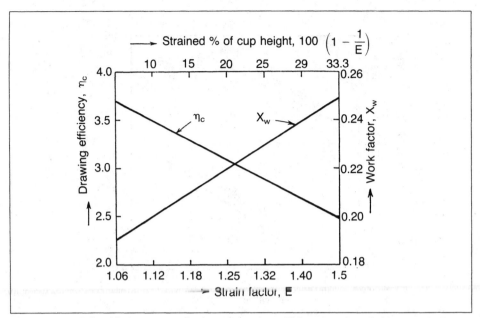

Fig. 10-15. Relationship between strain factor, drawing efficiency, and work factor.

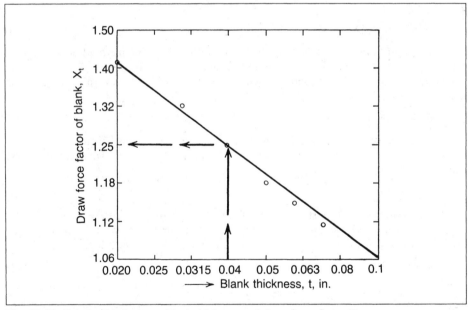

Fig. 10-16. Relationship between blank thickness and draw-force factor X_t.

DIE REQUIREMENTS FOR DEEP DRAWING OF CYLINDRICAL SHELLS

Draw and Punch Ratio. The draw radius of the draw die should be kept as large as possible to aid in the metal flow, but if it is too large, the material will be released by the blankholder too soon, and wrinkling will result. When the radius is too small, the material will rupture as it is bent around the draw edge.

The ratios of the die and punch radii to the metal thickness largely determine the maximum strain factor for a single deep drawing operation. The larger the strain factor, the larger the ratios of the die and punch radii to the metal thickness must be. For example, if the die radii are equal to the metal thickness, a strain factor not larger than $E_c = 1.25$ should be chosen to prevent rupture of the metal. If $E_c = 1.32$, the die radii must be increased to at least twice the metal thickness. For the strain factor $E_c = 1.5$, the draw-die radius must be between five and ten times the metal thickness, and the punch radius should be at least five times the metal thickness.

When cupping without a blankholder, it has been found that wrinkling will not occur if the flat contact distance between the blank and the die does not exceed three times the thickness of stock. The width between the die opening and the point of contact should not exceed twenty times the stock thickness. The shape of the inlet to the die opening may be a true radius or an elliptical curve or a taper of 45% to 60% measured from the horizontal. Figure 10-17 illustrates these points.

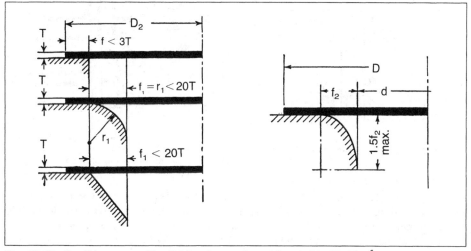

Fig. 10-17. Conditions to avoid wrinkling in cupping without a blankholder.[5]

The drawing, without a blankholder, of cupshaped parts of heavy-gage metals often requires draw radii of six to eight times the stock thickness. This reduces the width of the flat surface of the die upon which the blank lies. A taper or an elliptical curve may be used to increase the width of this surface and also to aid the flow of metal into the die. The minor diameter of the ellipse may be approximately one half the difference between the blank diameter and the shell diameter. The major diameter may be up to 1.5 times the minor diameter.

There is no set rule about how large the punch-nose radii should be for each successive die, except that each should be proportionally smaller than the succeeding shell. To prevent excessive thinning of the bottom of the cup, the punch-nose radius should be from four to ten times the metal thickness. When sharp radii have been used in the first draws, thinning often appears on the side wall of later operations as a line or a depression and is increasingly higher on the wall as the diameter is reduced. The nose radius and sides of the punch should be polished with vertical strokes, especially when drawing soft metals, to eliminate cross pockets into which the metal might flow and cause fracture when the metal is stripped from the punch.

Shells may be prepared with angular corners, as shown in Fig. 10-18. From the layout of the finished shell, the layout of each preceding shell is developed. The angle in the bottom of the last preliminary shell should start at a point equal to one-fourth of the bottom radius of the finished shell, measured inward from the inside of the shell. Angles for the other shells should start at a point equal to one-half the Y dimension. The angle Z measured from the horizontal should be 30° for stock up to 0.030 in. (0.76 mm) thick, 40° for 0.030- (0.76-mm) to 0.060-in. (1.5-mm) stock, and 45° for stock over 0.060 in. (1.5 mm) thickness. The radius r' at the intersection of the slope and the bottom and sides should be approximately $0.6Y$.

The relation of the punch-nose and die draw radii to minimize thinning of stock is shown in Fig. 10-19. The center point of the draw radius should be approximately 0.125 in. (3.18 mm) outside the previous cup, as illustrated at A. The center point of the punch-nose radius should be slightly inside the following shell, as at B. The center points of the punch-nose radii on the last two operations are about on the same line, maintaining the flat on the bottom of the cup, as at C.

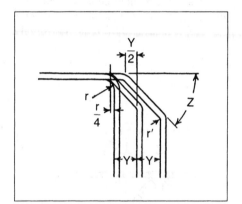

Fig. 10-18. Layout of bottom corners.

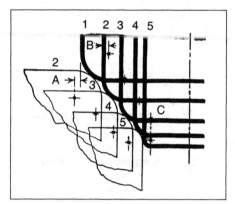

Fig. 10-19. Layout with sequence of punch-nose and draw-die radii for minimizing metal thinning.

Clearances. The die space usually allowed for drawing any metal should be proportional to the metal thickness plus an allowance to prevent wall friction. This allowance ranges from 7% to 20% of the metal thickness, depending upon the type of operation and the metal. As the shearing strength of the stock decreases, the allowance must be increased. Table 10-3 gives factors for determining typical die-clearance dimensions. The sizing draw clearance is used for straight-sided shells where diameter or wall thickness is important, or where it is necessary to improve the surface finish in order to reduce finishing costs.

TABLE 10-3
Draw Clearance to Blank Thickness Relationships*

Blank Thickness, in. (mm)	First Draws	Redraws	Sizing Draw†
Up to 0.015 (0.38)	1.07 to 1.09	1.08 to 1.1	1.04 to 1.05
0.016 to 0.050 (0.41 to 1.27)	1.08 to 1.1	1.09 to 1.12	1.05 to 1.06
0.051 to 0.125 (0.13 to 3.18)	1.1 to 1.12	1.12 to 1.14	1.07 to 1.09
0.136 and up (0.45)	1.12 to 1.14	1.15 to 1.2	1.08 to 1.1

* Expressed as a factor of blank thickness.
† Used for straight-sided shells where diameter or wall thickness is important, or where it is necessary to improve the surface finish in order to reduce finishing costs.

Air Vents. An air vent should be put in the punch and die to eliminate air pockets, which tend to collapse the cups when stripped from the die. Table 10-4 shows the practical air-vent diameters for different punch diameters when only one air vent is used in the tool. On noncircular shapes it might be advantageous to use two or more air vents. To prevent plugging of the air vents with drawing compound and dirt, they must be placed in such a position that they can be easily cleaned out.

TABLE 10-4
Air-vent Diameters[6]

Punch Diam., in. (mm)	Air-vent Diam., in. (mm)
Up to 2 (51)	0.188 (4.78)
2 to 4 (51 to 102)	0.250 (6.35)
4 to 8 (102 to 203)	0.312 (7.92)
Over 8 (203)	0.375 (9.52)

Blankholders. If used, a blankholder must be so aligned that the blank rim or flange is allowed to thicken when it approaches the die bore during cupping. Radial and circumferential forces (Fig. 10-1) shrink the blank rim and cause a temporary thickening of the metal until it is forced into the clearance between the draw die and punch. In this clearance, the thickening process is reversed, and the metal is strained (elongated). If the finished shell wall is as thick as the blank, the amount of thickening before entering the clearance is equal to the amount of subsequent straining in the clearance. The strain factor E_c [Eq. (40)] can therefore be used as the measure of thickening:

$$\frac{t_t}{T} = E_c = \frac{D/d + 1}{2}$$

(40A)

where t_t = the temporarily increased thickness of the blank
T = the initial thickness of the blank and thickness of the finished cup

If a rigid blankholder with a flat surface is used, the clearance between the holder and the upper surface of the blank must be carefully adjusted to allow the thickening and at the same time prevent development of wrinkles. So, the total space for the gap should be equal to, or slightly smaller than, t_t. The critical adjustment of a flat blankholder can be facilitated by using a tapered instead of a flat hold-down surface, permitting the progressive thickening of the blank. The widest opening of this taper does not need to be close to the theoretically required value of t_t and may vary plus or minus 50%. Furthermore, a tapered blankholder normally causes less friction than a flat one; consequently the additional draw force, necessary to overcome the friction of the blankholder, may be smaller with a tapered blankholder than with a flat one.

IRONING

Ironing, sometimes called "thinning," is a method of redrawing a tubular shell to reduce the wall thickness and assure a smooth, uniform wall surface with only minor reduction of the ID.

There are two reasons for ironing the side walls of a shell. One is to counteract or correct the natural thickening of the wall. This may be accomplished in the final redrawing operation by making the clearance per side between the punch and die equal to the original stock thickness. The second reason is to reduce its wall thickness considerably. This is done by making the clearance per side between the punch and die less than the thickness of the shell wall so the metal is thinned and the length of the shell is increased as it is forced through the die by the punch (Fig. 10-20). Seamless tubes and cartridge cases of steel and brass are examples of products which require ironing.

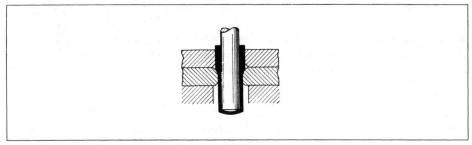

Fig. 10-20. Ironing operation partially completed.

Measure of Ironing. The principal measures of ironing, including shell-wall thickness reduction, strain factor, height of shell wall after ironing, and approximate working pressure required, are as follows.

The traditional measure of ironing is expressed by the per cent reduction of the shell-wall thickness:

$$R_i = 100 \frac{t_0 - t_1}{t_0} = 100 \left(1 - \frac{t_1}{t_0} \right) \tag{49}$$

However, this measure is impractical since it shows no simple relationship to the actual strains, and its usage should be discouraged.

The measure of strain caused by ironing is best expressed by the ratio of the wall thickness, before and after ironing.

$$E_i = \frac{t_0}{t_1} \tag{50}$$

The ratio must be in proportion to the ductility of the given material. Unless the metal being ironed is especially ductile, the ratio should be about 2:1 (50% reduction of wall thickness) for a single ironing operation. For most ductile and fully annealed alloys, a ratio of 2.5:1 (60% reduction of wall thickness) or higher can be utilized in the first ironing step.

Multiple Ironing Operations. Several ironing operations may be performed on a shell, but because of the resulting work (strain) hardening, there is a decreased ratio or strain factor for each subsequent ironing step. The resulting final strain of n ironing steps is the product of the n single ratios or strain factors:

$$E_{it} = E_{i1} = E_{i2} \times E_{i3} \ldots E_{in} \tag{51}$$

$$\frac{t_0}{t_n} = \frac{t_0}{t_1} \times \frac{t_1}{t_2} \times \frac{t_2}{t_3} \ldots \frac{t_n - 1}{t_n} \tag{51A}$$

The ability to become plastically deformed is a characteristic of a metal (depending upon the working conditions) and may be expressed by the maximum elongation factor for ironing $E_{max} = t_0/t_n$. When $E_{it,max}$ is reached, the metal must be annealed in order to restore its ductility by recrystallization or recovery. For metals which increase their ductility by extensive cold working and recrystallization, $E_{it,max}$ will grow larger within an extended working range. In relatively small working ranges, however, $E_{it,max}$ may be considered to be constant. It does not matter whether $E_{it,max}$ is reached in one or more processing steps. The total elongation factor t_0/t_n cannot be larger than $E_{it,max}$ without causing failure.

A total elongation factor larger than $E_{it,max}$ must be subdivided into suitable single elongations by the application of Eq. (51). For example, for a cup having a total elongation factor t_0/t_n of 3.15, the ironing is best done in two or more operations. The first operation is a large reduction in wall thickness; the second, a correspondingly smaller reduction. So, if $t_0/t_1 = 2.0$ is selected for the first operation, then, for the second operation, using Eq. (51A), $t_1/t_2 = 1.6$.

For convenient orientation, Table 10-5 shows some ironing strain-factor values, a statement of relative ironing ability, and the conventional percent reduction of wall thickness. The conditions or materials in the last column are in the row corresponding to their approximately maximum ironing value. The common ratio (1:12:1) of this series expresses equal strain intervals.

TABLE 10-5
Ironing Strain-Factor Values and Relative Ability of
Ironing of Various Materials

$E_{it.\max}$	Relative Ironing	Reduction R_i %*	Material (Example)
1.25	Extremely little	20.0	(For third draws)
1.4	Very little	28.6	
1.6	Little	37.5	(For second draws)
1.8	Medium	44.4	Steel
2.0	Medium-good	50.0	Stainless
2.24	Good	55.4	Copper
2.5	Very good	60.0	Aluminum
2.8	Excellent	64.3	
3.15	Extreme	68.3	

* $R_i = 100(1 - t_1/t_0)$.

It is often advisable to anneal the shell between the ironing operations, securing optimum efficiency for each ironing step. Greater ironing by one stroke is possible by using two-step (tandem) dies, such as shown in Fig. 10-21, rather than one single die, provided the two dies are spaced sufficiently to cause little or no interference with each other. The subsequent ratio t_1/t_2 of ironing by the second die must be smaller than the preceding ratio by the first die and depends upon the ratio s/d of the spacing s of the dies to the cup diameter and upon the ratio d/t of the cup diameter to the wall thickness t of the cup.

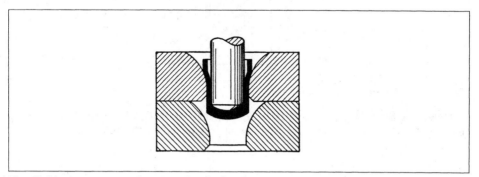

Fig. 10-21. Arrangement of tandem dies for ironing.

Very long thin parts are usually fabricated by first drawing a cup and then redrawing it without intentionally thinning the shell wall, until the cup diameter is sufficiently small to permit heavy ironing with minimum reductions of the inside diameter. This working mode produces a more uniform metal structure with a minimum of harmful residual stress in the finished part assures desirable even wall thickness and provides a smooth surface finish.

A die to reshape and iron the side walls of the cup in Fig. 15-11 is shown in Fig. 10-22. This die has an entrance angle of 45% per side with a radius blending it into the straight or bearing surface in the die cavity. The bearing surface is 0.250 in. (6.35 mm) wide, and the remainder of the die is relieved with a 0.5° angle per side. The part is pushed through the die and is stripped from the punch by three equally spaced spring-loaded stripping fingers. The punch and die are polished smooth and free from all grind marks, eliminating tiny pockets on the punch surface in the areas contacted by the part.

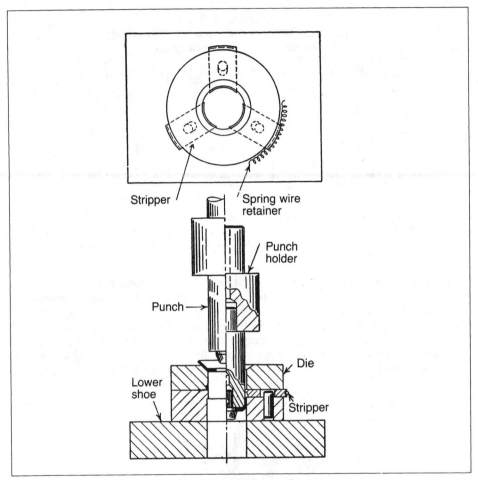

Fig. 10-22. Die to iron wall of cup to thickness. *(Coinex, Inc.)*

Height of Shell. By ironing the shell wall, the height h_0 of the shell is increased (elongated). The resulting height h_1 may be approximated by the equation

$$h_1 = \frac{h_0(D - d_1)t_0}{(D - d_0)t_1} \tag{52}$$

Ironing Force. An ironing operation is usually made without the use of a blankholder since the reductions in diameter are nil when the punch is fitted closely to the inside diameter of the shell or cup. The contour of the die used for ironing determines to a great extent the amount of draw force required for the operation. Most used are bell-mouthed dies with an included angle of 10° to 45°.

The force needed to iron a shell increases as the entrance angle of the die is decreased, increasing the length of contact of the metal with the die. The length of the straight or working face in the die slightly increases the force, in proportion to the length of the face. The force F_i for ironing, not considering die shape and friction, is approximated by

$$F_i = \pi d_1 t_1 S_{av} \ln \frac{t_0}{t_1}$$ (53)

$$F_i = \pi d_1 t_1 S_{av} \, 2.3 \log \frac{t_0}{t_1}$$

where d_1 = mean diameter of shell after ironing, in. (mm)
t_i = thickness of shell wall after ironing, in. (mm)
t_0 = thickness of shell before ironing, in. (mm)
S_{av} = average of the tensile strengths of the metal before and after ironing
ln = natural logarithm (base e)
log = common logarithm (base 10)

In the determination of the working pressure for ironing, some authorities recommend increasing the calculated value of F_i by 20% to compensate for friction between the metal and the die.

BLANK DEVELOPMENT FOR NONCYLINDRICAL SHELLS

Blanks for Rectangular Shells. The combination of drawing, side flow, and bending of metal during the drawing of seamless rectangular shells creates the problem of finding a shell blank that will assure the proper amount of metal for forming the shell corners without ears, folds, tears, or excessive thickness and, if possible, without the necessity for a trimming operation.

To start the layout for the blank, a rectangle $ABCD$ is drawn as shown in Fig. 10-23, having a width of $W - 2r$ and a length of $L - 2r$, where L, W, and r are the length, width, and bottom-corner radius, respectively. By continuing the sides beyond the points A, B, C, and D, for a length equal to $h + 1.57r$, where h is the height of the flat portion of the sides of the finished shell, and connecting the points by lines parallel to the outline of the original rectangle, the outline of a shell blank is obtained where bending only is done during the drawing operation. To this outline add the quadrants with a radius equal to R_c. The value of R_c is obtained by the equation

$$R_c = \sqrt{2Rh + R^2} + 1.41Rr$$ (54)

where R = corner radius, in. (mm)
h = height of the flat portion of sides, in. (mm)
r = bottom radius, in. (mm)

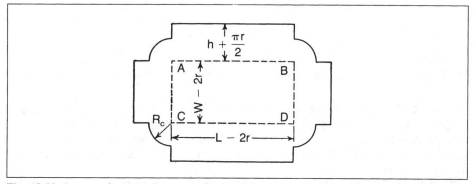

Fig. 10-23. Layout of a blank for a rectangular shell.

Theoretically, this blank outline (solid line) contains enough metal to draw the shell. Because the blank has sharp corners, slits, folds, and overlaps will occur at the points where the quadrants meet the side walls when the shell is drawn. The sharp corners must be blended by sloping curves by taking the metal from the side walls and adding it to the quadrants without any change in the amount of metal in the blank.

The method for blending the corners of rectangular blanks by sloping curves is illustrated in Figs. 10-24A to 10-24C. To develop the corners proceed in the following manner:

1. Bisect ab and cd. Through the dividing points e and f, respectively, draw gh and ij tangent to the quadrant arc.

Three cases may now occur as to the position of these tangents with respect to the quadrant are: these tangents may coincide (Fig. 10-24A), they may cross each other outwardly (Fig. 10-24B), or they may cross each other inwardly (Fig. 10-24C).

2. Draw arcs with a radius R_c, which will touch the outer edges of the side walls and the tangents gh and ij.

3. In case the tangents gh and ij cross each other before they touch the quadrant arc

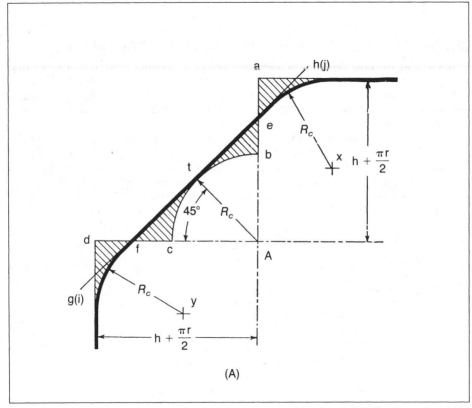

Fig. 10-24A. Corner development of blanks for rectangular draws: a straight line is produced if $R_c = 0.54(h + 1.57r)$.[7]

(Fig. 10-24C), an additional outside arc must be drawn for forming the sloping curve.

The development of the blank corners in this manner assures even distribution of the metal because the areas of the shaded curvilinear triangles outside the sloping curves in the side wall are equal to the areas of similar triangles added to the quadrant arcs inside the sloping curve.

Kidney-shaped Shell. Figure 10-25 shows a layout of a blank for an irregular kidney-shaped shell having one flat side with the corners of a radius r blending into the larger radius R and having a depth h. The diameter of the blanks containing the metal required to form shells having the radii R and r to a depth h may be calculated, using Eq.

10 - 24

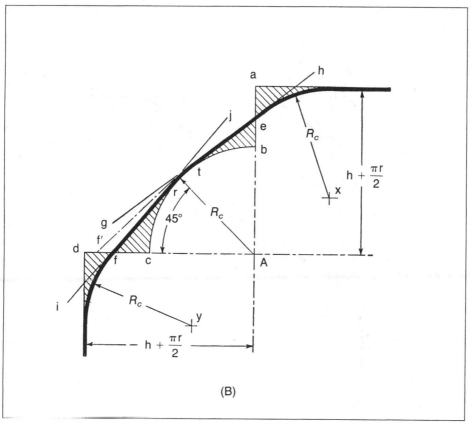

Fig. 10-24B. Corner development of blanks for rectangular draws: a convex sloping curve is obtained if $R_c > 0.54(h + 1.57r)$.

(1). Line F is drawn parallel to the straight side, a distance h below it. Arcs R' and r', radii of the above calculated blanks, are constructed from the appropriate centers. Blend the radii R' and r' and line F; the resulting outline will be a fairly accurate development of the blank shape and size necessary to produce the shell.

Elliptical Shell. Figure 10-26 shows a layout of a blank for an elliptical shell, with a minor ellipse axis AA' and a major ellipse axis BB'. The loci of radii c and R_2 are given as N and N', and the loci of radii d and R_1 are given as M and M'. The blank radii R_1 and R_2 are found by graphical and formula solution. To determine point O draw a horizontal line from A and a vertical line from B. The intersection is point O. The locus for R_2 is determined by drawing a line from O at right angles to the line AB', which connects A and B', and extending that line until it intersects the vertical center line at N'. This is the locus for R_2. The locus for R_1 is the point at which the drawn line ON' intersects the horizontal center line at M'. The blank radii R_1 and R_2 are found by application of the equations[8]

$$R_1 = \sqrt{(d - R_s)^2 + R_s(d - R_s) + 2R_s^2 + 2d(h - R_s)} \tag{55}$$

$$R_2 = \sqrt{(c - R_s)^2 + R_s(c - R_s) + 2R_s^2 + 2c(h - R_s)} \tag{56}$$

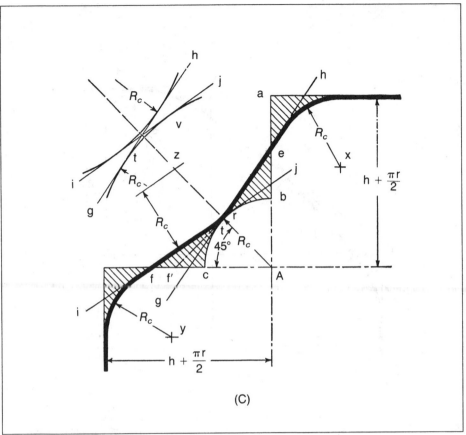

(C)

Fig. 10-24C. Corner development of blanks for rectangular draws: a concave sloping curve is produced if $R_c < 0.54(h + 1.57r)$.

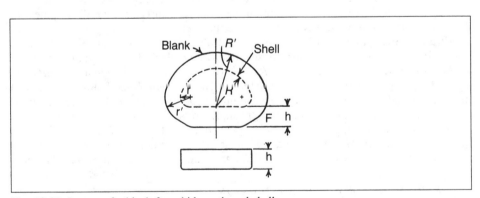

Fig. 10-25. Layout of a blank for a kidney-shaped shell.

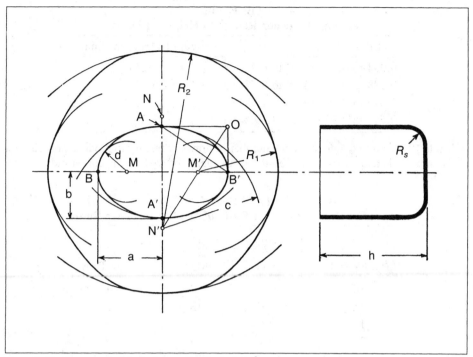

Fig. 10-26. Layout of a blank for an elliptical shell.[8]

DIE REQUIREMENTS FOR RECTANGULAR DRAWS

The number of drawing operations required to complete a rectangular-shaped part is governed by several factors, such as the quality of the material and its thickness, the corner radius, and the radius at the bottom edge of the part.

The limit of drawing a box can be expressed in various ways. Oval cups that are nearly cylindrical, and square or nearly square cups, can be drawn in one operation if the area of the blank is not four times greater than the cross-sectional area of the punch. For boxes with a length-to-width ratio between 1 and 9, the ratio of the blank area to the punch area may be somewhat larger than for square boxes; the area ratio increases to a maximum value of 4½ at a length-to-width ratio of 3 and decreases as the length-to-width ratio increases. Draws may be made to a depth of approximately 80% of the width in case of a square box with a rather small corner radius. Rectangular boxes can be drawn to a greater depth than can square boxes, and this maximum depth H increases with increasing ratio of length L to width W according to the following empirical equation:

$$\frac{H}{W} = C \sqrt{\frac{L}{W}} \text{ (in percent)} \tag{57}$$

The constant C in this equation varies slightly with the metal; use 80 for steels and such ductile metals as copper and brass, 70 to 75 for ductile aluminum alloys. The relation applies to boxes not longer than three times their width. For longer boxes, the maximum depth to which they can be drawn in a single operation is approximately 135% of their width.

It is also safe to assume that a part can be drawn in one draw to a depth of four to six times the corner radius. If the corner radius is greater than 0.50 in. (12.7 mm), the depth for the draw should be kept nearer the four times factor. Table 10-6 gives an idea of maximum depths that can be obtained from corners of a given radius.

TABLE 10-6
Relation of Corner Radius to Height of Draw[2]

Corner Radius, in. (mm)	Length of Draw, in. (mm)
0.094 to 0.188 (2.39 to 4.78)	1 (25)
0.188 to 0.375 (4.78 to 9.52)	1.5 (38)
0.375 to 0.500 (9.52 to 12.70)	2 (51)
0.500 to 0.750 (12.70 to 9.54)	3 (76)

Table 10-7 is used to determine the number of draws necessary to produce non-circular aluminum shells and is based on the ratio of the depth of shell to the corner radius (h/r).

TABLE 10-7
Number of Draws Based on Ratio of Depth to Corner Radius

Basic h/r Value	Allowable Range		No. of Draws
	Min	Max	
6	. .	7	1
12	7	13	2
17	13	18	3
22	18	24	4

Data from J. W. Lengbridge.

The amount of reduction between draws for a rectangular part depends upon the corner radius and diminishes as it becomes smaller. Where two or more draws are required, the length and width of each die can be determined by multiplying the corner radius by 3 and adding the product to the length and width, thus finding the length and width of the preceding die. This method should apply only to a corner radius of less than 0.50 in. (12.7 mm). For all radii over 0.50 in. (12.7 mm), use a constant of 0.5 instead of the radius. The corner radius of the first die may be as much as four or five times the radius of the succeeding die or finished part. The radii of the two dies should not be laid out from the same center but, as shown in Fig. 10-27A, with enough surface X between the two corners to provide a drawing edge. The reason for using a large corner radius for the first die is that, when the larger corners are reduced to the smaller radius in the second die, a large part of this compressed metal is forced out into the sides of the box. If the first die were laid out as in Fig. 10-27B there would be a comparatively large reduction at the corner and the metal would be more compressed. The drawing operation would therefore be made much more difficult, since the drawing action is confined to the corners when drawing rectangular work.

Punch-and-die Radii. The size of the draw and bend radii on the draw ring is generally the same for rectangular draws as for circular draws. Some designers prefer to make the radius between the corners smaller than the radius at the corners in order to equalize the stresses in the metal at the corners. To perform some very deep draws successfully, the radius on the draw ring may be from four to ten times the metal thickness. The top surface of the draw die and the draw radii should be polished smooth, free of grind marks and well blended together, to prevent localized retardation flow with consequent uneven drawing of the metal.

There may be bulged areas in boxes of thin stainless steels and other high-strength alloys, where the length of the flat area extends over fifty times the metal thickness, which may be made to deflect by snap action referred to as "oil canning." To eliminate this, such parts are formed in two operations using slightly different tools, with an annealing

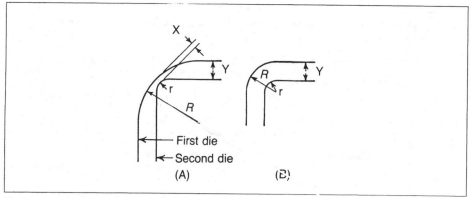

Fig. 10-27. Relation between corners of first- and second-operation dies for square and rectangular work.[2]

between operations. The second draw should stretch the metal into shape to eliminate the canning. To facilitate stretching in the center of the long walls of the box-shaped part, the nose radius of the first draw punch may be enlarged at these locations. The use of a constant radius on the nose of the redraw punch then provides the stretching action.

If the corner radius of a box-shaped shell is smaller than the punch-nose radius, a corner relief should be provided at the intersection of the punch-nose and corner radii to avoid tearing at these locations. Beveling the punch most at the corners permits the part to develop natural contours. The beveled portion should extend sufficiently far to permit the metal to curve at a radius equal to at least five times the metal thickness.

Draw Clearance. The clearance between the punch and die for a rectangular shell is about the same as for a cylindrical shell. There should be a little more clearance allowed at the corners than along the side walls. Some designers prefer to use the same corner radius on the die as on the punch to avoid ironing in these areas and to increase drawability.

Drawing Force for Rectangular Draws. Calculating the punch load for a rectangular shell should take into consideration the straight side-wall areas where only bending and friction are involved and stresses are low, and the corner areas where high compressive stresses are necessary to rearrange the metal. These areas are covered in the following equation:[9]

$$F_r = tS \, (2\pi r C_1 + L C_2) \tag{58}$$

where F_r = drawing punch force, lbf (N)
　　t = metal thickness, in. (mm)
　　S = nominal ultimate tensile strength, psi (Pa)
　　r = corner radius of the rectangular shell, in. (mm)
　　L = total length of straight sides of rectangular shell, in. (mm)
　　Constant C_1 = 0.5 for a very low shell, up to about 2 for a shell having a depth of five or six times the corner radius r
　　Constant C_2 = say 0.2 for easy draw radius, ample clearance, and no holding pressure; or about 0.3 for similar free flow and a normal blankholding pressure of about $P/3$; or a maximum of 1 for a metal clamped too tightly to flow

These values for C_1 and C_2 are roughly empirical, and judgment must be used in their application.

References

1. J.W. Lengbridge, "Theory and Practice of Pressing Aluminum," *The Tool Engineer* (1948-1949).
2. Frank W. Wilson and Philip D. Harvey, eds., *Tool Engineers Handbook*, 2nd ed. (Dearborn, MI: Society of Manufacturing Engineers, 1959).
3. Hjalmar Dahl, "How to Determine Exact Blank Diameter," *The Tool Engineer* (August 1943).
4. Anton F. Mohrnheim, "Strain Relations in Cupping and Redrawing of Tubular Parts," *The Tool and Manufacturing Engineer* (March 1963).
5. G. Sachs, *Principles and Methods of Sheet-metal Fabrication*, (New York: Reinhold Publishing Corporation, 1951).
6. F. Strasser, "Die Makers Kinks," *American Machinist* (June 23, 1952).
7. Serguis P. Brootzkoos, "How to Calculate Blanks for Seamless Rectangular Shells," *American Machinist* (July 28, 1949).
8. Ferenc J. Kuchta, "Blanks for Elliptical Shells," *American Machinist* (September 5, 1960).
9. E. V. Crane, *Plastic Working of Metals and Non-metallic Materials in Presses* (New York: John Wiley & Sons, Inc., 1944).

SECTION 11

PRODUCT DEVELOPMENT FOR DEEP DRAWING

Deep drawing requires dies that are carefully designed in order to produce high-quality stampings at a reasonable cost. Lack of attention to factors such as the depth of draw and shape of the blankholder can result in an operation that requires continuous adjustment of binder pressure and blank location—even with the finest quality drawing steel.

Analytical Tools. Analysis of drawability should start in the product design stage. The application of computer-aided formability analysis at this stage can provide vital information as to the manufacturability of a proposed part design.

If computer-aided formability analysis shows the part to have a good safety margin, the design of hard tool can begin. In cases where the part design is shown to provide marginal drawability, a soft tool trial may be required.

Soft Tool Development. In order to avoid costly delays, it is wise to build a temporary draw die to prove out the feasibility of a proposed deep-drawn stamping design. Section 21 provides information on such tooling.

An important, but often overlooked factor in determining stamping feasibility with soft tooling is the forming rate. It is normal to use low-speed hydraulic presses to operate soft tools. Hard tools are usually operated in much faster mechanical presses. Experience has shown that some types of fractures result from too slow a forming rate. One explanation is that the slow forming rate does not work-harden the metal rapidly enough, and it continues to thin rather than pulling metal from other areas.

Circle Grid Analysis is a press shop analytical technique used in many press shops throughout the world. This important investigative tool should be used to document the soft tool trial. This method of developing "safe" stampings is described in detail later in this section.

COMPUTER-AIDED FORMABILITY ANALYSIS*

Computer simulation of sheet metal forming processes to determine manufacturability of a part can be beneficial when applied early in the design process. With today's emphasis on improved quality, improved lead times, and simultaneous engineering, there is a great need for such predictive manufacturing analysis tools. Considerable progress has been made in the application of computer aided analytical methods to model sheet metal forming.

A Need for Manufacturing Analysis Tools. Figure 11-1 depicts the old sequential system. The progression of an automotive sheet metal stamping from conception to production is a segmented series of events. Styling, part design, material selection, die

*Material in this sub-section is based upon a paper presented by Mark R. Tharrett at an SME Die and Pressworking Tooling Clinic in Dearborn, Michigan, August 25-27, 1987. The analysis software was developed at General Motors Research Laboratories.[1, 2, 3, 4]

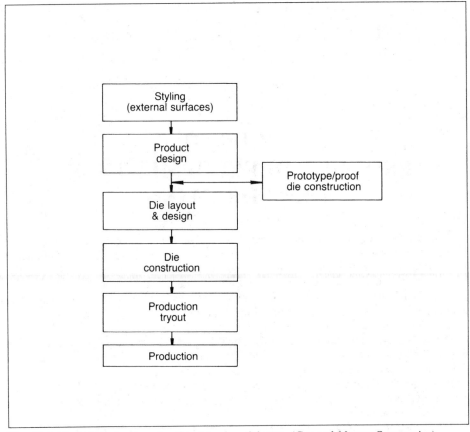

Fig. 11-1. Evolution of a stamping the old sequential way. (*General Motors Corporation*)

design, die construction, die tryout, and production were all performed in a set sequence. Each step was isolated from the other. Formability problems encountered during tryout sometimes resulted in a major redesigning of stampings resulting in costly production delays.

Simultaneous Engineering. In order to reduce product development time and cut costs, the trend is for mutual involvement of representatives from all units into a simultaneous engineering team (Fig. 11-2). Manufacturing engineers are involved in the earliest phases of a car program and work with product engineers and stylists to design for manufacturability as well as form and function. Manufacturing analysis is up-front, simultaneous with product design.

Analysis Methods. The use of analytical methods to predict success or failure in metal forming pre-dates the electronic computer. Such analysis is difficult because of the many nonlinear interactions that occur when forming metal. However, advances in analytical methods and growth in computer power now make computer-aided formability analysis a powerful tool that can be employed at the product design stage to prove out stamping feasibility.

Requirements of a Good Program. A good program must be modeled after actual press shop operations. Factors such as the plastic deformation of the sheet metal, effects of punch and die geometry, friction, properties of the material, forming limits, and binder configuration must be considered. The system must not only include an analysis of the above effects, but also account for their interactions.

Approach. A key issue in choice of analysis software systems for sheet metal forming is the trade-off between the use of many simplified programs (each tailored to a specific application) requiring short computing time versus a generalized method requiring relatively long computing time.

A number of programs are available that are tailored to a specific application. Such specialty analysis programs can be used predict formability in isolated critical areas of a panel. These programs are typically limited to a specific forming operation or deformation mode such as stretch flanging, punch progression analysis, sectional analysis, and springback analysis.

Since they are designed for a specific application, such programs are easy to use. The computing requirements are moderate; both personal computers and mainframes can be used.

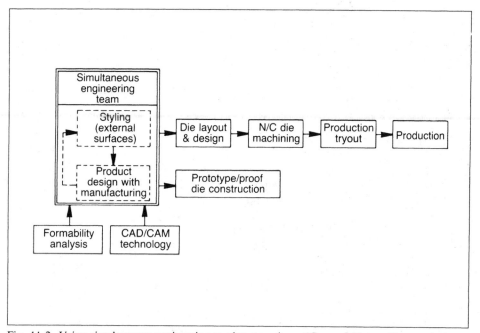

Fig. 11-2. Using simultaneous engineering to plan stampings. (*General Motors Corporation*)

SPECIALTY ANALYSIS APPLICATIONS

Stretch Flanging. Figure 11-3 illustrates two stretch flanged parts. The length of the free edge of the sheet metal is increased as a result of the operation. Splits will occur if the material is stretched beyond its limit. By using a stretch flanging program as a design tool, the designer can calculate the maximum strain at the free edge of the flange, the maximum flange length, as well as the maximum bend angle or minimum bend radius for a specified material.

Sectional Analysis. Figure 11-4 shows an application of sectional analysis. A die cross section is examined neglecting the deformation in the direction normal to the section line. This assumption is valid only when the cross section does not vary too much in the direction normal to the section line plane strain condition. The input requirements are: die geometry, material properties, friction between the punch and sheet (lubrication condition), and whether draw beads or lock beads are used. Strain distribution along the part cross section, amount of metal draw-in, and punch load estimate are given as output.

Figure 11-5 illustrates an example of a sectional analysis applied to a floor pan reinforcement. Splits were occurring in the wall section of the part during tryout of the soft tooling. A computerized sectional analysis was performed in the area of the failure location. This initial design is shown on the right. Only half of the part cross section is

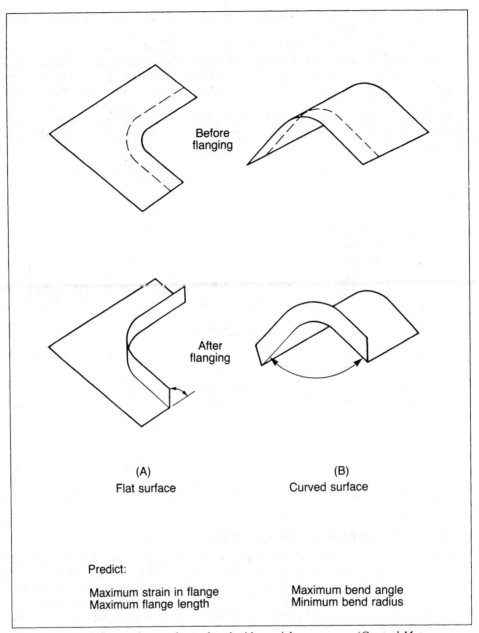

Before
flanging

After
flanging

(A)
Flat surface

(B)
Curved surface

Predict:

Maximum strain in flange
Maximum flange length

Maximum bend angle
Minimum bend radius

Fig. 11-3. Stretch flanges that can be analyzed with specialty programs. (*General Motors Corporation*)

shown due to symmetry. The analysis also predicts the failure in the wall section which was observed during tryout.

Using the predictive capabilities of the code, a series of calculations were made changing the lubrication conditions, material properties, and rail geometry.

Computerized Die Tryout. The proposed part design is shown on the left. The strain distribution is now below the material forming limit. Failure was eliminated by increasing the die radius and opening the wall angle.

A second example is shown in Figs. 11-6 and 11-7. In this example a splitting condition was predicted in the header region of a door inner panel. Allowing more metal

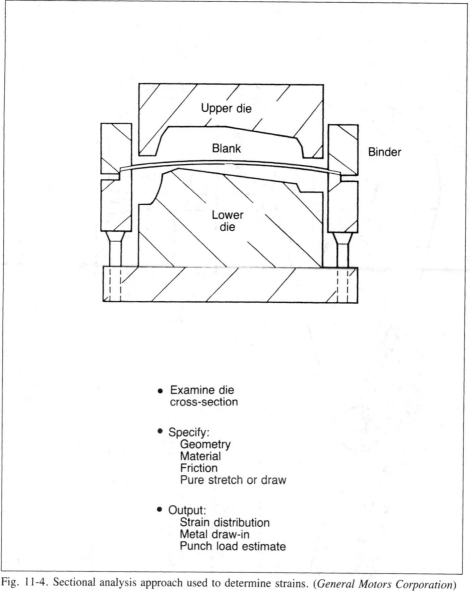

- Examine die
  cross-section

- Specify:
  Geometry
  Material
  Friction
  Pure stretch or draw

- Output:
  Strain distribution
  Metal draw-in
  Punch load estimate

Fig. 11-4. Sectional analysis approach used to determine strains. (*General Motors Corporation*)

draw-in did not reduce the strain in this area because of "locked metal" on the punch and die surfaces, at points A and B (Fig. 11-7), resulting in a pure stretch condition. Part design changes were required to relieve this condition and lower the strain in the header region.

These were just two examples of how the predictive capabilities of sectional analysis can have a significant influence in product development. To provide maximum benefit, analysis tools should be applied early in the design process, before tool construction.

With these tools the designer can evaluate design alternatives and select the optimum within the process capabilities.

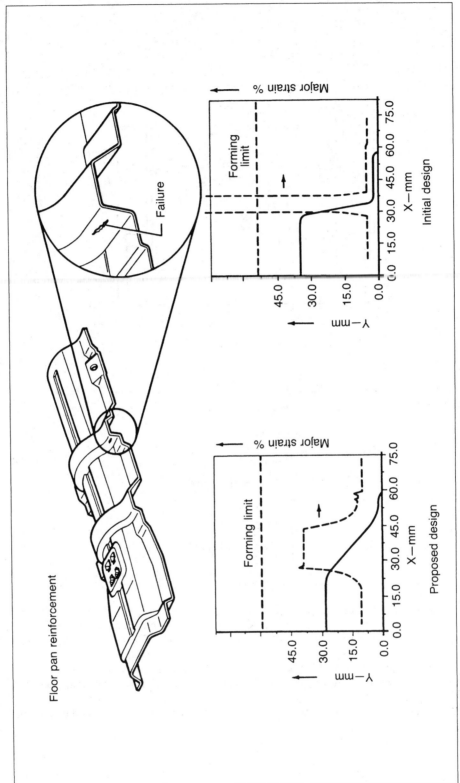

Fig. 11-5. Example of computerized die tryout. (*General Motors Corporation*)

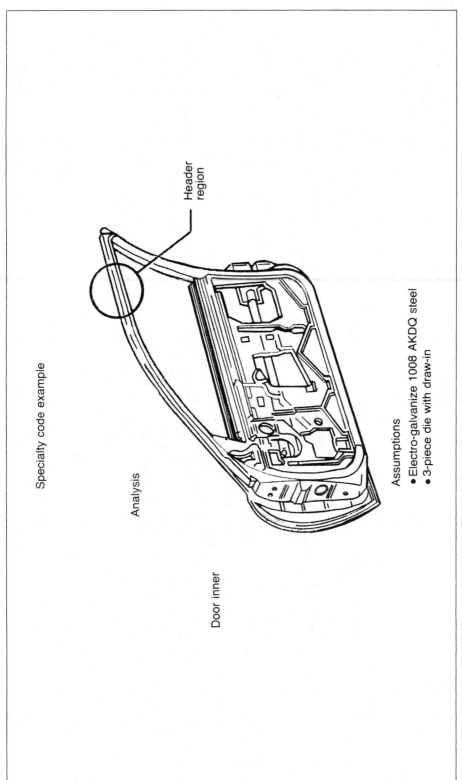

Specialty code example

Analysis

Door inner

Header region

Assumptions
• Electro-galvanize 1008 AKDQ steel
• 3-piece die with draw-in

Fig. 11-6. Door inner panel with fracture in header. (*General Motors Corporation*)

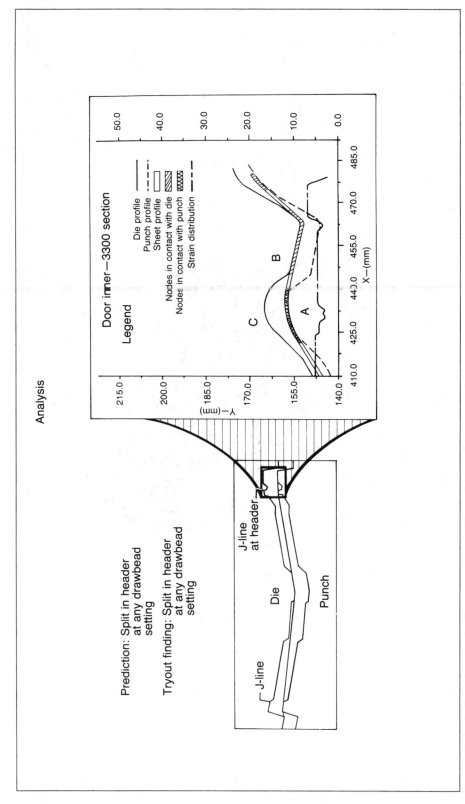

Fig. 11-7. Analysis of door header fracture problem. (*General Motors Corporation*)

11-8

GENERAL ANALYSIS

General analysis is a full-part modeling all-inclusive approach. Stamping of the entire panel is simulated in three dimensions with a single computer program. Because of the complex interaction of many variables during the stamping process, computing requirements can be quite extensive, requiring a mainframe, or even a super computer.

There are several general analysis programs in use in the industry, and more are being developed. These programs are usually based on the finite element or finite difference method.

An example of full part modeling using a finite element code developed by a large automotive manufacturer is shown in Fig. 11-8. The part is a stretch-formed roof outer panel. Since the part is symmetrical, only one half of the part need be analyzed. Strains are color coded showing magnitude and location on the stamping. Output of punch contact information, strain distribution, and die force information is available at any punch depth. Strains are compared to the material forming limit to determine the severity of the stamping. Figure 11-8 illustrates the calculated major strain distribution at final punch depth of the draw operation.

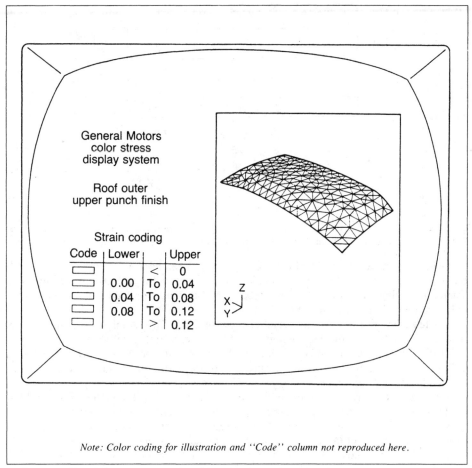

General Motors
color stress
display system

Roof outer
upper punch finish

Strain coding

Code	Lower			Upper
		<		0
	0.00	To	0.04
	0.04	To	0.08
	0.08	To	0.12
		>		0.12

Note: Color coding for illustration and "Code" column not reproduced here.

Fig. 11-8. Finite element analysis of roof panel. (*General Motors Corporation*)

MEASURING DEFORMATION WITH CIRCLE GRID ANALYSIS*

Measuring Deformation. The key to the Circle Grid Analysis technique is the ability to measure deformation. The basis of almost all deformation measurements is a grid on the surface of the blank. This grid deforms with the blank and allows point to point calculations of the deformation that occurred during the stamping operation.

Stretch Calculations. The simplest of deformation measurements is a series of grid lines which divide the blank into equal segments. One-inch lengths were popular several decades ago. The initial and the final lengths of the segments are measured. A simple formula is used to calculate percent stretch (percent elongation, increase in length of line, or other press shop terms for stretch). Fig. 11-9 shows a blank which elongated from 4 in. (101.6 mm) to 6 in. (152.4 mm) for an overall stretch of 50%. If the blank is stretched uniformly, then each 1 in. (25.4 mm) increment also will stretch 50%.

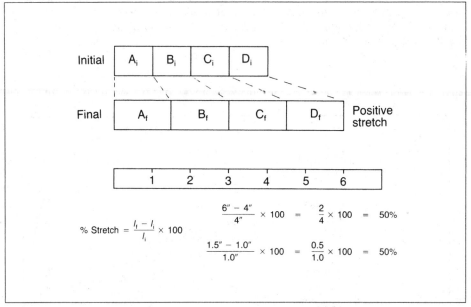

Fig. 11-9. Calculating stretch in drawn sheet metal. (*National Steel Corporation*)

Grid Placement. Within the blank, the grid lines need not uniformly cover the entire blank. A grid will measure the deformation of the blank within the confines of the grid. Therefore, a single pair of grid lines can be placed anywhere on the blank to monitor the deformation of a specific area (Fig. 11-10).

Stretch Distribution. An important element of Circle Grid Analysis is point-to-point measurements, because most stampings do not stretch uniformly. Deformation will vary as the blank is deformed around complex tools. In Fig. 11-11, the stretch varies from 0% to 100% and back down to 0%. If a measurement of the entire blank is made, all segments are averaged. Here the blank average is 40%.

Visualizing the string of numbers that represents the percent stretch in Fig. 11-11 is difficult. An easier, more graphic method is to plot the percent stretch vertically for each location across the stamping. Each data point represents the measurement of a single pair of grid lines. The resulting curve is called the *stretch distribution*.

The values of stretch calculated in Fig. 11-11 are plotted on Fig. 11-12 as a stretch

*The material in this subsection was taken from a booklet by Stuart Keeler, "Circle Grid Analysis (CGA)," published by National Steel Corporation's Product Application Center, Livonia, Michigan, 1986.[5, 6]

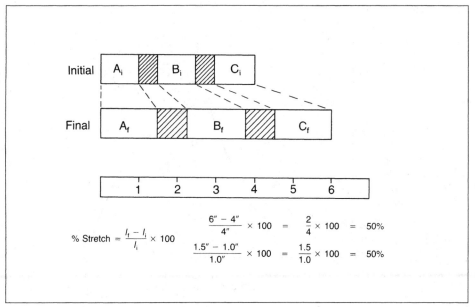

Fig. 11-10. Using grid lines to determine stretch. (*National Steel Corporation*)

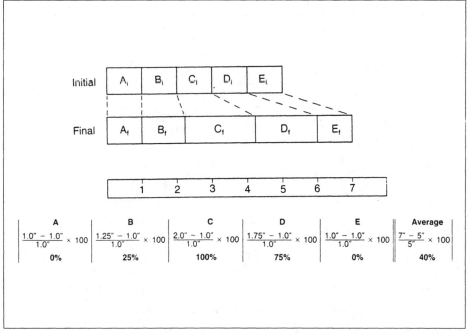

Fig. 11-11. Stretch distortion example. (*National Steel Corporation*)

distribution. The location of each grid is spaced uniformly on the horizontal axis because the original grid spacing was uniform. The vertical height of each data point is the percent stretch. The values obtained are typical of most practical stretch distributions with a rather sharp gradient.

Negative Stretch. Instead of elongating, some segments become shorter. Elongation of segments is called *positive stretch;* compression of the segments is called *negative stretch.* The formula to calculate the percent stretch remains the same. In Fig. 11-13, a

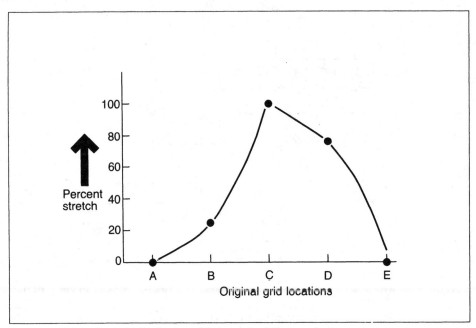

Fig. 11-12. Plot of stretch distribution. (*National Steel Corporation*)

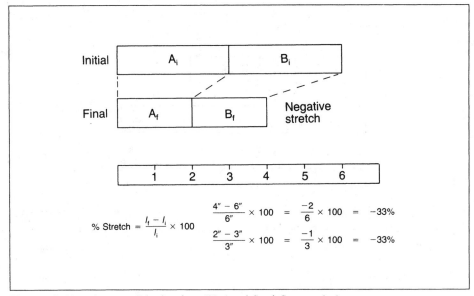

Fig. 11-13. Negative stretch in drawing. (*National Steel Corporation*)

negative 33% stretch is obtained.

When measured in a line from blank edge to blank edge, some stampings can have negative stretch on both edges and positive stretch in the center. In Fig. 11-14, the percent stretch varies from −50% to +100% to −50%. However, measuring the entire length of the blank shows no change or 0% stretch.

Two-directional Stretch. Sheet metal stampings have two surface dimensions—length and width. Therefore, to fully describe the stretch state of the stamping, both the length and width stretch components must be measured. For this purpose, a square grid is used. In Fig. 11-15, both *X* and *Y* stretch components are positive.

11- 12

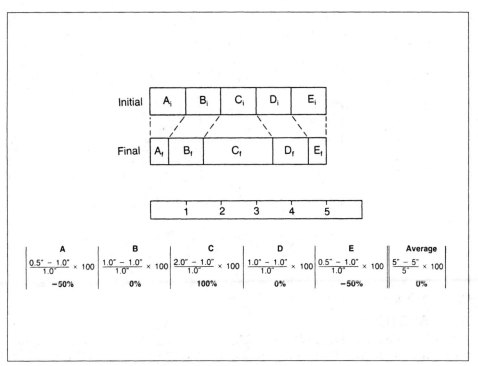

Fig. 11-14. Distribution of two directional stretch. (*National Steel Corporation*)

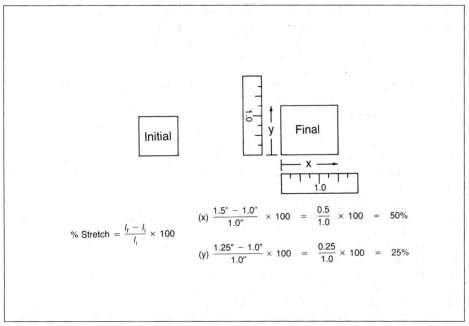

Fig. 11-15 Measuring two directional stretch. (*National Steel Corporation*)

Many sheet metal stampings have a positive stretch in one direction and a negative stretch in the other direction as shown in Fig. 11-16.

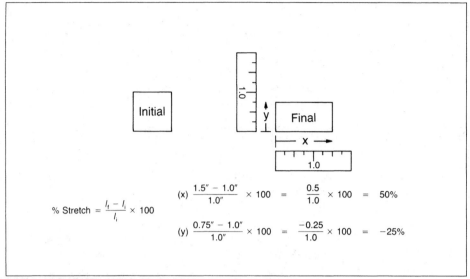

Fig. 11-16. Negative stretch. (*National Steel Corporation*)

GRID SYSTEMS

One-inch Square Grid. The one-inch (25.4 mm) square grid system is large compared to the localized stretch and rarely oriented along the direction of maximum stretch. A 1-in. (25.4 mm) square is often too large to provide the desired information.

The scribbling requires a long time to complete and must be done with care to avoid deep grooves which may cause preferential fracture along the scribe marks.

Scribbling of blanks for large body panels still has applicability to determine low strains on large fairly flat drawn shapes. Often a minimum strain must be developed to stiffen the part and remove roll marks. Circle grid analysis may not be as accurate a technique for determining strains under 3%.

Circle Grid. A better grid pattern for most stretch measurements is a small diameter circle. In response to the complex interaction of deformation forces, the circle transforms to an ellipse as shown in Fig. 11-17. The long axis of the ellipse is the direction of maximum stretch and is called the *major stretch*. The direction perpendicular to the maximum stretch is the minimum stretch and is called the *minor stretch*. The major stretch is always positive. The minor stretch can be positive or negative. Therefore, for each location on the stamping, two measurements must be made—the major stretch and the minor stretch.

Typical Circle Grid Patterns. Figure 11-18 shows several typical patterns. The pattern on the left is most commonly used. Each 0.1-in. (2.54 mm) circle stands alone, so that the resulting ellipse can be easily visualized and measured. The lines surrounding the circles can be used to determine overall metal flow patterns, while the circles detail the point to point variations.

Grid measurements are usually made with a transparent mylar tape with "calibrated, diverging, railroad tracks" (Fig. 11-19). The tape is flexible and can be laid around a tight radius or tucked into a tight corner. The calibration of the tape eliminates any calculation of stretch. The tape is used to measure the major axis of the ellipse first and then rotated to measure the minor axis.

Figure 11-20 shows a sketch duplicating the distribution of stretch in a bumper jack hook. Note that the amount of stretch and direction of the major stretch change from point to point on the stamping. Large changes are observed within a 0.3 in. (7.62 mm) length.

Electrochemical Etching. A practical grid marking system is electrochemical marking with an electrical stencil that contains 5,000 circles in a master pattern. The stencil is placed on the cleaned blank (Fig. 11-21), covered with a felt pad containing the electrolyte, and then etched using a variety of electrode configurations. The grid is permanently etched into the surface of the blank. The type of electrolyte used is determined by the type of metal being etched. The actual gridding takes about one minute per application.

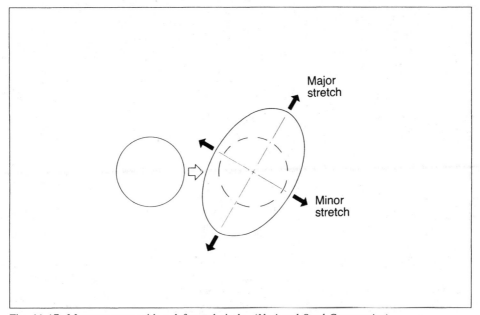

Fig. 11-17. Measurements with a deformed circle. (*National Steel Corporation*)

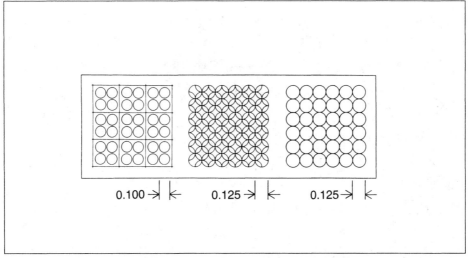

Fig. 11-18. Examples of circle grid patterns. (*National Steel Corporation*)

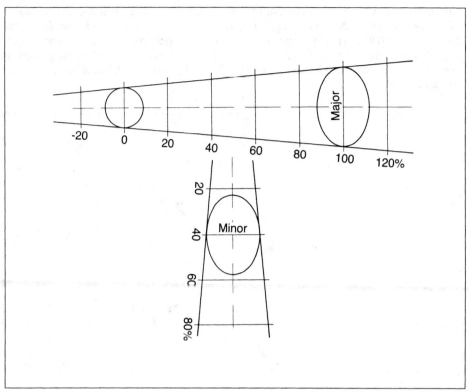

Fig. 11-19. Use of mylar tape measuring system. (*National Steel Corporation*)

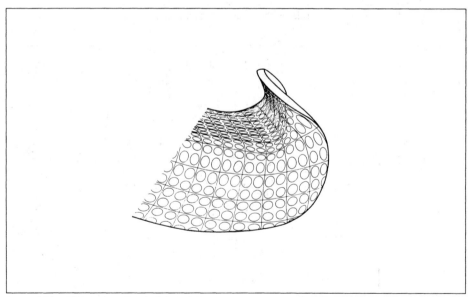

Fig. 11-20. Etched part after forming. (*National Steel Corporation*)

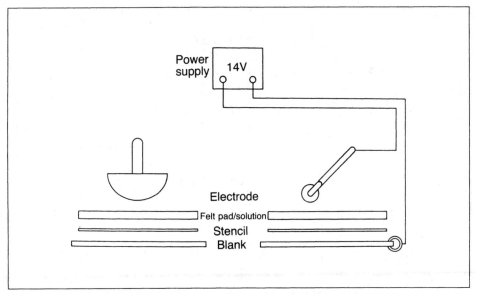

Fig. 11-21. Schematic of etching procedure. (*National Steel Corporation*)

DETERMINING FORMING SEVERITY

Distribution of stretch is useful information by itself. Location of high stretch concentrations and the direction of the maximum stretch often are sufficient information to suggest solutions to forming problems. However, some numerical rating of forming severity is desirable.

The system of rating forming severity is based on the circle grid measurements and the Forming Limit Diagram.

Stretch Combinations. An infinite number of combinations of major and minor stretch can be found on different stampings. One special case is shown on the right of Fig. 11-22, where both the major and minor axes of the ellipse are equal—the circle becomes a larger circle. This special case is called balanced biaxial stretch. The middle case also is special. Here the minor stretch component is zero. This is called *plane strain*. This stretch condition is found over edge radii or across character lines. Its importance will be shown later. Other combinations are also illustrated.

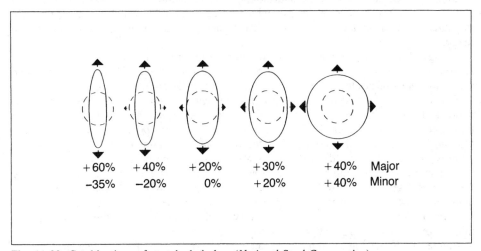

Fig. 11-22. Combinations of stretched circles. (*National Steel Corporation*)

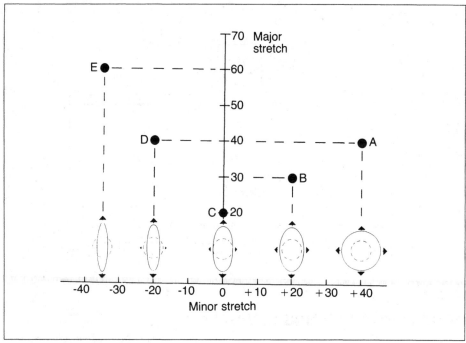

Fig. 11-23. Circles plotted according to type of stretch. (*National Steel Corporation*)

Plotting Circle Grid Measurements. With the variety of combinations, a method for plotting them on a single graph is necessary. The plotting technique used in Fig. 11-23 allows both the major and minor stretch for each circle to be plotted as a single point. The major stretch is plotted on the vertical axis, while the minor stretch is plotted on the horizontal axis. Circles which plot on the left side of the diagram have negative minor stretch, while circles which plot on the right side of the diagram have positive minor stretch.

The ellipses from Fig. 11-22 are plotted on this diagram. Note that the case of plane stretch (minor stretch equal to zero) is plotted on the vertical axis. The next step is to plot on the same diagram the values for all circles which fail.

Failed vs. Unfailed Circle Measurements. To determine the conditions for the onset of failure for a given material, circles which failed and circles which had not failed are measured on representative stampings. These are plotted on the major/minor stretch diagram.

In Fig. 11-24, a stretch condition representing a failure is plotted as an open circle. A stretch condition which does not fail is plotted as a solid circle. A line drawn between these two types of circles indicates the conditions for the onset of failure.

This diagram is called the *Forming Limit Diagram*. It separates the combinations of major and minor stretch which cause failure from those which do not fail. While the failure combinations are easily detected because of a certain percentage of failures, the proximity of failure of any point below the line can be ascertained only by plotting the point on this diagram and determining its relationship to the failure line.

The shape of the Forming Limit Diagram (Fig. 11-25) is constant for most low-alloy sheet steel used in the automotive, appliance, agricultural, container, and similar industries. The curve moves up and down the axis for different coils of steel. Thus, the location of the curve can be fixed by specifying the intersection of the curve with the minor stretch axis. This point is shown as FLD_0.

Figure 11-26 illustrates that the FLD can be raised or lowered for different steel sheets. The level of the FLD—as specified by FLD_0—is a characteristic of the sheet steel. For

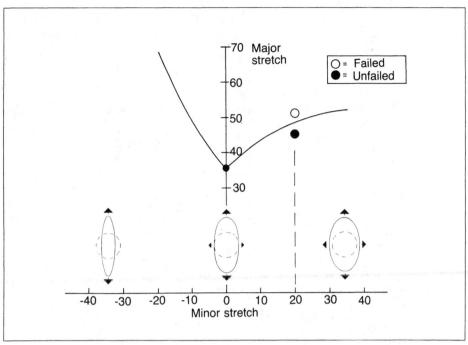

Fig. 11-24. Evolution of the forming limit diagram (FLD). (*National Steel Corporation*)

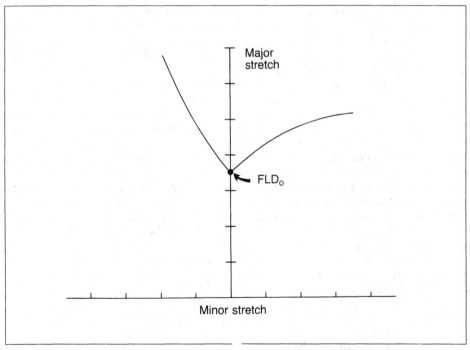

Fig. 11-25. Intersection of the FLD curve with the minor stretch axis (FLD_0). (*National Steel Corporation*)

example, a thinner sheet of steel would have a lower FLD than a thicker sheet of steel. Also, a higher strength steel would have a lower FLD than a lower strength steel.

Nomograph for FLD_0 Determination. Nomographs such as the one shown in Fig.

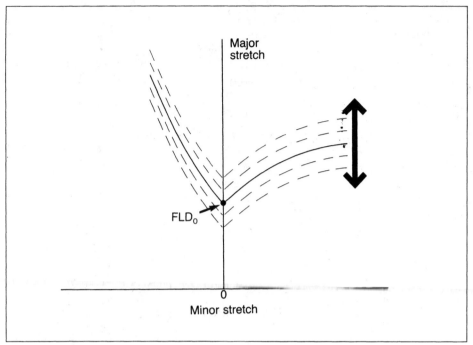

Fig. 11-26. Different steels move the FLD_0 point. (*National Steel Corporation*)

11-27 are used to determine FLD_0 for any combination of strength and thickness of low carbon steel. The yield strength is used to approximate the formability of the steel sheet. A more accurate curve uses the work hardening exponent, n, derived from a tensile test of the steel sample. For a first approximation, however, the actual yield strength can be used. The dotted lines illustrate the steps used.

The FLD separates conditions for onset of failure from those that will not fail. For practical purposes, a second line is drawn parallel to the FLD but offset by 10% stretch (Fig. 11-28). This line and the FLD define a zone which is marginal. While failure does not occur for stretch combinations in this zone, an inadequate safety margin exists. Here, safety margin is defined as the proximity to the FLD and is equal to the FLD value minus the major stretch value for the measured minor stretch. An example would the the FLD_0 point located on this diagram in Fig. 11-28. Its safety margin would equal zero.

Example of FLD_0 Determinations. Figure 11-29 shows the reduction in FLD_0 due to a reduction of stock thickness and use of a stronger material. In Fig. 11-30, the left diagram is for the thicker, low strength steel and the right diagram is for the thinner higher strength steel. These diagrams represent the stretching capability of the sheet. The actual major and minor stretch combinations have to be measured from the stamping in order to be placed on the diagram. When changing the thickness and strength of a sheet of steel, the values of the major and minor stretch also are likely to change. Therefore, the Circle Grid Analysis technique is a very useful post-mortem type of analysis. The severity of the stamping made with steels A and B can be assessed. However, the data from stamping A cannot be combined with the reduction in the Forming Limit Diagram to predict the severity of stamping B.

Severity as a Function of Minor Stretch. According to the Forming Limit Diagram, the circle with the largest major stretch may not be the most severe. Severity depends on the specific major/minor stretch combination. For example, in Fig. 11-31 a major stretch of 80% is in the safe zone with a negative minor stretch of -30%. However, a 40% major stretch in the plane stretch state (minor stretch $= 0$) is in the failure zone. It is obvious from the Forming Limit Diagram that the plane stretch condition permits the least amount

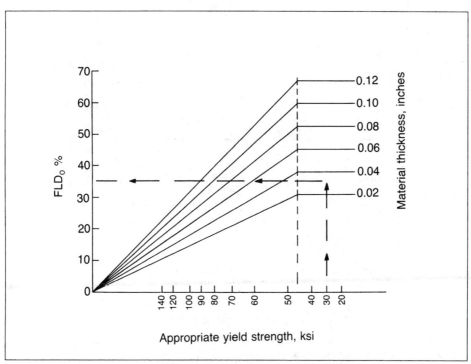

Fig. 11-27. Yield strength and thickness used to approximate FLD_0. (*National Steel Corporation*)

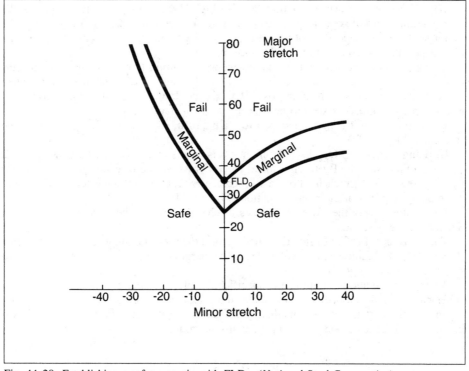

Fig. 11-28. Establishing a safety margin with FLD_0. (*National Steel Corporation*)

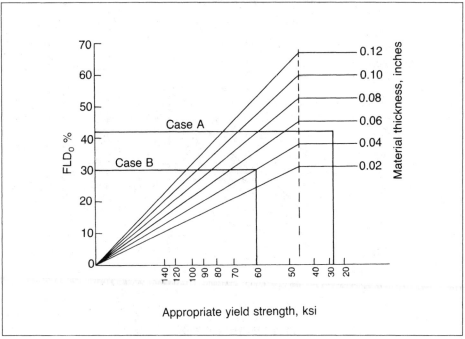

Fig. 11-29. Change of material thickness and strength affects FLD_0. (*National Steel Corporation*)

of major stretch before failure.

In many stampings, the major stretch cannot be changed because of design restrictions, character lines, etc. However, as illustrated in Fig. 11-32, a major stretch can be in the safe, marginal, or fail zone depending on the value of the minor stretch. The value of the minor stretch often depends, not on the design of the stamping, but on the design of the die—especially the contours of the draw die binder.

Changes in binder configuration can have a positive impact on stamping economy. By using circle grid analysis and the forming limit diagram as a development tool when making binder changes, it is often possible to reduce drawing severity enough to permit a less expensive grade of steel to be used for a given stamping.

The Forming Limit Diagram can be used like a map, showing possible avenues to the safe zone (Fig. 11-33).

Utilizing Negative Stretch. Very high levels of major stretch can be obtained by a negative minor stretch. By forcing metal to compress in one direction (negative minor stretch), the major stretch is allowed to increase without a damaging tensile (positive) force being required in the major direction (Fig. 11-34).

This factor is why the extreme changes of shape shown in the data and formulas in Section 10 are possible.

The FLD as a Traffic Light. One method of indicating stamping severity is by color (Fig. 11-35). The failure zone is colored red, the marginal zone is colored yellow, and the safe zone is colored green.

As another method to simplify the specification of safety margin, numbered severity zones have been developed. This eliminates the calculation of the safety margin. Plotting the major/minor stretch value immediately assigns a severity zone. In the example shown in Fig. 11-36, zones 9 and 10 are in the marginal zone.

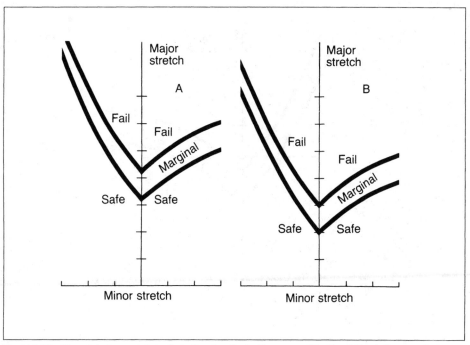

Fig. 11-30. Material change plotted as forming limit diagrams. (*National Steel Corporation*)

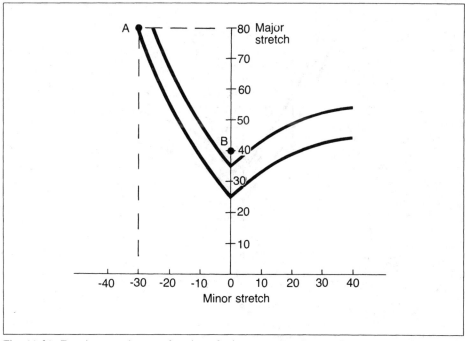

Fig. 11-31. Forming severity as a function of minor stretch. (*National Steel Corporation*)

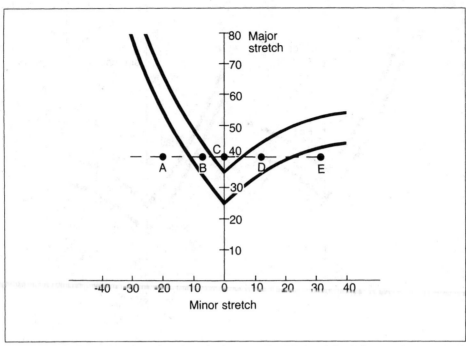

Fig. 11-32. Severity varies as a function of minor stretch. (*National Steel Corporation*)

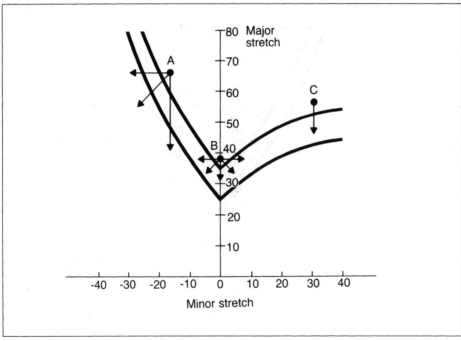

Fig. 11-33. Changing stamping severity affects the FLD. (*National Steel Corporation*)

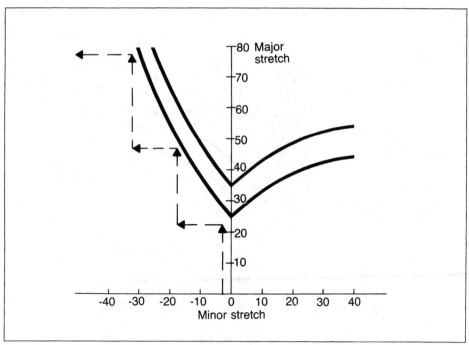

Fig. 11-34. Compression on one axis can aid drawing. (*National Steel Corporation*)

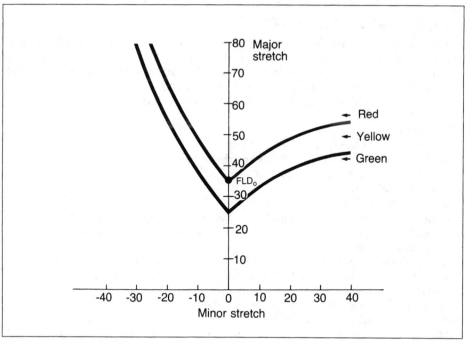

Fig. 11-35. Color coding forming severity on a FLD. (*National Steel Corporation*)

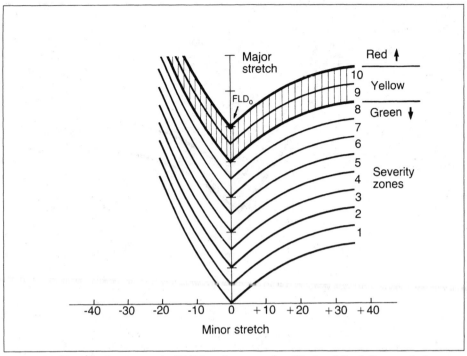

Fig. 11-36. Numerical zones on the FLD based on severity. (*National Steel Corporation*)

DISTRIBUTION OF STRETCH

The distribution of stretch is a measure of the efficiency of the deformation. A sharp gradient with only a few circles having high stretch is not efficient. Since depth of stamping is average stretch across the stamping, a few highly stretched circles can reach the failure level of stretch with a very low average stretch.

To determine the distribution of stretch, the analysis sequence should be as follows:

A. Grid the blank.
B. Form the stamping.
C. Measure the major/minor stretch values.
D. Plot the major stretch distribution in area of interest.
E. Plot the stretch limit from the FLD.
F. Determine the safety margin.

Steps A through D are illustrated in Fig. 11-37.

Figure 11-38 illustrates the maximum allowable stretch derived from the minor stretch values and the FLD (Steps E and F). In this example, the peak value of the major stretch is at the failure value from the FLD.

Rating Forming Severity. Four common conditions of severity are schematically shown in Fig. 11-39. In *A*, a number of the circles, sometimes called elements, are above the maximum allowable stretch limit. The peak stretch value is well into the fail zone. In this case, a high percentage of the stampings would fail. In *B*, the peak stretch is only slightly above the maximum allowable level. In this case, about 1 in 20 stamps would fail. However, a failure rate of 5% generally is not acceptable. When doing stamping tryout, a single gridded stamping can be used to indicate a finite failure rate. The usual 100 to 1000 stamping trial is not required to assess the stamping severity. In *C*, the stamping is just in the safe zone. Without the circle grid system, a severity could not be assigned to this stamping because no breakage would be encountered in the trial run, although material variations, die wear, and press adjustments could cause future problems. In *D*, an excessive safety margin is observed. In this case attention should be paid toward cost

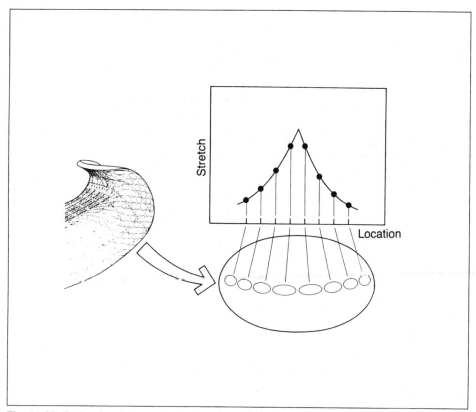

Fig. 11-37. Rating forming severity by location. (*National Steel Corporation*)

reduction through blank size changes, downgrading steel drawing quality, and lubricant cost reductions.

Importance of Stretch Gradient. In addition to the value of the peak stretch, the shape of the stretch gradient can be important in assessing forming severity. For example, in Fig. 11-40 the uniform gradient of Stamping A means that the stretch is well distributed among all the elements of the stamping. A change in the forming conditions will mean that the change is divided among all the elements and that only a slight increase in the peak stretch—and resulting decreases in the safety margin—will occur. For Stamping B, however, any change in the forming conditions will be carried by only a few elements; the peak stretch will rise very quickly and reach the maximum allowable stretch limit. Thus, Stamping A has a higher peak stretch and lower safety margin than Stamping B, but Stamping A has the potential of being more resistant to changes in severity.

In practice, stampings are not made deeper and deeper until they fail but are taken only to the design depth. In the schematic in Fig. 11-41, the average stretch (area under the curve) is equal for both A and B. Therefore, Stampings A and B are both at the design depth. The stamping that has the most uniform distribution of stretch (Stamping B) has the largest safety margin and the most favorable stretch distribution.

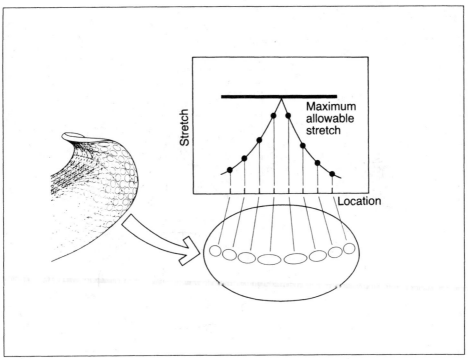

Fig. 11-38. Establishing maximum allowable stretch. (*National Steel Corporation*)

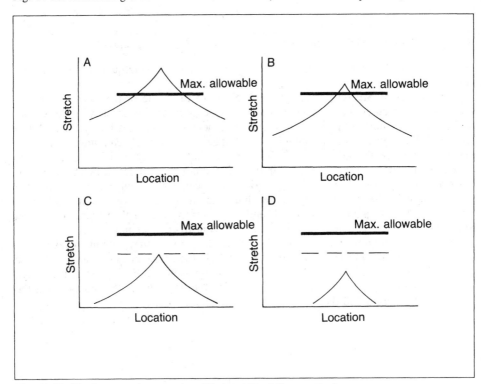

Fig. 11-39. Plotting stamping severity. (*National Steel Corporation*)

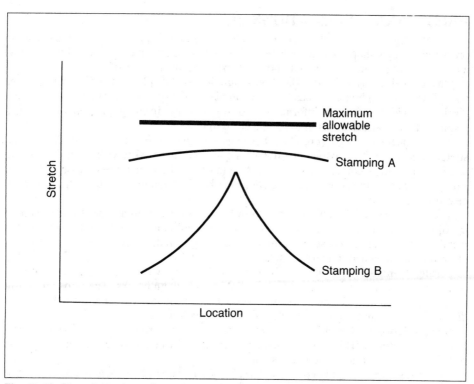

Fig. 11-40. Plot of stretch gradient. (*National Steel Corporation*)

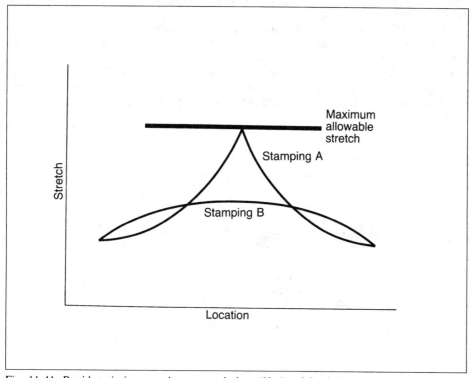

Fig. 11-41. Rapid strain increase due to part design. (*National Steel Corporation*)

PRESS SHOP APPLICATIONS

All of the press shop applications of Circle Grid Analysis presented here are currently practiced in press shops. An advantage of this technique is that everybody gets the same answer—independent of their tool and die experience or the lack of it.

Circle Grid Analysis Visual Display. A simple visual display of the deformed ellipses will provide the high stretch location, magnitude, direction and distribution. The visual display provides sufficient information to solve many forming problems. The grid displays the actual deformation experienced by the sheet.

This bumper jack hook (Fig. 11-20) shows a single, severely deformed ellipse with stretch being almost zero two circles away. A question can be asked whether this stamping is critical or not. A correct answer cannot be made until the measured major and minor stretch values are plotted on the correct Forming Limit Diagram and the resulting severity determined.

The major axes of the ellipses line up in a given direction in a panel for a reason. If the direction of the major axes is followed, it often will point directly to the cause of the metal flow restraint. These restraints include tight radii, insufficient punch-to-die clearance, mismatched drawn beads, and folded metal.

Stretch Comparisons. The stretch distribution patterns will show differences between mirror image stampings—especially when only one of them breaks. The grids are useful to show progress after forming system modifications. The last stamping made before the die is reworked can be gridded and saved for evaluation.

Stretch History. Stretch history allows analysis of the critical stretch conditions and how they developed. This history could be determined for punch travel within a single die or between dies of a multiple die system.

The grid system is rapid and allows simple monitoring with a single blank before and a single blank after any modifications. The amount of severity change indicates the sensitivity of the system to that variable. Documentation and press shop records can be developed using the numerical grid measurements.

Monitor Diesetting and Production. The grid measurements allow for one die set to be compared with another die set, as well as comparison of different press lines, press types, and punch/hold-down adjustments.

Production runs can be monitored to show the effects of die wear. The cause of sudden breakage can be determined by a system of reference steel samples and reference lubricants samples. The status (severity) of the dies at shutdown can be determined so that

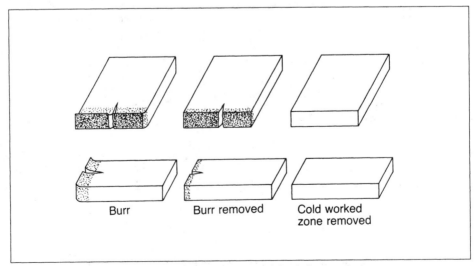

Fig. 11-42. Examples of blank edge damage. (*National Steel Corporation*)

necessary work can be scheduled before the next startup.

Assuring a Safety Margin. The correct specification of material and lubricant is the combination which provides a sufficient safety factor. Dies can be released (or purchased) when sufficient safety factor is obtained with the production steel and production lubricant.

All experimental changes in material, lubrication, die modification, and press adjustments can be tracked, documented, and evaluated using the circle grid system.

Tearing on a Blanked Edge. This is a common forming problem which can be analyzed with the circle grid method. The forming severity, location of severe areas, and the effect of blanking damage can be measured using the circle grid system.

The important question is whether the base metal has sufficient formability to satisfactorily make the stamping or whether the formability of the blanked edge is critically reduced by the blanking operation.

While a burr is an indication of blanking damage, removal of the burr will not remove all of the damage. Studies have shown that the damage can extend inward a distance equal of one metal thickness from the blanked edge. One quick technique to determine if blanking damage is the cause of breakage is to file back the metal on the edge a distance of one metal thickness (Fig. 11-42). If the fracture disappears, then the base metal formability is sufficient to form the stamping.

Blanking damage also reduces the stretchability of the base metal. By placing a grid on the stamping and plotting the stretch on Fig. 11-43, the amount of blanking damage (shown as burr height) before fracture can be assessed.

Another means of reducing forming severity on a blanked edge is to weaken metal surrounding the high stretch locations, thereby redistributing the stretch. The stretch distribution shows the exact location to weaken the metal, such as notching or scalloping the metal, at locations 1 and 2 (Fig. 11-44).

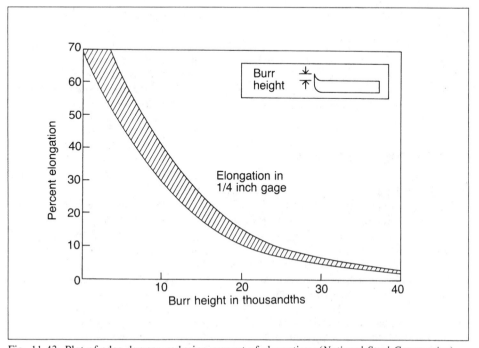

Fig. 11-43. Plot of edge damage reducing percent of elongation. (*National Steel Corporation*)

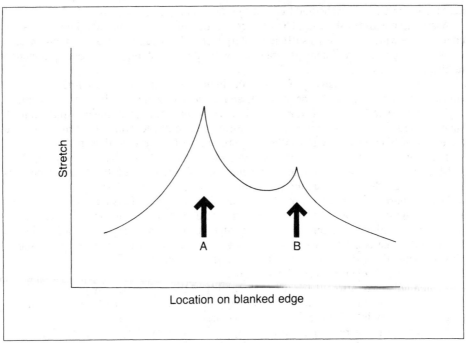

Fig. 11-44. Redistributed stress concentration. (*National Steel Corporation*)

USING CGA AS A TROUBLESHOOTING TOOL

Full understanding of the interaction between the tools, materials, lubricant, friction, temperature, speed, and all other forming variables would be ideal. However, the interactive system is so complex that final deformation severity cannot be predicted from the components of the forming system.

The most common press shop application technique is based on the "black box" procedure.

This technique provides for making a "best guess" modification of the forming system. If the severity of the system is reduced, then that modification was a good choice. If the system becomes more severe, the opposite modification is attempted. If that also causes an increase in severity, then no change is made to that variable. Proficiency in die tryout will reduce the time to find the correct variable, but does not change the general analytical technique.

THE FORMING SYSTEM

Once the concept of the black box analysis of the forming system is accepted, then the four primary inputs to the forming system can be studied.

The four primary inputs are material, lubrication, die design, and press adjustment.

Material Selection. The results of trials on two steels are shown in Fig. 11-45. Steel A has its maximum severity in the safe zone, while steel B is in the fail zone. A study of the forming characteristics of steel A compared to steel B will assist in specification of the correct steel for this application. The interesting aspect of circle grid analysis is that the specific stamping need not be identified in order to correctly analyze the data.

Lubrication. A major contributor to stretch distribution in many stampings is the lubrication.

In this example of an automotive fender (Fig. 11-46), lubricant A takes the forming system into the fail zone. Lubricant B not only creates a safe zone condition, but forming severity is so low that additional cost savings should be investigated.

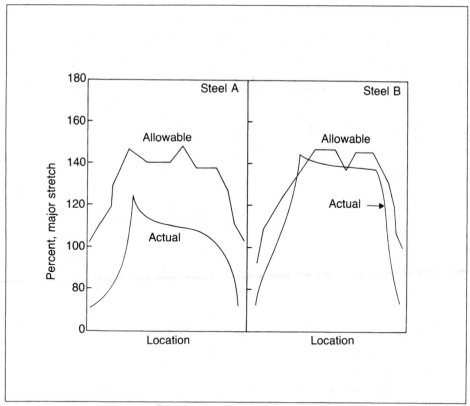

Fig. 11-45. Effect of steel selection. (*National Steel Corporation*)

Troubleshooting Multiple Die Operations. Die design, and the initial part design, provide the greatest opportunity for modification of the stretch distribution and forming severity.

In Fig. 11-47, breakage occurs in the seventh die of a multiple die set. The stretch distribution, however, is critical in the second die. The stretch gradient in the second die is the result of the nonuniform stretch distribution generated in the first die. Modification of the first die eliminated breakage in the seventh die.

Press Adjustment. Blankholder adjustments can affect stretch distributions. After a successful trial number one, breakage was encountered in trial number two (Fig. 11-48). Stretch in trial two was greater everywhere within the die opening. This meant less metal was flowing into the die from the binder area. An increase in binder pressure had created the reduced metal flow.

A good way to insure setup repeatability from run to run is to use a tonnage meter to make the binder adjustments to a tonnage setting that is known to produce good parts.[7]

To work most effectively, the whole forming system must be treated as an interactive system. Specific parameters may be effective or ineffective depending on the other components of the forming system.

The circle grid analysis system is excellent for training apprentices. By making tooling, lubricant, and material changes and then observing the metal deformation changes, cause-and-effect patterns can be readily discerned.

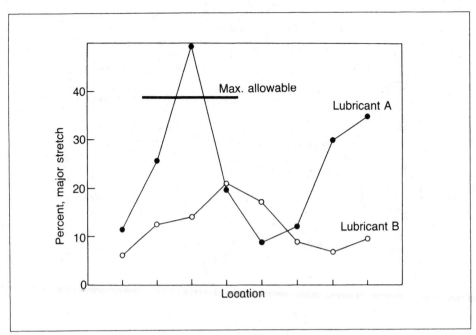

Fig. 11-46. Effect of lubricant selection. (*National Steel Corporation*)

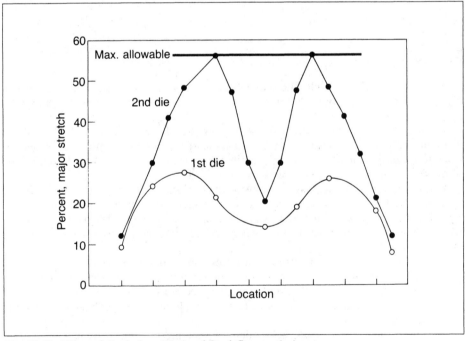

Fig. 11-47. Effect of die design. (*National Steel Corporation*)

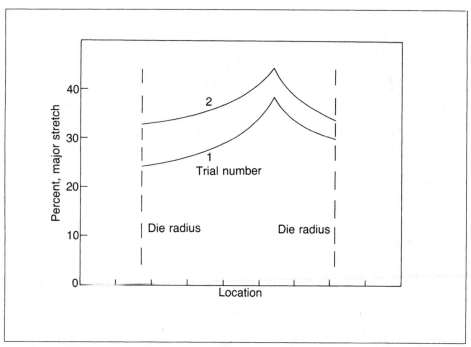

Fig. 11-48. Effect of blankholder adjustment. (*National Steel Corporation*)

References

1. N. M. Wang and M. L. Wenner, "Elastic-viscoplastic Analysis of Simple Stretch Forming Problems," *Mechanics of Sheet Metal Forming*, D. P. Koistinen and N. M. Wang (New York: Plenum Press, 1978), p. 367.
2. N. M. Wang and B. Budiansky, "Analysis of Sheet Metal Stamping by a Finite-Element Method," *J. App. Mech. 45* (1978), p. 73.
3. F. J. Arlinghaus, W. H. Frey, T. B. Stoughton, and B. K. Murthy, "Finite Element Modeling of a Stretch Formed Part," *Computer Modeling of Sheet Metal Forming Process*, N. M. Wang and S. C. Tang, eds. (Warrendale, PA: The Metallurgical Society, 1986).
4. N. M. Wang and M. L. Wenner, "An Analytical and Experimental Study of Stretch Flanging," reprinted from International Journal of Mechanical Science (England: Pergamon Press, 1974), Volume 16, pp. 135-143.
5. S. Keeler, "From Stretch to Draw," SME Technical Paper MF69-513 (Dearborn, MI: Society of Manufacturing Engineers, 1969).
6. S. Keeler, "Circle Grid Analysis (CGA)," National Steel Corporation Product Application Center, Livonia, Michigan, 1986.
7. D. A. Smith, "How to Improve Hit-To-Hit Time With a Tonnage Monitor," SME Technical Paper TE88-780 (Dearborn, MI: Society of Manufacturing Engineers, 1988).

SECTION 12

DRAW DIES

Draw dies are designed for use on many types of presses. They are used on single-action, double-action, and triple-action presses which are mechanical or hydraulically powered.

The simple single-action dies (Fig. 12-26) employ only a punch and die arranged so that they can be mounted in a press. As the shapes being drawn become more complex and difficult to fabricate, blankholders and pressure pads are added to the dies and the parts are developed in several operations rather than one. The blankholders may be permanently attached to the draw ring (Fig. 12-27) with spacers that allow insertion of the blank and its proper positioning over the die cavity. The movable blankholders and pressure pads are actuated by pressure arrangements built into or attached to the die or press.

Analysis of Drawn Shapes. The finished shape of a shell must be carefully analyzed to determine the shape of each redraw. The shape to which a shell is redrawn determines to some extent the number of redraws required to produce the finished shape. Figure 12-1 is an analysis of redrawing operations. In Case 1, the reduction is 55%, which is usually considered too much for most materials; therefore, two draw operations are recommended. Three draws are necessary in Case 2 because of the contour. The small diameter of the bottom of this shell creates a condition in which there is a substantial area out of control at the start of the draw, if attempted in one operation. Shapes similar to this must be drawn in two or more operations, and a suggested method is shown at D. Case 3 is a 10-in. (254-mm) -diameter blank drawn to a 5.5-in. (140-mm) -diameter shell. Although the reduction is only 45%, the thickness-diameter ratio is only 0.10, requiring a smaller reduction in the first draw as suggested at F.

Shells of the same maximum overall dimensions are shown in Fig. 12-2 but, because of their contours, require different shape and number of redraws. A represents a straight-sided shell 6 in. (152.4 mm) in diameter and 6 in. (152.4 mm) deep, drawn in two draws from a 13.75 in. (349.25 mm) -diameter blank, a total reduction of about 56%. B is a tapered shell of the same overall dimensions, drawn in five draws from a 12-in. (304 mm) -diameter blank. The illustration at B-1 shows the condition of the metal out of control if an attempt is made to redraw the tapered shell in the third operation. C illustrates a domed shell drawn in three operations because of the shape of the bottom. The blank for this shell in 11 in. (279 mm) in diameter.

A typical arrangement of redraws for a circular shell of 3003-O aluminum is shown in Fig. 12-3. This shell was drawn in three operations without any intermediate anneals.

A layout of drawing operations taken from an actual production job is shown in Fig. 12-4. The thickness-diameter ratio on this job was such that the 45% reduction on the first draw gave some trouble and an arrangement containing an additional draw would have been more satisfactory.

In Fig. 12-5 a tubular part is shown formed from type 316 stainless steel. The first draw is a 40% reduction, and the second is 20%. The third operation sizes and flattens the flange. The die material is aluminum bronze with a 0.50 in. (12.7 mm) draw radius. The

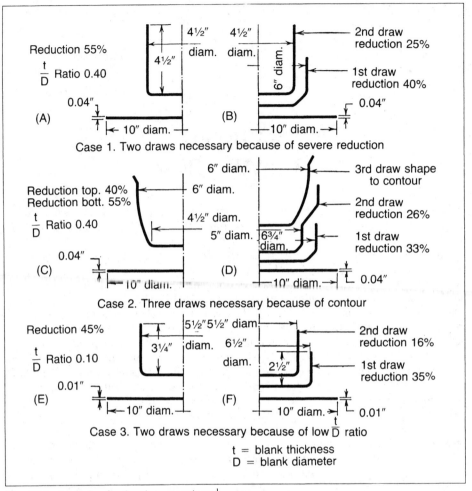

Fig. 12-1. Analysis of redrawing operations.[1]

punch is made of high-carbon high-chromium steel, chromium-plated.

These die materials are used for draw stainless steel, because they are less likely to scratch or gall the workpiece during the operation. For further information on die materials, refer to Sections 28 and 29.

A large stainless-steel shell (Fig. 12-6) with a four-draw sequence uses a two-step punch in the third draw and a shallow reverse contour at the bottom, using a mating die with a contour-bottom pad. The stock was type 302 stainless, annealed.

A shell which requires seven draw operations to reduce it to the final dimension is shown in Fig. 12-7. To facilitate the redraws, the shell is annealed after each operation. This part of type 347 stainless steel 0.091 in. (2.31 mm) thick drawn in alloy cast-iron tools on a double-action hydraulic press. The tools are repolished every 100 to 300 parts.

The development of a type 316 stainless-steel part, in which the bottom is pierced out and the remaining lower wall is straightened out, is shown in Fig. 12-8. High blankholding pressures were required on this operation, because of the large draw and punch-nose radii. This part is drawn with a hardened-tool-steel punch, die, and blankholder in a double-action press.

Cone Shapes. The drawing of cone-shaped shells requires additional draws to minimize the amount of material out of control. The tapered shell in Fig. 12-9 is completed in six draws, while the very severe cone in Fig. 12-10 required eight draws.

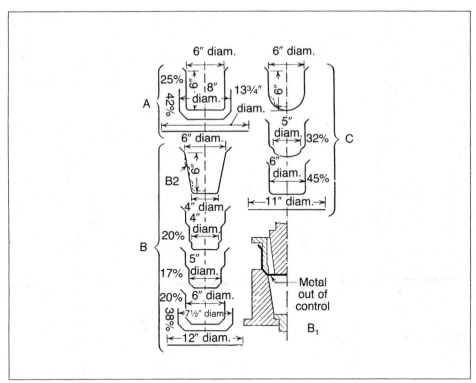

Fig. 12-2. Shell contours—a factor in determining the number of draws.[1]

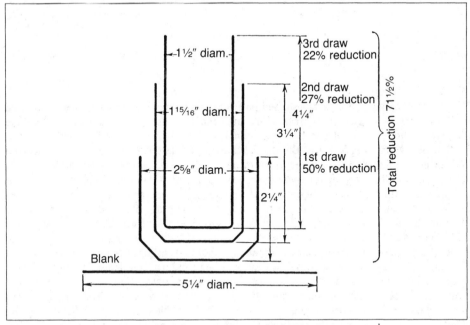

Fig. 12-3. An arrangement of draws for a circular shell of 3003-O aluminum.[1]

These shells are made of annealed aluminum, but sometimes fewer steps are possible with higher-strength metals such as stainless steel because of their ability to withstand high

12- 3

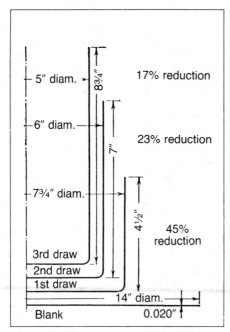

Fig. 12-4. Layout of drawing operations.

Key labels in figure 12-4:
- 5" diam. — 8¾" — 17% reduction
- 6" diam. — 7" — 23% reduction
- 7¾" diam. — 4½" — 45% reduction
- 3rd draw
- 2nd draw
- 1st draw
- 14" diam. — 0.020"
- Blank

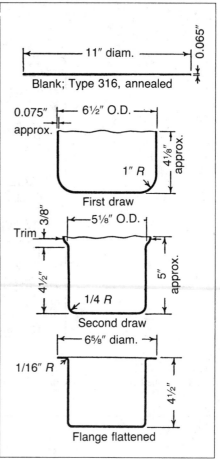

Fig. 12-5. Forming a tubular part of stainless steel.[2]

Key labels in figure 12-5:
- 11" diam.
- 0.065"
- Blank; Type 316, annealed
- 0.075" approx.
- 6½" O.D.
- 4⅛" approx.
- 1" R
- First draw
- Trim 3/8"
- 5⅛" O.D.
- 5" approx.
- 4½"
- 1/4 R
- Second draw
- 6⅝" diam.
- 1/16" R
- 4½"
- Flange flattened

blankholding pressures.

The development of a funnel-shaped part, drawn of type 302 stainless steel, is shown in Fig. 12-11. A single-action press with a die cushion is used with aluminum bronze die and blankholder and hardened-tool-steel punch for all the drawing operations.

Tubular Part with Vertical Bead. The draw schedule for a tubular part formed from a 5.25-in. (133.35-mm) -diameter blank of dead-soft sheet steel is shown in Fig. 12-12. The plan view B shows the relative position through which these sections are taken. The workpiece is first formed as a cup; then the beads in the side are formed. In three draws the stepped diameter is sunk into the bottom, and the cup is perforated and redrawn into the tubular shape. The redraw (third-draw) die (shown at B) to form the beads on the sides of the cup is an inverted-type die where the punch (D1), with inserts of the shape of the beads, is mounted on the die shoe, with the blankholder (D2) operated by the die cushion. The draw ring (D3) is mounted on the punch holder, and a spring-loaded stripper (D4) is recessed into the punch holder. In the die in view C, in which the previously perforated bottom is redrawn into a tubular shape, the shell is held between the draw ring (D1) and the blankholder (D2), then drawn over the punch (D3). Near the end of the stroke, the flange is trimmed to size. The positive knockout (D4) ejects the workpiece from the die.

Reverse-drawn Shell. A part in which the final shape required a reverse draw is shown in Fig. 12-13. For this type 347 stainless-steel part, the die and blankholder were tool steel and the punch was cast iron, mounted in a double-action press. The second and final draw

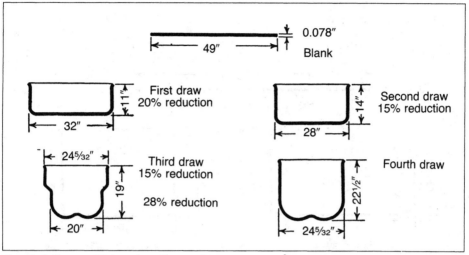

Fig. 12-6. Draw reductions for a large stainless-steel shell.[3]

pieces were reverse-drawn to dimensions and shapes shown.

Reverse-contoured Cup. The steps taken in press-forming a reverse-contoured cup are illustrated in Fig. 12-14. This part is drawn on a single-action press with an aluminum bronze die ring and blankholder. The cast-iron punch has a hardened-steel nose. The second draw piece is made in a regular redraw die, and the third draw piece uses a reverse draw to form the recess and outside contour. The part is sized, and a hole pierced and flanged in a three-station die on a single-action press. The bead is formed on rolls and crimped in a single-action press.

Reverse-bottom Cup. Figure 12-15 illustrates a type 304 stainless-steel part, 0.057 in. (1.44 mm) thick, drawn in a triple-action die on a double-action press. This operation is unique in that, since the limit of reduction is not reached in the cupping operation, the redraw starts while the cup possesses a considerable flange. During the first forming stage, the blank is held by the blankholder (D1) attached to the outer ram against the die (D3). The punch (D2) on the inner ram descends until it comes in contact with the recessing punch (D4). The second hold-down (D5), actuated by the die cushion, has been holding the metal against the punch and, when the recessing punch starts working, the metal flows in a reverse draw between the main punch and the second hold-down, also continuing to draw the flange metal from under the first hold-down, producing the form of the first draw piece. A second die completes the part to the finished shape.

The main punch and recessing punch are cast iron with inoculated cast-iron inserts. The die is cast iron with aluminum bronze inserts. The first and second hold-down are aluminum bronze.

Three-die Four-draw Shell. Figure 12-16 shows a deep-drawn cup in which four reductions were done in three dies. The first operation is a regular draw with a blankholder. The second operation is a regular redraw using a blankholding sleeve. In the third draw die, the first draw ring reduces the part form 7.496 to 6.256 in. (190.40 to 158.90 mm) diameter, and the second ring reduces it to 5.973 in. (151.71 mm) diameter. All the dies are the push-through type, and the second and third have spring-operated stripping fingers (D1).

Large-flange Cup-shaped Part. The drawing of a stainless-steel part with a large flange is shown in Fig. 12-17. The first die draws a part with a beveled bottom corner. The second die has a blankholding sleeve that has the same shape as the shell drawn in the previous die. The redrawing of a shell with beveled corners is easier than drawing one with radius corners, since the metal does not have to flow around two 90° angle corners. The third operation flattens the flange and bottom surface.

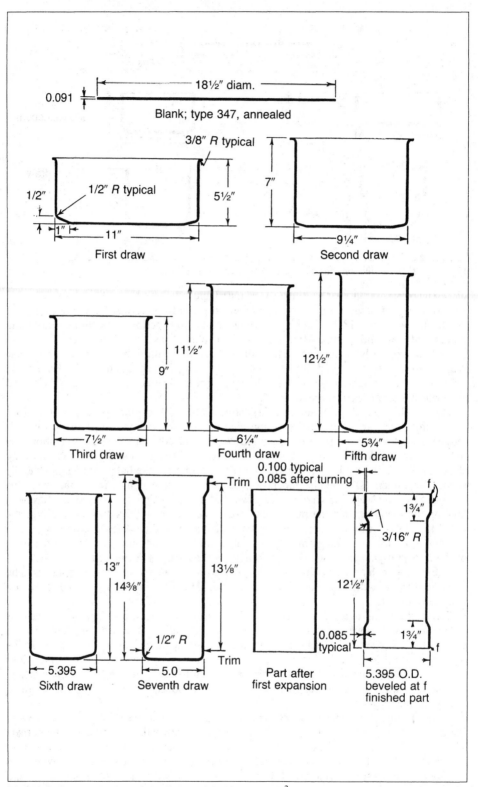

Fig. 12-7. Draw reductions for a deep stainless-steel shell.[2]

12-6

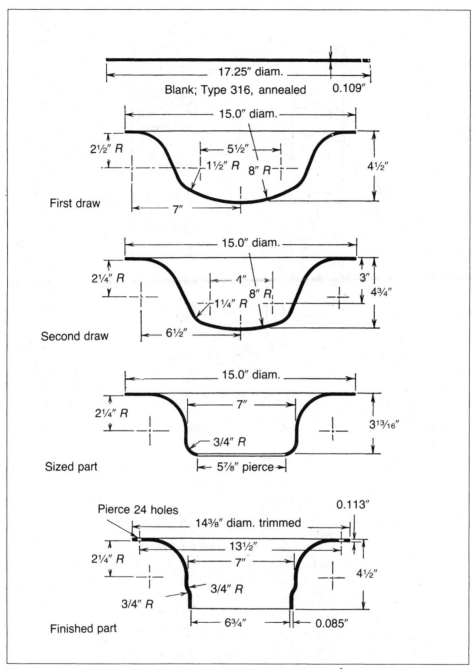

17.25" diam.
Blank; Type 316, annealed   0.109"

15.0" diam.
2½" R
5½"
1½" R   8" R
4½"
First draw
7"

15.0" diam.
2¼" R
4"
8" R
1¼" R
3"
4¾"
Second draw
6½"

15.0" diam.
2¼" R
7"
3/4" R
3¹³/₁₆"
Sized part
5⅞" pierce

Pierce 24 holes
0.113"
14⅜" diam. trimmed
2¼" R
13½"
7"
3/4" R
4½"
3/4" R
Finished part
6¾"
0.085"

Fig. 12-8. An example of a draw using a high blankholding pressure.[3]

Large-flange Tubular Part. The drawing operation for a tubular part with a large shaped flange is shown in Fig. 12-18. The 10-in. (254-mm) -diameter cup is sunk to final diameter in three operations. A bevel is used on the bottom corner of the cup in the second and third draws as an aid to the drawing operation. The bevel in the flange, remaining after the third draw, is a start in forming the large radius around the 10-in. (254-mm) -diameter tubular portion. After the third draw, the bottom is blanked out of the cup, the

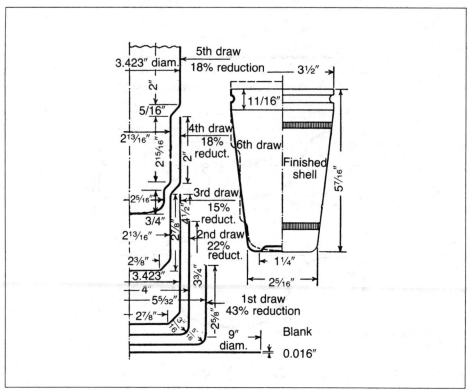

Fig. 12-9. Draw reductions for a tapered shell.[1]

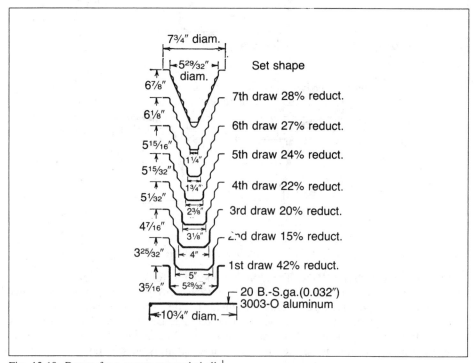

Fig. 12-10. Draws for a severe tapered shell.[1]

12-8

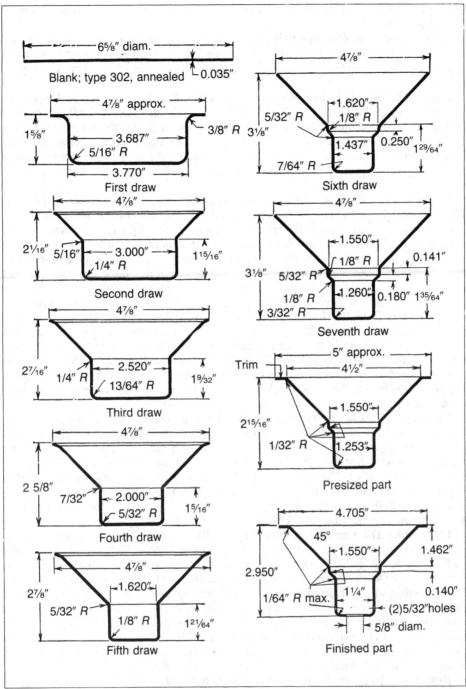

Fig. 12-11. Press-forming a funnel-shaped part.[2]

12-9

tubular portion completed, and the radius around the tube set. The outside contour is formed in the fourth draw.

Square Shell from Circular Blank. On square shells, it is sometimes possible to use round blanks and even make the first draw circular. The square shape is approached gradually, as suggested by the 2.25 in. (57.2 mm) square by 4-in. (101.6-mm) -deep radio shield (Fig. 12-19).

Square Flanged Shell from Square Sheared Blank. A box-shaped part drawn from a square sheared blank is shown in Fig. 12-20. Apparently, the blank shape is unimportant as long as the total surface area of the formed part does not exceed three to four times the cross-sectional area of the punch. However, a part formed from a rectangular blank usually deviates considerably from the finished contour and necessitates a large trimming allowance. In this part, note the change in material thickness in the various parts of the shell.

Drawing a Long Narrow Box. A long narrow box-shaped part is illustrated in Fig. 12-21. During the first draw, the preshaped blank was drawn to a flat elliptical shape and part of the flange was trimmed off before the second draw. In the second draw, the side of the shell was straightened, and to aid in eliminating canning, the punch-nose radius was increased from 0.250 in. (6.35 mm) at the ends to 0.50 in. (12.7 mm) at the center. The third draw shaped the shell to the finish dimensions and stretched the long flat side wall to eliminate the canning effect. This type 304 stainless-steel part is annealed after the first and second draws. This cast-iron double-action die is used in a double-action press.

Square Box from Preshaped Blank. A preshaped blank and the box to which it is drawn are shown in Fig. 12-22. This part is drawn in a conventional tool-steel die in a double-action mechanical press. The material was 1100-O aluminum 0.064 in. (1.63 mm) thick. The draw radius is 0.313 in. (7.95 mm), increased to 0.375 in. (9.52 mm) at the corners. This draw is rather severe, since the part is drawn to a depth 23 times the corner radius, and the area of the blank is 4.5 times the area of punch, but the depth is only 80% of the width.

Rectangular Box from Preshaped Blank. The rectangular box-shaped part shown in Fig. 12-23 is drawn of 0.064 in. (1.62 mm) 2024-O aluminum from a preshaped blank. The draw radius is 0.167 in. (4.24 mm), the punch-nose radius 0.50 in. (12.7 mm), and the punch corner radius 0.438 in. (11.12 mm). The part is drawn to a depth of about 6.6 times the corner radius. The blank area is 3.4 times the punch area, and the depth is 80% of the width. This part has a 0.093 in. (2.36 mm) offset in the vertical portion of the ends and a slop of 0.156 in. (3.96 mm) at the bottom. A conventional-type die in a double-action press is used for operation.

Eliminating Oil Canning in Rectangular Box Draw. An oil canning effect is experienced in the forming of the part shown in Fig. 12-24. The nose radius of the preform punch was gradually increased from 0.75 in. (19.0 mm) to 1 in. (25.4 mm) at the center along the longest sides. The finish form punch has a 0.75 in. (19.0 mm) radius the entire length, and the sharpening of the radius near the center of the long sides caused stretching of the center portions of the walls, eliminating the canning. Excess metal, or ears, is used at the corners to relieve the strain at those points, thus distributing it over the entire surface of the blank, avoiding corner breaks. The part is annealed between draws to aid in forming. A cast-iron double-action die is used in a double-action press. The shell material is type 302 stainless steel 0.050 in. (1.27 mm) thick.

Drawing an Irregular Contour. A contoured part of 0.040 in. (1.02 mm) thick 52SO clad aluminum is shown in Fig. 12-25. The tool-steel draw ring has a 0.50 in. (12.7 mm) draw radius with draw beads on the long sides. A 1 in. (25.4 mm) nose radius was used on the tool-steel punch. The blankholder is also made of steel.

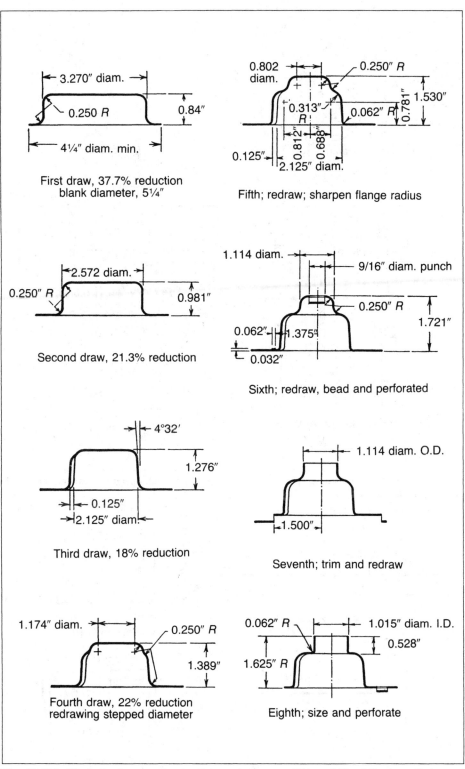

First draw, 37.7% reduction
blank diameter, 5¼"

Second draw, 21.3% reduction

Third draw, 18% reduction

Fourth draw, 22% reduction
redrawing stepped diameter

Fifth; redraw; sharpen flange radius

Sixth; redraw, bead and perforated

Seventh; trim and redraw

Eighth; size and perforate

Fig. 12-12A. Blank development for drawing a tubular part. (*General Metal Products Co.*)

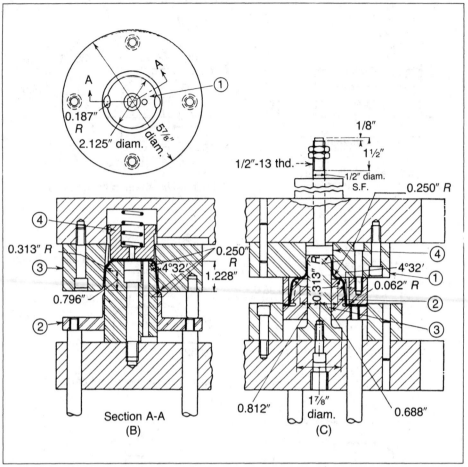

Fig. 12-12B and C. (B) Third-operation draw die, and (C) seventh-operation trim and redraw die for the part in Fig. 12-12A.

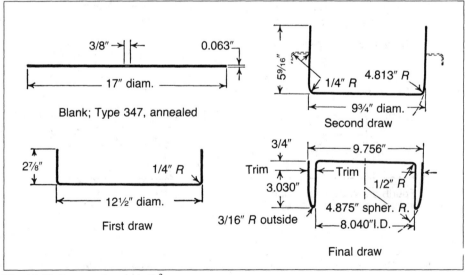

Fig. 12-13. Reverse-drawn shell.[2]

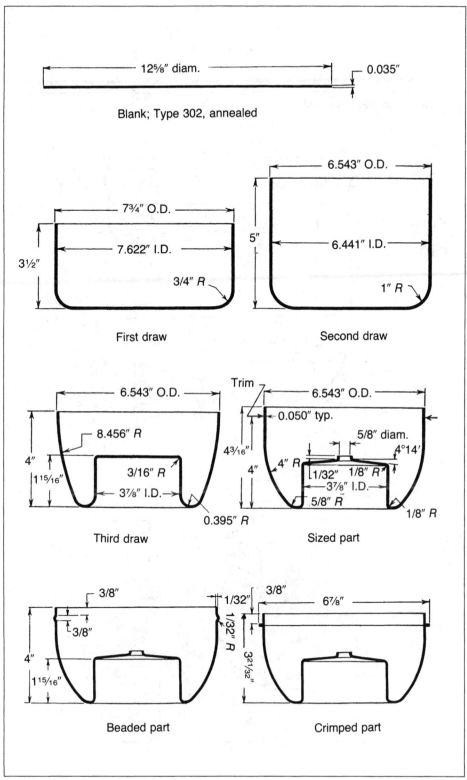

Fig. 12-14. Reverse-contoured cup.[2]

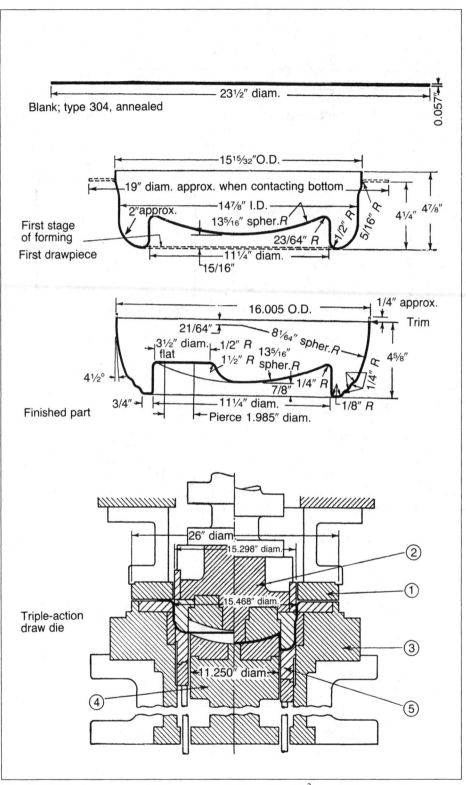

Fig. 12-15. Forming a reverse-bottom cup on a triple-action die.[2]

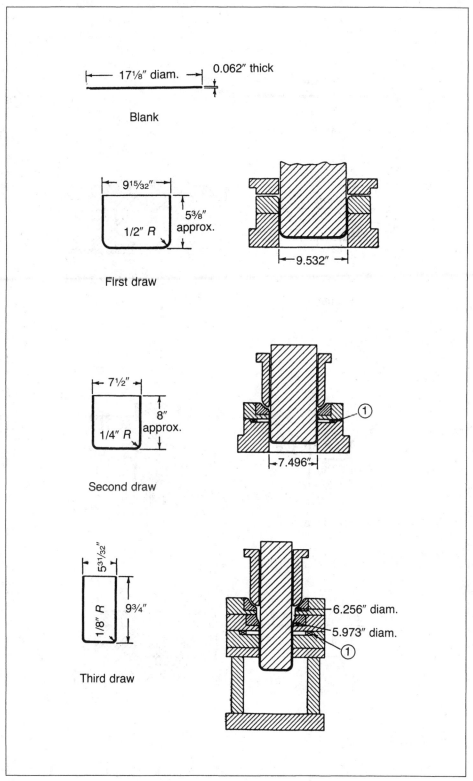

Fig. 12-16. Deep-drawn cup, four operations in three dies.

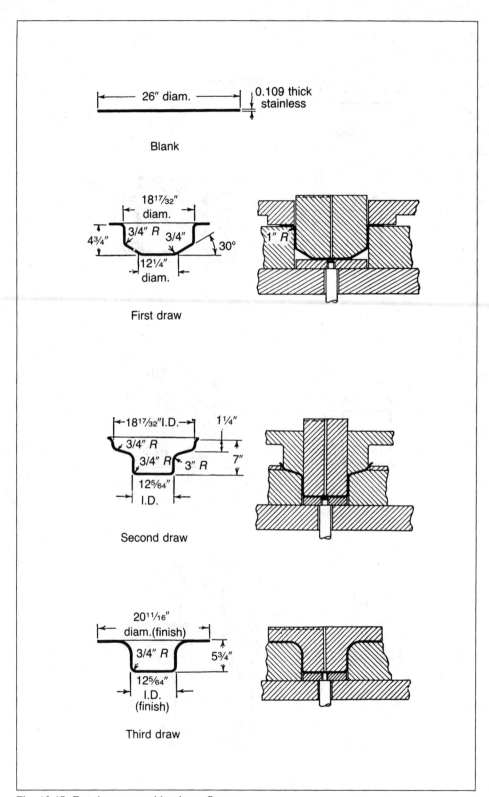

Fig. 12-17. Forming a part with a large flange.

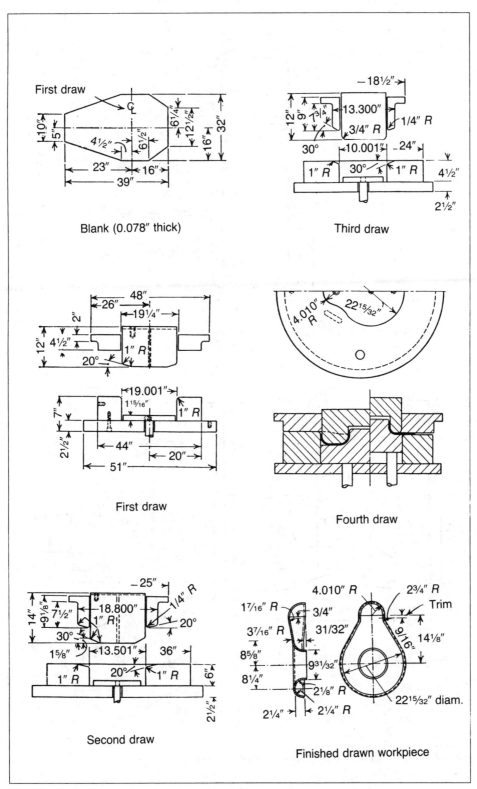

Fig. 12-18. A tubular part with a large, shaped flange.

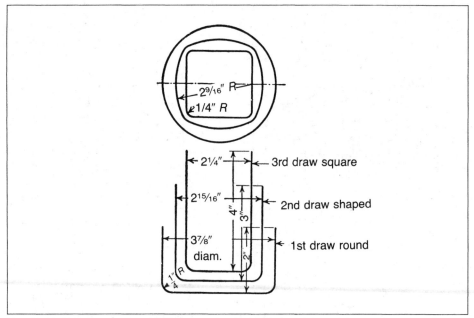

Fig. 12-19. Drawing a square shell from a circular blank.[1]

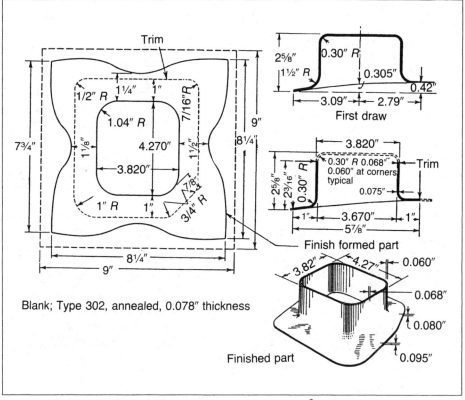

Fig. 12-20. Box-shaped part drawn from a square sheared blank.[2]

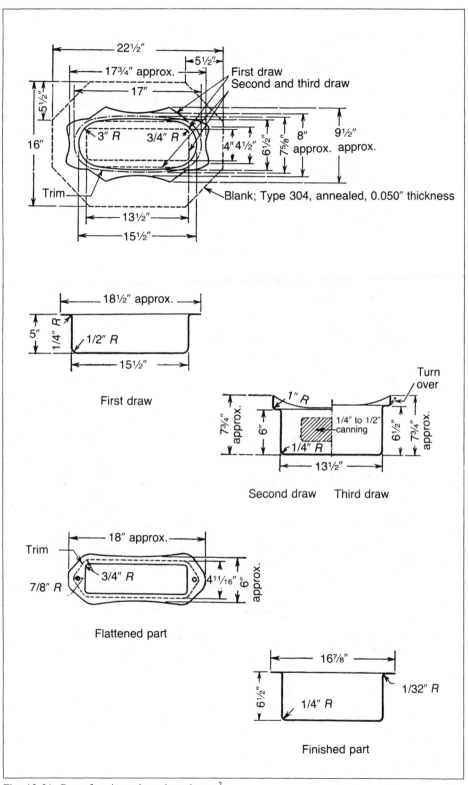

Fig. 12-21. Press-forming a box-shaped part.[2]

12- 19

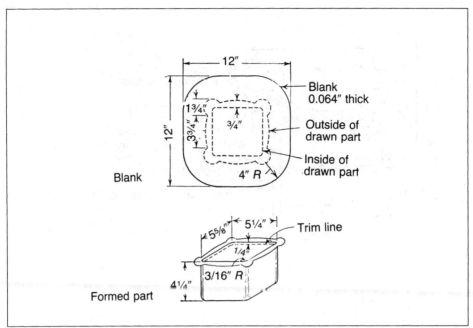

Fig. 12-22. Preshaped blank drawn into a box shape.[4]

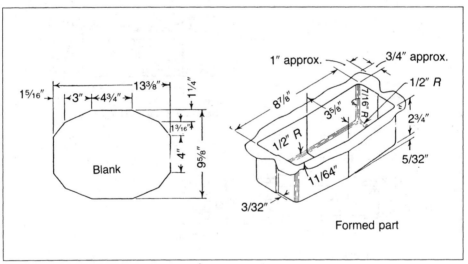

Fig. 12-23. Rectangular box-shaped part.[4]

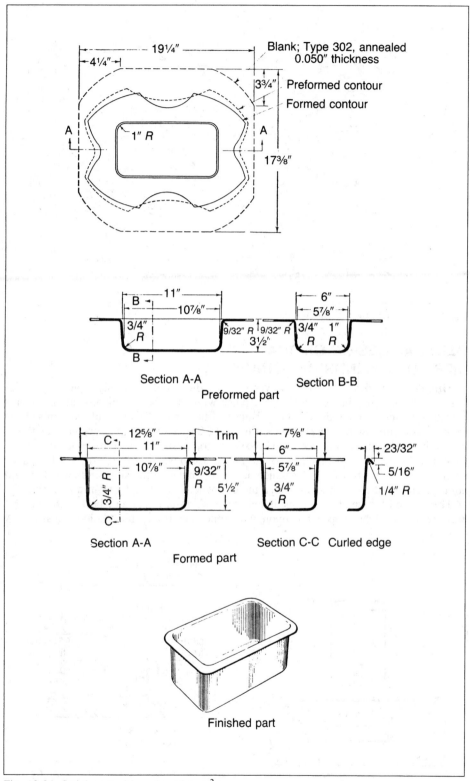

Fig. 12-24. Stainless-steel box-shaped part.[2]

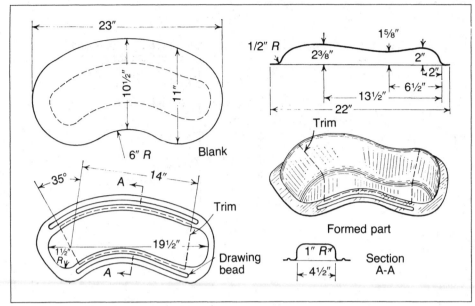

Fig. 12-25. An irregularly contoured part.[4]

GENERAL DESIGN OF DRAW DIES, DIES FOR CYLINDRICAL SHAPES

Single-action Dies. The simplest type of draw die is one with only a punch and die. Each component may be designed in one piece without a shoe by incorporating features for attaching them to the ram and bolster plate of the press. This type of die forms shells from blanks with low D/d or D/t ratios. Figure 12-26 shows a simple type of draw die in which the precut blank is placed in the recess on top of the die, and the punch descends, punching the cup through the die. As the punch ascends, the edge of the shell expands slightly to make this possible. The punch has an air vent to eliminate suction which would hold the cup on the punch and damage the cup when it is stripped from the punch.

The method by which the blank is held in position is important, because successful drawing is somewhat dependent upon the proper control of blankholder pressure. A

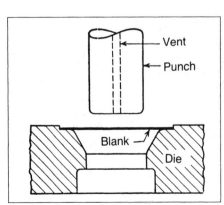

Fig. 12-26. Simple type of drawing die.

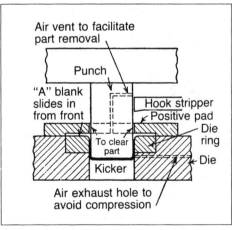

Fig. 12-27. Simple form of drawing die for use with heavy stock.

simple form of drawing die with a rigid flat blankholder for use with heavier stock is shown in Fig. 12-27. When the punch comes in contact with the stock, it will be drawn into the die without allowing wrinkles to form.

Another type of drawing die for use in a single-action press is shown in Fig. 12-28. This die is a plain single-action type where the punch pushes the metal blank into the die, using a spring-loaded pressure pad to control the metal flow. The cup either drops through the die or is stripped off the punch by the pressure pad. The sketch shows the pressure pad extending over the nest, which acts as a spacer and is ground to such a thickness that an even and proper pressure is exerted on the blank at all times. If the spring pressure pad is used without the spacer, the more the springs are depressed the greater the pressure exerted on the blank, thereby limiting the depth of draw. Because of limited pressures obtainable, this type of die should be used with light-gage stock and shallow depths.

A single-action die for drawing flanged parts, having a spring-loaded pressure pad and stripper, is shown in Fig. 12-29. The stripper also may be used to form slight indentations or reentrant curves in the bottom of a cup, with or without a flange. Draw tools in which the pressure pad is attached to the punch are suitable only for shallow draws. The pressure cannot be easily adjusted, and the short springs tend to build pressure too quickly for deep draws. This type of die is often constructed in an inverted position with the punch fastened to the lower portion of the die.

Deep draws may be made on single-action dies, where the pressure on the blankholder is more evenly controlled by a die cushion or pad attached to the bed of the press. The typical construction of such a die is shown in Fig. 12-30. This is an inverted die with the punch on the lower portion of the die.

Double-action Dies. In dies designed for use in a double-action press, the blankholder is fastened to the outer ram which descends first and grips the blank; then the punch, which is fastened to the inner ram, descends, forming the part. These dies may be a push-through type, or the parts may be ejected from the die with a knockout attached to the die cushion or some delayed action. Figure 12-31 shows a cross section of a typical double-action draw die.

The inverted-type redraw die in Fig. 12-32 has the punch (D1) mounted on the lower shoe and the draw ring (D2) mounted on the upper shoe. The pressure exerted on the metal between the blankholder (D3) and the draw ring is controlled by the stop pin (D4). The large stop pin (D5) controls the depth of the redraw.

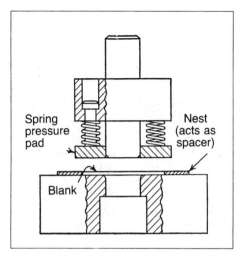

Fig. 12-28. Draw die with spring pressure pad.

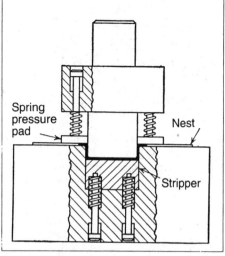

Fig. 12-29. A draw die with spring pressure pad and stripper.

The redrawing of cups can be made with less cold working of the metal by using a beveled bottom corner rather than a radius. If a radius is used, the metal is made to flow around two 90° angle bends. Figure 12-33 shows a redraw die for a cup with a beveled bottom corner. The blankholder sleeve and the mouth of the draw ring are shaped to fit the beveled bottom corner. In this die, the top of the cup has a tapered flange and may be flattened later. Cups also may be drawn into a die of this type leaving no flange on the part.

Figure 12-34 illustrates the typical construction of a double-action die for a cup with a radius at the bottom corner. The finished cup has a flange at the top; therefore, the bottom of the blankholding sleeve and top of the draw ring are flat. On this die, the ejector is shaped to indent the bottom of the cup at the end of the stroke. Note the air vents in the punch, die, and ejector, to eliminate air pockets.

Combination Draw-redraw Die. When the reduction of the first draw is limited because of a low diameter-thickness ratio, but the yield point of the material is still low, a shell may be drawn in a combination draw and redraw die as shown in Fig. 12-35. The blank is placed in the nest and held there by the blankholder (D1). The main punch (D2) draws the shell with the aid of the pressure pad (D3) into the draw ring (D4) and over to the reverse draw punch (D5) into the cavity in the main punch. The top surface of the redraw punch should be below the top of the draw ring so that the first draw has been well-started before the redraw starts. The left-hand part of the figure shows the cup partially drawn; the right-hand side shows the completed cup.

Combined Cylindrical and Elliptical Draw. A combination cylindrical- and elliptical-shaped draw is shown in Fig. 12-36, with details of the die design. The shape of the developed blank is shown with the position of the first draw. The stock is 0.120-in. (3.05-mm) deep-draw steel.

Operation 1 includes a combination blank and draw. D1 is a combination blanking punch with hardened steel inserts and draw ring; the blankholder (D2) is actuated by a die cushion. The blanking die (D3) is scalloped to provide shear. The draw punch is 9 in. (229 mm) in diameter, and the spherical crown is finished to size. This draw die is mounted on a heavy die set with guide pins and long guide bushings.

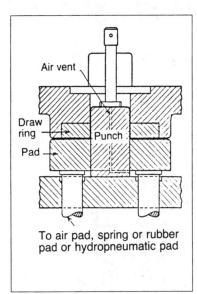

Fig. 12-30. Cross section of inverted drawing die for single-action press; die is attached to ram; punch and pressure pad are on bed of press.

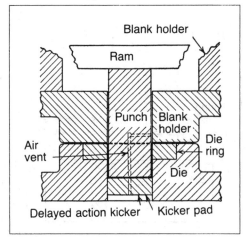

Fig. 12-31. Cross section of typical double-action cylindrical draw die.

Operation 2 is a regular redraw with the blankholder operated by the die cushion. The punch and draw ring are vented to allow the air to escape and prevent suction. The part can be ejected from the draw ring by a knockout operated by the press knockout bar.

In Operation 3, the elliptical portion is drawn, with a spring-operated sleeve (*D1*) sliding over the cylindrical portion to prevent its being pulled out of shape in this operation. In this inverted-type die, the holding sleeve is recessed into the die shoe and slides inside the draw ring (*D2*). The blankholder (*D3*) is operated by a die cushion. For ease of manufacture, the punch is made in two parts, one cylindrical (*D4*) and the other elliptical (*D5*).

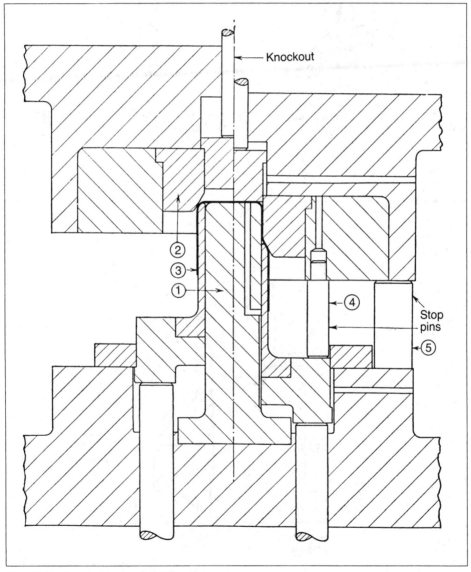

Fig. 12-32. Inverted-type redraw die.

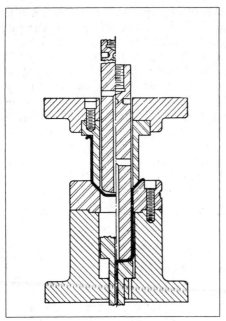

Fig. 12-33. Double-action redraw die for beveled bottom corner cups.[1]

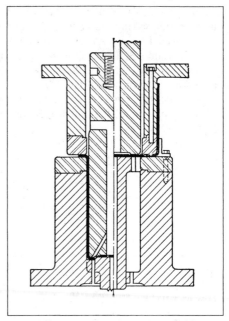

Fig. 12-34. Double-action redraw die for radius bottom corner cups.[1]

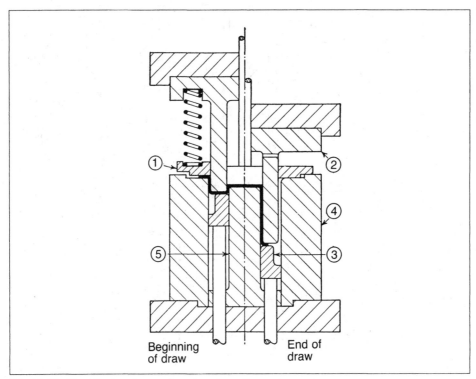

Fig. 12-35. Combination draw and reverse draw die.

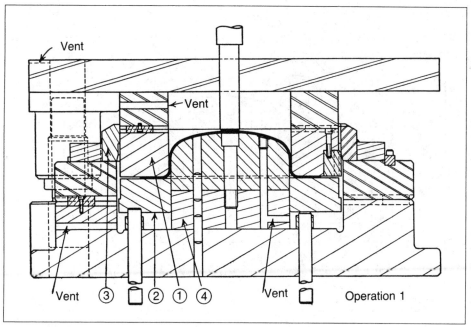

Fig. 12-36A. Combination cylindrical and elliptical draw die, operation 1. (*Right Tool & Die Co.*)

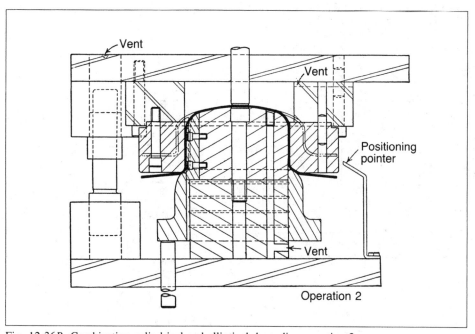

Fig. 12-36B. Combination cylindrical and elliptical draw die, operation 2.

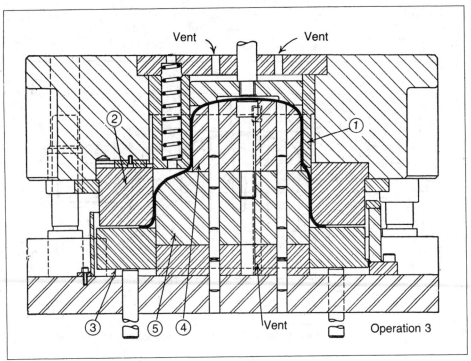

Fig. 12-36C. Combination cylindrical and elliptical draw die, operation 3.

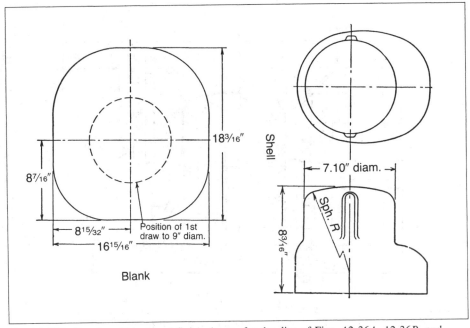

Fig. 12-36D. Developed blank and finished part, for the dies of Figs. 12-36A, 12-36B, and 12-36C.

DIES FOR BOX-SHAPED DRAWS

Combination Blank, Draw, Pinch-trim. The die shown in Fig. 12-37 is a combination die to blank, draw, and pinch-trim to height a rectangular box of 0.050-in. (1.27 mm) cold-rolled deep-drawing steel. The strip stock is fed into the die against the stop (D1) and is held in position by the spring-loaded pressure pad (D2), while the combination blanking punch and draw ring (D3) punches the blank in the die (D4). The box is formed over the draw punch (D5) with the aid of the blankholder (D6) and is pinched to height with the pinch punch (D7). The die is used in a single-action press with a die cushion and uses a positive knockout to eject the box from the draw ring.

Inverted Deep Draw. An inverted-type draw die for a single-action press with a cushion is shown in Fig. 12-38. The die is for a box of 0.031-in. (0.79 mm) cold-rolled steel. The preshaped blank is located on the blankholder by disappearing pins. Near the end of the stroke, the box is trimmed between the draw ring and pinch-trim block.

Draw Die for a Transmission Support. Figure 12-39 is a draw die for a transmission support made of 0.187-in. (4.75-mm) -thick AISI-SAE 1008 stock. A preshaped preflanged blank is placed on the locating pins (D1). As the die closes, the center portion of the punch (D2) and the pressure pad (D3) grip the metal, and the two end punches (D4) start drawing the metal into the die. Because of the thickness of the material, no blankholder is required in this die. The closing of the die sets the radii and flanges to dimensions.

Drawing an Oval Can. A die to blank and draw an oval-shaped can is shown in Fig. 12-40A. This is an inverted-type die for use in a single-action press with a die cushion.

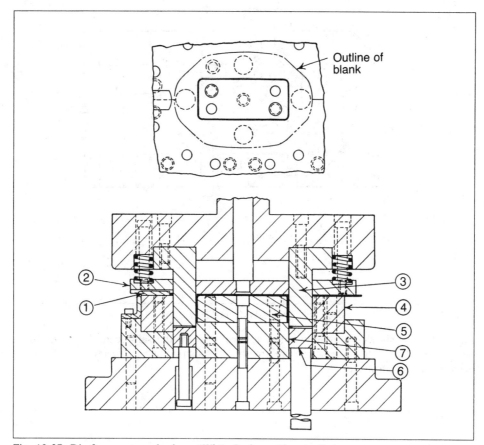

Fig. 12-37. Die for a rectangular box. (*White-Rodgers Electric Co.*)

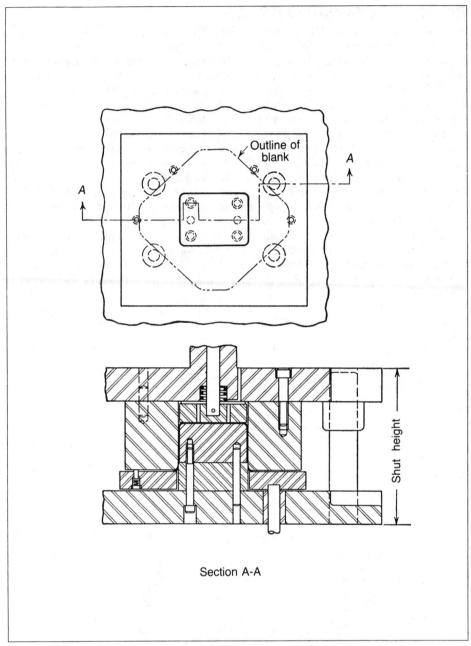

Fig. 12-38. Draw die for a deep box. (*White-Rodgers Electric Co.*)

Figure 12-40*B* shows the redraw and pinch-trim die for the same can. This die also is designed for use on a single-action press with a die cushion using a positive knockout for ejecting the part from the die.

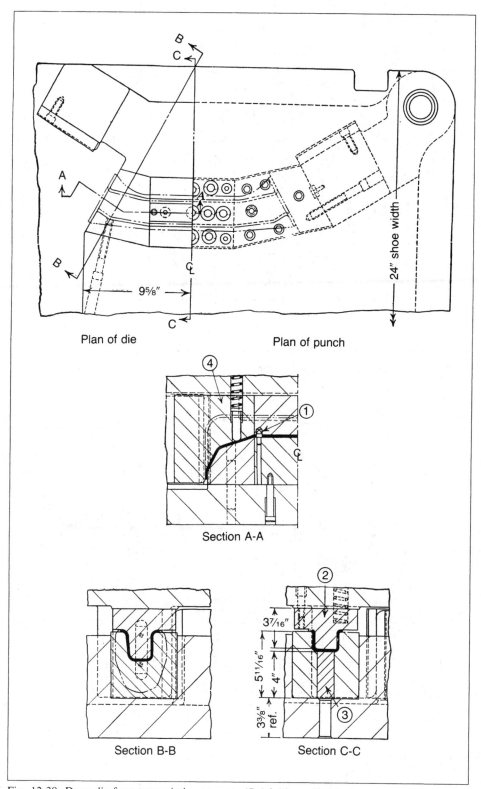

Plan of die   Plan of punch

Section A-A

Section B-B   Section C-C

Fig. 12-39. Draw die for a transmission support. (*Buick Motor Division*)

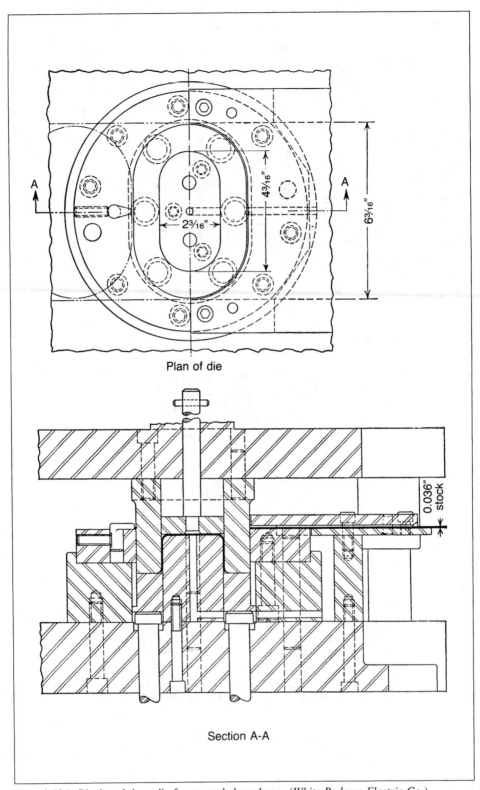

Plan of die

Section A-A

Fig. 12-40A. Blank and draw die for an oval-shaped can. (*White-Rodgers Electric Co.*)

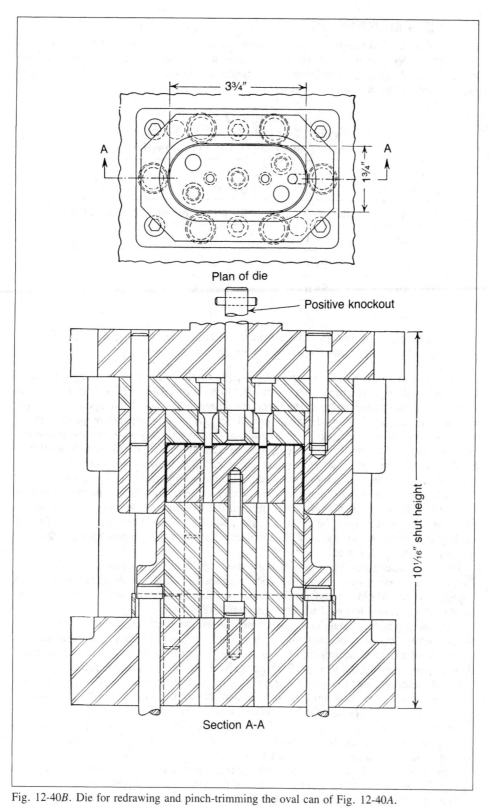

3¾"

1¾"

Plan of die

Positive knockout

10¹⁄₁₆" shut height

Section A-A

Fig. 12-40B. Die for redrawing and pinch-trimming the oval can of Fig. 12-40A.

DRAWING MAGNESIUM

At proper elevated temperatures, magnesium alloys have been drawn with reductions up to 70%. Deep draws, that require the use of multiple die sets and intermediate annealing in other metals, can be made in magnesium in one stage. Table 12-1 lists test data for one magnesium alloy.

TABLE 12-1
Drawability Test Data on AZ31B-0 Alloy at
Various Temperatures and Press Speeds*

Drawing Temperature	Percent Reduction		
°F (°C)	20 fpm (6.0 mpm)	50 fpm (15 mpm)	80 fpm (24 mpm)
75 (24)	14.3	14.3	
250 (121)	30.3	30.3	
400 (204)	58.6	57.1	53.8
500 (260)	62.5	61.3	53.8

*Test made on 0.064-in (1.63 mm) drawn into a 3-in. (76.2-mm) diameter cylindrical cup. Draw radius was 6*t* and punch radius 10*t*.

At room temperatures, magnesium alloys require rather large radii for simple bends or very shallow draws. At temperatures from 300° to 800° F (150° to 427° C), the rapid work hardening which occurs at room temperature disappears. Elevated temperatures of the annealed magnesium alloy sheet do not affect the room-temperature properties. If full-hard sheet is worked at too high a temperature for too long a period, the properties after forming will be similar to those of the annealed alloy. When full-hard material is to be formed, the proper time-temperature curve must be studied for that alloy. Rapid forming at relatively high temperatures can yield parts with good mechanical properties.

For low-temperature runs, the magnesium can pick up its forming temperature from the die itself. If heated blanks are fed into the die, the heating can be done on portable hot plates or ovens placed near the press. The method of die heating or the methods of blank heating is dictated by the type of job involved. For short-run or prototype work, hand torches sometimes are used.

Hydraulic and mechanical presses are used for drawing magnesium sheet plate. Double-action presses are desirable, since magnesium alloys at elevated temperatures require less blankholder pressure. The operator should have accurate controls on the blankholder pressure to avoid tearing and wrinkling.

For low production quantities, gas burners often are used. Black iron pipe may be formed to the contour of the draw die, as shown in Fig. 12-41. Holes are drilled in the pipe so that the flame will impinge on the draw ring. To provide space for an efficient gas flame, approximately 0.750 in. (19.05 mm) should be allowed between the gas burner ring and the draw ring. For longer-run operations, electric heat is usually applied to the die (Fig. 12-41) with automatic controls so that the temperature of the die components may be maintained very accurately without attention from the press operator. In certain configurations, the punch may tend to overheat. In such cases, water cooling may be used in the punch so that it may hold a given temperature. This is especially true when planning for hot forming on a continuous-production basis.

Care must be taken to ensure that the heating method employed does not cause the magnesium to ignite. Once ignited, magnesium burns in air. Water cannot be used to quench the fire because it reacts violently with burning magnesium resulting in the

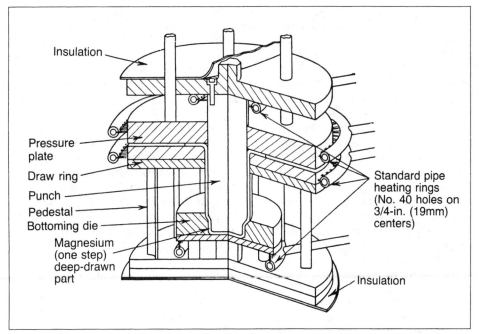

Fig. 12-41. General die construction for deep-drawing magnesium using gas ring burners for heating various die elements. (*The Magnesium Association.*)

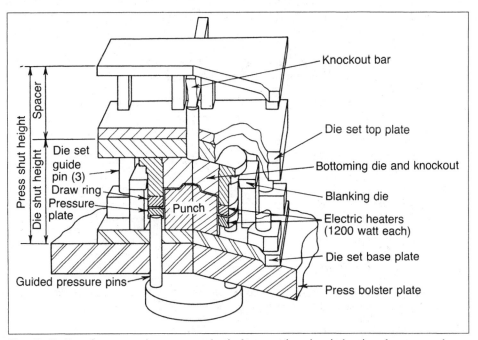

Fig. 12-42. Drawing magnesium on a mechanical press using electric heating elements on the drawing and pressure plate. (*The Magnesium Association.*)

release of hydrogen gas. Dry powder fire extinguishing agents formulated for use on magnesium fires should be available in case of an emergency.

When the draw die is placed in the press, insulating material should be used to keep the die from direct contact with the press ram and bolster plate. Compressed mineral fiber board often is used for this purpose. If the die is exceptionally large and a great deal of heat is given off, a water-cooled plate can be mounted between the die and the press. In this case, the insulation is still required so that the coolant plate does not take too much heat from the die while it is protecting the press.

As in other metals, the radius of the punch nose as well as the entrance radii of the drawing ring are important. Draw ring entry radii are normally from about four to seven times sheet metal thickness. Sharper radii can be used. However, if they become too sharp, metal fracture can result.

It is important that both the magnesium alloy sheet and the draw die surfaces be lubricated. Since most severe forming is done at temperatures over 500° F. (260° C) colloidal graphite suspended in a solvent often is used. It is best to use sheet which has no chemical treatment or, as it is commonly known, in the *as-rolled condition*.

Very mild forming, when done at temperatures up to 250°F (121° C), is being done with oils, greases, and waxes. Molybdenum disulphide is sometimes used for the higher temperatures, but it is doubtful if it has any practical advantages over colloidal graphite. Colloidal graphite tends to fill the pores of a cast-iron or Meehanite die so that the drawing condition improves with age. No attempt should be made to clean this graphite material from the pores of the die ring.

Following the outlined procedures, dies have been known to yield good parts after repeated heatings for many years.

When designing dies for drawing magnesium alloys, it may be necessary to consider the thermal expansion of the workpiece and die materials. The average values for the coefficients of thermal expansion in parts per million (ppm) per degree F (C) of different materials are as follows: aluminum alloys, 14.4 (26); cast iron, 6.8 (12.2); magnesium alloys, 16.1 (29); and zinc alloys, 15.4 (27.7). Table 12-2 lists conversion factors which can be used in designing dies. Typical blankholder pressures are shown in Table 12-3.

TABLE 12-2
Dimensional Factors for Steel Dies
Used for Hot-forming Magnesium

Forming Temperature °F (°C)	Dimensional Factor
300 (149)	1.00187
400 (204)	1.00270
500 (260)	1.00359
600 (316)	1.00450
700 (371)	1.00542
800 (427)	1.00642

The magnesium part dimension should be multiplied by the appropriate factor to obtain the correct steel die dimension for the temperature being used.

TABLE 12-3
Blankholder Pressures for Drawing a 1.250 in. diameter
Cup to a 63% Reduction On a Hydraulic Press

Drawing Temp., °F (°C)	Blankholder Pressure lbf* (kN)
400 (204)	575 (2.6)
425 (218)	425 (1.9)
450 (232)	350 (1.6)
475 (246)	300 (1.3)
500 (260)	250 (1.1)
525 (274)	200 (0.9)
550 (288)	175 (0.8)
575 (302)	150 (0.7)
600 (316)	125 (0.6)

Determined on 0.064-in. AZ31B-0 sheet at a drawing speed of 1.25-in./min. Radius of draw ring = 6t.

CARBIDE DRAW DIES

The use of sintered carbide for the draw ring and punch in a draw die greatly increases the life of these parts. The die set, punch plate, and die block should be made heavier and of higher-quality materials than are used in all-steel dies. The rules and practices used in designing steel dies apply in the design of carbide dies, but it should be remembered that carbide is not so easily machined as steel and must be designed to permit the use of grinding wheels to finish the contours. To facilitate finishing, inserts often are made in sections instead of in a solid piece. The draw radii are the same as for steel, but the bearing length may be shorter, because of the longer life of carbides. Liberal back relief on the exit side of the nib is recommended as a precaution against the possibility of flaking. Table 12-4 may be used to determine the various nib dimensions in reference to the inside diameter.

When a carbide die insert is subjected to high-impact loads and internal bursting pressures, it must be adequately supported externally by pressing or shrinking the carbide ring into a hardened steel case. Suitable steels for die cases include SAE 4140, 4340, and 6145, as well as AISI type H13 tool steel, hardened to R_C 38-48. The outside diameter of the steel case should be two to three times the outside diameter of the carbide ring when high internal pressures are involved. While carbide can be shrunk into steel successfully, shrinking steel into carbide is not advisable. With the thermal expansion of steel being about three times that of carbide, the steel can expand enough to break the carbide with only a moderate increase in temperature.

The shrink allowances listed in Table 12-5 are only general guidelines. Actual calculations with the formulas given in Fig. 12-43 are preferred. Calculations should be performed for any new or unusual designs, such as dies for forming parts to complex geometry, for drawing or ironing thin-walled cylinders, for operations at elevated temperatures, and for applications that exert high internal pressures on the dies.

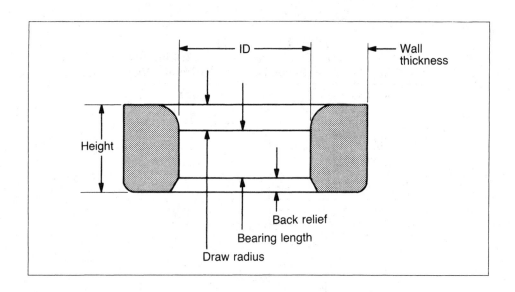

TABLE 12-4
Typical Dimensions for Carbide Draw Dies[5]

Inside Diameter, in. (mm)	Wall Thickness, in. (mm)	Bearing Length, in. (mm)	Back Relief Length, in. (mm)
to 0.50 (12.7)	0.312 (7.9)	0.125 (3.2)	0.062 (1.6)
0.50 to 1 (12.7 to 25.4)	0.375 (9.5)	0.188 (4.8)	0.094 (2.4)
1 to 1.50 (25.4 to 38.1)	0.50 (12.7)	0.25 (6.4)	0.125 (3.2)
1.50 to 2.50 (38.1 to 63.5)	0.562 (14.3)	0.312 (7.9)	0.125 (3.2)
2.50 to 5 (63.5 to 127)	0.675 (15.9)	0.375 (9.5)	0.156 (4.0)
5–10 (127 to 254)	0.75 (19.1)	0.438 (11.1)	0.188 (4.8)
10–15 (254 to 381)	0.875 (22.2)	0.50 (12.7)	0.25 (6.4)

TABLE 12-5
Approximate Shrink Allowances for
Carbide Cylinders Mounted Inside Steel Rings[6]

Outside Diameter of Carbide Cylinder, in. (mm)	Medium Diametral Interference, in. (mm)	Heavy Diametral Interference, in. (mm)
0.50 to 0.750 (12.7 to 19)	0.0020 (0.051)	0.0025 (0.064)
0.750 to 1 (19 to 25.4)	0.0025 (0.064)	0.0038 (0.097)
1 to 1.250 (25.4 to 31.7)	0.0035 (0.089)	0.0050 (0.127)
1.250 to 1.50 (31.7 to 38.1)	0.0040 (0.102)	0.0063 (0.160)
1.50 to 2 (38.1 to 50.8)	0.0050 (0.127)	0.0075 (0.190)
2 to 2.50 (50.8 to 63.5)	0.0070 (0.178)	0.0100 (0.254)
2.50 to 3 (63.5 to 76.2)	0.0080 (0.203)	0.0125 (0.318)
3 to 3.50 (76.2 to 88.9)	0.0100 (0.254)	0.0150 (0.381)
3.50 to 4 (88.9 to 101.6)	0.0110 (0.279)	0.0175 (0.444)
4 to 5 (101.6 to 127)	0.0140 (0.356)	0.0200 (0.508)
5 to 6 (127 to 152.4)	0.0165 (0.419)	0.0250 (0.635)
6 to 7 (152.4 to 177.8)	0.0200 (0.508)	0.0300 (0.762)

(Kennametal Inc.)

Carbide Draw Punches. The same general design used for steel punches should be followed for carbide punches. The carbide section of the punch should be long enough to cover the wear points. The carbide may be secured to the steel shank in various ways depending upon the diameter. If the punch is large enough and holes in the face are not objectionable, the carbide may be fastened as shown in Fig. 12-44A. With the development of EDM, holes may be tapped in the sintered carbide and the cap screws, running through the steel body, as shown in Fig. 12-44B, hold the tip to the body. Unless the pilot on the carbide is longer than the depth of the tapped hole, a counterbore should be made a little deeper than the length of the pilot to direct the stresses away from the thin section around the tapped hole. Punches which are too small in diameter for a tapped hole should be made of solid carbide or can be brazed as shown in Fig. 12-45C.

Mechanical retention of carbide wear rings to large punch bodies is illustrated in Fig. 12-46. View A shows holes tapped in the carbide for cap screws; the construction in view B utilizes a shoulder on the ring and a steel clamping plate.

A die with a carbide draw ring is shown in Fig. 12-47. This die produces a cup from 0.250-in. (6.35-mm) stock without the use of a blankholder. The part is ejected from the die with a delayed-action ejector and is stripped from the punch with a riding stripper. The employment of a riding stripper makes it possible to use a shorter punch for a long stroke. The stripper is supported by springs above the die and guided by headed guide pins in the lower shoe at a sufficient height to permit loading and unloading the die. The punch descends, punching the stripper down, to enable the short punch to form the cup. On the upstroke of the press, the stripper is raised by springs until it engages the heads of the guide pins, then strips the cup from the punch.

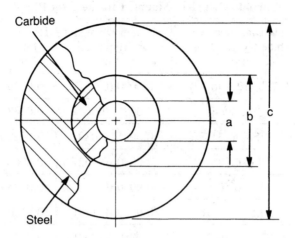

Carbide

Steel

a b c

δ — Diametral interference
P — Pressure between cylinders
E_s — Modulus of elasticity of steel
E_c — Modulus of elasticity of carbide
μ_s — Poisson's ratio of steel
μ_c — Poisson's ratio of carbide

$$\delta = \frac{bP}{E_s}\left(\frac{b^2 + c^2}{c^2 - b^2} + \mu_s\right) + \frac{bP}{E_c}\left(\frac{a^2 + b^2}{b^2 - a^2} + \mu_c\right)$$

If a steel ring is to be shrunk on a solid carbide cylinder, the diametral interference can be calculated by considering "a" to equal zero in the above formula. In the design shown, the tangential stress at the inner surface of the steel due to shrink is:

$$\sigma_t = \frac{P(b^2 + c^2)}{c^2 - b^2}$$

The maximum compressive prestress at the inner surface of carbide due to shrink is:

$$\sigma_t = \frac{-2Pb^2}{b^2 - a^2}$$

Fig. 12-43. Methods of determining diametral interference prestress when shrinking carbide rings into steel cases.[6] (*Kennametal Inc.*)

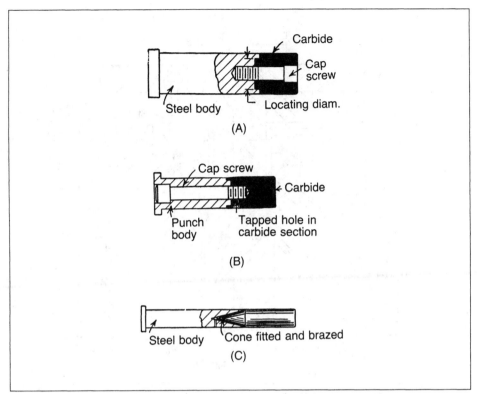

Fig. 12-44. Carbide-punch construction.

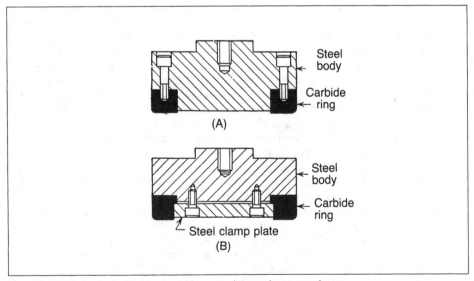

Fig. 12-45. Method of attaching carbide wear rings to large punches.

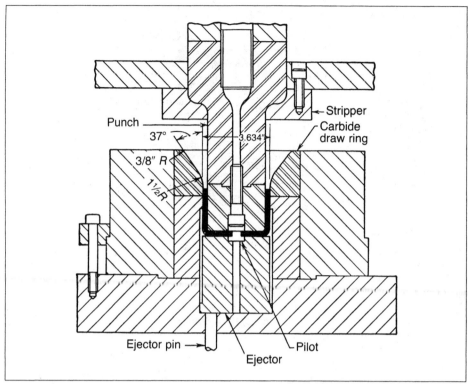

Fig. 12-46. Cupping die with a carbide draw ring. (*Worcester Pressed Steel Co.*)

DRAWING THIN OR FOIL STOCK

Die for Drawing Foil Pan. The die shown in Fig. 12-47 draws a foil pastry pan. A coil of flat stock aluminum foil is fed through the die when the die is in the "open" position. As the die closes, a blank is cut from the flat stock when the upper blanking punch ($D4$) enters the lower cutting edge ($D1$). The blanking punch ($D4$) grips the blank against the outer draw ring ($D2$) while it is being drawn over the male form punch ($D5$). The pressure actuating the draw ring ($D2$) is exerted by an air cushion under the die. The pan depth is established when the inner draw ring ($D6$) bottoms in the die shoe by the upper form punch ($D9$). The upper form punch ($D9$) is held firmly against the inner draw ring ($D6$) by air pressure above the air piston ($D7$). The bottom of the pan is also embossed at this time by the embossing pads ($D3$ and $D10$) when the set screw ($D8$) contacts the piston ($D7$). The blanking punch ($D4$) continues downward past the top surface of the inner draw ring ($D6$) to draw a wipe-down vertical flange around the periphery of the pan. As the die opens, this flange is rolled into a curl or bead around the periphery of the pan by the outer draw ring ($D2$). The pan is lifted from the die by the two draw rings ($D2$ and $D6$). It is rejected from the die by means of compressed air.

The die set is made of mild steel, and the remaining elements of oil-hardening tool steel, hardened and ground.

Draw Die for Pie Pan. The forming of 0.012-in. (0.30 mm) tin plate into pie pans is accomplished by the die shown in Fig. 12-48. Precut blanks are positioned by a nest (not shown) on the draw ring ($D1$) with the punch ($D2$) supporting the blank at the center. The combination die and blankholder ($D3$) and the draw ring have an offset which serves as a draw bead, thus keeping the sides of the pan smooth. The blankholder is supported by a die cushion during the forming operation.

A suction cup ($D4$) removes the pan from the punch; the positive stripper ($D5$) removes the pan from the suction cup and the die. Air holes through the punch and lower shoe

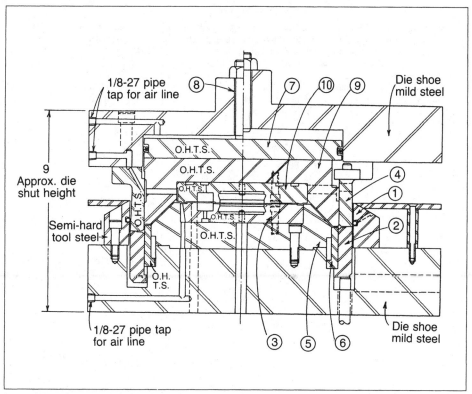

Fig. 12-47. Draw die for aluminum-foil pan. (*Ekco Containers, Inc.*)

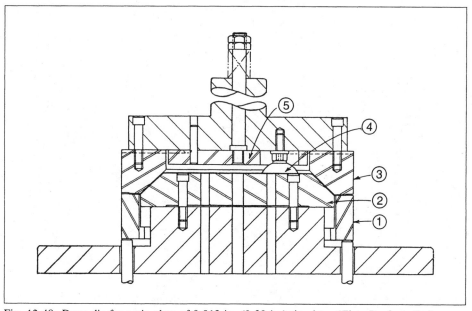

Fig. 12-48. Draw die for a pie plate of 0.012-in. (0.30-in.) tin plate. (*Ekco Products Co.*)

prevent the formation of vacuum under the formed pan.

The draw ring, die, and punch are made of oil-hardening tool steel, heat-treated to Rockwell C55.

References

1. J. W. Lengbridge, "The Theory and Practice of Pressing Aluminum," *The Tool Engineer* (June 1948).
2. "Forming of Austenitic Chromium-nickel Stainless Steels," The International Nickel Co., Inc. (1948).
3. L. F. Spencer, "Cold Forming Stainless Steels," *Iron Age* (March 31, 1949).
4. "Correlation of Information Available on the Fabrication of Aluminum Alloys," National Defense Research Committee (1943).
5. Charles Wick, John W. Benedict, and Raymond F. Veilleux, eds., "Die and Mold Materials," *Tool and Manufacturing Engineers Handbook*, 4th ed., Vol II (Dearborn, MI: Society of Manufacturing Engineers, 1984), p.2-26.
6. *Ibid*, p.2-27.

SECTION 13
DIES FOR LARGE AND IRREGULAR SHAPES

Most of the dies in this category are used to form the large sheet metal panels used in automotive and appliance assembly.

The materials used in dies for large irregular shapes depend upon the severity of the operation. Cast irons are commonly used, although cast steel die components are sometimes used for severe operations. Steel weldments are generally a more expensive form of construction, but can avoid the lead time required for pattern construction.

Computer-aided design (CAD) of large irregular-shaped dies has many important time and cost saving advantages. For example, CAD drawings of castings for the pattern shop can be plotted full size with the correct shrinkage allowance automatically provided. The cutter paths for computer numerical controlled (CNC) machine tools can be programmed directly from CAD data.

Safe Handling of Large Dies. Large dies present a special handling problem because of their weight. A method preferred over chain slings is the use of double eye cable slings in sets of four (Fig. 13-1). The die is provided with four captive self-locking pins (Fig. 13-2) in each die shoe to permit safe lifting, opening, turning, and closing. Spring loaded retention devices (not shown) are used to insure that the pins do not slip out of position during lifting and turning operations.[1]

GENERAL CONSTRUCTION PRACTICES

Placing Steel Inserts in Iron Castings. Iron die castings are usually a one of a kind item. The usual pattern material is styrofoam. This greatly simplifies both patternmaking and mold preparation, as there is no need to remove the pattern from the mold prior to pouring the iron.

Because the melting point of steel is several hundred degrees higher than cast iron, steel objects placed in the styrofoam pattern will become an integral part of the finished casting. For example, steel pipe used for heating or cooling a die can be cast directly into the die shoe.

Figure 13-3 shows a method for placing an eye bolt hole insert in a casting. The insert has a groove and flat which prevent the insert from turning or pulling out of the casting under extreme loading. Safe eye bolt holes are available for handling the casting as soon as it is removed from the sand. The cost per finished hole may also be lower than that of conventional drilling and tapping.

A finished hole with closer tolerances than could be obtained with a sand core can be provided with the use of the casting liner shown in Fig. 13-4. The liner material is round low-carbon steel, mechanical tubing. This style of liner provides holes in the die shoe to accommodate the cable sling lifting method shown in Fig. 13-2.[1]

Draw Beads. The main function of draw beads in the blankholder of a draw or redraw die is to provide additional resistance to metal flow, thus helping to control the movement

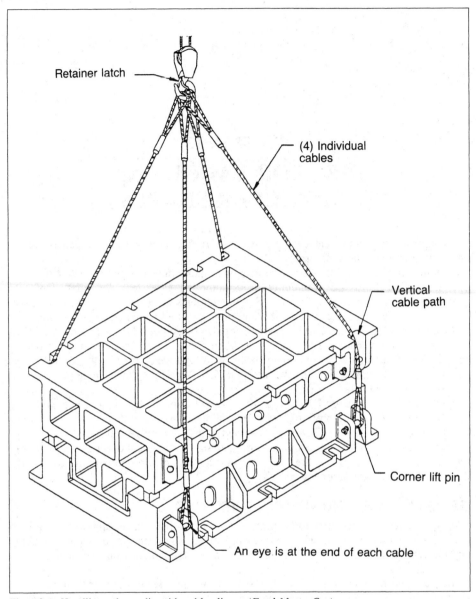

Retainer latch

(4) Individual
cables

Vertical
cable path

Corner lift pin

An eye is at the end of each cable

Fig. 13-1. Handling a large die with cable slings. (*Ford Motor Co.*)

of the metal into the die cavity. This greatly reduces the amount of blankholding pressure required for an operation.

In double-action presses it is possible to adjust or shim the blankholder to grip the blank more tightly in certain areas to control the metal flow. However, the fine adjustments required are very time consuming and do not permit quick die change (QDC).

Two or more beads may be placed in areas requiring greater control of the metal flow into the die cavity. While the location of the beads can be determined in the die tryout, dies for producing similar parts may be used as a guide. Often, a single bead is placed around the die cavity and additional beads are placed in local areas only as required. In tryout, it may be found that the single bead must be reduced in size or eliminated in some areas.

Commercial rolled-steel sections used for inserting draw beads into a binder are shown

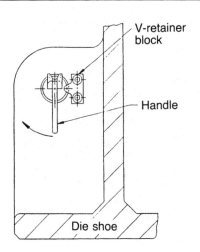

V-retainer block

Handle

Die shoe

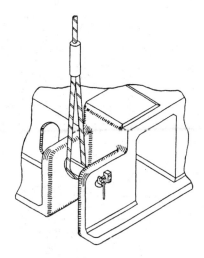

Lift pin is in position for lifting.
Handle is in down position: Therefore,
slot in pin is not lined up with point
of the V-retainer block.
Lift pin can not move out in this position.

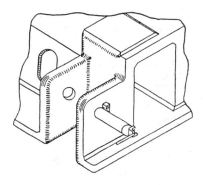

25 radius in all casting corners
most likely to be in contact with
lifting cables.

Lift pin has been pulled out to
accept the eye of the cable.
Pin is retained as the blind slot
surface in the pin contacts the V-
retainer block.

Operating Procedure

- Rotate handle toward the outside of the die until contact is made with the the
  V-retainer block.
- Slot in lift pin will be lined up with V-retainer block.
- Pull pin out of opposite casting wall.
- Blind slot in the pin prevents the pin from coming completly out of the casting.
- Insert the eye of the cable.
- Push pin thru opposite cast wall.
- Rotate pin so handle is in down position.
- Lift pin cannot move out in this position.

Fig. 13-2. A good method for safe attachment of a cable sling to a large die. (*Ford Motor Co.*)

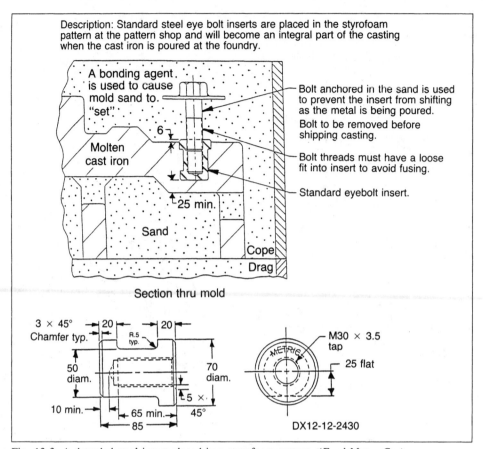

Description: Standard steel eye bolt inserts are placed in the styrofoam pattern at the pattern shop and will become an integral part of the casting when the cast iron is poured at the foundry.

A bonding agent is used to cause mold sand to "set"

Molten cast iron

6

25 min.

Sand

Cope

Drag

Section thru mold

Bolt anchored in the sand is used to prevent the insert from shifting as the metal is being poured.
Bolt to be removed before shipping casting.

Bolt threads must have a loose fit into insert to avoid fusing.

Standard eyebolt insert.

3 × 45° Chamfer typ.

20

20

R.5 typ.

50 diam.

70 diam.

5 ×

10 min.

65 min.

45°

85

M30 × 3.5 tap

25 flat

METRIC

DX12-12-2430

Fig. 13-3. A threaded steel insert placed in a styrofoam pattern. (*Ford Motor Co.*)

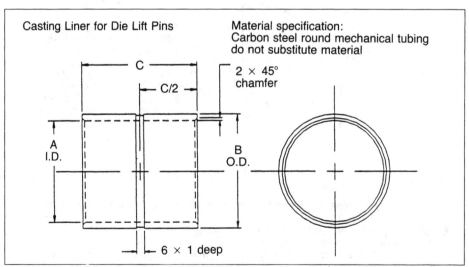

Casting Liner for Die Lift Pins

Material specification:
Carbon steel round mechanical tubing do not substitute material

C

C/2

2 × 45° chamfer

A I.D.

B O.D.

6 × 1 deep

Fig. 13-4. Steel mechanical tubing placed in a styrofoam pattern (provides holes for lifting pin shown in Fig. 13-2). (*Ford Motor Co.*)

in Fig. 13-5. Figure 13-6 illustrates another method of inserting a draw bead into a blankholder.

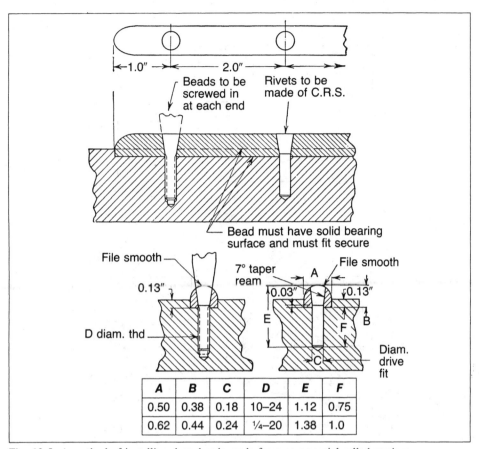

A	B	C	D	E	F
0.50	0.38	0.18	10–24	1.12	0.75
0.62	0.44	0.24	¼–20	1.38	1.0

Fig. 13-5. A method of installing draw beads made from commercial rolled sections.

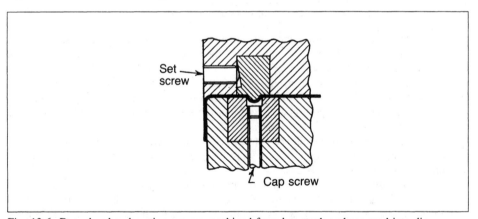

Fig. 13-6. Draw bead and mating groove machined from bar stock and recessed into die members.

Cast Draw Beads. In order to increase wear life and reduce metal pickup, the blankholder is often made of cast irons or steels having alloying constituents. These additives often enhance the ease with which the wearing surfaces can be flame hardened. For this reason, draw beads cast into the blankholder can outwear inserted beads, and are usually favored for most new die construction.

Another reason for cast beads being favored is that the working surfaces of dies for

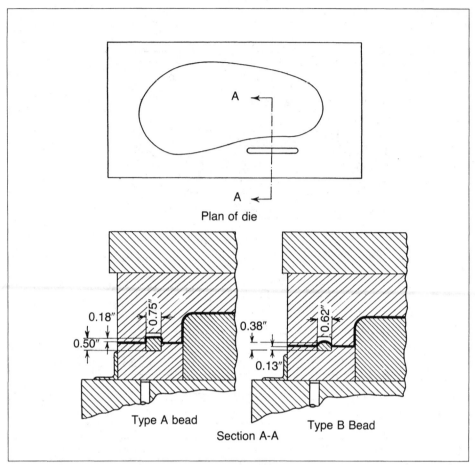

Fig. 13-7. Draw beads to restrict the flow of metal in local areas.

automotive outer panels are often hard chrome plated or ion-nitrided to increase wear life. These hard surfaces greatly reduce pick up of the zinc coatings now used to give such stampings corrosion protection.

Ion nitriding is considered to provide some important practical advantages over hard chrome, such as the ability to pop a slug mark with heat without destroying the coating. If inserted beads are used, there is a danger that the studs holding them will come loose in the nitriding furnace.

The size and location of beads are determined from past experience on similar parts. Where additional beads are required, they are built up with welding rods and ground or machined to shape and smoothness.

Short beads used to restrict the flow of metal in local areas have small radii on the corners to provide greater resistance to the metal flow (Fig. 13-7). The lower type of bead is adaptable to dies using springs, an air cushion or nitrogen cylinders to exert the pressure on the blankholder. The optimum height and corner radius of these beads must be determined by the results obtained at the tryout of the die.

The lock-type draw bead in Fig. 13-8 is installed in dies to provide maximum restriction to metal flow. The beads are rectangular in shape with minimum corner radius. While the figure shows inserted construction, they are usually cast and machined directly into the draw ring.

Figure 13-9 illustrates a half lock bead that requires a smaller blank than is the case when a full lock bead is used. Die tryout is simplified through this type of bead. If metal

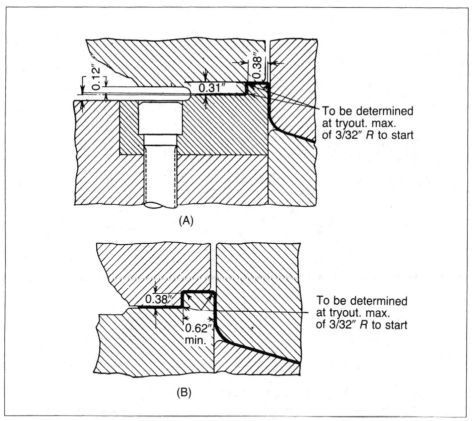

Fig. 13-8. Locking beads: *(A)* insert type; *(B)* cast integral type.

movement is required in a given binder area, the bead is simply reduced in size by grinding.[2]

Wear Plates. In addition to the regular leader pins and guide bushings, strong heel blocks are often provided to assure alignment. To assure a good wear life on long-run tooling, the heel blocks are fitted with hardened-steel or bronze wear plates. Wear plates are often fastened to the punch and blankholder rings to keep them in alignment and to avoid excessive wear.

An alternative to routing grease lines to the wear plates is to use self-lubricating wear plates which need no grease. Lubrication is provided by graphite composition plugs that are pressed into holes in the wear plates.

Hardened tool steel wear plates will outlast bronze wear plates in extreme pressure applications.

Air vent holes are often drilled into the punch and lower cavity insert, to permit the escape of air and excess draw lubricant. The holes in the punch are often provided with a short piece of copper or plastic tubing to prevent dirt from entering the die cavity when a vacuum is created as the die opens.

Positioning the Blank in Draw Dies. The amount of metal lying on the blankholder surface outside the draw radius is important. Too large a blank is wasteful and can retard metal flow. Too small a blank will not provide enough metal for proper gripping. Positioning of the blank can be achieved by nesting it between pins or blocks or by using full or partial holes in the blank. A means of adjusting the location of the gages should be provided. This is particularly important in die tryout where it may be necessary to reposition them.

Parts removed from the die by automatic means may require that the gages be retracted

13- 7

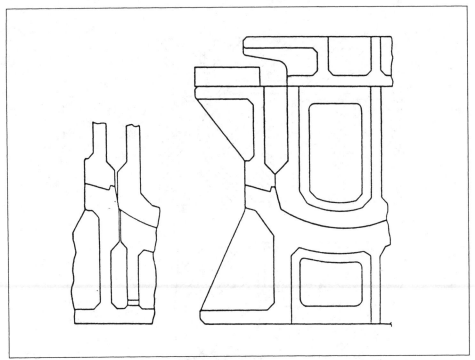

Fig. 13-9. Single step lock bead saves metal. (*Chrysler Corp.*)

during the time of part removal and then returned to position in time to catch and position the incoming blank. The air cylinders used to actuate the gage pins are usually controlled by an air valve system that is part of the press automation.

Part Lifters. In order for the automation or the operator to remove the part from the die a part lifter is normally built into the die. This can be as simple as a spring-loaded plunger in the lower die.

At one time, the in-die automation for most automotive stamping dies was tremendously overdesigned. High-pressure hydraulic-style cylinders of 4-in. (102-mm) bore size or larger were used as air cylinders to actuate the lifters. Many large levers and bellcranks were driven from 2-in. (51-mm) diameter steel jackshafts. The whole system required a lot of room under the lower die shoe. This required numerous risers of 10-in. (254-mm) or greater height. This construction style resulted in expensive dies that required presses with large shut heights.

In order to reduce die sizes and construction costs, the latter system has been largely replaced with spring-actuated-lifters, rack-and-pinion gear type lifters, or lifters actuated by servomechanisms built into the press bed.

Spring-actuated Lifters. The spring-actuated type lifter should be used wherever possible, regardless of production volume. The advantages include long dependable life, low cost, and simple installation.

Spring-actuated lifters have several disadvantages. The upper die must drive the lifter down with the blank or part in place, which can result in unacceptable damage to the part. The part will follow the upper die as the press opens, which can result in the part sticking to the upper die and being mislocated. The lifter action may be too rapid, resulting in part bounce.

Proper design and placement of the lifter can overcome these problems. Magnets can control part bounce. Spring plungers on the upper die can be used to prevent sticking to the upper die. Proper placement and/or the use of outriggers on the lifters actuated by the upper die can avoid problems with the lifters damaging the part.

Rack-and-pinion Lifters. While this type of lifter mechanism permits more compact die construction than the bellcrank system that it replaced, the new system is more prone to jam. Most jamming problems can be avoided in the design stage of the die.

It is very important to keep dirt and slugs out of the gear teeth. Measures that can help include placing the gear rack so the teeth face downward, and using shields or guards to keep foreign objects away from the mechanism.

A common design mistake is to attempt to operate three or more gearboxes in series from a common cylinder. A large amount of the initial force is lost each time that it is transmitted through a gearbox, due to mechanical losses. A good rule to follow is to transmit force through no more than two gearboxes. If this is not possible, more than one driving cylinder should be used to stay within the two-gearbox-maximum rule. Multiple cylinders can provide a more compact construction, as smaller cylinders can be used. In multiple-cylinder designs, the rack-and-pinion system serves to keep all of the lifter heads synchronized.

Servomechanisms in the Press Bed. Some large three-axis transfer presses have brushless, DC motor-driven servomechanisms located in the press bed that actuate a lifter plate under the bolster. The bolster has a pin pattern similar to that used for cushion pins. The die has sub-pins that line up with the bolster pin holes. The die sub-pins directly actuate the lifter heads in the die.

An important advantage of this type of lifter system is that one lifter drive can serve any number of dies. This greatly reduces the cost of the lifter mechanism in the die. A further advantage is that the motion of the DC servo drive can be controlled very precisely facilitating close timing in conjunction with the transfer rail motion.

PAD SHOCK AND IMPACT PROBLEMS

When the pad stops at the end of its return travel, all of the momentum is suddenly dissipated. This can result in severe shock and noise.

The result is rapid keeper wear as well as broken die shoes. In extreme cases, the bolts used to fasten the upper die to the press slide fail, permitting the upper shoe to become detached from the press.

Shock Absorber Characteristics. Shock absorbers designed for OEM use in heavy truck and bus applications are among those suitable for adaptation to control die pad shock problems. Heat buildup is usually not a problem at 16 strokes per minute, a normal rate for presses designed to stamp large and irregularly-shaped parts. Information on the extension force at a given linear velocity is available from the shock absorber manufacturer. For example, a shock absorber that is rated to provide an extension force of 1,000 pounds (4.45 kN) at 11.5 feet (3.5 m) per second, would permit a 1,000-pound (454 kg) weight to fall no faster than 11.5 feet (3.5 m) per second. This is similar to the upward slide velocity at the time of keeper impact of the presses used to produce large automotive body stampings.

Shock Absorber Selection. The total extension force of the shock absorbers should equal 1.5 times the pad weight. If air or nitrogen cylinders are being used, there should be at least one inch (25 mm) pad free travel to allow slow-down distance for the shock absorbers.

Since the shocks develop pressure upon die closure, the force that they develop will actually aid in maintaining pad pressure. When stem-mount style shock absorbers are used, the upper shoe and pad are provided with drilled and counterbored mounting holes (Fig. 13-10). The use of an impact wrench to tighten or loosen the stem nuts is advised to overcome any tendency of the shock absorber to spin.

Loop mounted shock absorbers can be used with angle iron mounting brackets (Fig. 13-11).

The pad must not hang on the shocks. There must be 0.25-in. (6.4-mm) or more unused travel at each end of the stroke.[3]

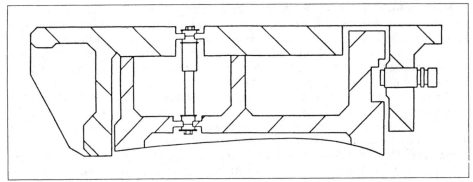

Fig. 13-10. Automotive stud mount pull rod shock absorbers used to reduce the keeper impact of a heavy pad. (*Smith & Associates*)

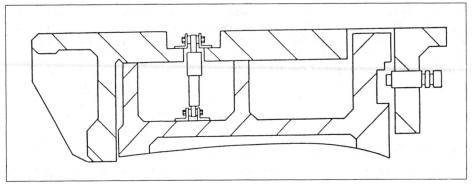

Fig. 13-11. Automotive loop mount pull rod shock absorbers used to reduce the keeper impact of a heavy pad. (*Smith & Associates*)

REPRESENTATIVE DIE DESIGNS

Roof-panel Draw Die. The roof-panel die shown in Fig. 13-12 produces a shape which requires careful positioning to enable the low point of the punch to contact the blank near the middle so that a uniform stretch is obtained in the blank as the operation proceeds. The shape and relatively shallow depth of the roof panel require the blank to be stretch drawn to retain the finished shape. The blankholder has lock beads that do not permit metal movement.

An elongation in the blank of 5 to 10% will generally produce a high-quality roof panel free from any oil-canning effect.

As a general rule, a certain amount of crown is needed in all automotive panels to provide strength. The minimum acceptable percentage of elongation required to stiffen an automotive panel is approximately 3%. Stretching the metal also serves to remove mill roll marks.

The actual percentage of elongation in this example can be varied. To increase the percentage of elongation, the cavity insert in the lower shoe can be lowered by milling metal from the bottom side.

This will permit greater punch travel.

Forming Parts by Stretching and Drawing. Shallow parts with contours in either one or both directions, i.e., longitudinally and laterally, often require that the dies grip the metal rather tightly so that it can be stretched to a certain extent rather than being drawn to shape. Stretching the metal in this manner can practically eliminate springback (Section 21).

The front panel for a space heater (Fig. 13-13, view *A*) is an example of this type of part. The blankholder at the ends, view *B*, has a curvature similar to the finished outline

13- 10

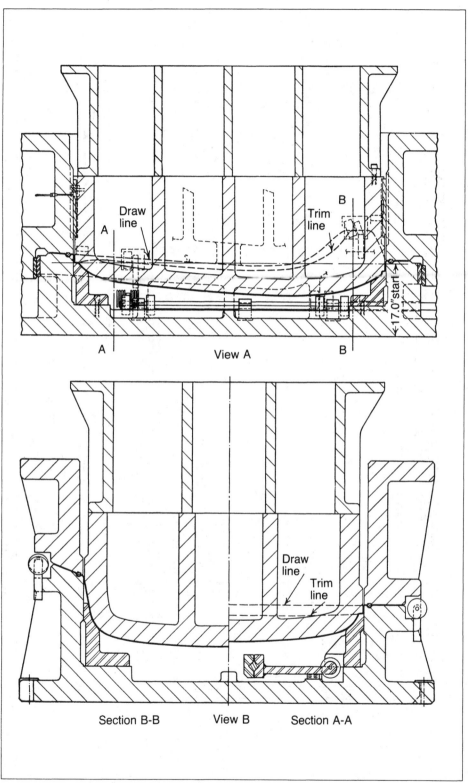

Fig. 13-12. Cross section through a roof-panel draw die: view *A*, section at center line; view *B*, half sections at front and rear of roof panel. (*Ford Motor Co.*)

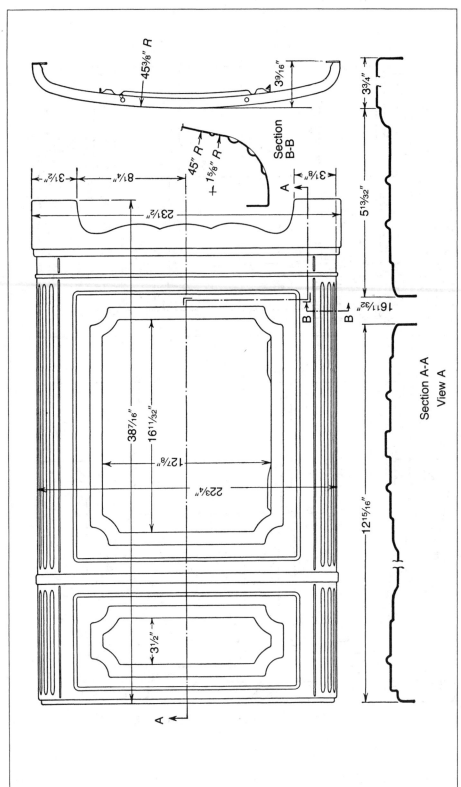

Fig. 13-13A. View A, front panel of a space heater. (*Motor Wheel Corp.*)

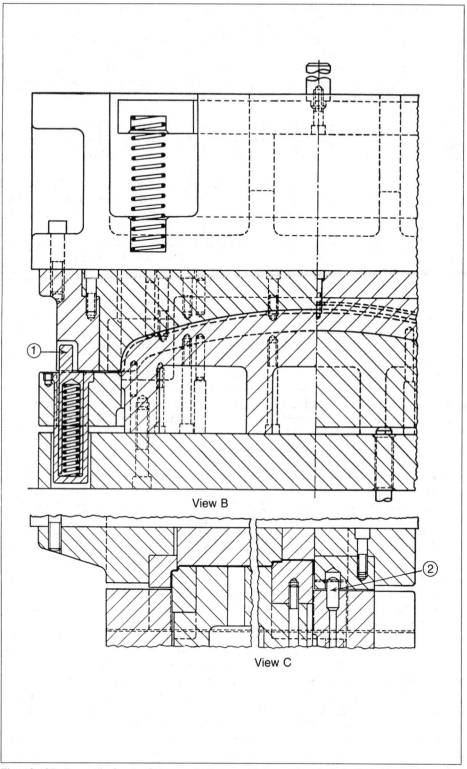

View B

View C

Fig. 13-13B. Draw die for the front panel of a space heater: view B, lateral cross section, and view C longitudinal cross section. (*Motor Wheel Corp.*)

to preform the sheet before the actual forming takes place. Along the sides and lower edge, the blankholder is straight and contains beads to help control metal flow. Positioning of the blank in the die is accomplished by two spring-loaded locating pins (*D1*) on each side and a fixed pin (*D2*) at the lower end. The blankholder, which is actuated by the die cushion, grips the blank while a flange at each end and the contour are formed. This die closes tightly on the workpiece to emboss or coin the sharp corners of the offsets as shown in view *C*. Additional operations on this front panel will be discussed later in this section.

Development of a Fender Draw Die. The section through the lowest point of an automotive-fender draw die (Fig. 13-14, view *A*) illustrates the tipping at an extreme angle to draw the nose end. The blank for this fender is trimmed, preformed, and welded into a cone shape.

The preshaping of the blank reduces the amount of metal drawn into the die, reduces the actual depth of draw, and facilitates control of the metal in the die during drawing. The shape of the blankholder ring was determined by the periphery of the finished part, to permit use of a minimum-sized blank.

The acute slope of the blankholder surface at the nose end of the fender has been eliminated by using steps. These steps act as locking beads and add strength to the steel insert. View *B* shows the use of a locking bead (*D2*) on the high side of the die.

In addition to offering greater resistance to the metal movement, the steps provide horizontal surfaces to facilitate spotting. The actual pressure on the blank is increased, since the resultant force is at right angles to the metal blank.

The excess material in the nose end of the fender is pulled into the head-lamp area, which is blanked out in a succeeding operation, by a small hook-shaped extension (*D3*) on the punch.

Alignment of the blankholder and lower die member is assured by the guide surfaces (*D4*). An air-operated lifter is incorporated into this die to facilitate gripping and removal of the drawn part by a mechanical ejector. On the production line, this die was scheduled to operate in a 120-in. (3.2-m) double-action press. In the die shop, the largest die tryout press available was a 108-in. (2.74-m) press. For this reason the blankholder was designed with two detachable channel-shaped extensions, as shown in the illustration, to be used during the production run.

Use of a Third Action. Figure 13-15 illustrates a die with a lower punch actuated by a third-action press.

Construction of this type should be avoided wherever possible through good product design. A third-action press is much more costly to purchase and maintain than a double action drawing press. In automotive drawing applications, approximately one second is added to the cycle time per stroke, due to the need for the inner slide to go into dwell while the third action is working. This loss of speed is a problem even though the third action is not being used.

If product designs that would normally require a third-action are unavoidable, the die can be designed with a cam-actuated lower forming member driven by the inner slide of a double-action press.

Lancing to Gain Metal. There are cases when so much stretching or drawing of the metal must take place in local areas, that the required shapes can not be drawn before fracture is certain to occur. A controlled fracture in a scrap area of the panel can be accomplished by lancing the metal in the scrap area. Figure 13-16 illustrates the construction of the lancing elements. The part being formed is the inner panel of a luggage-compartment lid for an automobile. The cutters are placed in several of the lightening holes which are blanked out later. This device is used wherever the metal is stretched completely over the face of the punch and into the draw-ring face, and more stretch or metal movement is required locally. The cutting operation is timed to lance metal before the occurrence of an uncontrolled fracture that could extend into the part area. The stock is cut on only three sides, so that a hard-to-eject blank is not left in the

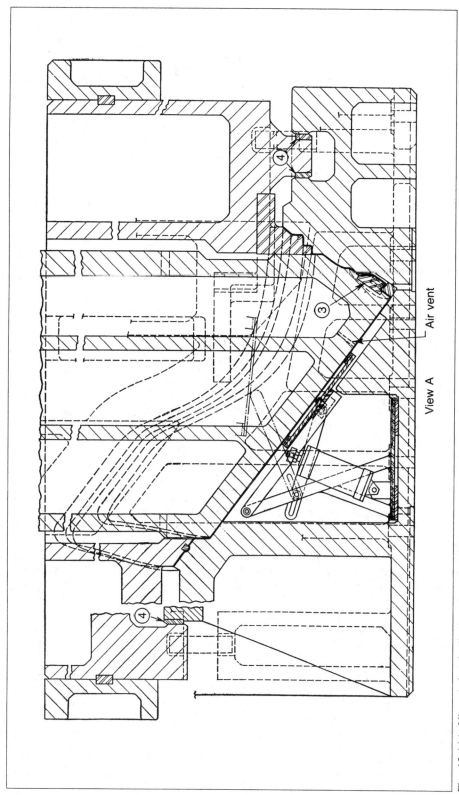

Fig. 13-14A. View A, longitudinal section of a draw die for an automobile fender, showing tip angle. (*Ford Motor Co.*)

View A

Air vent

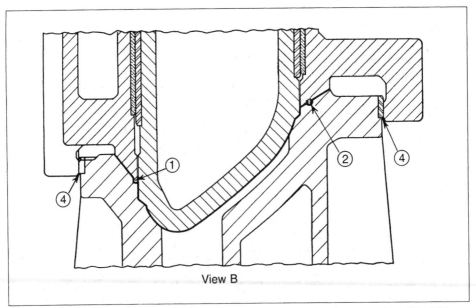

Fig. 13-14*B*. View *B*, cross section of fender die showing draw bead construction. (*Ford Motor Co.*)

die but necessary metal movement is still allowed. The utilization of a third action of the press to lance and form local areas is shown in Fig. 13-17. This die forms the outer panel for an automobile door, and the third press action lances and draws the recess at the window opening. The male portion of the lancing punch (*D*1) is mounted in the draw punch, and the female portion (*D*2) is mounted in the third-action drawing punch (*D*3). A locking bead is installed in the blankholder to grip the metal for stretching around the draw punch. The die cavity has an open center since it is not necessary to squeeze the metal between the two surfaces to set it to shape. The die is designed in such a manner that the blankholder is guided by the draw ring to maintain their proper alignment during the operation.

Another method of lancing a stamping while it is being drawn is to provide a sharp die steel with no mating half as the lance. It can be mounted either in the die cavity, or on an area on the punch into which metal is drawn by a reverse contour in the lower die. The lance is timed to make contact with the stamping, in an area that will be subsequently trimmed away, just as enough strain is developed to initiate a fracture. This lance produces a controlled fracture. The advantages of this method are that only one steel is needed, and that the guiding of the punch in the upper blankholder is not critical.

All types of lances must be maintained to avoid the generation of slivers which can cause marks on the finished stamping.

Combination Forward and Reverse Draw. The drawing of an inner-panel, upper lift gate on a station wagon is an example of combination forward and reverse drawing in the same operation. The blank, with a rectangular hole in the center, is positioned in the die (Fig. 13-18) and gripped by the blankholder (*D*1) against the surface of the draw ring (*D*2). The punch (*D*3) draws the stock into the cavity from the outside over the regular draw ring and from the inside over the punch (*D*4). The pressure pad, which is in the die cavity, holds the blank against the draw punch, thus controlling the amount of metal flowing into the die cavity. Since the metal being drawn into the die from the inside of the blank is under tension, no blankholder is required on the inner portion of the blank. The pressure pad also lifts the finished part out of the die.

Air-cushion Drawing. Single-action presses with an air cushion are less expensive and can produce drawn parts faster than a double-action press.

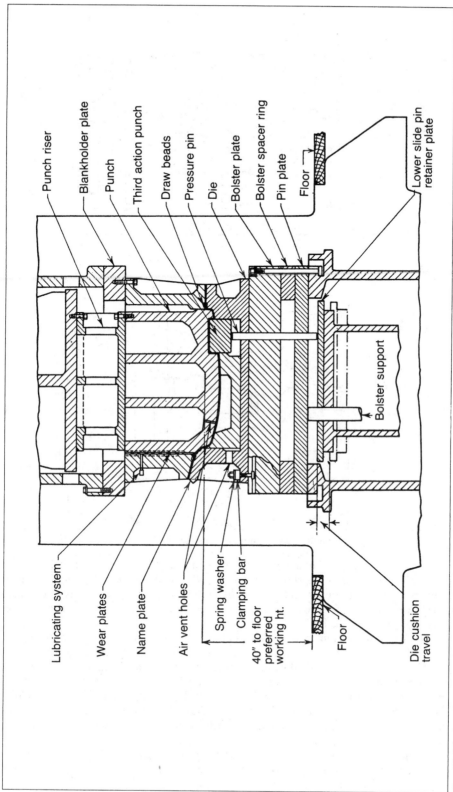

Punch riser
Blankholder plate
Punch
Third action punch
Draw beads
Pressure pin
Die
Bolster plate
Bolster spacer ring
Pin plate
Floor
Lower slide pin retainer plate

Bolster support

Lubricating system
Wear plates
Name plate
Air vent holes
Spring washer
Clamping bar
40" to floor preferred working ht.
Floor
Die cushion travel

Fig. 13-15. A press third-action used for form a complex drawn part.

13-17

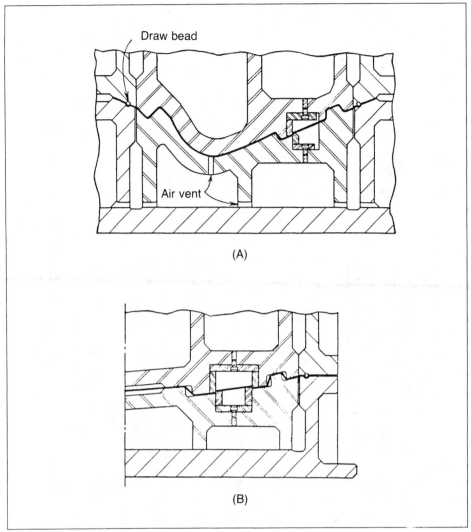

Draw bead

Air vent

(A)

(B)

Fig. 13-16. Lancing a blank while drawing to gain metal. (*Ford Motor Co.*)

The die for drawing a pair of automobile front-seat side shields on a single-action press equipped with an air cushion is shown in Fig. 13-19. The area of actual draw of this part is uniform even though beads and flanges are formed into the part as it is drawn and stretched over the punch. Drawing a complex shape by any means is easier if the shape and depth of the draw are fairly uniform.

The blank is positioned in the die by the piloting pins ($D1$), and the closing of the upper portion of the blankholder preforms the blank to a U-shape over the center portion of the blankholder. This is accomplished by the arch-shaped part of the blankholder as shown in view A, which fits into a recess in the punch as shown in view B. Continued descent of the press slide results in the forming of the part over the punch. The punch and die close tightly to emboss the sharp contour. The air cylinder ($D2$) lifts the ejector pad ($D3$) and the part to such a height that the jaws of the mechanical ejector (view B) are able to grip the part and remove it from the die.

Lateral movement of the blankholder is restricted by a wear plate on an overhanging flange of the upper die. As a safety measure, a flange is cast onto the base plate surrounding the blankholder.

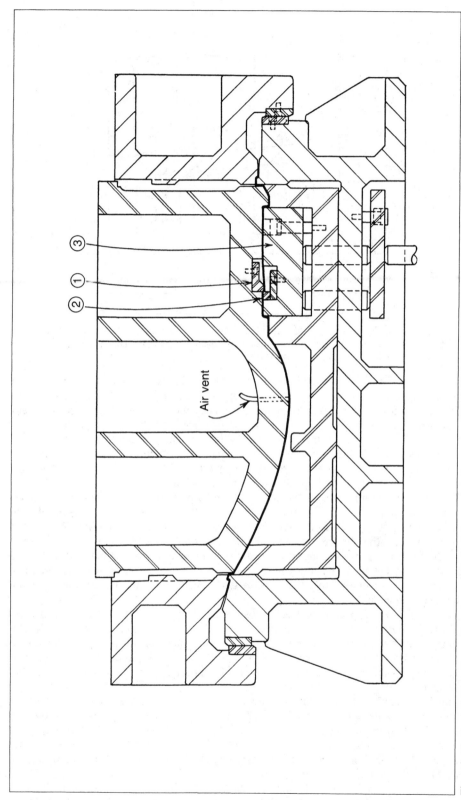

Air vent

③ ① ②

Fig. 13-17. Press third-action used to lance and final-form a panel. (*Ford Motor Co.*)

13- 19

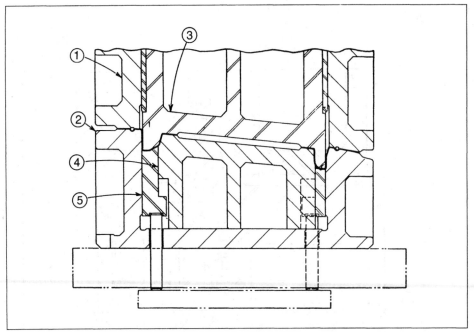

Fig. 13-18. Drawing metal from both the outside and inside edges of a blank to produce a station-wagon lift-gate inner panel. (*Ford Motor Co.*)

This type of die can be adapted to a double-action press with a few minor changes, in case either a single-action press is not available, or a satisfactory part is not produced.

Die for Range Top and Backsplasher. The one-piece electric-range top and backsplasher shown in Fig. 13-20, view *A*, is produced in a series of operations from 0.059-in. (1.50-mm) thick vitreous enameling steel sheared into rectangular sheets. A blanking die trims this sheet on three sides to obtain the developed blank shown in view *B*. The two half holes on the sides are used to position the blank in the first two forming dies. View *C* shows the part after the first forming operation.

The die to preform the corner flanges, face, and top of the backsplasher and the front edge is shown in Fig. 13-21. The forming punch on the upper shoe is comprised of four sections. The portion *D*1 forms the front edge and the two punches (*D*2) are for preforming the corner flange. The blankholder functions between these three punches, since the forming operations are performed on the outer edges of the blank. Pressure is applied to the blankholder by four round rubber springs held in place by center locating pins. The blankholder is also used to bend the sheet to form the backsplasher and to emboss several areas.

The two foolproof locating pins (*D*4) position the blank on the lower die. The part of the die (*D*5) containing the locating pins is elevated by an air cushion to a sufficient height to permit the flat blank to be placed in position. When the die is open, the upper blankholder drops below the lower edges of the fixed members and the air cushion raises the lower blankholder to grip the stock. The foolproof locating pins recede with the upstroke of the press, permitting quick removal of the stamping.

The die for drawing the side flanges, finish-forming the front edge and top rear flanges, and embossing three switch-bracket openings of the part shown in Fig. 13-20 is shown in Fig. 13-22. A spring-loading pressure pad (*D*1) on the upper die holds the preformed blank on the punch (*D*2) and the locating pins (*D*3) during the drawing operation. This die is made of Meehanite iron castings with hardened-steel inserts at the points of wear and areas where the embossing is performed.

The die in Fig. 13-23 is for piercing the four holes for the heating units, the small holes

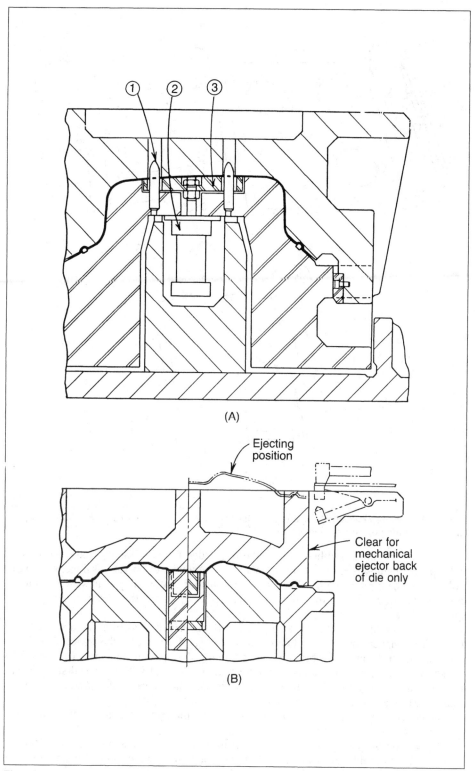

(A)

Ejecting
position

Clear for
mechanical
ejector back
of die only

(B)

Fig. 13-19. Double-action die with a press-cushion actuated blankholder used to form a front
seat side shield. (*Ford Motor Co.*)

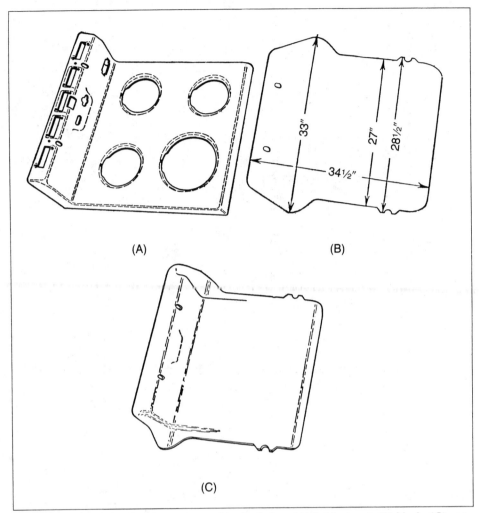

(A)

(B)

(C)

Fig. 13-20. One-piece electric-range top and backsplasher: (A) finished part; (B) blank; (C) part after first forming operation. (*General Electric Co.*)

for securing the units, and holes along the front-edge flange. This die and other dies for subsequent operations employ cams to increase the number of operations done in each die. All exposed cam-return springs are fitted with tubular steel spring guards attached to the die members (not shown) to contain the springs in the event that the center rod should unscrew.

The piercing of the holes in the front-edge flange is achieved by a fixed vertical cam (D1) on the upper shoe, and a horizontally sliding cam (D2) on the lower shoe which transmits the motion to the block (D3) containing the piercing punches. Commercially standard punches, punch retainers, and die buttons are used with hardened-steel backup plate behind the punches. The die rings for the larger holes are made from air-hardening tool-steel tubing and are recessed into a machine-steel subplate.

The upper die shoe has been fabricated from steel plates as a weldment; the lower shoe is a casting.

The next operation performed on this part is piercing five rectangular slots and twelve holes in the top of the backsplasher and, at the same time, two irregular-shaped holes for instruments in the face of the backsplasher.

The part is positioned in the die (Fig. 13-24) so that the two holes in the face of the part

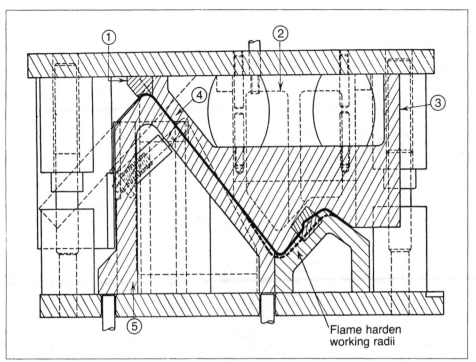

Fig. 13-21. Section through die to produce part shown in Fig. 13-20. (*General Electric Co.*)

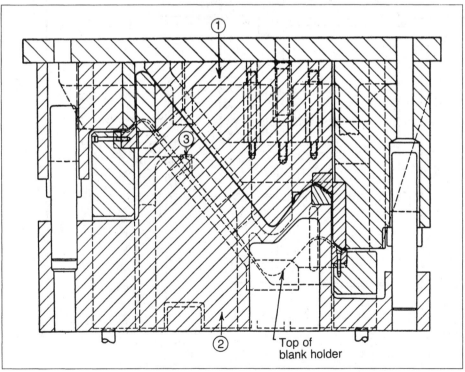

Fig. 13-22. Die to draw and emboss range top and backsplasher shown in Fig. 13-20. (*General Electric Co.*)

13-23

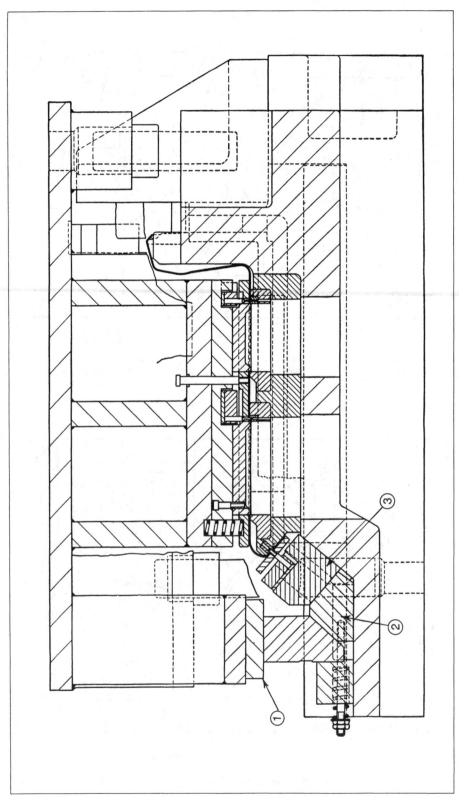

Fig. 13-23. Combination cam-actuated and conventional type piercing die (workpiece shown in Fig. 13-20). (*General Electric Co.*)

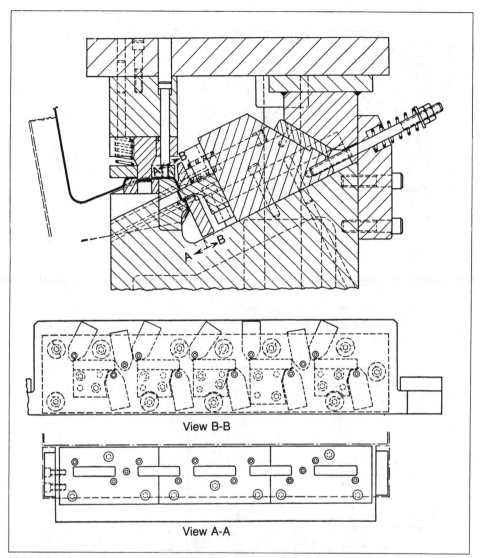

View B-B

View A-A

Fig. 13-24. Cam-actuated slotting die for backsplasher (workpiece shown in Fig.13-20). (*General Electric Co.*)

are pierced in the conventional manner, while the remainder of the holes are pierced by a cam-actuated slide. Spring-loaded pads precede the punches and clamp the workpiece to the die, also serving as punch strippers on the upstroke of the die. Springs are utilized to return the slide for loading and unloading the die.

Additional piercing and embossing operations are performed on the part by the die shown in Fig. 13-25. The conventional vertical punch cuts a hole in the face of the backsplasher; the cam-actuated punch sitting at about a 45° angle pierces a rectangular hole at the radius, and the other cam-actuated sliding block embosses a circular indentation around four of the instrument-mounting holes.

A commonly used type of cam flanging die is shown in Fig. 13-26, but incorporated in the die are additional cams for other die functions.

The cam-actuated flanging punch ($D1$) is mounted on the plate ($D2$) which is supported by pins extending the die cushion. Also on this plate is a sliding form block ($D3$) around which the flange is formed. The raising of this plate by the air cushion engages a set of

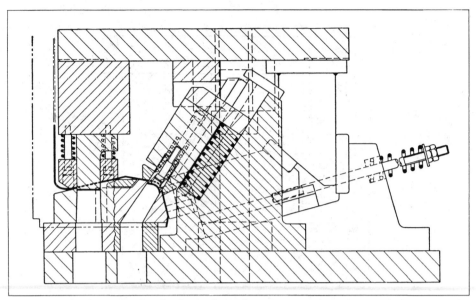

Fig. 13-25. Piercing and embossing backsplasher by cam-actuated punches (workpiece shown in Fig. 13-20). (*General Electric Co.*)

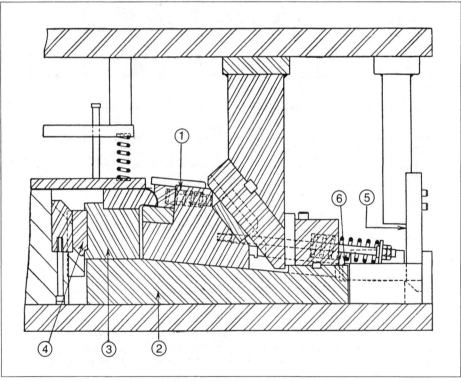

Fig. 13-26. Cam-operated flanging die for workpiece shown in Fig. 13-20. (*General Electric Co.*)

cams (*D*4) to withdraw the form block to facilitate removal of the part, since the flanges are being formed on three sides by the die. Before the plate can be raised by the air cushion, the upper shoe must ascend carrying the cam with it (*D*5) and releasing the latch plate (D6).

Not shown but included in this die are cam-operated punches for forming a vertical flange on each side of the backsplasher, and a conventional punch and form block to flange downward the rear edge of the top of the backsplasher.

A section through the die for drawing the recesses for the heating units in the range top is shown in Fig. 13-27. The die is an inverted type with the draw punches and air-cushion-actuated blankholder on the lower shoe, and the draw rings attached to the upper shoe. Spring-loaded pressure pads are installed in the die cavities to control metal movement. These details bottom at the low point of the press stroke to flatten the flanges and sharpen the corners.

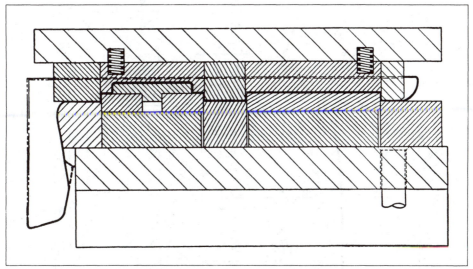

Fig. 13-27. Draw die for recesses in range top of Fig. 13-20. (*General Electric Co.*)

Inverted Draw Die. This type of die construction is very popular for drawing such automotive body panels as roofs, fenders, hoods, deck lids, and inner doors. In a tandem line there is an advantage in that a part turnover is not needed to get the drawn panel into position to load onto the trim die. This often permits higher line speeds. Other advantages include possible use of a smaller blank than would otherwise be required.

A disadvantage may be potential press damage because substantial tonnage is required some distance from the bottom of the press stroke. This is especially troublesome if the die is driven by the blankholder slide of a double or triple action drawing press. Section 27 has detailed information on this problem.

A typical inverted draw die construction for drawing automotive roof panels is shown in Fig. 13-28. A nitrogen manifold (D1), which also functions as the lower die shoe, has twenty-four 6 ton (53.4 kN) nitrogen cylinders (D2) to provide a total of 144 tons (1,281 kN) of pressure at 1,500 psi (10.3 MPa) nitrogen pressure for the draw ring (D3). Cylinder travel is 4-in. (102-mm). Spacer blocks attached to the draw ring (D4) are carefully machined so that the draw ring is even with or slightly below the height of the draw punch (D5). The exact dimension is established in tryout. The spacer blocks are stepped in height to permit gradual engagement of the nitrogen cylinders, to reduce press loading high in the stroke. Usually only 4 to 8 of the 24 cylinders are needed for pressure at initial contact of the blankholder (D6). The working surfaces of the punch and blankholders are ion nitrided for wear resistance and freedom from pickup of the electrogalvanized zinc coating used on the blanks.

Floating-draw-ring Die. In this type of die construction, the draw ring can be elevated to a height above or nearly above the high point of a reverse contour in the part. This

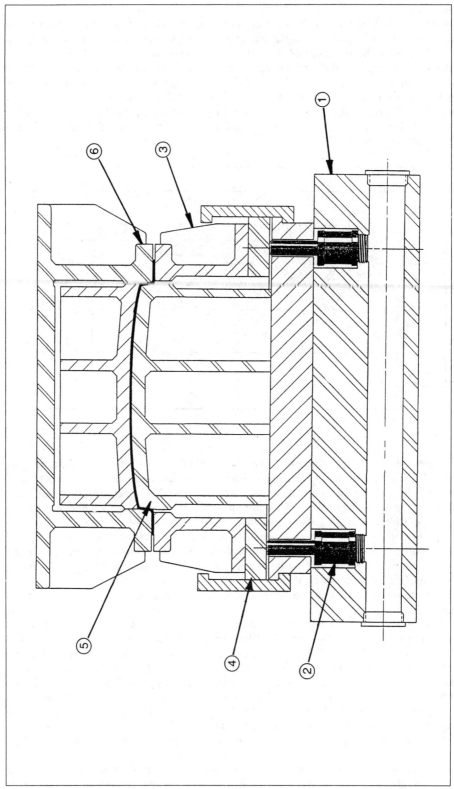

Fig. 13-28 Inverted automobile roof draw die using a nitrogen manifold to provide draw-ring pressure. (*Forward Industries*)

permits the blank to be gripped with a natural deflection that would not be possible with the blankholder line in the low position. With a fixed draw line at the higher position, a deeper draw would be required. But a part with a high center and high sides may be rather difficult to draw satisfactorily, because of a possibility of being unable to flow the metal in the reverse direction around the punch nose. The shape of some parts does not require the higher drawn sides; therefore, a smaller blank is used in these cases. This type of die is used on a regular double-action press, with the floating draw ring actuated by a die cushion in the press bed or nitrogen cylinders built into the die.

The floating draw ring grips the metal against the surface of the blankholder, and the simultaneous descent of these two die members draws and stretches the blank over the die. With the blankholder in dwell position, the punch continues downward, forming the part to the contours of the punch and die.

This type of die construction has a higher first cost than a regular double-action die, but the savings in blank size reduction and the use of a lower-grade steel than would be required in a conventional draw die can more than offset the higher initial cost.

An example of the floating draw ring is shown in Fig. 13-29. This die forms an automobile dash panel, which is the panel between the passenger and engine compartments and the sloping portion of the floor board. It has a hump in the center to clear the transmission and the engine.

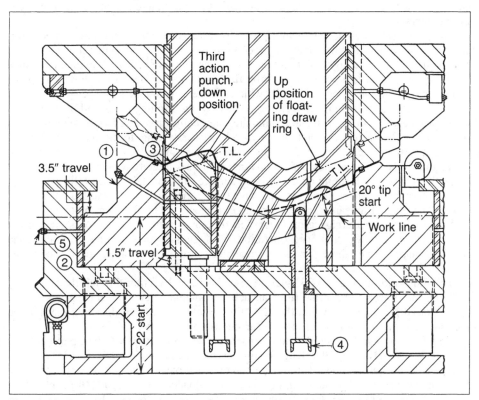

Fig. 13-29. Draw die with floating draw ring to form a dash panel. (*Ford Motor Co.*)

The floating draw ring (*D*1) is elevated by a group of air or nitrogen cylinders (*D*2) and has a travel of 3.5 in (89 mm). The reentrant contour of the part necessitates the use of a third-action punch (*D*3). The ''up'' position of the draw ring and the ''down'' position of the third-action punch are shown by the phantom lines and indicate the interferences given to the blank by the reverse contour in the die.

Incorporated in this die is a lifter mechanism (*D*4) to raise the drawn part out of the die so that the jaws of the mechanical ejector can grip and remove the part from the die. Also shown are grease fittings (D5) at holes and tubing leading to the wear plates for lubrication purposes.

Die to Form Stiffening Beads. A die to form crossed stiffening beads, a large-radius contour, and flanges along two sides is shown in Fig. 13-30. The floating punch (D1) is of open construction, contacting the sheet only at the beads, through the center, and around to the outer edges. The floating punch is supported by springs to form the contour and flanges before the beads are formed by the stationary punch (D2).

Die to Form Shallow Recesses. The liner top and bottom panels of a refrigerator are formed together in one piece by the die shown in Fig. 13-31, view A. Springs lift the blankholder above the top surfaces of the forming punches and provide the pressure for gripping the blank while forming. In addition to guide pins, the upper die is aligned with the punches by heel blocks on each end of the die. View B is a section through the die to separate the two panels, trim the periphery, and perforate holes in the bottom panel. The pressure pad (D2) in the upper die holds the part against the punch (D3) to be cut by the die (D4). The scrap is stripped from the punch by the stripper pad (D5). The parting punch (D6) and the die steels (D7) separate the two panels. The punch and die units (D8) pierce 15 holes in the liner bottom.

Die to Form Refrigerator Bottom Pan. The die in Fig. 13-32 forms the spherically curved surface, two beads, a rectangular-shaped indentation, and flanges on the four sides in one operation.

The punch (D1) for the spherical segment is fastened solidly to the lower shoe (D2). The form block (D3) for the remaining operations is spring-loaded so that it can be raised to a height equal to the high point on the punch. The pressure pad (D4) on the upper shoe is spring-loaded and is constructed to drop flush with the surface of the draw ring (D5).

Closing of the die grips the blank, which has been positioned by six spring-loaded disappearing gage pins (D6) between the surfaces of the forming block and pressure pad. Continued descent draws the metal around the spherically curved surface of the punch. As the form block reaches the end of its travel, the draw ring forms the four flanges. Bottoming of the pressure pad against the upper shoe embosses the beads and rectangular-shaped indentation This die is mounted on a large four-post die set. To limit deflection caused by forming on an inclined surface, a heavy heel block (D7) with wear plates is provided. Sheet-metal shields surrounding the form block and fastened to the lower shoe prevent unwanted objects from getting underneath the form block.

Dies to Produce Refrigerator Doors. Refrigerator doors have contours and corner radii which require the combined operations of stretching and forming to produce the required shapes. Since these panels are enameled, care must be taken not to scratch the outer surface.

The illustration in Fig. 13-33, view A, is a horizontal section through the draw die, showing the draw punch, draw ring, air-actuated blankholder with draw heads, and a leather-faced pressure pad within the draw ring. View B is a vertical section through the die, showing the forming of an offset near the top of the door.

A section through the trimming and piercing die is shown in view C. The part is placed over the block, (D1), with the flange resting on the trimming die (D2). The periphery is trimmed by the punch (D3). Continued downward travel of the punch straightens the flange into the side wall of the stamping. The spring-loaded pressure pad (D4) controls the flow of the metal over the draw radius on the inner edge of the punch.

The punch and die assembly (D5) blanks out a hole for the door handle. This assembly is interchangeable with the block on the opposite side of the die to facilitate the blanking of right- and left-hand doors. Small holes are pierced near the top and bottom of the door in this die with commercial punches and dies.

Flanging of the door panel is accomplished in the die shown in view D. The actuating cam (D6) for the flanging slides (D7) is secured to the lower die shoe. Likewise is the cam (D8) for actuating the sliding form block (D9). The unique feature of this flanging die is

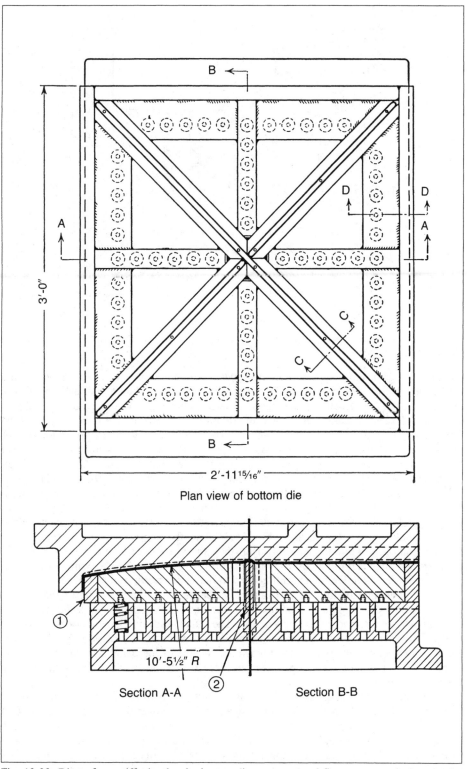

Fig. 13-30. Die to form stiffening beads, large radius contour, and flanges on an air duct panel. (*Pullman Standard Car Mfg. Co.*)

Plan view of bottom die

3'-0"

2'-11¹⁵⁄₁₆"

Section A-A

Section B-B

10'-5½" R

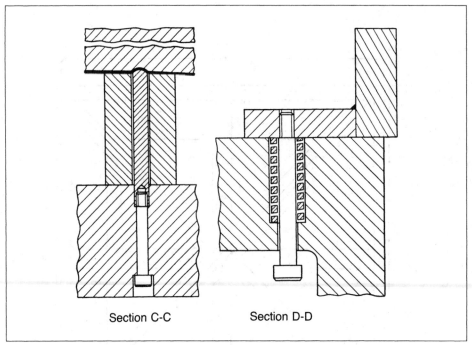

Section C-C   Section D-D

Fig. 13-30. (*Continued.*)

that the sliding form blocks and flanging slides are mounted on a vertically moving plate (*D*10) actuated by the die cushion. The upper die serves as a blankholder and the closing of the die lowers the floating plate, forcing the form blocks outward to a dwell position. Continued descent of the upper die and the floating plate permits the flanging slide to be forced inward by its actuating cam. Cams incorporated in the die assure positive return of the sliding members upon the raising of the floating plate by the press cushion. To prevent the form block from being tipped upward by the flanging operation, a slot has been machined in the side of the block to accommodate the overhanging block (*D*11).

Dies for Space-heater Front Panel. Dies for performing additional operations on a space-heater front panel (Fig. 13-13) are shown in Fig. 13-34. The die section shown in view *A* is taken through the various cutting operations of the die. The flanges along the sides are held against the locating block (*D*1) and the segmental punch (*D*2) by the pressure pad (*D*3), while the die (*D*4) trims the excess stock. The punch (*D*5) blanks out the door opening, leaving stock for flanges formed in a later die. The smaller piercing punch (*D*6) pierces holes for mounting trim to the upper portion of the panel.

The straightening of the flange trimmed in the previous die, forming of the flanges around the door opening, and forming of the numerous embossments are performed in the die of which a cross section is shown in *B*. The punch-and-die elements of this tool are made with several inserts because of the beads and indentation embossed on the panel. View *C* is a section taken along the vertical center line of the panel showing the various inserts used in the die construction. The panel is lifted from the die by pads actuated by the press cushion.

A section through a cam-action die for flanging the vertical sides of the panel is shown in view *D*. The panel is placed face down in a nest in the lower die. The form block (*D*7) is forced into position by the cam (*D*8) and the flanging punch is pushed inward by the beveled surface on the upper shoe. The form block and flanging punch are carried on a spring-loaded pad attached to the upper die shoe. To assist in setting the bend radius, the sliding block (*D*9) on the lower shoe is pushed against the panel by the same beveled surface that actuates the flanging punch. The sliding members are all returned to their

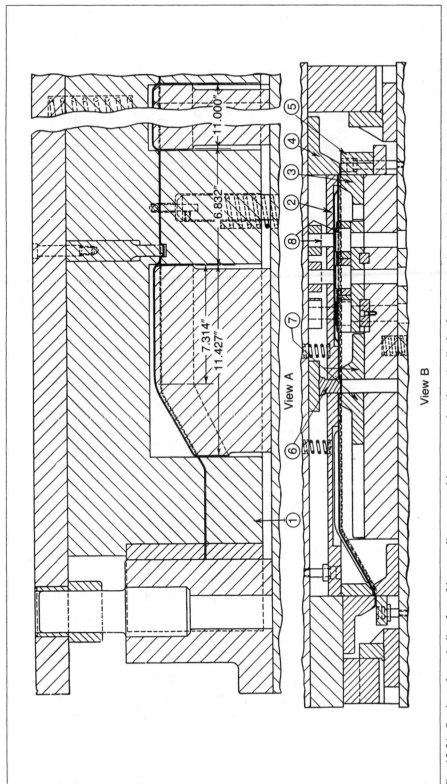

Fig. 13-31. Sections through dies for refrigerator liner top and bottom panels: view A, die for forming panels; view B, die for trimming and parting panels. (*Ready Machine Tool & Die Co.*)

13-33

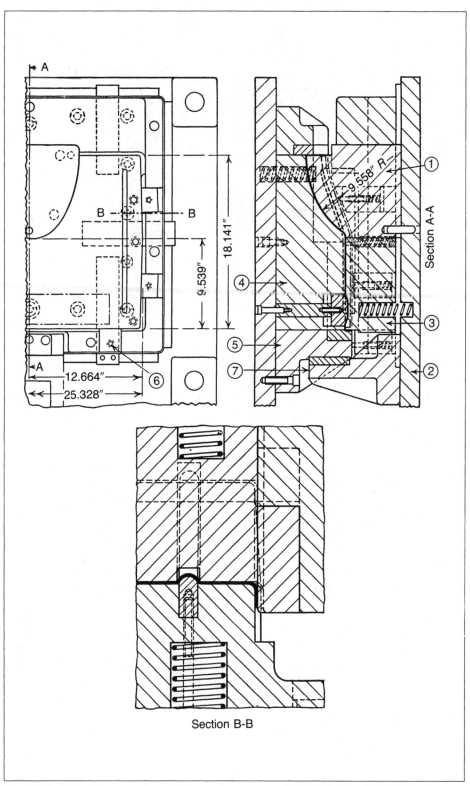

Section A-A

Section B-B

Fig. 13-32. Die to form complete refrigerator bottom pan. (*Ready Machine Tool & Die Co.*)

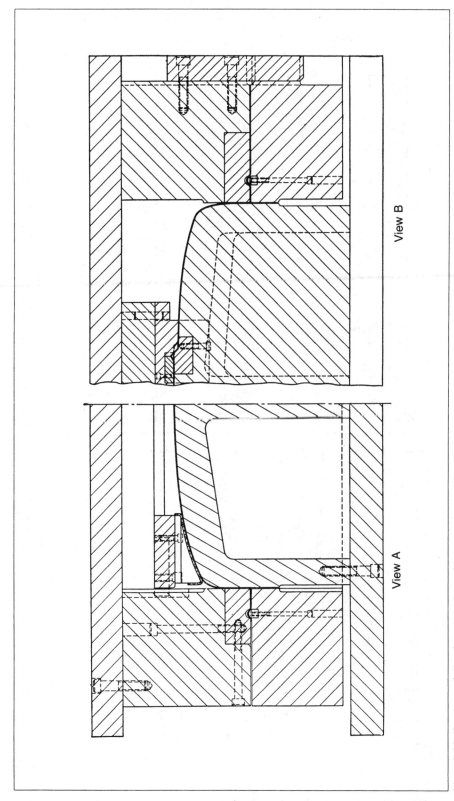

Fig. 13-33. Sections of dies for producing refrigerator doors. View A, horizontal section through the draw die. View B, vertical section through the draw die near top of door. (*Ready Machine Tool & Die Co.*)

View A

View B

13- 35

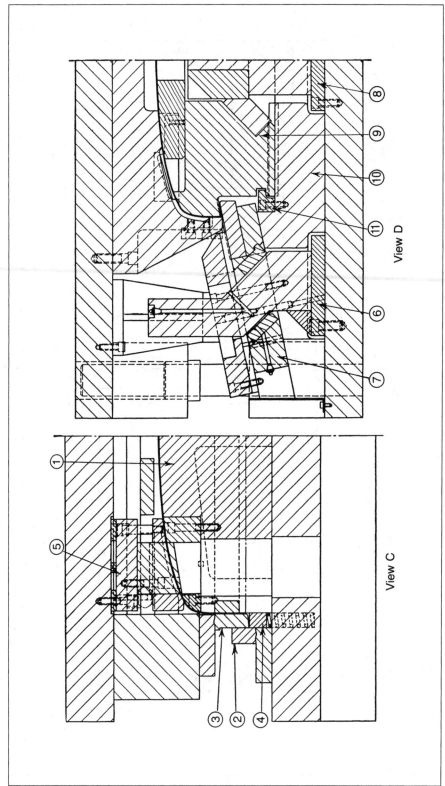

View C

View D

Fig. 13-33. (Continued.) View C, section of die for trimming and straightening the sides. View D, section of the cam-actuated flanging die. (Ready Machine Tool & Die Co.)

13- 36

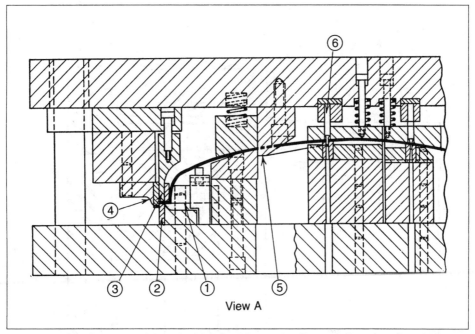

Fig. 13-34. Dies for a space heater front panel. View *A*, section through trimming and blanking die.

neutral positions by springs. Safety guards (not shown) are used over the exposed springs.

Pierce, Form, and Trim Die for a Large Shell. Trimming of the top edge and piercing and forming the bottom of a washing-machine tub are shown in Fig. 13-35. The drawn tub is placed in the locator (*D*1) and held in place by two spring-loaded hold-downs (*D*2 and *D*3). The trimming of the top edge is performed by four cam-actuated trimming punches (*D*4). The combination piercing and forming punch (*D*5) blanks out the bottom and forms a flange around the bottom. The part is raised for removal by a spring-actuated lifter. Another spring lifter in the die elevates one side of the scrap blank for ease of removal.

Die to Draw Cylinder-head Cover. The cylinder-head cover produced in the die shown in Fig. 13-36 is basically rectangular in shape but has some irregular curves in the top. On each side are three semicircular recesses to provide clearance for mounting bolts. Steel inserts (*D*1) are placed in the punch in these areas. Four circular indentations are formed in the cover by the inserts (*D*2). Material is provided for the recesses by piercing with the nail punch (*D*3), thus avoiding fracture by over-stressing the the material at the nose of the punch. A die-cushion-operated pressure pad holds the material against the nose of the punch during the drawing operation. The die is constructed of cast iron and used in a double-action press.

Draw Die for an Automobile Oil Pan. Sections of the draw die for forming a body oil pan having an uneven depth of draw are shown in Fig. 13-37. The material for this part is 0.060-in. (1.52-mm) SAE 1008 steel. View *A* is taken along the longitudinal center line of the die, showing the die construction and the shape of the part. The partial sections *A-A* and *B-B* show the relative difference in the depth of draw at each end of the oil pan. Hardened-tool-steel wear plates guide the punch within the blankholder and the blankholder over the lower die. The die is designed for use in a double-action press with a die-cushion-actuated pressure pad to assist in the control of the metal within the die cavity.

Draw Die for Bumper Part. A front bumper impact bar made from 0.120-in. (3 mm) high-strength steel is shown in Fig. 13-38. The irregular shape of the draw line is shown

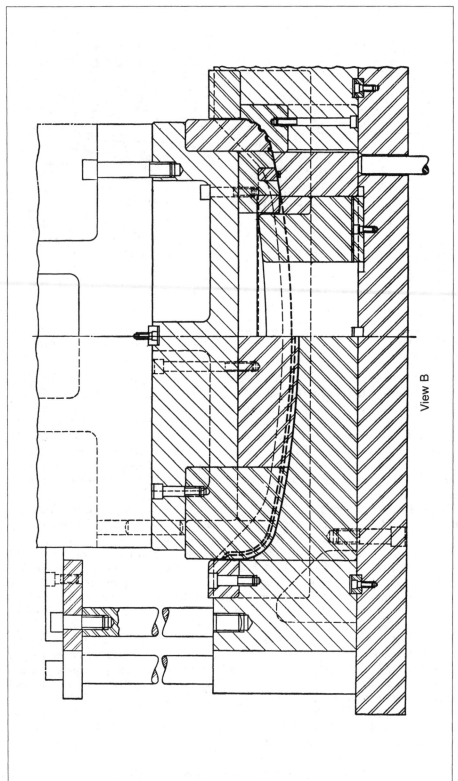

View B

Fig. 13-34. (Continued). View B, cross section of the flanging and embossing die.

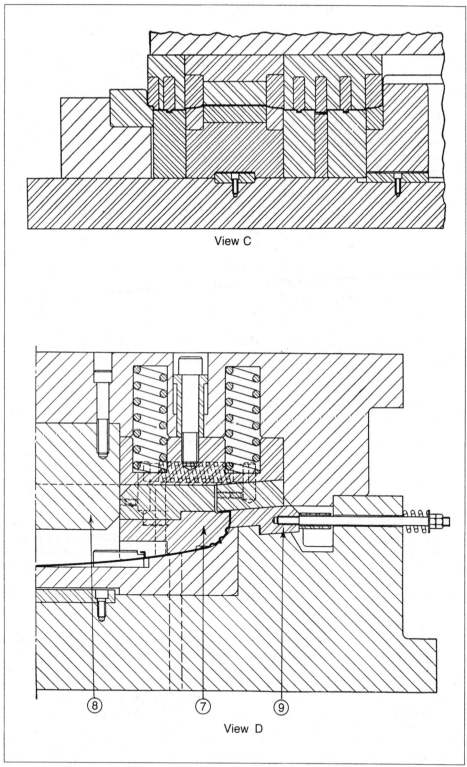

View C

View D

⑧   ⑦   ⑨

Fig. 13-34. (*Continued*). View *C*, partial vertical section showing inserts for embossing. View *D*, cam-action flanging die.

13- 39

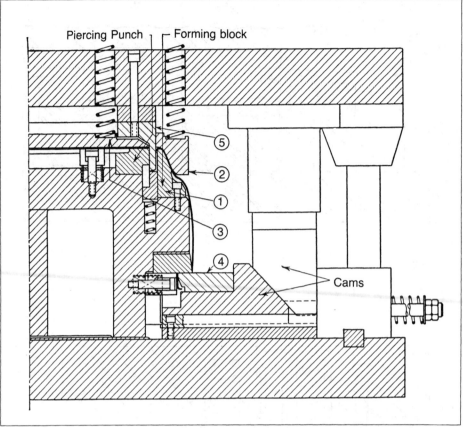

Fig. 13-35. Combination die performing pierce, draw-neck, and edge-trim operations on a large drawn shell. (*Maytag Co.*)

by the plan view of the die. Sections *A-A* and *B-B* illustrate the shape to which the part is drawn. A spring-loaded lifter mechanism shown in section *B-B* is incorporated in the die to raise the part so it can be removed from the die by automation.

Dies for Soft-drink-vending-machine Cabinet. The front door of the cabinet shown in Fig. 13-39 is made from 0.042-in. (1.07-mm) deep-drawing quality, prelubricated, cold-rolled steel rectangular blanks. The first drawing operation forms the vertical recess through the center, the offset across the face of the panel near the top, and the flanges on the sides and bottom to the depth of the offset. Sections through the die showing its simple construction are shown in Fig. 13-40. The section *A-A* is taken along the vertical center line of the die, the partial section *B-B* shows the construction above the vertical recess, and section *C-C* is through the vertical recess.

A second draw operation forms the flanges around the door. To facilitate the final draw, the flanges are trimmed to width across the top and along the sides for a distance of about half the height of the door. The top corner areas are blanked to a developed shape so that an even flange height is produced in the final draw die.

The door panel was finish-drawn and the side walls pinch-trimmed to height in the die from which the sections in Fig. 13-41 were taken. Punch-and-die steels are incorporated in this die to blank out the openings in the face of the panel.

The spring-actuated blankholder (*D1*) surrounds the punch and grips the flange of the part against the face of the draw ring (*D2*). The pinch-trim steel (*D3*) is inserted into the cast-iron punch. The draw ring of the die is also cast iron with hardened-steel inserts.

After completion of the operations, the door panel is removed from the draw ring by

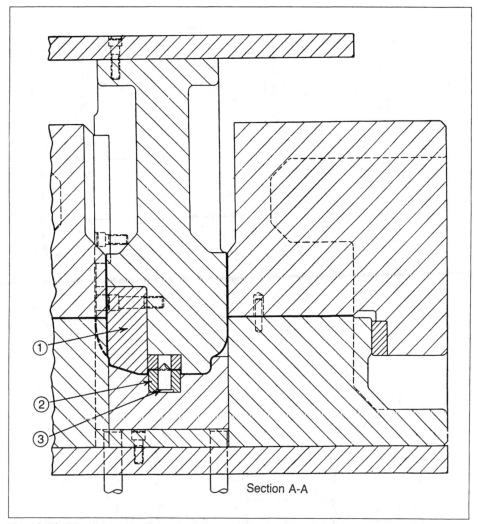

Fig. 13-36. Die to draw a cylinder head cover. (*Chrysler Corp.*)

the positive stripping arrangement. The stripper consists of a pad (*D*4) fitting the inside of the draw ring, the extension bars (*D*5), and the headed studs (*D*6) attached to the lower shoe.

The smaller slugs are permitted to drop through the die. The larger cut outs are lifted to the surface of the die by individual pressure pads.

The forming of the return flange along the top edge is achieved by a cam-action die of conventional design. The flanges along the sides are formed in the dies shown in Fig. 13-42. Shoes and spring-loaded pads hold the door panel firmly against the form block while the flanging operations are performed.

Dies for Automatic Washer Top. The top and backsplasher for the machine cabinet are made in one piece from 0.041-in. (1.04-mm) thick deep-drawing-quality steel. A rectangular draw to form the backsplasher is made from a developed blank which is later cut into two pieces to make two parts. The section of the draw die shown in Fig. 13-43 illustrates a split draw ring which permits the insertion of shims. This construction facilitates adjusting the draw pressure. The ability to localize the draw pressure helps avoid an oil-can effect in the large flat surface surrounding the drawn portion. This type of construction also greatly reduces spotting time.

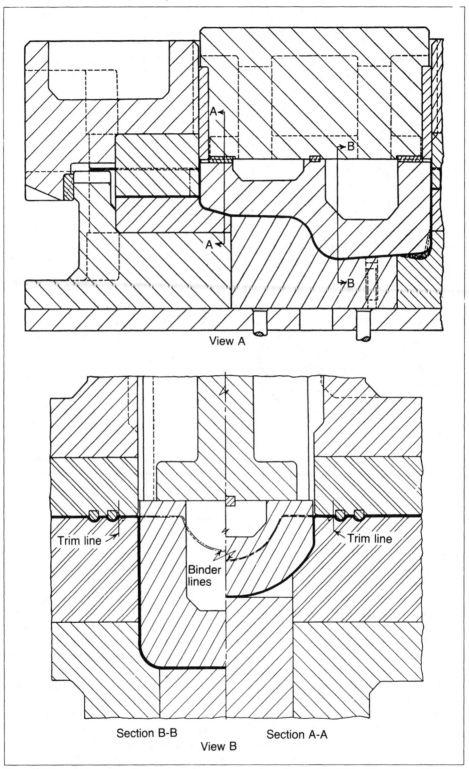

View A

Trim line

Binder
lines

Trim line

Section B-B   Section A-A

View B

Fig. 13-37. Double-action die for drawing an automotive oil pan. (*Cadillac Motor Car Division, General Motors Corp.*)

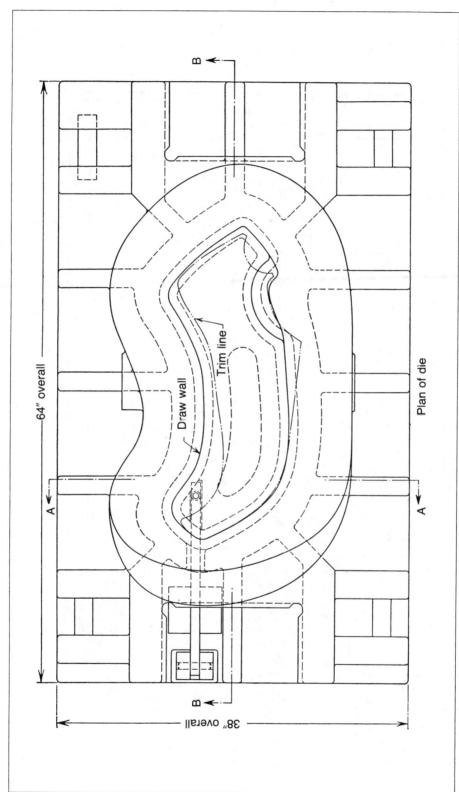

Fig. 13-38. Draw die for forming a bumper part of 0.120-in. (3.04-mm) high-strength steel.

13-43

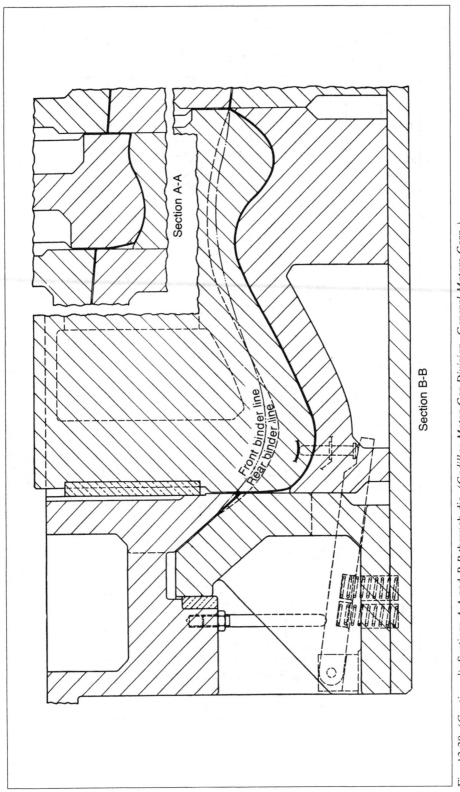

Section A-A

Section B-B

Front binder line

Rear binder line

Fig. 13-38. (Continued). Sections A-A and B-B through die. (Cadillac Motor Car Division, General Motors Corp.)

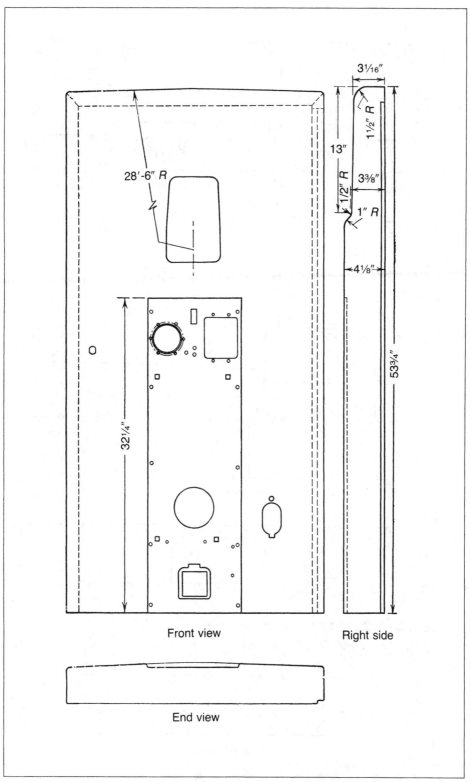

Front view

Right side

End view

Fig. 13-39. Front-door panel for a soft-drink vending machine cabinet. (*Mills Industries, Inc.*)

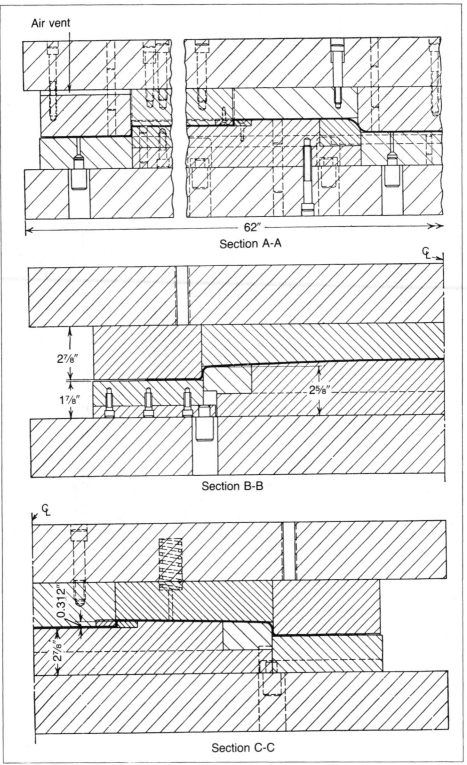

Air vent

62″

Section A-A

$2\frac{7}{8}″$

$1\frac{7}{8}″$

$2\frac{5}{8}″$

Section B-B

0.312″

$2\frac{7}{8}″$

Section C-C

Fig. 13-40. Sections through the first draw die for the part shown in Fig. 13-39. (*Mills Industries, Inc.*)

13- 46

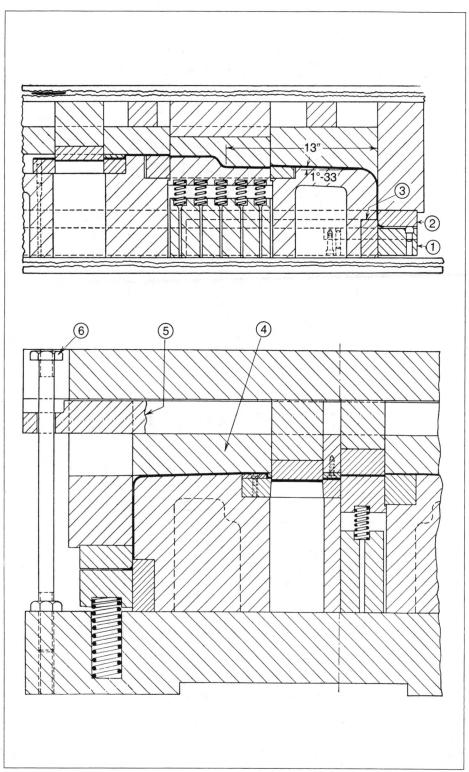

Fig. 13-41. Final draw, pinch trim, and blanking die for vending machine door. (*Mills Industries, Inc.*)

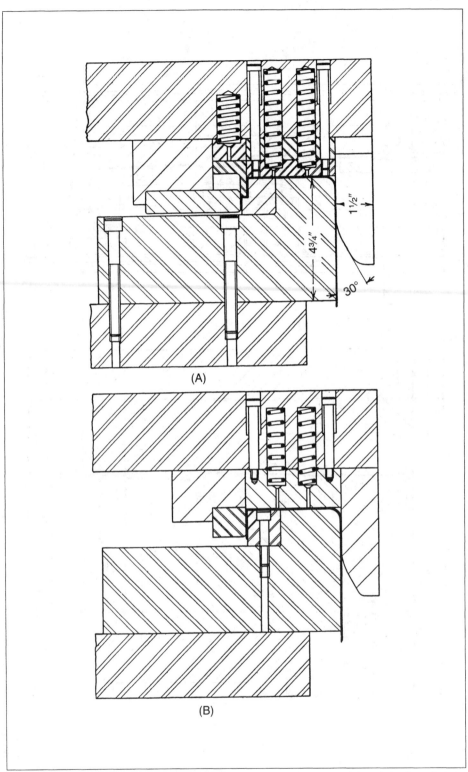

(A)

(B)

Fig. 13-42. Sections through dies for forming the return flanges on the sides of the front-door panel. (*Mills Industries, Inc.*)

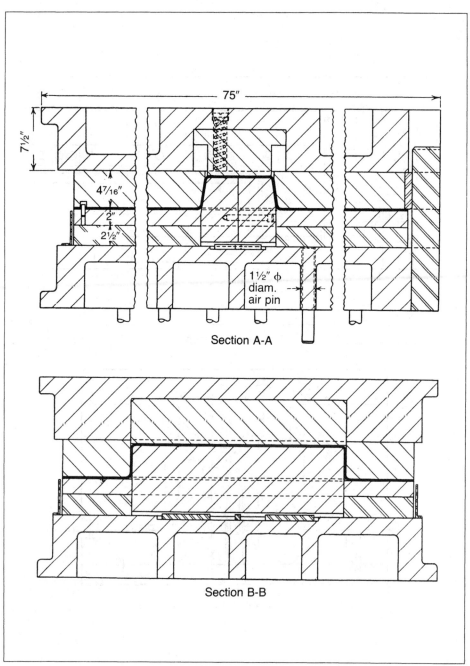

Section A-A

Section B-B

Fig. 13-43. Draw die for automatic washer top. (*General Electric Co.*)

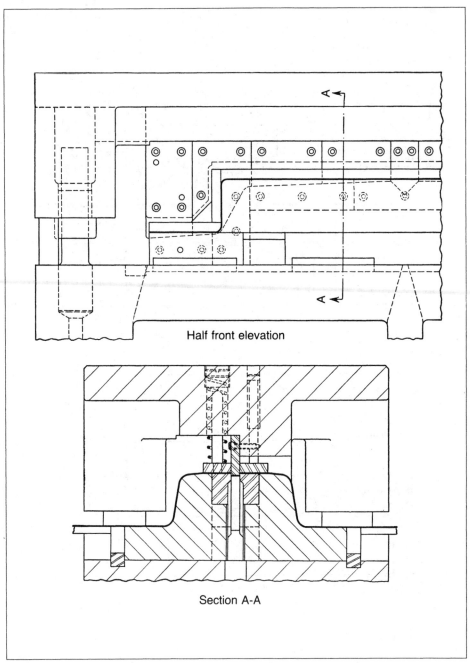

Half front elevation

Section A-A

Fig. 13-44. Parting die for the part drawn in the die shown in Fig. 13-43. (*General Electric Co.*)

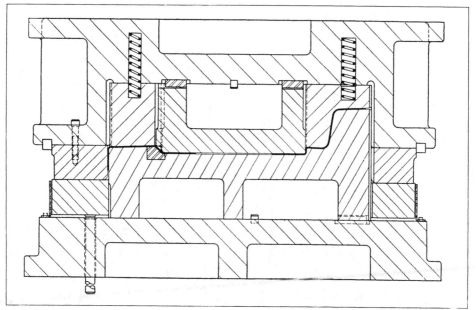

Fig. 13-45. Die to draw flanges and restrike recess in washer top. (*General Electric Co.*)

The parting die (Fig. 13-44) uses a sectional punch that incorporates shear to reduce the load. A chisel-shaped lance in the center of the punch cuts the scrap into two pieces for easier disposal. A spring-loaded pressure pad surrounding the punch holds the part securely against the die members. Guide pins and heel blocks are utilized to maintain die alignment. The die is of cast iron construction with tool-steel inserts.

Subsequent operations rough-pierce and draw a recess in the center of the top, then trim three sides of the blank to a developed shape prior to drawing the flanges and restriking the center recess. The later operations are performed in the die shown in Fig. 13-45. A spring-loaded pressure pad holds the blank against the draw punch while a press-cushion actuated blankholder grips the metal in the flange area.

The return flanging of the three horizontal and two vertical sides is accomplished in the cam-action die shown in Fig. 13-46. The cam (*D*1) attached to the upper shoe pulls the slide (*D*2) (to which the form block is attached) outward to a dwell position. This action also raises the form block for the backsplasher to position. During the dwell period of the form-block slide, another surface on the cam engages the flanging slide (*D*3), forcing it inward to form the flanges. This arrangement is typical for the three horizontal flanges on the top, and two additional cams actuate slides to form the vertical flanges on the backsplasher.

A flange around the center opening is drawn in the die shown in Fig. 13-47. The workpiece is placed over the locating ring (*D*1) fastened to the lower spring-loaded pressure pad. The closing of the die grips the part by the upper pressure pad (*D*2) against the draw ring (*D*3), while the punch (*D*4) draws the metal into the die. The next operation forms a curl on this flange as shown in Fig. 13-48.

Dies for Forming Serving Trays. The dies for forming, curling, and flattening the curl around the edge of a serving tray are shown in Fig. 13-49. A developed blank is cut to shape in a blanking die and placed in the die as shown at view *A* for forming and curling. The section of the die at view *B* is at the mid-point of the closing position showing the flange formed. View *C* shows the die completely closed to form the recess and flange. The section at view *D* is taken through the flattening die where the flange is folded in and flattened.

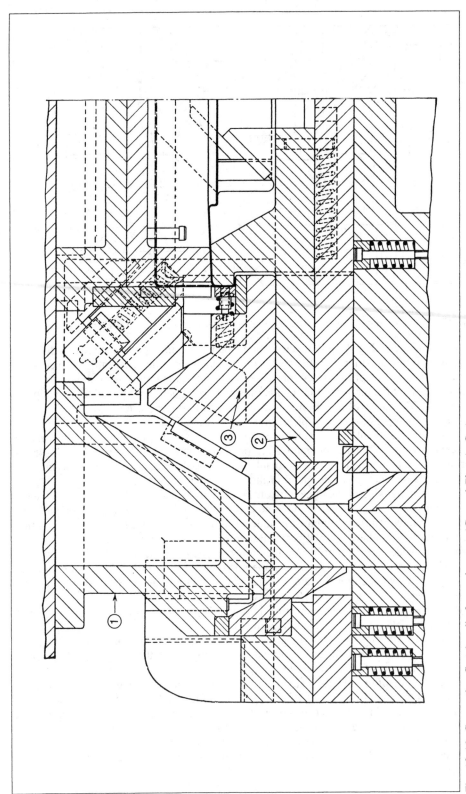

Fig. 13-46. Cam-action flanging die for washer top. (*General Electric Co.*)

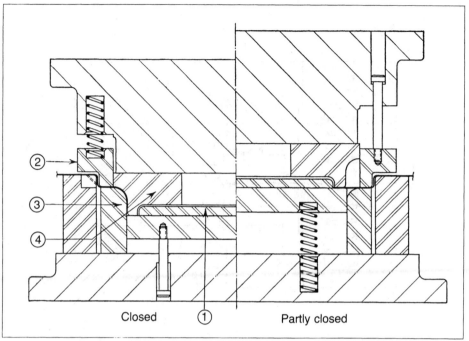

Closed ① Partly closed

Fig. 13-47. Die for drawing flange around center opening. (*General Electric Co.*)

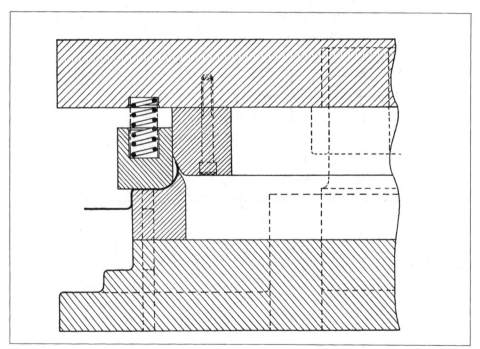

Fig. 13-48. Curling die for flange around center opening of washer top. (*General Electric Co.*)

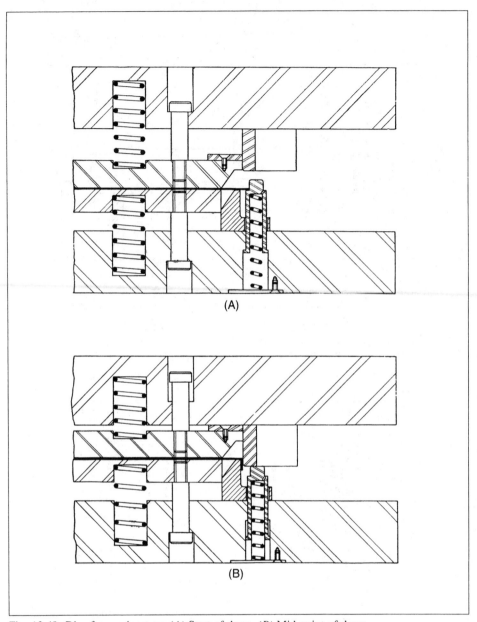

Fig. 13-49. Dies for serving tray: (A) Start of draw. (B) Mid-point of draw.

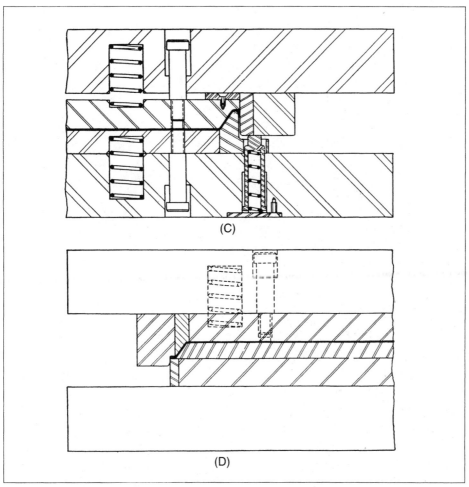

(C)

(D)

Fig. 13-49. *(Continued)*. *(C)* Closed position for form and curl of tray. *(D)* Section through flange-flattening die. *(Davey Products Co., Inc.)*

References

1. Ford Motor Company, Stamping Engineering, ''CT-20 J/M Die Design Standards,'' Body and Assembly Operations, Dearborn, Michigan, 1989.
2. Chrysler Corporation, Stamping and Assembly Division, ''Die Engineering Standards,'' Detroit, Michigan, 1989.
3. D.A. Smith, ''Back to the Basics: Techniques for Improving Stamping Quality and Lowering Costs,'' Technical Paper presented at an SME Die and Pressworking Tooling Clinic, Schaumburg, Illinois, August 29, 1989.

SECTION 14
RUBBER-PAD AND HYDRAULIC-ACTION DIES

The use of rubber in conjunction with press tools takes advantage of a property it possesses in common with fluids, viz., its *ability to flow.* If a quantity of rubber is placed in a cylinder and pressure is brought to bear upon it by applying a force to a ram, any such force must set up a resultant reaction on every surface with which the rubber comes in contact.

Rubber also possesses the property of *cohesion,* or resistance to flow, not exhibited by a fluid, which plays a vital part in the working of materials. The cohesive property of rubber has its limits, and great care must be taken in the design of tools not to expect too much of the material; also, every assistance should be given to the rubber to enable it to maintain its form unbroken and so preserve its life.

The several rubber-die processes possess a common characteristic—only the male portion or punch is made. This lowers the tool cost considerably. The rubber pad is attached to the ram of the press and, as it is lowered, the rubber is made to flow around the form block, forming the blank to the shape of the form block.

GUERIN PROCESS

The Guerin process employs a rubber pad on the ram of the press and a form block to be placed on the lower platen. This process is the oldest and most widely used but is limited to the forming of relatively shallow parts in light materials, normally not exceeding 1.25- to 1.5-in. (31.8- to 38.1-mm) deep. Parts with straight flanges, stretch flanges, and beaded parts formed from developed blanks are most suitable for this process.

Rubber Pressure Pads. These are held in a steel or cast-iron container with walls of sufficient cross section to withstand the pressure exerted upon it by the rubber. The thickness of the pad may vary from 6 to 12 in. (152 to 305 mm). The pads may be a solid type or a laminated type. The solid type is produced in two ways: (1) by curing as a homogeneous mass, or (2) by curing in sheets or slabs, usually 1 in. (25.4 mm) thick, then curing these sheets together in a single slab of the required thickness. The laminated type includes all pads made up of individual sheets placed one on top of the other in the press, but not cured into a solid mass. Laminated pads have the advantage that the working surface can be restored when worn by merely reversing the top layer or exchanging it with one of the others. Most users favor a pad hardness of 55 or 65 durometer hardness for forming work. When shearing or blanking is done with a rubber pad, a throw sheet with a 90 durometer hardness is sometimes placed over the blanks or cemented directly on the face of the pad. Figure 14-1 shows the arrangement of the rubber die and lower platen in the press.

One way to hold the solid pad in place is to make it larger than the container in which it is to be used. The pad is then forced into the box by placing it on the bolster of the press

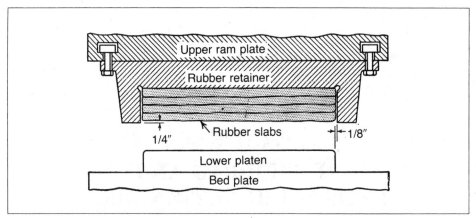

Fig. 14-1. Arrangement of rubber-pad die in press.[1]

and closing the press. One percent minimum oversize on the length and width has been used successfully.

A framework whose sides are Z-shaped (Fig. 14-2) retains the rubber slabs without them being cemented together. The retaining strip may be part of the framework or a separate strip fastened on with cap screws. The dimensions of the lower platen should provide 0.062- to 0.125-in. (1.59- to 3.18-mm) clearance per side between it and the inside of the rubber retainer.

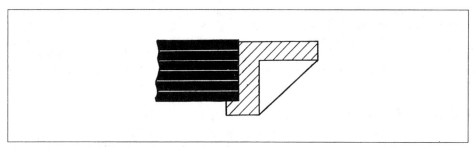

Fig. 14-2. Z-shaped frame for rubber retainer.

Form Blocks. The form block (Figs. 14-3 and 14-4) is the primary tool used in rubber forming. It is a contoured flat plate of sufficient height to accommodate the part. A minimum of two pins are placed in the form block to locate the blank in the proper position. The height of these pins should be kept at a minimum to prevent damaging the rubber pad. Most of the parts are formed on a one-piece block in one operation. The more complex shapes require more than one piece to the form block or more than one operation. C- or Z-flanges on parts usually fall in this category.

Cross sections of form blocks for rubber forming are shown in Fig. 14-3. View A illustrates how the block is undercut to allow for springback in the flanges. A trap, as shown in B, is a means of directing an increased amount of pressure against the flange. A roll, wedge, or hinged wiping plate as shown in views C, D, and E, respectively, increases the pressure against the flange and helps to eliminate wrinkles. The cover plate shown clamps the blank on the form block to prevent its slipping and reduces the distortion of the web. A wiping plate which extends over a peripheral segment of more than 180°, and therefore may be left floating, is called a draw ring (Fig. 14-3F).

Form blocks are made from a wide range of materials. Masonite die stock or impregnated fiber will successfully stand the pressures and is relatively inexpensive. Masonite will break down rapidly if the block is made with sharp corners or has an overhanging section. In either of these two cases, or for a large number of parts, the block

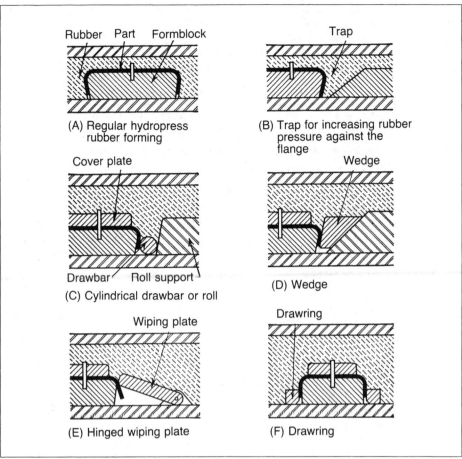

Fig. 14-3. Auxiliary rubber-forming tools.[2]

should be made of cast iron or steel. Kirksite, magnesium, and aluminum are also popular form-block materials. When hot-forming magnesium blanks, Kirksite has a tendency to flow at a temperature of over 450°F (232°C). Magnesium form blocks are easy to machine, light to handle, and have the same coefficient of expansion as the material being formed.

To increase the life of the rubber, form blocks should be kept as low as possible, and all corners that the rubber must flow down over should be rounded off as much as the part permits. When determining the height of the form block, consideration must be given to the fact that rubber will not form itself into sharp internal recesses or corners. For example, it will not form itself into the 90° corner between the side of the form block and the platen on which the block rests, but will form a natural radius between the two surfaces. Additional height must be added to the form block so that the bottom edge of the formed flange will be higher than the tangency point of this natural radius formed by the rubber. This distance is usually a minimum of 0.125- to 0.187-in. (3.18- to 4.75-mm) higher than the greatest height of the part.

Bend radii, springback, and bend allowances for rubber forming are the same as for die forming (see Section 8). The minimum flange widths for different metals and thicknesses T are shown in Table 14-1. These limits apply to simple rubber forming, but auxiliary devices may be used to form narrower flanges.

The simplest of all rubber forming is the forming of a straight flange by means of a single bend over a form block. Figure 14-2 illustrates a form block for forming a straight

14-3

flange along one side of the blank. This form block is designed to flange a right- and left-hand part and uses a cover plate to reduce the distortion in the web.

TABLE 14-1
Minimum Flange Width for Rubber Forming[3]

Material	Minimum Flange Width
2024-0 and 7075-0	0.062-in. (1.57mm) + 2.5T
2024-T3	0.125-in. (3.18 mm) + 4T
Annealed stainless steels	0.187-in. (4.75 mm) + 4.5T
Quarter-hard stainless steels	0.625-in. (15.9 mm)

The second-operation block, for forming flanges in the opposite direction in the part in Fig. 14-4, is shown in Fig. 14-5. The parts are placed on the form block, and a cover plate (not shown) is used to prevent the rubber from damaging the flange already formed. Flanges are formed along one side and an end of this part. The center portion of the form block is also removable to facilitate removing the finished part from the form block.

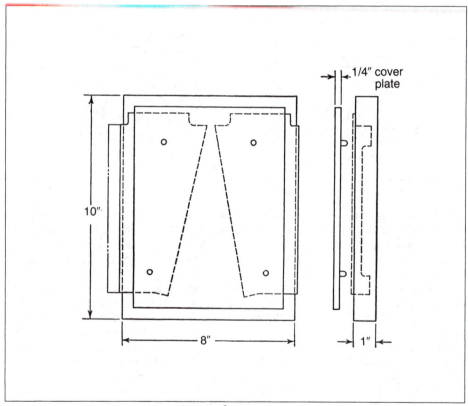

Fig. 14-4. Form block for straight flange part.[2]

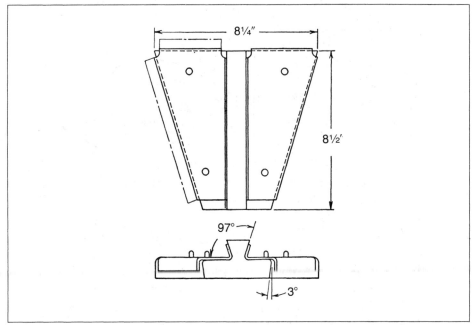

Fig. 14-5. Second-operation block for reverse flanges on part shown in Fig. 14-4.[2]

RUBBER-FORMED STRETCH FLANGES

Parts with stretch flanges are well suited for rubber forming. Often such parts can be produced more economically and more accurately by this process than by die forming. Stretch flanges may either be along concavely contoured portions of the part edges or around holes. There is no difference in forming requirements between the two types. The length of the outer flange is usually limited to a segment which is usually not larger than 180°, while hole flanges extend over the entire periphery of the hole.

Stretch flanges are obtained by being bent over a form block in the same manner as straight flanges. The width of the flange and the radius of the contour affect the bending of a stretch flange. Such a flange, having a large radius, can be formed in soft materials quite readily, but forming of sharply contoured flanges requires the use of pressure-increasing devices.

Aluminum alloys in either the annealed or as-quenched state are stretch-flanged in thicknesses up to 0.125 in. (3.18 mm). Austenitic stainless steels up to 0.050 in. (1.27 mm) thickness in the annealed condition, and quarter-hard stainless-steel parts between 0.016 in. (0.41 mm) and 0.030 in. (0.76 mm) thick, can be stretch-flanged. Very thin quarter-hard stainless-steel parts develop an irregular, wavy flange edge. A severely stretched flange of thick and hard material does not touch the block at all, or possibly contacts it only at the flange edge. The forming limits of stretch-flanging different materials are discussed in more detail in Section 2.

A part with a stretch flange of type 302 annealed 0.030-in. (0.76-mm) thick stainless steel is shown in Fig. 14-6. The base of the form block is zinc alloy. A steel cover plate is required to avoid distortion of the web, and a steel trap was used to obtain sufficient pressure for forming the stretch flange. The joggle was partially formed on this block but required a hand operation to finish-form.

Rubber forming of various types of lightening-hole flanges is shown in Fig. 14-7. These flanges are usually formed from prepunched blanks without the aid of pressure-increasing devices.

The J-flange is another type of stiffener around a lightening hole. This flange is formed in two operations. The first uses regular rubber forming which provides enough

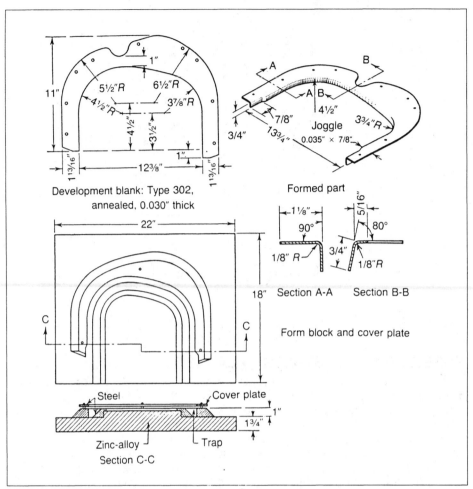

Fig. 14-6. Stainless-steel part with stretch flange.[2]

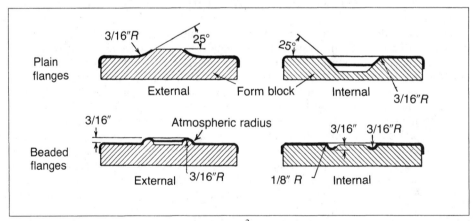

Fig. 14-7. Rubber-formed lightening-hole flanges.[2]

impression to locate an insert (Fig. 14-8). In the second operation, this insert accumulates the rubber pressure exerted on the whole area of the top surface of the insert and bridges it over so that this pressure forces the flange into its final position.

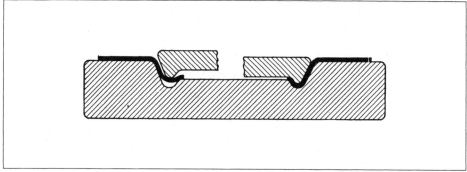

Fig. 14-8. Forming J-flanges.[1]

RUBBER-FORMED SHRINK FLANGES

Many rubber-formed parts contain shrink flanges; however, they can be accurately produced only within narrow limits. A shrink flange on a part that was rubber-formed without any auxiliary mechanical device will tend to exhibit wrinkles. The extent of this wrinkling depends upon the intended shrink, which increases with increasing flange width, increasing bend angle, decreasing contour radius and, up to a certain limit, increasing flange length. These wrinkles, if in soft metal and not too deep, can be worked out by hand.

To eliminate excessive shrinking at corners, a corner cutout entirely removes the otherwise severely shrunk flange corner. Flutes are often used as a means of controlling the wrinkles. Figure 14-9 shows a part which has flutes and its form block.

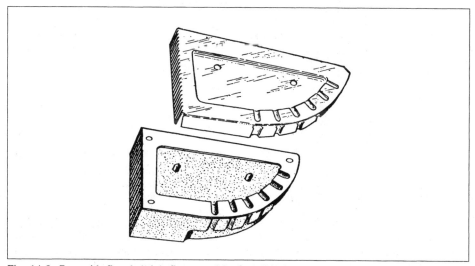

Fig. 14-9. Part with fluted shrink flange and its form block.

A part with a shrink flange having a large contoured radius is shown in Fig. 14-10. This part has a stretch flange which is formed in the first operation. The part is then placed on the second-operation form block with the cover plate to protect the already formed flange. The positioning of the two blocks on its base plate provides a trap which increases the pressure exerted by the rubber on the blank, and aids in forming the short return bend on the shrink flange.

Parts that contain both stretch flanges and shrink flanges can be accurately formed by a special two-operation procedure termed "slip" forming. A part with both types of flanges and an extended shallow recess is shown in Fig. 14-11. To avoid distortion to the

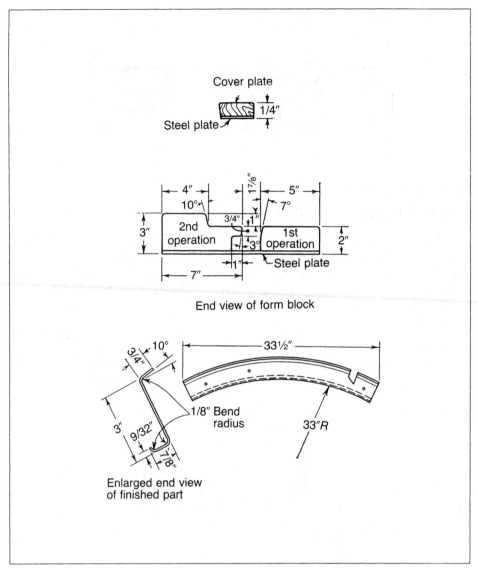

Fig. 14-10. A large-radius shrink-flange part.

web, the recesses are formed first in the flat. Recesses near the stretch flanges allow some draw-in of the metal from the periphery. Recesses close to the shrink flanges are prevented from drawing metal from the periphery by a suitably shaped pressure pad. The flanges are turned over in the usual manner by a second form block. The edges of the blank in the area of the shrink flange are notched to aid in the shrinking.

A mechanical means of forming shrink flanges is shown in Fig. 14-12. The blank is placed on the form block and pressure pad; then cover plates are placed over the blank. The cover plate on the pressure pad acts as a floating draw ring as the rubber die descends.

The use of a female form block and a spring-loaded male pressure pad is shown in Fig. 14-13. The cover plate aids the rubber to act as a punch. As the ram descends, the blank is drawn into the die, forming the shrink flange without wrinkling. The straight flange across the end is formed downward in the die with the aid of the trap. The first operation also blanks a hole in the web. The second-operation form block flanges the lightening hole and forms the internal beads.

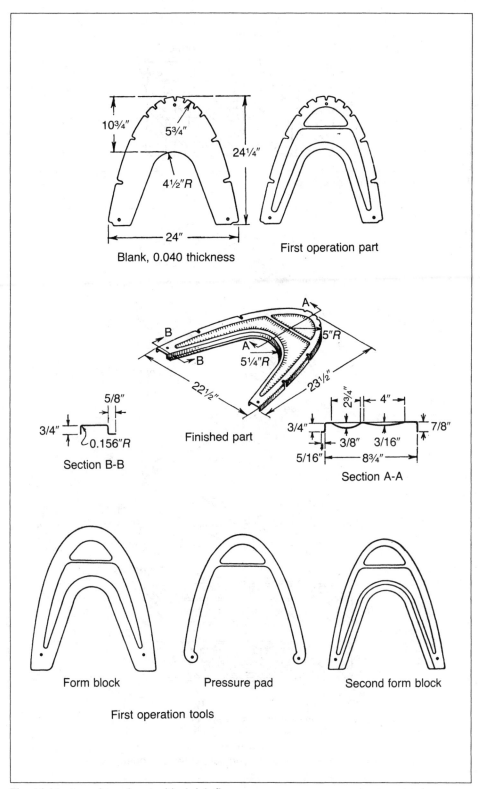

Fig. 14-11. An embossed part with shrink flanges.

14-9

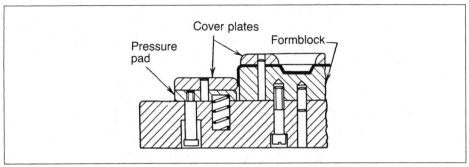

Fig. 14-12. Die to form shrink flange.[1]

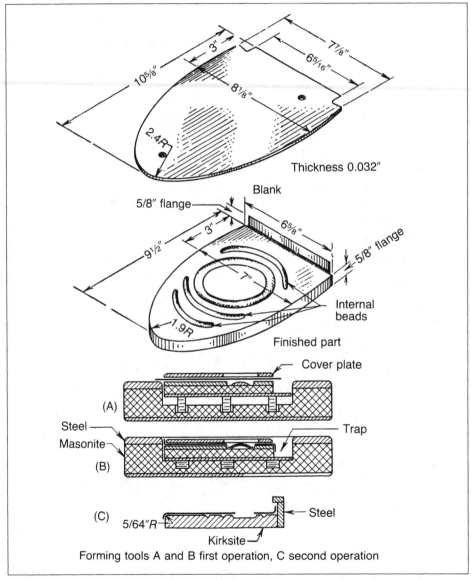

Fig. 14-13. Combination die to form shrink and straight flanges.

DRAWING OF SHALLOW PARTS

The rubber-die process may also be used to draw shallow recessed parts. To prevent the flanges from wrinkling, the metal must be held firmly yet be allowed to move in the same manner as drawing in a mechanical press. The flange portion of the metal can be lubricated with paraffin or pressure-relieved by means of an undercut protecting block.

The part shown in Fig. 14-14 is a typical draw by the rubber-pad method. Without a lubricant or pressure-relieving block, the rubber locked the metal to the form block, causing the metal to crack almost continuously in the radius on all four sides of the pan. This was because the amount of metal required to fill the radius was greater than the amount of material allowed to flow, plus the elongation of the material. The undercut-protecting block allows the material to flow without wrinkling. Careful design of this block is essential. When the base is increased at the expense of the undercut, the rubber exerts more pressure to hold the protecting block firmly against the form block, and less pressure to hold the protecting block firmly against the flange of the blank. When the undercut is increased at the expense of the base, the opposite is true and more pressure is exerted against the flange of the blank. The height of the undercut is also important and should be 0.003 to 0.006 in. (0.08 to 0.15 mm) greater than the thickness of the metal blank.

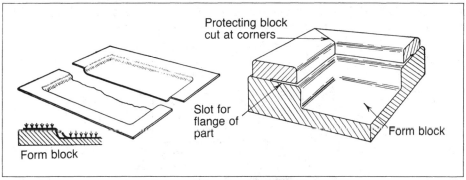

Fig. 14-14. Design of form block to relieve pressure on wide flange.[1]

When designing tools for rubber forming, it should be remembered that parts can be formed by external pressure that cannot be formed satisfactorily by internal pressure; that is, forming should take place over a projection rather than into a recess. For example, in Fig. 14-15 where a shape is to bend through a 90° bend and, if external pressure is applied with a form block as shown at A, relatively little pressure would be required, because of the even distribution of the load, and a close fit will be obtained at the apex, irrespective of how large or small the radius may be at that point. It may not be possible to obtain a sharp corner at B with the internal form, because of the localized pressure and the locking of the metal to the flat surface of the form block by the rubber.

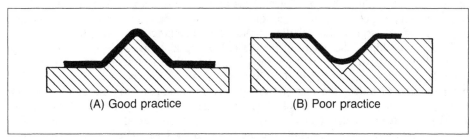

Fig. 14-15. Bending with external pressure vs. internal pressure.

BLANKING WITH A RUBBER DIE

Blanking in rubber produces an edge that is better than that obtained by band sawing and almost as good as that obtained by routing. Since blanking requires higher pressures than forming, the thickness of stock possible to blank depends upon the unit stress exerted by the rubber die. The rubber die is capable of blanking up to 0.032-in. (0.81-mm) 2024-O aluminum without difficulty, and in some cases up to 0.040-in. (1.02-mm) stock. Long curves and straight sides are easily blanked; also cutouts and reverse curves, if not too small, can be made. The minimum hole diameter or width of cutout possible is about 2 in. (50 mm).

The usual amount of metal wasted in rubber-die blanking is greater than for conventional steel blanking dies. A minimum edge distance of about 1.5 in. (38 mm) is required between the edge of the cutout and the edge of the part or sheet. When cutting several blanks from a large sheet, the blocks should be spaced about 3 in. (76 mm) apart to ensure a good blanked edge.

To shear a blank or to trim a part, the form block is provided with sharp cutting edges. The sharp cutting edge required can be machined onto most metal form blocks, but the nonmetallic blocks require steel inserts at the cutting edge. Figure 14-16 illustrates the use of a bead as a locking ring around the blanking die to create a recess and to localize the pressure of the rubber at the cutting edge. The lock ring also prevents slippage of the workpiece before it is cut. The upper view shows the blank in place on the block, and the lower view shows the shape of the stock before the fracture occurs.

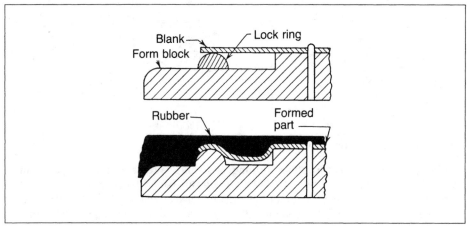

Fig. 14-16. Shearing (trimming) a rubber-formed part.[3]

When blanking heavier-gage metals with a rubber die, an edge radius up to the thickness of the material can be obtained. A grip plate shown in Fig. 14-17 provides two places to grip the material: on the grip plate and on the die. Between these two points is an unsupported section which is subject to the pressure of the rubber. Before the stock has had time to rupture at the cutting edge, the material has been drawn down over the cutting edge, giving the large radius peculiar to rubber-die blanking of thicker stock.

The relationship of the height of the die to the grip plate has little effect on the edge radius, but may have some effect on the pressure required to fracture the material. The grip plate should be about 0.5 in. (12.7 mm) wide and 0.125 in. (3.18 mm) thick. It is not necessary to follow the intricate details of the contour of the die block. Pins for locating the blank on the die block should be used if possible. The die block may be positioned on the base plate by loosely fitting dowel pins so that other blocks may be used with the same base plate. Fastening the die block to the base plate is not necessary.

Combination Rubber-pad Dies. Combination rubber-pad blanking and forming dies are frequently used for lightening holes (Fig. 14-18). A preshaped blank is placed on the form block, which has a cutting edge for the cutouts, and is positioned by locating pins. As the rubber die descends, the center portion of the lightening hole is cut out, and the bead around the hole is formed at the same time as the flanges on the outside contour. Cover plates should be used where necessary to avoid distortion of the web.

Combination dies are frequently used on parts similar to the one shown in Fig. 14-11 for cutting the inside contour, to avoid handling flimsy blanks.

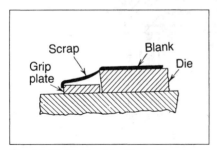

Fig. 14-17. Rubber blanking with grip plate.

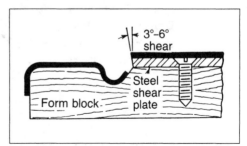

Fig. 14-18. Combination blanking and forming die.

DEEP DRAWING WITH RUBBER-PAD PROCESS

Methods have been developed for deep-drawing shells using the inexpensive tools possible with the rubber-pad process. The parts produced are comparable in quality to those produced in all-metal dies.

The Marform process and the Hidraw process employ a deep rubber pad on the ram of the press with a stationary punch on the bed of the press. A blankholder plate actuated by a specially controlled die cushion controls the pressure on the blank as it is drawn around the punch.

The descending platen makes contact with the blank and confines the forming metal to prevent wrinkling. As the platen continues to descend, pressure is generated in the rubber pad, which grips the blank between its face and the blankholder plate. The rubber acts as a fluid pressure, forcing the blank over the punch so that it conforms to the contour of the punch. Pressure exerted by the cushion controls the forming pressure developed in the rubber pad, and is adjustable to a preset stroke pattern as required by the form of the part being drawn and the tensile strength of the metal.

The components of these two processes are shown schematically in Fig. 14-19. The punch is fixed, and the upper platen containing the rubber pad moves downward to meet the blank. Constant control of pressure on the blankholder provides smooth forming and eliminates wrinkles. The pressures used range from 5,000 to 15,000 psi (34.5 to 103 MPa). The blankholder plate has about 0.062 in. (1.57 mm) clearance between it and the rubber-pad holder to prevent the rubber from squeezing out of the holder.

The rubber-pad process of deep drawing usually allows a greater reduction percentage than by conventional drawing dies. It is apparent that, when one side of the material is gripped by the rubber and the other side by the steel blankholder, work hardening does not take place so rapidly as when the material is gripped and drawn between the two hard surfaces of a steel die. Thus the material will flow more readily, permitting the use of a larger blank. The variable draw radii in a rubber pad permit the material to draw more easily than the fixed radii of a steel die.

The maximum blank size for aluminum and steel cups drawn by the rubber-pad process may be found by multiplying the punch diameter by 2.34.

In the forming of square or rectangular-shaped boxes, it is not necessary to use a developed pattern; in fact, some tests have shown that best results are obtained by using a sheared blank. The blanks must be large enough to allow for trimming, since a portion of the draw radius is left on the end of the box or can.

The minimum corner radii for square and rectangular boxes drawn from the softer

14- 13

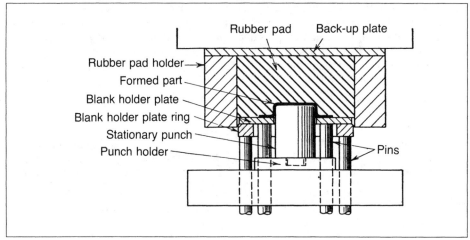

Fig. 14-19. Schematic view of deep drawing by the rubber-pad process.[4]

aluminum alloys and deep-drawing quality steel are normally not less than 25% of the box width. The safe depth of draw for the same materials ranges from 75 to 100% of the width.

The forming or draw radius is affected by the blank size, the depth of draw, the material and its thickness, the flange left on the part after forming, and the shape of the finished part. For a round cup, the radius is smaller when the final flange is narrow. In drawing square or rectangular-shaped boxes, the forming radius at the corners will be larger than along the straight sides.

The minimum flange radii for round cups and at the corners of square or rectangular boxes vary according to the hardness of the metal. For the aluminum alloys 1100-O and 3003-O, the radius is about 1½ times metal thickness t; for 5052-O and 6061-O it is $2t$; for 2024-O and 7075-O it is $3t$. When the stock is deep-drawing steel, $4t$ should be the minimum radius.

When the parts being made require a sharper radius than is practical to produce by rubber forming, a restrike operation is necessary using a plate with the required radius.

The punch-nose radius depends upon the material, its thickness, and the ratio of the depth of draw to the diameter of the cup. The deeper the draw with reference to the diameter, the larger punch-nose radius required to prevent fracture to the part. The safe radius relative to material thickness for the softer aluminum alloys ranges from t to $4t$ for depth-to-diameter ratios ranging from 1:4 to 1:1, respectively. For steel the radius is $0.5t$ for the 1:4 depth-to-diameter ratio, and $2t$ for a ratio of 1:1.

According to tests made at 6,000 psi (41.4 MPa), in forming cups from the softer aluminum alloys with a minimum diameter of 1.5 in. (38 mm), the minimum stock thickness should be approximately 1% of the cup diameter to obtain the maximum depth of draw. A minimum stock thickness of 0.016 in. (0.41 mm) is recommended for cups smaller than 1.5 in. (38 mm) in diameter.

In the forming of tapered shapes, there is a limitation as to the size of the gap A (Fig. 14-20) between the top of the punch and the edge of the blankholder. If the metal is thin and the gap too large, the rubber pad will tend to shear the metal in the gap that is not supported by the punch or pressure plate. The maximum gap A varies as a percentage of the punch diameter d varies for different thicknesses of aluminum and steel. This percentage of punch diameter varies from 7½ to 30% for 0.025- to 0.125-in. (0.64- to 3.15-mm) thick aluminum alloys, and 10 to 55% for similar thicknesses of deep-drawing steel. The percentages increase proportionally as the stock thicknesses increase.

When the maximum gaps shown are used, the depth of draw for the softer aluminums is equal to the punch diameter and about 75% of the punch diameter for the harder

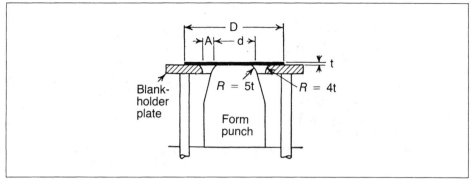

Fig. 14-20. Forming of tapered caps by the rubber-pad process is possible by proper design of punch and blankholder plate.

aluminums and SAE 1010 deep-drawing steel. The values have been determined by using 6,000 psi (41.4 MPa). When using 10,000 psi (68.9 MPa), cups and boxes can be drawn approximately 20% deeper.

HYDRAULIC-ACTION DIES

In the deep-drawing process called "Hydroforming" (Fig. 14-21), the ram is specially designed to act as a forming chamber. This is accomplished by hydraulic fluid contained in a cavity in the ram, which exerts pressure on a rubber diaphragm. The forming tools consist of a punch machined to the required shape, and a blankholder ring with a hole which closely matches the shape of the punch. The pressure pad is built up and is clamped on the bolster which is fastened to the press bed. The punch is fastened to a hydraulic-ram assembly located under the bed of the press. The top of the punch in the lowered position is usually flush with the top of the pressure pad.

The forming chamber is lowered to clamp the metal blank between the rubber diaphragm and the blankholder ring. Hydraulic pressure is then exerted against the diaphragm to prevent wrinkling of the metal during the forming operation. The punch moves upward causing the metal to flow around it and, as forming progresses, the compression of the fluid in the forming chamber causes higher forming pressures to develop against the top, radius area, and side walls of the formed part. The deep-drawing operations are carried out under pressures ranging up to 15,000 psi (103 MPa). Figure 14-22 illustrates some parts formed by the Hydroform process on Cincinnati Hydroform machines.

When deep drawing with the hydraulic-action processes, the rubber diaphragm has a tendency to clamp the part as it is formed to the punch surfaces, thus preventing further stretching or straining of these formed areas, and at the same time causing the metal to flow in around the punch. Hydraulic-action forming also has a variable draw radius, and the high local strains introduced in the initial phase of the forming operation are minimized by the ability of the rubber pad to change its radius to suit the forming cycle. With the lower pressures at the beginning of the cycle, the draw radius is large, and it decreases as the forming pressures increase. The rubber pad does not produce any marks or scratches on the outer surface of the shell, as a metal draw ring sometimes does.

The Wheelon forming process uses a method of applying direct hydraulic pressure to the rubber forming pad. The blanks are placed over simple male dies similar to those used in the Guerin process, positioned in the press frame, and forming pressure is applied by hydraulically inflating a rubber bladder mounted in the immobile roof of the press. Figure 14-23 shows a cross section of the press frame with a die and blank in place, with the rubber bladder in the released and the forming positions. This method is limited in depth of draw to about the same as the Guerin process but, with pressures of 6,000 to 10,000 psi (41.4 to 68.9 MPa) available, practically all wrinkling is eliminated.

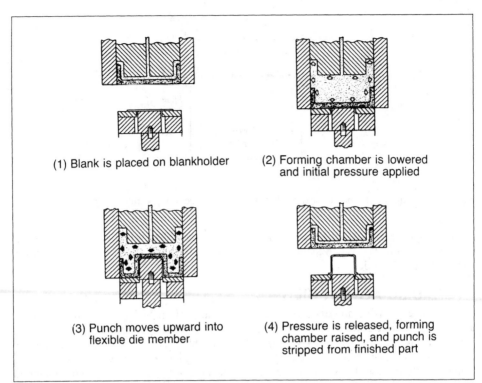

(1) Blank is placed on blankholder

(2) Forming chamber is lowered and initial pressure applied

(3) Punch moves upward into flexible die member

(4) Pressure is released, forming chamber raised, and punch is stripped from finished part

Fig. 14-21. Operational sequence for the Cincinnati Hydroform process. (*Meta-Dynamics Division, Cincinnati Milacron*)

A method that is particularly adaptable to the forming of shallow shapes and the drawing of cone-shaped and tapered stampings is called the "Hydrodynamic" process (Fig. 14-24).

The Hydrodynamic process was so named because it uses the power of a fluid under pressure to perform drawing and embossing operations. A die only is required, since the fluid under high pressure is substituted for the punch. A predetermined size of blank is placed on the top face of the spring-loaded pressure pad, nested between spring-pin gages. The container, pressure pad, and blank are then raised by the hydraulic ram until the underside of the die comes in contact with the top face of the container. High-pressure fluid is then admitted through the intake opening in the container. The fluid is forced up through the hole in the center of the pressure pad and acts as a fluid punch, exerting a uniform pressure over the entire surface of the blank, thus forming it to the desired shape. Air-vent holes are provided in the cavity of the die to permit the air to escape from the die cavity as the part is formed. Parts of tapered or conical shapes can be drawn in one operation by this method, whereas two or three operations are necessary by the conventional die methods.

The SAAB fluid-form method, illustrated in Fig. 14-25, uses a flexible (fluid) punch which assumes the shape of the die. At the same time, the punch serves as a blankholder. Enclosing of the fluid within the punch is accomplished by use of a diaphragm of such relative hardness that it can practically be regarded as a fluid at the high pressures used.

The punch and blankholder of the conventional draw-die press are thus replaced by a steel cylinder which contains hydraulic fluid and is exposed to pressure by a telescoping piston. The die for the SAAB fluid-form method can be conventional, with escape holes for the air enclosed in the die cavity.

The process can be adapted to a conventional single- or double-acting press by the installation of a fluid-form unit. This unit can be removed, thus permitting the press to be

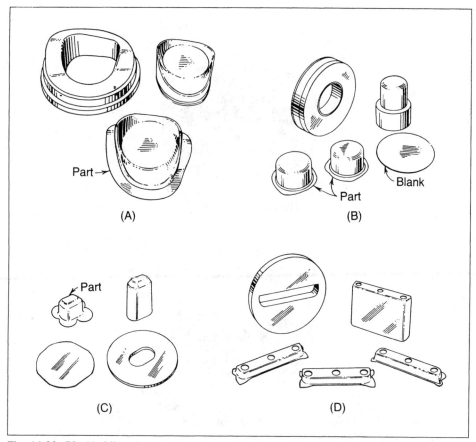

Fig. 14-22. Blankholding rings and punches used and parts formed by the Hydroform process: (A) automobile-headlight bevel; (B) cooking utensil; (C) pump body; (D) escutcheon cover. (*Meta-Dynamics Division, Cincinnati Milacron*)

used alternately for fluid-form presswork and conventional presswork. The fluid-form unit consists of two main parts, the punch or press chamber and the die holder. The die holder consists of a steel cylinder containing a telescoping piston which is sealed against the cylinder wall by a rubber or bronze packing and is retained in its upper position by a spring. The lower end of the cylinder is closed by a rubber diaphragm which is retained by an annular slot in the cylinder wall. The size of the unit (diameter of the diaphragm) depends primarily on the available press power. Presses between 100 and 3,000 tons (890 kN to 26.7 MN) or more are considered to be well suited for the process.

A particular application of the process is suitable for forming parts with deep cavities. With the aid of a hydropneumatic cushion on the press bed of a single-action press, the diaphragm is used as a *flexible die* which draws the blank over a rigid punch. The hydropneumatic cushion should afford a counterpressure corresponding to at least 10% of the press capacity. A control valve permits variation of the counterpressure during the downward movement of the cushion. This means that the blankholder pressure can be varied, enabling adjustment of this pressure to the lowest required at any time. The blank is wrapped gradually over the rigid punch by the flexible die diaphragm, which is in continuous contact with the blank, thus allowing a larger reduction than is possible with the conventional method.

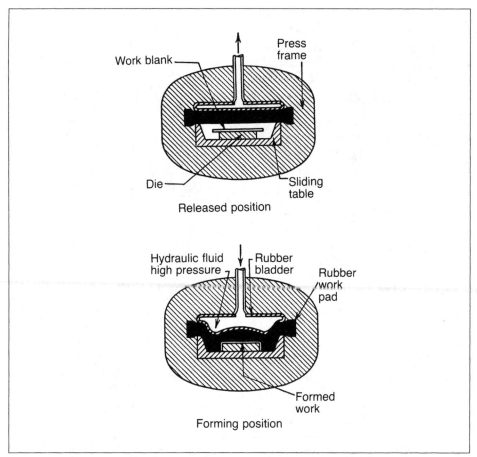

Fig. 14-23. The Wheelon forming process. (*Verson Allsteel Press Co.*)

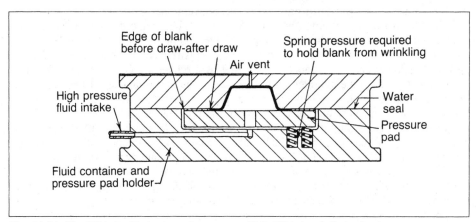

Fig. 14-24. The Whistler Hydrodynamic process. (*S.B. Whistler & Sons, Inc.*)

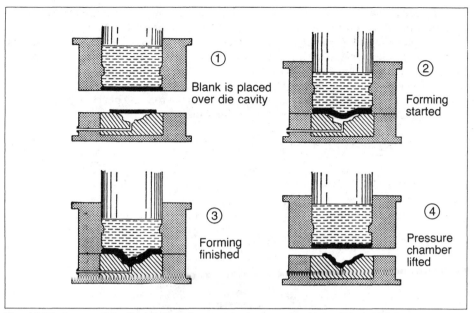

Fig. 14-25. Stages of the SAAB fluid-form method of pressworking. (*SAAB Fluid Form Division, McMahon and Co., Inc.*)

URETHANE-PAD DIES

Like rubber, urethane as a metalforming medium can be characterized by its ability to flow under pressure and to regain its original shape when subjected to cyclic forces. Urethane can be cut or machined into various forms and generally results in simpler, less expensive tooling.

Bulging Die Using Urethane Punch. In Fig. 14-26 a combination of a horizontally split female die and urethane punch is used for bulging a deep-drawn shell. As the upper die descends, the pressure exerted on the elastic punch forces the shell to expand in all directions where the flow is unrestricted. At the end of the stroke, both the punch and the shell assume the shape of the female die. During the upstroke of the press, the urethane punch regains its original shape, permitting the insertion of another workpiece.

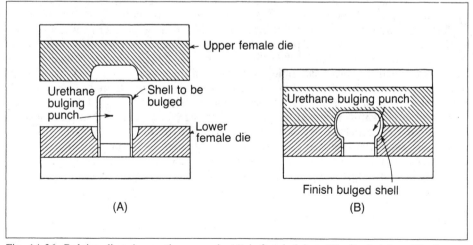

Fig. 14-26. Bulging die using urethane punch: (*A*) before bulging; (*B*) after bulging. (*Kaufmann Tool & Engineering Corp.*)

U-forming Die. The die shown in Fig. 14-27 has advantages for forming heavy gages of sheet metal. When the forming punch advances between the forming blocks, the urethane pressure pad is deflected and keeps the steel plate forced against the work. This is an effective aid in maintaining the flatness of the bottom of the workpiece. The pressure pad acts like a back-up punch. During the return stroke of the press, the pressure pad assumes its original shape, therefore acting as a knockout. When the wedge is moved in the direction shown, the finished part is released from the forming punch. Changeover for different gages can be done by shimming behind the forming blocks.

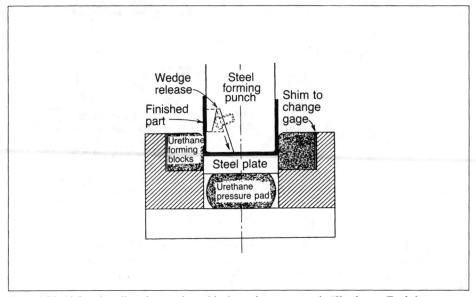

Fig. 14-27. U-forming die using urethane blocks and pressure pad. (*Kaufmann Tool & Engineering Corp.*)

Urethane Draw Die. Figure 14-28 shows a combination of steel die and urethane punch for drawing a cup-shaped part. This simple setup may be used for such difficult drawing as indicated only if the definition at the bottom of the shell is not critical. The blank which is placed on the urethane punch initially is drawn up into the die cavity by this punch as the ram is lowered. Notice that the urethane functions here as draw punch and backup punch. To permit the air to escape from the die cavity, an air vent is added.

Bulging Annular Grooves in Tubing. Figure 14-29 shows another application of the urethane tooling. Here both the punch and the split dies are stationary. When the ram of the press descends, the length of the urethane punch is shortened by the plunger. As the urethane is practically incompressible (the volume of the punch is constant), the material seeks to escape into the cavities of the die block, thus forming the desired shape. To allow the finished parts to be removed, the die is split into segments.

Combination Rib-forming and Cavity-embossing Die. Operations, such as shown in Fig. 14-30, can be handled by urethane tooling. The die indicated consists of a steel female die and a flat urethane pad, the latter functioning as a forming and embossing punch. Depending upon the amount of deflection of the pad, adhesive or double-faced clamping tape may be applied instead of the mechanical retainer shown.

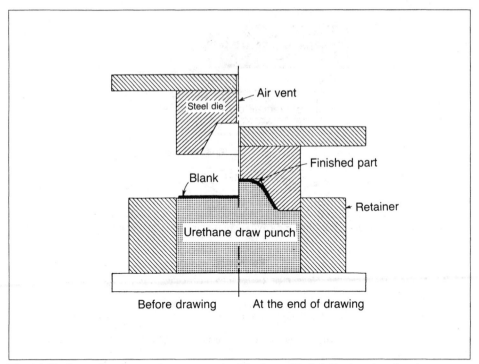

Fig. 14-28. Draw die using urethane as punch. (*Kaufmann Tool & Engineering Corp.*)

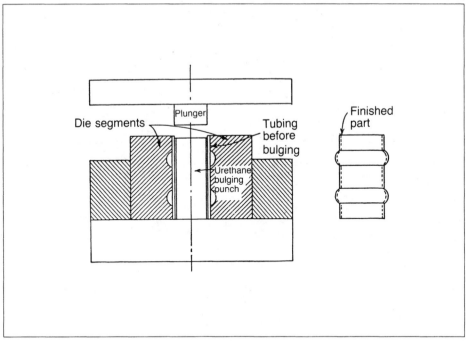

Fig. 14-29. Split die for bulging annular grooves in tubing. (*Kaufmann Tool & Engineering Corp.*)

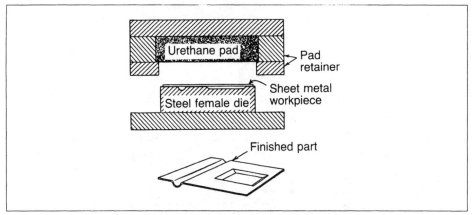

Fig. 14-30. Combination rib-forming and cavity-embossing die. (*Kaufmann Tool & Engineering Corp.*)

References

1. R.B. Schulze, *Aluminum and Magnesium Design and Fabrication* (New York: McGraw-Hill Book Company, 1949).
2. "Forming of Austenitic Chromium-nickel Stainless Steels," The International Nickel Co., Inc., 1948
3. G. Sachs, *Principles and Methods of Sheet-metal Fabricating* (New York: Reinhold Publishing Corporation, 1951).
4. W.M. Stocker, Jr., "Controlled Pressure Aids Deep Drawing," *Am. Machinist* (Dec. 10, 1951).

SECTION 15
COMPRESSION DIES

Compression or squeezing dies change the form of a metal slug or blank by plastically deforming it through the directed application of compressive forces. Metal, strained by compressive stresses, behaves like a viscous liquid. For practical purposes it is incompressible, so that a lessened volume in one direction will result if a volume is expanded in another direction. Very small increases or decreases occur in the total volume of the compressed metal depending upon its temperature, type, and condition as well as the amount of applied force. The applied force, or total pressure needed for a successful compression operation, depends upon the area to be squeezed, the extent and speed of the squeeze, and the freedom of flow of the metal. It is therefore difficult to compute working pressures. Pressures needed for coining can amount to five times the compressive strength of metal, or up to 3,000 times the Brinell hardness of the metal.

The nomograph (Fig. 15-1) may be used as an aid in estimating compression-die pressures. It is based primarily upon theoretical yield points and does not take into account degrees of restriction in metal flow (dependent on part and die contours), nor does it include allowances for resistance to flow due to strain hardening. Any given metal does not have a yield point that is fixed. The nomograph is used as follows:

Given: An area of 5 sq in. (32 sq cm.) of yellow brass, having a theoretical yield point of 40,000 psi (276 MPa).

Connect point 5 on the A scale and point 40,000 on the S scale with a straight line, intersecting P scale, giving an approximate value of 100 tons (8,896 kN) for the press capacity required.

CLASSIFICATION OF COMPRESSION DIES

Compression operations on metals may be divided into four general classifications, according to the relative total amounts of all resistances to metal flow:

1. The *sizing*, or the flattening and smoothing, of areas of forgings, castings, and stampings by squeezing the metal to a desired dimension. There is little if any resistance to metal flow, and the volume of metal moved is relatively small compared with the volume of the workpiece.

2. *Swaging* (swedging) is somewhat more severe than sizing, since the shape of the blank or slug is considerably altered as part of it flows into the contours of the die; the remaining metal is unconfined and flows generally at an angle to the direction of applied force. Compared to sizing, there is greater resistance to metal flow, although more metal is moved, as in such operations as the swaging of small gears, cams, or other small parts of irregular contour. The upsetting of heads on bolts and many cold- and hot-forging operations are classified as swaging operations.

3. *Coining* operations usually force metal to flow within a die, but not out from it, so that all work surfaces are confined. The distances through which the metal flows are comparatively short, but most or all of the metal flows to form new surface contours, thus necessitating high pressures. Embossing of sharply defined, but relatively shallow,

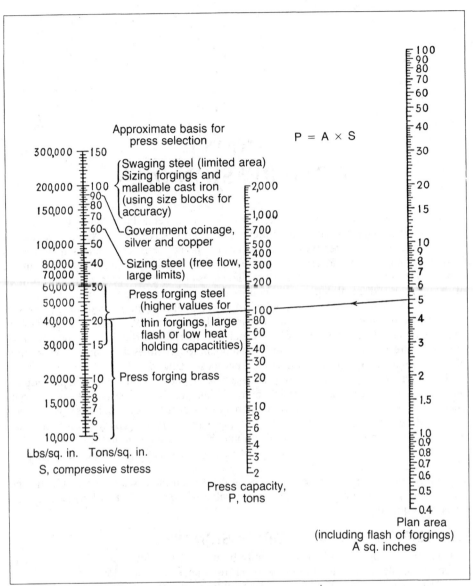

Fig. 15.1 Nomograph for determining compression die pressures.[1]

indented or raised letters, lines, or designs in thin metals with theoretically no change in metal thickness, may require only moderate pressures; this is classified as a type of coining.

4. *Extruding* operations compress and force metal to flow plastically through a die orifice, generally into a continuous length of uniform cross section.

Sizing. Surfaces of bosses, for example, on castings and forgings are squeezed to a dimensional tolerance as close as ±0.001 in. (±0.03 mm).[2] The resulting finish is comparable to a milled surface; excessive squeezing reduces surface quality.

Figure 15-2 shows a typical operation in which a casting or forging is compressed to size. The thickness to which the part is compressed is controlled by the depth of the die cavity or stop blocks placed in the die.

Blanked parts, such as the one shown in Fig. 15-3, are placed in compression dies to flatten certain areas to size. Steps machined in the surface of the die produced the

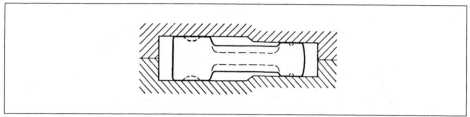

Fig. 15-2. Typical sizing die operation.[2]

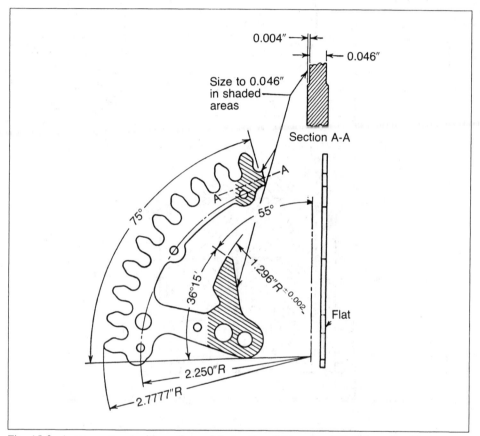

Fig. 15-3. A gear segment with portions of the surface flattened and sized in a die. (*National Cash Register Co.*)

0.004-in. (0.10-mm) offsets in the surface of the part.

Swaging. A gear-swaging die is shown in Fig. 15-4 with the part produced in the die. The flange integral with the gear is trimmed to size in a later die. To produce this part, the blank is placed over the locating pin, and the press is tripped. The punch or swaging block forces the metal to flow and fill the cavities or tooth spaces in the die. The ejector block is carefully machined to conform to the outline of the die cavity and to such a height as is required by the thickness of the gear. The press is set so that the punch bottoms hard against the die to produce uniform parts. The size of the blank in this case is not critical as long as it contains sufficient metal to produce the part, since the cavity in the punch is large enough to allow free flow of all surplus metal. A positive mechanism for ejection is built into this die, although any method of positive delayed-action ejection may be used.

Die for Swaging Type Segment. A die for swaging the bevel and two welding

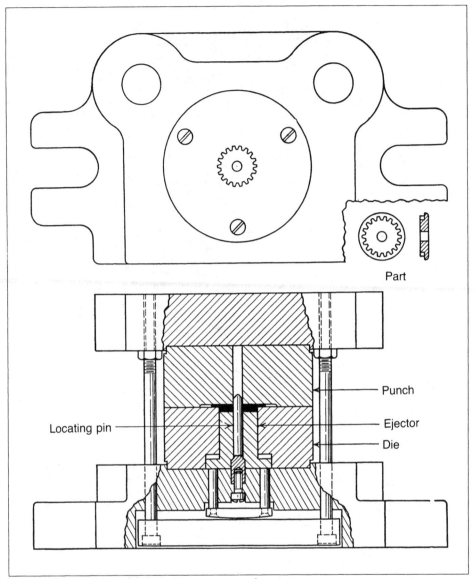

Part

Punch

Locating pin

Ejector

Die

Fig. 15-4. Die to swage a gear with integral flange.[3]

projections on a printing group-type segment reinforcing plate is shown in Fig. 15-5. The preblanked segment is placed in the die and, as the slide descends, the bevel and welding projections are swaged into the part. For ease of manufacture, the die is made in sections. The impression is machined into the insert ($D1$) which is set into the main die block ($D2$). This block is stepped to serve as a part locator. End stops ($D3$) are also fastened to the main die block. The hand-operated ejecting lever ($D4$) lifts the pin ($D5$) to eject the segment from the die cavity.

Part Swaged from Screw-machine Blank. A part which has been swaged on both sides is shown in Fig. 15-6. The blank at view A was prepared in a lathe or screw machine. The outline produced by the first swage operation is shown at view B. The plastic flow of the metal deformed the OD of the blank to some extent. The part was next semiturned in an automatic lathe to prepare a blank to be finish-swaged as shown at view C. This operation formed the gear teeth to shape, with the excess material flowing out into

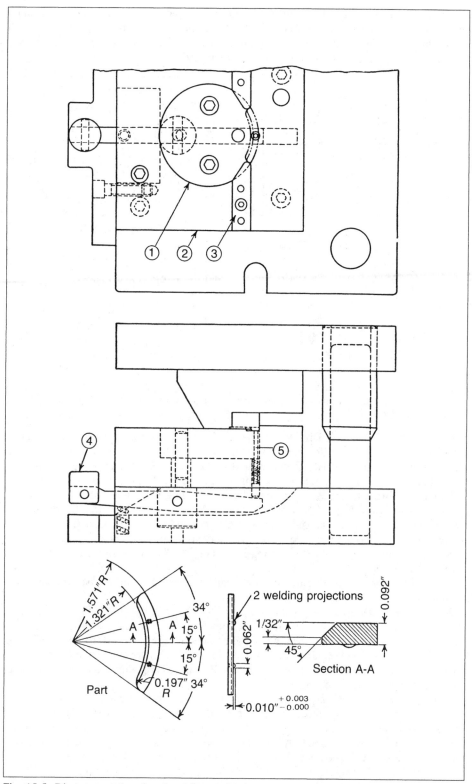

Fig. 15-5. Die to swage bevel and welding projections on a type segment. (*National Cash Register Co.*)

a scalloped outline as shown. To ensure integrally complete teeth on the gear, the blank was made with a rim on the outer edge higher than in the center, as shown by the sketch of the blank. After the final swage operation, the part was returned to the automatic lathe for finish turning and facing of the gear as shown at D.

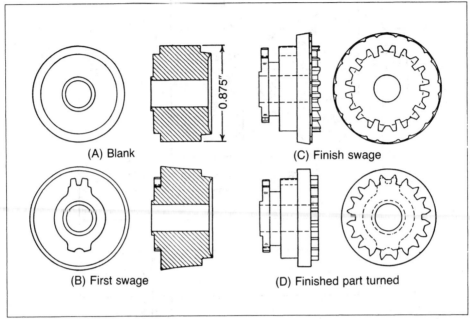

(A) Blank

(C) Finish swage

(B) First swage

(D) Finished part turned

Fig. 15-6. A mild-steel part produced from a screw-machine blank in two swaging operations. (*Pitney-Bowes, Inc.*)

Cam-operated Swaging Die. A cam-operated swaging die for a phonograph needle is shown in Fig. 15-7. A needle which has been cut to length and pointed is placed in this die to swage the flat area. The needle is held in the die by the spring-loaded hold-down (D1), while the cam (D2) forces the sliding swaging dies (D3) to close.

Universal Swaging Die. A universal die for swaging both sides of a blank in one operation is shown in Fig. 15-8. The die holders (D1) are identical so that the die inserts (D7 and D8) may be interchanged between the upper and lower positions. The die-insert shedders (D2 and D3) are actuated by positive knockouts to ensure positive ejection. The blank is positioned in the die by the spring-loaded pin (D4) which is retracted by the fixed pin (D5) as the die closes.

Timing of part ejection from the lower die can be adjusted to suit the part thickness by the threaded adapters (D6) in the upper shoe.

Swage Die for Dado Blade. The swaging of 0.125- or 0.250-in. (3.18- to 6.35-mm) -thick dado blades is accomplished in the die shown in Fig. 15-9. A 0.125-in. (3.18-mm) -thick spacer (D1) is removed from the die when swaging the 0.250-in. (6.35-mm) -thick blade. The sliding block (D2) is actuated by the cam (D3) to clamp the blade in the vertical position. The spring-actuated plungers (D4 and D5) position the blade endwise and hold it down in the die.

The fixed die blocks (D6 and D7) support the swaging punch (D8) during the operation. These blocks are recessed into a mounting plate, and this assembly is recessed into the lower die shoe. Safety stops prevent overtravel of the die. Alignment of the upper and lower shoes is maintained by two guide posts engaging long, shouldered guide bushings. The part is turned end for end in the die for swaging the second surface.

Swaging and Coining Type Bar. The single-piece type bar shown in Fig. 15-10A is made in five operations. The blank is produced from strip stock in a conventional blanking

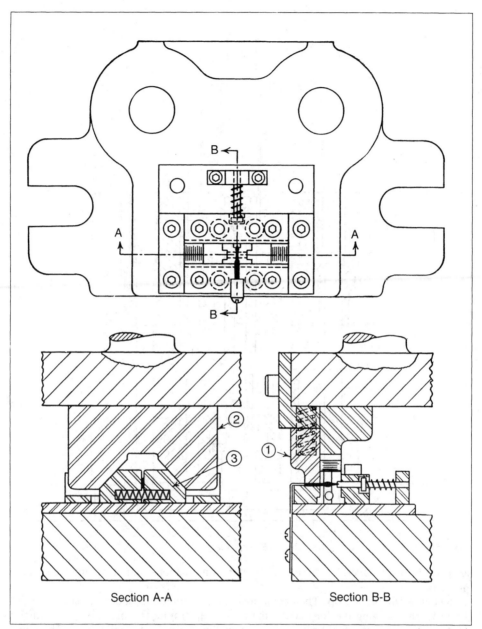

Section A-A   Section B-B

Fig. 15-7. Double swaging die for a phonograph needle. (*Harig Mfg. Corp.*)

die in the first operation. The notched section of the blank is the area in which the characters are formed and required considerable experimentation to develop. The blank is 0.050 in. (1.27 mm) thick, and the swaged surface upon which the 0.021-in. (0.53-mm) high characters are coined is trimmed to 0.124 in. (3.15 mm) wide.

The operation is performed in a split die using a spring-loaded wedge cam to actuate the sliding block shown in Fig. 15-10*B*.

In order to obtain sharp characters of uniform contour, the first operation coins the characters with an included angle of 67° and a height of 0.031 in. (0.79 mm), and the second coining operation produces characters with an included angle of 82° and 0.021 in. (0.53 mm) high. This required a set of readily interchangeable punches. The type bars are

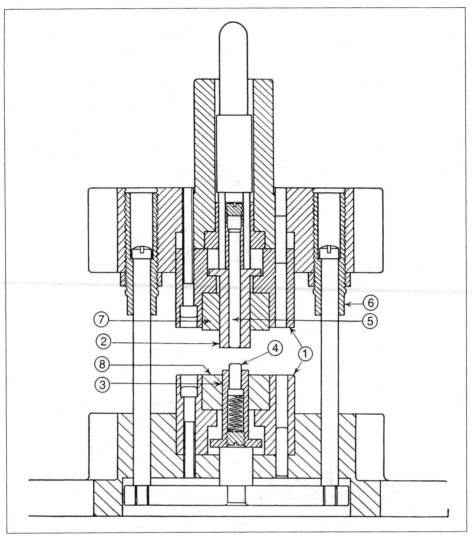

Fig. 15-8. Universal swaging die. (*Pitney-Bowes, Inc.*)

also made with several different character combinations requiring interchangeable punches.

Swaging Die for a Cup. The operations required to produce a cup are shown in Fig. 15-11A. The swaging die for making the flared cup from a flat blank is shown in Fig. 15-11B. The washer-type blank is placed over the spring-loaded pilot (D1). The punch (D2) is used to size the center hole while the die (D3) is swaging the flange to shape and thickness around the punch (D4). The knockout sleeve (D5) bottoms in the upper die at the end of the downstroke to flatten the bottom of the cup. The ironing of the side wall of this part is described in Section 10 and the coining of the bottom later in this section.

Dies for Producing Generator Pole Shoe. The group of dies shown in Fig. 15-12 produces the part as shown. It is first bent from a flat blank. In the bending of the part, depressions were produced at the junction of the wings and body by concentrated pressures.

The final operations were: (1) prebend the blank (view A); (2) pierce and countersink one side of the screw hole; (3) swage the wings surrounding the body of the part (view B); (4) coin the center section to thickness (view C); and (5) drill and tap the pierced hole.

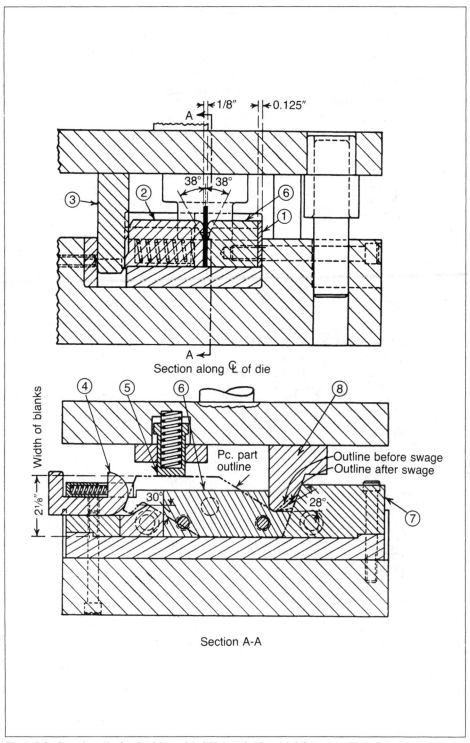

Fig. 15-9. Swaging die for 0.125- and 0.250-in. (3.18- and 6.35-mm) -thick dado blades. (*Harig Mfg. Corp.*)

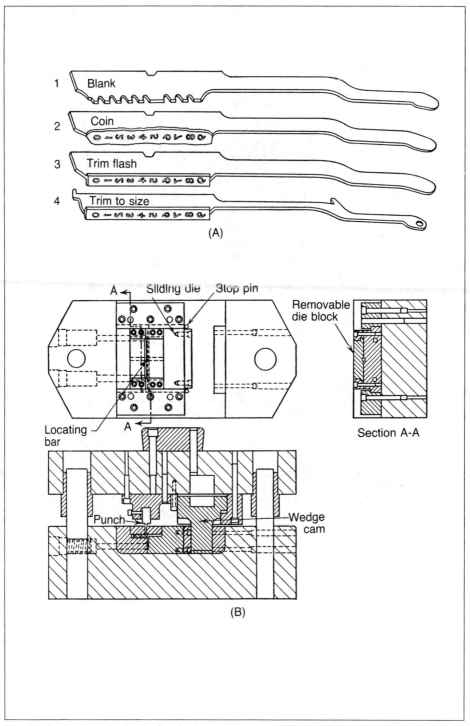

1 Blank

2 Coin

3 Trim flash

4 Trim to size

(A)

A — Sliding die   Stop pin

Locating bar   A —

Removable die block

Section A-A

Punch   Wedge cam

(B)

Fig. 15-10. Swage and coin die for single-piece type bar. (*Monroe Calculating Machine Co.*)

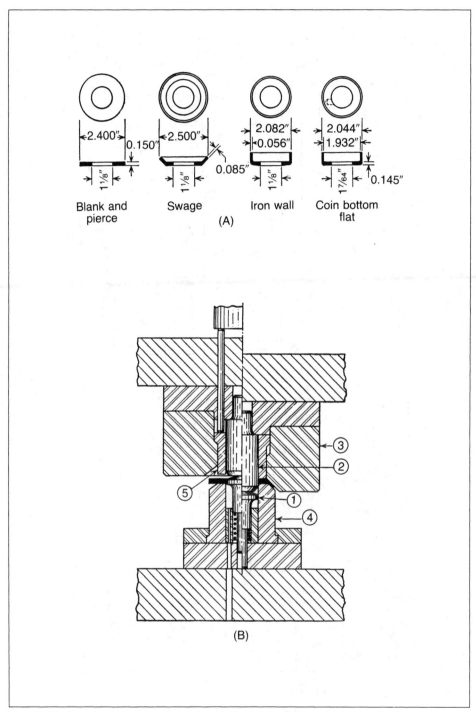

Fig. 15-11. Part development and swaging die for thin-walled part. (*Coinex, Inc.*)

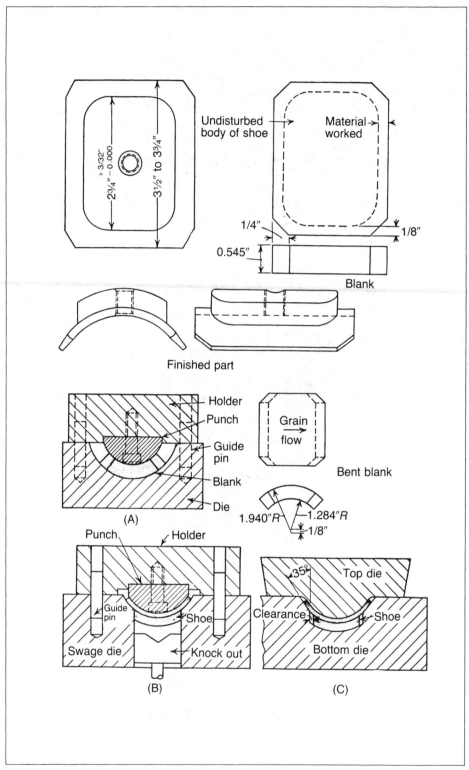

Fig. 15-12. Dies for producing generator pole shoe: (*A*) preform blank; (*B*) swage wings; (*C*) coin center section to thickness. (*Delco-Remy Division, General Motors Corp.*)

COINING

Coining is a very severe pressworking operation, since the flow of the metal is entirely confined in the die cavity. Portions of a blank or workpiece may be coined: corners of previously formed or drawn cups may be built up or filled in, or indented or raised sections of a blank may be formed by coining dies.

Die to Coin Indent. The die to coin the indent in the bottom of a cup (Fig. 15-11A) is shown in Fig. 15-13.

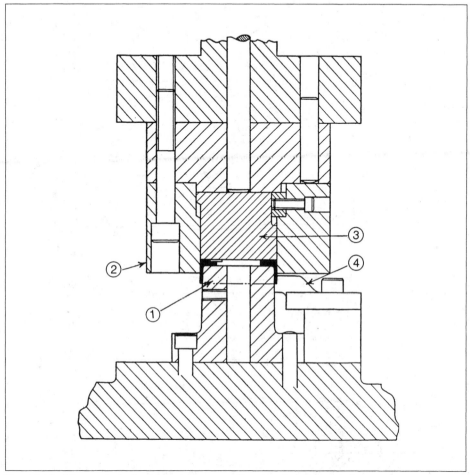

Fig. 15-13. Die to coin indentations in bottom of a cup. (*Coinex, Inc.*)

The part is placed on the post (*D*1), and the coined area is confined by the die ring (*D*2), while the coining punch (*D*3) indents the bottom of the cup. The coining punch is designed to slide up and down inside the die ring and is actuated by the positive knockout bar to eject the part from the die. The locator (*D*4) properly positions the part in a previously cut notch. The parts of this die are made of oil-hardening tool steel mounted on a standard two-post die set.

Die to Coin Serrations. The coining of serrations on both sides of a 0.060-in. (1.52-mm) thick part is shown in Fig. 15-14. The serrating dies are inserted in hardened-steel plates. A nest on the lower die locates the part accurately while it is being serrated. The depth of the serrations is controlled by two safety stop blocks. To assist in the removal of the part after the operations, a small pick-off slot is machined in the lower die block. The insert mounting block on the upper shoe is machined to the contour of the

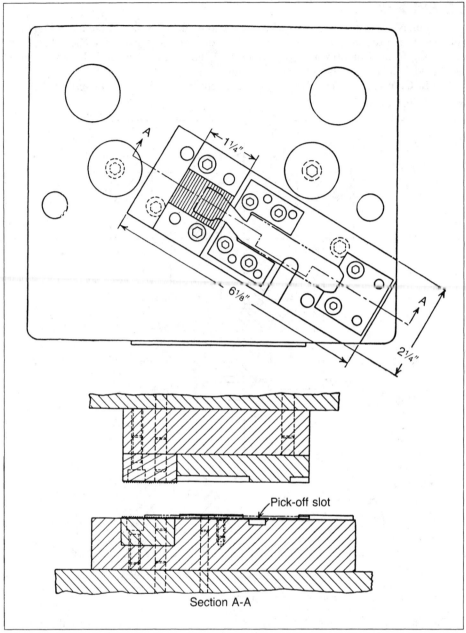

Fig. 15-14. Die to coin serrations on both sides of a part. (*National Cash Register Co.*)

part to clear the nesting plates which are thicker than the part.

Coining Serrations in Two Operations. The part shown in Fig. 15-15 has the serrations coined in two operations. The first coining die shown upsets the rim of the segment and forms the crowned section. This area is serrated in a succeeding coining die.

The die is made in three pieces inserted into the lower die shoe. The center section is movable in order to eject the workpiece from the die. The die for coining the serrations is of similar construction, the punch being so designed as to coin proper-shaped serrations.

Coining Die for Pipe Union. The die for coining a part of a pipe union is shown in

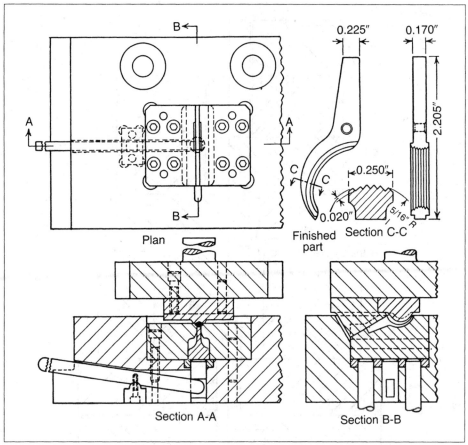

0.225" **0.170"**

2.205"

A   A

B   B

C   C

0.250"

0.020"   5/16" R

Plan   Finished   Section C-C
part

Section A-A   Section B-B

Fig. 15-15. Part coined in two operations and the first-operation coining die. (*Barth Corp.*)

Fig. 15-16. The preliminary operations produce the sleeve from 0.249-in. (6.32-mm) thick and 6.5-in. (165-mm) wide hot-rolled, pickled, and annealed SAE 1008 stock. The 400-ton (3,600 kN) coining pressure required to produce this part calls for a heavily constructed die.

The left-hand portion of the illustration shows the die open, with the ejector in the "up" position and a blank in place. The closing of the die allows the ejector plug to slide down, forming part of the cavity into which the metal in the blank is forced to flow. The right-hand portion shows the die closed with the part formed. The heavy ring surrounding the die on the lower shoe also acts to confine the portion of the die attached to the upper shoe, preventing it from expanding during the operation and producing out-of-tolerance parts.

Die for Coining an Expansion Disk. The part shown in Fig. 15-17, view *A*, prior to being fabricated in dies, had been produced as a screw-machine part at a very high cost and with considerable difficulty.

The first operation is a conventional combination blanking and drawing operation. The second, a swaging operation, reshapes the flange and puts a small chamfer around the outside edge to help prevent a fin from forming in the following operation. Operation 3 coins the part to fill out the corner and to produce the initial taper on the inside of the flange. The die for this operation is shown in view *B*. The forming punch (*D*1) is recessed into the lower die shoe. The forming ring (*D*2) coins the metal into the contour of the punch, and the knockout (*D*3) bottoms at the end of the stroke against a hardened plate inserted in the upper shoe, to form the top of the part.

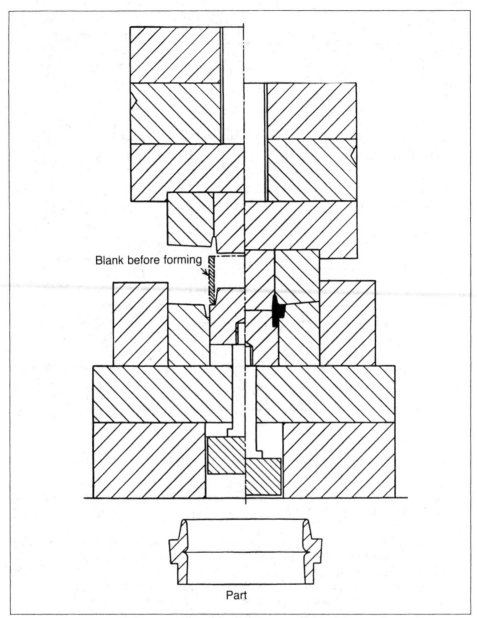

Blank before forming

Part

Fig. 15-16. Die for coining part for pipe union. (*Worcester Pressed Steel Co.*)

In the final coining die shown at view *C* (operation 4), the flange is produced as well as coining the other dimensions to size in the confined die space. The die cavity is made in two pieces with the center part (*D*4) acting also as an ejector. The top flange is made oversize in this die to ensure a fully formed part, and the excess material is removed in a trimming die. The part is made of 0.053/0.057-in. (1.35/1.45-mm) thick cold-rolled steel annealed to Rockwell B70.

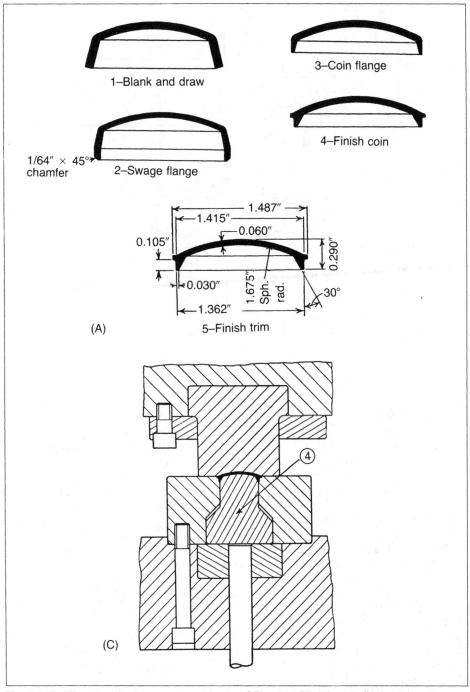

1–Blank and draw

3–Coin flange

1/64″ × 45° chamfer

2–Swage flange

4–Finish coin

1.487″
1.415″
0.060″
0.105″
0.290″
0.030″
1.675″ Sph. rad.
30°
1.362″

(A)

5–Finish trim

④

(C)

Fig. 15-17. View *A*, coined cup produced in die of Fig. 15-17*B*; View *C*, final coining die.

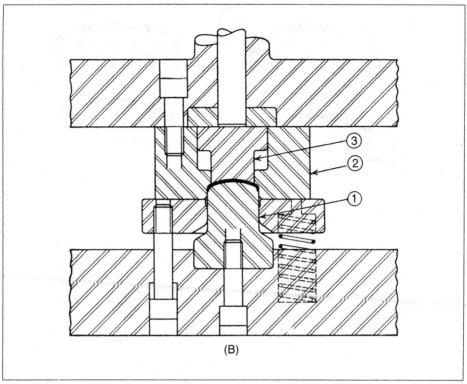

Fig. 15-17. *Continued*. View *B*.

EMBOSSING

An embossing die forms raised letters, areas, or designs in relief on the surface of sheet-metal parts. Embossing differs from forming in that the designs are comparatively small or shallow, and usually in the nature of relief work upon a surface.

Embossing differs from coining in that the latter usually has different designs on each side of the part. An embossment has the same design on both sides, one being the reverse of the other; that is, one side h⌐ ⌐he depressed design and the other side the raised design.

There is a local stretching and compressing of the metal in all embossing operations; the amount depends upon the design and how much it extends above or below the surface.

Dies to Emboss Bowl of a Coffeepot. The forming of four small rectangular, one larger rectangular, and one circular embossment in the bottom of a shell is done in the die in Fig. 15-18. The open end of this coffee pot bowl is smaller in diameter than the diametral position of the four small rectangular embossments, thus requiring that their embossing punches be retractable so that the shell may be placed in working position. These embossing punches (*D*1) are pivoted and expanded into position by the spring-loaded center portion of the die (*D*2). Because of their size and shape, the bottoms of the die cavities for these embossments are inserts (*D*3). The embossing punches (*D*4 and *D*5) for the other two embossments are also inserts in the upper portion of the die. The position of the embossments and the relative location of the section through the die are shown in the bottom view of the part.

A die for embossing another area in the coffeepot bowl is shown in Fig. 15-19. This operation embosses water-level marks in the side of the coffeepot while it is supported on a horn. The embossing and stamping punches are inserted into the punch-holder plate and retained by setscrews.

The above-described dies perform operations on the lower bowl of a coffeepot while the die in Fig. 15-20 embosses a circular bead in the upper bowl of the same coffee maker.

15-18

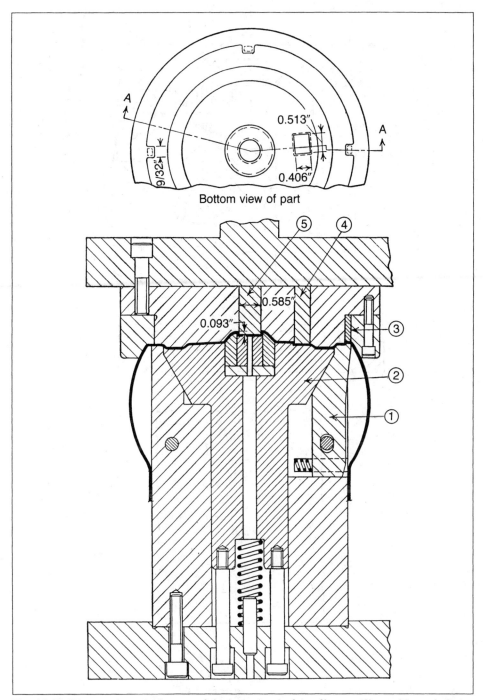

Bottom view of part

Fig. 15-18. Die to emboss the bottom of the lower bowl of a coffeepot. (*Knapp-Monarch Co.*)

The bowl is placed upside down in the die and is positioned by the center locating pin while the bead is being formed.

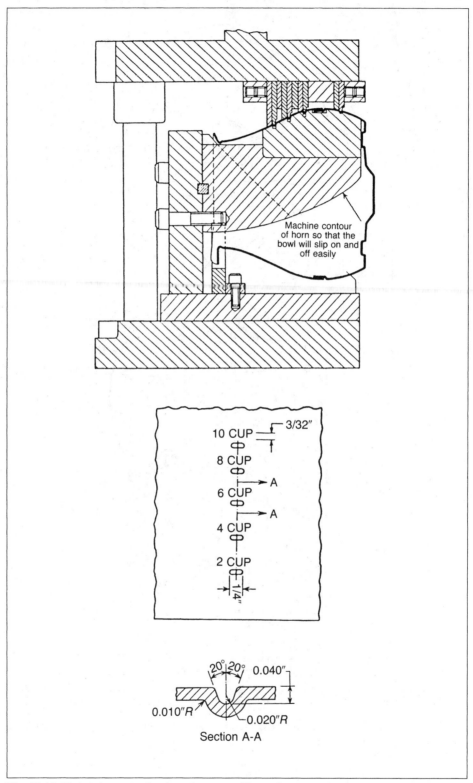

Fig. 15-19. Embossing water-level marks on a coffeepot lower bowl. (*Knapp- Monarch Co.*)

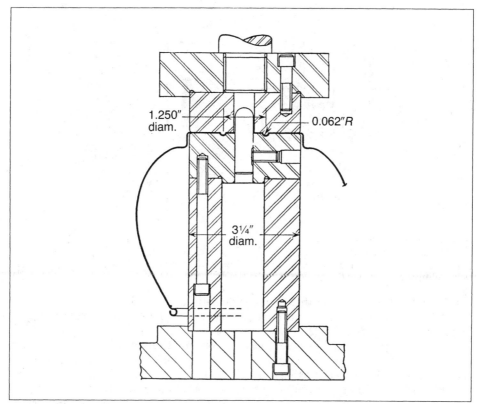

Fig. 15-20. Embossing die for a circular bead in the bottom of an upper bowl of a coffeepot. (*Knapp-Monarch Co.*)

EXTRUDING

One of the most severe of press operations is the cold-extrusion method of shaping metal. Articles produced by this method range from collapsible tubes and continuous shapes in the softer metals to heavy artillery shells, cold-formed from steel billets. The cross-sectional shapes which can be cold-extruded are square and rectangular as well as round shapes. The so-called extruded items are often a product of a series of operations which may combine coining, backwards and forward extruding, ironing, and embossing. The success of cold extruding depends upon product design, raw materials, lubrication, tool design, and heat treating, either individually or in combinations. What appears to be a lubrication failure can be due to other factors such as impractical product design, tooling, annealing, or defective raw materials.

Less expensive metals may be used for cold extrusion, because of the excellent strength developed in the low-carbon steels from the high deformations attained in the process. There is a considerable saving of raw material, since almost all of the metal present in the original blank is present in the final extruded article. Additional savings can be realized by using blanks from hot-rolled bar stock instead of from hot-rolled plate.

An extruding process is defined as either forward or backward extrusion, corresponding to the direction of metal flow with respect to the direction of the applied force (Fig. 15-21). Backward-extruding dies force metal to flow between the punch and die in a direction opposite to the direction of the force on the punch. Forward-extrusion dies force the metal to flow ahead of the punch through the die orifice.

Any lack of symmetry in the side walls of a shell creates lateral pressures which force the punch out of alignment. For this reason, ribs or similar designs on the inside or outside walls of a shell should be symmetrical. Bosses, indentations, or cavities on the inside or

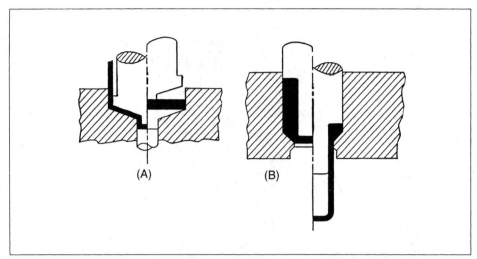

Fig. 15-21. Schematic representation of extrusion process: (A) impact or backward extrusion; (B) uniform pressure or forward extrusion.

outside of the bottom of cup shapes, not symmetrically located, create irregular metal flow which tends to force the punch out of alignment. Typical shapes illustrating these points are shown in Fig. 15-22.

Reductions up to 85% of cross-sectional area on cold steel billets by extrusion in one operation have been successfully made. Under normal conditions, it is good practice to limit the first reduction to approximately 60%, and subsequent operations to about 40%. The average hardness of steels before extrusion should not be over Rockwell B60.

The pressure required in extrusion work depends upon the yield point of the material before and after extrusion, and the varying amounts of friction or resistance to flow depending upon the size and contour of the die. Higher operating speeds create additional

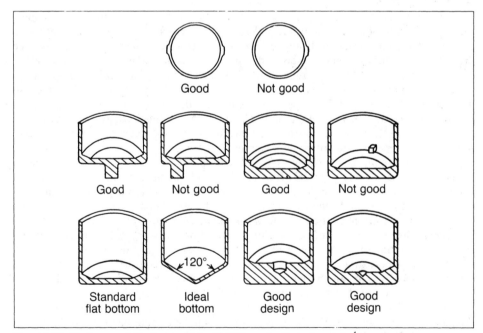

Fig. 15-22. Design considerations which affect the extruding of metals.[4]

loading on the tools. When there is little restriction to flow, and speed is low, the press load can be based on the yield point after work hardening. The pure metals can be worked at lower pressures than the alloys. Table 15-1 lists extrusion pressures for common metals. The pressures listed cover a considerable range depending upon the alloy, its microstructure, the restrictions to flow, and the severity of work hardening.

The pressures required for extrusion also depend upon the percentage of reduction in area. For parts made of various aluminum alloys, when the percentage of reduction of area is known, the required pressure may be taken from Fig. 15-23. Although the alloys mentioned are usually extruded at room temperature, a reduction in press pressure can be achieved by extruding at elevated temperatures. Closely related to, and a factor in establishing, reduction of area is part wall thickness. The increasing effect on punch pressure required as the wall becomes thinner is illustrated in Fig. 15-24.

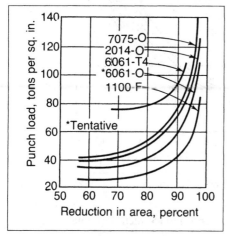

Fig. 15-23. The effect of reduction in area to the punch load in extruding.[5]

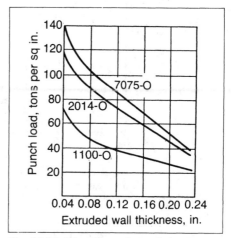

Fig. 15-24. The effect of extruded wall thickness on the punch load.[5]

TABLE 15-1
Extrusion Pressures for Common Metals[1]

Material	Pressure	
	psi ($\times 10^3$)	MPa
Pure aluminum, extrusion grade	80–140	552–965
Brass, soft	60–100	414–689
Copper, soft	50–140	345–965
Steel C1010, extrusion grade	100–330	689–2,275
Steel C1020 (spheroidized)	120–400	827–276

The relationship between reduction of area and extrusion pressures for a series of plain carbon steels is shown in Fig. 15-25. The steels referred to in the different curves have carbon content in the range of 0.05 to 0.50%, and less than 0.03% each of sulfur and phosphorus. Steel 11 contained 0.58% chromium, 0.11% carbon, and 0.36% manganese, and 0.03% each sulfur and phosphorus.

Correct lubrication in extrusion work considerably lowers the pressures required. Ordinary die lubricants break down because of the high pressures and excessive surface heat attending extruding operations. A bonded steel-to-phosphate layer overlaid with a bonded phosphate-to-lubricant coating is satisfactory for the heat and pressures encoun-

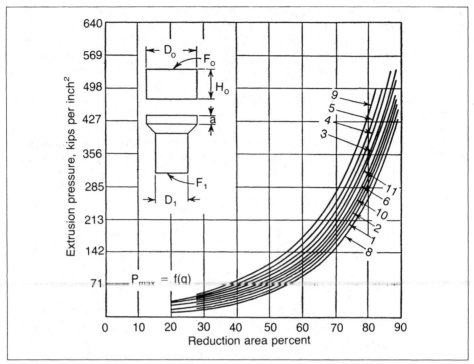

Fig. 15-25. Extrusion pressures: reduction relationships for the forward extrusion of a series of steels with carbon contents in the range 0.005 to 0.50%.[6]

tered. The phosphate layer serves as a parting layer to reduce metal-to-metal welding or pickup, and as a carrier for the lubricant.

Tool Design. The magnitude of the forces in extruding is very large, since the workpiece is stressed to its elastic limit in compression, resulting in lateral pressure on the die walls which is in addition to the pressure caused by the actual plastic deformation. Under these high pressures, there is danger of local collapse and plastic deformation of the tool surfaces. To prevent this, the surface hardness must be approximately Rockwell C60.

Resistance to bursting is frequently achieved by shrinking the tool-steel die ring into a massive die shoe so that the ring is normally in compression. These compressive forces must be overcome before the die ring is stressed in tension.

Various tool steels are suitable for extrusion dies, the choice depending partly upon the required die life. A steel which, in heat treating, acquires a hard wear-resisting case supported by a softer but strong and tough core is useful for cold-extrusion dies. Since there is a possibility of extruded parts leaving the die at a temperature of nearly 550°F (290° C), the steel should resist softening due to tempering.

Careful heat treating of the die elements is necessary because high working stresses are encountered. Proper grinding wheels and procedures must be used to avoid grinding checks and surface cracks that will become more prominent when the part is stressed in use. Die steels that are especially sensitive to grinding cracks and checks should not be used.

The length of a backward-extrusion punch is governed by the punch diameter and the yield strength of the workpiece material. The practical limit of the length of the punch for extruding steel should be about three times its diameter, four times the punch diameter for 7075-0 aluminum, and six times for 6061-0 aluminum. These proportions may be increased by using guide bushings. The die cavity should be 0.250 to 0.50 in. (6.35 to 12.7 mm) longer than the desired part, since the extrusion does not come out with uniform length and may require trimming or sizing. It is not necessary that the blank closely fit the die cavity; it should be of such a shape as to make it impossible to place it off center and thus lose concentricity in the finished part.

A typical backward-extrusion die is shown in Fig. 15-26, in which the metal flows in the opposite direction to punch movement. The carbide insert and its ring are tapered in the holder which is built up of two members shrunk together. The taper should be about 1° per side. The holder ring requires an adequate number of holddown screws to prestress the carbide ring in order to minimize its expansion and fatigue failure. The carbide insert and its ring are supported by toughened steel plates to distribute the high local loads. The extruding punch ($D1$) is guided by a spring-loaded guide plate ($D2$) which is positioned by piloting in a ring ($D3$) on the lower die. Ejection of the part from the die is by a delayed-action stripper which lifts the bottom portion of the die cavity.

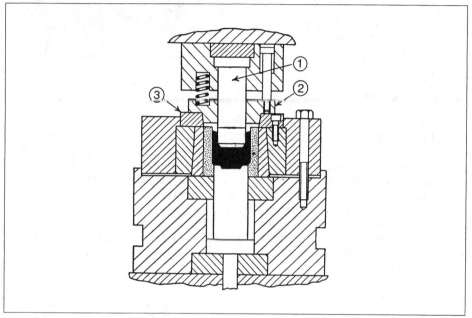

Fig. 15-26. A typical backward-extrusion die with a carbide insert.[7]

A backward-extrusion die with the extrusion die ring extended to serve as a guide for the punch is shown in Fig. 15-27. The punch has a tapered shank and is retained in the upper shoe by a bushing within a threaded nut. This design facilitates interchangeability and reduces die costs.

The outside retainer ring of cast or forged steel is shrunk on the OD of a hardened insert to provide some compression on the insert. The hardened insert is ground to a 1° to 2° taper per side on the ID corresponding to the taper on the OD of the die ring. The retainer assembly is pulled tightly over the die ring by means of clamps or cap screws to compress the die ring. This compression preloads the die ring so that it can better withstand the forces tending to expand and crack the die.

Included in the illustration are enlarged views of the extrusion die and punch. The striking or working end of the punch is developed as a cone with a side angle of 5° to 7° for steel, and 1° to 2° for aluminum. The corners have small radii of 0.016 to 0.047 in. (0.40 to 1.19 mm), and the punch is relieved above a short bearing to assist in part removal.

The radius at the outer edge of a punch should be kept as small as possible because a large radius tends to create a wedging action. The inside corner radius (on parts requiring a large outside radius) may be reduced by using an angular fillet of a height equal to the radius on the outside bottom of the shell and at an angle of 15° from the bottom.

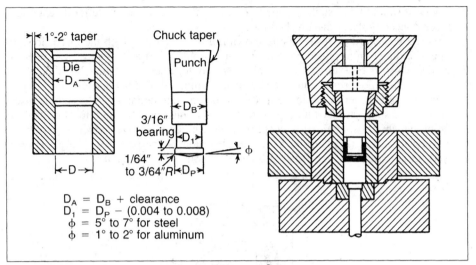

Fig. 15-27. Backward-extrusion die for forming a cup.[8]

A forward-extrusion die, in which the metal flows in the same direction as the punch travel but at a faster rate because of the change in cross-sectional area of the part, is illustrated in Fig. 15-28. The stroke of the press is of sufficient length to allow a preformed slug to be inserted in the nest ($D1$) above the die ring ($D2$). This nest also serves as a guide for the punch during the operation. The die ring and a secondary guide ring ($D3$), added to help produce a straight part, are encased in a pair of tapered rings ($D4$). These rings are retained and clamped to the lower shoe by an outer ring ($D4$) so that only the inner ring is bearing on the lower shoe. This mounting ensures that the die and guide rings will always be held in a shrunk-in state. The taper of the mandrel is calculated to suit the elongation and the desired taper on the inner wall of the extrusion.

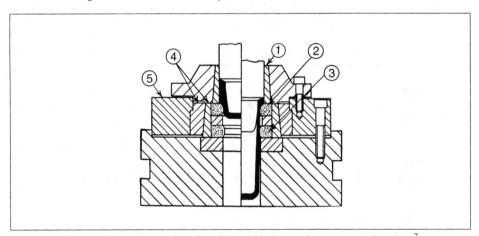

Fig. 15-28. Typical forward-extrusion die with carbide inserts in a compression ring.[7]

The forward-extrusion-die insert in Fig. 15-29 is placed in a retainer assembly similar to the one shown in Fig. 15-27. The punch can be attached to a taper shank with a through bolt so that a similar upper shoe may be used. The punch shown is made of a lower portion and an upper portion; the lower portion is subject to wear. The diameter D_1 of the die insert is such that it properly positions the preformed cup or slug. The bearing diameter D produces the desired OD of the extruded section. Below this bearing, the die is relieved to reduce frictional resistance, and a guide is placed on the lower end to maintain

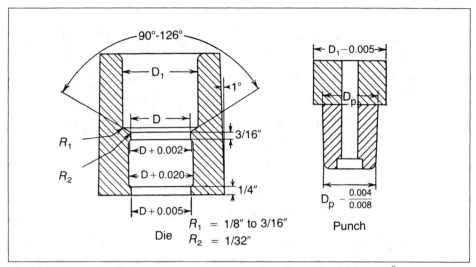

Fig. 15-29. Extrusion-die insert and punch elements for a forward-extrusion die.[8]

straightness as the extruded section flows out ahead of the punch.

The radius R_1 should be fairly large to aid metal flow and carry lubrication around this point. The included angle of the shoulder should be between 90° and 126°. Although the wider die angles increase tonnages to some extent, they are of great advantage in reducing radial forces and in maintaining lubrication at the work areas. The die angle also influences the amount of work hardening imparted to the work; the wider the included angle, the larger the amount of work hardening. The radius R_2 is very small. The bearing is short and has a slight back rake to provide clearance for the extruded metal.

The stepped dies in Fig. 15-30A and the die with small included angle in view B impose limitations on the amount of cold work practical in one operation. The top bearing of a stepped die cannot be relieved with back rake because of the compressive forces imposed by the approach to the second step; the metal would tend to upset into the relief area, making stripping impossible. The principles of good design are shown in view C.

Two methods of obtaining stepped diameters on a part by forward extrusion are shown in Fig. 15-31. At view A is a slug with a diameter of D_1. It is placed in a die and it is forward-extruded to diameter D_3. The workpiece is then placed in another die using the same punch and it is extruded to diameter D_2. In view B a slug is "sized" to the diameter

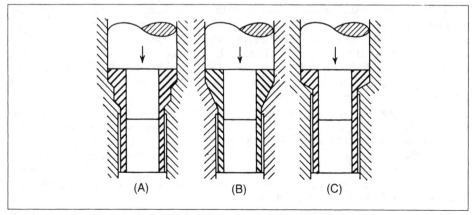

Fig. 15-30. Shoulder designs for forward-extrusion dies: (A) stepped die, not recommended; (B) long approach, not recommended; (C) short approach, good design.[8]

15-27

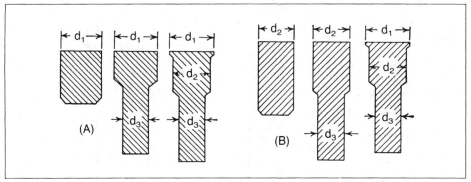

Fig. 15-31. Two methods of obtaining stepped diameters by forward extrusion.[8]

D_2 in a coining die, and is then placed in an extruding die to form the diameter D_3. The final operation upsets the head to diameter.

Extruding Die for a Generator Frame. The die in Fig. 15-32 forward-extrudes a portion of an automotive generator frame. The blank for the frame has a tongue and slot for locking purposes. After extrusion, the seam does not open, thus eliminating a conventional welding operation. After the 0.375-in. (9.53-mm) thick stock is grit-blasted to remove scale and provide pockets to hold lubricant, the cylindrical blank is produced in five operations on a transfer press. The operations are: cutoff, notch, form to an oxbow shape, V-form, and finish form.

This die is constructed with a floating arbor which allows it to move forward with the flow of the metal, which may be faster than the speed of the press stroke. The punch in this die surrounds the arbor and forces the metal to flow through the carbide extrusion ring. Since the punch is attached to the stripper assembly, it acts to strip the part from the arbor, and a knockout attached to the die cushion ejects it from the die cavity.

Impact Extrusion of Magnesium Alloys. Certain magnesium alloys can be impact extruded. The extrusion process is basically the same as for other metals except that heating elements must be added to the die. Depending on the alloy, the extrusion temperature will vary from 450° to 750° F (230° to 400° C). Because of the additional working of the metal, the properties of impact extrusions normally average slightly higher than those for the material from which the blank was made.

The die for impact extruding magnesium should have highly polished cavities as well as the punch surface. The selection of the proper tool steel for the punch and die is governed by the run expected.

The side wall of the die cavity should have a draft of approximately ⅛ degree. This prevents the impact extrusions from sticking in the cavity. In normal operating procedures the part adheres to the punch and is stripped from it on the upward stroke.

The success of impact extruding magnesium alloy depends largely on a uniform thickness of a lubricating film on the blank. If the film is uneven, or is left off part of the blank a product with uneven wall thicknesses will result. Tumbling or spraying of heated magnesium slugs with colloidal graphite has been successful. Dipping slugs has not proved successful because of uneven lubricant films.

One advantage of magnesium alloy impact extrusions is that they can be finished after the forming operation. The impact extrusion is a wrought product readily accepting any chemical finish and can be anodized or electroplated.

Since the impact extrusion process is a high-speed process, automatic heating is required for the blank. This can be accomplished by adding a resistance heating element to the die chute or feed bottom. Gas or electric radiant heat can be adapted to a small oven, which is integral with the feed mechanism.

In addition to heating the blanks for forming, the dies must be heated to approximately the same temperature. The dies may be heated electrically or by gas heaters. Die cavities

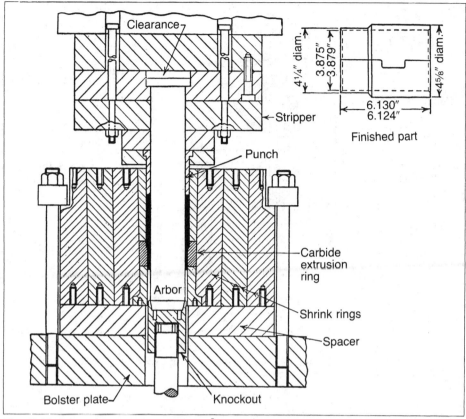

Fig. 15-32. Die to extrude generator frame.[9] (*Delco-Remy Division, General Motors Corp.*)

must be made oversize to compensate for part shrinkage upon cooling.

Figure 15-33 illustrates the use of heating elements for the die sections and insulating materials to protect the press. Typical extrusion pressures are listed in Table 15-2.

TABLE 15-2
Impact Extrusion Pressures for Magnesium Alloys*
psi ($\times 10^3$) (MPa)

Alloy	Temperature, °F (°C)						
	450 (230)	500 (260)	550 (290)	600 (320)	650 (340)	700 (370)	750 (400)
AZ31B	66 (455)	66 (455)	60 (414)	54 (372)	52 (358)	50 (345)	46 (317)
AZ61A	70 (482)	68 (469)	66 (455)	64 (441)	62 (427)	60 (414)	58 (400)
AZ80A	72 (496)	70 (482)	68 (469)	66 (455)	64 (441)	62 (427)	60 (414)
ZK60A	68 (469)	66 (455)	64 (441)	62 (427)	58 (400)	54 (372)	52 (358)

* Values based on 85% reduction in area.

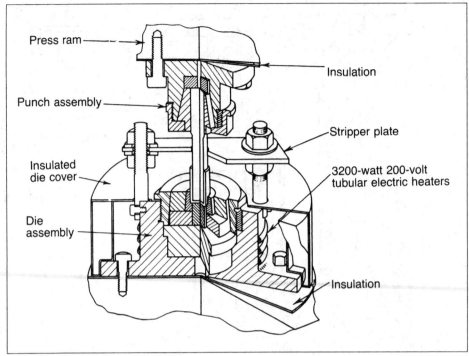

Fig. 15-33. Sketch of tooling setup for the production of magnesium impact extrusion. (*The Magnesium Association.*)

References

1. *Bliss Power Press Handbook*, E.W. Bliss Co., Toledo, Ohio, 1950
2. E.V. Crane, *Plastic Working in Presses*, 3rd ed. (New York: John Wiley & Sons, Inc., 1945).
3. K.L. Bues, "Gear Swaging Die," *Western Machinery & Steel World* (May 1949).
4. *Alcoa Aluminum Impact Extrusions*, Aluminum Co. of America, 1949.
5. J.D. Shoemaker, "How to Make Impact Extrusions of High-strength Aluminum," *American Machinist* (July 6, 1953).
6. D.V. Wilson, *Metallurgical Requirements of Steel for Cold Extrusions*, Sheet and Strip Users Association, 1953.
7. *Computations for Metalworking in Presses*, E.W. Bliss Co.
8. J.F. Leland and J.W. Helms, *The Influence of Proper Lubrication on the Design of Cold Extruded Components* (Society of Automotive Engineers, Inc., 1954).
9. R.L. Kessler, W.A. Fletcher, and W.P. Bowman, "How Delco-Remy Cold Forms Metals," *American Machinist* (July 20, 1953).

SECTION 16
PROGRESSIVE DIE DESIGN

A progressive die performs a series of fundamental sheet-metal operations at two or more stations during each press stroke in order to develop a workpiece as the strip stock moves through the die. Each working station performs one or more distinct die operations, but the strip must move from the first station through each succeeding station to produce a complete part. One or more idle stations may be incorporated in the die to locate the strip, facilitate interstation strip travel, provide maximum-size die sections, or simplify their construction.

The linear travel of the strip stock at each press stroke is called the "progression," "advance," or "pitch" and is equal to the interstation distance.

The unwanted parts of the strip are cut out as it advances through the die, and one or more ribbons or tabs are left connected to each partially completed part to carry it through the stations of the die. The operations performed in a progressive die could be done in individual dies as separate operations but would require individual feeding and positioning. In a progressive die, the part remains connected to the stock strip which is fed through the die with automatic feeds and positioned by pilots with speed and accuracy.

When parts are made from individual blanks moved from die to die by mechanical fingers in a single press, the dies are known as transfer dies.

Selection of Progressive Dies. Whenever the total production requirements for a given stamping are high, a progressive die should be considered. The savings in total handling costs by progressive fabrication compared with a series of single operations may be great enough to justify the cost of the progressive die.

The present application of computer-aided die design, together with the general use of wire burn EDM for making die sections, has greatly simplified the designing and construction of progressive dies.

The quality of stampings made on progressive dies is often higher than that produced on individual dies. There is less chance for off-gage conditions due to part locating problems. The human factor has less influence on part quality. Often the savings in labor costs together with the more consistent quality of progressive die stampings have been the deciding factors in justifying the greater material cost of coil stock over producing automotive stampings from offal in recovery dies.

The fabrication of parts with a progressive die should be considered when:

1. Stock material is not so thin that it cannot be piloted.

2. Coil handling equipment, stock straighteners, and feeders are available or can be justified.

3. The overall size of the die, which is determined by part size, number of stations, and strip length, is not too large for available presses.

4. The total press tonnage capacity required is available.

5. The press is level and in good condition. Problems with worn gearing, loose gibbing, and worn bearings can result in alignment problems that can damage precision tooling.

6. Quality and part consistency requirements are high.

7. Quick die change and flexible manufacturing requirements exist.

STRIP DEVELOPMENT FOR PROGRESSIVE DIES.[*]

The individual operations performed in a progressive die are usually simple, but when they are combined in several stations, the most practical and economical strip design for optimum operation of the die requires careful analysis. The sequence of operations on a strip and the details of each operation must be carefully developed to ensure that the design will produce good parts without production or maintenance problems. The following strip development sequence is applicable to both manual and computer-aided design.

A tentative sequence of operations should be established and the following items considered as the final sequence of operations is developed:

Step 1. Analyze the part.
    a. What is the material and thickness?
    b. What hole dimensions—size and location—are critical?
    c. What surfaces are critical?
    d. What forms are required?
    e. Where can carriers be attached?
    f. Is direction of material grain important?

Step 2. Analyze the tooling required.
    a. What production is required per month, per year, total?
    b. What presses are available? It is important to know bolster area, shut height, feed height, strokes per minute, air cushion, etc.
    c. What safety conditions must be met?

Step 3. Make dummy drawings. A part dummy drawing shows the finished part and all positions in which the part will be formed in order to achieve the final form. Provide for proper metal movement for drawing or forming operations.
    a. Tips for dummy drawings:
        1. Show the complete part.
        2. Show all positions necessary to form part in plan and elevation views.
        3. Show over-bend positions if they are critical.
        4. Show all necessary views to achieve clarity.
        5. Show work lines and set up lines.
        6. Provide vertical and horizontal center lines for measuring when assembling the strip layout.
        7. Show the strip carriers, if known, before strip layout is assembled.
        8. Trace dummy from part print, if part print is dimensionally accurate (less time and less errors).
        9. Use design aids:
            a. Wax for sample parts and carriers.
            b. Rubber skins.
            c. Plastic skins.
            d. Models.
      10. Check accuracy.

Step 4. Make strip layout. A strip layout for a progressive die is a series of part dummy drawings marked up to indicate the die operation to be performed in each station of the die. How to construct a strip layout:
    a. Determine the proper progression for the part.

* Arnold Miedema, President, Capitol Engineering, and Edwin A. Stouten, Retired Vice President, Capitol Engineering, contributed to this section.

b. Tape on a drawing board a series of prints of the plan view of the part in die position using the horizontal center line for alignment.

c. Apply clear tape over prints in order to hold prints together after removal from the drawing board.

d. Mark all die operations on each station that will be performed in the die. (Use red color pencil for cutting operations and green or blue for forming operations.)

e. Mark operations directly on prints or use overlay-sheet for preliminary layout, but mark directly on the prints for final layout to prevent something being missed during die design.

Step 5. Discuss proposed process with another person to check for errors. Use a checklist when analyzing the preliminary strip layout.

a. What stations can be eliminated by combining with another station?

b. Are good die steel conditions maintained?

c. Does movement of part between stations require a stretch web?

d. Are idle stations provided to permit "breathing" of strip, if stretch web is not feasible?

e. Provide for pitch notch(s), if possible, to maintain proper progression.

f. Avoid sight stop for first hit, if possible.

g. If possible, pierce in first station and pilot in second station to establish pitch control.

h. Provide adequate pilots for all subsequent stations.

i. Is there room for stock lifters to permit free flow of strip during feed?

j. Are close tolerance holes pierced after forming to eliminate development of hole location?

k. Use an overlay sheet to run a simulated strip through the die to check each operation and to spot any loose pieces of scrap which might be left on the die.

Step 6. Draw plan views by tracing proper part positions for each station from the strip layout.

Step 7. Draw the plan of the punch over the plan of the die to permit tracing as much as possible and to reduce scaling errors.

Step 8. Make views and notes to communicate properly to the die maker (Remember: to assume is to blunder).

Step 9. Problem areas to watch:

a. Part lifters.

b. Part gages.

c. Part control—pilots, etc.

d. Pad travels.

e. Scrap ejection.

f. Part ejection.

g. Poor die steel conditions.

h. Will the die fit the press?

i. Will the die fit production requirements?

Step 10. Receive preliminary design approval.

Step 11. Finish design layout.

Step 12. Detail as required.

COMPUTER-AIDED DESIGN AND MACHINING[*]

The main advantage of computer-aided progressive die design and machining is the ability to build precision tooling in less time and at a lower cost. Integrating the part and

[*] Jeffrey L. Smolinski, Engineering & Technologies Manager, LA-Z BOY Chair Co.

die design process with the generation of cutter path information also greatly reduces the chance for errors.

Traditional Design Procedure. In order to understand the full ramifications of being able to cut tool steel directly from an electronic drawing data base, the major faults of traditional methods need to be examined. Below is an outline of the major tasks that occur in order to move a progressive die project from conception to machine:

1. An engineered part drawing is given to the die designer to begin work on the development of a new die.

2. The die designer then begins the detailed task of laying out the design of the progressive die that will in the end produce the finished part that is defined by the part drawing.

3. A detailer takes the finished die design and produces the detailed drawings of each of the die parts, i.e., punches, punch holders, die steels, etc.

4. The finished details are then passed on to an NC programmer who determines the tool path coordinates of each line, circle, and arc that defines the die part geometry.

5. The NC programmer then writes out the complete scripted program including all tool-path coordinates and any "G" or "M" codes that will be required by the machine to perform such tasks as offset cutting compensation, feed-rate settings, coolant on/off switching, etc.

6. The written program must then be typed directly into the machine controller, or into an input device that will transfer the cutting instructions to the controller, or onto a medium such as paper tape that will be read by the machine controller.

7. The machine is now ready to cut the die part that will be required to manufacture the designed part from step one.

Error Sources. Close examination of each step of the die design procedure as outlined above clearly shows that there are many opportunities for errors to be introduced into the machine instructions. During the course of steps one through seven, the original design data of the part has gone through five transformations.

The first transformation occurs when the die designer transfers the dimensional data from the new part drawing to the drawing board when developing the design of the new die. It is in this transformation that most geometric errors occur. Complex shapes can make calculations of offset arcs and angular lines very difficult when designing clearances between punches and die steels. Such a miscalculation can lead to scrap parts and unworkable dies.

Transformation of the data happens again, when the detailer takes information from the die design drawing and begins drafting the die part details. Even if the die design had been created flawlessly, there is still a good chance of an error occurring in the drawing of the die part, as the detailer reads the design drawing and transfers the drawing data for the second time.

The third transformation of the drawing data begins as the NC programmer starts to calculate all of the endpoint coordinates of the lines and arcs of the geometry on the die part drawings. In most NC applications, accuracy can be carried out to three or four decimal places. Complex shapes, again, may make for some very difficult calculations of precise endpoint or tangency point locations. Miscalculation of a single point could mean the difference between a workable and a nonworkable die.

Transformation number four occurs when the NC programmer transfers the coordinate information from the drawing to a written program.

Example of Error Generation. The most frequent occurrence of errors happens in step four. To understand why errors occur easily, consider the amount of data that must be entered into a CNC machine tool program just to machine ten progressive die sections.

On each detail drawing, there are 100 points that define the tool path for the section, and each of these points must be written into the NC program. Each point consists of an X-coordinate and a Y-coordinate, and precision is to four decimal places. An example of a point on the drawing might be:

$(X, Y) = (1.3574, -3.3468)$

and the written line in the NC program might look like this:

G01 X13754 Y-33468

Notice that two numbers in the NC program have been transposed, creating an error. The transposing of numbers can occur in the misreading of a point coordinate from the drawing or the miswriting of a point coordinate in the program, and the possibility of error increases with the greater number of points that have to be transferred. The number of point coordinates that must be transferred is as follows:

(10 drawings) × (100 points) × (2 coordinates) = 2,000

and the total number of digits would be:

(2,000 point coordinates) × (5 digits) = 10,000

In order to machine ten die sections without error, 10,000 number digits must be read and written without transposing a single one of them.

In addition to the possibility of transposing the point coordinate digits, there is still another potential for error in transformation number four. Besides the point coordinate information, the NC machine is going to need some other instructions in order to properly cut the part. These instructions are generally handled in the "G" and "M" codes that are written into the NC program along with the point coordinates. The codes tell the machine such things as which side of the line to offset to compensate for cutter kerf, or which direction to travel on an arc, i.e., clockwise or counterclockwise. The misplacement or misuse of one of these commands could produce erroneous cutting instructions.

For example: "G02" tells the cutter to travel along an arc in a clockwise direction and "G03" in a counterclockwise direction. If the NC programmer were to inadvertently use a "G02" where a "G03" should be used, the cutter would travel off the desired tool path.

The fifth and final data transformation is the keypunching of the NC program into the machine controller or any other input device that serves as the communicator of the cutting data to the NC machine. The possibility of error here is obvious, as all the lines of the written NC program, including all point coordinates, machine codes, and cutting comments, have to be keyed in. Errors once again may be a result of the transposing of digits, either in reading or writing the point coordinates or by inadvertently striking the wrong key while inputting the data.

In any one of the above five described data transformations, there is a chance for error. Some errors may be detectable before machining, others may not show up until the part has already been cut wrong. In either case errors expend both energy and time, and may result in the production of unsalvageable die details. Reduction of the number of data transformations will greatly reduce the possibility of error.

Integrating CAD with CNC Machining. One proven solution to the reduction in data transformations is the integration of computer-aided design and CNC programming. The concept-to-machine procedure can be simplified through the use of computer-based aids. For example:

1. A CAD engineered part drawing is given to the die designer to begin work on the development of a new die.

2. The die designer electronically copies the geometry of the part drawing onto his or her die design drawing and adds any additional geometry to complete the die drawing.

3. The detailer electronically dissects the die design drawing into the necessary drawings of each of the die parts.

4. The finished CAD detail drawings of the die parts are then tool-pathed by an NC programmer.

5. A NC post-processor program on the CAD system automatically transforms the tool-path information into the NC program.

6. The NC program is then electronically uploaded to the machine controller.

7. The machine is now ready to cut the die part that will be required to manufacture the designed part from step one.

The reason one can be confident that the program in the machine is correct is that the program itself is generated from the data that originated in NC form in step one. Through all of the steps, the exact same CAD data that was used in the part design is used in the die design.

When the die designer specifies die clearances for punches, it is not necessary to calculate offsets for complex shapes with arcs and lines at odd angles. The only instruction that the computer requires is the offset clearance dimension.

The detailer no longer has to transfer the dimensional data of the die design to create die section drawings, but simply pulls the geometry and dimensional data off the die design drawing electronically. The detail drawing is also a by-product of the same geometry that defines the part drawing.

Calculation of the endpoint coordinates is virtually nonexistent for the NC programmer, because the endpoints are automatically defined to the CAD system, by the presence of geometry. The NC programmer needs only to define to the CAD system, the order of the coordinates he or she wants programmed, and does this simply by tracing a tool path on the CAD drawing itself by selection the geometry in the order of desired cutting. ''G'' and ''M'' codes can be added to the tool path at this time as well, and some commands will be automatically determined by the computer.

Since there is no manual writing of the program, there is very little opportunity for the errors to occur that were possible with the manual programming methods. A program, which can reside as part of the CAD system, post-processes the NC tool path information, and generates the NC program automatically and electronically.

Finally, keypunching the data into the controller or any other input device is eliminated. The electronic NC program can be directly uploaded to the NC machine controller, without the keying in of a single point coordinate or machine code.

Cost Effectiveness. The following are actual times at one company before and after the CAD and CNC integration for wire EDM machining:

	Task Description	Time
	Drawing of the die design	16 hrs
BEFORE	Detailing of ten parts	20 hrs
	NC point calculations	10 hrs
	NC programming	10 hrs
	NC program keypunching	10 hrs
		66 hrs
	Task Description	**Time**
	Drawing of the die design	8 hrs
AFTER	Detailing of ten parts	5 hrs
	NC tool path defining	5 hrs
	NC postprocessing	1 hrs
	NC program uploading	1 hrs
		20 hrs

No allowance has been made for the savings resulting from machining error reduction. If records of the cost of machining errors is available, this data can be used to further justify the cost of integrating CAD with a CNC machining system. Shortening the cycle time for getting a progressive die from the design concept into production also enhances flexible manufacturing capability.

GOOD DESIGN PRACTICES

Designing for Strong Die Blocks. Figure 16-1 illustrates the use of a three-station die to avoid weak die blocks. At *A*, the pierced hole is near the edge of the part where it is cut off, thereby weakening the die block at this point. If an idle station is added so that the piercing operation is moved ahead one station, the die block is stronger and there is less chance of breaking. At *B*, the pierced holes are centered on the strip but close together. In this case the holes should be pierced in two stations to avoid thin sections in the die block between the holes. The addition of an idle station also permits a larger, more robust punch retainer to be used.

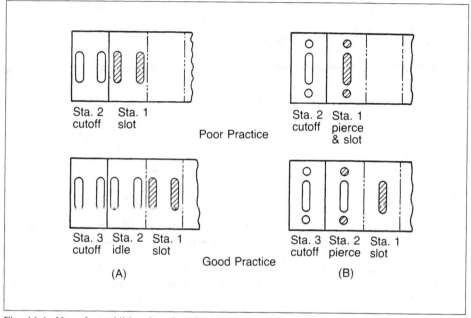

Fig. 16-1. Use of an additional station to avoid weak die blocks: (*A*) pierced hole close to the edge of part; (*B*) pierced holes close together.

Layout for a Flanged Channel. The layout of the strip for a flanged channel shown in Fig. 16-2 illustrates several of the points to be considered in preliminary layouts of progressive dies. A hole is pierced in station 1 which can be used in any succeeding stations as a pilot hole. Notching the strip with two punches in the first station and trimming the rounded edges of the part in the second station avoided the use of a delicate J-shaped punch. The first forming operation is done in station 5, with two idle stations on each side to allow the strip to drop below the level of the other stations to form the flanges upward. Final forming is done at station 8 and a shear cutoff is achieved at station 9. The carrier strip is through the center of the strip development.

The need to have the grain across the width of the stock strip means that the die cannot be fed from a coil and that strips no longer than available coil width must be fed by hand. This layout is not recommended when reproduction requirements are high. An improvement in steel quality may permit the use of coil stock with the grain rotated 90°.

Forming U-channels. The strip development for a channel-shaped part, and a section through the die, is shown in Fig. 16-3. Slots are cut in station 1, of sufficient length to allow the flanges of two adjoining parts to be formed at one time in station 3. The strip

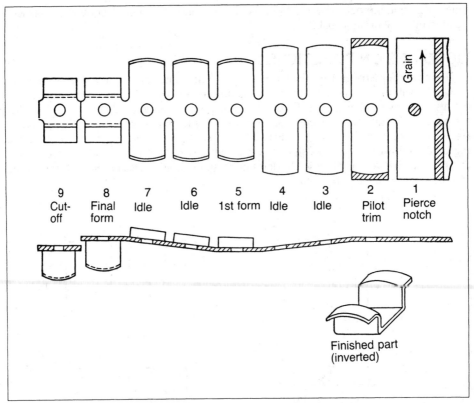

Fig. 16-2. Strip development for a flanged channel. (*C. R. Cory.*)

is lanced and one flange partially formed at one time in station 2. The part is cut from the two side carrier strips in station 4. The use of a lance rather than a notching punch conserves material.

Part Orientation Can Affect Die Cost. The layouts of Fig. 16-4, using wide and narrow stock widths, illustrate the use of oblong commercial punches as compared to special heeled punches. Commercially available piercing punches should always be considered where possible to avoid the cost of special punches. Figure 16-4A shows a part made from a wide stock strip using two special heeled punches, and the same part made from a narrow strip using a purchased oval punch. In both cases a shear-type cutoff punch is used. Figure 16-4B shows a wide strip using two heeled rectangular notching punches and a slug-type trim and cutoff punch. The narrow strip development for the same part uses a square piercing punch, heeled trimming punches and a shear-type cutoff punch. Costs of various stock widths and the greater number of pieces that can be produced between coil changes may outweigh the savings of using commercially available punches.

Providing for Carrier Strip Distortion. The strip development for drawing in progressive dies must allow for movement of the metal without affecting the positioning of the part in each successive station. Figure 16-5 shows various types of cutouts and typical distortions to the carrier strips as the cup-shaped parts are formed and then blanked out of the strip. Piercing and lancing of the strip around the periphery of the part as shown at A, leaving one or two tabs connected to the carrier strip, is a commonly used method. The semicircular lancing as shown at B is used for shallow draws. The use of this type of relief for deeper draws places an extra strain on the metal in the tab and may cause it to tear. The carrier strip is distorted to provide stock for the draw. A popular cutout for fairly deep draws is shown at C. This double lanced relief suspends the blank on narrow ribbons, and no distortion takes place in the carrier strips. Two sets of single rounded

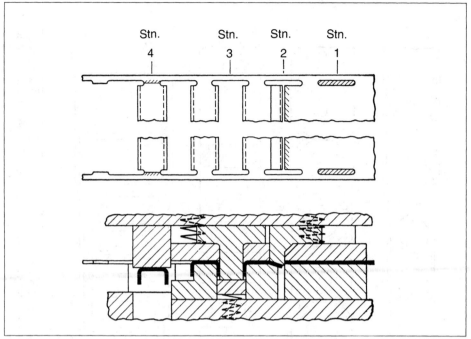

Fig. 16-3. Channel forming progressive die. (*C. R. Cory.*)

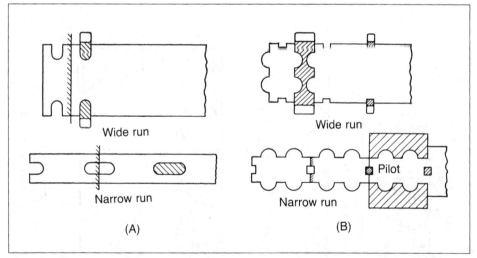

Fig. 16-4. Strip layouts for using commercial punches: (*A*) oval punch; (*B*) square punch.

lanced reliefs of slightly different diameters are placed diametrically opposite each other to produce the ribbon suspension. The hourglass cutout in *D* is an economical method of making the blank for shallow draws. The connection to the carrier strips is wide, and a deep draw would cause considerable distortion. An hour-glass cutout for deep draws is shown in *E*, which provides a narrow tab connecting the carrier strip to the blank. The cupping operations narrow the width of the strip as the metal is drawn into the cup shape.

The hourglass cutout may be made in two stations by piercing two separate triangular-shaped cutouts and lancing or notching the material between them in a second station. The cutouts shown at *F* and *G* provide an expansion-type carrier ribbon that tends to straighten out when the draw is performed. These cutouts are made in two stations to

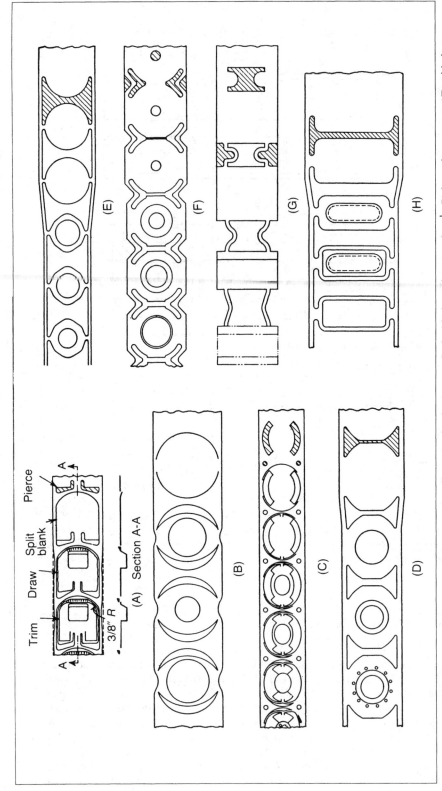

Fig. 16-5. Cutout relief for progressive draws: (A) lanced outline; (B) circular lance; (C) double-lanced suspension;[1] (D) hourglass cutout; (E) wide base hourglass cutout; (F) cutout providing expansion-type carrier ribbon for circular draws; (G) cutout providing expansion-type carrier ribbon for rectangular draws;[2] (H) I-shaped relief for rectangular draws.

allow for stronger die construction. Satisfactory multiple layouts may be designed using most of the reliefs by using a longitudinal lance or slitting station to divide the wide strip into narrower strips as the stock advances. The I-shaped relief cutout in *H* is a modified hourglass cutout used for relatively wide strips from which rectangular or oblong shapes are produced.

Straight slots or lances across the stock are sometimes used on very shallow draws or where the forming is in the center portion of the blank. On the deeper draws, this type of relief tends to tear out the carrier strips or cause excessive distortion in the blank and is not recommended.

Round Button Construction. Figure 16-6 illustrates the use of round buttons and punches for all piercing and forming stations. Many of the punches and buttons shown are standard commercially available items.

An advantage of this system is that pre-hardened, tool steel button blanks can be kept on hand and quickly finished by means of wire-burn EDM as needed.

Multi-operation Layout. A layout of the sequence of operations for producing an actuator bracket is shown in Fig. 16-7. In the first station the center hole is pierced. The hole is used to pilot the strip for each successive operation. In the second station, a circular stretch flange is formed around this hole to part print dimensions, and the strip is notched to a partial outline of the left leg. In the third station similar notching for the right leg and cutting of three slots in the left leg occur. In station 4, similar slots are cut out and the central portion, as well as the completed outline of the left leg. The right leg outline is completely cut out and the central portion is trimmed to length at station 5. The bulges in the sides are formed up and tabs on the center section are formed down in the sixth station. In station 7 the part is severed from the strip, the carrier tab on the left side is cut off, and the legs are formed downward.

The die for this part is made of high-carbon, high-chrome steel with hardened backup plates for the punches and die blocks. An air-operated ejector removes the part from the form block. The die operates at a speed of 75 strokes per minute producing about 150,000 parts per sharpening.

Push Back into Carrier Strip. The nature of the forming or the size or shape of a part sometimes requires that the blank be completely severed from the strip before forming. The blank is then pushed back into the strip so that it can be advanced properly to the succeeding stations. A strip development of this type is shown in Fig. 16-8. This is sometimes referred to as a cut-and-carry operation.

The strip is pierced and notched in the first two stations and the blank is severed from the strip in station 3; then, by the action of a spring-loaded pressure pad, it is pushed back into the strip. In station 4, the blank is spanked to flatten and secure it into the strip. The first forming is done at station 5, and finish forming and removal from the strip are achieved at station 7. Station 6 is idle to add strength to the die. In addition to the piloting hole, a notch is cut along the edge in the first station. The purpose of the notch is to engage a locator which operates a limit switch. The switch is connected to the electrical circuit controlling the press clutch to prevent press operation if the strip is improperly positioned.

The strip development and part drawing for a camera diaphragm plate are shown in Fig. 16-9. The stock is 0.020-in. (0.51-mm) thick by 1.875-in. (47.62-mm) wide 6061-T6 aluminum. The strip development allows the periphery of the part to be trimmed in several steps to simplify punch shapes and also to prevent concentration of stresses which would tend to warp the workpiece. This arrangement facilitates fabrication of the die sections. The dies for the triangular-shaped cutouts are round inserts made in two pieces for ease and accuracy of grinding.

Avoiding Pitch Growth. During normal die operations, die components are deflected and compressed as defined by the laws of physics. For example, a 1.0 in. (25.4 mm) steel cube subjected to a force of 14.5 tons (129 kN) will compress 0.001 in. (0.02 mm), or one part per thousand. This is based on a modulus of elasticity of 29,000 ksi (200 MPa). The

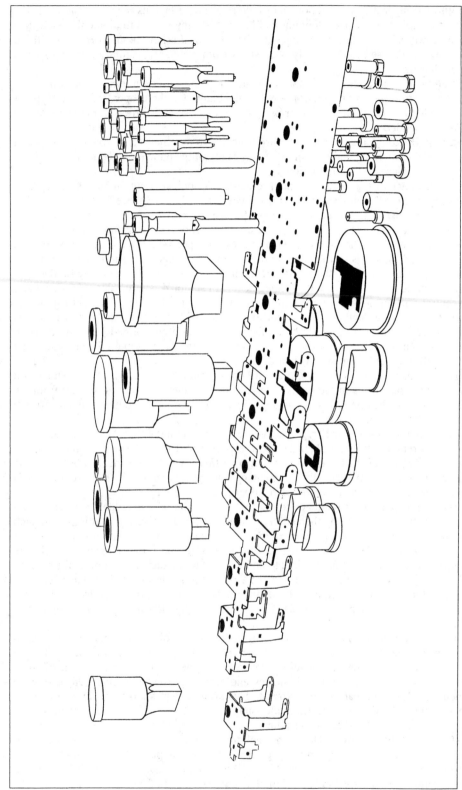

Fig. 16-6. Round button construction maximizes utilization of commercially available components. (*Dayton Progress Corp.*)

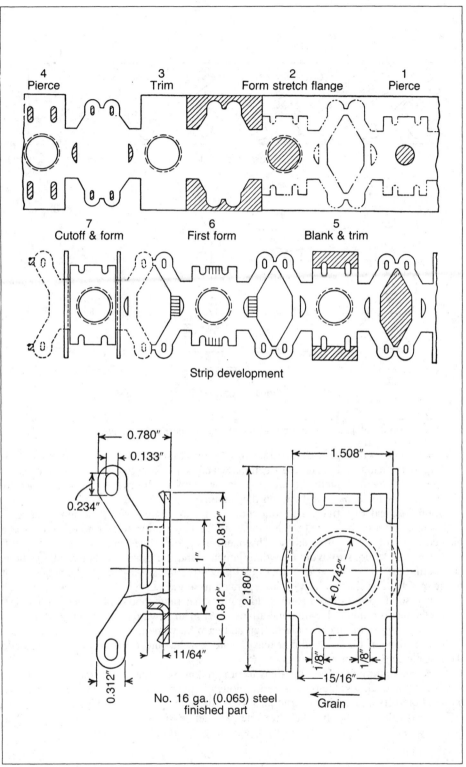

Fig. 16-7. Strip development and part drawing for actuator bracket. (*The Emerson Electric Mfg. Co.*)

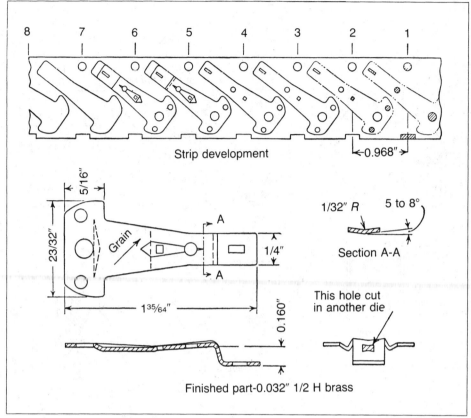

Fig. 16-8. Strip development with a push-back blank. (*White-Rogers Electric Co.*)

sides of the cube will expand 0.0003 inch (0.008 mm) based upon a Poisson's ratio of 0.3.

This size change of steel die sections when subjected to pressure can cause a troublesome growth of pitch. Figure 16-10 shows how pitch growth can be lessened by providing slots for expansion between the stations.

Good General Practices. The following are good progressive die design practices.

1. Pierce piloting holes and pitch notches in the first station. Other holes may also be pierced that will not be affected by subsequent noncutting operations.

2. Use idle stations to avoid crowding punches and die blocks together. Idle stations also permit the use of larger die blocks and punch retainers. An added advantage is that future engineering changes can be incorporated at low cost.

3. Use solid spring-loaded stock guide rails rather than spool lifters where possible.

4. Plan the forming or drawing operations either in an upward or a downward direction, whichever will assure the best die design and strip movement.

5. The shape of the finished part may dictate that the cutoff operation should precede the last noncutting operation.

6. Locate cutting and forming areas to provide uniform loading of the press slide. If this is not practical, and the press is large enough to permit the die to be offset, determine the required offset and have instructions to the diesetter placed on the die.

7. Design the strip so that the scrap and part can be ejected without interference. The best way to eject the part is to cut it off and drop it through the lower die shoe.

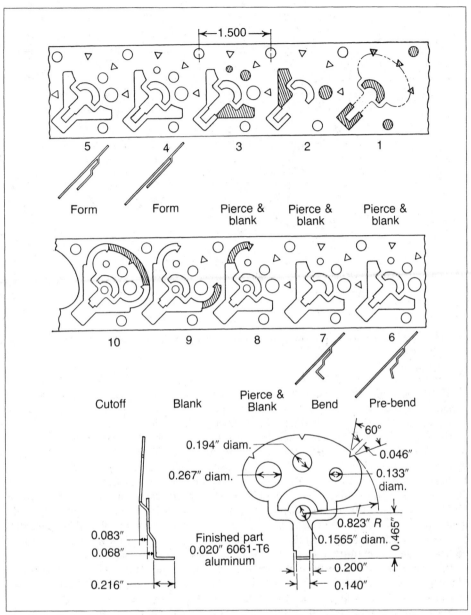

Fig. 16-9. Strip development for a die to pierce, form, and trim a diaphragm plate. (*Argus Cameras, Inc.*)

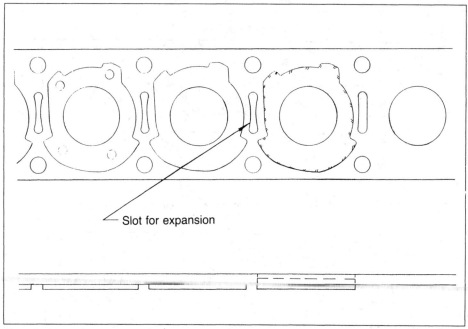

Slot for expansion

Fig. 16-10. Providing a slot for expansion to avoid pitch growth when cutting heavy stock.[3] (*Capitol Engineering Co.*)

PILOTS

Since pilot breakage can result in the production of inaccurate parts and jamming or breaking of die elements, pilots should be made of good tool steel, heat-treated for maximum toughness and to hardness of Rockwell C57 to 60.

Retaining Pilots. Pilots are usually retained by conventional heads as well as standard ball lock retainers. Several other methods are shown in Fig. 16-11. A threaded shank, shown at view A, is recommended for high-speed dies; thread length X and counterbore Y must be sufficient to allow for punch sharpening. For holes 0.75 in. (19 mm) in diameter or larger, the pilot may be held by a socket-head screw, shown at B; dimensions X and Y again must provide for sharpening. A typical press-fit type is shown at C. Press-fit pilots, which may drop out of the punch holder, are not recommended for high-speed dies but are often used in low-speed dies. Pilots of less than 0.25 in. (6.4 mm) diameter may be headed and secured by a socket setscrew, as shown at D.

Indirect Pilots. Designs of pilots which enter holes in the scrap skeleton are shown in Fig. 16-12. A headed design, at A, is satisfactory for piloting in holes of 0.2 in. (5 mm) in diameter or less. A quill is used to support the pilot shown at B.

Spring-loaded pilots are desirable for stock exceeding 0.062 in. (1.57 mm) in thickness. A bushed shouldered design is shown in Fig. 16-12C. A slender pilot of drill rod shown at D is locked in a bushed quill, which is countersunk to fit the peened head of the pilot.

Tapered slug-clearance holes through the die and lower shoe should be provided, since indirect pilots generally pierce the strip during a misfeed.

Combined Piloting and Stock Lifting. Figure 16-13A illustrates how a pilot guide bushing can be used to support the stock at the feed level to aid pilot entry. Provision should be made for slug clearance as shown at B in the event of a misfeed.

Electrical Die Protection. Figure 16-14 illustrates the use of an electrical limit switch to stop the press in the event that the spring loaded pilot ($D1$) and limit switch ($D2$) detect a misfeed. In order for this system to be successful the press must stop very quickly, which limits speeds at which this method is applicable. The pilot should be as long as

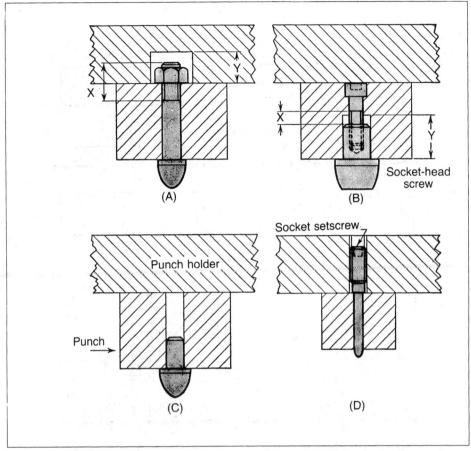

Fig. 16-11. Methods of retaining pilots: (A) threaded shank; (B) screw-retained; (C) press-fit; (D) socket setscrew.

possible to provide extra stopping time. Section 26 has additional information on die protection systems.

Using a Pilot Release. If the feeder is equipped with a pilot release, at least one pilot must be long enough to engage the stock before the pilot release is activated. Figure 16-15 illustrates the use of a slight amount of overfeed in conjunction with a pilot release system. The pitch stop shown in Fig. 16-14A provides for 0.010 in. (0.25 mm) overfeed, which is easily corrected by the pilots when the pilot release is actuated.

If a spring-loaded pilot is used in conjunction with a limit switch, the longest pilot should be so equipped. The two pilots shown at (B) provide gradual cam-like action in moving the progression strip into precise location prior to die closure.

Miscellaneous Piloting Methods. The pilot need not always be affixed to the upper die shoe. Figure 16-16 illustrates a pilot (D1) attached to the lower die shoe. A spring ejector (D5) assists in lifting the stock off of the pilot upon the ascent of the ram. Pilots may also be attached to pressure pads provided that the pads are precisely guided. A pilot (D2), attached to a pad (D3) which is guided (D4) by pins and bushings is much stronger than a longer pilot attached to the upper die shoe.

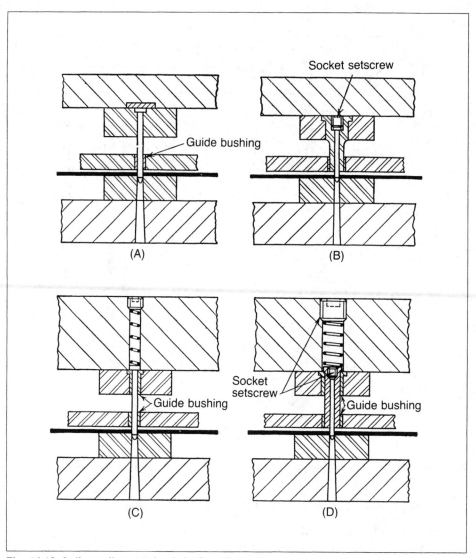

Fig. 16-12. Indirect pilots: (A) headed; (B) quilled; (C) spring-backed; (D) spring-loaded quilled.

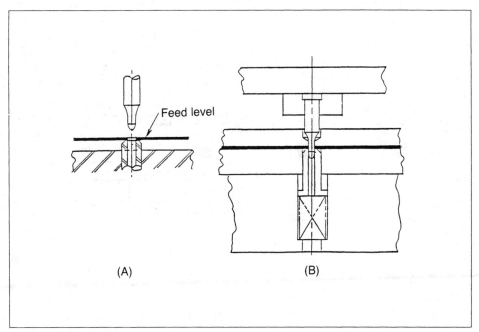

Fig. 16-13. Using a spring-loaded pilot bushing to support stock: (A) stock supported at feed level; (B) view showing spring and slug clearance. (*Capitol Engineering Co.*)

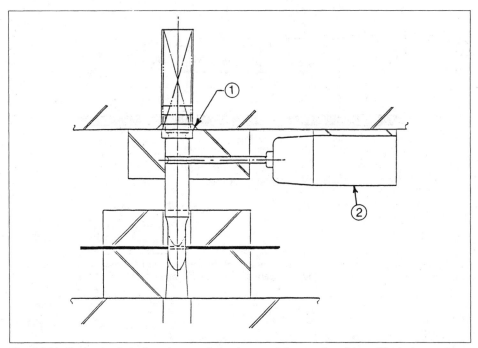

Fig. 16-14. Spring loaded pilot trips a switch when stock is out of position. (*Capitol Engineering Co.*)

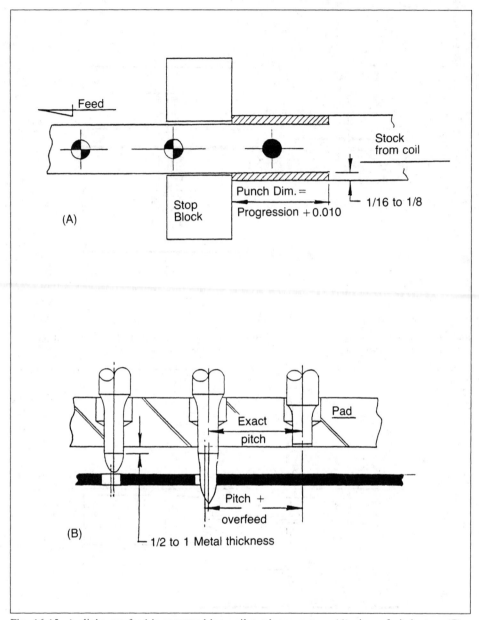

Fig. 16-15. A slight overfeed is corrected by a pilot release system: (*A*) view of pitch stop; (*B*) pilot action returns the stock to correct location. (*Capitol Engineering Co.*)

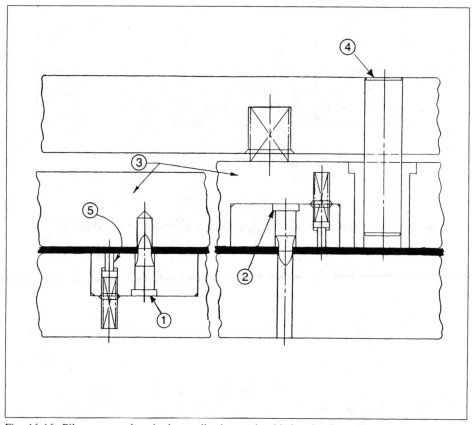

Fig. 16-16. Pilots mounted to the lower die shoe and guided pad. (*Capitol Engineering Co.*)

STOPS

Solid blocks are often used with final blanking and cutoff operations to position the end of the stock or workpiece. Solid stops may be incorporated in the final cutoff station of a progressive die to gage one end of a completed part which is attached to the advancing strip. The part is then severed to the desired dimensions by a cutoff punch. The finished part drops through the die.

Another design incorporates a cutoff punch having a heel which bears against the vertical surface of a stop, gaging the part and confining it (attached to the strip) to the station. The punch prevents the part from cocking while being sheared.

Pin Stops. One design incorporates a small shouldered pin, pressed in the die block, to engage an edge of blanked-out portions of manually fed strip stock. Since the operator must force the stock over the shoulder to secure a desired feed length, this stop is not suitable for high production or high speed dies.

Another design that may be used with a double-action blank-and-draw die, or a die in which very little or no scrap is left between blanked-out areas in strip stock, also has a small pin pressed in the die block, but the pin has a sharp edge which faces the incoming stock strip. The edge cuts through and thrusts aside the thin fins of stock left between successive blanked-out openings as the strip is fed into the die. The pin functions as a stop when it engages the trailing edge of a blanked out area before, and only before, the next succeeding area is cut out.

Latch Stops. A latch pivoting on a pin, as shown in Fig. 16-17, is held down by a spring. The latch is lifted by the scrap bridge and falls into the blanked area as the stock is fed manually into the die. It is then necessary to pull the stock back until the latch bears

16-21

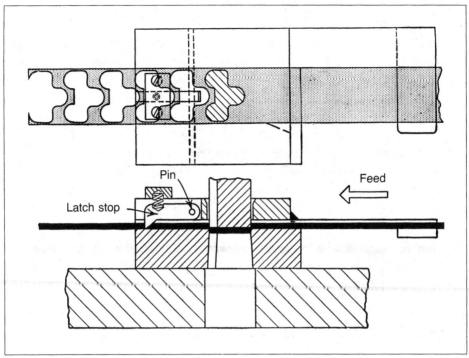

Fig. 16-17. Latch stop.[4]

against the scrap bridge. This design is suitable for low production only.

Starting Stops. A starting stop, used to position stock as it is initially fed to a die, is shown in Fig. 16-18A. Mounted on the stripper plate, it incorporates a latch which is pushed inward by the operator until its shoulder (D1) contacts the stripper plate. The latch is held in to engage the edge of the incoming stock; the first die operation is completed, and the latch is released. The stop will not be used again until a new strip is fed to the die.

The starting stop shown at view B, mounted between the die shoe and die block, upwardly actuates a stop plunger to position incoming stock. Compression springs return the manually operated lever after the first die operation is completed.

It is important to plan the starting sequence of a strip or coil to avoid damage. In some cases when a partial cut is made, excessive punch deflection caused by an unbalanced load results in a sheared die section. Figure 16-19A illustrates an acceptable starting sequence. The stock is first positioned against a stop pin (D1) and the press cycled to pierce the pilot hole (D2). The second hit does not produce loose pieces of scrap. The production of a loose piece of scrap is shown at (B).

Trigger Stops. Trigger stops incorporating pivoted latches (D1, Fig. 16-20A and B) at the ram's descent are moved out of the blanked-out stock area by actuating pins (D2). On the ascent of the ram, springs (D3) control the lateral movement of the latch (equal to the side relief) which rides on the surface of the advancing stock and drops into the blanked area to rest against the cut edge of the cutout area.

Stops for Double Runs. The stops shown in Fig. 16-21 were designed for double runs in the same direction. The stock is turned over for the second run. A conventional automatic trigger stop (D3) functions continuously after the starting stop (D2) for the first run is pushed in to engage the notch in the strip. The rough starting stop (D4) for the second run and accurate stop (D5) are actuated by handle (D1) for the cutting of the first blank of the second run; the automatic trigger stop then functions for the remainder of the run.

Another stop design for double runs in the same direction is shown in Fig. 16-22, in

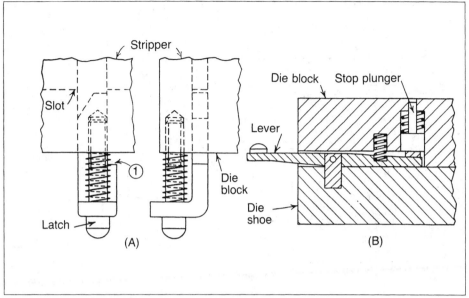

Fig. 16-18. Starting stops.[5]

which flat spring-returned pin stops (D1) are actuated by a rod terminating in handle (D2), which is swung to either of the two positions for corresponding stock runs.

Trim Stops. Trimming or notching stops bear against edges previously cut out of strip edges. These stops are also known as pitch stops. A trimming punch cuts the strip to the exact width desired and to a length equal to the feed distance or pitch. The punch length is slightly greater than the feed advance so that no scrap can remain attached to the strip to impede proper stock travel.

Double trimming stops are shown in Fig. 16-23; stop (D1) is a starting stop as well as a running stop. It bears against the trimmed edge cut by the first notching or trimming punch (D2).

The second punch (D3) trims the stock to width and provides a cut stock edge for the second stop (D4) to ensure proper stock positioning for severing the workpiece from the scrap skeleton.

Preferred Pitch Notch Design. In cases where a notching punch is combined with the punch used to cut a pitch notch, there is a chance that the scrap may lift with the punch and jam the incoming strip. Figure 16-24A illustrates the preferred method of accomplishing the combined operations in a single station. A poor method is illustrated at B.

If the condition illustrated at B cannot be avoided, the portion of the punch that cuts the pitch notch should be sharpened as shown at C to curl the slug and break the oil film suction.

Providing for Strip Removal. A solid pitch stop block will not permit the progression strip to be withdrawn from the end of the die where the parts exit. In some cases, the strip cannot be withdrawn from the other end either because the partly formed parts jam in the stock guides. A swing-away stop (Fig. 16-25) permits the strip to be withdrawn from the discharge end of the die.

Pitch Stop Limit Switch. The pitch stop shown in Fig. 16-26 pivots a small amount to permit actuation of a limit switch when the feed advance is completed. It is important to provide a long beveled lead on the edge of the stop to avoid shearing the pitch notching punch when the press is cycled in the event of a short feed or no feed.

Detecting Both Under and Over Feed. The stop shown in Fig. 16-27 has several advantages when compared to a conventional pitch stop. The small semicircular cut-out on the edge of the strip saves stock, and both under and over feeds are detected. The stop

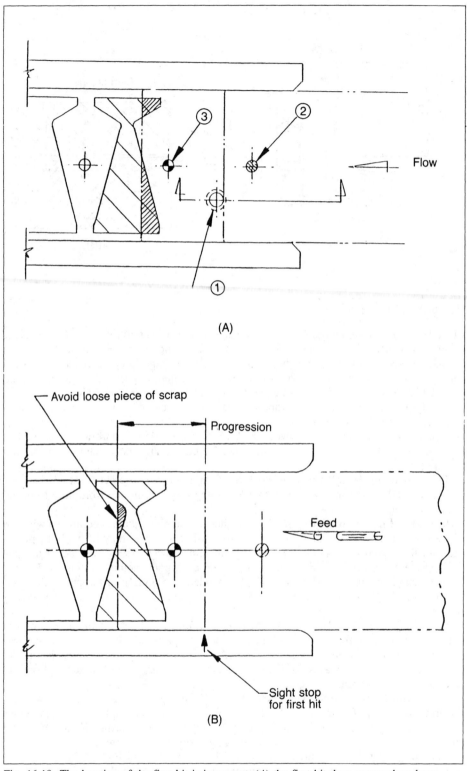

Flow

(A)

Avoid loose piece of scrap

Progression

Feed

Sight stop
for first hit

(B)

Fig. 16-19. The location of the first hit is important: (A) the first hit does not produce loose pieces of scrap; (B) the production of a loose piece of scrap.

16-24

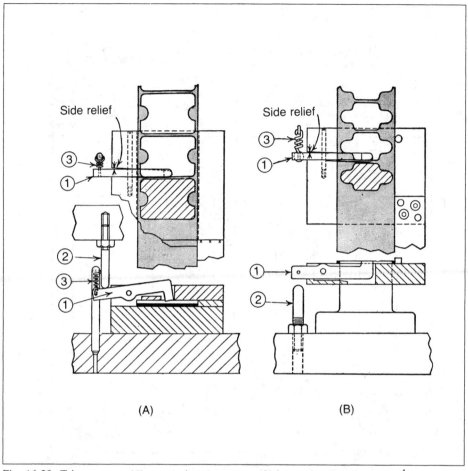

Fig. 16-20. Trigger stops: (*A*) top stock engagement; (*B*) bottom stock engagement.[4]

acts as a detent to retain the stock in position. Another advantage to this type of protection system is that a mis-feed is detected much sooner than is possible with a pilot-actuated limit switch.

The usual practice with a die protection system of this type is to stop the press during that portion of the stroke between completion of forward feed until just after feeder pilot release. This will serve to detect both proper forward feed and stock slippage upon release by the feeder.

Since a partial cut is made by the notching punch, some method to heel up the punch is required.

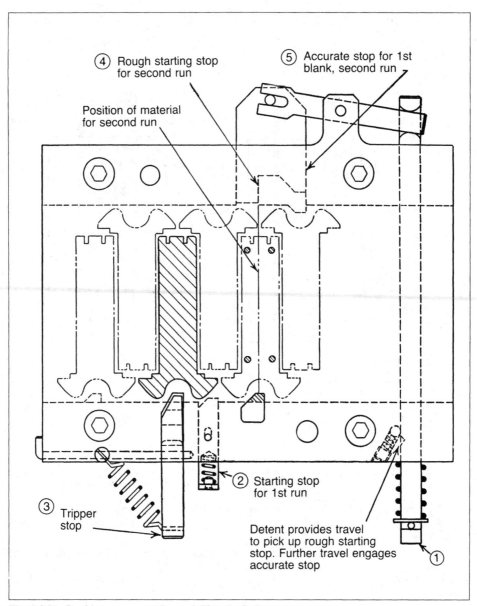

④ Rough starting stop for second run

⑤ Accurate stop for 1st blank, second run

Position of material for second run

② Starting stop for 1st run

③ Tripper stop

Detent provides travel to pick up rough starting stop. Further travel engages accurate stop

①

Fig. 16-21. Double run stops. (*General Electric Co.*)

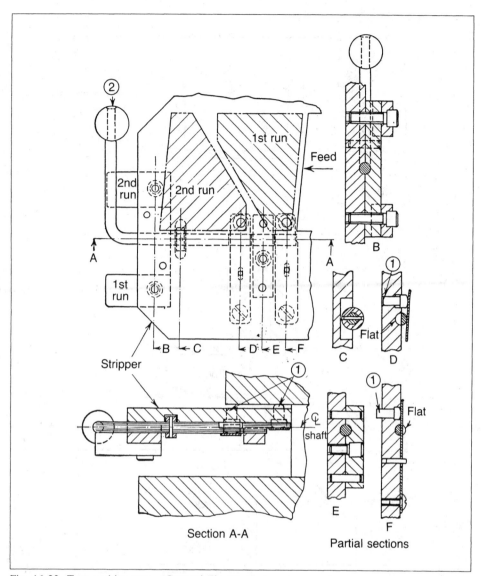

Fig. 16-22. Two-position stop. (*General Electric Co.*)

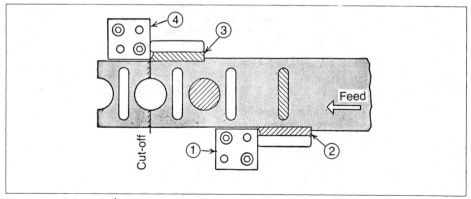

Fig. 16-23. Trim stop.[4]

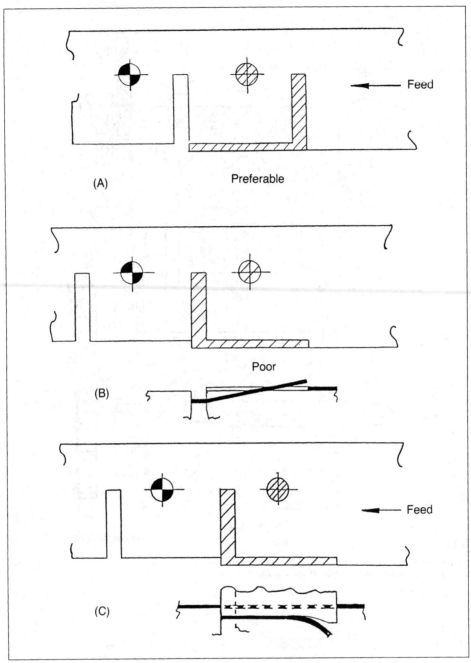

Fig. 16-24. Avoid slug tipping when cutting pitch notch: (A) preferred method; (B) poor method; (C) curved surface on punch used to break oil suction. (*Capitol Engineering Co.*)

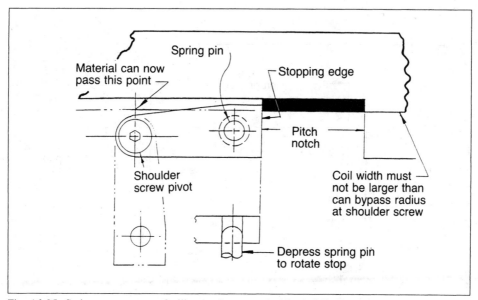

Fig. 16-25. Swing-away stop to facilitate strip removal. (*Capitol Engineering Co.*)

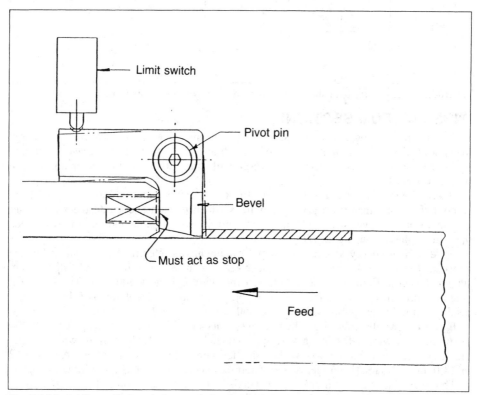

Fig. 16-26. Adding a limit switch to a pitch stop to detect proper feed. (*Capitol Engineering Co.*)

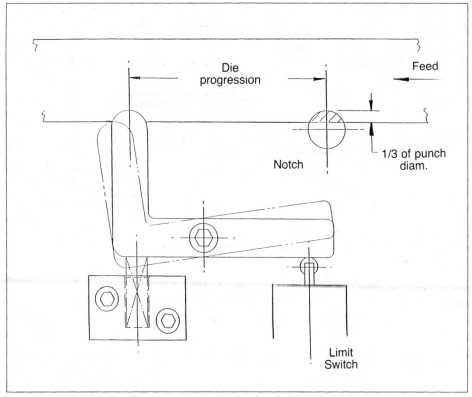

Die progression

Feed

Notch

1/3 of punch diam.

Limit Switch

Fig. 16-27. Checking for both short and over feed. (*Capitol Engineering Co.*)

WIRE CUT EDM SECTIONS

Production of progressive die sections by means of wire cut electrical discharge machines (EDM) has eliminated much of the tedious surface and jig grinder work required to build most progressive dies. Figure 16-28A illustrates the principle of operation of a wire EDM machine.

Brass wire ranging from 0.004 to 0.012 in. (0.10 to 0.30 mm) in diameter is fed from a spool (D1), and through a series of pulleys which tension and guide the wire. After passing through the workpiece (D2), several pulleys, which guide and pull the wire, feed the wire to the take-up spool (D3).

An electrical energy source (not shown) maintains an electrical potential difference between the wire (D4), and the workpiece (D5). A reservoir, pump, and filter system supply dielectric fluid to the gap between the wire and the workpiece. The electrical potential difference between the wire and workpiece causes a breakdown of the dielectric resulting in multiple sparks which erode both the wire and the workpiece.

Figure 16-28B illustrates how the wire (D7) cuts a kerf (D8) in the workpiece. A 0.004 in. (0.10 mm) wire will leave a kerf approximately 0.006 in. (0.15 mm) wide. Larger diameter wire will cut more rapidly than smaller wire and leave a wider kerf. A wire 0.12 in. (3.0 mm) in diameter will leave a kerf approximately 0.018 in. (0.46 mm) in diameter.

The worktable is driven in the X- and Y-axis by a numerically controlled servo system. The wire path can also be tilted automatically to provide tapered relief. In addition to the wire path information needed to produce the desired die detail, the servo system also controls the rate of feed based on cutting conditions sensed by the control system.

Wire EDM Speed and Accuracy. A large section can take in excess of 8 hours to cut. The main factor is not how complicated the wire path is, but rather the length of cut and thickness of the section.

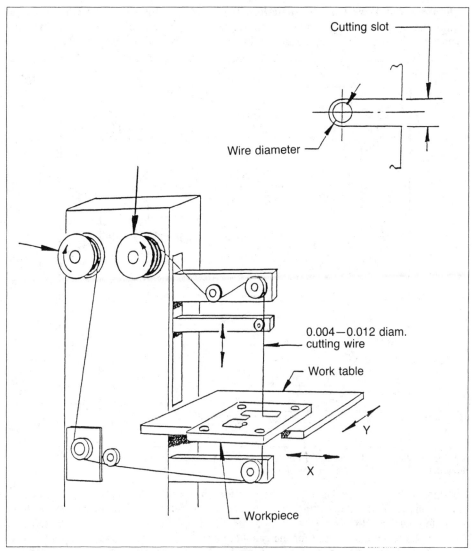

Cutting slot

Wire diameter

0.004—0.012 diam. cutting wire

Work table

Y

X

Workpiece

Fig. 16-28. Wire electrical discharge machining: (A) view of machine and workpiece; (B) cutting wire and resulting kerf.[3] (*Capitol Engineering Co.*)

The usual practice is for one specially trained operator to set up and run a number of machines. Some shops operate these machines unattended during the night shift.

Should the wire break, some models will re-thread the wire automatically. Automatic cutting of pilot holes is available on some machines. In the event of a malfunction that the machine is unable to repair, an annunciator system summons the operator—by telephone or pager in some cases.

Accuracies of ±0.0002 in. (±0.005 mm) are routinely achieved, with higher accuracy attainable. A factor in the accuracy of the finished die detail is the amount of residual stress present in the blank prior to cutting. This stress is partly relieved when the blank is cut, resulting in movement. This problem can be largely overcome by taking a roughing cut followed by a finishing cut.

As a general rule, tool steel types and heat treating processes that result in the least size change during heat treatment, are best suited for the production of accurate sections by means of wire cut EDM.

Examples of Wire Cut Construction. Figure 16-29 shows how the wire is started within the slug area (*D*1), and a small rectangular test cut made to check machine operation. The dowel holes are also wire cut as shown at *D*2 to achieve complete

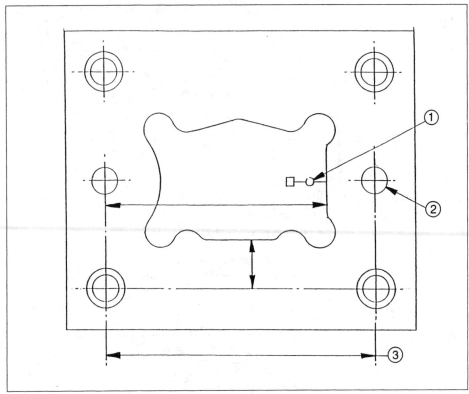

Fig. 16-29. Typical wire cut EDM die steel with wire cut dowel holes to facilitate interchangeability.[3] (*Capitol Engineering Co.*)

interchangeability of replacement steels. Reference lines (*D*3) are used to establish dimensions from the die cutting path to insure that future duplications of steels have accurate dowel locations.

A similar technique is used for the production of a punch steel as is illustrated in Fig. 16-30. The wire is started and a test cut is made (*D*1). The dowels are wire cut on standard locations (*D*2) to provide future interchangeability. The screw holes can also be wire cut, but the counterbores will require conventional EDM machining.

Wire starting holes can either be cut by a small accessory conventional EDM electrode or provided by drilling prior to hardening. Screw holes having proper clearances can be drilled on location prior to hardening if reasonable care is exercised.

An advantage to EDM machining the detail completely out of a solid hardened block is that replacement parts can be produced without delays for heat treating.

Dealing with Corner Radii. Just as a perfectly square inside corner cannot be ground with a surface grinder because of normal wheel breakdown, a square corner cannot be cut with a round wire. Figure 16-31*A* illustrates how this radius in the die steel causes a lack of proper clearance in the corner which results in rapid breakdown of the corner of the punch. One solution would be to stone a small radius on the punch steel, but this will not produce a square corner in the part.

An easy solution to the problem of obtaining a square corner is shown at Fig. 16-31*B*. This simple wire path deviation requires very little additional time and provides the extra die clearance needed to reduce the load on the punch corner.

16- 32

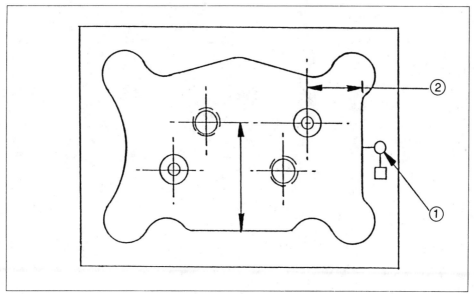

Fig. 16-30. Typical wire cut EDM punch from rectangular block with wire cut dowel holes.[3] (*Capitol Engineering Co.*)

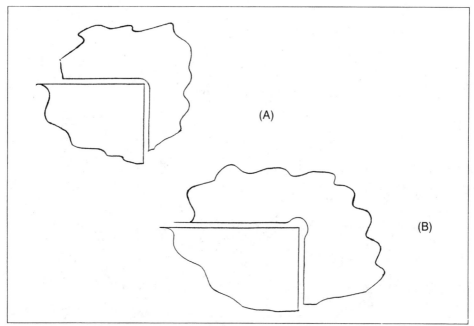

(A)

(B)

Fig. 16-31. Extra die clearance at a corner provides increased punch life.[3] (*Capitol Engineering Co.*)

References

1. R. Le Grand, "Scrap Suspension Maintains Lead in Progressive Die," *Am. Machinist* (May 14, 1951).
2. R. Le Grand, "Progressive Dies Make Precision TV Parts," *Am. Machinist* (Dec. 11, 1950).
3. A. Miedema, "Proven Principles and Applications of Progressive Dies," The Society of Manufacturing Engineers, Nov. 7-9, 1989, Grand Rapids, Michigan.
4. P. Balsells, "Stops for Dies," *Am. Machinist* (December 8, 1952).
5. P. Balsells, "How to Design Strippers and Ejectors for Progressive and Forming Dies," *Am. Machinist* (March 30, 1953).

SECTION 17
PROGRESSIVE DIES

GENERAL DIE DESIGN

A progressive die should be heavily constructed to withstand the repeated shocks and continuous runs to which it is subjected. Precision or antifriction guide posts and bushings should be used to maintain accuracy. The stripper plates (if spring-loaded and movable), when also serving as guides for the punches, should engage guide pins before contacting the strip stock. Lifters should be provided in die cavities to lift up or eject the formed parts, and carrier rails or pins should be provided to support and guide the strip when it is being moved to the next station. A positive ejector should be provided at the last station. Where practical, punches should contain shedder or oil-scale-breaker pins to aid in the disposal of the slug. Adequate piloting should be provided to ensure proper location of the strip as it advances through the die. For more die-design considerations, see Sections 3 and 16.

Scrapless Die. An elementary pierce, pilot, and cutoff die is shown in Fig. 17-1. A hole is pierced in the strip stock at the first station by the punch (*D*1). The strip is accurately located in station 2 by the pilot (*D*2) and the finished blank is cut off from the stock strip in station 3 by the cutoff punch (*D*3) shearing the metal against the edge of the die (*D*4). The part slides off the die along the inclined surface. The use of a shear-type cutoff die saves stock since there is no scrap. The parts, however, are not so accurate in width as those from a blanking die since the width of the strip stock varies. This inaccuracy could be remedied by adding trimming punches to trim the sides of the stock strip.

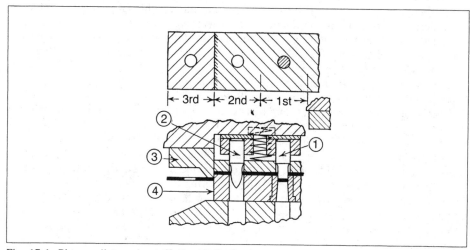

Fig. 17-1. Pierce, pilot, and cutoff die. (*C.R. Cory*)

Forming Dies. Parts with straight edges may be formed in a shear-type cutoff die as shown in Fig. 17-2. The strip is pierced in station 1, then moved against the stop (*D*1), sheared by the cutoff punch (*D*2), and formed over the inverted V-shaped punch (*D*3). A spring-loaded pressure pad prevents the strip from moving while the part is cut off and acts as a stripper for the piercing punch.

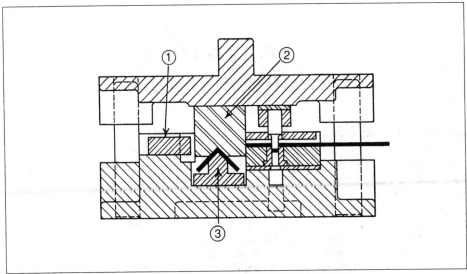

Fig. 17-2. Progressive forming die with shear-type cutoff.[1]

When the edges of the blank are not straight, a slug cutoff die as shown in Fig. 17-3 may be used. The shearing forces in a slug-type cutoff punch are balanced, since metal is cut by both edges. This type of die uses a little more stock but the scrap falls through the die out of the way. In this die, the strip is pierced in station 1 and moved against the stop (*D*1) to be cut off by the punch (*D*2) and formed by punch (*D*3). The die can be used on an inclined press and the part ejected at the rear by a jet of air.

The die shown in Fig. 17-4 can produce the same part shown in Fig. 17-3 and use a little less stock, since only two triangular-shaped scrap pieces are left. With this die the scrap is left on top of the die and must be removed before the stock is advanced.

In dies where the blank might shift or the bend line is not straight, a blankholder is used. The die shown in Fig. 17-5 uses a spring-loaded blankholder pad (*D*1) to hold the blank during the forming of the flange, and to eject the part from the die after forming. The one-piece, shear-type cutoff punch also forms the flange; an inserted adjustable shear blade in a combination cutoff and form punch eliminates regrinding the form radius after resharpening the shearing edge.

Cam-action Forming Die. The die shown in Fig. 17-6 performs three forming operations. The first forms a slight offset, the second forms a crown, and the third forms a flange of over 90°. A hinged forming punch (*D*1) engages a 45° cam surface at the lower part of the stroke, which forces it against the right-angle flange to set it to 110°. The amount of forming of this flange and the offset are controlled by adjustable tapered wedges (*D*2 and *D*3). The strip is pierced and trimmed to width in station 1, positioned in station 2 by a pilot (*D*4), and on the upstroke of the press the strip is lifted above the surface of the offset forming block (*D*5) by the spring-loaded lifter (*D*6). The part is cut from the strip in station 3 by a slug-type cutoff punch, just prior to the last two forming operations in station 4. The small spring-loaded pressure pad (*D*7) holds the part while it is being cut off and formed. The part is ejected from the die by the slanting spring-loaded ejector (*D*8).

Progressive Die to Pierce and Form a Lens Retainer. The die in Fig. 17-7*B* trims,

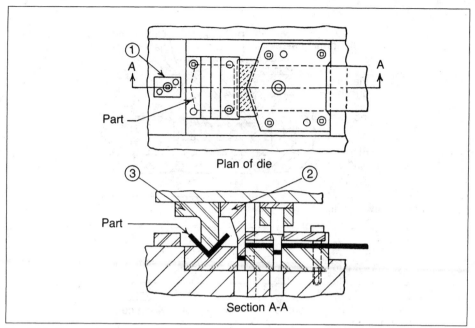

Fig. 17-3. Progressive forming die with slug-type cutoff.[1]

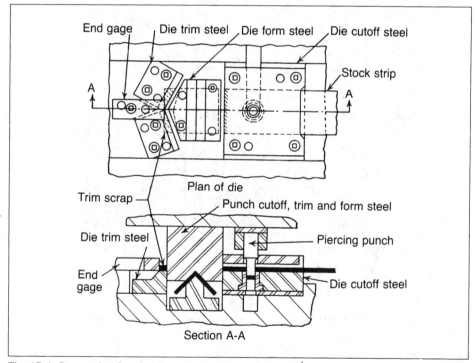

Fig. 17-4. Progressive forming die with blanking-type cutoff.[1]

forms, pierces, and blanks a lens retainer of 0.010-in. (0.25-mm) spring-temper brass strip 1.188 in. (30.18 mm) wide. A center and two side piloting holes are pierced in the first station and the area around the tabs is trimmed in the next three stations by six rectangular-shaped punches. Station 4 shears the ends of the tabs from the strip and

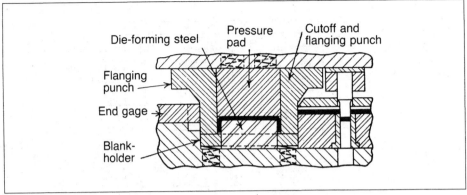

Fig. 17-5. Progressive forming die with blankholder.[1]

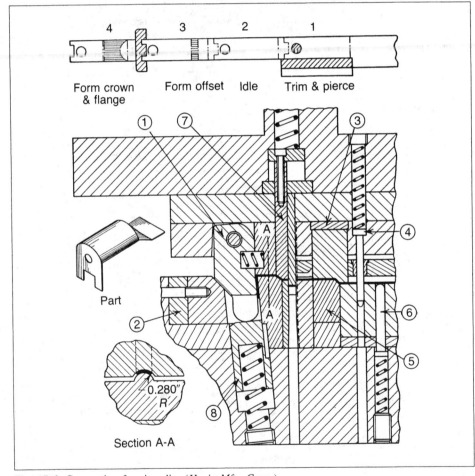

Fig. 17-6. Cam-action forming die. (*Harig Mfg. Corp.*)

partially forms them, and the following operation sets them to shape. The center hole is pierced in station 6, and in station 7 the part is blanked from the strip and drops through the die. The unnumbered idle stations are provided to increase die strength; actually the first part is completed on the eleventh press stroke. The strip is piloted in the forming stations by the center holes, and in the cutting stations by the side piloting holes.

Spring-loaded stock lifters elevate the stock to avoid interference of the formed tabs with the die. The punches and die inserts are held in position by flats machined on the inserts to fit recesses in the retainers. Recesses are machined deep enough behind all cutting elements to allow them to be resharpened and spacers to be placed behind them to bring them back to height. The punches are guided by the spring stripper which is in turn guided by leader pins.

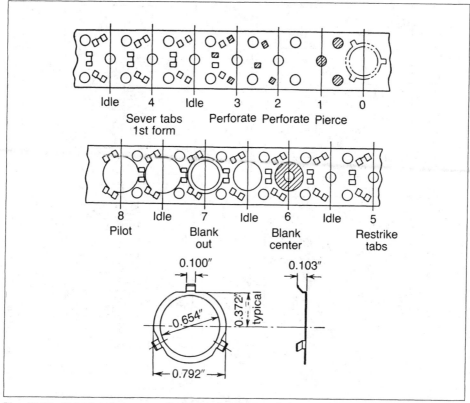

Fig. 17-7A. Strip development and part drawing for lens retainer. (*Argus Cameras, Inc.*)

Forming and Embossing after Cutoff. The strip in Fig. 17-8 is notched, pierced, and cut off before being pushed into the forming and embossing station. The blank is confined by a tunnel-type stock guide and stripper. The advance of the strip stock pushes the blank against solid end gages and under the forming and embossing punch. The section of the die shown is through the last station with the stroke partially down and the U-shape formed in the part. The remaining portion of the downward stroke embosses the part over the pin (*D1*) and into the recess in the punch (*D2*). On the upstroke of the press the part is ejected from the die by the ejector (*D3*) and stripped from the punch by (*D4*). The stock is fed from front to back, and the inclined position of the press enables the part to fall out the back.

Cupped Washer Made in Progressive Die. The small part of Fig. 17-9 is carried on a side web through the die. The contour of the first perforating punch allows the subsequent forming of the flange with its sharp corner. Then the notching punch trims the metal around the two adjacent tabs. Cutting of this outline using two punches in two operations simplifies punch design by eliminating the sharp corners required on one punch for single-operation cutting. The perforating of the hole follows. The punch (*D4*) pilots, cuts off, and draws the washer to shape. The washer is stripped off this punch by two spring-operated stripper arms (*D1*) mounted in the die backup plate (*D2*). A scrap cutter

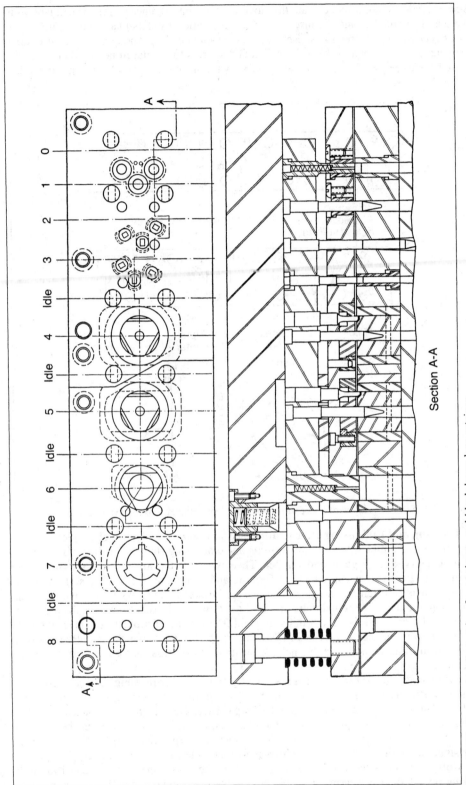

Section A-A

Fig. 17-7B. Progressive die to trim, form, pierce, and blank brass lens retainer.

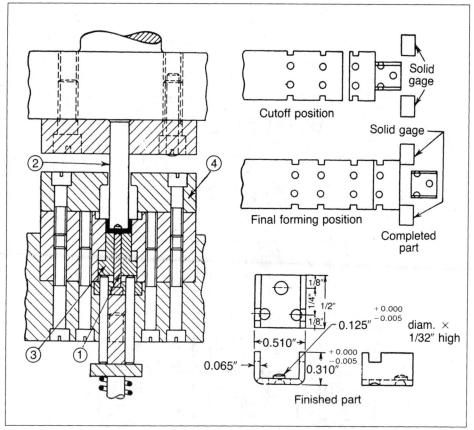

Fig. 17-8. Progressive die to notch, pierce, and emboss U-shaped part. (*Wilson-Jones Co.*)

(D5) is incorporated. An inexpensive yet rigid way of mounting all punches in the punch pad or retainer (D3) is provided by the use of low-melting-point alloy.

Progressive Die Fabrication of Small Brass Connecting Links. The brass strip is conventionally perforated and trimmed at the first two stations (Fig. 17-10). Section *A-A* shows the use of a heeled first forming punch. The spring-loaded combination pressure pad and ejector contains a pilot for positioning and holding the strip during forming.

Section *B-B* shows the positive-return cam-action slide for final forming of the part. The final operation blanks the connecting link from the carrier tabs, allowing it to fall through the die. The strip development indicates that there are several idle stations to add strength and rigidity to the die. The form block or mandrel at the forming stations and the die section at the cutoff station are inserts to simplify construction and replacement.

In Station 1 of this die, the pilot hole is perforated and the strip trimmed to width on either side. The trimming of the stock provides two projections which serve as stops during its progression through the die. The projections require a little more material, particularly when they are cut on both sides, but they do allow the thin strip to be pulled straight through the die without shifting. These projections are commonly called "shear stops." In station 2 the six small holes are perforated. Normally, it is a bad practice to have such small punches so close together, but it ensures accurate hole location and spacing. The area around the part, except for a connecting tab along the axis of progression, is trimmed in station 3. Station 4 is an idle station. The carrier tab falls through the die as the part is severed by the slug-type cutoff punch. This cutoff and a forming operation are combined in station 5. Ejection is by air blast.

Double-V Bending Die. The die shown in Fig. 17-11 makes two brackets having two

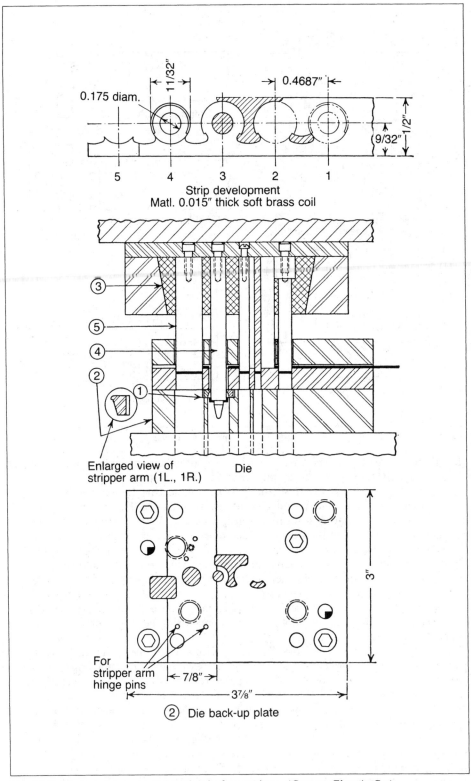

0.175 diam.

11/32"

0.4687"

1/2"

(9/32")

5 4 3 2 1

Strip development
Matl. 0.015" thick soft brass coil

③

⑤

④

②

①

Enlarged view of
stripper arm (1L., 1R.)

Die

3"

For
stripper arm
hinge pins

7/8"

3⅞"

② Die back-up plate

Fig. 17-9. Die to produce cupped washer in four stations. (*Century Electric Co.*)

17-8

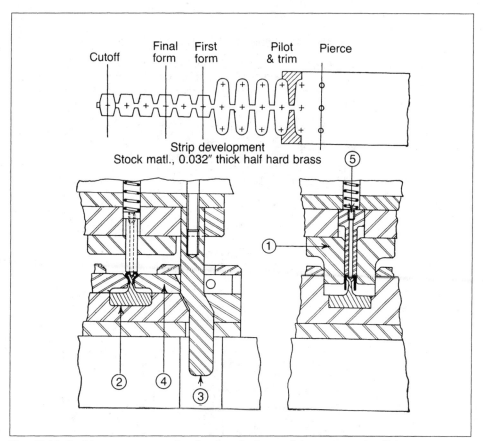

Fig. 17-10. Positive-return cam-action forming die for small brass connecting link. (*Harig Mfg. Corp.*)

right-angle bends. The material is 0.36-in. (9.1-mm) by 2-in. (51-mm) wide cold-rolled-steel strip. In the first station, the strip is notched to the end contour of the brackets, and a 0.375-in. (9.52-mm) diameter hole is punched for use as a pilot in stations 2 and 3. Four small holes are also punched in this station. Additional slotting, notching, and punching are performed in station 2. In station 3, the left-hand ends of the brackets are trimmed to shape, and the leading parts are sheared from the strip. In station 4, the strip is located by a positive stop before any cutting or forming is done. The two right-angle bends, or a Z-shape, are formed in each bracket at this station. Spring-loaded pins lift the parts from the die surface while an air blast ejects them from the die. Since the strip has a tendency to close in, a separator is incorporated in station 4 to assure that the bends are square with the edges.

Two-station Pierce, Cutoff, and Form Die. A progressive die to pierce, pilot, cut off, and form a small channel shape with a semicircular cutout at the edge of each flange is shown in Fig. 17-12. The stock, 0.075-in. (1.90-mm) hot-rolled steel, is confined in a closed stock guide and stripper. A hole is pierced in the first station, and the strip is piloted, cut off, and formed in station 2. Instead of having a separate piloting station, the semicircular pilot (*D*1) is inserted adjacent to the combination cutoff and form punch (*D*2). The pilot is stepped beneath the cutoff and form punch to enable its full diameter to position the metal correctly before it is cut off. The strip is firmly gripped between the cutoff punch and the pressure pad (*D*3) during the cutoff and form operation. The part is flattened between the form block (*D*4) and the positive knockout pad (*D*5) at the end of the forming strokes. Ejection from the form block is done by the pressure pad and from

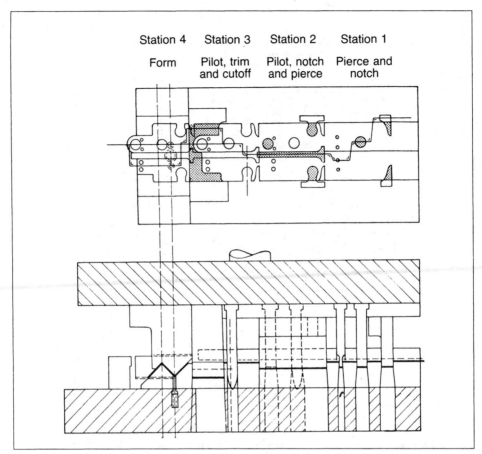

Station 4	Station 3	Station 2	Station 1
Form	Pilot, trim and cutoff	Pilot, notch and pierce	Pierce and notch

Fig. 17-11. Progressive die for stainless-steel part. (*White-Rodgers Electric Co.*)

the punch by the positive knockout. This die can be fed by hand or by an automatic feeder. The forming punch heel block (*D6*) forms a positive stop for the strip.

Progressive Die with a Cam-operated Final Forming Station. The blank for the part shown in Fig. 17-13 is cut off from the strip by a slug-type cutoff and pushed progressively along the top of the die to the forming stations by the advancing stock strip. The drawing shows a section through the last forming station. The forming punch (*D1*) is mounted on a spring-loaded pad and forms the open U of the part. The part is further formed as the continued downward motion of the ram closes the dies (*D2*) which are spring-returned by cams (*D3*). Finished parts are pushed off the die by the blanks moving to the left.

Overforming by a Hinged Cam. A small beryllium-copper part has its center piloting hole pierced in the first station of the die of Fig. 17-14. In the same station the part is trimmed to width and the ends cut to a partially rounded outline, leaving connecting tabs. A small hole in the left-hand end of the workpiece is pierced in station 2; an embossment in the opposite end is formed in station 3. In station 4, the part is cut off by a slug-type cutoff punch prior to forming around the form punch (*D1*). The hinged member (*D2*) is actuated by a descending cam (*D3*), backed up by a heel block (*D4*). The finished part is stripped from the forming punch by stripper pins (*D5*) actuated by a positive knock-out bar.

Adjustable Cams Control Bending. The 0.036-in. (0.91-mm) thick half-hard brass part shown in Fig. 17-15C is made in seven stations in a progressive die. Cams threaded into the upper shoe and secured with locknuts provide a means of adjusting the travel of

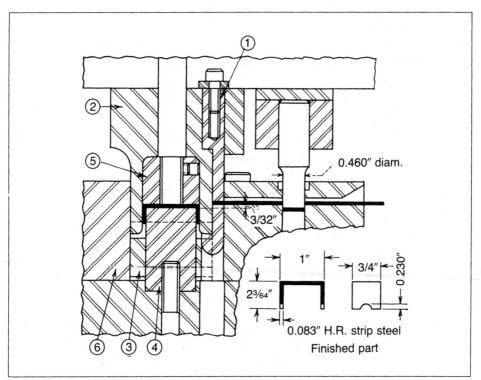

Fig. 17-12. Two-station die for channel-shaped part. (*Harig Mfg. Corp.*)

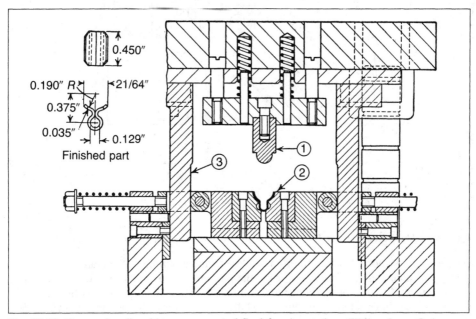

Fig. 17-13. Progressive die with a cam-operated final forming station. (*Wilson-Jones Co.*)

the forming slides to control spring-back due to different hardnesses, tensile strengths, and thicknesses of material.

In the first station, the six holes in the part are pierced in addition to two strip piloting holes. Two slots are blanked in station 2; station 3 is idle to provide space for stronger die

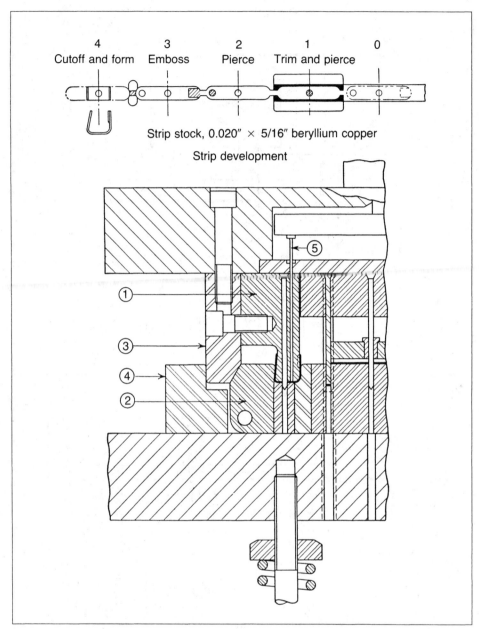

Fig. 17-14. Hinged cam to overform part in a progressive die. (*Harig Mfg. Corp.*)

construction. Stations 4, 5, and 6 progressively trim the outline of the blank. A slug-type cutoff punch cuts off the blank and trims the partial contour on the rear tab in station 7. The forming of five right-angle bends to complete the part is also accomplished in this station. Three slides actuated by cams bend the two side lugs and the tongue. Two slides, one on each side of the blank (Fig. 17-15E) bend the two lugs while an end slide (see Fig. 17-23D) forces the tongue against the upper punch for its first right-angle bend. This is done at the same time that the forming punch is pushing the blank into the die cavity, bending the rear projection at a 90° angle. As the forming punch reaches its final depth, the recess in the punch forces the tongue into a second 90° bend over the end forming slide. An air cushion applies pressure to a pad under the punch to keep the part flat and

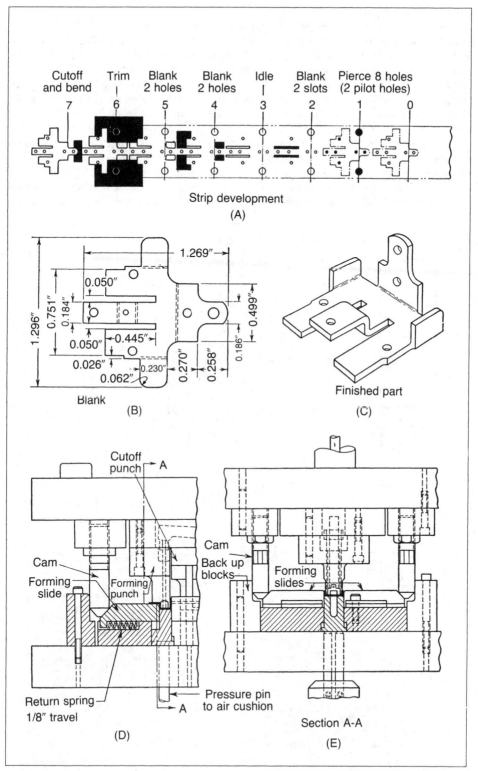

Fig. 17-15. Adjustable cams control bending: (A) strip development; (B) blank; (C) finished part; (D) end forming slide; (E) side forming slides.[2]

eject the finished part from the die cavity.

Die for Leadwire Anchor Staple. A small part made of 0.010-in. (0.25-mm) thick brass is blanked and formed in a three-working-station progressive die. (Fig. 17-16). The first station trims the stock to width and pierces a piloting hole. Station 2 blanks the contour of one side of two adjacent parts. Station 3 blanks the part from the carrier strip and forms it to shape. At this station the carrier strips are cut into slugs and fall through the die. Small shedder pins are incorporated in these punches to ensure the ejection of the slugs. Two pilots (*D*1) position the strip. Four idling stations are incorporated because the part is very small.

Progressive Die for Fabricating a Drawer Saddle. A progressive die which embosses, notches, and forms a drawer saddle is shown in Fig. 17-17. The strip is embossed in station 1, and four round-end notches and two square-end notches for piloting are cut at the second station. The metal between the square notches serves as a carrier tab. The flanges are formed down in station 3. The cutoff and final forming

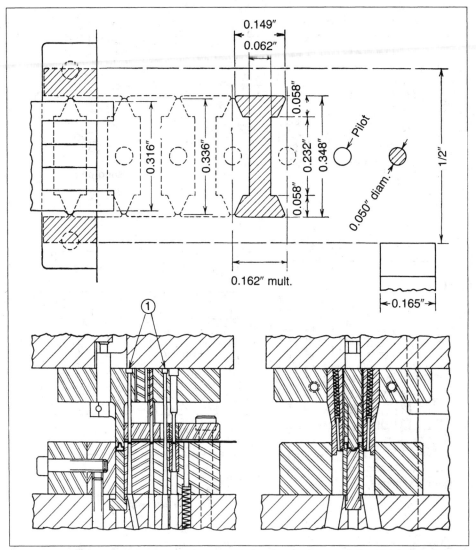

Fig. 17-16. Strip development and section through forming station to produce lead-wire anchor staple. (*Harig Mfg. Corp.*)

17- 14

operations are performed at the fourth station. The punch and die are made in sections for ease of construction and replacement. A forked stripper removes the part from the forming punch.

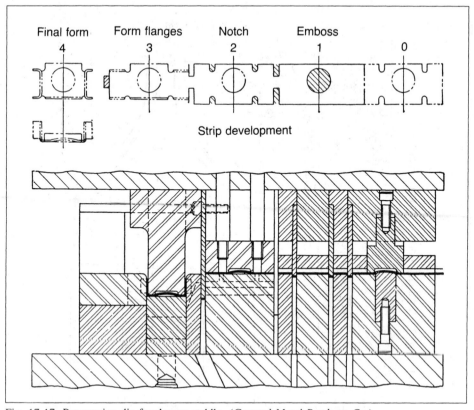

Fig. 17-17. Progressive die for drawer saddle. (*General Metal Products Co.*)

Progressively Piercing and Forming Spring Bronze. A switch arm is made of spring-temper bronze 0.0159-in. (0.404-mm) thick as shown in the layout in Fig. 17-18. Two pilot holes are pierced in the scrap area of the carrier strips to be used for piloting at all stations. The die plate is sectioned for ease of sharpening and maintaining relationship between cutting and forming stations. The punches are divided into several simple shapes to avoid thick and thin sections and sharp corners. Inserts are used for the embossing operations.

A feature of this die is the forming of the tab to a 15° closed angle. The first forming operation produces a groove across the part. The next station has a flat punch which hits home on the groove, causing the tab to bend up from a horizontal plane to approximately 100°. The following station operation pushes the tab down to 165° from its starting point. In the next operation the formed piece is blanked through the die. The scrap is cut into small pieces by a shear blade at the left-hand end of the die. The die is operated with an automatic slide feed and a stock straightener at a speed of 75 strokes per minute.

Die for Internal Gear Segment. The die shown in Fig. 17-19 is used to stamp an internal gear segment for a can opener from cold-rolled-steel strip stock. A channel-type stripper guides the strip through the die. A spring-loaded pressure pad holds the stock down during the stamping operation and guides the pilots and piercing and extruding punches.

In station 1, two pilot holes are punched along each edge of the stock. Stations 2, 4, 7, and 9 are idle stations. In station 3, a hole is punched containing the tooth profiles. At

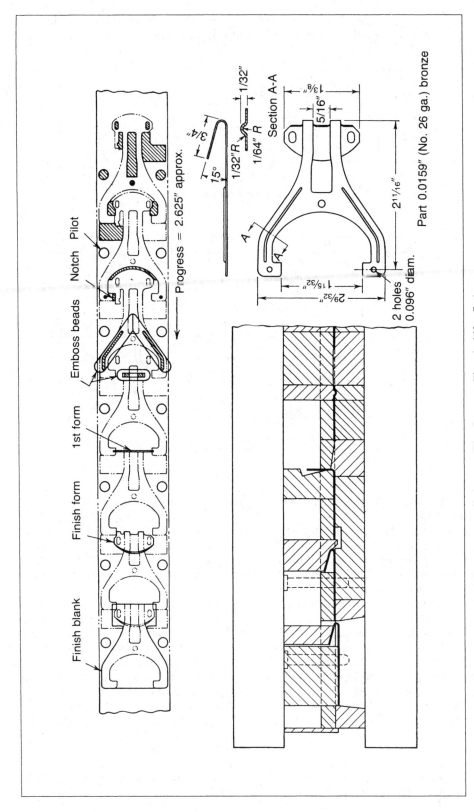

Fig. 17-18. Progressively piercing and forming a bronze switch arm. (*The Emerson Electric Mfg. Co.*)

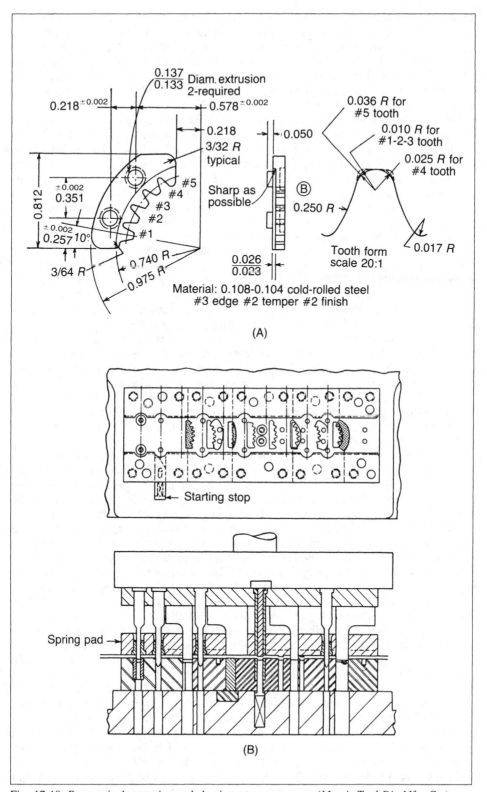

Fig. 17-19. Progressively swaging and shaving a gear segment. (*Morris Tool-Die Mfg. Co.*)

station 5, the step is swaged. The two bosses are extruded in the next station. The teeth are shaved to the correct profile in station 8, and the part is blanked out of the strip at station 10. To assure an accurate part, the strip is piloted in stations 2, 4, 6, 9, and 10.

The die block is made in sections to facilitate machining the required profiles. The swaging is done by an insert in the die block. A spring-loaded pin lifts the extruded bosses from the die and succeeding clearance holes. The die is mounted in a standard 10 × 5 in. (254 × 127 mm) two-post die set.

Die for Pinion Gear. A 12-tooth, 20-diametral pitch pinion gear, which mates the internal gear segment made in the die of Fig. 17-19, is stamped in the die of Fig. 17-20.

The oval-shaped hole in the gear is punched in station 1 and shaved to size in station 6. The gear is blanked halfway through the strip in station 4 and is shaved down through the blank in station 7. At station 1, pilot holes are punched along each edge of the strip. To ensure an accurate gear, the strip is piloted in station 2, 5, and 6.

A spring-loaded pressure pad holds the strip during the stamping operations while guiding and stripping the punches and pilots. A spring-loaded pad is used to eject the partially blanked gear from the die in station 4.

The gear is made of 0.083 in. (2.11 mm) thick by 0.906 in. (23.01 mm) wide cold-rolled steel.

Progressive Die for Coining, Piercing, and Forming a Flange. The part, strip development, and view of sectional die blocks are shown in Fig. 17-21. This part is a rewind pawl for a camera and is made of 0.032-in. (0.81-mm) cold-rolled stainless steel. Serrations are coined in the flange in station 2 and the scrap around the flange is blanked out in station 3. In station 4, this flange of 90° is formed. A concave radius is also formed in this flange by a punch which has the identical serrations as previously coined in the strip. In station 5, final blanking of the part is completed. The positioning of the cutouts in the strip simplifies the construction of the sectional die blocks shown.

Nine-station Die for a Contact Spring. The die in Fig. 17-22 produces contact springs made of grade D phosphor bronze strip, 0.006 in. (0.15 mm) thick by 1.000 in. (25.40 mm) wide, at the rate of 5,000 to 6,000 per hour. The strip layout shows the steps which are taken to produce the part.

The punches are held in a split plate which is hardened and ground to suit the profile of the individual punches. The two halves are fastened together and screwed and doweled to the upper die shoe. The nine die stations are made of sectional ground dies held together in a hardened and ground U-shaped channel by end plates. The die used for piercing the circular segment in station 3 consists of a hardened and ground shouldered plug (D1) surrounded by a bushing (D2). One side of the plug is ground to form the inner half of the circular segment, while the contour of the outer half is ground on the inside of the bushing. These two pieces are pressed together in the correct relation, and set into the die. The assembly is doweled in place to prevent shifting or turning. The trimming operations in stations 4 and 5 are performed by the combined action of the ground inserts (D3 and D4). The strip is piloted in stations 2 and 6 through the hole pierced in station 1. The forming is done in stations 6, 7, 8, and 9, with cutoff also accomplished in station 9.

Progressive Curling of a Hinge. Stainless-steel strip can be manually or automatically fed into the die of Fig. 17-23. Three rivet holes and two holes for piloting holes are cut at the first station by quilled punches (D1), as well as an edge notch (used by the automatic stop). The central M-shaped hole is also cut at the first station. Forming punches (D3, D4, and D5) and forming dies (D6, D7, and D8) respectively, first-form, second-form, and finish a cylindrical curl in the part. Four pilots (not shown) are used for positioning the part at the second and third stations; at the last station the part is bowed to a 3.094-in. (78.59-mm) radius and cut off from the scrap skeleton. Production in a 50-ton (445-kN) OBI press is 900 parts per hour (manual feeding) and 1,200 parts per hour with an automatic feed. Ejection of part and scrap is by air jet.

Progressive Production of a Brass Bracket. Four perforating punches (D10) and four notching punches (D4 and D5) cut the holes and the outline of the part in the die of Fig.

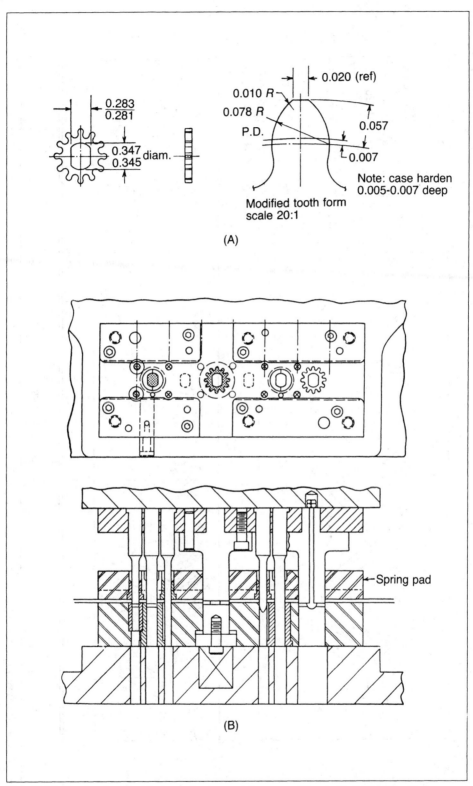

0.283 / 0.281

0.347 / 0.345 diam.

0.020 (ref)

0.010 R

0.078 R

P.D.

0.057

0.007

Note: case harden
0.005-0.007 deep

Modified tooth form
scale 20:1

(A)

←Spring pad

(B)

Fig. 17-20. Progressive blanking and shaving a pinion gear. (*Morris Tool-Die Mfg. Co.*)

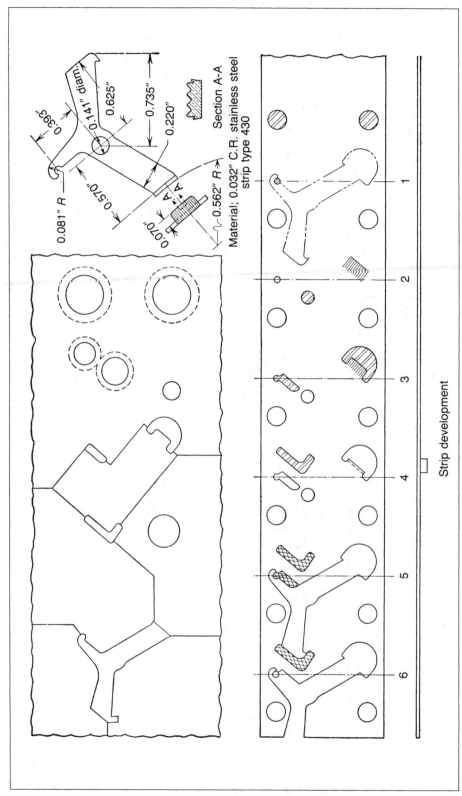

Section A-A

Material; 0.032" C.R. stainless steel strip type 430

Strip development

Fig. 17-21. Progressively coining and forming a flange on a stainless-steel part. (*Argus Cameras, Inc.*)

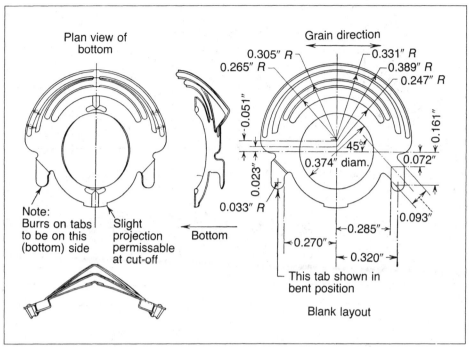

Plan view of
bottom

Grain direction

0.305″ R
0.265″ R

0.331″ R
0.389″ R
0.247″ R

0.051″
0.023″

45°
0.374″ diam.
0.072″
0.161″
0.093″

Note:
Burrs on tabs
to be on this
(bottom) side

Slight
projection
permissable
at cut-off

Bottom

0.033″ R

0.285″
0.270″
0.320″

This tab shown in
bent position

Blank layout

Fig. 17-22. Phosphor bronze contact spring produced in a nine-station progressive die, and layout of the blank. (*John Volkert Metal Stampings, Inc.*)

17-24. Six pilots (D3) and a pressure pad (D9) align and clamp the part (a bracket) as the legs are bent down and around a form block (D8). Two punches (D2) slide inwardly in holders (D1) actuated by cam blocks (D7) to form the part. A cut-off punch (not shown) severs the part from the strip. Stock is slide-fed into this die of 10-in. (254-mm) shut height, which is used in a press having a 3-in. (76-mm) stroke.

Progressive Die for Cover. A U-shaped cover of 0.036-in. (0.91-mm) cold-rolled steel is pierced, notched, trimmed, flanged, and formed in three progressive stations as shown in Fig. 17-25. Heeled trimming punches in station 1 trim the strip to the required outline and notches are cut, forming the carrier tab and freeing the metal for flanging in the next station. Station 1 also contains piercing punches for the piloting hole in the carrier tab and punches for the two holes. These punches are guided by the stripper. Individual die buttons are installed in the die block for each of these punches. In station 2, the flanges are formed along each side of the strip. In station 3, the blank is parted from the strip, the carrier tab is sheared off, and the U-shaped part is formed. The right-hand side of view A is through station 1 and the left-hand side is through station 2. View B shows the stripping action in the final station. The lever (D1) is actuated at the top of the stroke by a positive knockout. The lowering of this lever ejects the part from the die by the knockout (D2). This lever also depresses the shedder pins (D3), which causes the bell-crank-shaped stripper (D4) to rotate, allowing the part to fall out.

Die to Produce a Sprocket Chain. Sectional hardened-tool-steel punches and die elements progressively score, lance, notch, form, curl, and assemble a series of connected links for a sprocket chain from 0.062-in. (1.57-mm) cold-rolled steel strip. The strip is conventionally fed until it reaches the cutoff station. In order to move the chain to the next station which assembles the links, a sprocket (not shown) actuated by the press slide pulls the chain into the curling or assembling station (Fig. 17-26).

Die with Variable-length Cutoff. This die was designed to produce two memo-book-back binders in lengths, containing an even number of rings from 6 to 18. The strip is 0.025-in. (0.64-mm) thick, 1.25-in. (31.8-mm) wide cold-rolled steel. The strip

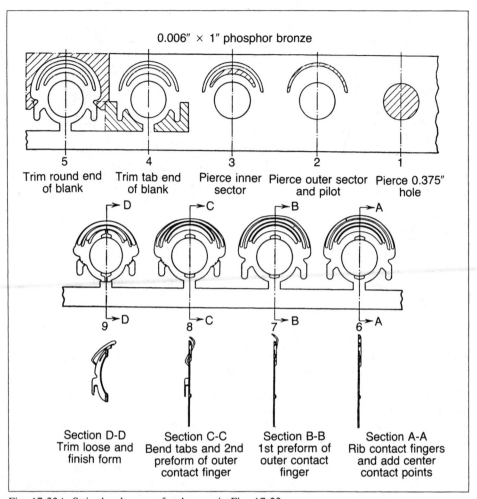

0.006" × 1" phosphor bronze

5	4	3	2	1
Trim round end of blank	Trim tab end of blank	Pierce inner sector	Pierce outer sector and pilot	Pierce 0.375" hole

D → C → B → A →

9 → D 8 → C 7 → B 6 → A

Section D-D	Section C-C	Section B-B	Section A-A
Trim loose and finish form	Bend tabs and 2nd preform of outer contact finger	1st preform of outer contact finger	Rib contact fingers and add center contact points

Fig. 17-22A. Strip development for the part in Fig. 17-22.

development, sections through the forming stations, and the cutoff control are shown in Fig. 17-27. The control (Fig. 17-27B) is effected by ratchets (D1) with different numbers of teeth and a cam (D2) with two or four lobes, which contact the top of the cutoff punch. Ratchet rotation is accomplished by contact of the lever arm (D3) with the stud (D4) at the top of the press stroke. The number of teeth picked up on each stroke can be varied by moving a stop stud (D5) to any one of four positions. The cam shown has four lobes, being made in two pieces with two lobes each. By removing a dowel pin and rotating one cam 90°, a cam with only two lobes is obtained. The versatility of the cutoff control is such that, by using a 36-tooth ratchet and a two-lobe cam, and by placing the stop stud in position 3, skipping three teeth on the ratchet, a binder with 12 rings is cut off. Using the same position of the stop stud, in conjunction with a four-lobe cam, a 6-ring binder will be cut off by the die. Figure 17-27C shows the first forming operations on the ring; Fig. 17-27D shows the final forming operation. The slide (D6) moves across the die to form the second quarter of the ring and the upward motion of D7 completes the form. The form block (D7) is actuated by the rod (D8) and the lever (D9).

Progressive Blank and Draw Die. The strip development and a section through the die for an automotive front-door armrest are shown in Fig. 17-28. The stock-strip is 0.0345-in. (0.876-mm) cold-rolled steel 12.313-in. (312.75-mm) wide. The blank is trimmed at station 1 and piloted at station 2. The strip is slit to provide two-point

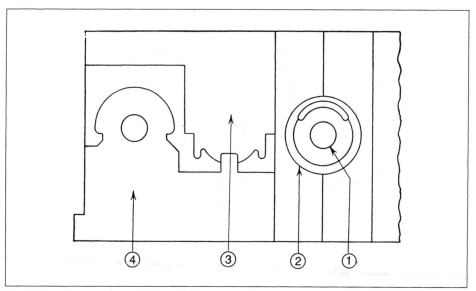

Fig. 17-22*B*. Layout of die in Fig. 17-22 at the cutting stations 3, 4, and 5.

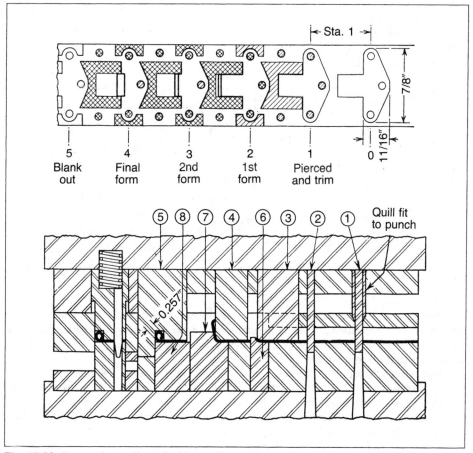

Fig. 17-23. Progressive curling of a hinge. (*Knapp Monarch Co.*)

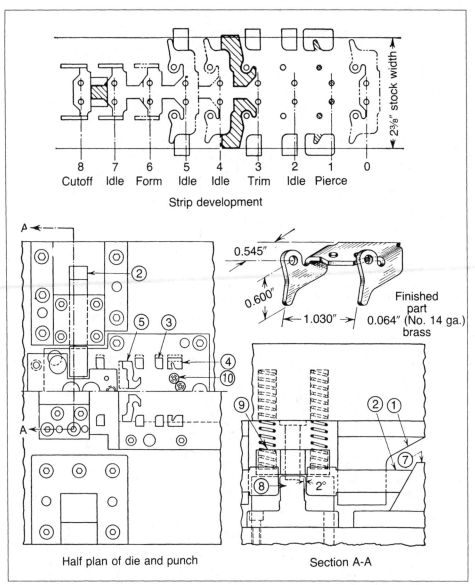

Fig. 17-24. Strip development and die for progressively producing a brass bracket. (*The Emerson Electric Mfg. Co.*)

suspension for the blank and the bead is formed in station 3. Station 4 accomplishes the first draw operation, station 5 the trim operation for the excess stock, and station 6 the cutoff and final draw operation. Spring-loaded strippers remove the part from the punch, allowing it to drop through the die.

Progressive Die for a Lock Part. This four-station die (Fig. 17-29) notches, pierces, embosses, forms, and shears off a part of 0.065-in. (1.65-mm) thick cold-rolled steel. The stripper is a fixed type attached to the die block and contains four spring-loaded stops for hand-feeding strips into the die in order to get as many pieces as possible from short lengths of stock.

Progressive Louvering Die. Sectional dies are used in the two piercing stations shown in the plan view of the progressive die in Fig. 17-30A. The strip is positioned by the pilots (*D2*, Fig. 17-30B) and guided by spring-loaded guide pins (*D4*). The punch and die

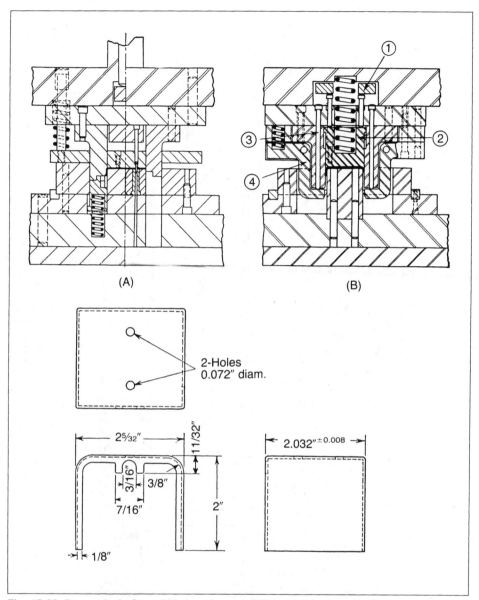

Fig. 17-25. Progressively formed U-shaped cover. (*White-Rodgers Electric Co.*)

sections (*D5*) are cut away to facilitate strip movement in and out of the louvering station which contains six sets of mating forming members (*D3*). This station incorporates spring-loaded lifter pins (*D1*). The completed part is blanked out and drops through the last station (not shown). The part is shown in view *C*.

Nineteen-station Die for Brass Terminal. This progressive die (Fig. 17-31) to make a brass electrical terminal of 0.040-in. (1.02-mm) brass has 18 stations plus a notching station which trims the edge of the strip and thereby creates a notch for locating purposes. A spring-loaded stripper plate upwardly strips the part from the drawing punches mounted in the lower punch retainer plate. The stripper plate raises the stock strip above the ends of the drawing punches to provide unimpeded stock travel through the die. Shedder pins (*D2* and *D3*) push the part down and out of the die cavities in the upper draw die (*D1*). A spring-loaded pressure pad and ejector (*D4*) clamps and ejects the part in the final cutoff

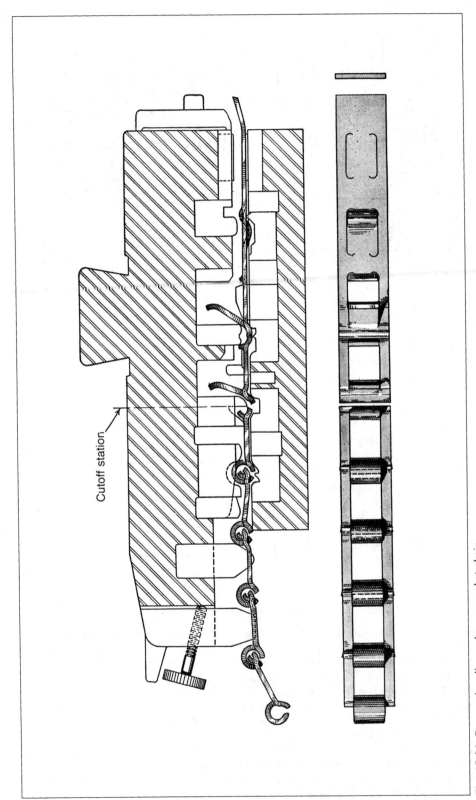

Cutoff station

Fig. 17-26. Progressive die produces sprocket chain.

17-26

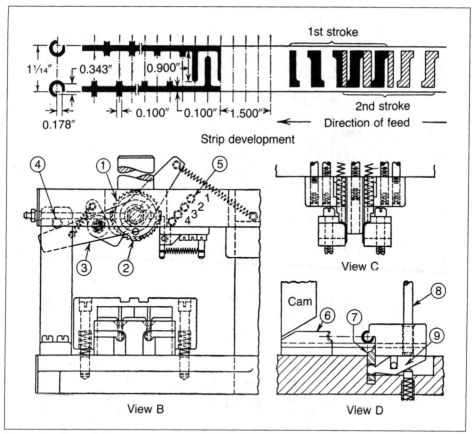

Fig. 17-27. Progressive die with cutoff control. (*National Blank Book Co.*)

and flanging station.

The small diameter of the cup allows it to be pierced in air (without die support) at station 12. A blast of air through the punch assures that the slug will drop down through the die instead of clinging to the punch. After forming the cylindrical portion, the strip is trimmed and a flange is formed. At the last station, the part is cut off and an additional flange is formed with the aid of a spring pressure pad which also ejects the part from the die.

Die to Draw, Pierce, and Flange a Plate. The part and sections through the die for producing the parts are shown in Fig. 17-32. This die pierces an hourglass cutout in the strip at the first station; in the second station two stepped-diameter cups are drawn. Station 3 pierces a hole in the bottom of each cup, seven 0.149-in. (3.78-mm) diameter holes and one 0.5-in. (12.7-mm) diameter hole in the face of the plate. In station 4, the sides of the 0.503-in. (12.78-mm) diameter portion and of the 1.031-in. (26.19-mm) diameter portion are set to depth and diameter. At station 5, a side of two adjacent plates is trimmed and in station 6 the part is blanked from the side carrier strips and the tab bent downward. All punches are guided by a pin-guided stripper plate. The cup-sizing punches are bushed in the stripper plate. All the stations use inserted die buttons where space permits.

A previous but not satisfactory design produced the 1.031-in. (26.19-mm) diameter cup in several draw operations. The bottom of the cup was pierced and the tubular portion formed. The operation was too severe and caused excessive thinning and cracking.

Progressive Production of an Automobile Generator Fan. The strip development in Fig. 17-33A shows the results of the die operations that produce the part shown in view B. A flanging punch (*D1*, view *C*) forms the 12 blades downward against a forming ring

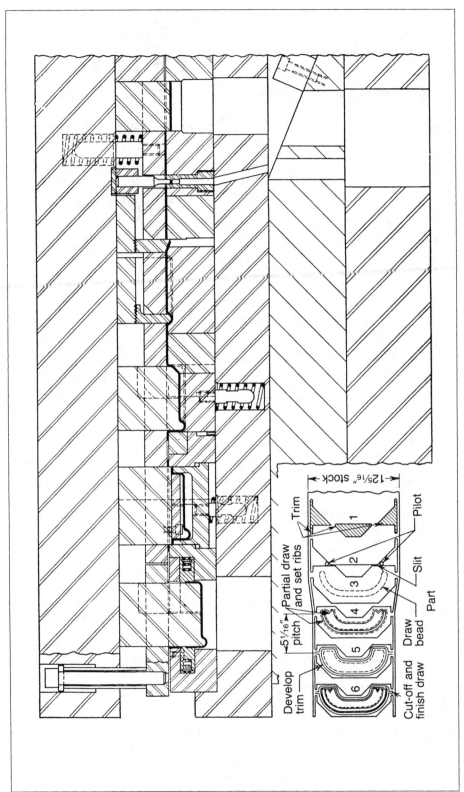

Fig. 17-28. Progressive blank and draw die for armrest.

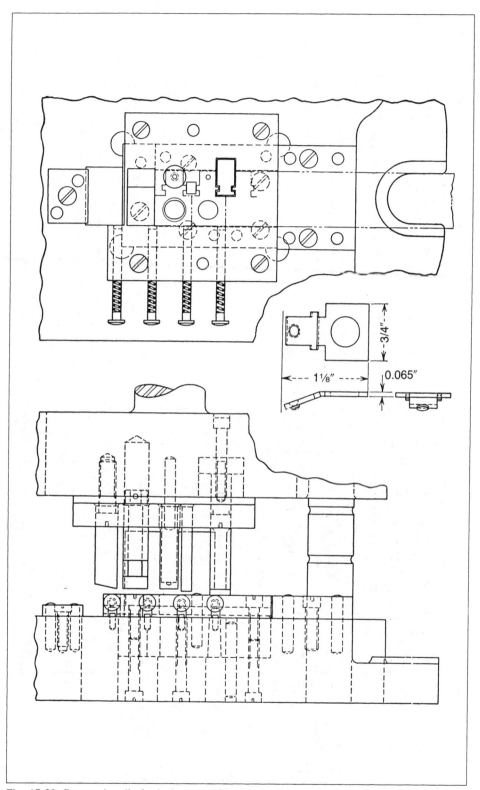

Fig. 17-29. Progressive die for lock part. (*Wilson-Jones Co.*)

17- 29

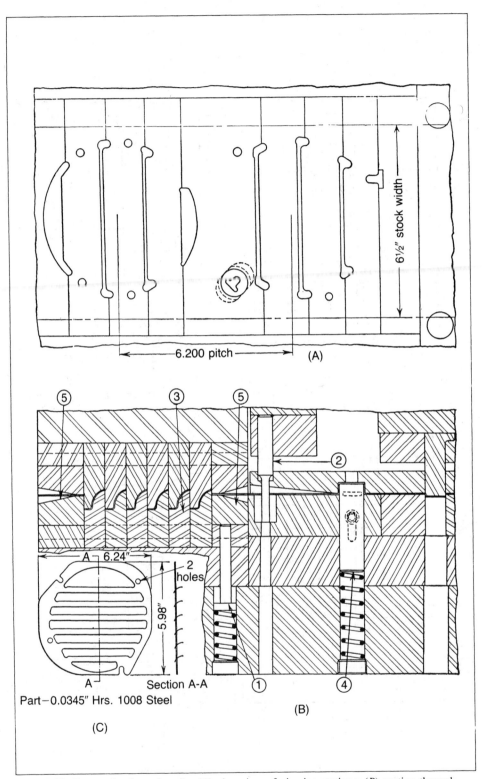

Fig. 17-30. Progressive louvering die: (A) plan view of piercing stations; (B) section through louvering station; (C) part. (*Buick Division, General Motors Corp.*)

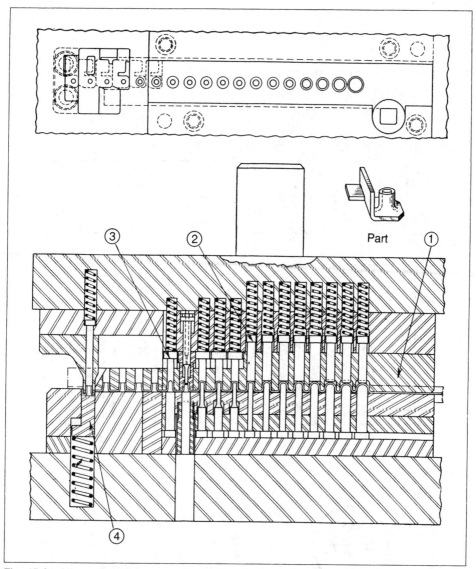

Fig. 17-31. Nineteen-station progressive die to make brass electric terminals. (*Harig Mfg. Corp.*)

(*D*3). Ejector pins (*D*4) push the part up and off the ring on the upstroke. A hold-down ring (*D*2) actuated by a positive knockout rod (*D*5) strips the part from the punch.

Progressive Die to Produce a Tank Flange. A progressively drawn, pierced, flanged, and trimmed gasoline-tank-filler neck flange, produced in a 10-station die (one idle station) is shown is Fig. 17-34. The strip development is shown in Fig. 17-34A. The strip remains horizontal during the operations. Because the tubular portion position slopes from back to front, a cam-operated punch is needed to pierce the bottom of the cup prior to the tube-sizing operation, which is also cam-operated. The strip is carried through the die on a spring-loaded strip carrier which travels upward 1.75 in. (44.4 mm) to enable the cups to clear the die cavities. The first station uses a double-concave or hourglass-shaped punch to produce a partially circular blank with carrier strips on each edge. The first draw is performed in station 2 producing a half-spherical cup. Figure 17-34B is a section through station 3, the second drawing operation which reshapes the cup. The blankholder (*D*1) and

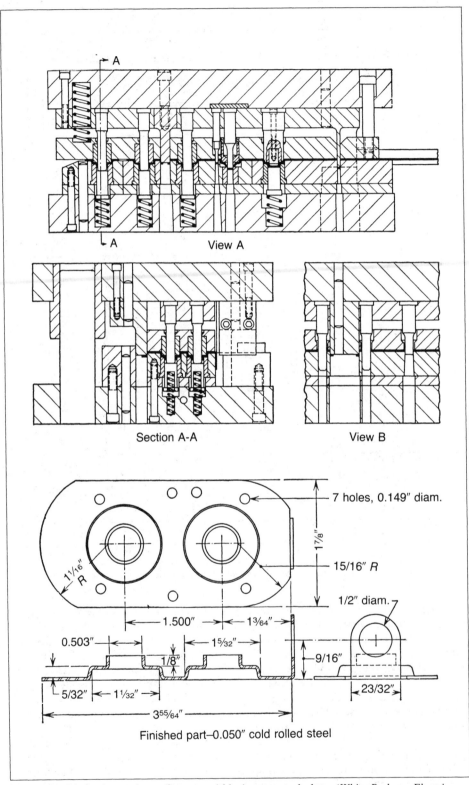

Fig. 17-32. Die to draw, pierce, flange, and blank out a steel plate. (*White-Rodgers Electric Co.*)

17-32

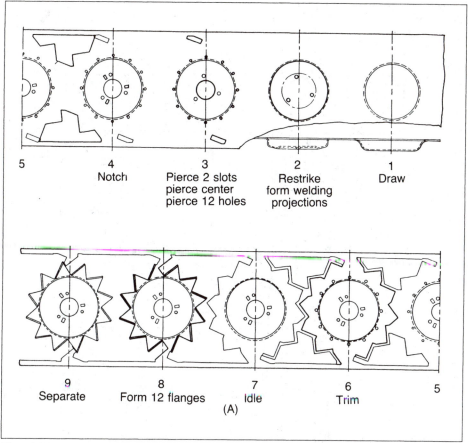

5	4	3	2	1
	Notch	Pierce 2 slots pierce center pierce 12 holes	Restrike form welding projections	Draw

9	8	7	6	5
Separate	Form 12 flanges	Idle	Trim	

(A)

Fig. 17-33. Progressive die for producing an automobile generator fan: (A) strip development.

the die (D2) are composed of individual sections in each station, a good practice from the standpoint of maintenance and possible change in design. The punch (D3) in all forming stations is made of two pieces of hardened tool steel to enable a smaller piece to be changed in case of wear or design change. A spring-loaded ejector (D4) is used in all forming stations to lift the cup out of the die cavity. The tubular section is further shaped, the bottom flattened, and the bead formed in the third draw at station 4 (Fig. 17-34C).

The bottom is pierced out of the cup with the inclined cam-operated punch shown in Fig. 17-34C. Stations 7 and 8 straighten and restrike the tubular portion to size. The forming punch (D5) shown in Fig. 17-34D is mounted in a sliding holder. The downward movement of the ram moves the sliding holder along the inclined surface (D6). A spring-loaded blankholder (D7) holds the strip firmly during the descent of the ram and strips the part from the punch on the ascent. The piercing operation at station 6 uses the same sliding toolholder arrangement. The spring-loaded die (D8) is split on its longitudinal center line and remains open while the ram is up so that the strip can be raised vertically. The cam (D9) attached to the punch holder locks the die closed for the forming operations. Two holes are pierced in the flange in station 9. At station 10, the flange is blanked to shape and the part drops through the die and bolster plate. This die is mounted in a heavy, cast-alloy-iron die set with four guide posts. The general construction of the die is heavy, using good steel and hardened parts to withstand the high production requirements.

Die for Fabricating a Spark-plug Seat Gasket. This progressive die produces five seat gaskets at a press stroke. The strip pattern shows alternate rows of three and two

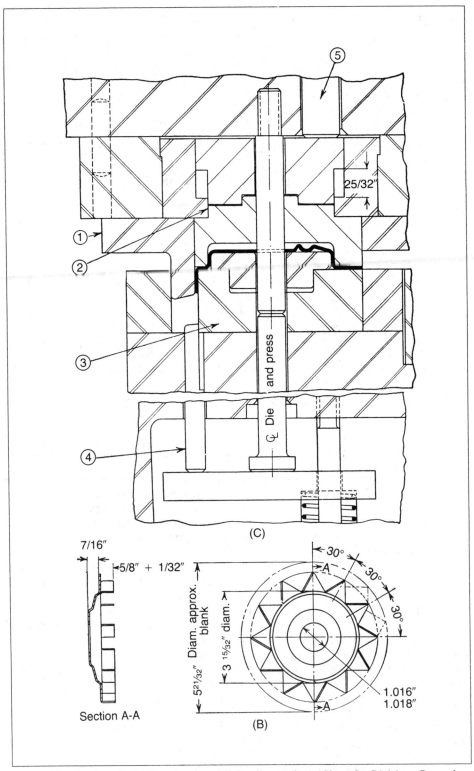

Fig. 17-33. (*Continued*). (*B*) Part drawing; (*C*) flanging station. (*Chevrolet Division, General Motors Corp.*)

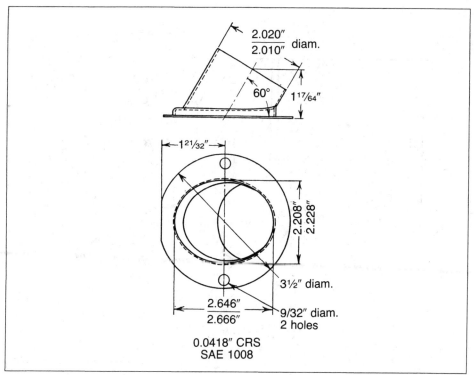

Fig. 17-34. Gasoline-tank-filler neck flange. (*Chevrolet-Flint Division, General Motors Corp.*)

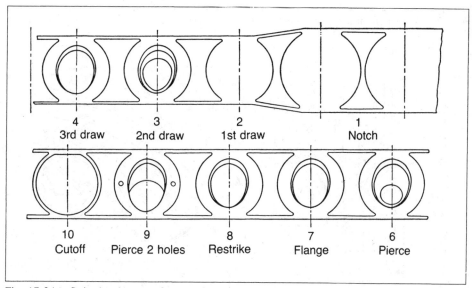

Fig. 17-34A. Strip development for tank-filler neck flange.

gaskets across a strip of 0.020-in. (0.51-mm) thick SAE 1008 sheet steel 5.75-in. (146-mm) wide. The part and a section through the die showing the development are shown in Fig. 17-35 (see page 17-38). The blank advances approximately 1.25 in. (31.8 mm) at each stroke of the press and there is an idle station between each working operation to allow more space in the die for heavy working stations. Station 1 shears the blank from the strip except for a small tab on each side to carry it through the die. Station 2 draws

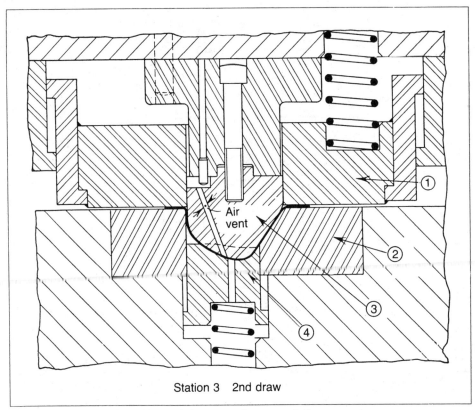

Station 3   2nd draw

Fig. 17-34*B*. Section through second-draw operation on neck flange.

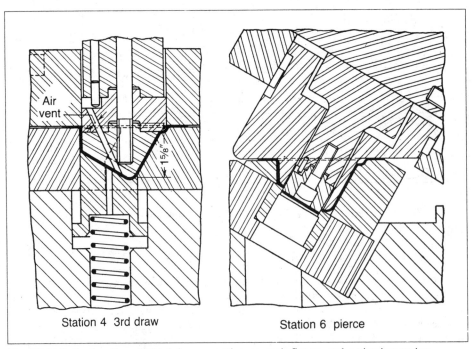

Station 4   3rd draw   Station 6   pierce

Fig. 17-34*C*. Section through third-draw operation on neck flange, and a piercing station.

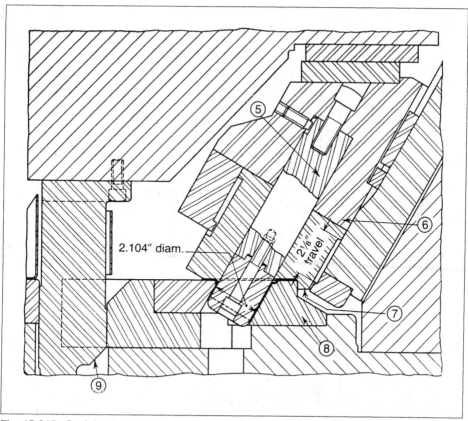

Fig. 17-34D. Straightening and restriking the tubular portion of neck flange.

a cup; station 3 forms a bead around the cup, and station 4 pierces the flat section from the bottom of the cup. The remaining radiused portion of the bottom is formed into a straight tube in station 5. The curling operation is done in station 6. The part is closed tightly in station 7, and the part is blanked and dropped though the die in station 8. All the punches and the die inserts are backed up by a 0.375-in. (9.52-mm) thick hardened and ground steel plate. Spring-loaded lifters push the part out of the die cavities, and spring-loaded stock carriers hold the strip up while advancing to the next station. Limit switches stop the press in case of misfeeding.

References

1. C.R. Cory, *Die Design Manual*, 1940.
2. G.H. De Groat, "Adjustable Cams Time Bending," *American Machinist* (June 9, 1952).

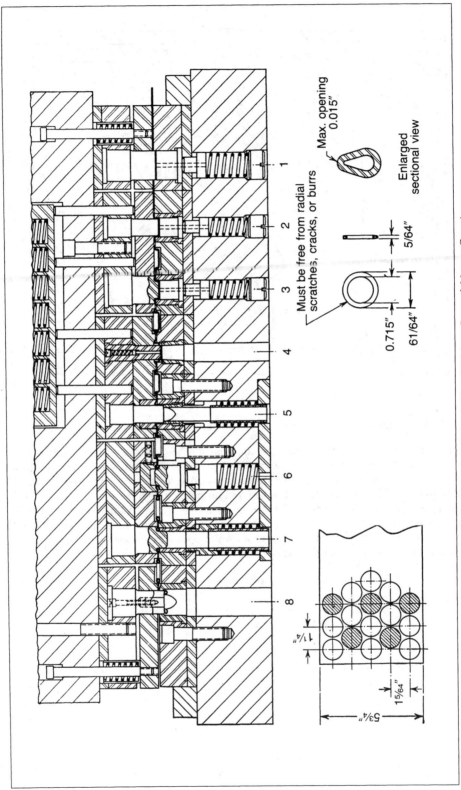

Fig. 17-35. Progressive die for producing spark-plug seat gaskets. (*AC Spark Plug Division, General Motors Corp.*)

17-38

SECTION 18
COMPOUND AND COMBINATION DIES

The terms *compound* and *combination* have been interchangeably used to define any one-station die, the elements of which are designed around a common center line (usually vertical), in which two or more operations are completed during a single press stroke. The dies described in this section are classified as follows:

1. *Compound Dies*. Press tools in which only cutting operations are done, usually blanking and piercing
2. *Combination Dies*. Press tools in which a cutting operation (usually blanking) is combined with a shaping or deforming operation (bending, forming, drawing, coining, etc.)

COMPOUND DIES

A common characteristic of compound-die design is the inverted construction, with the blanking die on the upper die shoe and the blanking punch on the lower die shoe. This construction commonly calls for the pierced slugs to pass through the lower die shoe.

Blank-and-pierce Dies. Compound dies are particularly useful for producing pierced blanks to close dimensional and flatness tolerances. Generally, the sheet material is lifted off the blanking punch by a spring-actuated stripper, which may be provided with guides to feed the material and a stop to position it for the next stroke. The blank tends to remain in the die, from which it is removed by a spring stripper or by a positive knockout. A positive knockout is most satisfactory when blanking relatively hard or heavy materials that remain flat without the use of a holddown or pressure pad. A combination spring-actuated blankholder and knockout is used for blanking thin and springy materials when flatness and accuracy are required. It also is used when the press has no positive-knockout attachment, or when the blank is too large to eject properly. Ejection of the blank from the die by spring or positive knockouts makes angular die clearance unnecessary, assuring constant blank size through the entire life of the die.

A typical example of a compound (blanking and piercing) die is shown in Fig. 18-1. During the cutting cycle, the stock is held flat between the faces of the stock stripper and the blanking die. The blanking die makes contact with the stock slightly before the piercing punch, which pierces a hole in the center of the piece after it is blanked out of the strip. As the piece is blanked out, the strip is carried below the cutting edge of the blanking punch (Fig.18-1B) brought back slightly above the punch level by the lower stripper.

A compound die for blanking and piercing, a clutch disk, is shown in Fig. 18-2A. The clutch disk (Fig. 18-2B) is made of 0.072-in. (1.83-mm) half-hard cold-rolled sheet steel. The blank is produced from a 10-in. (254-mm) wide strip, and a 6.5-in. (165-mm) diameter hole is pierced in the center. Subsequent operations in other dies pierce 12 small holes in the disk and bend up the ears on the five tongues. In this two-section compound die, the blank is cut from the strip and forced downward into the die by the punch.

The piercing punch (D1) is a solid block and fits a counterbore in the die shoe. The

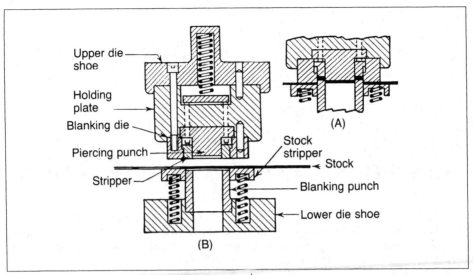

Fig. 18-1. Typical compound blank-and-pierce die.[1]

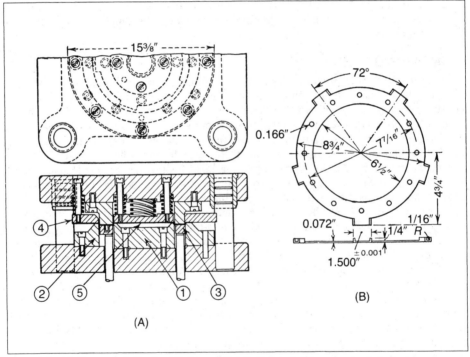

Fig. 18-2. (A) Compound die that blanks out clutch disk and pierces center hole in one operation. (B) Finished clutch disk. (L.L. Locke)

blanking die (D2) also is fitted in a counterbore in the die shoe. A pressure pad (D3), located between the die sections, is supported and operated by four pressure pins. Two stripper plates are operated in connection with the punch, one (D4) being located on the outside of the outer cutting edge, and the other (D5) on the inside of the inner cutting edge. Helical springs furnish stripping pressure for the plates. The construction of this compound tool is substantially the reverse of that shown in Fig. 18-1, because the blanking die and piercing punch are supported by the lower die shoe.

A compound die for making a pierced blank for a washer is shown in Fig. 18-3. One press stroke punches the center hole and blanks the piece from 0.015-in. (0.38 mm) cold-rolled-steel strip. The piercing punch is attached to the upper die shoe, and the blanking punch is attached to the lower die shoe. (The piercing punch contacts the material slightly ahead of the blanking die.) The part is stripped from both the blanking die and piercing punch by a positive knockout. The blanked strip is lifted off the blanking punch by a spring-loaded pressure pad.

In the blanking and piercing die (Fig. 18-4) the blanking die is made in three pieces whose cutting edges form the outside shape of the part. The part is blanked from 0.050-in. (1.27-mm) cold-rolled-steel strip. A blanking punch mounted on the lower die shoe mates with the sectionalized blanking die mounted on the upper shoe. Piercing punches mounted in the upper shoe pierce two small holes in the part as it is being blanked from the strip. A stripper plate removes the work from the blanking punch, and a shedder pin strips the blank from the small punches.

Three washers and a slug are produced at a single stroke of the press by the die shown in Fig. 18-5. Three concentric punches are attached to the upper shoe, and two concentric sleeve dies are attached to a special combination die block and lower shoe. Two concentric ejector sleeves fit between the punches and two concentric strippers for the blanking dies, one between the dies and one outside the outer die. The outermost of the three upper punches functions as a blanking die, cutting on its ID only. It is seated firmly in a groove in the punch holder and held in place with a screw-on ring. The intermediate blanking punch and the solid center piercing punch are integral and are screwed to the bottom of the punch holder. The ejector sleeve and knockout assembly slide freely between the punches, and gravity holds the ejectors down when the die is open. The knockout ejects the two washers when the press ram reaches the top of its stroke. The spring is intended to balance the weight of the ejector knockout so the washers will not drop out accidentally.

The strippers, which hold the stock and remove both the pierced stock and the intermediate washer, are actuated by a die cushion through pressure pins. The solid slug

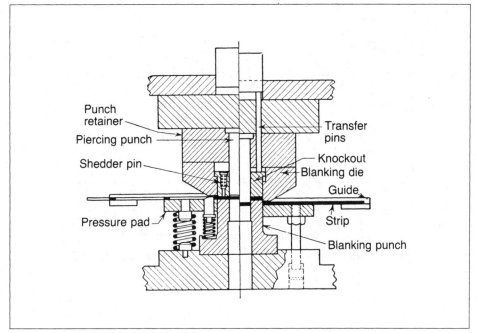

Fig. 18-3. Blank-and-pierce die for a washer. (*Cousino Metal Products Co.*)

18- 3

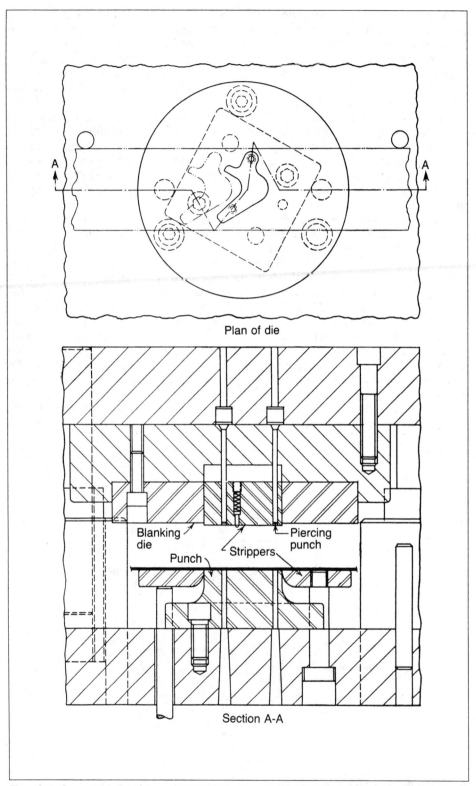

Plan of die

Blanking
die

Punch

Strippers

Piercing
punch

Section A-A

Fig. 18-4. Compound die with sectionalized blanking die. (*The National Cash Register Co.*)

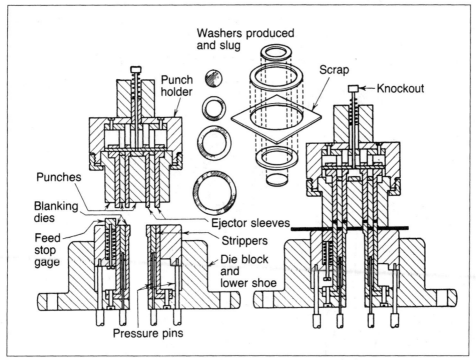

Fig. 18-5. Compound blank-and-pierce die makes three concentric washers at each stroke of the press.[2]

falls through the center die. All punch and die edges are sharp except the OD of the outside punch. This die cuts cardboard washers from 0.0625- (1.59-mm) and 0.125-in. (3.18-mm) stock but also could cut thin metal. A progressive die could be used to make these washers and automatically sort them.

Blank, Pierce, and Notch Die. In the compound die shown in Fig. 18-6, for producing gray fiber spool heads, the 0.094-in. (2.39-mm) sheet fiber stock is fed by hand and is located by a finger stop. This is an inverted die with the blanking punch mounted on the lower shoe. To simplify machining the contour of the cavity in the blanking die (D1), the projections into the cavity to cut the notches on the periphery of the blank are inserts in the die. These inserts (D2) are keyhole-shaped, the circular portion having a tapped hole to use in securing it to the die plate. The knockout (D3) is made in two pieces to facilitate the construction of slots to guide the four slender piercing punches (D4). The piercing punch for the center hole is also guided by the knockout bushing. The round and rectangular punches are secured in the die by the same retainer (D5). The knockout is actuated by a positive-knockout bar in the press through shedder pins.

The blanking punch (D6) is made in two sections. The outer profile of the outer section has the same shape as the blank. The inner profile of this section is circular, with notches or keyways the size of the small perforations. The inner section is circular, with flats to serve as the inner edge of the die for the piercing punches. A spring stripper contains the stock guides and strips the blanking punch.

Trim-and-pierce Dies. Most drawn shells must be trimmed. If the trimming is to be performed in a regular single-action press instead of a special flat-edge trimming press, and if other cutting operations are required after the draw, it usually is economical to combine these with the trimming operation. Often shells must be pierced, and the trimming and piercing are readily performed by a compound trim-and-pierce die, two examples of which are shown in Fig. 18-7. A die that pierces holes in a shell flange at the same time the edge is trimmed is shown in Fig. 18-7A. After trimming, in case the shell

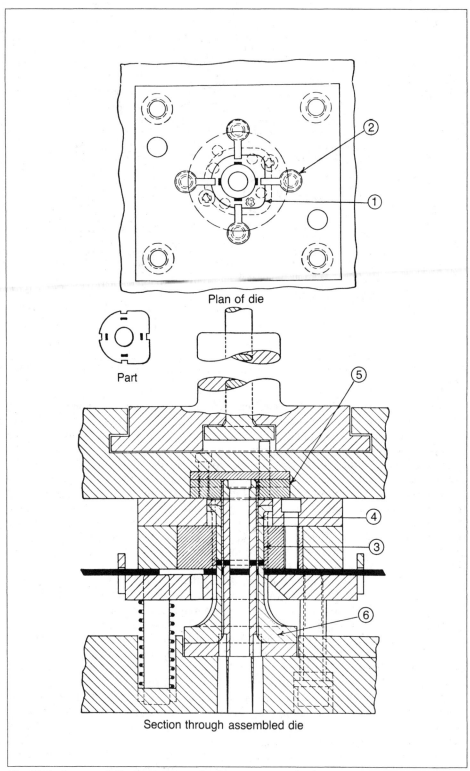

Plan of die

Part

Section through assembled die

Fig. 18-6. Compound die for blanking, piercing, and notching fiber spool head. (*Harig Mfg. Corp.*)

18 - 6

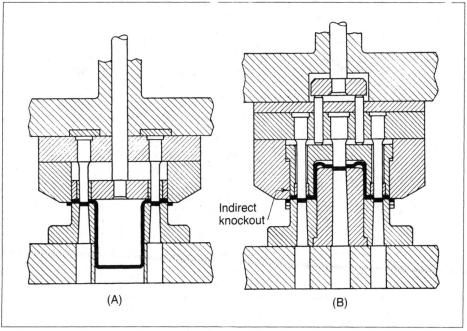

Fig. 18-7. Compound trimming and piercing dies: (A) for piercing holes in flange and trimming flange of deep shell; (B) for piercing holes in flange and bottom and trimming flange of shallow shell.[3]

is carried up by the upper die, it is stripped from the die and the punches by a knockout. An inverted die for trimming and piercing shallow shells is shown in Fig. 18-7B. Punches pierce holes in the flange, and a center punch pierces a hole in the shell bottom. An indirect knockout must be used because of the centrally located punch. Bushings in the knockout support the long slender punches that pierce the flange holes. The central punch is shorter and does not need support. Oversize parts would stick in these dies, and means of ejection should be provided.

The scrap material beneath the trim die and around the punch can be slit by a chisel-point scrap cutter for easier removal.

Shave-and-pierce Dies. A compound shave-and-pierce die (Fig. 18-8) shaves around the slot and teeth and pierces two holes in a key. A knockout is provided to prevent the part from remaining in the shaving die and a stripper plate strips the part from the shaving punch. A shedder pin prevents the part from adhering to the knockout.

In a compound die for shaving and piercing a signal stop yoke arm (Fig. 18-9), the shaving die (D1) is made up of three sections whose cutting edges form the outline of the part. The shaving die and piercing punch are attached to the upper shoe and the shaving punch (D3), in which the piercing die is incorporated, is attached to the lower shoe. A knockout containing an oil-seal breaker pin is provided to eject the part from the shaving die and piercing punch. A stripper plate is provided to free the shavings from the shaving punch. This part was previously blanked and pierced in the compound die shown in Fig. 18-4.

Broach, Cutoff, and Pierce Die. A broach, pierce, and cutoff die (Fig. 18-10) is used to produce a notched bar from cold-rolled-steel stock 0.188 in. (4.78 mm) thick and 0.50 in. (12.7 mm) wide. Holes 0.188 in. (4.78 mm) in diameter are located near each end of the part, and there are three notches along one edge. The notches are 0.125 in. (3.18 mm) wide by 0.032 in. (0.81 mm) deep and must have square edges all around with a good finish. A groove machined in the stripper plate accommodates the stock and guides it through the die. Base blocks support the stock during the broaching operation and act as

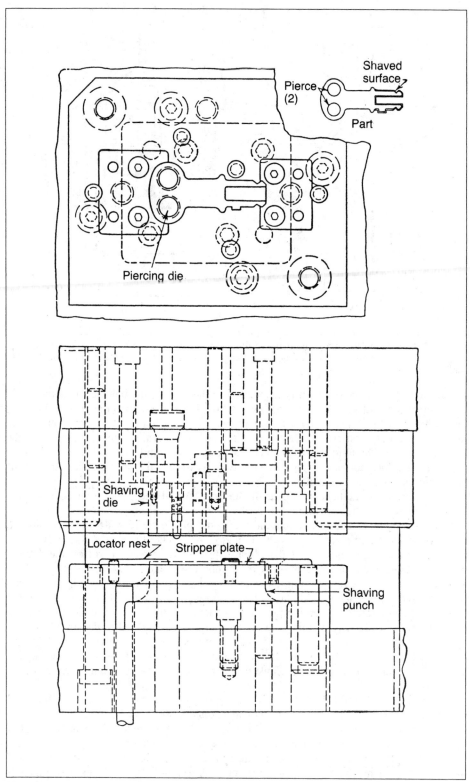

Fig. 18-8. Compound die for shaving and piercing a key. (*The National Cash Register Co.*)

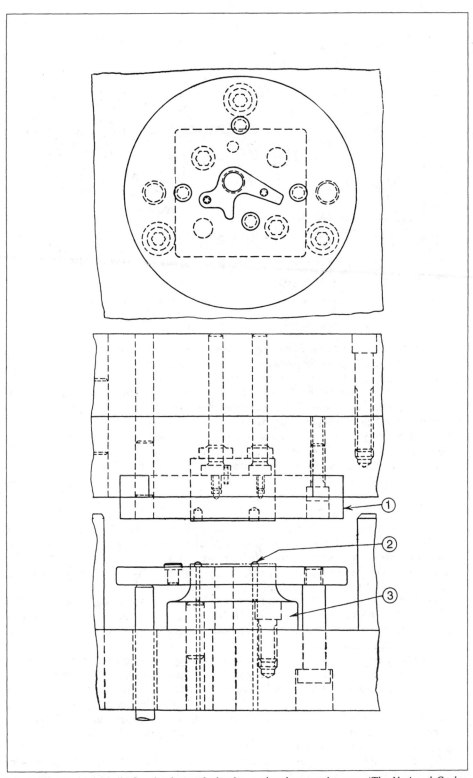

Fig. 18-9. Compound die for shaving and piercing a signal stop yoke arm. (*The National Cash Register Co.*)

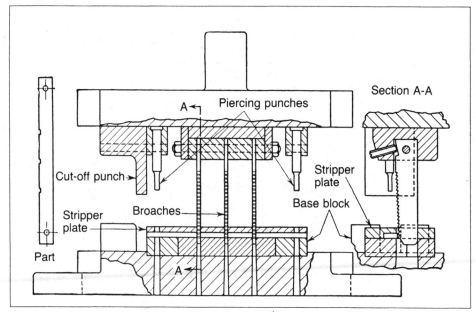

Fig. 18-10. Compound broach, cut-off, and pierce die.[4]

guides for the three broaches. A punch attached to the upper shoe cuts off the stock to length. A clearance hole in the punch blocks and cutoff punch permits removal of the broach-adjustment shaft when repairs are needed on the broach assembly.

When the die is in operation, the broaches engage the work almost immediately at the start of the press downstroke. When the broaches have completed two-thirds of their stroke, the cutoff punch contacts the stock and cuts it to length. The part is still held securely in place by the broaches, locating the part accurately, so the piercing punches, coming into contact with the work immediately after it is cut off, pierce the two holes in correct relation to the broached notches. After the press has completed its cycle, the stock is hand-fed against the outboard stop.

COMBINATION DIES

The range of types of combination dies is much greater than in the case of compound dies. Two or more operations such as forming, drawing, extruding, and embossing may be combined with each other or with the various cutting operations such as blanking, piercing, trimming, broaching, and cutoff. Much of the success of such dies depends on the provisions for stripping and ejecting finished parts.

Cutoff-and-form Dies. The die illustrated in Fig. 18-11 was designed to form clips (from hot-rolled low-carbon sheet steel), used as clamps in the assembly of a wire product. A design requirement is that the sides contact each other at the formed center section with a specified pressure.

Two punches are attached to the upper shoe, one for cutting off the strip and one for forming the clip. Strip stock is fed through a grooved guide block until it contacts a stop. As the ram descends, the strip is cut to length just before it is contacted by the forming punch. A pilot pin in the forming punch enters a prepierced hole in the piece to locate it accurately. The two forming blocks pivot around pins within openings in a die block with a U-shaped depression. The U-forming die is free to slide vertically, but is restrained by a spring until the forming punch descends.

After the workpiece has been cut to length, the forming punch bends it between the upper inside corners of the forming blocks. This temporary bend is removed as the work is forced into the U-shaped groove in the center of the die block. This die block is

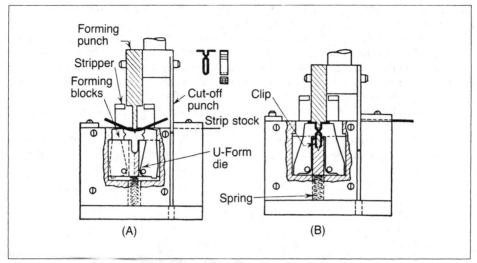

Fig. 18-11. Die for cutting off and forming a spring clip: (A) relative positions of die members at beginning of operation; (B) positions of die members at completion of forming operation.

restrained from sliding downward by a spring, until the center of the workpiece is forced to the bottom of the groove. Continued descent of the press ram forces the die block downward until the lower ends of the forming blocks contact the bottom of the base. This causes the forming blocks to pivot toward each other, forcing the workpiece into a rectangular opening in the forming punch, as seen in Fig. 18-11B.

On the upward stroke of the press, the forming blocks return to their original positions. This permits the formed clip to be carried upward until it is removed from the punch by the stripper. A blast of air then blows the clip clear of the die. Speed of operation for this die is 200 strokes per minute.

The combination die shown in Fig. 18-12 forms six bends in one press stroke to produce a spacer clip from SAE 1020 strip steel, No. 4 temper, 0.062 in. (1.57 mm) thick and 0.250 in. (6.35 mm) wide. Average production is 5,000 pieces per hour, with

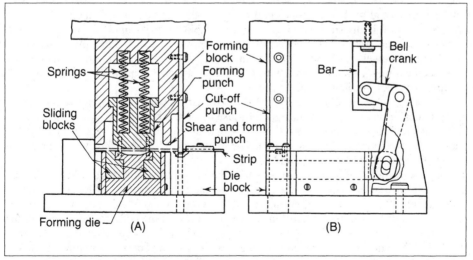

Fig. 18-12. Combination cutoff and forming die for spacer clip: (A) and (B) die with parts shown in position for beginning of operation.

the punch press operating at 150 strokes per minute. The clips are used as spacers on tubular framework and are spot-welded to the tubes at assembly.

In operation, stock is fed through a die block and guided by pins to a stop. As the press ram begins to descend, a cutoff punch enters the die block and shears the strip. Springs under pressure from the descending ram force a forming block to mate with a forming die and form the center depression in the clip. As the ram continues to descend, slight excess material is cut off by a sharp corner on one edge of the forming block, which shears the piece against the die block. At this stage, the lower ends of the forming block form right-angle bends near the ends of the piece over the corners of two sliding blocks (Fig. 18-13*C*). These two blocks slide to the rear to provide clearance so the inward projections of the forming block can bend the piece over the edges of the forming die thus completing the part (Fig. 18-13*D*).

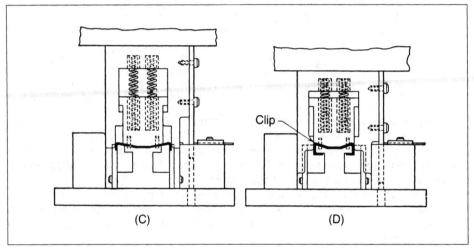

Fig. 18-13. Die of Fig. 18-12: (*C*) partial stroke, four bends completed; (*D*) end of stroke, sliding blocks withdrawn, all bends completed.

The sliding blocks are operated by a bell-crank lever. The longer arm of the bell-crank lever is attached to the blocks by a pin. The shorter arm is attached to a pin passing through a slot in a bar bolted to the ram. As the ram falls, the bell-crank remains stationary until the upper end of the slot contacts the pin. Further descent causes the bell-crank to pivot about its fulcrum and withdraw the sliding blocks from underneath the forming block. As the ram ascends, the bell-crank lever is not pivoted until the lower end of the slot in the bar comes in contact with the pin. At this point, the forming block and punch have been entirely withdrawn, allowing the sliding blocks to advance to their original position and eject the finished part.

The combination die shown in Fig. 18-14 is for lancing and forming louvers in a side panel of a cabinet. This die was designed for use in a press brake. Two louvers 6 in. (152 mm) long are formed at each press stroke. The spacing of the louvers is controlled by two pins (*D*1) in the spring pressure pad (*D*2). The sheet is held by the spring pressure pad and the upper die while being lanced and formed.

Die to Produce a Radiator Fin. Two longitudinal beads and three rows of flanged slots are formed at one stroke of the press in the die shown in Fig. 18-15. There are 57 slots spaced at 0.373 in. (9.47 mm) in each of the three rows. The stock used to make this radiator fin is 0.004-in. (0.10-mm) thick by 2.76-in. (70-mm) wide copper strip. On the downstroke, the stripper plate (*D*1) is depressed, carrying the stock over the shear-form punches (*D*2). The concave-shaped punch lances the stock and extrudes it upward into the die (*D*3). The punches are 0.086 in. (2.18 mm) wide by 0.766 in. (19.46 mm) long and held in the die by the keys (*D*4). Shedder pins (*D*5) actuated by the spring pad (*D*6) eject

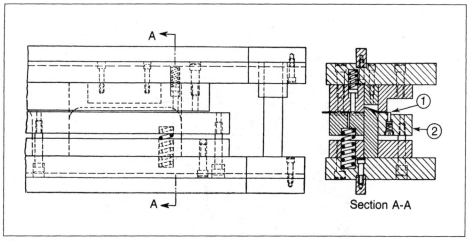

Fig. 18-14. Louvering die. (*Mills Industries, Inc.*)

the radiator fin from the die. Four guide pins assure proper alignment of the upper and lower sections during the operation.

Lance-and-form Die. In designing dies for simultaneously forming lugs equally spaced around the outside of a shell, the forming members must be the receding type to allow the shell to be removed from the die after the lugs are formed. An example of this type of die is illustrated in Fig. 18-16. Three lugs are formed equidistant around a circular shell. To form the lugs, the shell is centered on the pilot end of the holder. As the ram descends, the cam forces a slide radially inward, bringing two die inserts (one a shearing member for lancing, the other a forming member that forms the lug) into position against the shell. At the same time, the upper end of the shell is centered by a spring pad. While the ram descends, the pointed end of the actuating plug forces three pivoted punches radially outward and into the walls of the shell and thus the punches lance and form the lugs in the inserts.

The lower end of the cam, acting through an adjustable link, forces a punch-control ring downward, allowing the punches to move outward. As the ram returns, the spring at the bottom raises the cam and control ring, causing the forming punches to recede from the lugs and allowing the slide to move away from the shell, thus permitting removal of the finished shell.

The plug floats, so equal pressure is transmitted to all three punches. The plug is prevented from turning by a triangular section, free fit in a hole in the plate and secured to the punch block. Plug thrust is taken by a spherical shoulder, which normally is held against its seat by a coil spring.

Blank, Draw, and Pinch-trim Die. High production is achieved in the die shown in Fig. 18-17 by combining blanking from a strip with drawing and trimming operations. A blanking punch blanks and draws the shell over the plug, and the shell is pinch-trimmed by a punch in the lower member. A knockout ejects the shell at the top of the stroke, and a spring stripper removes the strip from the blanking punch. The pressure pad supplies pressure for drawing and strips the scrap ring from the trimming punch.

Blank, Draw, Form, and Pierce Die. A spice-box cover, part of a two-piece cover assembly, is made on the combination die shown in Fig. 18-18. Material for both the cover and "slide" which is used to close the openings in the cover is tin plate 0.010 in. (0.25 mm) thick. Both parts are produced on 28-ton (250-kN) double-action inclinable presses equipped with air-blast aided ejection. Metal strips are fed into the dies by automatic vacuum-feed attachments. Speed of operation (for the cover) is 135 strokes per minute.

At each stroke of the press the stock is fed in, blanked, drawn, and formed, characters

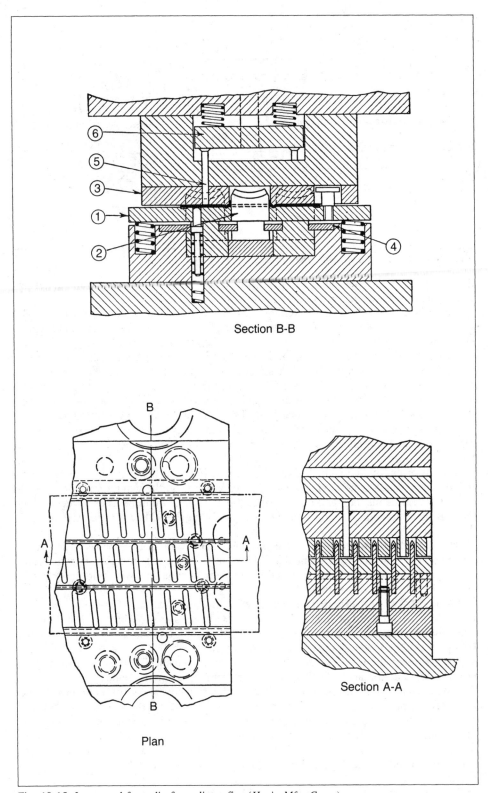

Section B-B

B

A——————A

B

Plan

Section A-A

Fig. 18-15. Lance-and-form die for radiator fin. (*Harig Mfg. Corp.*)

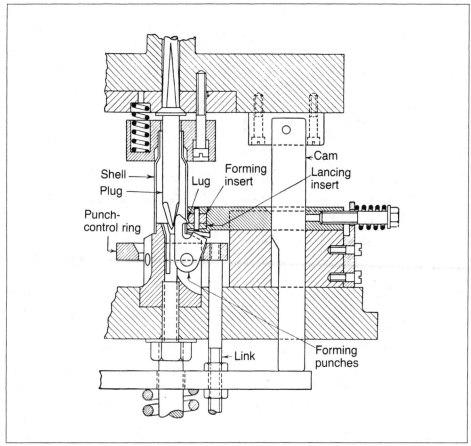

Fig. 18-16. Die for forming three lugs simultaneously around a shell. Only one die section is shown, since the other two lug-forming sections are similar. (*E.L. Soltner*)

stamped, spoon entry hole pierced, slug returned to the blank, and the finished cover ejected. Near the bottom of the stroke, the guides for the slide are formed, and the spoon entry hole is pierced and stamped with the letters "Punch In." The rubber pad in the upper die eliminates the necessity for minute coordination between the piercing and stamping dies and the forming punches. This rubber pad also provides the pressure for returning the slug to the blank. A positive knockout ejects the part from the die.

Pierce, Blank, Lance, and Emboss Die. Production of the slide for the spice-box cover (die shown in Fig. 18-18) is at the rate of 150 strokes per minute (four pieces per stroke) on the die shown in Fig. 18-19. This die also is sectional, and the four units are arranged to distribute pressure evenly over the die face. It is ejected by a positive knockout.

On the finished cover assembly, after the slide has been assembled to the cover plate, movement of the slide is performed by hooking a small U-shaped projection with a spoon or fingernail. The lancing punch used to form this projection is sharp on the sides but has rounded ends, and slits the two sides and draws out the projection without breaking the ends.

The die completes the piece in six virtually simultaneous steps: feed in, pierce small holes, blank slide, lance projection, emboss, eject. The holes are pierced with 0.076-in. (1.93-mm) diameter, long high-speed steel punches held in place with socket-head setscrews.

Cutoff, Form, and Curl Die. A valve for an air-control device is completed by

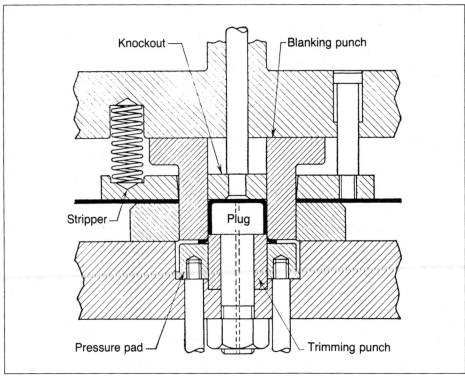

Fig. 18-17. Blank, draw, and pinch-trim die for producing shells.[3]

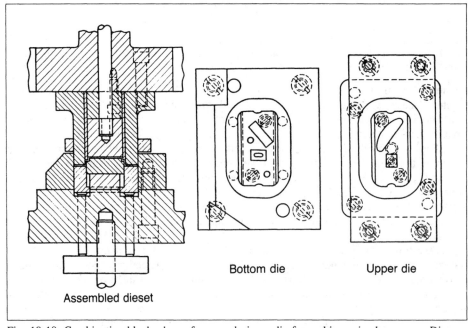

Fig. 18-18. Combination blank, draw, form, and pierce die for making spice-box cover. Die components are arranged in sections for adjustment after sharpening.[5]

combining three operations in one die (Fig. 18-20). Two of the operations (cutoff and form) are performed during the downstroke and the third (curling) is performed during the

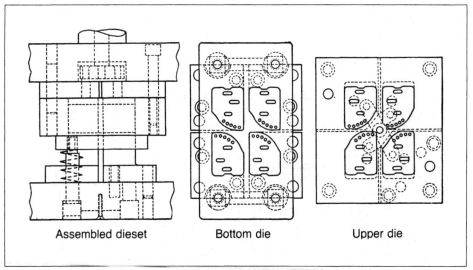

| Assembled dieset | Bottom die | Upper die |

Fig. 18-19. Die for producing slide, which is assembled to spice-box cover.[5]

upstroke. Steel 0.018 in. (0.46 mm) thick is used, and the parts range in length from 5 to 14 in. (127 to 355 mm).

As the punch plate descends, the material is held in place by the pressure pad. A cutoff punch shears the stock against the cutoff die as the ram continues downward. At a predetermined point, the punch plate contacts four pins in the bottom die, which compresses the die cushion, counteracting the upward pressure of the forming and curling dies. Forming punches force the metal down to the necessary depth, where half curls are produced by bottoming of the forming and curling dies.

As the forming punches rise with the upstroke of the press, the pressure pad continues to hold the part while the two curling dies are forced up by the die cushion, curling the two edges into the finished diameter.

Combination Blank, Form, and Pierce Die. The die shown in Fig. 18-21 blanks, forms, and pierces the pronged collet in one press stroke. This die is designed for use on a double-action press. The blank is cut by the punch ($D1$) and die ($D2$) on the outer press slide. On the inner slide, the spring-loaded drawing punch ($D3$) forms the blank into a shallow cup. Continued descent of the inner slide causes the draw punch to act as a blankholder, while the nail-point-shaped punch ($D4$) pierces the cup and forces the four prongs into the ring ($D5$). The ejector ($D6$) is actuated by the die cushion and lifts the part up to be blown out through the opening in the blanking die.

Cutoff, Form, and Pierce Die. Figure 18-22 shows a punch and die used to cut off, form, and pierce the part illustrated. In operation, strip stock is fed through the die until it contacts the end stop. After the press is tripped, the ram descends and a pressure pad comes down onto the strip stock to hold it securely. After the stock is cut to length, the swing punch contacts the material, carrying and forming it to suit the form in the die. Since the swing punch is free to swing on its center, it follows the die contour as it forms the bend in the part. Die construction is such that the left-hand angle-form operation is completed at the same time as the right-hand radius forming. Just prior to completion of the form operations, the punches pierce the two small holes in the part.

Blank, Draw, Form, Trim, and Pierce Die. A die that combines the five operations of blanking, drawing, forming, trimming, and piercing a ferrule is shown in Fig. 18-23. At the left of the vertical center line, the various parts of the tool are shown as they appear at the top of the press stroke; while the right-hand portion of the drawing shows relative positions of the parts at the bottom of the stroke. The ferrule is made of 0.020-in. (0.51 mm) thick cold-rolled steel, and the work is done in a single-action press.

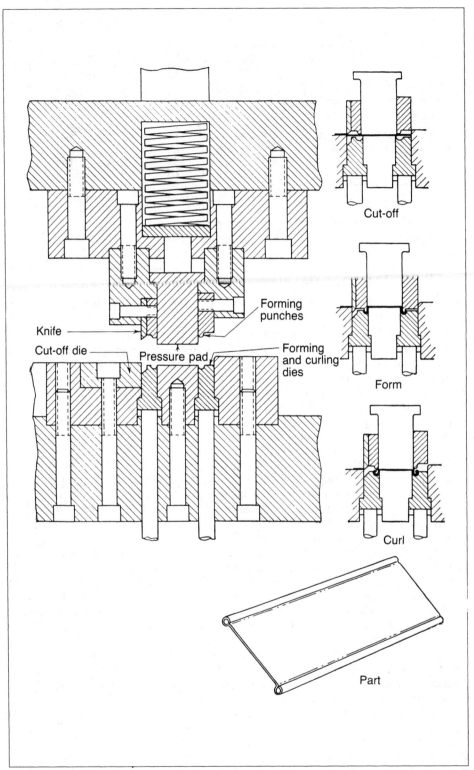

Fig. 18-20. Combination die for cutting, forming, and curling operations on a valve for air-control device.[6]

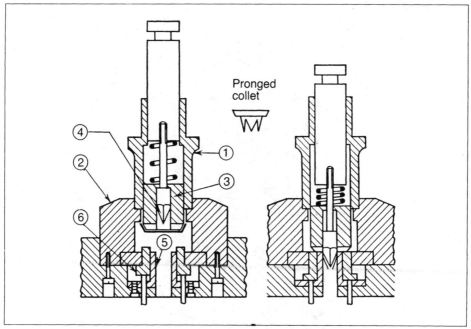

Fig. 18-21. Combination die to blank and form pronged collet.[7]

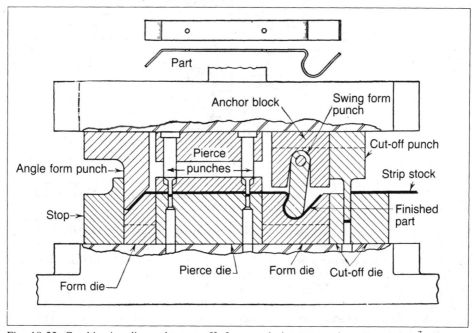

Fig. 18-22. Combination die used to cut off, form, and pierce a part in one operation.[7]

The blanking punch (D1) is screwed into a ring, above which is heavy coil spring (D2), designed to deliver ample pressure upon the punch to blank out the stock, yet allow it to recede as the press continues its downward movement. Blanking die (D3) has about 0.062-in. (1.59 mm) shear, which reduces cutting pressure required by about 50%.

The blanking die is held in place by a stripper, secured to the die shoe (D4) by screws. The combined drawing die and trimming punch (D5) is secured to the punch holder by a

18 - 19

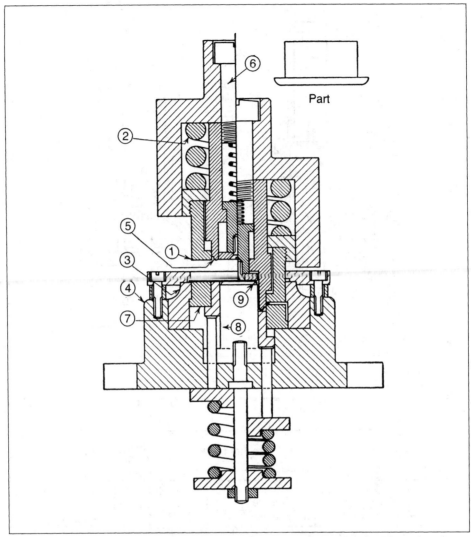

Fig. 18-23. Combination die to blank, draw, form, trim, and pierce a ferrule. Illustration shows relative positions of die components at both ends of stroke.[8]

large screw (*D6*) passing through the shank of the holder. Drawing ring (*D7*) inside the blanking die is made in two pieces and rests on plungers, by which the support is transferred to a washer above a heavy coil spring below the press bed. Drawing punch *D8* sets in a recess in the die shoe and is held down by a shouldered stud upon which the lower spring mechanism is assembled.

In operation, the strip of material is fed through the die from the right. As the ram descends, a blank is cut and gripped between the punch face and drawing ring in the die. As downward movement continues, the blank is drawn over punch *D8* by die *D5*. After the blanking punch motion is stopped, the turning motion of the drawing tool forms the flange of the ferrule.

As the downward movement nears the end of the stroke, the grip of the tools prevents further drawing of the stock, and the center of the shell is pierced out by the sharp upper corners (*D9*) of the drawing post. On the upstroke of the press, the ferrule is ejected by the drawing ring and is carried out of the tools by the next advance of the stock.

Blank-and-draw Die (Thin Stock). Dies for stock under 0.020 in. (0.51 mm) thick

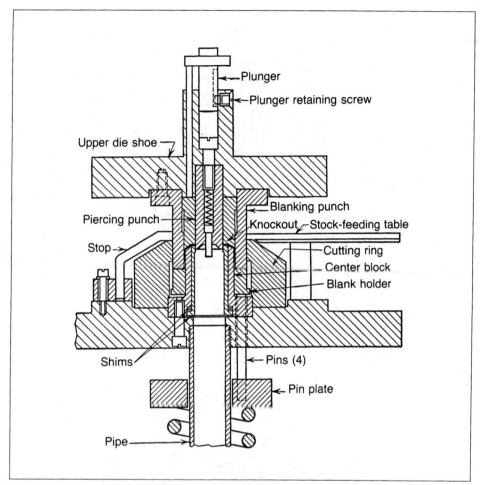

Fig. 18-24. Die combining center piercing with blanking and drawing of thin stock. The stock is blanked and drawn to shape before the center punch cuts to avoid distortion of the hole.[9]

differ from thick-metal dies in such construction details as clearance, die hardness, and various features of die design. Figure 18-24 illustrates a die for center piercing, combined with blanking and drawing. On such a die, the piercing punch should not cut until the stamping has been drawn to its full depth. This is especially true when the pierced hole in the bottom of the part is relatively large, and the blankholder consequently does not grip a sufficient area of stock to prevent enlargement of the hole. The center piercing punch on this die has a spring-loaded shedder pin to ensure ejection of the slug.

The bushing, which is the center blanking die, should be provided with shims which may be removed and built up under the bushing to raise it when it is shortened by sharpening. The bushing is given 1/2° taper and hardened. The center piercing punch, however, is semihard and forced through the bushing to avoid clearance and ensure a clean cut of thin metal.

Blank, Form, and Pierce Die. A combination blanking, forming, and piercing die (Fig. 18-25) produces a circular flanged part from 0.013-in (0.33-mm) tin plate. The punch blanks the stock to size as it passes through the die. To avoid wrinkling during the forming process, the stock is held by the upper blankholder against the blanking die and the lower blankholder against the forming punch.

The forming die is held in position by an auxiliary die cushion in the position shown in Fig. 18-25A, until the work is completely formed. Then, as the ram travels downward,

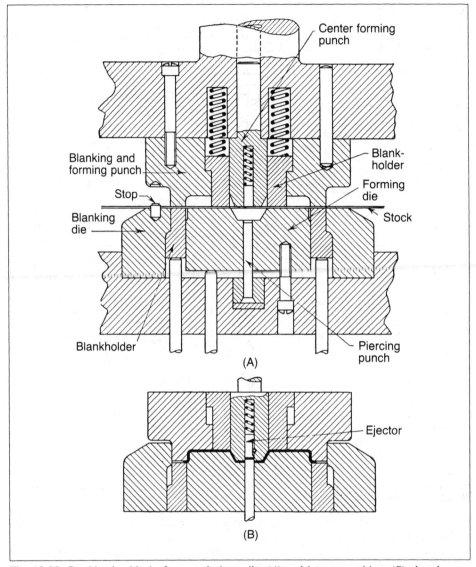

Fig. 18-25. Combination blank, form, and pierce die: (*A*) tool in open position; (*B*) closed position. (*F.W. Curtis*)

the forming die is forced down, allowing the piercing punch (which is held stationary in the die shoe) to pierce a 0.188-in. (4.78-mm) hole in the part. When the punch is raised, the ejector removes the slug from the die. The work itself is ejected by action of the pressure ring against the flange.

Pierce, Serrate, Countersink, and Blank Die. Figure 18-26 shows a cross-section of the punch-and-die assembly to produce the part illustrated. This part is made in extremely large quantities and must be held to fairly close tolerances. It has two pierced holes which must be countersunk to allow riveting of two studs. (Refer to Section 2 for permissible tolerances on stampings.)

The center hole on the part must be serrated to prevent a hub from turning under radial load after it is staked into location. Because the hub staking is a secondary operation, the serrated indentations around the edge of the hole are required to be consistently and accurately located.

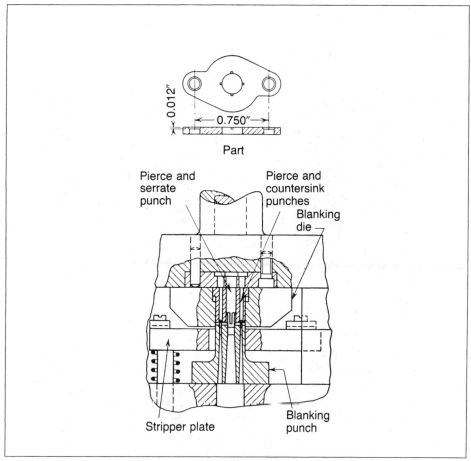

Fig. 18-26. Combination pierce, serrate, countersink, and blank die.[10]

Since the parts are not carried through the die but forced back out after blanking is complete, walls of the die hole are straight, not tapered as is conventional for clearance. A knockout rod assures positive action of the ejector. A stripper plate carries the stock up and off of the punch.

Punches must be very accurate, so that the 0.012-in. (0.30 mm) depth to be countersunk can be attained in both holes. Stop blocks are mounted to punch and die holders at opposite corners to limit the ram downstroke and punch travel, thereby ensuring the desired depth.

Blank, Draw, and Pierce Die. A 60-pitch, 24-tooth gear is made from 0.030-in. (0.76 mm) nickel silver on the combination die shown in Fig. 18-27. As the press slide descends, the strip is held between the faces of the spring stripper ($D3$) and the blanking die ($D4$), while the cup is drawn into the spring-loaded draw die ($D5$) by the combination perforating die and draw punch ($D2$). As the press slide continues downward, the center hole is pierced, the OD is blanked by $D1$ and $D4$, and the bottom of the cup is flattened between $D2$ and $D6$. The pierced slug falls through the hole in the perforating die.

On the upstroke, $D5$ strips the part from the blanking die and forces it back into the strip, from which it is easily removed when desired. A round-nosed pin ($D7$), which fits the center hole in the work, positions the strip for the next stroke.

Shear-and-form Die. Figure 18-28 illustrates a die that shears and forms five blades on a ventilating fan from a precut blank. The shearing punches ($D1$), which have shear ground on them, cut the blades to shape, then form a 90° bend on the heel end of each

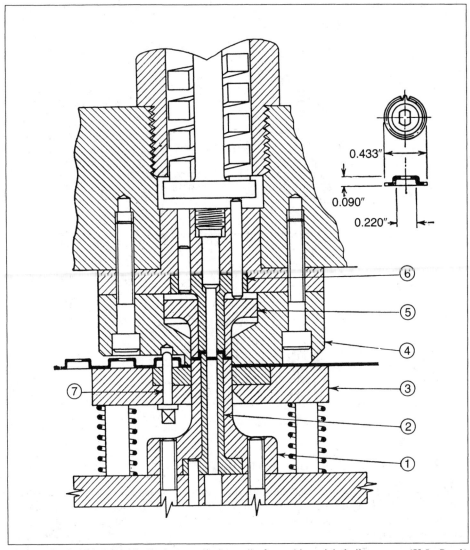

Fig. 18-27. Combination blank, draw, and pierce die for making nickel-silver gear. (*H.L. Barth*)

blade. These punches are individual and mounted on the lower shoe. Shedder pins actuated by the positive knockout remove the part from the punches.

Pierce-and-form Die. Center pierce and tab-forming operations are performed on a ball-bearing retainer by the combination die illustrated in Fig. 18-29. This is a secondary operation, and the preformed part is located by a portion of a ball-shaped locator (*D*1). The tabs are formed between the forming punch (*D*2) and forming die (*D*3). The forming punch is supported by the die cushion and moves downward to allow the retainer to be pierced by the punch (*D*6) ejects the scrap from the die and actuates the shedder pins (*D*7) to remove the retainer from the form die. The spring stripper (*D*8) lifts the part from the forming and piercing punches.

Redraw, Pinch-trim, and Pierce Die. The redrawing, pinch-trimming, and piercing of an oval-shaped can are shown in the combination die in Fig. 18-30. The blank holder is supported by the die cushion and retracts with the closing of the die to reduce the can to size. Near the end of the press stroke, the open end is pinch-trimmed to height and the bottom is pierced by two punches. A positive knockout removes the shell from the die,

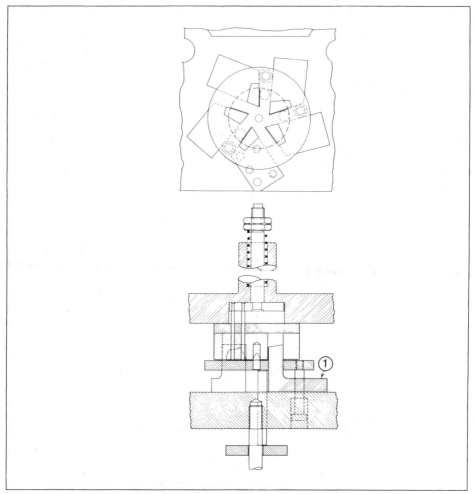

Fig. 18-28. Combination shear and forming die for five-blade fan made from 0.015-in. (0.38-mm) steel. (*Harig Mfg. Corp*)

and a slight expansion of the open end of the shell enables the blankholder to lift it from the draw punch. Air holes are provided to eliminate a vacuum or back pressure in the die.

Pierce, Blank, and Emboss Die. A combination die (Fig. 18-31) blanks a gear to shape and pierces holes in the blank, and embosses welding projections on the surface of the gear blank. An embossing punch, projecting upward from the blanking punch, forms a projection by forcing metal into a hole in the knockout. A second punch, in the hole in the knockout, forms a flat top on the projection. A shedder pin is used to break the oil seal between the face of the stripper and the finished part.

Combination Die for 12-gage Steel Shell. The combination die in Fig. 18-32 blanks, draws, pinch-trims, pierces, and embosses a symmetrical shell of 12-gage (0.1046 in./2.65 mm) cold-rolled deep-draw-quality steel. The blanking die has shear ground in it to reduce shock and blanking load on the press. After the bottom is pierced, eight radial indentations are embossed in the bottom of the shell. The pinch-trim operation on this gage stock left a chamfer around inside the open end of the shell, which was desirable in this case, since it is chamfered more in a later operation. Ejection from the punch is accomplished by the air-cushion and from the die by a positive knockout. If the part is not immediately removed from the punch, the cooling of the part will shrink it enough to freeze to the punch. The success of the operation is largely dependent upon the proper

18 - 25

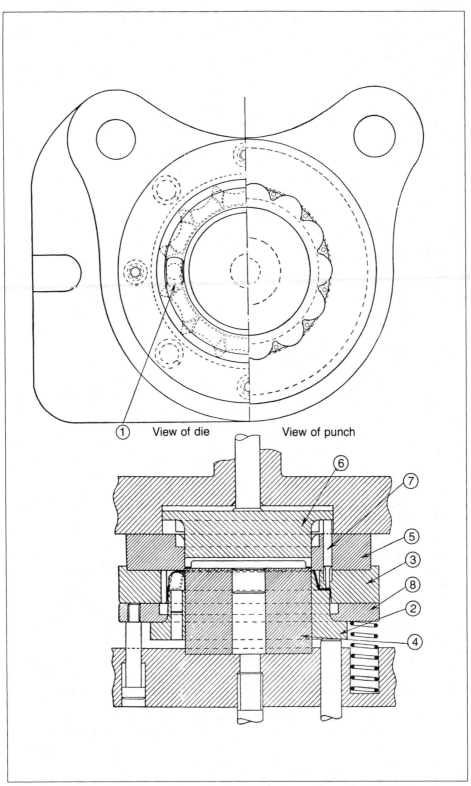

① View of die   View of punch

Fig. 18-29. Combination pierce-and-form die for ball-bearing retainer. (*Harig Mfg. Corp.*)

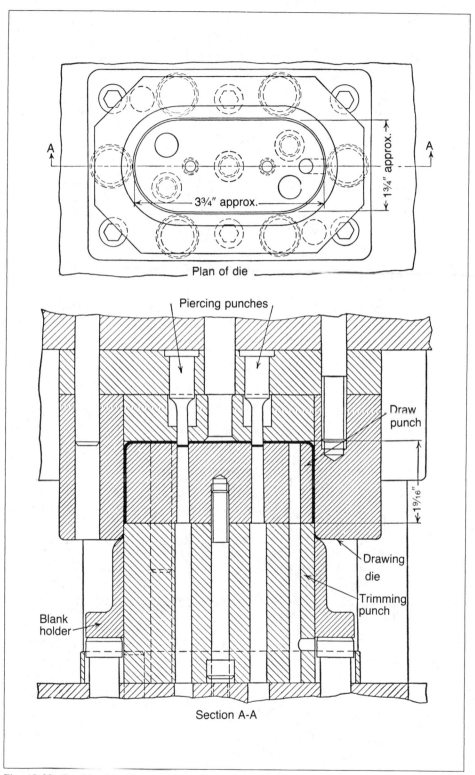

Plan of die

Piercing punches

Draw punch

Drawing die

Trimming punch

Blank holder

Section A-A

3¾" approx.

1¾" approx.

1⁹⁄₁₆"

Fig. 18-30. Combination die to redraw, pinch-trim, and pierce two holes in steel cover. (*White-Rodgers Electric Co.*)

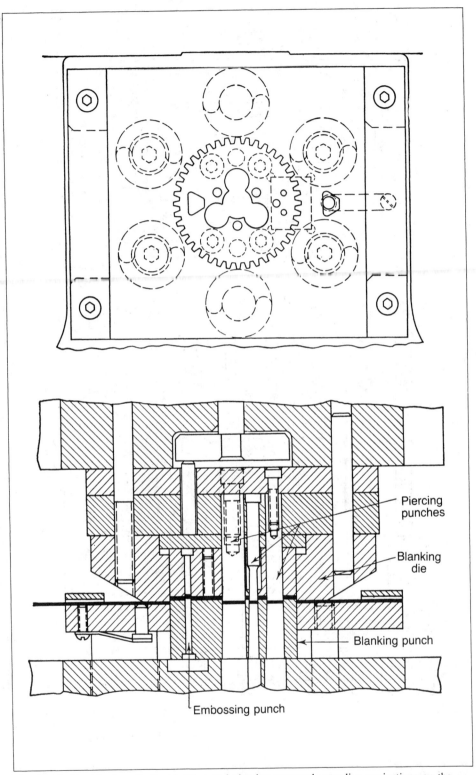

Fig. 18-31. Combination die for blanking and piercing gear and extruding projections on the surface. (*Barth Corp.*)

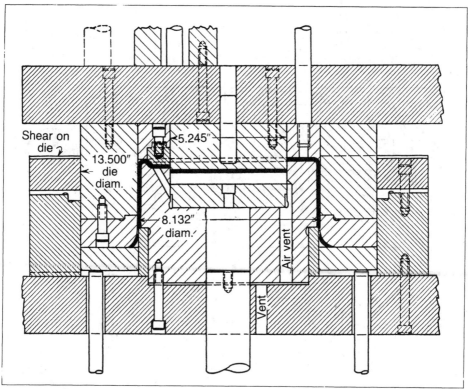

Fig. 18-32. Combination die produces symmetrical shell of 12-gage (0.1046-in./2.65-mm) stock. (Davey Products, Inc.)

clearance between the pinch-trim ring and the draw ring. Clearance, per side, should be 10% of stock thickness, so the metal will pinch to the point of fracture and then tear clearly, leaving a pinched-off scrap ring that will overlap the shell periphery and lift it from the punch on the return stroke. A mirror finish on the draw radius of the draw ring, and a speed of the 150-ton (1,330 kN) hydraulic press suitable to the drawing compound used, are prime factors in the operation.

COMBINATION DIE USING NITROGEN MANIFOLD SYSTEMS.

High pressure nitrogen systems, compared to springs, have important advantages. Generally, much higher pressures per unit volume are possible. The pressure is available at the start of the cylinder stroke.

The combination die in Figure 18-33 produces the faucet housing shown in Figure 18-33A in a single press stroke. Cutting steels, (D1) attached to the lower die and (D2) attached to the upper shoe, cut the blank from coil stock fed into the die. As the die closes further, (D2) acts as an upper blank holder as the stock is sequentially drawn and wiped over the post (D3). (D4) is the binder for the draw and wipe sequence. Binder pressure is supplied by a nitrogen manifold unit (D5) which is placed in a milled opening in the lower shoe. Finally, the three holes are pierced and (D2) pinch trims the part against the sharp post steels (D6).

The upper shoe also functions as a nitrogen manifold (D7), which is used to apply pressure to the pad (D8) used to hold the stock in place for blanking.

Drilled passages in (D5) and (D7) serve as surge tanks to prevent excessive pressure increases as the cylinders are stroked. Large port plugs with O-ring seals are used to close the ends of the drilled passages, as well as to seal the threaded cylinder to the manifold joints.

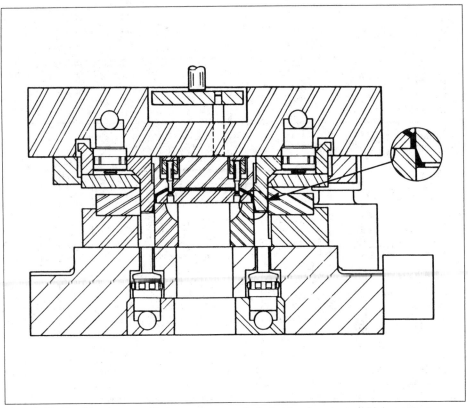

Fig. 18-33. Combination cut-off, draw, pierce, and pinch trim die uses two high pressure nitrogen systems.[11] (*Forward Industries*)

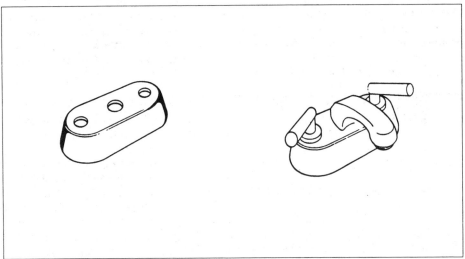

Fig. 18-33A. Faucet housing.[11] (*Forward Industries*)

Nitrogen manifold systems operate at gas pressures as high as 1,500 psi (10.3 MPa). The individual cylinders, which screw into the manifold, are available in several sizes with pressure ratings from 1 to 6 tons (8.9 to 53.4 kN).

An additional advantage of using a nitrogen manifold system is that the pressure is adjustable to allow compensation for stock variations.

References

1. J.W. Lengbridge, "Types and Functions of Press Tools," *The Tool Engineer* (June 1949).
2. B. Menkin, "Compound Punch and Die Makes Three Washers," *American Machinist* (October 23, 1947).
3. J.R. Paquin, "How to Choose Trim Dies for Drawn Shells," *American Machinist* (October 30, 1950).
4. K.L. Bues, "Die-Grams," *Western Machinery & Steel World* (March 1950).
5. H. Dahl, "Rotating Punch Bends Small Rings," *American Machinist* (June 9, 1952).
6. W.M. Stocker, Jr., "Compound Dies Blank, Pierce and Form Multiple Parts," *American Machinist* (July 21, 1952).
7. B. Menkin, "Practical Ideas," *American Machinist* (November 10, 1952).
8. K.L. Bues, "Die-Grams," *Western Machinery & Steel World* (July 1949).
9. J.A. Strama, "A Multiple-operation Press Tool," *American Machinist* (February 11, 1926).
10. W.C. Mills, "Thin Stock Demands Special Die Design," *American Machinist* (February 11, 1926).
11. G.A. Stauch, "Solving Die Design Pressure Problems Using Nitrogen Die Cylinder Systems," SME Technical Paper TE87-759 (Dearborn, MI: Society of Manufacturing Engineers, 1987).

SECTION 19

DESIGNING PRESSTOOLS FOR FINEBLANKING*

Fineblanked Part Characteristics. Although fineblanked parts look much like conventionally stamped parts, close examination will show that they have many characteristics of a precision machined part. Generally, the fineblanking process can produce closer part-to-part tolerances at a higher production rate than is possible with any machining process having equal precision capabilities.

Fineblanking is a hybrid stamping process that produces parts essentially by cold-extruding them from strip or sheet material. As shown in Fig. 19-1, fineblanked parts

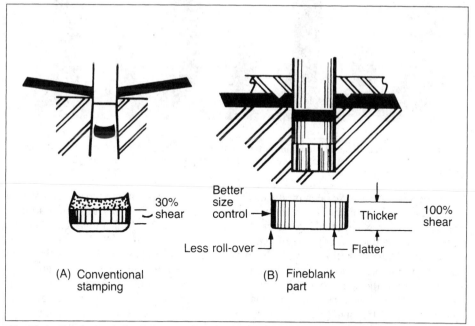

Fig. 19-1. Comparison of a fineblanked part (A) to a conventionally stamped part (B). (*Feinblänking Ltd.*)

have near-perfect square edges free from breakage; the process has the additional advantage of being able to perform many forming and sizing operations in a single station.

While the smooth edges are the most apparent difference when comparing a fineblanked part with a conventional stamping, the elimination of secondary operations,

*Contributed by Karl A. Keyes, President, Feinblänking Ltd., Fairfield, Ohio.

such as drilling, reaming, milling, hobbing, embossing, coining, and shaving, is the chief advantage. The other advantages include:

1. Flat parts practically free from rolled edges.
2. Holes, external tabs, and slots can be smaller than stock thickness.
3. Predictable cut-to-form relationships at the design stage.
4. Good mechanical properties due to cold-working.
5. A wide range of three-dimensional shapes can be cold-formed.
6. Use of materials to 0.625 in. (15.88 mm) or possibly greater thickness.
7. Break-free cutting edges up to 100% of metal thickness.

FINEBLANKING TOOLS

Function of the Impingement Ring. The secret of fineblanking is rigid stock control. Counter-blanking and material clamping pressures are combined with the restriction of lateral movement by a V-shaped impingement ring placed around the periphery of the part on the blank-punch side. Figure 19-2 illustrates the V-shaped impingement ring.

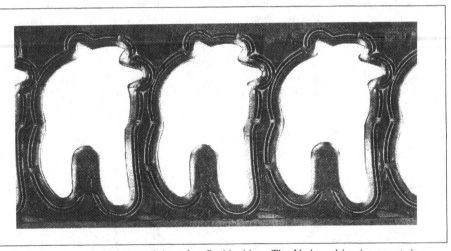

Fig. 19-2. The scrap skeleton remaining after fineblanking. The V-shaped impingement ring restricted lateral movement as it clamped the material rigidly against the matrix plate. (*Feinblänking Ltd.*)

While some parts have been produced as fineblanked parts without the use of a V-ring, it is a requirement for true fineblanking quality. Also, a V-ring may be required on the matrix (die) side when blanking materials over 0.200 in. (5.08 mm) thick or when it is necessary to produce an edge free from roll-over.

Sequence of Operations. With the press open, material is fed into position (Fig. 19-3A). As the press closes, the V-ring (B) starts imbedding into the stock, clamping it tightly against the surface of the matrix plate. As the die continues to close, the blank-punch makes contact with the material. At the same time, the counter-punch, a mirror-image of the blank-punch, resists the forward movement of the blank-punch.

At this point, the blank-punch must overcome both the shear resistance of the material and the force being applied by the opposing counter-punch (C). By controlling the material in this way, and using tight punch to matrix (die) clearances, extremely close tolerance parts with break-free edges can be produced.

At the end of the stroke, the press starts to open (D) leaving the fineblanked part secure in the matrix cavity.

The material is stripped off the blank-punch (E). Slugs, if any, are ejected. Both the part and slugs are evacuated from the press (F) where they are separated.

To properly accomplish this sequence of operations, a triple-action press is required.

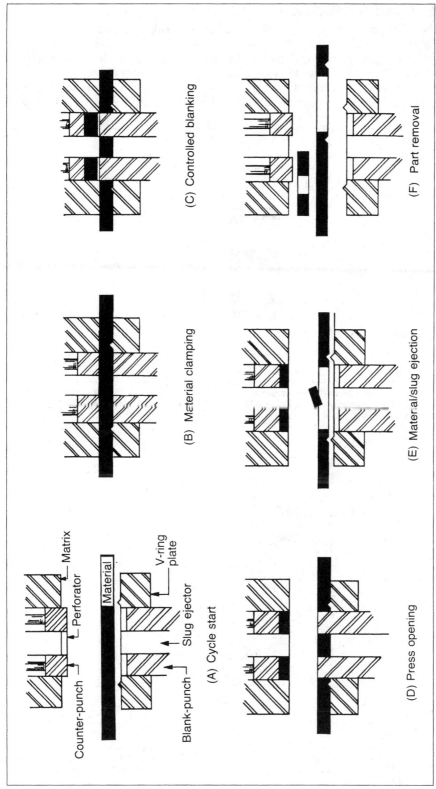

Fig. 19-3. Sequence of fineblanking operations; material is fed into position (A); the V-ring (B) imbeds into the stock; the blank-punch opposed by the counter-punch shears the material into the matrix (C); stroke ends and press opens (D); the material is stripped away and the slug ejected (E); part evacuated from the matrix (F). (*Feinblänking Ltd.*)

19-3

FINEBLANKING PRESSES

Fineblanking presses are built in both mechanical and hydraulic types. The V-ring and opposing pressures are actuated by controlling the exhaust oil leaving the working cylinder, or by oil pressure being sent directly to the side opposite where the work is being done, or both. Figure 19-4 illustrates a typical hydraulic fineblanking press.

Figure 19-5 illustrates a type of mechanical fineblanking press. Generally such presses are suited for smaller fineblanking work due to the limited tonnages available. Mechanical

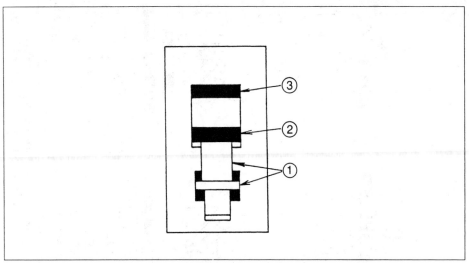

Fig. 19-4. Principle of operation of a triple-action hydraulic fineblanking press: hydraulic pressure raises the main ram (1); the lower table (2) clamps the material; the upper table (3) controls the part being blanked. (*Feinblänking Ltd.*)

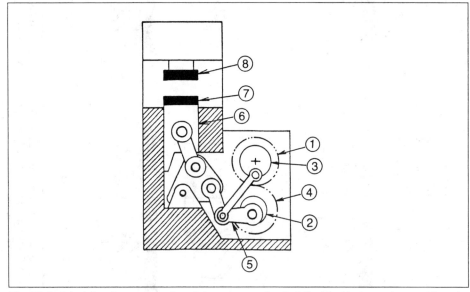

Fig. 19-5. Principle of operation of a triple-action mechanical fineblanking press: gears (1) and (2) are driven by a variable-speed power transmission (not shown); crankshafts, (3 and 4) actuate a double knuckle joint mechanism (5) which operates the slide (6); as the slide moves upward, hydraulic pressure in the lower table (7) clamps the material; hydraulic resistance in the upper table (8) controls the part being blanked. (*Feinblänking Ltd.*)

presses have an advantage of having higher stroking rates than hydraulic presses (up to 160 spm) and higher energy efficiency.

Many fineblanking presses are placed in pits to bring the presstool height to a convenient level. The pit can also be used to collect metalworking fluids for reuse.

Since most presses operate from the bottom up, presstool placement can be completed quickly and support rings (inserts) changed with ease.

Some of the smaller presses provide for mounting the blank-punch downstairs in a manner similar to a conventional compound die. This arrangement has the advantage of providing gravity-assisted part ejection.

Transfer of Force. V-ring and opposing forces are obtained in one of two ways depending upon the press design.

Large presses have pressure pins which supply the force. Support posts can then be added to reduce the load span thereby reducing presstool deflection. Generally, the use of support posts will improve part quality and reduce the frequency of tool regrinding. Figure 19-6 illustrates a tool having support posts.

In smaller hydraulic presses, these pressures are taken directly from the end of the piston. This requires a presstool of robust construction in order to bridge the load span without excessive deflection. Figure 19-7 illustrates a tool having no support posts.

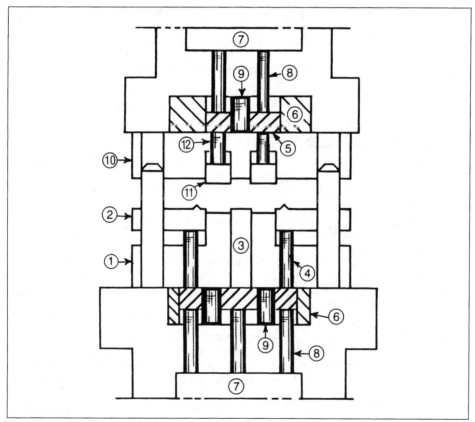

Fig. 19-6. A fineblanking presstool having support posts: (D1) blanking-punch assembly; (D2) V-ring plate; (D3) slug ejector; (D4) pressure pin (V-ring); (D5) pressure pad; (D6) support insert; (D7) hydraulic piston; (D8) pressure pin; (D9) support post; (D10) matrix assembly; (D11) counter-punch; (D12) pressure pin (counter). (*Feinblänking Ltd.*)

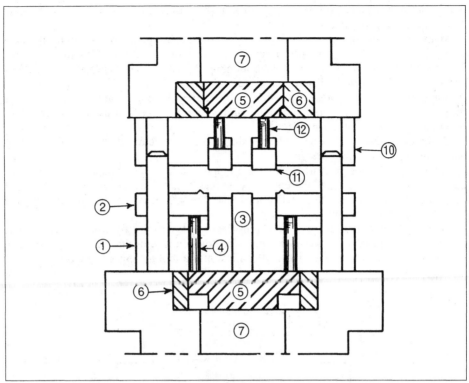

Fig. 19-7. A fineblanking presstool having no support posts (see Fig. 19-6 for key to numbers). (*Feinblänking Ltd.*)

FINEBLANKING PRESSTOOLS

Two basic presstool designs are used to produce fineblanked parts: sliding or moving blank-punch and fixed or stationary blank-punch.

The determination for when to use which design is based on one or more of the following considerations:

1. Press design (sliding, fixed, or combination).
2. Part thickness and/or size.
3. Ease of construction.
4. Type of presstool (unistation or multistation).
5. Personal preference.

Figure 19-8 illustrates a sliding or moving punch presstool. A fixed or stationary presstool is illustrated in Fig. 19-9.

Unistation Presstools. Sliding and fixed blank-punch presstools are further designed into unistation (compound) or multistation (progressive or transfer) presstools.

Unistation presstools are limited to producing only certain types of part characteristics. These characteristics include but are not limited to: blanking, round and shape perforating, coining, semi-piercing, lancing, chamfering, internal shaping, offsetting, and forming prior to blanking.

A major limitation of unistation presstools is the inability to work both sides of the part. Being able to work both sides can further reduce or eliminate other types of costly secondary operations.

Figure 19-10 illustrates a typical unistation presstool layout.

Multistation Presstools. Multistation presstools can produce all of the characteristics possible in unistation tools along with others which require more than one station to develop. All conventional progressive or transfer die operations can be duplicated with

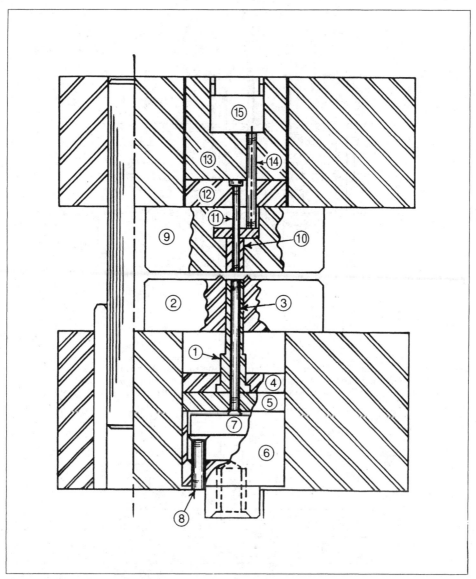

Fig. 19-8. An example of a sliding or moving blank-punch presstool: (D1) blank-punch; (D2) V-ring plate; (D3) slug ejector; (D4) retainer; (D5) backup plate; (D6) sliding base; (D7) bridge; (D8) bridge pin; (D9) matrix plate; (D10) counter-punch; (D11) perforator; (D12) retainer; (D13) backup plate; (D14) pressure pin; (D15) pressure pad. (*Feinblänking Ltd.*)

one major difference. The multistation presstool can be designed with very useful compound stations. This design feature makes it possible to maintain relationships between external shapes and internal openings and/or forms while assuring part flatness and restricting roll-over to one side of the part. Even though several cutting stations are used in the tool, the material can be controlled to produce parts with excellent accuracy. Figure 19-11 illustrates a multistation presstool layout.

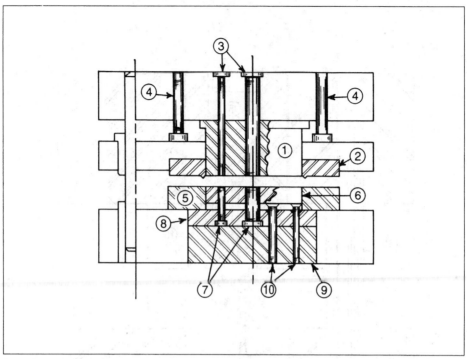

Fig. 19-9. An example of a stationary or fixed blank-punch presstool: (D1) blank-punch; (D2) V-ring plate; (D3) slug ejector; (D4) pressure pin (V-ring); (D5) matrix plate; (D6) counter-punch; (D7) perforator; (D8) retainer; (D9) support; (D10) pressure pin (counter). (*Feinblänking Ltd.*)

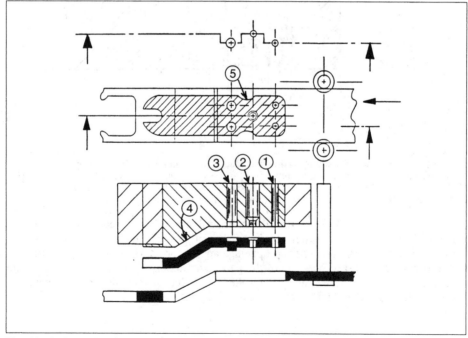

Fig. 19-10. Example of a unistation presstool: (D1) hole smaller than stock thickness; (D2) hole thru with chamfer on one side; (D3) semi-pierced locator; (D4) 30° offset form; (D5) radius notch in edge. (*Feinblänking Ltd.*)

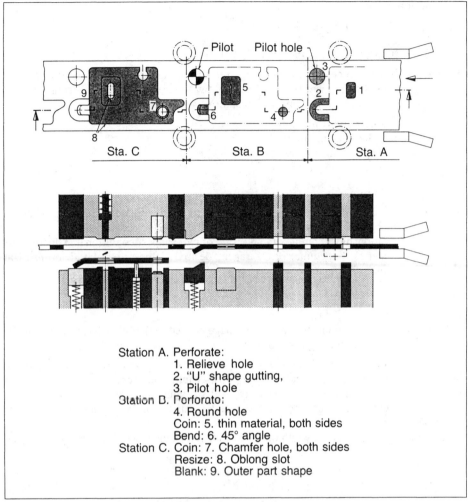

Fig. 19-11. Example of a multistation presstool. (*Burkland Inc.*)

ENGINEERING THE PRESSTOOL

Press Size. Part size and thickness, the height of forms or the number of stations in the tool can all contribute to determining the proper press size. However, press tonnage, a combination of blanking force, V-ring force, and counter force, is the minimum requirement in press selection.

While shear strength is used in force calculations for unistation tools, tensile strength is better for calculations for multistation tools due to the unknown effects of material workhardening as it passes from station to station. Besides offering a 20% safety factor when blanking mild steel, tensile strength data is readily available from all of the leading material producers. Tensile strength (S_t), in ksi (Pa), has been substituted for shear strength (S_s) in the following calculations. Blanking force:

$$F_b = (L_e + L_i) \times T_s \times S_t/2 \qquad (1)$$

where F_b = blanking force, the tonnage required to produce the piece part, US tons (N)
L_e = external shape length, in. (mm)
L_i = internal shape length, in. (mm)
S_t = tensile strength
T_s = stock thickness, in. (mm)

V-ring force:

$$F_v = (L_v \times H_v) \times S_t \times [1.5 \ (\log S_t/25) + 1.2] \tag{2}$$
$$L_v = L_c/0.948 \tag{3}$$
$$H_v = 0.134 \ (T_s) + 0.008 \ \text{(when } T_s > 0.020) \tag{4}$$

where F_v = V-ring force, the tonnage required to impinge the material which restricts
lateral movement (some presses allow for the reduction of V-ring force by
40% prior to blanking), US tons (N)

H_v = V-ring height, in. (mm)
L_f = feed length, in. (mm)
L_v = V-ring length, in. (mm)

Counterforce:

$$F_c = 0.20F_b \tag{5}$$

where F_c = counter force, the tonnage required to rigidly support the material being
blanked, US tons (N). The final counter force selection, influenced by size
relationships, complexity and style of part shapes, condition and ductility of
material, punch/matrix clearance, desired flatness, etc., is determined at
setup. This equation is used to establish a starting point.

Press tonnage:

$$PT = F_b + F_v + F_c \tag{6}$$

where PT = total press tonnage, the total tonnage needed to fineblank the part, equals the
minimum size press required.

Strip width (W_s) and feed length (L_f): Shape of the piece part and stock thickness are
the most influential factors in determining strip width and web between parts; other
considerations are tensile strength, ductility, and size and shape of the V-ring.

Piece parts with round sides require less material outside of the cut-edge than do parts
with straight sides.

Strip width (W_s) equals the width of the part (W_p) plus the stock thickness (T_s) factored
for proper impingement area.

$$W_s = W_p + 2.4T_s \quad \text{for round or rounded-side part} \tag{7}$$
$$W_s = W_p + 3.0T_s \quad \text{for straight sided part} \tag{8}$$

Feed length (L_f) equals the part length (L_p) with the strip plus stock thickness (T_s)
factored for proper impingement area, 0.040 in. (1.02 mm) minimum.

$$L_f = L_p + 1.5T_s \quad \text{for round or rounded-side part} \tag{9}$$
$$L_f = L_p + 2.0T_s \quad \text{for straight sided part} \tag{10}$$

Center of Load. The force of the press ram must align as closely as possible with the
blanking load in the press tool. While this condition is desirable with unistation tools, it
is a must for multistation tools. Off-center loading can result in poor production,
unpredictable piece part distortion, and excessive tool failure through wear or breakage.

The center of load can be determined by calculating the center of gravity of the piece
part or strip layout using the cut lengths of outside and inside shapes.

For cut lengths symmetrical about a center line, the center of gravity will be located on
the center line. To find the exact location, take moments with reference to any axis at right
angles to the center line. Divide the cut lengths into the geometrical shapes, (A, B, C,
etc.), where the center of gravity can be easily found. Calculate the distance (a, b, c,
etc.), from the axis to the center of gravity of each shape. The distance from the axis to
the center of load is found from the equation

$$\text{Center of Load} = \frac{Aa + Bb + Cc}{A + B + C} \tag{11}$$

For cut lengths not symmetrical about a center line, the center of gravity must be found by using moments with relation to two axes (X and Y) established at right angles to each other. The center of load is determined from the equations

$$X-\text{axis} = \frac{Aa_x + Bb_x + Cc_x}{A + B + C} \tag{12}$$

$$Y-\text{axis} = \frac{Aa_y + Bb_y + Cc_y}{A + B + C} \tag{13}$$

V-ring Size and Location. Stock thickness (T_s) is the primary factor in determining the size and location of the V-ring, however in some cases tensile strength and alloy elements in the material must be taken into consideration. Figure 19-12 illustrates the sizing and location of the V-ring as a function of stock thickness.

Although a V-ring around the matrix is normally recommended when blanking heavy materials, where increased roll-over or slight edge inclusions do not affect part function, it may not be necessary. In such cases, the V-ring size specified for the matrix should be applied around the blank-punch.

Punch-to-matrix (Die) Clearance. Considered as the total gap that exists between the blank-punch or internal shape punch and the matrix opening so at any given location the distance between the punch and matrix equals 1/2 the published clearance. Clearance as it affects the blank-punch is designated C_p, the matrix is C_m.

The larger the clearance, the greater the production between grinds. However, with larger clearances the fineblanking quality of the piece part suffers. Roll-over and break-out are directly affected as the clearance increases.

Optimum clearance between the blank-punch and matrix is 0.5% of stock thickness (T_s) when producing parts from fineblanking quality material. As a general rule, materials with poor formability require tighter clearances to maintain plastic deformation than do materials with good formability.

The clearance should be doubled where shapes re-enter the part, such as around the root and crest of gear teeth.

Perforators and other internal shapes may require various clearances depending on the relationship of the opening to the stock thickness. The optimum blanking clearance can be doubled for openings which are less than 5 times stock thickness; openings smaller than stock thickness require even greater clearances.

Punch-to-matrix clearances for mild steel (50 ksi (345 MPa) tensile strength, with unit elongation of 20% to 25%) with stock thicknesses of 0.020 to 0.500 in. (0.51 to 12.70 mm) can be determined as follows:

External shapes

Stock thickness (T_s)	Clearance
0.020 to 0.156 in. (0.51 to 3.96 mm)	$C_p = 0.0038T_s$
0.157 to 0.500 in. (3.99 to 12.70 mm)	$C_p = 0.0051T_s$

Hole Diameters

$$C_m = 0.0096T_s + 0.0005 \text{ in. } (0.013 \text{ mm}) \text{ for under } 65\% \ T_s \tag{14}$$
$$C_m = 0.0078T_s + 0.0003 \text{ in. } (0.008 \text{ mm}) \text{ for } 65\% \text{ to } 2T_s \tag{15}$$
$$C_m = 0.0048T_s + 0.0002 \text{ in. } (0.005 \text{ mm}) \text{ for over } 2T_s \tag{16}$$

Internal Shapes

$$C_m = 0.0037T_s \text{ for area greater than } 5T_s \tag{17}$$

Shut Height. The height from the bottom to the top of the presstool, with part material in place between the face of the blank-punch and the matrix, is the shut height of the presstool. Shut height is used to establish the press setup position from which all cutting and/or forming operations start.

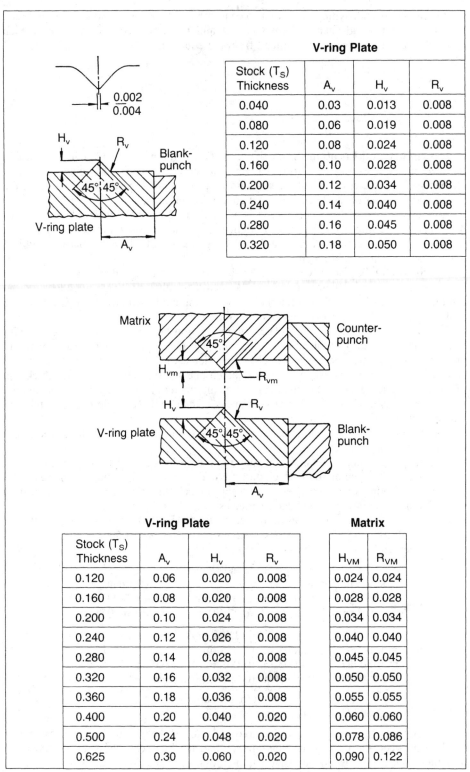

V-ring Plate

Stock (T_S) Thickness	A_v	H_v	R_v
0.040	0.03	0.013	0.008
0.080	0.06	0.019	0.008
0.120	0.08	0.024	0.008
0.160	0.10	0.028	0.008
0.200	0.12	0.034	0.008
0.240	0.14	0.040	0.008
0.280	0.16	0.045	0.008
0.320	0.18	0.050	0.008

V-ring Plate				Matrix	
Stock (T_S) Thickness	A_v	H_v	R_v	H_{VM}	R_{VM}
0.120	0.06	0.020	0.008	0.024	0.024
0.160	0.08	0.020	0.008	0.028	0.028
0.200	0.10	0.024	0.008	0.034	0.034
0.240	0.12	0.026	0.008	0.040	0.040
0.280	0.14	0.028	0.008	0.045	0.045
0.320	0.16	0.032	0.008	0.050	0.050
0.360	0.18	0.036	0.008	0.055	0.055
0.400	0.20	0.040	0.020	0.060	0.060
0.500	0.24	0.048	0.020	0.078	0.086
0.625	0.30	0.060	0.020	0.090	0.122

Fig. 19-12. V-ring size and location as a function of stock thickness: (*A*) V-ring plate; (*B*) matrix; (*C*) typical blend radius. (*D*) V-shaped impingement. (*Feinblänking Ltd.*)

DESIGNING PRESSTOOL COMPONENTS

Components for fineblanking presstools are very similar to those used in conventional compound and progressive dies. However, the designer should keep in mind that the load requirements are much greater in fineblanking so more massive details are required.

The method used to retain the blank-punch, perforator, and other internal shape components must be strong enough to withstand stripping and ejection forces. A safety factor of 1.5 to 2 times is recommended with special attention given to the blank-punch which is subjected to both forces occurring almost simultaneously.

$$F_s = 0.10F_b \quad \text{for common shapes} \tag{18}$$
$$F_s = 0.15F_b \quad \text{for complex external and/or internal shapes} \tag{19}$$

where F_s = stripping force, the tonnage required to strip the material from the blank-punch, US tons (N).

$$F_e = 0.10F_b \quad \text{for common shapes} \tag{20}$$
$$F_e = 0.15F_b \quad \text{for complex external and/or internal shapes} \tag{21}$$

where F_e = ejection force, the tonnage required to eject the part from the matrix plate as well as slugs and other scrap, US tons (N)

Blank-punches. The blank-punch is the main working component in the tool. It can be directly attached to the retainer plate, constructed as one piece with a pedestal mounting, or designed with a head or other protruding shape to be secured by the retainer plate. Commercial head-type punches purchased for this purpose can be used as is or altered as required. Figure 19-13 illustrates various styles of blank-punch mounting techniques.

Whichever method is used the blank-punch is fastened to the moving member in sliding punch tools. The moving member must be keyed in cases where the outside of the blank punch is round and the internal openings are either shaped or located off center. To ensure accurate radial alignment, the opposing component must be keyed in a similar fashion. Fig. 19-14 illustrates the use of a round key to fix the radial location of the blank-punch.

In fixed punch presstools, the blank-punch is fastened directly to the punch holder (Fig. 19-15) or to the load-bearing block (Fig. 19-16) which extends through the punch holder to the surface of the press. During blanking, the stripping and ejection pins push back against hydraulically controlled pressure plates. Stripping and ejection pins which are also used for V-ring or counter-punch pressure should be slightly longer than other pins.

Bridges. Most sliding punch tools will require an internal bridge for ejecting slugs, partial cuts, and forms. Beam strength of the bridge must be substantial enough to overcome the ejection force of the internal shapes with a minimum safety factor of 3.

Bridge travel should exceed twice stock thickness plus the expected tool life. Figure 19-17 illustrates the bridge in a sliding punch tool.

Fineblanking Perforators. Perforators and other internal shape punches used in unistation presstools and the compound station of multistation tools are secured by or fastened directly to the retainer plate and guided by a close fit in the counter-punch which in turn fits snugly in the matrix plate. These close fits allow the fineblanking of holes 50% or less of material thickness in mild steel as illustrated in Fig. 19-18.

Perforators and other internal shape punches should be dimensioned in the upper half of the part tolerance to allow for wear and changes in mating fits.

Counter-punch. The counter-punch, a mirror image of the blank-punch, serves to keep the material under control as it enters the matrix cavity. Counter-punch retention can be insured by peening the back edges, using a head, or affixing a backing plate as illustrated in Fig. 19-19.

Counter-punch pressure pins should always be larger in diameter. All pins should be the same size and properly spaced to evenly support the load of the counterforce (F_c).

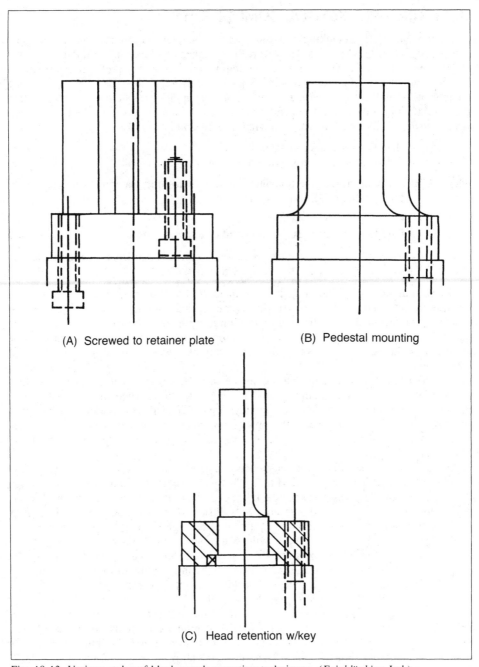

(A) Screwed to retainer plate   (B) Pedestal mounting

(C) Head retention w/key

Fig. 19-13. Various styles of blank-punch mounting techniques. (*Feinblänking Ltd.*)

Pin strength for 70 ksi (483 MPa) material can be determined as follows:

$$S_p = 0.7854D^2 \times 35, \text{US tons (N)} \tag{22}$$

To find the required number of a given size pin

$$\text{Number of pins} = F_c \div S_p \tag{23}$$

Matrix Plates. The matrix (die) plate produces the outer shape of the piece part and as such should be dimensioned between the nominal and the lower half of the part

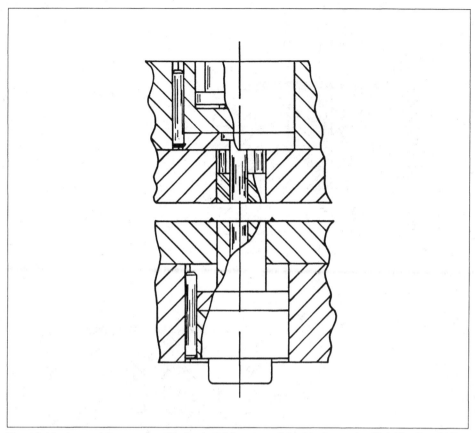

Fig. 19-14. A round key used to fix the radial location of the blank-punch. (*Feinblänking Ltd.*)

tolerance, thus allowing the opening to wear with little detrimental effect. When blanking low tensile materials up to 50 ksi (345 MPa), which are thicker than 0.375 in. (9.52 mm) or medium tensile materials above 0.250 in. (6.35 mm), the matrix plate may need to be encompassed by a shrink ring. Shrink rings may also be used to support thin or weak cross sections. Made from tool or alloy steel, the ring is normally machined 1.500 in. (38.10 mm) larger in diameter and about 0.250 in. (6.35 mm) thinner than the matrix plate, with a 0.001 to 0.0015 in. (0.02 to 0.038 mm) per diametric inch shrink allowance.

V-ring Plates. The V-ring plate, a mirror image of the matrix minus the cutting clearance, is the guiding source for the blank-punch, and should be positively aligned with the matrix plate by locating pins. These pins must be large enough to resist any lateral loads caused by off-center cutting or forming that might be taking place between the two plates. The V-ring plate substitutes as a stripper plate when the tool opens, removing the material from the blank-punch. V-ring protection pads (Fig. 19-20), machined on the edges of the plate, keep the ring from being crushed when the die is closed without material in place.

A V-ring may be necessary around the opening on the matrix when blanking heavy material or to reduce roll-over on the piece part.

Tool Steel Selection. The complexity of the piece part as well as expected production volume plays an important role in selecting the material to be used for presstool components. Figure 19-21 lists recommended steels for presstool construction by intended application and production volume.

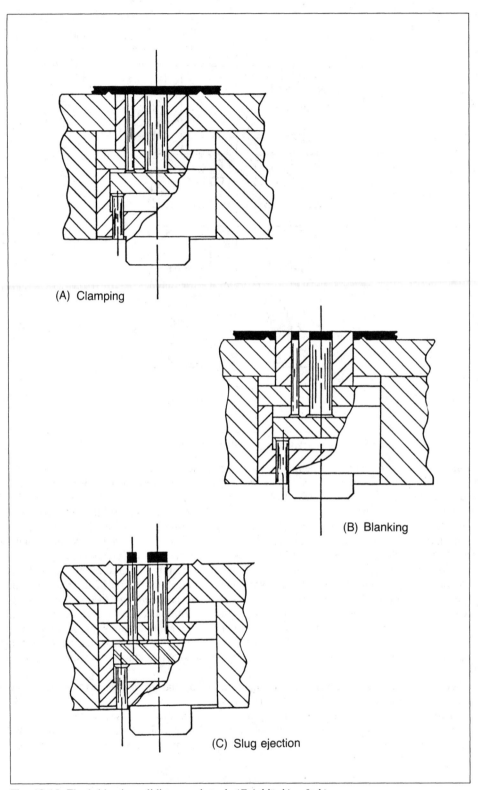

(A) Clamping

(B) Blanking

(C) Slug ejection

Fig. 19-15. The bridge in a sliding punch tool. (*Feinblänking Ltd.*)

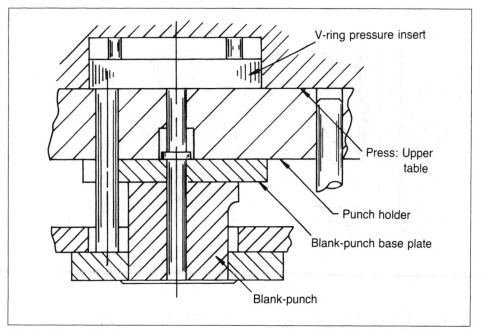

Fig. 19-16. Method of mounting the blank-punch to the punch holder. (*Feinblänking Ltd.*)

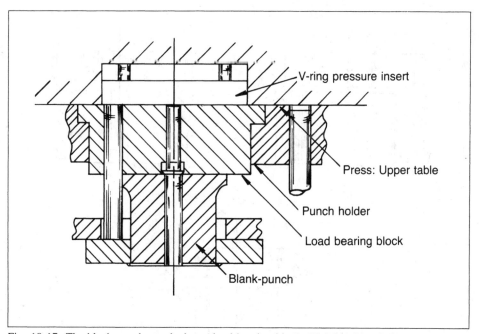

Fig. 19-17. The blank-punch attached to a load bearing block. (*Feinblänking Ltd.*)

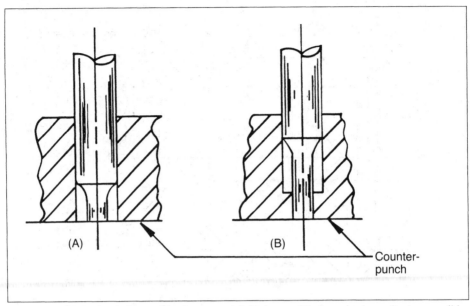

Fig. 19-18. Producing holes smaller than material thickness: (A) the point is as short as possible and guided on shank; (B) both point and shank are guided. (*Feinblänking Ltd.*)

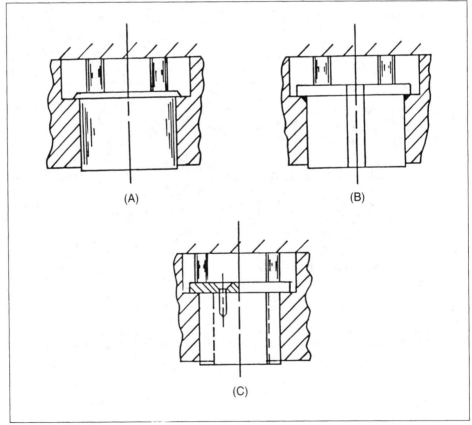

Fig. 19-19. Methods for retaining counter-punches: (A) peened edges; (B) head-type construction; (C) backing plate retention. (*Feinblänking Ltd.*)

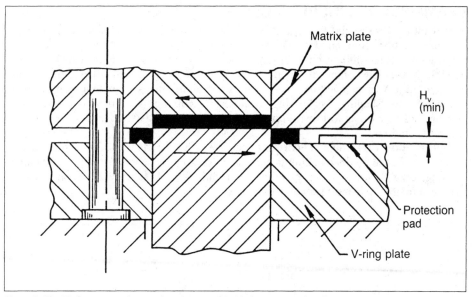

Fig. 19-20. V-ring protection pads, shown with V-ring to matrix alignment pin. (*Feinblänking Ltd.*)

	PRODUCTION		
	Low	Medium	High
	AISI	AISI	AISI
Blank-punch	A2	D2/M2	CPM M4/10V
Perforator > 1.5T$_s$	A2	M2	CPM 10V
Perforator < 1.5T$_s$	M2	M2	CPM 10V
Inner-shape Punch	A2	D2/M2	M2/CPM 10V
Matrix Plate	D2	D2	M2/CPM M4
Counter-punch	A2	A2	A2
V-ring Plate	A2	A2	A2
Backup Plate	8620	8620	A2
Retainer Plate	8620	8620	S7
Blank-punch Base	A2	A2	A2
Pressure Plate	8620	8620	S7
Shrink Ring	1045	1045	4140

Fig. 19-21. Tool material chart. (*Feinblänking Ltd.*)

EXAMPLES OF FINEBLANKING PRESSTOOLS

Unistation Presstools. Because it is economical to produce and simple to maintain, this single station, compound type presstool is used in more than 90% of all fineblanking applications. Three-dimensional forming prior to cutting is one feature which makes it possible to complete complicated piece parts in a unistation presstool. Also, other features can be used in various combinations, such as combining blanking with any or all of the following: inter-shaping, perforating, chamfering, coining, semi-piercing, embossing, lancing, and even offset forming. Figure 19-22 illustrates a unistation presstool combining several operations.

Multistation Presstools. This type of tool, which consists of two or more stations, expands the scope of fineblanking to include counterboring, countersinking, trepanning, bending, drawing, re-shaping, beveling, extruding, etc. Figure 19-23 illustrates some features possible in a multistation presstool.

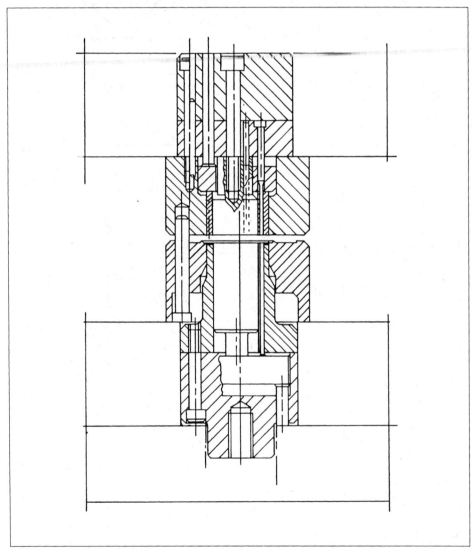

Fig. 19-22. Unistation presstool: (A) with sliding blank-punch; blanking, perforating, and inter-shaping. (*Feinblänking Ltd.*)

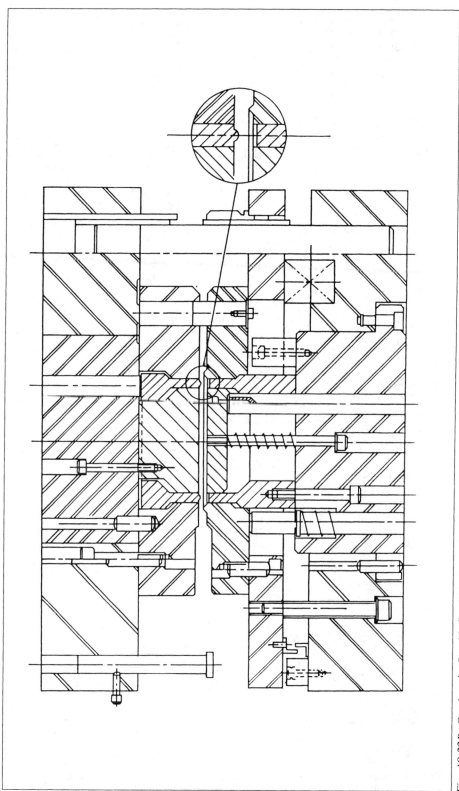

Fig. 19-22B. *Continued.* (*B*) with fixed blank-punch; coining, blanking, and inter-shaping. (*Feinblänking Ltd.*)

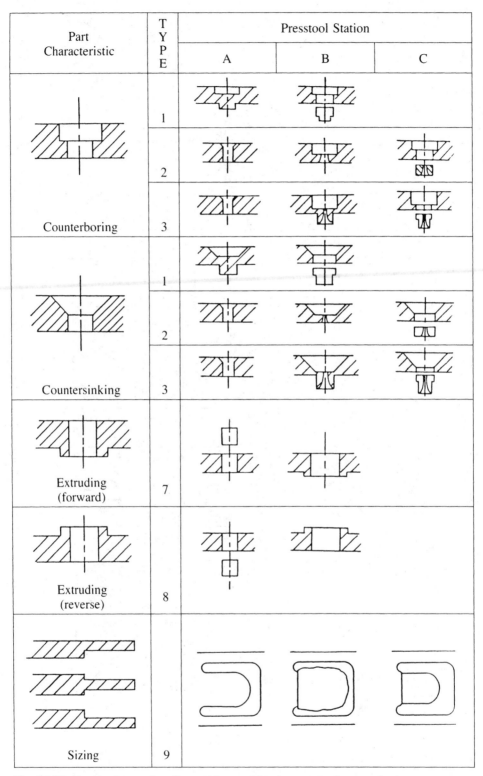

Fig. 19-23. Combined operations featured in a multistation presstool: counterboring; countersinking; extruding, forward and reverse; thinning (free coining). (*Feinblänking Ltd.*)

19- 22

SECTION 20

TOOLS FOR MULTIPLE SLIDE FORMING MACHINES*

Machine Evolution. Slide forming machines were originally built to produce wireforms. Wire was fed into the machine, severed to length and formed into the desired configuration. From the initial dedicated machines evolved universal machines capable of producing a variety of wireforms simply by changing tooling.

Strip or ribbon stock was introduced into the slide machine and processed the same as wire: feed, cut-off, and form. A conventional press was added to the machine, as an integral member, and ribbon material could then be formed in a progressive die prior to the cut-off and forming sequence of the slide machine.

The misnamed "four slide" machine was introduced to the metal stamping industry for processing parts that required more than could be achieved from conventional power presses and the progressive dies that complemented them. The so-called four slide machine had more than four sliding members if the cut off slide, the stripper slide, the feed slide and the press slide were included. The machine had four main driving shafts arranged in a rectangle where synchronous power was transmitted from one to the next by bevel gears at the corners. The shafts carried cams of various configurations for various functions including four prominent cams, one on each shaft for forming, hence the machine was labeled a "four slide". Later machine versions included a provision for a sliding kingpost or mandrel and a twin front slide primarily utilizing one slide for blankholding and the other for forming.

A bracket known as the kingpost holder was fastened to the top platen of the machine and it housed the kingpost plus the stripper shaft. The kingpost, or forming mandrel, suspended from the kingpost holder was the center of all forming activity. Single or multiple stages of forming were feasible with the stripper slide being utilized for transferring the blank from one stage to the next or ejecting the finished part through a large hole in the machine platen located directly under the kingpost. Figure 20-1 is a plan view of a typical four slide machine looking down.

Further enhancements to four slide machine versatility included multiple press attachments and the ability to reverse individual presses 180° to accommodate burrside requirements or for forming or extruding in the press section toward the desired direction. Mechanical blankholders were also devised as a standard feature to facilitate tooling.

A vertical four slide machine is essentially a horizontal four slide machine tilted on edge. The press sections can no longer be reversed but all slugs from gutting and piercing have gravity to assist their ejection. The kingpost or forming mandrel is relocated to the rear of the machine along with the stripper slide. The formed parts are now transferred and ejected forward toward the front of the machine rather than through the base platen. The

* Larry Crainich, Design Standards Corporation, Bridgeport, Connecticut contributed to this section.

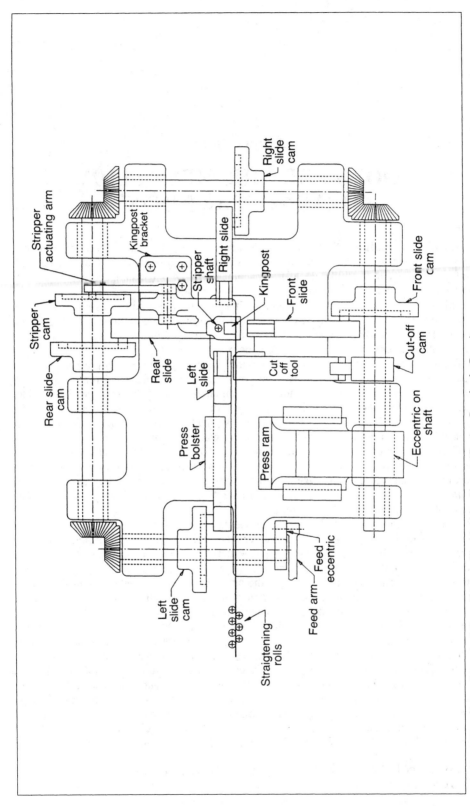

Fig. 20-1. Plan view of a typical four slide machine looking down. (*Design Standards Corp.*)

Labels in figure:

Right slide cam

Stripper actuating arm

Kingpost bracket

Stripper shaft

Right slide

Kingpost

Front slide

Front slide cam

Stripper cam

Rear slide cam

Rear slide

Left slide

Cut off tool

Cut-off cam

Press bolster

Press ram

Eccentric on shaft

Left slide cam

Straigtening rolls

Feed arm

Feed eccentric

major advantage of the vertical four slide machine is convenience of tool setup as well as service and maintenance of the tooling in the machine. The need to bend over a machine to service tooling was eliminated and visibility was improved due to vertical proximity and the relocation of the kingpost holder to the face or rear of the platen.

The vertical shaft driven slide machine progressed further with the addition of auxiliary slides for angular orientation toward the forming center of the machine. The press section was located further to the outside to permit added positioning and flexibility for the forming slides themselves. The kingpost became the center mandrel and the number of motions available, for transfer or ejection from the rear of the machine into the working area, increased.[1, 2]

EXAMPLES OF CONVENTIONAL MULTI-SLIDE FORMING

Forming a Cylinder in One Stage. A round cylinder shown in Fig. 20-2A can be formed on a multi-slide machine in one stage. Figure 20-2B illustrates how the correct size of blank is cut from the strip stock fed in at the left by a cut-off die and punch. A shelf is provided on the front slide to keep the severed blank level.

The front slide then moves inward, forming the blank into a U (Fig. 20-2C). While the front slide is in dwell, both the left and right slides move inward (Fig. 20-2D) completing more of the form around the arbor.

In Fig. 20-2E, the rear slide moves inward and finishes the form. All slides now are retracted, and a stripper removes the completed part from the arbor.

Forming a Cylinder in Two Stages. The cylinder shown in Fig. 20-2A can also be formed in two stages by transferring the partly formed shape to a second position on the arbor.

Figure 20-3A illustrates how the blank is cut off and partly formed around an oval arbor.

The side slides next move inward and preform the cylinder around the oval arbor (Fig. 20-3B). The stripper arm, illustrated in Fig. 20-4, next drops the preformed blank onto a shelf and the rear slide begins the curling sequence. (Fig. 20-3C).

The form is finished when the tool operated by the rear slide bottoms out on the arbor (Fig. 20-3D). The part is stripped from the arbor by the next part being transferred from the first to the second stage.

Stripper Arm and Actuating Cam. Figure 20-4 illustrates a sectional view of a four slide machine similar to that shown shown in Fig. 20-1. The stripper cam and stripper arm are shown together with the kingpost bracket and front and rear slide tools. View Z-Z illustrates the attachment of the stripper arm to the stripper slide.

Forming an Overlapped Box. The part shown in Fig. 20-5A is cut-off and U-formed. To limit arbor deflection, the rear slide moves forward to support the arbor during forming (Fig. 20-5B).

Next the rear slide is withdrawn, (Fig. 20-5C) and the right hand slide tool forms the right hand leg of the part to a 90° bend while the front slide continues to dwell, holding the part and limiting arbor deflection. The left slide next pre-bends the left hand leg of the part as shown (Fig. 20-5D).

Finally, the rear slide again moves forward and finishes the form by closing the box tight (Fig. 20-5E). A special double-throw cam is required in order for the rear slide to perform a dual function.

Forming an Electrical Hardware Component. As was shown in the previous example, special cams enable a slide to perform more than one function. The part is illustrated in Fig. 20-6A.

In this example, the front slide performs a dual function by first forming the outside U after the blank is cut off (Fig. 20-6B). The two moving arbors are activated by the stripper slide.

The side slides move inward folding the wings over. The side form tools are then spanked by the rear slide (Fig. 20-6C). The rear slide, side slides, and the stripper arbors

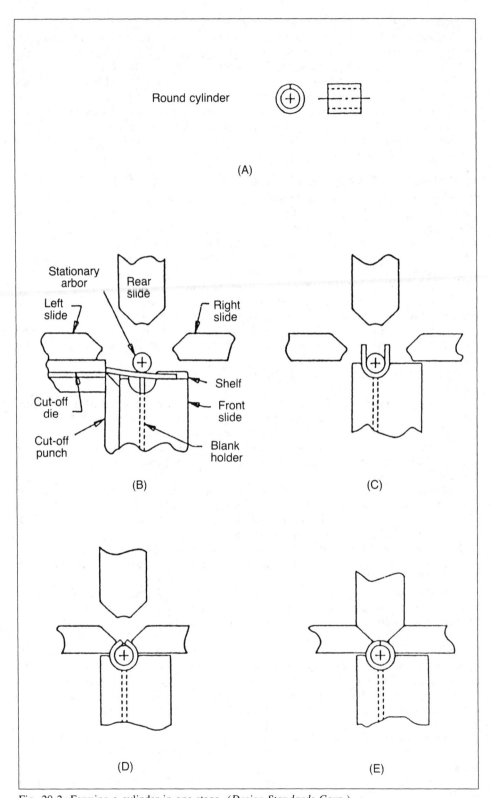

Round cylinder

(A)

Stationary arbor
Rear slide
Left slide
Right slide
Shelf
Cut-off die
Front slide
Cut-off punch
Blank holder

(B)

(C)

(D)

(E)

Fig. 20-2. Forming a cylinder in one stage. (*Design Standards Corp.*)

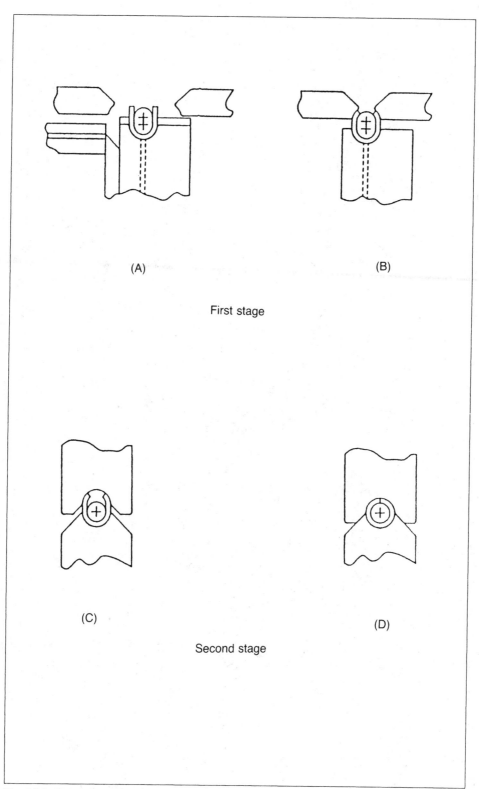

(A)   (B)

First stage

(C)   (D)

Second stage

Fig. 20-3. Forming a cylinder in two stages. (*Design Standards Corp.*)

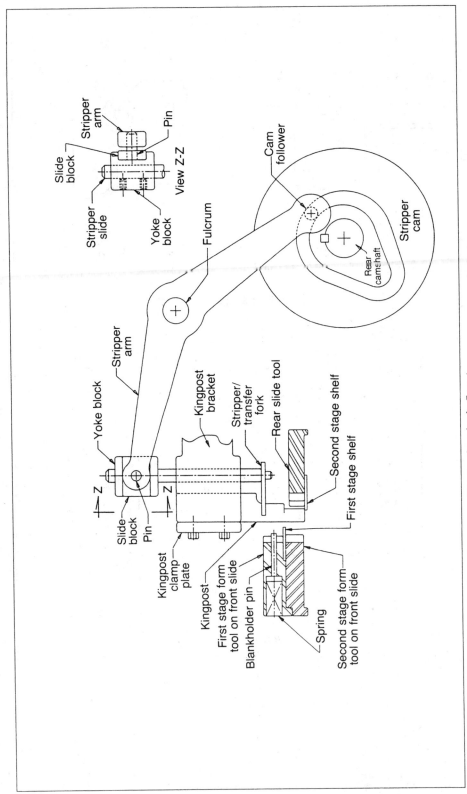

Fig. 20-4. Four slide sectional view showing stripper arm. (*Design Standards Corp.*)

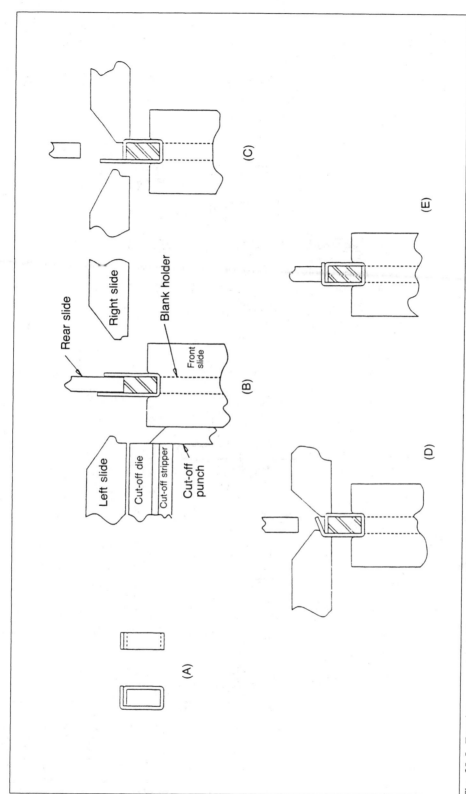

Fig. 20-5. Forming an overlapped box on a multislide machine. (*Design Standards Corp.*)

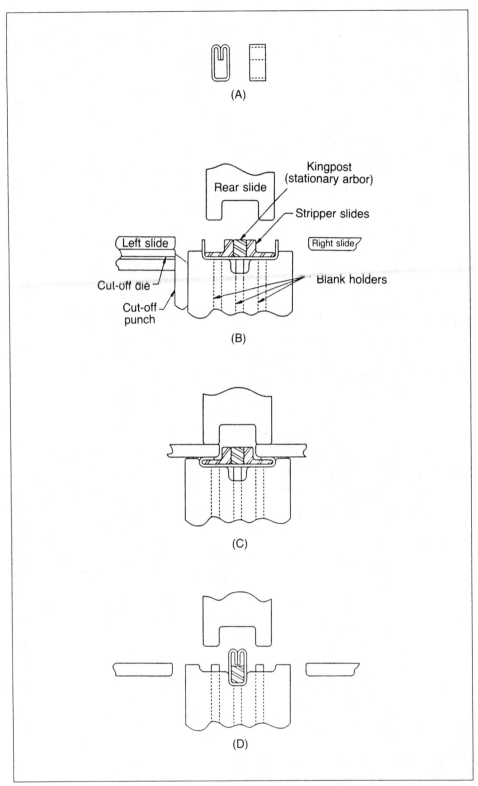

Fig. 20-6. Forming an electrical hardware component. (*Design Standards Corp.*)

are then retracted, and the front slide performs its second function by advancing (Fig. 20-6D) and forming the inner U.

Forming a Brass Contact. The contact shown in Fig. 20-7 is made of brass strip stock 1 in. (25 mm) wide and 0.024 in. (0.61 mm) thick. The die operations to make the part are shown in the strip layout of Fig. 20-8A. At the first station, the hole is pierced, and at the next a pilot enters the hole and locates the stock at the correct pitch of 1.34 in. (34 mm). At the third station the rectangular hole is punched and is followed at the fourth by the complete severing of the blank from the strip.

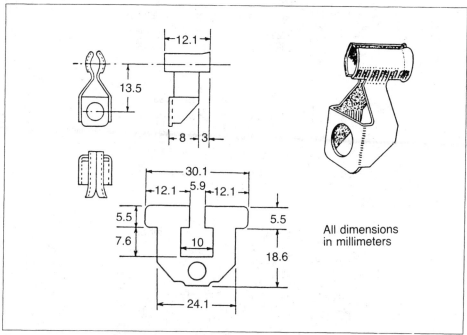

Fig. 20-7. Brass contact made on a multiple-slide machine.[3] *(Heenan & Froude Limited)*

As soon as the blanking punch (*D1*) has retracted, the transfer fingers (*D2*) are brought into position in front of the die. The blank which has been held in the die aperture is pushed out and loaded into the fingers by the plunger operated from a cam from the rear shaft.

When the transfer finger loaded with a blank has been moved to the forming position, it dwells to allow the front tool (*D4*) to enter the finger's aperture and push the blank out against the square section mandrel (*D5*). Continued movement of the tool forms the blank to a U-shape.

Upon being unloaded, the transfer finger is returned to the die set for another blank. Meantime, the two side tools advance and form the two contact portions of the part against a formed mandrel which is also fixed to the rear slide. When the tools retract, the finished component is ejected from the machine.

The contact could also be produced by progressive tools, as shown by the strip layout of Fig. 20-8B. The cutting of the blank to the required outline is made by three punches and dies, which leave adjacent blanks joined by a tab which is cut away prior to forming. The part is bent to a U-shape in the forming station, and the contact faces are embossed in the die set.

Figure 20-8C shows a strip layout for producing two contacts per cycle.[3]

Tooling for a Toy Crane Hook. The 0.08-in. (2.0 mm) wire crane hook shown in Fig. 20-9F is made on tooling where the anvil acts to bend the wire. This method can be used where the wire is of small diameter and little force is required to bend it.

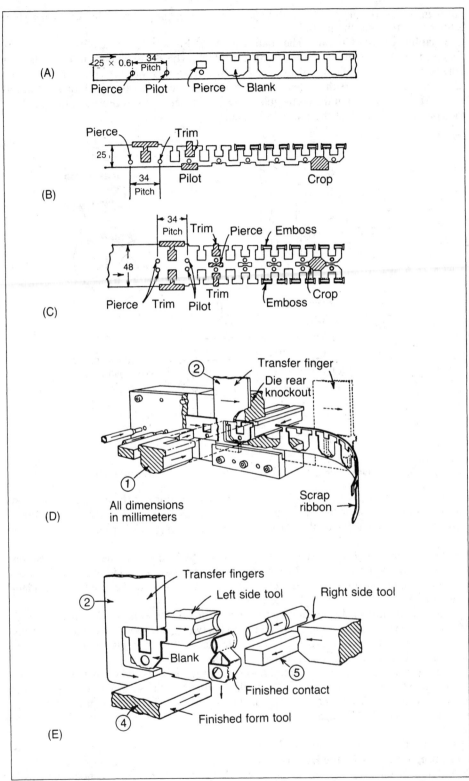

Fig. 20-8. Multiple slide tooling for contact of Fig. 20-7.[3] *(Heenan & Froude Limited)*

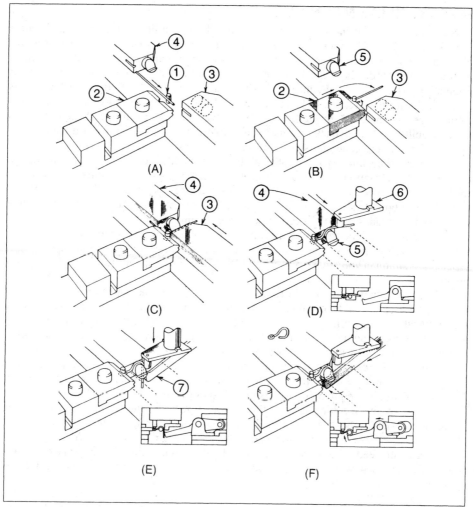

Fig. 20-9. Forming a toy crane hook from wire on a multi-slide machine.[3] *(Heenan & Froude Limited)*

A small diameter peg ($D1$, Fig. 20-9A) is supported in a tubular holder in the center tool post, and the eye is formed around it by the combined action of the three forming tools ($D2$, $D3$, and $D4$). The forming tool ($D2$) has a semicircular groove cut in it to hold the eye portion securely on the peg while further bending takes place.

Side tool ($D4$) is fitted with a shaped peg ($D5$) as shown at views B and D and is used as a former for the large loop of the hook. The opposite side tool ($D3$) is drilled out to receive the peg ($D1$).

The right side tool ($D3$), timed to move inward slightly in advance of tool ($D4$), bends the short leg of the small eye. As the tool ($D4$) moves in, the tapered former ($D5$) slides under the wire which is forced upward by the inclined surface of the peg. Any tendency for the wire to move to the right past the center line of the component is prevented by the flat face of tool ($D3$), while excessive upward movement of the wire is prevented by the vertical form tool ($D6$).

When the two side tools have completed their movement, the vertical form tool moves down to bend the wire down as shown in view E. Completion of the hook portion is done by the forward and then upward movement of the rocking lever ($D7$), which is pivoted on a small bracket bolted to the rear slide and tilted by a cam plate fixed to the bed.

ROTARY SLIDE MACHINES

Since the advent of the rotary slide machine the versatility of function has become virtually unlimited. The rotary slide machine is powered by a large sun gear that drives planetary gears on a common shaft with cams on individual slides for custom, linear motion. The slides can be located almost anywhere around the sun gear and at any angular orientation. The axis of the sun gear is the center of forming activity with form tools available from any direction. Placement and subsequent withdrawal of forming arbors from the rear of the machine facilitates progressive forming operations. Added to that is the availability to "piggyback" cams for multiple motions from common directions for intricate forming or sequential stage forming operations.

The rotary slide machines have the ability to feed from either or both ends of the machine and the location of press attachments at either end also. Accessories are available for feeding from the rear or from the front of the machine if so desired. Tapping, welding and vibratory bowl feeding of parts for automatic assembly have been available on slide machines for decades, and their refinements are gaining prominence.

Developments adjunct to rotary machines are machines whose slides are powered by timing belts or worm gear drives that facilitate the "stacking" of entire machine bases in series, for simultaneous make-and-assemble production capabilities limited only by imagination.

A Typical Rotary Slide Machine. As was the case with the conventional multi-slide machine, the rotary slide machine derives its power from a mechanical gear, so all synchronization is essentially absolute. As shown in Fig. 20-10, the slides can be placed at any location around the periphery of the sun gear and at any angular orientation. Great versatility of motion is made possible by custom-made cams.

An additional press and feed mechanism can be fitted to the left side of the machine. To increase versatility even further, feeding from the rear of the machine is also common. Tapping and welding equipment are also common options.

The base of the machine houses the motor, flywheel, clutch, brake, and central lubrication system. The stock is pulled through a stock straightener and fed into the machine by a reciprocating feed arm driven by an adjustable feed-eccentric.

To detect misfeeds, both conventional and rotary multi-slide forming machines are commonly equipped with proximity, optical, and strain-gage based misfeed detection systems.

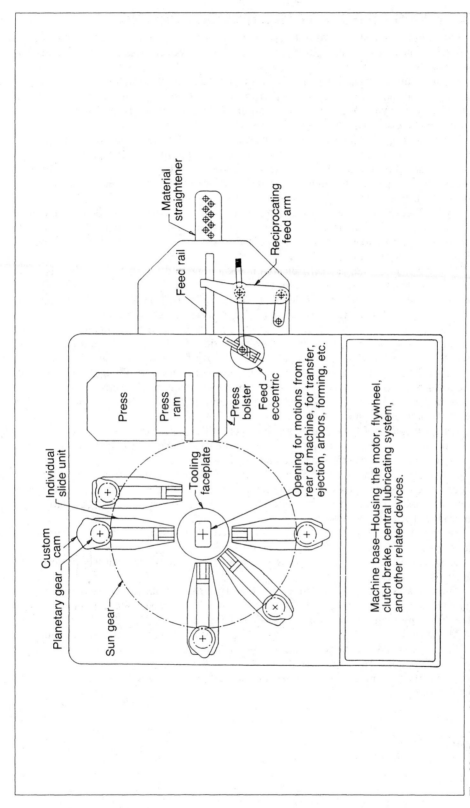

Fig. 20-10. A typical basic rotary slide machine. (*Design Standards Corp.*)

20- 13

EXAMPLES OF ROTARY MULTI-SLIDE FORMING

Forming a Complex Terminal. The terminal-line prong shown in Fig. 20-11A is an example of how a very complicated part can be formed on a single machine at efficient speeds.

The incoming material is first pierced, notched, and embossed in the press section (not shown) of the machine before being preformed and cut off. The top form tool and top cut-off blade shown in Fig. 20-11B are stationary. The action involved is rather complex, but the drawing is self-explanatory.

Figure 20-11C illustrates the preforming of the tail. After the cut-off blade mounted to the bottom slide severs the part, the top slide advances a form tool to preform the tail.

The slides then retract (Fig. 20-11D) to permit the strip to advance. The severed, partially formed part is also advanced to the next form station where the remainder of the operations begin.

Figure 20-11E illustrates a rather complex sequence of events. A blade (D1) mounted to the center mandrel holder advances and is positioned under the blank after the feed is complete. A spring-loaded pad (D2) mounted to the bottom slide advances to provide a platform for forming the part.

A spring-loaded plunger (D3) for both blankholding and forming moves into position on an angle, clamping the part. Form tool (D4) is actuated by the same slide as D3 and bends the tail of the part around the blade (D1). Slide (D5) then further forms the tail.

A small round mandrel (D6) driven by a motion source separate from the center mandrel moves forward to provide a forming arbor. Form punch (D7) is mounted to the bottom slide and is also hinged to provide side motion. First it forms the part up and then, as it contacts cam X (recessed behind the part) it overforms the part onto plunger (D3).

Figure 20-11E illustrates the remaining sequence of operations. Plunger (D3) and tools (D4 and D5) retract while a piggyback slide mounted on the top slide closes the form. The slides and center mandrel retract and the part is ejected out the front of the machine by a blast of compressed air.

An Example of Make-and-Assemble Versatility. The rotary slide machine is so versatile that one machine can make and assemble worm gear hose clamps complete. The only part that is not made in the machine is the worm screw.

Figure 20-12A illustrates the finished part. The shroud is made from strip stock on a progressive die in the press station on the left-hand side of the machine. The right-hand side of the machine also has a press section used to stamp, pierce, and slot the strip stock to produce the hose clamp band.

In stage I (Fig. 20-12B) the strip is embossed, preformed, cutoff, and inserted into the shroud.

Stage II (Fig. 20-12C) clinches the shroud to the band and sequentially preforms the curl in the band.

The screw is oriented and fed by a vibratory bowl and an in-line feeder. It is then automatically inserted (stage III) into the expanded shroud and the correct thread gap established by the tool on the left-hand side of Fig. 20-12D.

The assembly is completed in stage IV (Fig. 20-12E) by completing the forming and curling of the band. At the fourth stage, the screw is again rotated to engage the serrated band and complete tightening for shipment.

If desired, the shroud could also be spot-welded to the band in the same machine with off-the-shelf equipment.

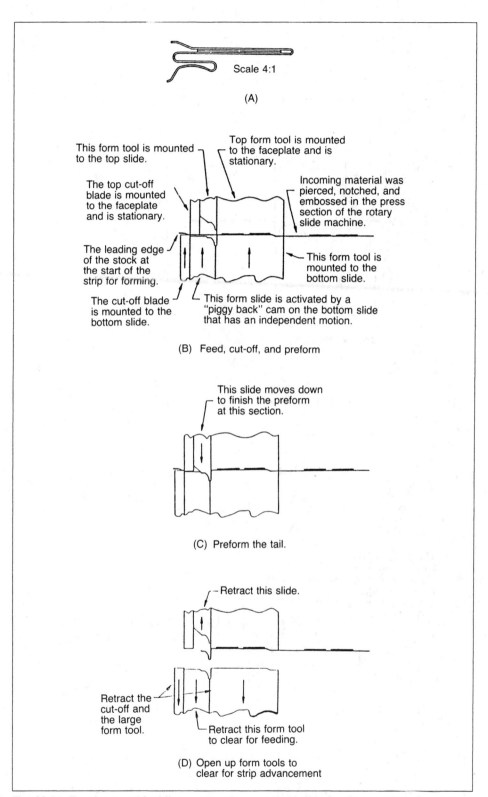

Scale 4:1

(A)

This form tool is mounted to the top slide.

Top form tool is mounted to the faceplate and is stationary.

The top cut-off blade is mounted to the faceplate and is stationary.

Incoming material was pierced, notched, and embossed in the press section of the rotary slide machine.

The leading edge of the stock at the start of the strip for forming.

This form tool is mounted to the bottom slide.

The cut-off blade is mounted to the bottom slide.

This form slide is activated by a "piggy back" cam on the bottom slide that has an independent motion.

(B) Feed, cut-off, and preform

This slide moves down to finish the preform at this section.

(C) Preform the tail.

Retract this slide.

Retract the cut-off and the large form tool.

Retract this form tool to clear for feeding.

(D) Open up form tools to clear for strip advancement

Fig. 20-11. Forming a terminal line prong on a rotary multi-slide forming machine. (*Design Standards Corp.*)

20-15

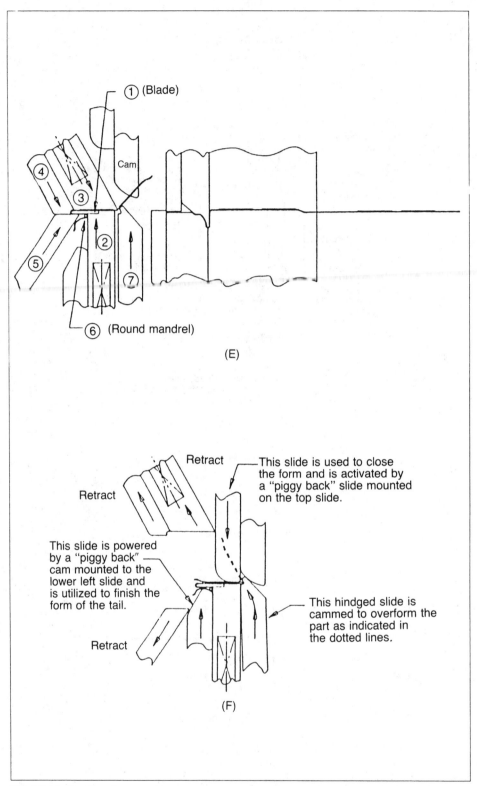

(1) (Blade)

Cam

(4)

(3)

(2)

(5)

(7)

(6) (Round mandrel)

(E)

Retract

Retract

This slide is used to close the form and is activated by a "piggy back" slide mounted on the top slide.

This slide is powered by a "piggy back" cam mounted to the lower left slide and is utilized to finish the form of the tail.

This hindged slide is cammed to overform the part as indicated in the dotted lines.

Retract

(F)

Fig. 20-11. *Continued.*

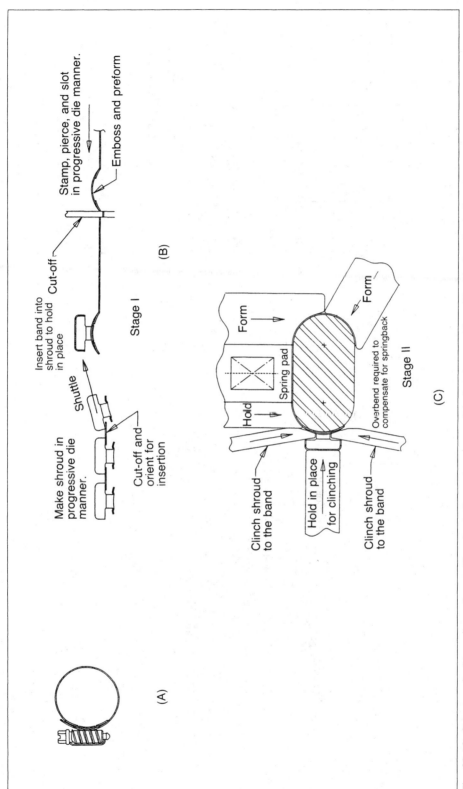

Insert band into
shroud to hold
in place

Stamp, pierce, and slot
in progressive die manner.

Cut-off

Emboss and preform

Shuttle

Make shroud in
progressive die
manner.

Cut-off and
orient for
insertion

Stage I

(B)

Form

Hold

Spring pad

Form

Clinch shroud
to the band

Hold in place
for clinching

Clinch shroud
to the band

Overbend required to
compensate for springback

Stage II

(C)

(A)

Fig. 20-12. Steps required to make and assemble a hose clamp on a rotary slide forming machine. (*Design Standards Corp.*)

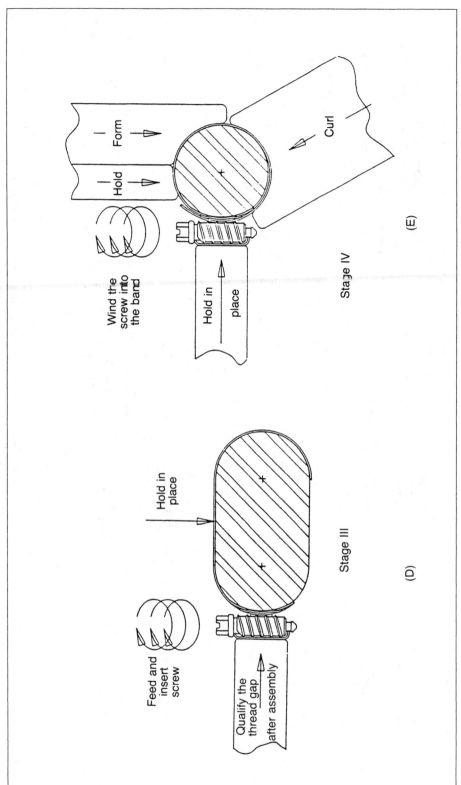

Fig. 20-12. *Continued.*

20-18

Advantages of Slide Machine Utilization. The advantages are:
1. The ability to produce intricate forms from coil stock in a single machine.
2. Multiple stage forming for complex products.
3. Standard tapping equipment availability.
4. Standard welding equipment availability.
5. The ability to feed and assemble alien parts to stampings in progress.
6. Simultaneous fabrication and automatic assembly of multiple parts in a single machine.
7. Individual timing and adjustment of all slides.
8. Virtually unlimited versatility.

Disadvantages of Slide Machine Utilization. Many design and construction sources lack the tool designers and builders who are versed in this machine tooling technology. Also, the lack of universal standards has resulted in tooling being dedicated to a machine of a specific make and size.

References

1. Charles Wick, John W. Benedict, and Raymond F. Veilleux, eds., *Tool and Manufacturing Engineers Handbook*, 4th ed., Vol 2 (Dearborn, MI: Society of Manufacturing Engineers, 1984).
2. Engineering and design standards of The Design Standards Corporation, Bridgeport, Connecticut.
3. *The Multiform Manual*, Heenan & Froude Limited, Worcester, England, 1958.

SECTION 21
LOW-COST AND MISCELLANEOUS DIES

Various types of dies used primarily for large-scale production have been presented in previous sections. There is also widespread demand for another class of tooling where short-run stampings are involved, requiring relatively low die costs. To reduce lead time and cost, such tools must usually be designed and made with techniques which are not applicable to the making of more costly tooling.

Development of dies in the low-cost, low-production category must combine, in addition to a low design and fabrication cost, the following related factors:

1. Accuracy. In general, close tolerances should be avoided.
2. Facilities. The tool designers should have a thorough knowledge of the facilities available in the tool room and in production and maintenance areas.
3. Product design. Complicated geometric product design should be avoided.
4. Tool storage. Tool setup cost vs. inventory cost should be evaluated. Determine whether the tools will be stored for future use or whether one production run will produce the required total number of parts.
5. Product obsolescence. This factor may determine a specific choice of low-cost tooling within alternative design ranges.
6. Safety. Compliance with all safety rules and regulations is required.

Dies with Interchangeable Elements. A die set constructed of conventional materials and by conventional methods is a major factor in the cost of low-requirement parts having a variety of shapes, hole sizes, and patterns. This cost can often be reduced by using dies with interchangeable or reusable elements.

The dies illustrated in Figs. 21-1 and 21-2 feature a master die set which accommodates interchangeable punch and die units made for specific parts. To assure accurate alignment between any unit and the master die set, a drill jig is used to transfer all dowel, screw, and stripper bolt holes into an interchangeable punch and die unit.

Incorporated in the master die set ($D1$) of Fig. 21-1 are spring stripper units, punch and die mounting screw holes, locating dowel holes, a punch backup plate, and a slug clearance slot. The punch and die plates ($D2$) are drilled to locate and support commercial punches ($D3$) and die buttons ($D6$). The stripper plate ($D9$) is tapped to accommodate the stripper units ($D4$ and $D5$) in the upper shoe and has clearance holes for the punches. Where necessary, support buttons ($D7$) are used to minimize deflection of the die plate. A precut blank is located by locating pins ($D8$).

The punch and die plates may be removed and stored as a unit or the punches, die buttons, and support buttons may be removed for use in other units.

A blanking-die arrangement in which the punch and die are attached to relatively thin plates that can be readily attached to a master die set is shown in Fig. 21-2. Dowel pins permanently pressed into the die shoes position the mounting plates in which coordinating holes are jig-drilled and reamed. Pieces of rubber, cork, or polyurethane may be cemented around the punch and in the die cavity to function as strippers. Stock guides and stops are permanently fastened to the mounting plates.

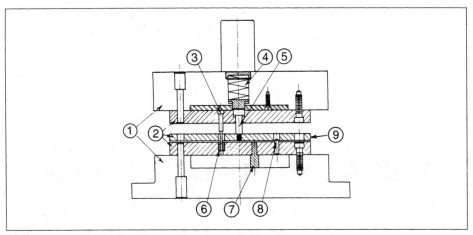

Fig. 21-1. Die set with interchangeable punch and die units.

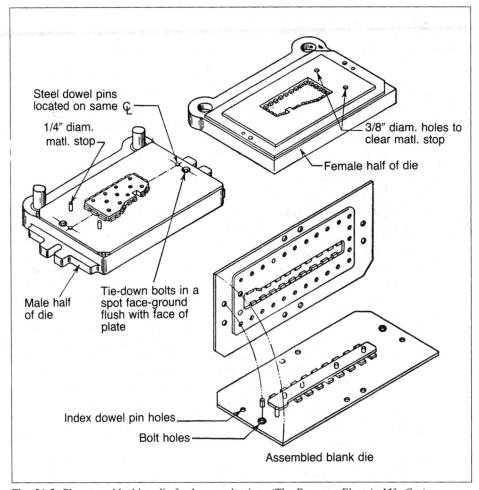

Steel dowel pins located on same ℄

1/4″ diam. matl. stop

3/8″ diam. holes to clear matl. stop

Female half of die

Male half of die

Tie-down bolts in a spot face-ground flush with face of plate

Index dowel pin holes

Bolt holes

Assembled blank die

Fig. 21-2. Plate-type blanking die for low production. *(The Emerson Electric Mfg Co.)*

Die sets incorporating interchangeable punches, button dies, and strippers are commercially available. The die set of Fig. 21-3 incorporates T-slots in both upper and lower shoes (*D1* and *D2*) to which slotted interchangeable retainers (*D3* and *D4*), for interchangeable punches (*D5*), button dies (*D6*), and stripper units (*D7*), are mounted. A slotted stock gage (*D8*), mounted to correct T-slots, provides for the exact positioning of various sizes of blanks or strip.

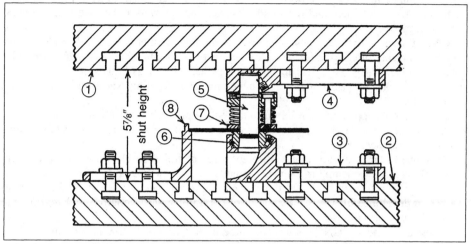

Fig. 21-3. Interchangeable units for punching sheet metal. *(S. B. Whistler & Sons, Inc.)*

Another commercial design (Fig. 21-4) consists of a one-piece unit which also may be set up on T-slotted plates in presses or press brakes. Interchangeable punches (*D2*) and dies (*D1*) of various sizes are available. Stripping pressure is provided by cylinders (*D3*) containing oil under pressure.

A cam-actuated, side-piercing unit which may be fastened to T-slotted plates is manufactured with interchangeable punches and dies of various sizes and shapes. A number of self-contained punching units of various sizes and shapes, designed to be bolted to templates on the upper and lower shoes, are also available commercially.

Reusing Obsolete Dies. Many shops reclaim obsolete dies that would otherwise be scrapped, and retrofit them for short run jobs. The die sets as well as many of the components can be reused.

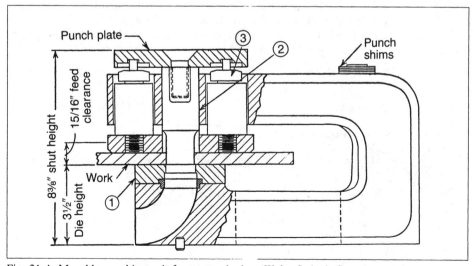

Fig. 21-4. Movable punching unit for a press brake. *(Wales-Strippit Corp.)*

Die shoes, steel plates, composite cutting sections, air cylinders, nitrogen pressure systems and commercial cam slides are usually easy to recycle into temporary tooling on short notice.

Quick construction practices that are not acceptable for long-run tooling can be used to good advantage to build a die needed on very short notice. Such practices may include mounting punch retainers and die blocks by welding, as well as hot shearing irregular cutting edges.

CNC Turret Punch Presses. Automatic computer numerical control (CNC) turret punch presses are often used for flexible manufacturing applications to produce small lot runs of sheet metal parts. These machines, together with press brakes or CNC bending machines, make up work cells for punching and forming short-run sheet metal components.

Large irregular openings can be cut out by the appropriate selection of conventional punches in the turret. High-speed nibbling stations are also available for use in CNC turret punching machines.

Developed blanks having many different sizes of holes on close location are easily produced on turret punching machines. An advantage to the use of developed blanks that are subsequently formed is eliminating the cost of a complex pierce die. A drawback to the use of developed blanks is the amount of cut-and-try work required to determine the correct hole locations.

A real advantage of CNC equipment is the ease with which the machine instructions can be modified to speed up the trial-and-error hole moves that blank development work requires.

Once the correct hole pattern is determined, the required number of blanks can be produced on the turret machine. The hole punching instructions can be stored for later production runs, or if the volume is high enough, used to correctly determine punch and die opening locations for a permanent pierce die.

The *X-Y* hole coordinate locations can be exported directly from the CNC production equipment to the CAD equipment used to design the permanent tooling.

Punching. Emphasis has usually been placed on the improvement of die designs which reduce the cost of the die. This patented method has simplified die construction and also has reduced the number of operations required to build a die.[1]

The dies shown in Fig. 21-5 are suitable for all low-requirement punching operations with the die life sufficient to produce approximately 30,000 to 100,000 parts, depending upon the material and thickness. They are used in punching cold-rolled steel, aluminum, copper, brass, and many other materials. Accuracy can be held to within ±0.003 in. (±0.08 mm) on material up to and including 0.125 in. (3.18 mm) thick.

Tool construction is relatively simple. The thin die plates are held apart by springs, guided by liner pins and guide bushings, and held together by snap retaining rings. The

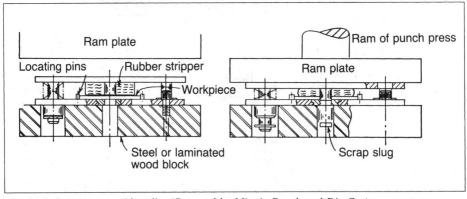

Fig. 21-5. Low-cost punching die. *(Patented by Minnie Punch and Die Co.)*

quick-change snap retaining rings also permit rapid assembly and disassembly of the tool. The rubber stripper, necessary to remove the blank from the perforators, is attached to the punch plate.

In setup, the die plate is mounted on a steel or laminated wood block that is secured to the bed of the punch press. The ram of the press is provided with a flat-bottomed ram plate. In operation, the ram plate descends and makes contact with the punch plate. As the ram descends further, the springs and the rubber stripper are compressed and the holes are pierced. The blank then is stripped from the perforating punches by the rubber stripper.

These self-contained units eliminate the standard die set and all unnecessary screws and dowels which are normally utilized in punch-press tools. They can be quickly set up in any punch press, regardless of the shut height. Other inherent advantages are reduction in setup time and conservation of storage space.

Floating-punch Die. An inexpensive method of diemaking produces the punch and the die from the same piece of tool steel. A die of this construction is shown in Fig. 21-6 with the punch (D1), punch-positioning plate (D2), spacers (D3), and die plate (D4). This die is used in a press equipped with a die set having hardened and ground steel faces and sturdy leader pins and bushings. The minimum shut height of this die set is determined by the thickness of the floating-punch die.

In operation, the stock is fed into the die; the punch is located in the positioning plate and on top of the strip, and the die is then slid in between the two surfaces of the hardened and ground faces of the die set. After tripping the press, the die is slid out of the die set, allowing the punch and blank to fall out.

The method of making the punch and die from one piece of tool steel is illustrated in Fig. 21-7. The outline of the finished part is laid out on both sides of the plate. The outline is then sawed out by tilting the blade at such an angle that it cuts on the inside of the layout on the top side and outside the layout on the lower side. The die block and punch are then finished by filing to the outline perpendicular to their respective faces. The cold-rolled-steel, punch-positioning plate is fastened to the die block with spacers between to provide a tunnel stock guide.

The same sawing technique can be used to make punches and dies, that are to be

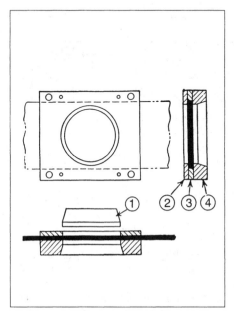

Fig. 21-6. Floating-punch die using the continental method. *(The DoAll Co.)*

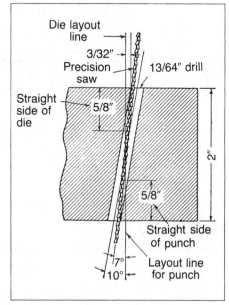

Fig. 21-7. Method of sawing punch-and-die block from same piece of steel.

mounted in a die set, from one piece of steel.

Low-cost Wire Burn EDM Dies. The time required to complete a cut on a wire burn electrical discharge machine is mainly a function of the thickness of the material being cut. The complexity of the cut, once programmed, does not reduce cutting speed. Thin hardened tool steel blanks can be cut into complex punch and die sections in a relatively short time. These thin sections are usually backed up with fairly thick shoes or sub plates.

The piece cut out of the die section can often be used as the punch. The width of cut, typically 0.015 in. (0.64 mm) would provide an excessive cutting clearance for thin stock. To overcome this problem, the wire path can be tilted in much the same manner as the band saw blade shown in Fig. 21-7. Generally, wire burn machines can be programmed to automatically provide tapered die clearance with no loss of cutting speed.

The wire kerf often can be located in a non-critical area of the punch or die opening. A sharp corner often is a good location. If this is not possible, the kerf can be carefully welded shut.

General-purpose Cutoff Die. The cutoff die shown in Fig. 21-8 may be used in shops where production does not justify the purchase of a shearing machine. The die is designed to utilize the four cutting edges on each cutoff blade before being removed for resharpening. To sharpen the die it is necessary to grind the face of the blades only and, on replacing in the die, to add shims to compensate for the stock ground off. These shims may be placed behind one or both of the blades. Shims can also be used to adjust the clearance for shearing thick or thin materials. Spring-loaded pressure pads under each blade grip the stock and assure a cleanly cut blank. An easily adjusted stop is provided for gaging the length of blanks. This die is mounted in a four-post die set to maintain the desired clearance between the two blades.

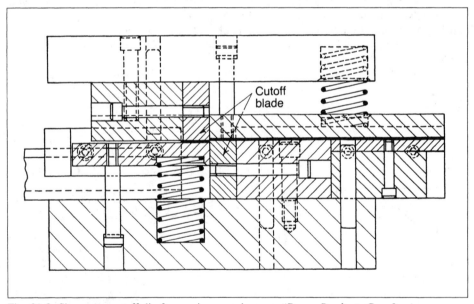

Fig. 21-8. Shear-type cutoff die for use in a punch press. *(Davey Products Co., Inc.)*

FORMING DIES

Roll Forming Punch and Die. Roll forming dies (Fig. 21-9) may be run in punch presses or press brakes. It is possible to hold close tolerances on bends since slippage is eliminated.

The prime feature of this tool is the method of forming the material around the form punch. Any angle up to 60° can be formed with the pivoting action of the two rolls in the die block. The rolls are spring-loaded to return them to their original position. To form

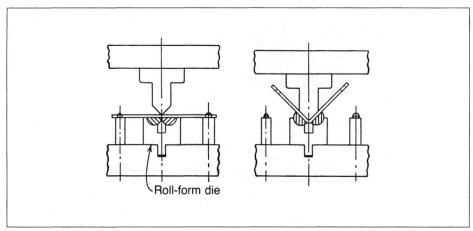

Fig. 21-9. Roll forming die.

angles of varying degrees would normally require several sets of V-type dies.

Springback is easily compensated for, since the required amount of overbend can be provided with ease. Less force is required to make bends than would be the case if the dies were bottomed to control springback. There are none of the usual scuff marks that can be a problem when a V-shaped die is used.

After the blank is perforated and clipped, it is placed over the two locating pins which locate it over the forming die. As the punch descends and the movement continues downward, the half-round inserts in the roll forming die pivot and fold the material around the form punch. This continues until the forming cycle is completed. A nest may also be used to locate the blank.

The forming members consist of a roll forming die, a form punch, and two pin-type locating gages. These details are hardened and mounted into a commercial die set. However, where short-run jobs are common, a more economical tool can be built by using a commercial form punch (not shown in Fig. 21-9). This form punch has a 60° angle, which allows forming of material from 0.005 to 0.250 in. (0.13 to 6.35 mm) thickness and angles from 0 to 60°. Radius bends from 0 to 0.375 in. (9.52 mm) can also be formed by changing the radius insert.

A similar type of tool with swiveling die members for inward flanging is illustrated in Fig. 9-12.

V-type Forming Die. The tool illustrated in Fig. 21-10 incorporates a punch and die used to bend a blank into a V-shape. It can be designed so that a variety of parts may be formed in the same tool by adding additional nests. The form punch and die block are made of tool steel and are mounted into a commercial die set. The forming members remain soft, and if production increases they can be hardened. A tool-steel nest, which locates the work, is mounted on the die block. In operation, the tool is attached to the ram and bed of a conventional press.

After the blank has been cut to the desired length, it is placed on top of the die, where it is located in a suitable nest. The punch then descends, and the blank follows the contour of the punch, thus completing the bending operation on the part. The size of the blank is a major factor in the size and weight of the tool. For additional bending dies, see Section 7.

Low-cost Forming Die. The making of a low-cost punch and die for an irregular shaped part is shown in Fig. 21-11. The punch is machined or cast to shape, depending upon the tolerance. From the finished punch a die is produced by pouring a zinc-based or bronze alloy around the punch contour. Clearance between the punch and die is obtained by shearing thin stock in a hydraulic press. As the clearance increases, thicker stock can be cut. A small amount of hand fitting may be required before the die is ready for

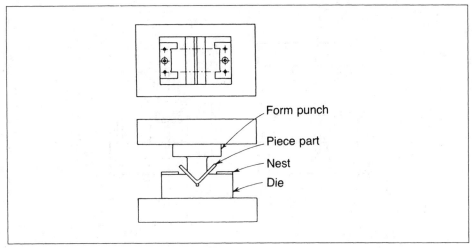

Fig. 21-10. V-type forming punch and die.

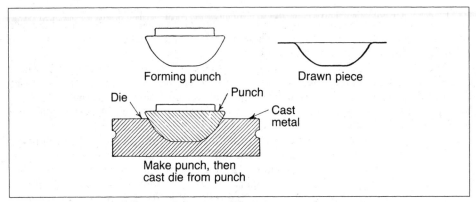

Fig. 21-11. Low-cost forming die for an irregular shape.

production. To complete the die, the punch is assembled to a mounting plate and a spring-loaded pressure pad surrounding the punch.

Knurling on a Punch Press. The knurling of all or part of a pin may be done on a punch press with the die shown in Fig. 21-12, where the die is shown partially closed. As the punch descends the piece is rolled between the knurl dies and finally drops through the

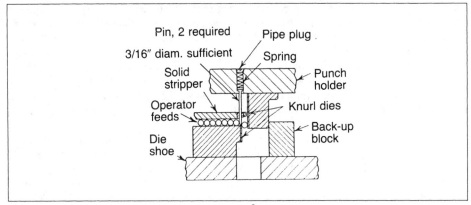

Fig. 21-12. Die for knurling pins on a punch press.[2]

21-8

die shoe. Two spring-loaded pins permit the feeding of only one part at a time. For best results, this die should be operated in an inclined press.[2]

Plastic Dies for Low Production. Experimental and low-production stampings can be produced at low cost by using lightweight, easily made plastic dies. These dies can be used to draw or form practically all grades of sheet steel and aluminum from which sheet-metal parts are fabricated.

The basic design of a plastic draw die is the same as for a conventional draw die; the punch, die, and blankholder or binder ring mount in the press in the same manner as conventional dies. The draw die shown in Fig. 21-13 has a metal container to support the plastic components. These metal containers can be made of cast iron, steel weldments, or cast zinc alloy. The working faces are solid cast plastic or are built up of layers of

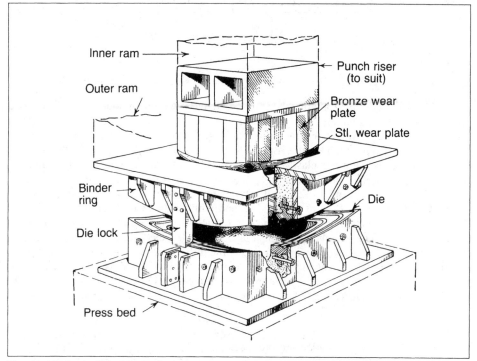

Fig. 21-13. Draw die with plastic working surfaces for low-production parts.[3]

plastic-impregnated glass cloth. The working surface is backed up by a plastic core or other filler material.

The cross section of a flanging die with plastic working faces is shown in Fig. 21-14. The locating member (D1) of the die has a cast epoxy surface with a polyester core on a steel base plate. The punch (D2) uses a cast-iron base for the cast epoxy surface. The steel backing is required, because the plastics have a compressive strength of approximately 11,000 psi (76 MPa) and an approximate tensile strength of 5,000 psi (34 MPa).

Dies of this type can be built by using a conventional die model with a developed binder line to make a plaster splash mold for the punch. The die model does not need to be compensated for shrinkage; the plastic material shrinks an insignificant amount in most cases. A plaster model of the binder ring is developed to the draw line on the die model. Draw beads, if considered necessary, may be worked into the face of the model so that they may be cast into the binder ring from the plaster splash mold. The edge of the draw beads should be about 1.25 in. (31.8 mm) from the die cavity to prevent the edge from breaking away during the operation. Metallic wear plates may be fastened to the punch and binder ring to prevent wear on the plastic surfaces.

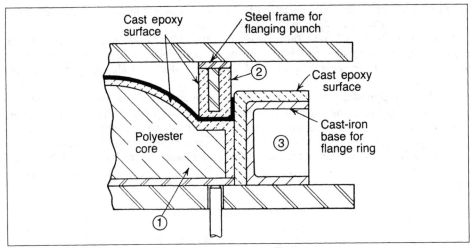

Fig. 21-14. Cross section of a flanging die with plastic working surfaces. *(Kish Industries.)*

The finished binder ring and punch are assembled to use as the model for the die cavity. The surfaces of the punch and binder ring are built up, equal to stock thickness, by using either patternmaker's wax or a spray-paint build-up. From this model, a plaster mold for the die is made, from which the die itself is finally cast. The curing of the material in the plastic die depends upon the plastic itself. Some plastics are self-curing; others require heating in an oven at a controlled temperature for a period of time.

The die construction in Fig. 21-15 illustrates the use of a zinc alloy container which was cast to the rough outline of the product and with flanges for attaching to the press. Because of its nonshrinking characteristics, the thickness of the plastic does not need to be held constant. The simplicity of the shape of the finished part allows the use of a partially-open-die construction where the punch and die mate only at the areas of severe forming. Holes were drilled at angles in the sand mold for the zinc alloy casting to provide the lugs to anchor the plastic face. The binder-ring faces and beads for this die are made of plastic. An alternate type of die construction may use a more uniform thickness of

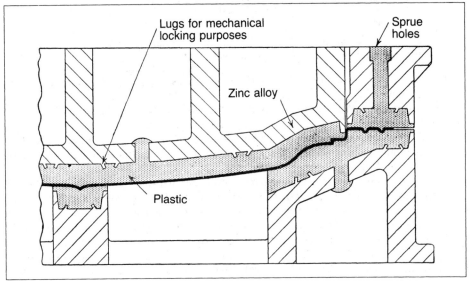

Fig. 21-15. Typical cross section of a plastic-faced die on a zinc-alloy base. *(Richard Brothers Division, Allied Products Corp.)*

plastic and zinc alloy for the beads and binder-ring surfaces. Cast iron may be used in place of the zinc alloy where the foundry facilities are available.[3]

STRETCH FORMING

The basic stretch-forming process consists of gripping a metal sheet, an extruded shape, or a brake or roll-formed section at each end with a pair of jaws, then stretching or wrapping it over a die of the desired contour.

The design of stretch dies usually need not include springback allowance. The reason why very little springback occurs after forming is illustrated in Fig. 21-16. The upper part of the diagram shows the effect of bending a sheet of metal. This cross section shows that the fibers in the upper portion are stretched, while the fibers in the lower portion are

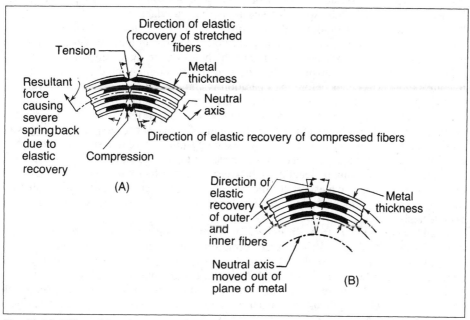

Fig. 21-16. Effect of forming on metal fibers: (A) by bending; (B) by stretch forming.[4] (*Reynolds Metals Co.*)

compressed. The lower part of the diagram represents a sheet formed by a stretch die. All the fibers have been stretched beyond the yield point and put in tension. The neutral axis, or the pivot point about which springback usually occurs, has been moved out of the plane of the metal so that all the residual stresses are in tension.[4]

Parts may not be stretch-formed to shapes that have a reentrant curvature. The shape must be so designed that a string, pulled by the jaws across the block, will follow the block from end to end without leaving the surface.

The typical contours which may be readily formed are shown in Fig. 21-17. As indicated by the arrows, the direction of stretch should be at right angles to the radii of curvature.

In most cases, stretching work-hardens the material to a higher strength uniformly throughout the part. This work hardening is done with a minimum reduction of stock thickness.

The primary cause of part failure is tearing at the point of maximum metal elongation. A sheet with a large curvature will usually fail by tearing between the jaw and form block. A sharp curvature causes failure at the crown or about halfway between the jaws. The point of failure in saddleback-shaped parts is at at the edge of the sheets where the

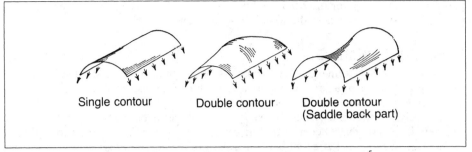

Single contour   Double contour   Double contour (Saddle back part)

Fig. 21-17. Typical contours to which sheets may be readily stretch-formed.[5]

maximum stretch is located. Smooth or polished edges on sheets will permit greater elongation of the parts than when the sheet has sheared edges.

The blank sheets should be of uniform width in order to distribute the stresses uniformly throughout the sheet. Cutouts and holes required in the finished part must be made after forming, to prevent distortion of the holes and to avoid stress concentration. When severe forming is involved, blanks with imperfections should not be used as a fracture failure will occur before the operation is completed.

Wrinkling or puckering of the sheet rarely occurs, except in the case of double-curvature parts, because stretch forming holds the sheet under tension during the operation. Stretch forming requires an extra length of material between the die and jaw to minimize the stress in the sheet caused by changing from a curve to a straight line. When forming saddleback parts, the metal has a tendency to slide toward the center of the die resulting in wrinkles running lengthwise between the jaws. Increasing the sheet width and extending it over the edges of the die will help to prevent the sheet from sliding. Increasing the stock thickness or using a material or alloy of greater tensile strength may also decrease wrinkling.

In addition to freedom from springback problems and the improvement of the physical properties of the metal, stretch or wrap forming has the following advantages:

1. Die cost is reduced since only one die is required, whereas a punch and die are necessary in drop-hammer forming and in single- and double-action presses.

2. Labor cost is lowered in making the die without the refinement of spring-back compensation. Also, low-cost die materials may be used, such as concrete, plastics, masonite, and wood.

3. Considerably less operator training is required than is the case with roll forming or hand finishing.

4. Parts formed have a minimum of residual stresses.

Stretch-form Machines. There are three basic types of stretch-forming machines (Fig. 21-18): the moving-ram type, the rotary-table type, and the rotary-arm type.

Moving-ram Type. These machines are of two designs: (1) moving die, and (2) moving jaw. The moving-die machine has stationary jaws, trunnion-mounted, to grip the stock while the die, mounted on a ram, is pushed into the sheet. The moving-jaw machine has a stationary die, the jaws being on trunnion-mounted hydraulic cylinders to stretch the sheet over the die.

Rotary-table Type. The machine has the die fastened to a rotary table. One end of the material is gripped in a jaw on the table; the other end is gripped in a jaw on the machine framework. As the die is rotated, the part is stretched and wrapped to the die contour.

Rotary-arm Type. On the rotary-arm machine, the jaws are fastened to the rotary arms and the die is stationary. The stock is gripped by the jaws tangent to the die contour and the rotation of the arms wraps the material to the die contour.

Each type of these stretch-forming machines is classified by tonnage, i.e., the pulling force that is exerted on a workpiece. A moving-ram machine is classified by the ram tonnage. The actual stretching force that the machine would be capable of applying to the

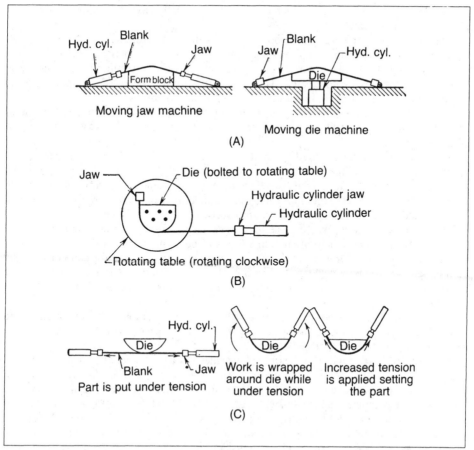

Fig. 21-18. The three basic types of stretch-forming machines: (A) moving-ram machine; (B) rotary-table machine; (C) rotary-arm machine.[5]

part would therefore be one-half its rated tonnage.

Machines designed primarily for parts up to approximately 20 in. (500 mm) wide are referred to as "extrusion-type" machines. Those capable of forming wider parts are called "sheet machines."

Radial-draw forming, employing the rotary-table type of stretch-forming machine, forms metal progressively as the die is rotated in a plane. Three forces are employed: stretching, bending and wiping, and side compression.[5, 6]

Metals That May Be Stretch-formed. Most metals may be stretch-formed; they need only to have a definite workable range between the yield point and the ultimate strength of the material.

The low-carbon steels (SAE 1010, 1020) are readily formed and will give excellent results. Austenitic stainless steels given excellent results in the annealed state, forming becoming increasingly difficult as the hardness of the material increases. Full-hard stainless steel is nearly impossible to form except under ideal conditions.

Aluminum is widely used for stretch forming, and most of its alloys are readily formed. Forming is done cold and may be accomplished by using heat-treated stock in many cases. When this is possible, the subsequent heat treatment, distortion, and restretch may be eliminated. The 7075-T6 aluminum alloy is difficult to form; 7075-O is readily formed.

Magnesium alloys lend themselves to very successful stretch forming. The punch (die) on stretch machines may be heated either with gas burners or electric elements. Common alloys such as AZ31B are usually stretched in the 400° to 600° F (204° to 316° C) range.

The temperature-resisting alloys such as HM 21 are usually stretched in the 500° to 700° F (260° to 371° C) range.

The jaws of the stretch press are operated at room temperature so that the magnesium being held has higher properties than the heated alloy which is being formed. After the stretch is complete, the jaws can be set to maintain a constant strain on the metal while secondary operations are completed. This secondary forming (beads, flutes, indents, etc.) should be done with a rawhide hammer or a hardwood blank.

Lubrication for stretch-forming magnesium is usually colloidal graphite. Very mild stretch forming at temperatures under 300° F (149° C) can utilize laundry soap, wax, or oil as a lubricant.

The amount of force required to stretch-form a part is defined as the product of the cross-sectional area of the part perpendicular to the force and the yield strength of the metal being formed. Since the yield strength of a metal is usually stated as the manufacturer's guaranteed minimum and work hardening is not considered, the calculated value should be increased by 25% to assure sufficient machine force.

In forming extrusions, where close tolerances are specified, annealed material should be used. When a heat-treated aluminum alloy is to be formed, the part is first formed in the annealed condition. It is then heat-treated and placed on the same form die, and a final setting tension is applied. Stretch forming after heat treating eliminates springback; therefore die correction is unnecessary. It also removes distortion arising in the heat-treated cycle.

The degree of success with which heat-treated aluminum alloys in either the quenched or aged condition can be stretch-formed depends upon the severity of the contour and the tolerance required in the finished shape.

Stretch-die Materials. The dies may be made of various materials, depending upon the production requirements. Common materials used are cast iron, wood, zinc alloys, masonite, cloth-based phenolic resin-bonded materials, and cast plastic resins. If the dies overhang the machine platen, sufficient strength must be built into them to prevent excessive deflection under the heavy stretching load. Any deflection in the die block will result in a like amount of change from the intended shape in the formed part.

The type of workpiece material, its physical condition, and its thickness have great influence on the selection of material from which the forming blocks are to be made. The thickness of the material specification determines the load which will be exerted upon the form block during the stretching operation.

Stretch dies made of well-seasoned wood are satisfactory for a limited production. White pine, straight-grained birch, and mahogany have been used with excellent results. Where there are sharp edges or shapes to be formed that result in points of very high pressure, the wood will wear excessively. At these points it is good practice to use zinc alloy, cast iron, or plastic inserts. Wooden dies should be a minimum of 4 in. (102 mm) thick. On very large dies, it may be necessary to reinforce the block with a subbase made of structural-steel members.

The zinc-based alloys will give longer service than wood, although points of extreme pressure may require inserts of a harder material. The inherently greasy nature of zinc alloys is desirable since it reduces lubrication requirements. A further advantage is that the material from obsolete dies can be reclaimed and reused. To minimize finishing operations, high-quality pattern construction and careful foundry practice are required.

Gray cast iron is good from the standpoint of inherent strength of the die blocks themselves. Because of the relatively high strength of cast iron, the die blocks may be cored out underneath the solid top to reduce weight and conserve material. As compared to zinc alloy dies, considerable labor is required to finish the working surface of a cast-iron die, but the long die life may justify the cost.

Plastic resin dies are light in weight and present virtually no problems when cast because no shrinkage allowance is necessary. This material gives good results from the standpoint of wear resistance, workability, and stability of contour regardless of changes in temperature and humidity. The glass-like surface that may be obtained eliminates practically all the galling or other die marks that frequently appear on the parts made with metal stretch dies. The die in Fig. 21-19 is a typical cross section of a plastic stretch die with a foam phenolic core and a solid phenolic working surface. The die has a steel backing plate with I-beams welded to it for rigidity. Unless the plastic form blocks are supported throughout their underside by a rigid structure, the lower edges will tend to chip off. The use of expanded metal lath in the compressive flange of the block can serve to prevent the chipping induced by the operational loading on the lower surface of the block. It is a good practice to strengthen plastic die blocks by placing reinforcing bars where tension loads are expected, while relying on the compressive strength of the material to withstand compressive forces.

Masonite should be used only for the more simple shapes, with no hammer work on the sheet while it is being stretched, as masonite tends to disintegrate under impact on its edge grain.

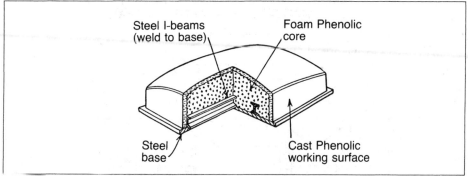

Fig. 21-19. Cross section of a typical plastic stretch-form die.[3]

ASSEMBLING DIES

Assembling dies fasten parts together by either of two methods: (1) fastening them together by means of a third part, such as a rivet, which is plastically deformed by the die, the parts being subjected to little or no deformation; (2) joining them by plastically deforming mating areas in either or both parts in operations of staking, folding, crimping, curling, seaming, or press-fitting. Conventional rivets may have one end upset in a die to fasten parts together; or pins, rods, bushings, or projections of the part itself may be used instead of a rivet.

Ram-coin dimpling is an assembly operation involving a type of riveting; pertinent data is found in Section 8 (Figs. 8-5 and 8-6 and Eq. (11)).

Riveting Die. A riveting-die design (Fig. 21-20) incorporates a slide (D3) which is pulled out by the operator for loading stator laminations. Four pilots (D4) and a post (D6) enter pilot holes and the armature hole, respectively, as the laminations are stacked on them and on four copper rivets. Outward slide movement is limited by a spring-loaded stop pin (D8); inward movement and positioning of the slide are controlled by a spring-loaded detent ball (D5). Four riveting punches (D1) upset the copper rivets. A shedder pin (D2) is mounted in the punch retainer plate between each pair of riveting punches.

Riveting and Skewing Die. Motor rotor laminations are stacked on a center pilot (Fig. 21-21, D4) and 19 copper pins are inserted vertically in holes circularly spaced in the laminations to project into undercut holes similarly spaced in an upper floating punch plate (D7) and in a lower rotating punch plate (D18). The holes contain 19 upper and 19 lower riveting punches (D12) which bear lightly against the ends of the copper pins. The operator depresses a foot lever (D1) until the floating punch plate is held down by a hook or latch (D3) to the hook bar (D9), which is adjustable for height by means of a pressure

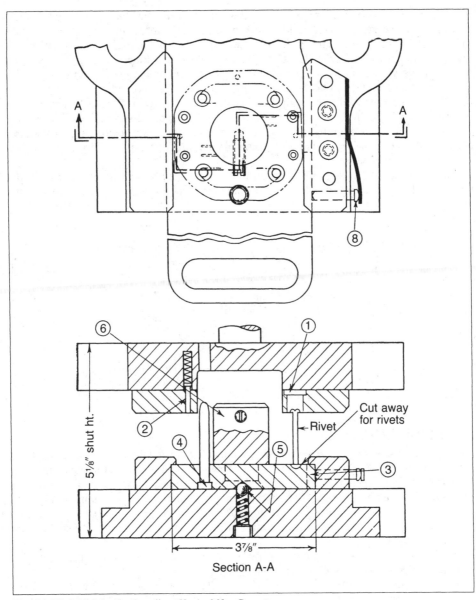

Fig. 21-20. Multiple riveting die. *(Harig Mfg. Corp.)*

plug (D10) and a socket-head screw (D11). The latch engages one of the three circular grooves in the hook bar corresponding to the desired assembly of three different heights of stacked laminations. The operator then pulls the skewing lever (D8) to the right, which rotates the lower punch plate and tilts the copper pins held in the undercut holes. This angular change of the pins from the vertical corresponds to a setting of an automatic indexing stop (D2). Its position, selected by an adjustment of a screw (D6), determines the amount of skew in a stack of laminations. Three notches in the indexing ring (D13) engage the indexing stop corresponding to three lengths of rotor assemblies and control the amount of skew desired in a completed assembly. The press is tripped after the skewing operation, allowing 38 riveting punches to upset both ends of the copper pins to complete the assembly of a rotor. On the downstroke a trip bar (D16) is depressed to release the hook bar from its latching position. The indexing stop, pivoting on its pin

(D17), is raised above the indexing ring during the last portion of the upstroke by a pickup pin (D14) contacting the edge of a flat on a post (D15). Upward movement of the stop allows the spring-loaded skewing lever to return to its starting position against its stop pin.

Riveting Magneto Laminations. A stack of laminations is placed between stationary and movable nest plates (Fig. 21-22, D7, D4); the latter is swung to the right by the operator's handle (D6) to align and clamp the stack as the toe of the handle bears against a clamping post (D8). Three riveting punches (D1) held in a punch holder (D2) upset three rivets as the stack is compressed by an insert (D3) shaped to the contour of the laminations. On the upstroke the handle is swung to the left, and a knockout (D5) is actuated through the linkage shown and returned by a spring (D9).

Die for Riveting Lever to Bushing. A lever is securely riveted to a bushing by the die shown in Fig. 21-23. The bushing is placed in a support block (D1) which is attached to a slide (D2). The slide enables the parts to be loaded and unloaded out from under the die. The lever is supported by the square head screw (D3) and located radially by two pins (D4). The spring pad (D5) holds the lever tightly on the bushing while the punch (D6) rivets the assembly. A pilot on the punch prevents the hole from closing in during the operation.

The slide is held in the retracted and working position by a spring detent (D7). To assure proper location and rigidity, a pilot pin (D8) pressed in the upper shoe enters a hole in the slide on the downstroke. The use of an electrical limit switch, to prevent the press being tripped when the slide is not in the correct position is advisable.

Staking Dies. The hub of a gear (Fig. 21-24A, D1), splined in a previous operation, is fastened to a wheel (D2) by a staking punch (D3). The four splines (D4), forced outward and against the washer by the four chisel edges of the staking punch, provide rigidity in the assembly. In view B, a stripping fork (not shown) prevents the staking assembly from sticking to the punch (D1) and its pilot (D2). The punch forces metal from the hub into the chamfer in the plate. In view C, two segmented staking rings (D1) inwardly force metal of a wheel (D3) against the serrated portion of a shaft (D2).[8]

A 0.062 in. (1.57 mm) diameter ball is staked into a 0.125 in. (3.18 mm) diameter shaft as shown in view D. The amount of taper on the punch provides enough metal displacement to hold the ball securely.

Staking-punch Design. Data for the design of four-bladed punches for staking hubs, tubes, and similar parts are given in Fig. 21-25.

Hydraulic Compensating Staking Die. A nest (Fig. 21-26, D7) positions aluminum hubs which are staked into holes in aluminum sheet stock. The hub is placed tenon side up in the nest which is bolted to the die top (D4) by cap screws (D5). Enough oil is contained in the well to maintain the 0.188 in. (4.78 mm) dimension between the die top and the cylinder (D6) as shown. O-rings (D2 and D3) seal the oil in the well in the base (D1). Upon actuation, the press forces the hub and cylinder down. Oil pressure forces the top up to clamp the stock between it and a rubber pressure pad on the ram. Further pressure compresses the rubber pad to allow the staking punch to enter the hole in the stock and stake it to the hub. This design compensates for hub-length variations by always allowing the stock to lie flat during the staking operation. The die assembly is placed in an adapter which can be permanently fixed to the press bed or attached to a sliding fixture for loading and unloading outside the punch.

Staking tips (Fig. 21-27A) are screwed into a holder (view B). The holder is mounted in an adapter (not shown) which is screwed in the ram of the press.

Commercial Assembling Die Sets. A design for fastening sheet-metal sheets together (Fig. 21-28A) incorporates a die which expands slightly under the impact of the lance-form punch. At B, a nut which does the piercing is held magnetically to the bottom of the punch and is staked to the workpiece. Interchangeable ball-lock punches and die buttons for staking various sizes and shapes of shouldered nuts are available.

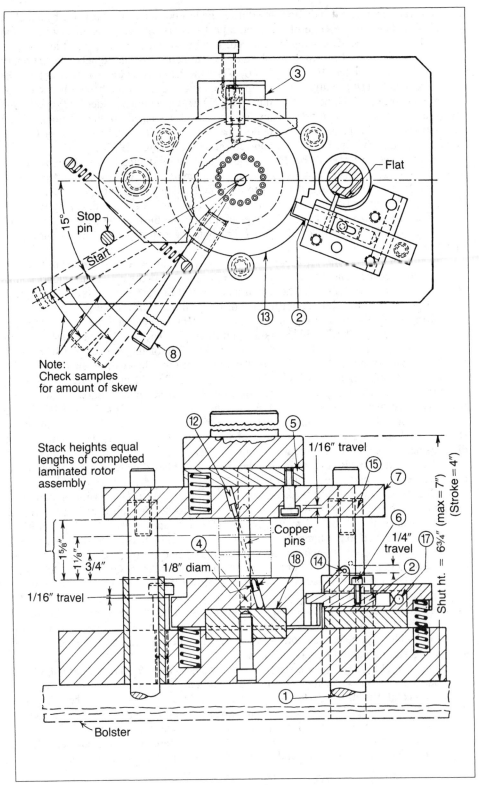

Fig. 21-21. Riveting and skewing die. *(Harig Mfg. Corp.)*

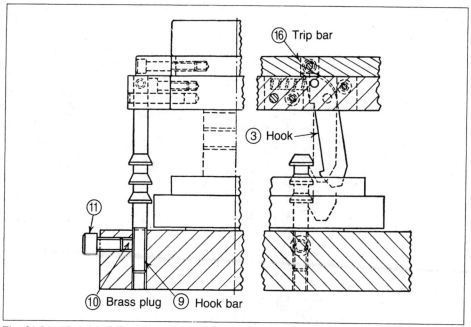

Fig. 21-21. (*Continued*). Latching fixture for laminations. (*Harig Mfg. Corp.*)

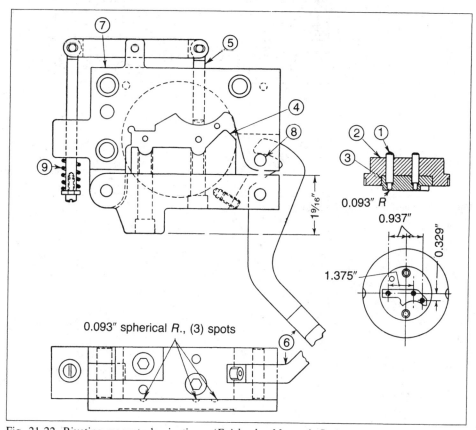

Fig. 21-22. Riveting magneto laminations. (*Fairbanks, Morse & Co.*)

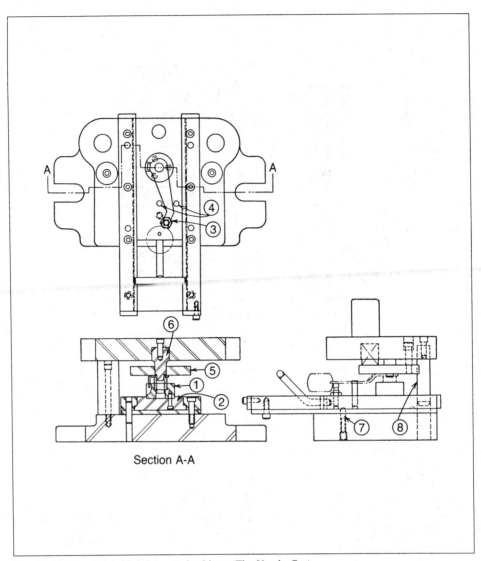

Section A-A

Fig. 21-23. Die for riveting lever to bushing. *(The Vendo Co.)*

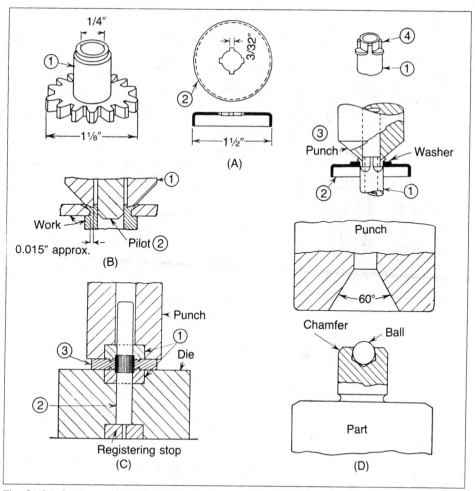

Fig. 21-24. Staking operations: (*A,B*) staking outwardly (*F. W Curtis*); (*C*) staking inwardly;[7] (*D*) staking a ball.[8]

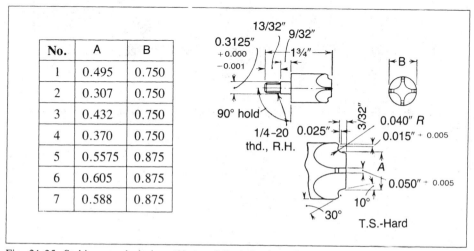

No.	A	B
1	0.495	0.750
2	0.307	0.750
3	0.432	0.750
4	0.370	0.750
5	0.5575	0.875
6	0.605	0.875
7	0.588	0.875

Fig. 21-25. Staking-punch design. (*National Cash Register Co.*)

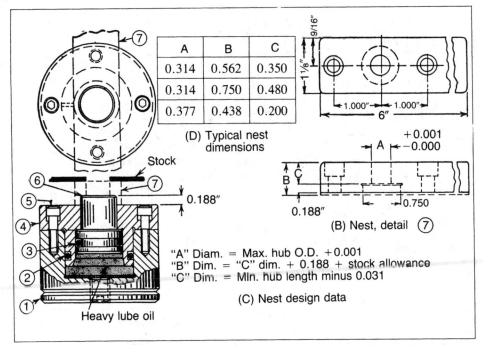

A	B	C
0.314	0.562	0.350
0.314	0.750	0.480
0.377	0.438	0.200

(D) Typical nest dimensions

Stock

0.188″

(B) Nest, detail ⑦

"A" Diam. = Max. hub O.D. +0.001
"B" Dim. = "C" dim. + 0.188 + stock allowance
"C" Dim. = Min. hub length minus 0.031

(C) Nest design data

Heavy lube oil

Fig. 21-26. Hydraulic compensating staking die: (*A*) assembly; (*B*) nest (*D*7); (*C*) nest design data; (*D*) typical nest dimensions *(National Cash Register Co.)*

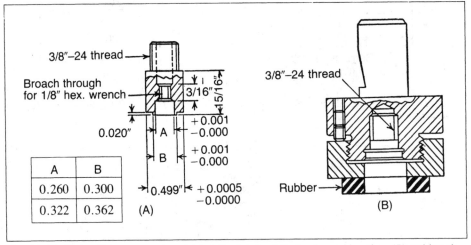

3/8″–24 thread

Broach through for 1/8″ hex. wrench

3/16″
15/16″

0.020″

A +0.001 −0.000
B +0.001 −0.000

A	B
0.260	0.300
0.322	0.362

0.499″ +0.0005 −0.0000

(A)

3/8″–24 thread

Rubber

(B)

Fig. 21-27. Hydraulic compensating staking die details: (*A*) circular staking tips; (*B*) staking-tip holder.

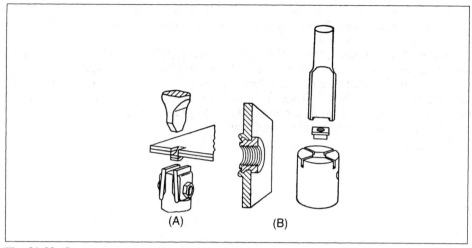

Fig. 21-28. Commercial assembling dies: (A) Metalace type (*Crockett Engineering Co.*); (B) clinch-nut type (*Richard Brothers Division, Allied Products Corp.*)

FOLD-OVER AND CRIMPING DIES

A die for assembling a doorknob part is shown in Fig. 21-29; the cap is folded over and around the inner knob (or shank) so that it will not slip when a torque of 50 in.-lbf (5.7 N·m) is applied. The closing or upper die incorporates a spring-loaded pilot for centrally locating the shank; another pilot in the lower die locates the cap. The stock is 0.056-in. (1.42-mm) thick quarter-hard brass. The forming of the shank is described earlier in this section.[2]

Assembling by Curling. A split beaded cylinder is placed on a spring-loaded pilot (Fig. 21-30, *D2*) and one end is curled over the edge of a hole in a shallow cup. There are no wrinkles in the curled flange in the cylinder, since it was previously fabricated with a longitudinal split. Sliding nests (*D5*), actuated by cams (*D3* and *D4*), slide inward to

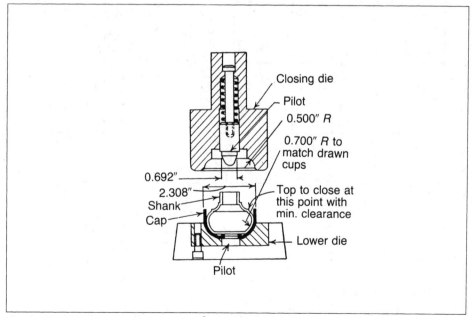

Fig. 21-29. Folding over doorknob parts.[2]

21-23

position and hold the cylinder during the curling operation. This die can be classified as a combination die, since horizontal punches (*D6*), actuated by 40° cam surfaces, cut narrow slots in the cup's rim.

Assembling by Folding Over. Three tabs of a metal clip (Fig. 21-31*A*, *D1*) are folded over to assemble it to a small bell crank (*D2*). A form block (Fig. 21-30, *D1*) mounted on a manually operated slide (*D6*) also functions as a nest. A spring-loaded stop pin (*D4*) disappears as the punch (*D2*) descends to bend the tabs over the crank's edges. Inward travel of the slide is limited by a stop pin (*D5*). A spring stripper (*D3*) also functions as a pressure pad to hold the parts in alignment.

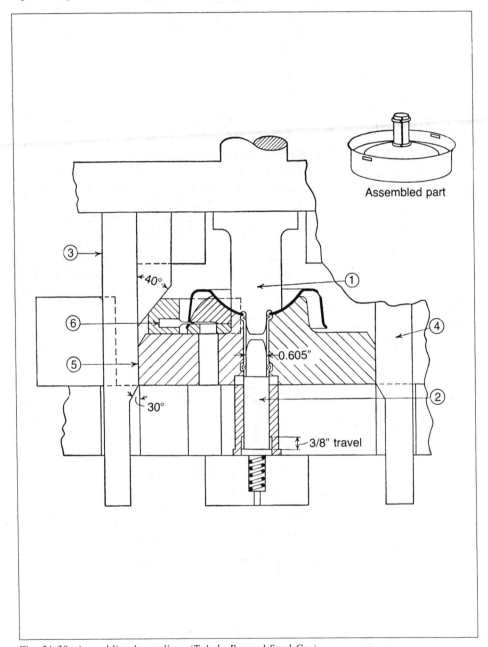

Fig. 21-30. Assembling by curling. *(Toledo Pressed Steel Co.)*

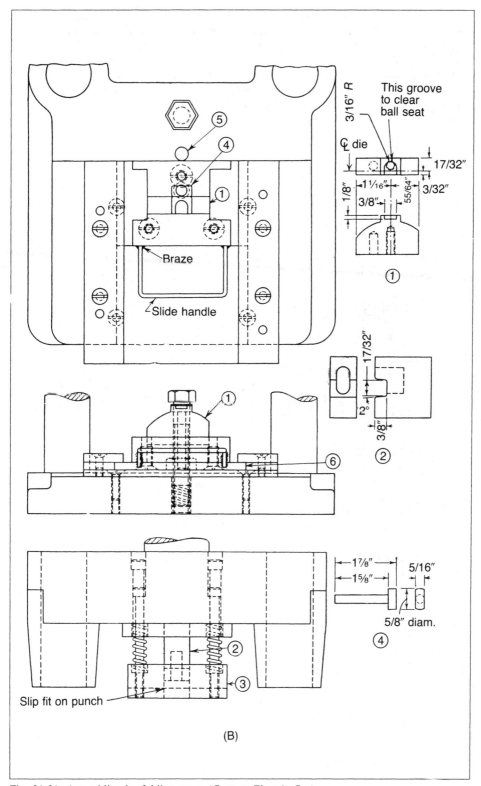

Fig. 21-31. Assembling by folding over. *(Century Electric Co.)*

21- 25

Die for Wire-to-eyelet Assembly. A metal eyelet or grommet is placed around the central post of the lower die (Fig. 21-32, *D*1), and the stripped end of an electrical wire is placed in the front notch (*D*6), bent around the post and down in the bottom of the rear notch (*D*7) as shown in view *B*. A notch (*D*4) in the revolving sleeve (*D*3) straddles the wire on the downstroke, wraps it around the post, and cuts it off as shown in view *C*. The clockwise revolution of the sleeve is actuated by ram pressure on the pin (*D*2) in its helical groove. A curling punch (*D*5) within the sleeve curls the grommet over and pinches it tightly on the wire as shown in view *D*.[9]

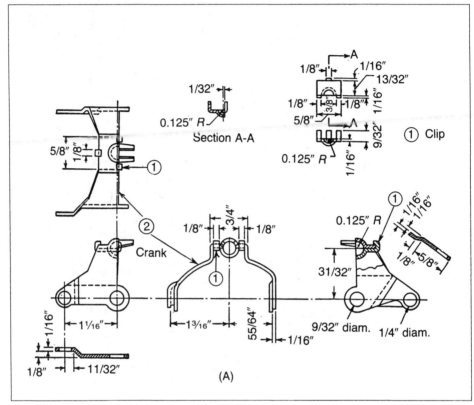

Fig. 21-31*A*. Parts assembled in die of Fig. 21-31.

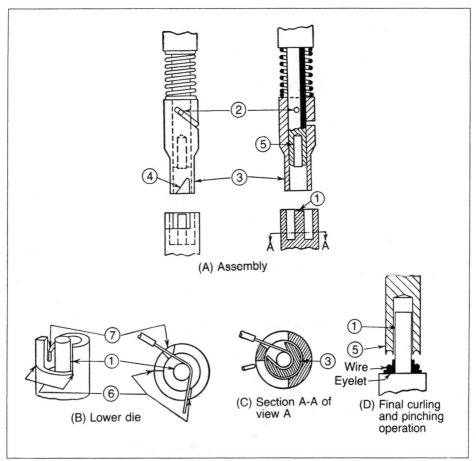

(A) Assembly

(B) Lower die

(C) Section A-A of view A

(D) Final curling and pinching operation

Fig. 21-32. Die for wire-to-eyelet assembly: (*A*) assembly; (*B*) lower die; (*C*) Section A-A of view *A*; (*D*) final curling and pinching operation.[9]

LASER CUTTING APPLICATIONS

The advent of the high-powered CNC laser cutting machine provides several important advantages for short run and prototype parts producers.

1. Small lots of irregularly shaped blanks can be economically produced without building a die.

2. Blank development work can be done much faster and with greater accuracy than can be done with a bandsaw or hand shears.

3. Drawn panels can be trimmed and irregular cutouts produced by a laser combined with a 5-axis positioning gantry.

4. Laminated pancake die sections can be cut with a laser.

Metal Cutting Lasers. Lasers cut by concentrating light energy on the surface of a workpiece, heating it so that it melts, vaporizes, or decomposes. An assist gas, delivered through a nozzle coaxial with the laser beam, removes the material (Fig. 21-33).

The important variables in the process are the laser power, the spot size, the focal position, the travel speed, and the assist gas. Dross and rough edges as by-products of laser cutting can be eliminated by the use of the right assist gas or water.

Cutting Carbon Steel. Most of the laser cutting done for metal fabrication is in low carbon steel of 0.125 in. (3.18 mm) thickness or less, with dimensional tolerance of ±0.005 in. (±0.13 mm) or more.

A 500-watt laser can cut 0.25 in. (6.4 mm) carbon steel, although it is most effective

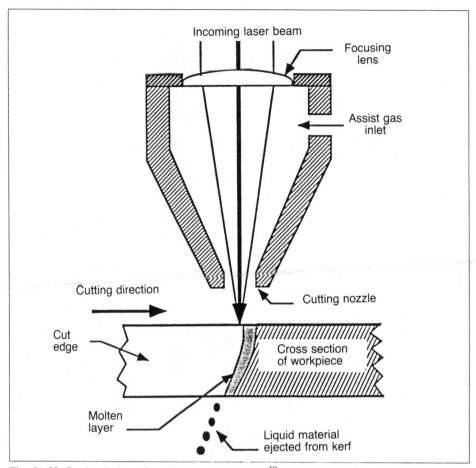

Fig. 21-33. Sectional view of metal cutting laser head.[10] (*Amada Laser Systems*)

below 0.125 in. (3.18 mm). More powerful lasers can cut thicker stock, but there is a problem with heat buildup. When laser cutting carbon steel, oxygen is generally used as an assist gas. The exothermic reaction of iron and oxygen adds energy and allows faster cutting and better edge quality than with an inert assist gas. If the workpiece temperature rises, the reaction can occur away from the laser beam. This causes a rough, wide kerf often called "burnout." A good method for reducing burnout problems is to apply water as a fine spray to the workpiece to cool the metal so that it will not react with oxygen other than at the laser spot. This allows intricate cuts to be made in steel up to 0.5 in. (13 mm) thick. This can be accomplished with a cutting head having an annular water nozzle surrounding the assist gas nozzle.

Materials Other Than Carbon Steel. When stainless steel is laser cut using oxygen as an assist gas, an adherent oxide scale remains on the cut edge. This scale must be removed by mechanical or chemical means before it is possible to weld the material. When laser cutting titanium, oxides are even more harmful, since oxygen dissolves in the metal and embrittles it.

These problems are eliminated by using an inert assist gas. Nitrogen or argon assist gasses yield clean, weldable edges with minimal heat-affected zones. Proper selection of gas pressure, travel speed, and lens focus produces burr-free parts from stock up to 0.25 in. (6.4 mm) thick.

Inert assist gasses are also used for cutting aluminum and nickel-base superalloys.

Surface Finish and Accuracy. Lasers regularly achieve 125 μ in. (3.18 μm) finishes on carbon steel 0.125 in. (3.18 mm) or thinner. Thicker steel has striations on the cut edge which read as 250 μ in. (6.35 μm) or rougher.

The spacing between the work and the nozzle must be maintained within close limits. In many cases, active control of this clearance is required. This can be accomplished with a servo-controlled axis driven by a capacitive, inductive, or contact sensor.

For any material, there is a range of speed which produces the best finish. Whenever there is a change in direction, the motion system must maintain this speed or cut quality will suffer. Smooth interpolation is required to prevent a "stair-step" effect on contours.

If accuracies of 0.001 in. (0.03 mm) or better are desired, the positioning accuracy of the motion system must be better than 0.0005 in. (0.013 mm). This accuracy must be achieved at the high travel speeds required to produce good edge quality. Using the latest servos, high speed CNC, and a light but rigid table, some newer laser cutting machines have a positioning accuracy of 0.0002 in. (0.005 mm). This, in itself, does not insure high accuracy. Thermal expansion occurring during laser cutting can cause distortion of the workpiece in excess of these tolerances. For the greatest accuracy, the part program must be modified to compensate for this effect.

Traditional Blank Development. Any blank development process involves a certain amount of cut-and-try work. The usual procedure is to build the draw or form die first, and then make a blank that will result in the desired part configuration after the subsequent metalworking operations.

This procedure for blank development starts with a master template which is used to scribe a layout onto the metal stock. The blanks are then cut out by hand using nibblers, bandsaws, drill presses, and files. Enough blanks are required for trial runs large enough to determine the stability of the process.

The resulting parts are compared to the part print, and any deviations from the print dimensions are used to make a revised master template. This procedure is repeated as often as necessary to obtain the required finished part. The final master template is then used as the pattern for building the blank die.

Laser Cut Blank Development. There is no physical master blank template required for laser cutting prototype blanks. The trial blank pattern is developed with computer aided design (CAD) equipment and the cutting information is downloaded into the CNC laser cutting machine which then cuts the required number of blanks under automated control. Irregular shapes including odd-shaped cut-outs are easily produced without the need for filing.

After the trial run, any corrections are simply made on the CAD equipment and the changes again downloaded to the CNC laser cutting machine. Once the final blank configuration is determined, the digitized blank pattern can be used to design the blank die. Often, this information can be used directly to determine wire burn EDM cutting paths.

A method for directly programming different types of wire burn EDM machines from a large mainframe computer is discussed in Section 16.

Laser Cutting Irregular, Three-dimensional Shapes. Trim dies for irregularly shaped drawn parts are both expensive and time consuming to build. Often the cost is greater than that of a short run, zinc-alloy draw die. Pierced holes and irregular cut-outs add to the cost.

Drawn panels can be trimmed and irregular cutouts can easily be produced by a metal cutting laser combined with a 5-axis positioning gantry. An advantage of this method over the use of a developed blank is that there is no distortion of the holes in the drawn panel. The break even point for building a die to avoid laser cutting is often over several thousand parts.

Laminated Blank Die Construction. Electrical discharge machining (EDM) is currently a favored method for the manufacture of tooling such as blanking dies for sheet metal fabrication. This method gives excellent quality and accuracy but is slow. Since lasers do not cut material as thick as most die cutting steels, the approach has been to cut several thin layers and laminate them.

An example of a laminated pancake die using this method of construction is shown in Fig. 21-34. The thin layer of prehardened steel is either 4130 or A2 tool steel. The other laminations are mild steel. The sheets can be held together with rivets or screws. This construction method has the incidental advantage of allowing the renovation of the die by replacing only the top lamination.

Figure 21-35 shows a compound blanking die of similar construction. In both designs, urethane rubber is used as a spring stripper.

Dies can be built by this technique in less than 8 hours provided that prebuilt universal shoes are on hand. This technology is useful for production quantities intermediate between those that can be economically laser cut and those produced on high-quality wire burn EDM built dies.[10, 11]

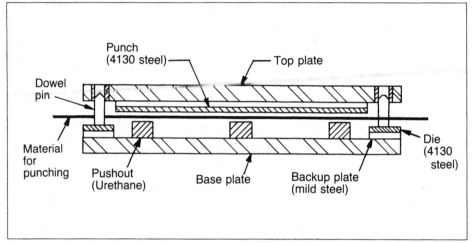

Fig. 21-34. Pancake die made of laser cut laminated sections.[11] (*Amada Laser Systems*)

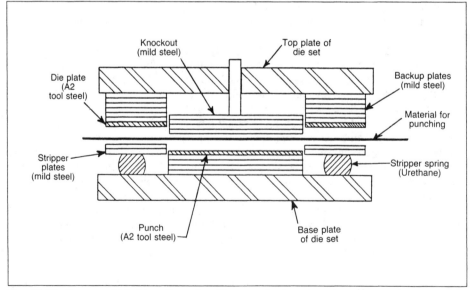

Fig. 21-35. Compound blanking die made of laser cut laminated sections.[11] (*Amada Laser Systems*)

References

1. Edward Bennett, "A New Short-run Piercing Die Set," *SME Technical Paper 62-43* (Dearborn, MI: Society of Manufacturing Engineers, 1962).
2. L.W. Montgomery, "Knurling on a Punch Press," *The Tool Engineer* (December 1951).
3. G.C. Adams, "Plastic Dies Move into Regular Production Service," presented at 22nd Annual Meeting of American Society of Tool Engineers, Philadelphia, Apr. 26-30, 1954.
4. *Aluminum Forming*, Reynolds Metals Co., Louisville, 1952.
5. K. Drone, "Principles of Stretch-wrap Forming," presented at 22d Annual Meeting of American Society of Tool Engineers, Philadelphia, Apr. 26-30, 1954.
6. K. Drone, "Stretch-wrap Forming," *Western Machinery and Steel World* (May 1956).
7. C.W. Hinman, *Pressworking of Metals,* 2nd ed. (New York: McGraw-Hill Book Co., 1950).
8. L.H. Austin, "Improved Staking Punch," *Am. Machinist* (June 22, 1953).
9. M. Conklin, "One Stroke Trims, Loops, Eyelets in Special Die," *Can. Machinery* (July 1954).
10. D. Hoffman, "Today's Laser Cutting—Fast and Clean," *Modern Machine Shop* (April 1989).
11. D. Hoffman, "A Tool For Making Tools," *The Fabricator* (March 1989).

SECTION 22

DIE SETS AND COMPONENTS

A punch-and-die set consists of an upper and a lower shoe, on which the die elements for doing the desired work may be mounted, and guide or leader pins and bushings for holding members in alignment. The number of guide pins and bushings varies from two to four, according to the accuracy required and the size of the die set.

Several styles of die sets are commercially available, some of which are shown in Fig. 22-1. Die shoes are made of cast iron, cast steel, and steel plate. Cast iron is used primarily for smaller dies and light-duty applications. Steel die shoes deflect less and have much greater strength than cast iron shoes.

Guide Pins and Bushings. Bushings are often retained in the upper die shoe by a press fit. If bushings are sized for a press fit, the pin hole should be made oversize by approximately the amount of press fit interference to allow for normal hole closure. A better method that provides ease of replacement is to make the bushing hole in the shoe a close slip fit and to retain the bushing with several small toe clamps.

The guide or liner pins are usually a press fit in the lower die shoe. In cases where it is desirable to remove the pins for die sharpening, the pins are provided with a flange and made a slip fit into the pin holes, and retained with toe clamps or a retaining ring. Figure 22-3D illustrates a type of removable pin. The location of the pins and bushings may be reversed with respect to the upper and lower shoe if required. The main reason that the bushings are located in the upper shoe is to lessen the possibility of dirt or foreign objects entering the bushing.

An accurate slip fit is established between the bushing and pin. Standard or purchased die sets have guide-pin diameters which depend on the size of the die set. Larger guide pins should be used where there are extreme alignment requirements. The diameters of guide pins for the average range of die sets vary from 1 to 4 in. (25 to 100 mm). To avoid striking the press ram, guide pins must not extend beyond the top surface of the upper shoe when the die is in the "closed" position (Fig. 22-2).

A mathematical analysis of guide-pin size is presented in Section 3, Die Engineering, which takes into consideration such design factors as vertical surfaces, side thrusts, and pin deflection.

Figure 22-3 illustrates some commonly used types of guide pins. Removable guide pins simplify die construction and sharpening. Shoulder pins simplify the in-plant manufacture of die sets by permitting through-boring of pin and bushing holes. The shoulder of the pin is made the same diameter as the OD of the bushing. Pins are also made with extra lead to facilitate assembly of the die set.

Most guide-pin bushings are honed on the ID and the OD is ground. Bushings are available in a variety of types, as shown in Fig. 22-4. Shoulder bushings are available made from aluminum phosphorus bronze, bearing bronze, steel, or steel with the ID bronze-plated. Steel bushings are generally AISI-SAE 1022 or 1117, carburized and hardened to Rockwell C58 to 62. Bushings are available finished to size or with stock for finish grinding. These are wring-fitted into the upper shoe and held in place with clamps.

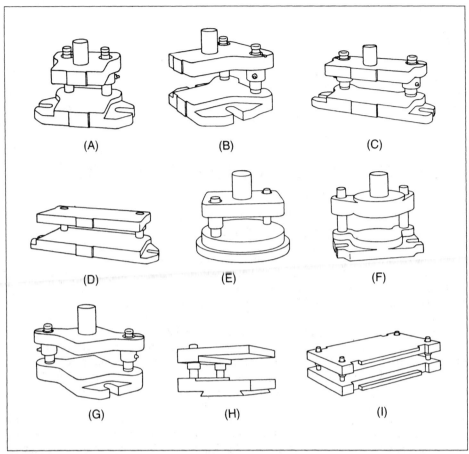

Fig. 22-1. Commercial die sets: (A) rear two-pin regular; (B) rear two-pin reverse; (C) rear two-pin narrow; (D) flange-type progressive rear two pin; (E) combination diagonal pin; (F) round rear pin; (G) diagonal pin; (H) multislide two pin; (I) special four-pin. (*Producto Machine Co.*)

Several types of ball-bearing (antifriction) bushings and pins are also manufactured.

The antifriction bushing and pin assembly is used on high-precision tools. There is a negative clearance or press fit of balls between the pin and bushing which prevents lateral movement of the pin in the bushing. Because of the rolling action of the ball bearings, no sliding action exists, and the initial preload is maintained throughout long die runs.

Self-lubricating bronze bushings are also available. Lubrication is provided by graphite composition plugs pressed into a number of holes drilled through the sides of the bushing.

Fool-proof Assembly. Some die sets are symmetrical and can be assembled in more than one way. To avoid this problem, the normal practice is to fool-proof the die set by either making one pin diameter differ from that of the other pin(s), or by offsetting pin locations.

These measures may not suffice in the case of die sets that may have one half removed from the press for work. Stenciling both of the shoes to indicate the front, or painting opposite sides a distinctive color, can prove helpful. Another method is to fasten the set-up blocks to the upper shoe on one corner and the lower shoe in the diagonally opposite corner.

Shanks. Shanks may be attached to, or made as an integral part of, the punch shoe. They are available in integral cast, welded, or inserted (screw-in). An example of the latter type is shown in Fig. 22-5.

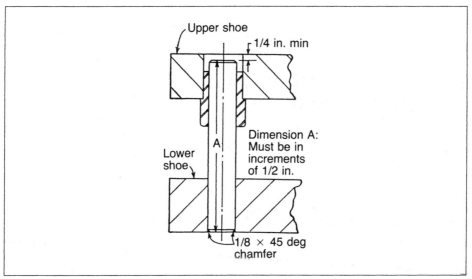

Fig. 22-2. Guide pin length (die in closed position).

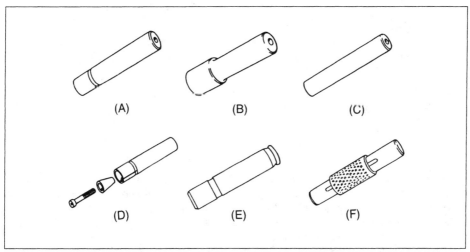

Fig. 22-3. Die-set guide pins; (*A*) regular; (*B*) shoulder; (*C*) commercial; (*D*) removable; (*E*) quick fit (*Producto Machine Co.*); (*F*) ball-bearing (*Lempco Industrial, Inc.*).

Inserted shanks, when furnished as a part of a purchased die set, should be locked in place, to prevent turning, by a setscrew with its center bisecting the junction of the shank thread and the punch holder of the die set. This locking device is called a "dutchman." Cast die sets are available with shanks cast integral, with inserted shanks, or with no shanks; steel die sets are available with inserted shanks, welded shanks, or no shanks.

Bosses. Demountable bosses, such as the one illustrated in Fig. 22-6*A*, are used as bushings or guide-pin supports to add rigidity to die sets which have thin shoes or extreme center distances. Bosses are available with screw-and-dowel construction. Welded-steel bosses are frequently provided on steel die sets.

A boss may be cast on the punch shoe to enable the guide pin to enter the bushing before the die begins to function (Fig. 22-6*B*). Bushings with long heads may be used to serve the same purpose.[1]

Standard Punches and Die Buttons. Specialized die component manufacturers produce a wide variety of standard punches and die buttons and stock them as catalog

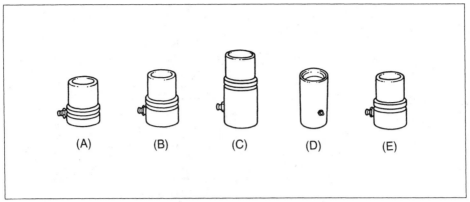

Fig. 22-4. Die-set guide-pin bushings: (*A*) short shoulder; (*B*) regular shoulder; (*C*) extra long shoulder; (*D*) long straight; (*E*) medium shoulder. (*Producto Machine Co.*)

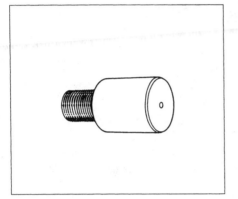

Fig. 22-5. Inserted-type (screw-in) punch shank.

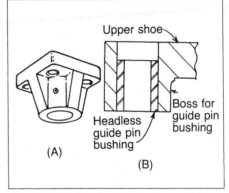

Fig. 22-6. Guide-pin bushing bosses; (*A*) demountable; (*B*) cast or welded integral.

items available on short notice. These are available as standard in many popular tool steels with less common steels available on special order. The manufacturers' catalogs also list many types of quick-change ball and roll-lock punch and die button holders.

Some manufacturers offer special processing such as chrome plating, ion nitriding, titanium nitride (TiN), and titanium carbide (TiC).

Metric Punch Conversion. The catalogs issued by punch manufacturers increasingly list metric size punches, buttons, and retainers as their standard or preferred product line. In order to take advantage of the savings possible through metric punch standardization, changes in product and tool design practices should be considered by every company.

The vast majority of punched holes in fabricated metal products are anything but critical in diameter and tolerance. Many are simply clearance holes for screws. These hole sizes can be standardized and consolidated to a great extent without loss of screw retention.

The key to efficiency is, of course, building hole size consolidation programs around standard, readily available metric punch point diameters. When hole-size requirements are reduced to a minimum and are based on standard tooling, and when product designers and tool designers have a mutual understanding of hole requirements, real progress in tool economy can be made.

Utilization of standard perforating tooling has resulted in substantial reduction in tooling costs for many users. In many cases, the initial cost of converting to metric tooling can be outweighed by the savings made possible through reduced inventories of replacement tooling.

Methods of Mounting and Securing Punches. Figure 22-7*A* illustrates how a low melting temperature eutectic alloy composed mainly of bismuth is used to anchor

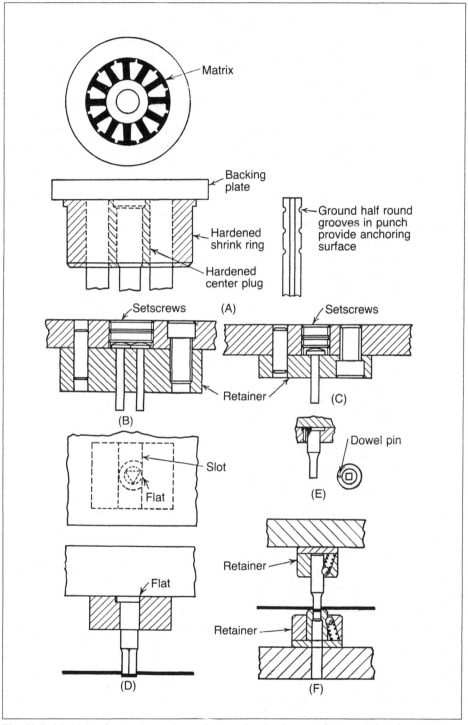

Fig. 22-7. Methods of mounting and securing punches by the use of: (*A*) a shrink ring (*Sterling Tool Co.*); (*B*, *C*) setscrews;[2] (*D*) a flat shoulder;[3] (*E*) dowel pin; (*F*) ball lock retainers.[1]

22-5

lamination die sectional punches within a shrink ring. The soft alloy also serves as a shock absorber.

Setscrews are used to retain two closely spaced punches at *B*, and a single punch at *C*. This method expedites changing punches from the top of the die shoe.

A flat on the punch shoulder, at *D*, bears on a slot in the punch retainer to prevent punch rotation; a pin, or "dutchman," extending through the punch shoulder, at *E*, accomplishes the same purpose.

The prevention of punch rotation and the easy replacement of both the punch and the die button are features of the design shown at *F*. The punches and dies, held in their respective retainers by spring-loaded balls or rollers, are commercially available for the piercing of round and irregular holes of many sizes.

Punch fabrication and sharpening are facilitated by the assembly of punch sections within a ring (Fig. 22-8*A*); a hardened button takes the thrust. At *B*, lifter keys (*D*1) fit the keyways in the side of the carbide punches (*D*2) to allow the removal of one punch without disturbing the others.

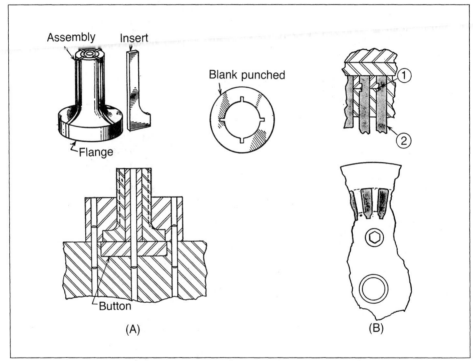

Fig. 22-8. Sectional punches: (*A*) split design;[4] (*B*) carbide rotor slotting.[5]

A forming punch (Fig. 22-9, *D*1) is bolted to a parting punch (*D*2) and is provided with a slot to allow for adjustment after both punches are ground.

Floating or Gag Punches. A cutoff blade (Fig. 22-10, *D*1) is held up in the nonoperating position by a spring (*D*2) but functions when a lever (*D*4) is swung clockwise to move a slide (*D*3) inward to the operating position shown.

Two slides (Fig 22-11*A*) control the operation of two punches; a similar design with a larger number of slides and punches, simple to set up to pierce holes in various combinations of size and location, is adaptable for short runs. A simpler arrangement, at *B*, for punching light stock, incorporates setscrews which are backed off to render the punch inoperative.

An operator-controlled pneumatic cylinder moves a gag bar having rack teeth along one side to engage the teeth of a rotary-face cam shown in Fig. 22-12*A*. A punch is

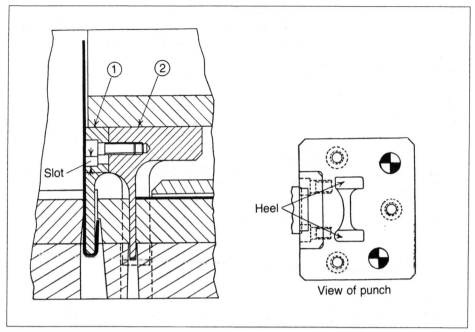

Fig. 22-9. Mounting a sectional punch for grinding adjustment. (*Harig Mfg. Corp.*)

mounted in the rotary cam which mates with a fixed-face cam, at *B*. The rotary cam and the punch move 0.67 in. (17.0 mm) up or down in 80° of rotation when the rotary cam is actuated by the toothed gag bar. Individual punches in several rows can be made operative or nonoperative in succession.

Gag punches can also be actuated by fast-acting electrical solenoids for such applications as cutting off a predetermined number of parts or rejecting misformed parts detected by in-die sensors.

Piloted Punches. Removable pilots facilitate the grinding of punches. A spring-loaded pilot, centrally located inside a punch, is suitable for engaging thin stock. If the hole in the stock is irregular in outline, a pin or thin dowel can be pressed through the body of the punch. The pin bears against a flat ground on the pilot, allowing it to move freely in a vertical direction, but preventing it from turning.

Flanging-punch Design. Combination punch design is shown in Fig. 22-13; a punch (*D3*) blanks and draws the outside of the cup. Another punch (*D1*) aids in drawing the cup and in forming the flange around the hole. A stepped punch (*D2*) enters a previously pierced hole to form the flange. A lower stripper (*D4*) is actuated by springs (not shown) through pins (*D5*).

Automotive-body Punches. A lance-and-bend punch commonly used in the automotive industry for producing anchoring tabs in automobile bodies is shown in Fig. 22-14.

Punch-and-die Design for Producing Holes for Self-tapping Screws. Punch-and-die button design (press-fit type) for producing holes in mild-steel sheets to be fastened to another sheet with self-tapping screws is shown in Fig. 22-15. These elements are also made in the ball-lock type.

The hole in the lower sheet is punched with a radial slit, and the surrounding metal is formed into a spiral cone conforming to the pitch of the screw thread. For maximum rigidity, use the smaller hole sizes in combination with the heavier stock thicknesses. Development of the pierced hole to a specified formed-hole diameter is a function of bottoming pressure and stock thickness.

Commercial Sleeved Punches. A punch commercially available (Fig. 22-16*A*) incorporates two intermeshing segmented sleeves inserted, respectively in the punch and stripper plates, providing excellent alignment and preventing buckling of the punch.

Another design, at *B*, is straight ground and incorporates a die-cast sleeve of soft metal to reduce punch vibration, a common cause of punch breakage. Both types are available in many point sizes and contours.

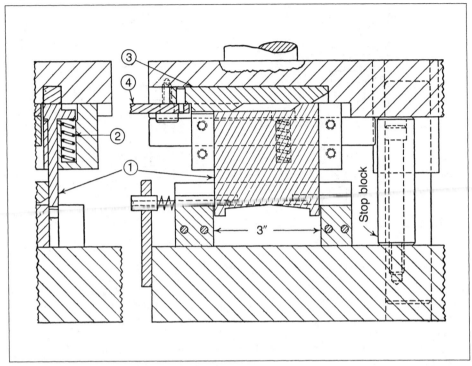

Fig. 22-10. Floating cutoff punch. (*Harig Mfg. Corp.*)

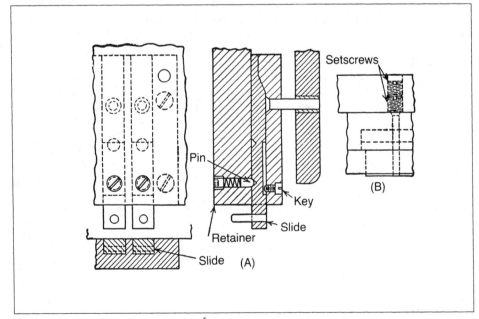

Fig. 22-11. Control of floating punches.[6]

22- 8

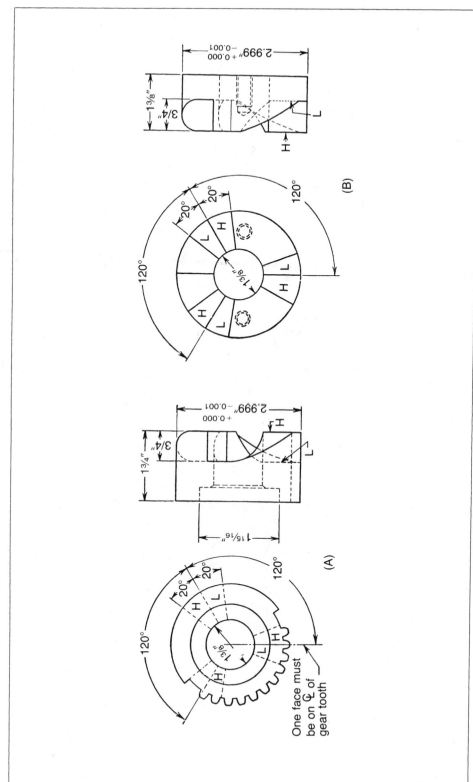

Fig. 22-12. Gag-punch cam design; *H* and *L* designate high and low faces.[7]

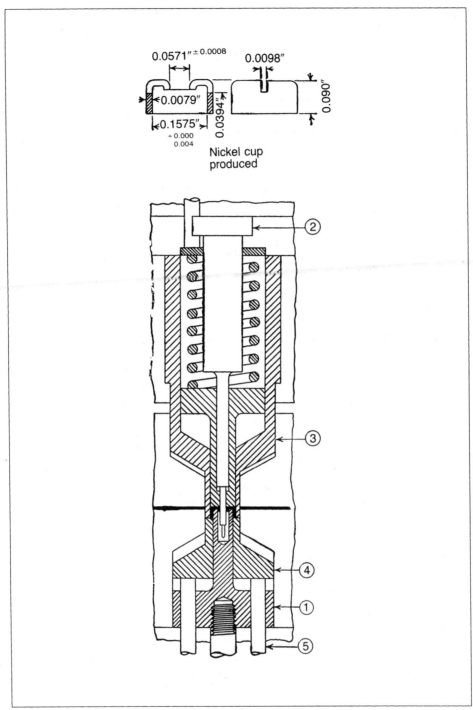

Fig. 22-13. Combination flanging-punch design.[8]

22-10

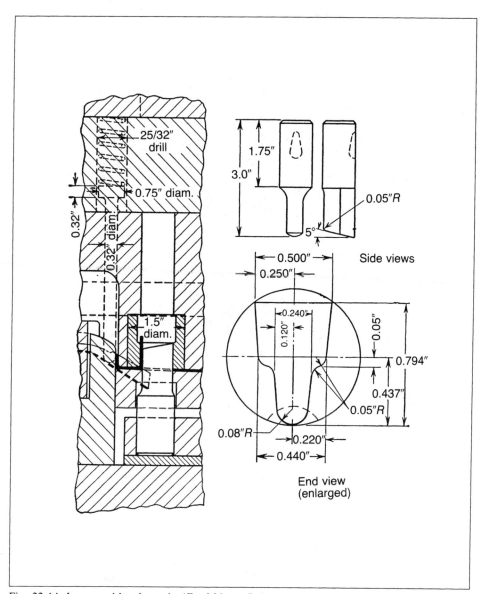

25/32" drill

0.75" diam.

0.32"

0.32" diam.

1.5" diam.

1.75"

3.0"

0.05"R

5°

Side views

0.500"

0.250"

0.240"

0.120"

0.05"

0.794"

0.437"

0.05"R

0.08"R

0.220"

0.440"

End view
(enlarged)

Fig. 22-14. Lance-and-bend punch. (*Ford Motor Co.*)

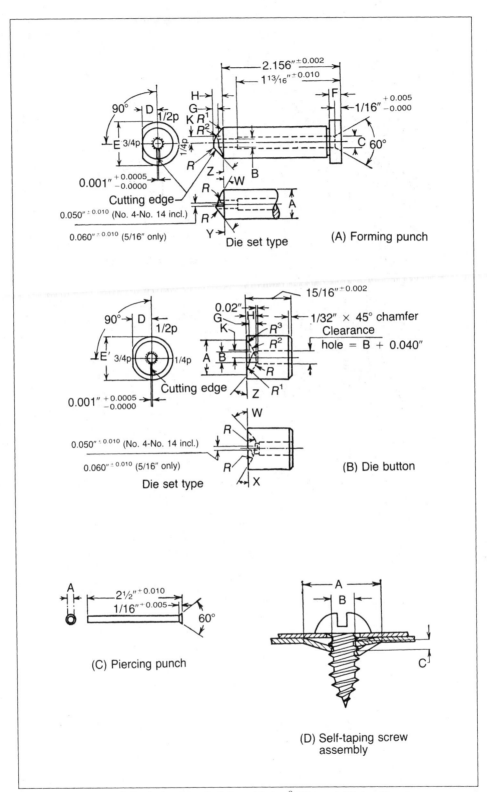

Fig. 22-15. Punch-and-die design for spiral slotted holes.[9] (*General Motors Corp.*)

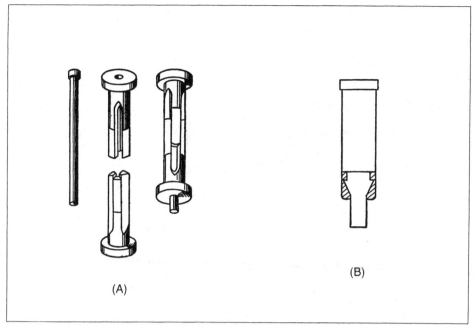

Fig. 22-16. Sleeved types, commercial punches ((A) *Durable Punch and Die Co.*;[10] (B) *Pivot Punch and Die Corp.*).

STOCK GUIDES OR GAGES

Stock may be solidly guided by suitable slots in a stripper, by stock rails or, as shown in Fig. 22-17, by pins, buttons, or angle iron. Solid guides may or may not require spring guides or spring pushers for optimum stock guiding. These may be of the preliminary type (Fig. 22-18A) used until permanent gage locations are determined, or of the typical adjustable-spring type, shown at B. Rollers may be used instead of pins or buttons to position the stock.

Stock guides are not always mounted to the die shoe; the types shown in Fig. 22-19, views A and B, are mounted on the stripper plate and can be called stock pushers although they do guide the stock. The guide pins (D3, view C) project through the pressure plate (D2) which holds the stock as it is severed by the cutoff punch (D1). A sliding plate (D2, view D) pushes and guides the stock between its edge and a slot in the stripper plate and is actuated by a cam (D1) mounted on the upper shoe. Cam adjustment is varied by a setscrew (D4) mounted in a plate (D3) secured to the upper shoe.

A spring-loaded disappearing guide or gage (Fig. 22-20) is entirely depressed after the die operation is completed, avoiding any interference with a stripper plate or other upper moving die components. The slot in the hollow stop clears the straight portion of the eyeleted rod which supports the spring.

Movable gages shown in Fig. 22-21 hold the blank level for a square cutoff. The punch forces the blank into the stacker, and the projecting lips of the gages prevent the blank from emerging from the stacker opening.

A light constant pressure is maintained on pusher slides (D1, Fig. 22-22); this compact design is practical for large stock areas in progressive and cutoff dies.

Adjustable Stops. An adjustable type of solid block (Fig. 22-23, D1) can be moved along a support bar (D2) in increments up to 1 in. (25 mm) to allow various stock lengths to be cut off.

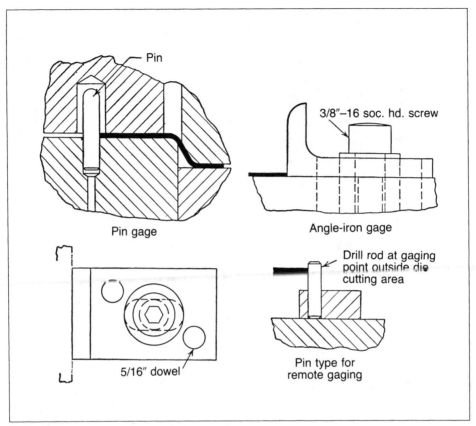

Fig. 22-17. Solid gage design. (*Ford Motor Co.*)

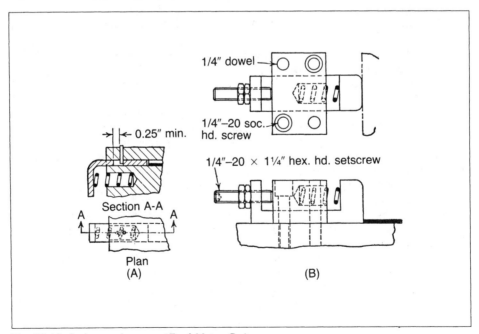

Fig. 22-18. Spring stock gages. (*Ford Motor Co.*)

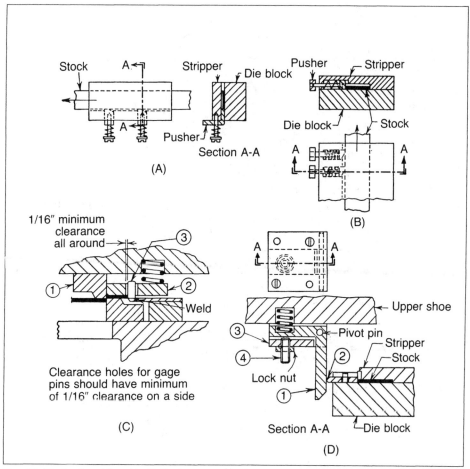

Fig. 22-19. Stock guides: (*A, B*) Pusher types;[11] (*C*) pin type (*Ford Motor Co.*); (*D*) lever actuated pusher type.[11]

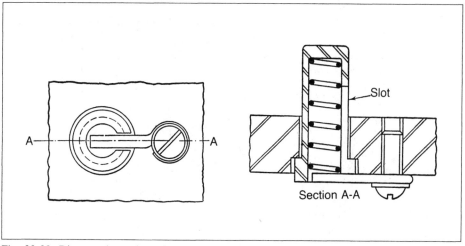

Fig. 22-20. Disappearing guide. (*Ford Motor Co.*)

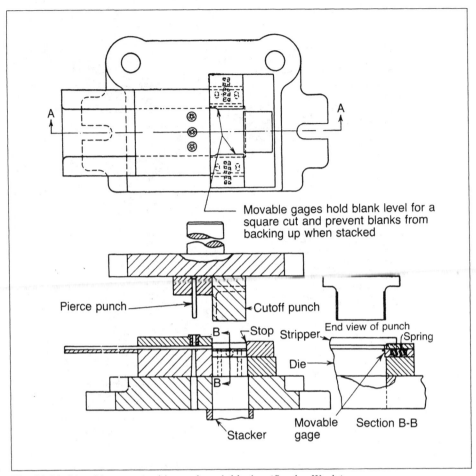

Movable gages hold blank level for a
square cut and prevent blanks from
backing up when stacked

Pierce punch

Cutoff punch

End view of punch

Stop
Stripper
Spring

B

Die

B

Section B-B

Stacker

Movable
gage

Fig. 22-21. Movable gages position and stack blanks. (*Stanley Works*)

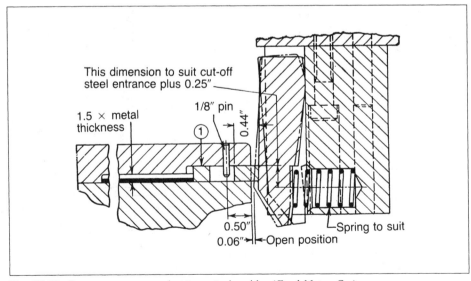

This dimension to suit cut-off
steel entrance plus 0.25"

1/8" pin

0.44"

1.5 × metal
thickness

①

0.50"

0.06"

Open position

Spring to suit

Fig. 22-22. Constant-pressure pusher-type stock guide. (*Ford Motor Co.*)

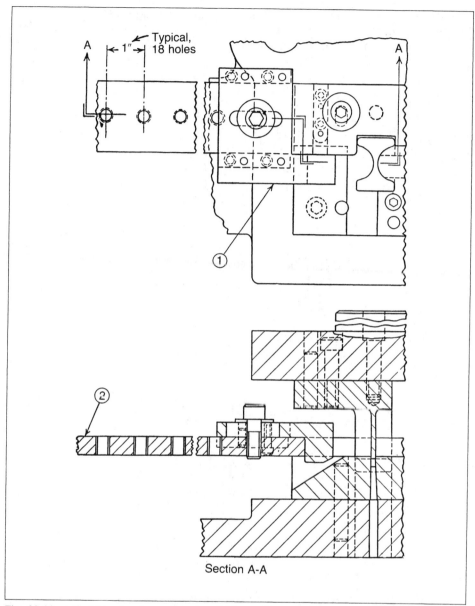

Fig. 22-23. Adjustable stop for parting die. (Chicago Chap. 5, SME)

STRIPPERS

Strippers are of two basic types, fixed or spring-operated. The primary function of either type is to strip the workpiece from a cutting, bending, forming, drawing, or coining punch or die. Coining includes extrusion, swaging, or any squeezing or compressing action that causes metal flow in the workpiece. A stripper that forces a part out of a die may also be called a knockout, an inside stripper, or an ejector. Besides its primary function, a stripper may also hold down or clamp, position, or guide the sheet, strip, or workpiece.

Fixed or Positive Stripper. A common type incorporates a slot in the stripper plate; such a "tunnel" also guides but does not hold the incoming stock.

Spring-operated Strippers. Typical spring-actuated strippers which also hold the stock down are shown in Fig. 22-24. At *A*, the springs surround the stripper bolts; at *B*, the springs are retained in pockets in the upper shoe and stripper plate. A guided design, at *C*, provides alignment and support for slender punches.

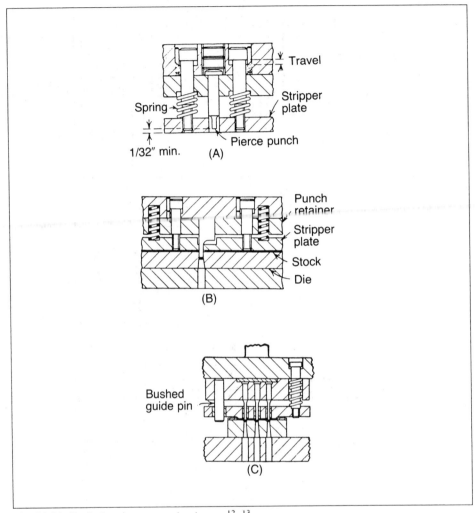

Fig. 22-24. Typical spring-operated strippers.[12, 13]

Interchangeable Punch and Stripper Design. Ball-lock punches and retainers (Fig. 22-25, *D*7, *D*8, *D*9, and *D*10) in various sizes are used with matching stripper assemblies, permitting quick changes of the assemblies through the use of spring-loaded locking devices for both stripper inserts (*D*14, *D*15, *D*16) and punches. To facilitate punch removal without having to remove the stripper plate, a hole may be drilled in the stripper plate, as indicated in lower view of Fig. 22-25, for inserting a releasing tool.

Double Stripper Design. An upper stripper (Fig. 22-26, *D*1) functions as a pressure pad on the downstroke to hold the work in the die block (*D*3); on the upstroke it strips the work from the forming punch (*D*4). The lower stripper (*D*2) strips or ejects the work from the die by forcing a plug (*D*5) up. The die block has a slot machined across the bottom to allow for the cross bar of the stripper (*D*2) which supports the plug (*D*5).

Strippers Using Latches. Spring-loaded latches are depressed as the part is formed in the die opening (Fig. 22-27, view *A*). After they clear the part, they snap up against the bottom of the locator block to strip the part from the ascending forming punch.

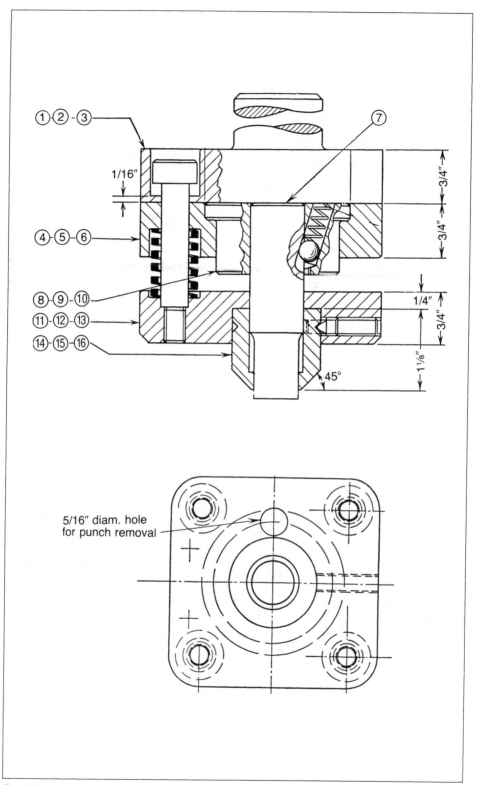

Fig. 22-25. Interchangeable stripper and punch design. (*General Electric Co.*)

22- 19

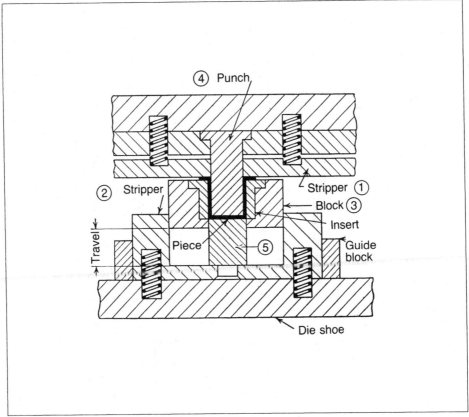

Fig. 22-26. Double stripper design.[12]

A single latch or spring-loaded hook, shown at view *B*, slides over an edge of a double bent part and strips it from the shedder-form block, as the form punch ascends. The shedder-form block functions as a shedder on the upstroke and as a pressure pad and form block on the downstroke.

A latch to strip a part from a mandrel is shown at *C*. On the downstroke the latch snaps upward as it contacts the sliding plate, but on the upstroke it forces the plate to the left. The bushing moves to the left and strips the part from the mandrel.

Stripping Fork. Figure 22-28 illustrates a fork (*D2*), actuated by an air cylinder (*D1*), which reciprocates horizontally to strip the part from the form block (*D3*) on the upstroke. This action takes place after the part is pushed out of the form punch (*D5*) by a spring stripper (*D4*). The spring-loaded stripper also serves as a hold-down and has a pilot (*D6*), to locate the part.

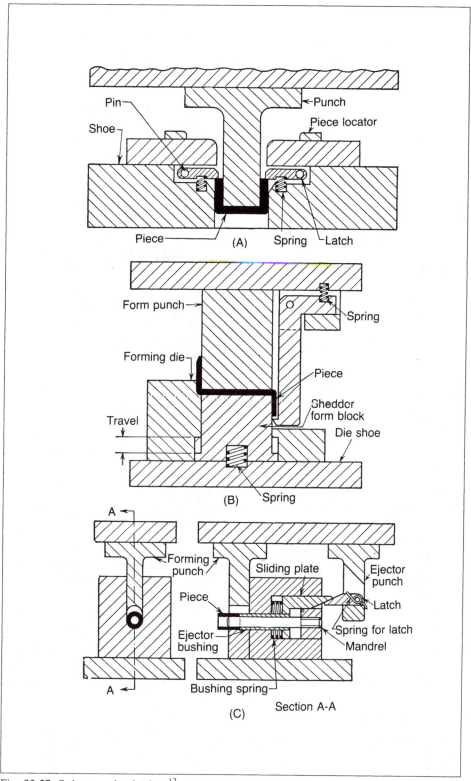

Fig. 22-27. Strippers using latches.[12]

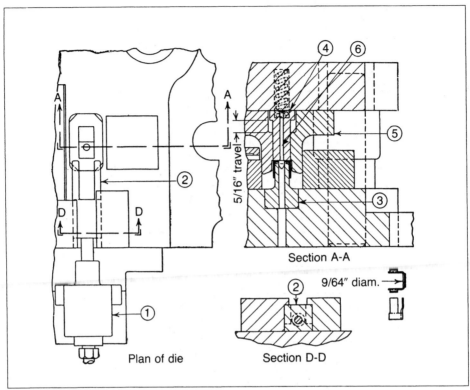

Section A-A

9/64" diam.

Plan of die   Section D-D

Fig. 22-28. Fork stripper. (*Brighton Division, Advance Stamping Co.*)

KNOCKOUTS

Positive Knockouts. A commonly used knockout design (Fig. 22-29, view *A*) incorporates a knockout rod (*D1*) which forces a knockout plate (*D2*) to strip the part from an inverted compound die. The part is prevented from adhering to the plate by the oil-seal breaker pin (*D3*). A stop collar (*D4*) is incorporated rather than a pin (*D1*) shown at *B*, which may easily shear off under heavy stripping pressures.

Indirect Knockout. Knockout plates with three and four auxiliary rods, as shown at view *B*, Fig. 22-29, are suitable for small dies. A stop pin, such as shown at *D1*, should not be used; a stop collar should be used instead, since extreme pressure may be applied by the upper or main knockout bar if it is set too low.

Spider-shaped knockout plates are used in large dies. The recess to accommodate the spider is not so apt to weaken the punch shoe as would be the case if a recess were provided for a large rectangular knockout plate.

Liftout Plates. A liftout plate (*D2*, Fig. 22-30*A*) is pulled up by lift bolts attached to the upper shoe. On the upstroke the flanged part is stripped from the lower form die (*D3*) to allow it to be blown off the die set with an air jet, or to slide off if the press is inclined.

Pressure-pad Liftouts. A pressure pad (*D2*, Fig. 22-30*B*) actuated by four springs (*D3*) also functions to lift the part from the combination cutoff-trim-form die (*D1*). The part is lifted to a position flush with the top of the die and then blown off with an air jet.

METAL SPRINGS

Springs for strippers, pressure pads, and other spring-operated die components can be selected from the ratings given in terms of pressure per units of travel as listed in manufacturer's catalogs.

Springs should be carefully selected in relation to the service for which they are

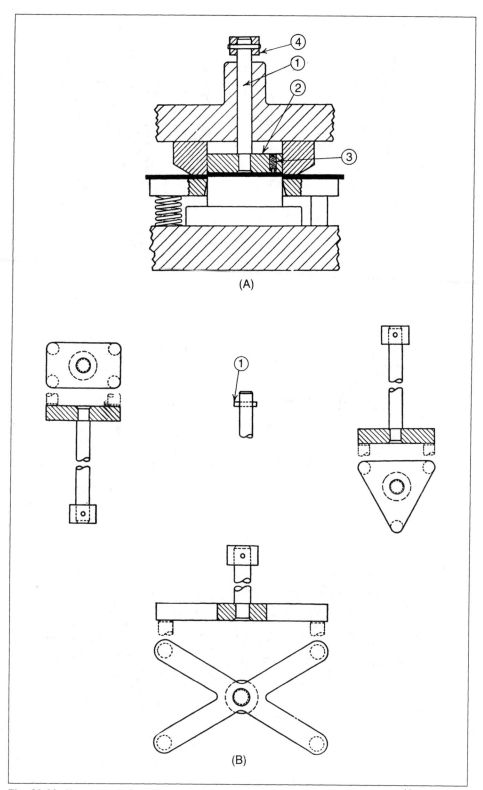

Fig. 22-29. Knockout design: (A) for inverted compound dies; (B) knockout plates.[14]

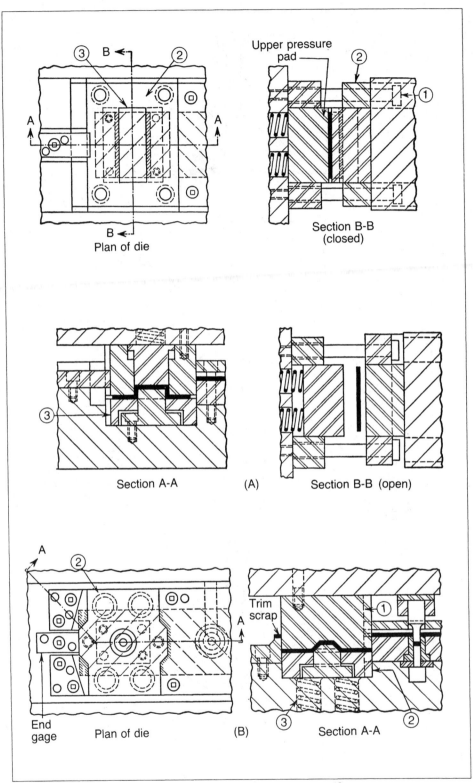

Fig. 22-30. Liftout design: (A) liftout plate; (B) pressure-pad liftout.[15]

intended; in addition to required pressures and deflections, space limitations and production requirements must be considered. In addition to helical die springs, washers (Belleville type) or rubber may be used.

Oval wire springs are available in three load ratings—30%, 37%, and 50% deflection—and in two types of steel—high-carbon and electric-furnace chrome-vanadium. Their popularity is widespread because the oval design avoids failures resulting from high stress concentrations.

Carbon steel springs of this type are suitable for low deflection applications and can be run for a considerable time before failure. Chrome-vanadium springs, while slightly higher-priced than those of carbon steel, last three or more times as long. When selecting die springs, best performance and reduced down time are ensured by use of chrome-vanadium-steel springs, and derating the travel from the maximum ratings.

Round wire springs are not recommended for pressure pad applications because of their low load ratings. They are still a good choice for such applications as latch return springs and for use in starting stops.

Spring Mounting and Care. The locating and mounting of springs in pockets around pilots or bolts, in tubes, or by other methods are determined by space available, service requirements, and malfunctioning through slug interference, misalignment, or other causes.

Even the best spring can fail if improperly used. The illustration in Fig. 22-31 provides examples of right and wrong ways to use springs.

Springs should be sufficiently supported in a hole or over a rod or stripper bolt, as

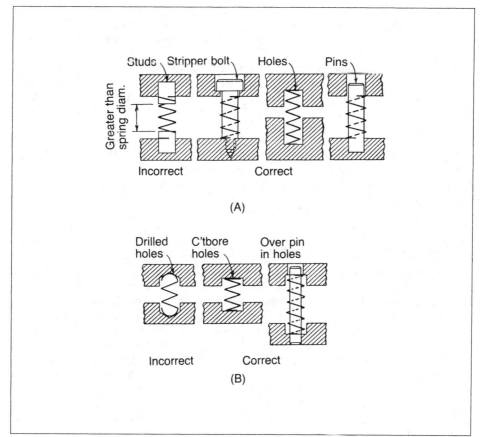

Fig. 22-31. Correct and incorrect ways to mount die springs.[16]

shown in view A, to guide them adequately under stress. Lack of support can result in crushing, twisting, binding, or surface wear.

When springs are set in holes, squaring the bottom of the hole (Fig. 22-31B, center) provides a flat seat and eliminates the possibility of crushing or deforming the ends, as shown in Fig. 22-31B, left, where holes have been left as drilled. Leading edges of holes should have a small chamfer to prevent interference with the movement of springs. However, if the unguided length of the spring is greater than the diameter, a center guide rod should be used, as shown in Fig. 22-31B, right. The guide rod also serves to retain the broken pieces should the spring fail. Tubular steel cans around the springs (not shown) are also used to contain any broken pieces of failed springs.

AIR SPRINGS

Pneumatic cylinders are favored over die springs for applications where long strokes and adjustable forces are required. A wide range of mounting styles and configurations are listed in component suppliers' catalogs. Commonly available piston diameters range from 1.0 in. (25.4 mm) to 10 in. (254 mm) in diameter. Stroke lengths range from 0.5 in. (12.7 mm) to over 10 in. (254 mm), with special types available by order.

The initial force available is piston area times the applied shop air pressure. Typical shop air pressures range from 75 to 100 psi (520 to 690 kPa).

To obtain higher pressures upon die closure, one-way check valves are sometimes used on the inlet line. Most commercial valves for this purpose are of the self-bleeding type. This design exhausts the pressure in the cylinder when the air supply source is removed. This feature is important if a pad or cam must have the pressure removed for maintenance.

Most types of die air springs have hollow pistons which act as surge tanks. The amount of pressure built up as a function of piston travel is available from manufacturers data sheets.

A few types are intended to be used with external surge tanks. The manufacturers' recommendations should be carefully followed as to the correct size and type of surge tank to use. Piping diameters must be be large enough to allow for unrestricted flow. Operations that require carefully balanced pressures under dynamic conditions may require hoses or piping of equal length with individual ports in the surge tank to insure even pressure.

Designs have been tried which attempted to provide very high pressures at the end-of-stroke by means of a high compression ratio. The pressures involved can exceed the working limits of the components. An often unanticipated catastrophic failure mode that can occur in such cases is a "diesel" ignition of the lubricating oil vapor. Nitrogen die pressure systems should be used in such cases.

PIPED NITROGEN DIE PRESSURE SYSTEMS

Piped Nitrogen Cylinders. From 5 to 15 times the force available in an air spring can be obtained from nitrogen die cylinders of the same size. The original designs of these systems were largely based upon hydraulic cylinder systems. Many of the hoses, packing and fittings used are available as generic hydraulic parts from hydraulic suppliers.

There are two basic types of cylinders in use: low and high pressure. The 500-psi (3.4-MPa) low pressure systems are seldom used in new designs. High pressure systems use an initial charging pressure of up to 1,500 psi (10.4 MPa). The forces available in high-pressure systems at full rated pressure range from 0.5 to 6 tons (4.4 to 53 kN). The smaller sizes are sometimes used as cam return cylinders. A variety of mounting styles are available. These piped systems are designed to be used with an external surge tank. The tank, cylinders, charging console, hose, and all fittings are usually ordered as a package from the manufacturer.

As the system wears, leaks can develop. An accepted figure for nominal leakage is 25 psi (172 kPa) per week. Whenever the packing in one cylinder starts to fail, it is best to rebuild the entire system.

As an aid to the designer, overlays of the system components are available from the manufacturers. CAD users can store these as macros.

Nitrogen Manifold Systems. Manifold systems have the advantage of few external fittings to leak. The manifold itself can serve as both a surge tank and a structural component such as a die shoe. Figures 13-28 and 18-33 illustrate the use of such systems.

SELF-CONTAINED NITROGEN CYLINDERS

Figure 22-32 illustrates a cross-section of a self-contained nitrogen cylinder. This type of cylinder features a special, semi-plastic seal material ($D1$) tightly contained between U-shaped packing cups. Up to 2,000 psi (13.8 MPa) of nitrogen gas is charged into the cylinder through fill valve ($D2$). The effective piston diameter is the actual rod diameter. $D3$ serves only to guide and retain the rod, as the gas pressure is equal on both sides of the container.

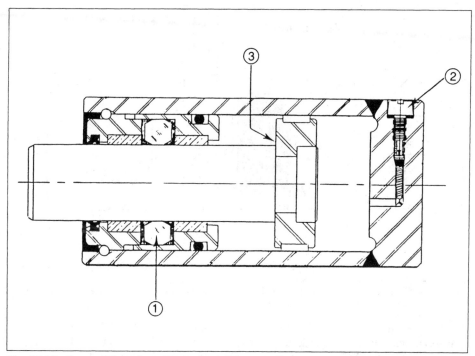

Fig. 22-32. Cross-section of a self contained nitrogen cylinder: (1) seal; (2) fill valve; (3) rod retainer. (*Dadco Inc.*)

Popular available sizes range from 0.75 to 5 tons (6.7 to 44 kN) in a wide range of stroke lengths. Many mounting styles are available including the no mount type which is placed in a drilled or cored hole.

The leakage rates of these cylinders is extremely low. The normal failure mode is a blown seal after the packing material becomes worn away. This type of nitrogen cylinder is favored by many die designers because of long life and low initial cost per ton as compared to air springs and other types of nitrogen cylinder systems.

Inverted Draw Die Application. The cost to tonnage ratio of the larger sizes of self-contained cylinders compared to manifold systems favors the use of self-contained cylinders in large inverted draw die applications. Figure 22-33 illustrates the use of 5 ton (45-kN), self-contained nitrogen gas springs used to provide draw ring pressure in an

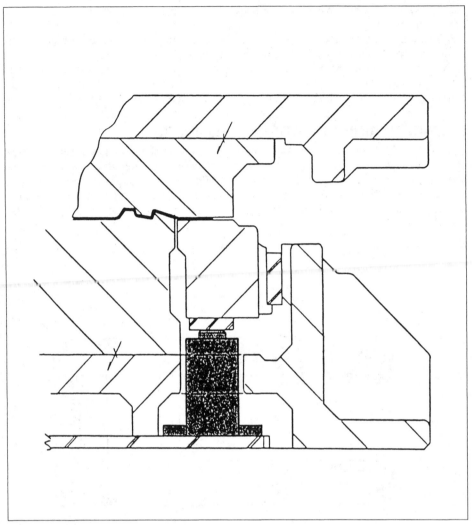

Fig. 22-33. 5,000 kg (11,000 lb) self-contained nitrogen gas spring used to provide draw ring pressure in an inverted draw die. (*Dadco Inc.*)

inverted draw die.

In such applications it is wise to provide cored inspection windows in the lower die shoe to permit a periodic visual check for failed cylinders. Failure will be evidenced by a collapsed cylinder. Normally, several extra cylinders are specified to compensate for the loss of any one cylinder to avoid interrupting a production run.

Cam Return Applications. Self-contained cylinders are increasingly used for cam return applications because of the large forces available and simplicity of installation. Figure 22-34 illustrates a self-contained nitrogen gas cylinder used to return an aerial cam.

Floating Draw Punch. Figure 22-35 illustrates self-contained nitrogen gas cylinders being used used to float a draw-die punch used in a double action press. For large automotive body panel dies having punches weighing up to 8 tons (7,300 kg), six 3 ton (27 kN) cylinders are used. Four cylinders would be sufficient. The two extra cylinders permit the production run to continue should any one cylinder fail. Cylinders last in excess of 500,000 strokes between repackings in this application.

Cylinders are usually used without mounts. They are placed in holes bored in the blankholder adaptor plate or bull-ring. The punch adaptor plate should be inlaid with hard

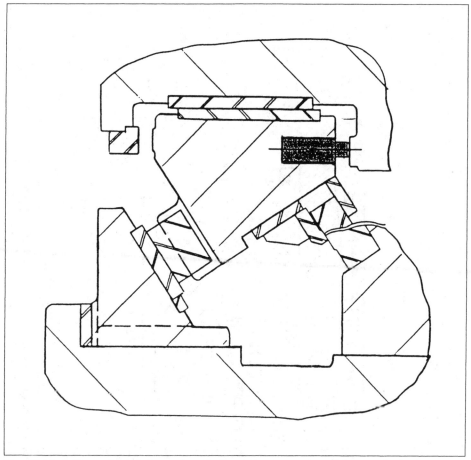

Fig. 22-34. Self-contained nitrogen gas cylinder used to return an aerial cam. (*Dadco Inc.*)

weld where the cylinder rod makes contact. The use of hardened steel contact blocks is not advised because they are apt to come loose and enter the die.

In addition to the obvious advantage of eliminating a diesetter task when setting the die, the floating punch draw-die is not affected by lateral movement of the press slides caused by worn or mis-adjusted presses. The chances of a punch or blank-holder wear plate loosening and falling into the die cavity is also practically eliminated.

Cylinder Filling. Self-contained nitrogen cylinders are filled from portable nitrogen tanks as shown in Figure 22-36. The tank, regulator, and other equipment are available from suppliers of industrial gases and nitrogen cylinders.

Pressure Adjustment. In cases where it is desirable to adjust the pressure of a self-contained cylinder, it can be converted to a piped system with miniature stainless-steel hydraulic hose and fittings. Figure 22-37 illustrates this system used with an external fill-valve to permit pressure adjustment.

In cases where it is desired to operate different parts of a system or even individual cylinders at different pressures, it can be done with a modular control panel illustrated in Fig. 22-38. This unit permits monitoring and adjusting each part of the system individually.

Fill Fittings. Figure 22-39 illustrates the use of a hydraulic diagnostic fitting that is widely used to fill high-pressure nitrogen die systems. The male fill fitting (*A*) is installed on the die while the female filling assembly (*B*) is installed on the hose from the regulator. These fittings can be easily hand-connected at full system pressure.

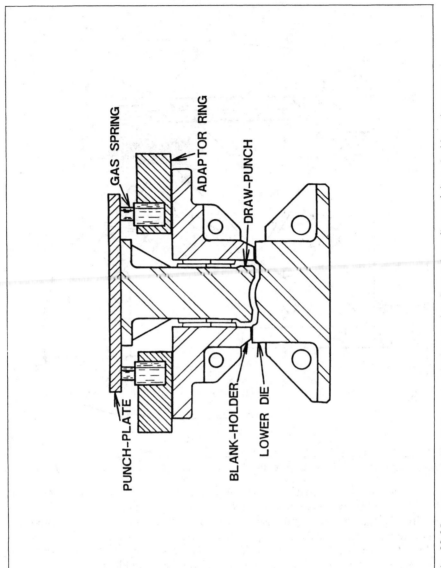

Fig. 22-35. *Self-contained nitrogen gas cylinders used to float a draw-die punch in a double action press.*

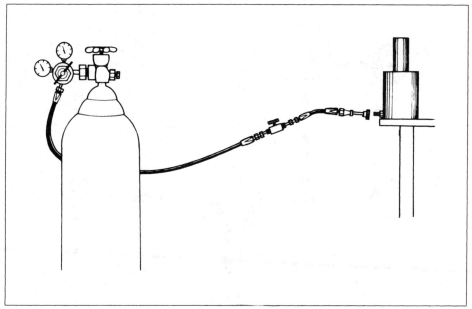

Fig. 22-36. Filling a self-contained nitrogen gas spring. (*Dadco Inc.*)

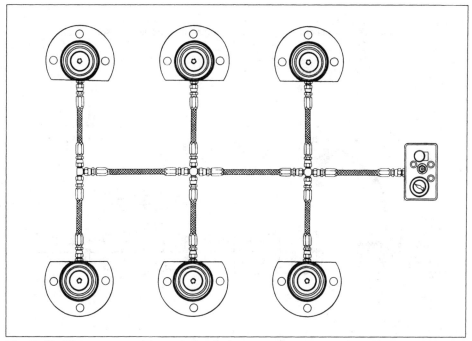

Fig. 22-37. High-pressure miniature hydraulic hose used with external fill-valve to permit pressure adjustment. (*Dadco Inc.*)

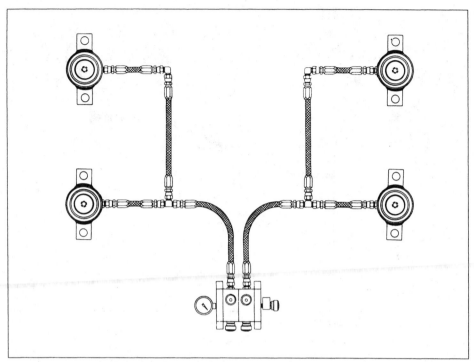

Fig. 22-38. Modular control panel for checking and adjusting nitrogen pressure. (*Dadco Inc.*)

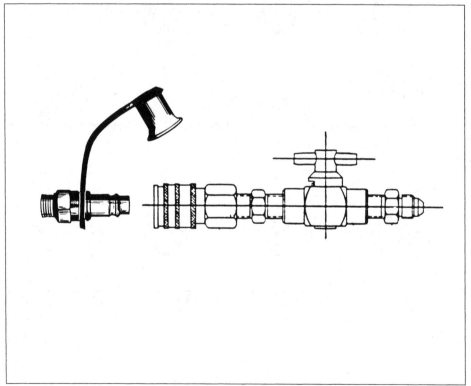

Fig. 22-39. Quick snap-on nitrogen fill fitting (*A*) and female filling assembly (*B*) for charging high-pressure systems. (*Dadco Inc.*)

RACK AND PINION LIFTERS

Rack and pinion lifter assemblies are used to actuate the part lifters in large dies. This type of lifter mechanism permits more compact die construction than the bellcrank system that it has largely replaced. The design considerations for this lifter system are in Section 13, Dies for Large and Irregular Shapes.

References

1. J.R. Paquin, "Choose Die Bushings Carefully," *American Machinist* (July 20, 1953).
2. P. Balsells, "Mounting of Piercing Punches," *American Machinist* (March, 16 1953).
3. J.S. Brozek, "A Notebook on Die Design," *The Tool Engineer* (May 1951).
4. C. Bossman, "Corner-grinding Troubles Avoided with Sectional Punch," *American Machinist* (June 30, 1949).
5. G. Eglinton, "Some Factors in Carbide Die Design," *The Tool Engineer* (May 1949).
6. F.A. Adams, "Retractable Punches Work on Alternate Jobs," *American Machinist* (October 30, 1950).
7. J.A. King, "Progressive Die Produces Safety Runway," *American Machinist* (November 6, 1947).
8. P. Wilmus, "Progressive Die Makes Tiny Cups," *American Machinist* (June 11, 1951).
9. "General Motors Standards," General Motors Corp., Detroit, Mich.
10. "Reference Book," Vol. 4, Durable Punch & Die Co., Chicago, Ill.
11. P. Balsells, "Stock Pushers for Dies," *American Machinist* (January 19, 1953).
12. J.R. Paquin, "Select Spring Strippers According to Die Requirements," *American Machinist* (November 17, 1949).
13. J.R. Paquin, "Positive Knockouts for Dies,"*American Machinist* (Oct.20, 1949).
14. C.R. Cory, *Die Design Manual,* Part 1 (Dearborn, MI: Society of Manufacturing Engineers, 1961).
15. P.R. Marsilius, "Talking About Die Sets—How to Select Die Springs," *Die Design Digest*, Vol. 1, No. 3 (Spring 1956), The Producto Machine Co., Bridgeport, Conn.
16. J.R. Paquin, "Pilots for Progressive Dies," *American Machinist* (May 20, 1948).

SECTION 23

DESIGNING DIES FOR AUTOMATION

WORKHANDLING AUTOMATION

True automation in press feeding and unloading implies not only the automatizing of the several different operations involved, but also their integration. Successful automation should plan for, and achieve, the following goals:

1. Maximum safety to operators and equipment.
2. Minimum damage to dies, presses, and automation equipment.
3. Higher productivity.
4. Improved quality of the product.
5. Less scrap.
6. Lower finished part cost.

Where safety is not the reason for considering automation, then the automation planning must be studied for cost effectiveness. For example, based upon projected volume, it may be less costly to feed blanks manually and to automate only the parts unloading.

Press automation requires close cooperation between the die designer and the process engineer. The shape and position of the part before and after each operation must be carefully studied to determine whether design changes, such as providing tabs or extra stock allowance on the blank for parts-gripping fingers, will facilitate automation.

To achieve flexible manufacturing, it is important to design the part, the process, and the required dies to permit successful application of automation. Some automation such as lifters, stock deflectors, and movable gaging is often built into the die and used in conjunction with computerized automation. There is extensive literature on the automation external to the die.[1, 2, 3]

Eliminate Turnover of Parts Between Operations. The turnover of a part between operations should be eliminated wherever possible because such operations add cost and are often the limiting factor in press or line speed. A study may show that part cost can be lowered by the use of a more complicated die, or even adding another die operation, instead of turning a part over. For example, an inverted draw die using high-pressure nitrogen cylinders for draw ring pressure may cost slightly more to build than a conventional die for a double action press, but the drawn panel can be loaded directly onto the post of a trim die without being turned over.

Use Shortest and Best Travel. Often it is best to feed the blank or part in a direction perpendicular to its longest dimension. There is usually an advantage in feeding the blank into one side of the press and withdrawing the part from the opposite side. This eliminates confusion and production delays due to stacks of blanks and finished parts located on the same side of the press. A shorter operation cycle will result since the blank can be fed into the press while the stamping is being withdrawn.

An obvious exception to this rule is feeding and unloading a press with a single programmable robot.

Install Gages Properly. Whether of fixed or retractable type, gages must assure correct positioning of the blank in the die.

Fixed gages should be so designed and located that they will not impede removal of the stamping. *Retractable* or *disappearing* gages may be spring-supported, and retract as the upper die or blankholder closes down upon them.

Mechanical limit or electronic proximity switches are often incorporated in the gages to indicate proper blank or part location. Often several such switches are used in series. Proximity switches are available with light emitting diodes (LEDs) built in to indicate actuation. These are a useful aid when setting up or troubleshooting an operation.

Pneumatic cylinders are sometimes incorporated in the die to retract gages upon press closure and to hold them down until the part is removed. They are frequently used as back gages when the part is to be removed out the rear of the press (Fig. 23-1).

Lifting Devices. The principal use of lifters in dies is to lift the part up after the press has cycled to permit either manual or automatic part removal. A lifter can be as simple as a spring-loaded latch or plunger to lift one corner of a drawn part to permit manual removal on a complex system of rack-and-pinion automation driven with multiple air cylinders (Sections 13 and 22).

Dies on automated operations often have limit switches or proximity sensors to indicate proper part nesting. Should the part fail to nest, the lifter can be quickly lifted and lowered by automatic means in an attempt to establish correct nesting.

Feeding Small Parts into Gap Frame Presses. These presses are usually inclinable, and parts can be fed by employing gravity or slides. The press frame may be inclined, or a chute may be installed at such an angle that the part can slide into position. Gravity, employed alone or assisted by an air blast, can be used to eject the part into a container or conveyor located at the rear of the press. Feeding can also be done by placing the part in the nest of a slide and then pushing it into the working area.

Scrap Handling. Scrap must be automatically removed from the die area by dropping

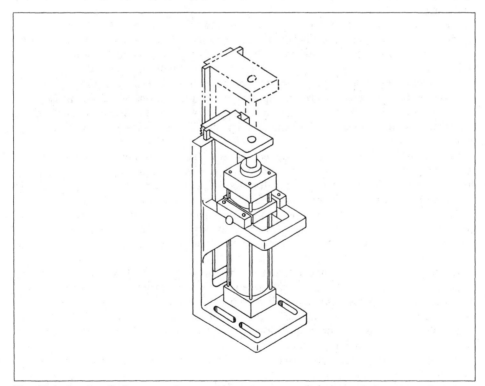

Fig. 23-1. An air cylinder-actuated retractable gage.

it off the die and bolster plate into containers or holes in the floor. Wide bolsters and/or low shut heights may not provide enough fall for proper shedding of scrap. In such cases, discharging some or all of the scrap through a hole or holes in the bolster is an excellent method.

Small, in-die scrap conveyors can be used to discharge scrap, but their use should be avoided. Their initial cost is much higher than a chute, and the die can be damaged should they fail to discharge scrap. If conveyors must be used, sensors should be installed in the die to stop the press in the event of a scrap jam-up.

Enough scrap cutters should be provided in trimming dies to ensure that the scrap will be cut into into small enough pieces for easy disposal.

Standardized Automation Equipment. The development of standardized pieces of automation equipment, engineered to perform specific operations, makes it possible to plan a gradual or building-block type of progression from very simplified, single operations to fully automated lines without obsolescence of equipment.

In general, automatic handling equipment can be divided into the following categories: (1) feeding equipment, (2) unloading equipment, (3) transfer equipment between press operations, and (4) stacking equipment.

With this variety of equipment it is possible to set up a press line by feeding the first press automatically and feeding the remaining presses manually. To increase the degree of automation, unloaders and standard conveyors may be added between the presses. The conveyors are later replaced with transfer and feed machines; finally, stackers are installed, thus making a fully automated line. The equipment can be installed at the outset to achieve a fully automated line.

When planning the automation of a press or press line, consideration should be given to the size of the equipment and future parts, especially when using the building-block concept. Since part size and configuration may change from year to year or from model to model, the equipment must be adjustable in stroke, height, and width to accommodate different-sized parts, if necessary.

In order to achieve flexible manufacturing and quick die change, the speed with which the equipment can be changed over is an important consideration.

FEEDING EQUIPMENT

The feeding of presses can range from fully manual to automatic devices for parts, automatic coil-stock feeders, and semiautomatic and automatic feeders for strip stock, sheets, and large blanks.

Manual Feeding. Low production may not warrant the expense of automatic feeds. Considering the savings in set-up time, an experienced operator following safe procedures can outproduce a mechanically fed press on some short runs.

The manual feeding of parts into a die can be accomplished by the use of a simple chute. The parts are pushed forward in the chute against a suitable stop in position for the stamping operation. The press can be inclined so that the formed parts can fall out the back of the press, or a jet of air can eject the finished parts. This die should be provided with a barrier enclosure for complete protection of the operator.

Hand-operated Slide Feed. The pusher in Fig. 23-2 can be used to feed a blank under the punch and withdraw it after the operation is performed. The pusher can have a nest machined into it to fit the shape of the part. Locating pins in the punch are used to position the pusher and blank in the die properly. If the part does not drop through the die or is not ejected by other means, it can be withdrawn by the pusher and allowed to drop through a hole provided in the bottom plate of the slide. This die should be provided with a barrier enclosure for complete protection of the operator. An electrical interlock should be provided to ensure that the press cannot be tripped unless the slide is in the correct position.

Magazines for Push Feeds. Manually operated push feeds may be supplied with parts

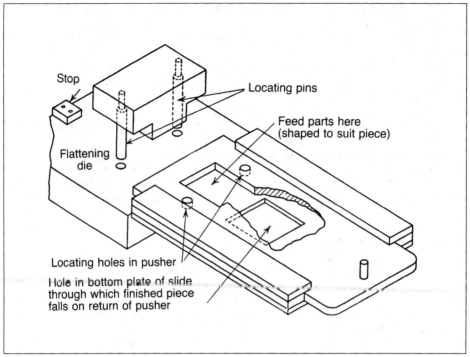

Fig. 23-2. Hand operated slide for loading and unloading die.[4]

from inexpensive magazines which can be attached to the slide rails. For best results, blanks must be flat and free from burrs. These push feeds should be interlocked with the press-tripping mechanism to ensure that the blanks are in proper position to prevent damage to the die and press.

Gravity-chute Feeding. Maintenance of a gravity-chute feed is usually required only if an escapement is incorporated. This type of feed is commonly used with press operations on shells. In a chute-feed die (Fig. 23-3A), the shell is fed by gravity into a locating nest. This cup is redrawn and pushed through the die. In case the shell sticks on the punch, a small recess is machined beneath the draw ring to strip the punch. At B, the design utilizes vacuum cups to pull the workpiece up and out of the die, and a knockout rod releases it from the cups. At C, an inclined chute on a flat die feeds flat blanks into a bending die. A spring pad is incorporated in the die to lift the part out of the cavity to be ejected by an air jet. When the press can be inclined, a straight chute can be used to load the die.

As is the case with all such designs, full barrier guards or other proper means must be employed to safeguard the operator.

Semiautomatic Gravity Feeding. A single-piece feeder for a gravity-feeding device is illustrated in Fig. 23-4. The illustration shows the trip arm at the top of the press stroke with the blank in the nest. As the press slide moves downward, the trip arm frees the lever on the end of the feed shaft. This allows the counterweight to tip the feeder so that the piece held by the upper leg is released and moves down to rest against the lower leg. As the slide continues downward, the part is formed and ejected. As the press slide approaches top center on its upward stroke, the trip arm attached to the slide contacts the lever on the end of the feeder shaft, tips the feeder to a position whereby one piece is released by the lower leg and, at the same time, the upper leg of the feeder catches and holds back the next piece.

Air-chute Feeding. Gravity assisted by a continuous weak air blast drives two balls down a tube (Fig. 23-5, D6) after a selective mechanism allows only two balls per press

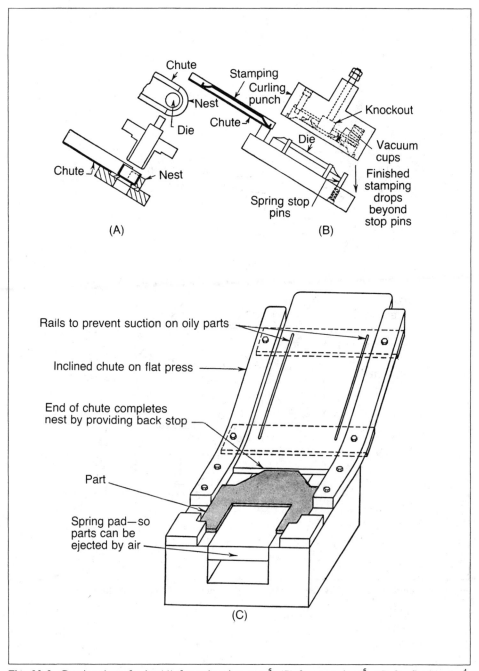

Fig. 23-3. Gravity-chute feeds: (A) for redrawing cups[5]; (B) for stampings[5]; (C) for flat blanks.[4]

stroke to be released from the hopper. The selective device comprises a revolving ball carrier (D1) with 12 equally spaced, 0.120 in. (3.05 mm) holes rotated, on the downstroke, by a spring-actuated, pawl-engaging ratchet. On the upstroke, the arm (D10) is forced to the left by the pivoting and linking members (D2, D3, and D4). The ball carrier passes under a metal plate (D8) and an inspection window (D5), and between the ends of the copper tubes (D6 and D7). The balls, released on the downstroke, drop through the tube into the die. A slide feed (Fig. 23-9) inserts the balls into the workpiece.

23-5

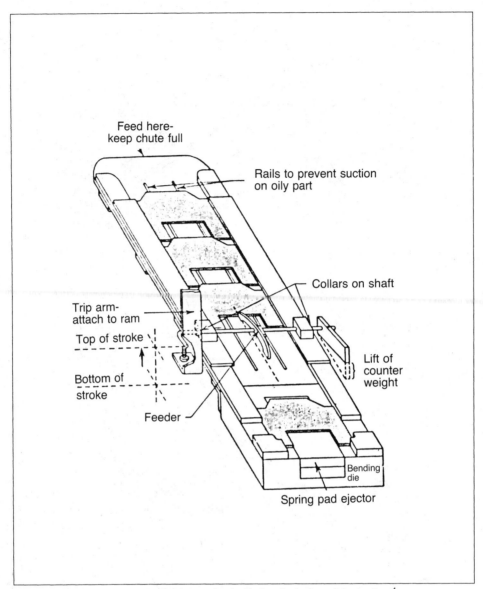

Feed here-
keep chute full

Rails to prevent suction
on oily part

Collars on shaft

Trip arm-
attach to ram

Top of stroke

Bottom of
stroke

Lift of
counter
weight

Feeder

Bending
die

Spring pad ejector

Fig. 23-4. Single piece gravity feeder mechanically interlocked to slide motion.[4]

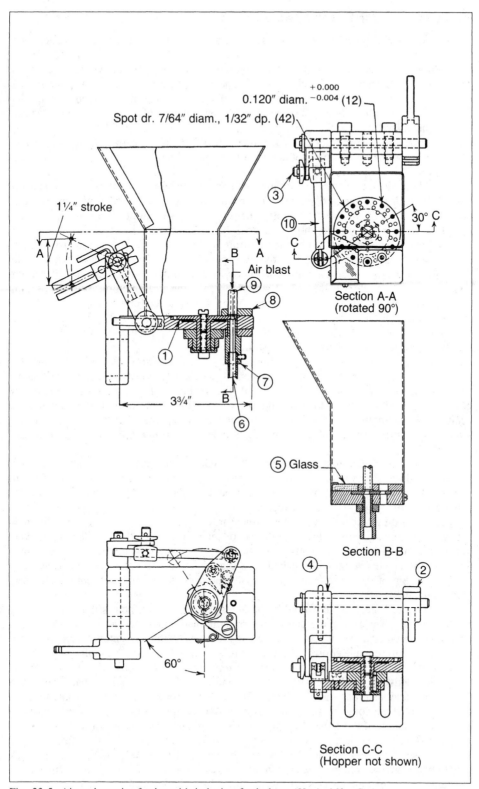

Fig. 23-5. Air and gravity feeder with indexing feed plate. (*Harig Mfg. Co.*)

AUTOMATIC MECHANICAL FEEDS

The most used types include roll feeds (slide, grip) or hitch feeds (hoppers, chutes, and magazines).

Roll Feeds. Basically this type of feed consists of two rolls, one above the other, often driven through a friction or ratchet clutch. The former provides an infinite number of stock feed lengths while the latter provides definitely set amounts of feed lengths that cannot be fractionally altered. Backward movement of the rolls during the non-feeding cycle is prevented by a brake.

Roll feeds may incorporate either a single pair of rolls (Fig. 23-6) which pull or push stock across the die, or two pairs of synchronized rolls with one pair pushing and the other pair pulling the stock across the die.

Roll release devices are frequently used when such feeders are used to feed progressive dies. The stock is momentarily released upon pilot entry to permit the pilot to correctly align the stock. A slight overfeed is often helpful when using pilot release feeders as it is usually easier for the pilot to shove the stock back than to pull it forward.

Good feeding practice is to separate the straightening rolls from the feeding rolls unless it is absolutely necessary to combine the two operations. Feeding in itself is naturally an intermittent operation, but removing coil curvature in a straightening machine will require less power and will do a better job of straightening if the straightening rolls are running at a uniform velocity. This is accomplished by providing space for a material storage loop that maintains a relatively constant supply of material between the feeding and the straightening operations. Progressive-die operations that require pilot pins provide further reasons for separating the feeding and straightening.

Upper feed rolls are easily and automatically lifted to allow pilot pins to shift the material, but the material may not have the desired flatness when the upper-straightener rolls are lifted after each progression of the material. Floor-mounted, self-driven coil-feeding machines fall into two general types; the rack-and-pinion machine is usually limited to maximum feed lengths of about 60 to 80 in. (1.5 to 2 m), but the material-measuring type will feed any desired length when employing digital control counters.

Basically, most rack-and-pinion types of machines are alike in that they all have self-driven drive shafts which are timed to turn slightly faster than the press crank-shaft and to which an adjustable throw block is mounted. The feeding function is about the same as that of the press-mounted feeds.

Measuring-type feeders have material-measuring rolls that are separate from the main feeding rolls so that accurate feed lengths can be maintained even though the material may slip occasionally on the feed rolls. The measuring rolls turn a digital pulse generator. When a preselected length is reached, the digital control initiates the slowdown and stopping action of the main feed rolls. Close feed accuracies and excellent repeatability can be obtained.

The latest roll feeds are powered by brushless direct current motors that provide very precise control. The feed pitch can be pre-programmed and stored as a job number in a on-board computer that can also preset press functions such as counterbalance pressure, shut height, and tonnage limits. The correct information for set-up of both the press and feeder can be done by inputting bar-coded information from the die or dieset work-order.

Hitch Feeds. A hitch feed consists essentially of a reciprocating head carrying a gripper unit and a similar stationary unit. On the downstroke of the press, a cam attached to the press slide contacts the cam roller on the reciprocating head. The continued downward motion of the press slide pushes the reciprocating head outward, compressing a spring. During the downward press stroke, the gripper plate on the stationary head prevents the stock from moving backward. On the upward stroke of the press, the stock is held by the gripper plate as the head moves inward propelled by the compressed spring. The amount of feed advance is set by a feed-length adjustment nut.

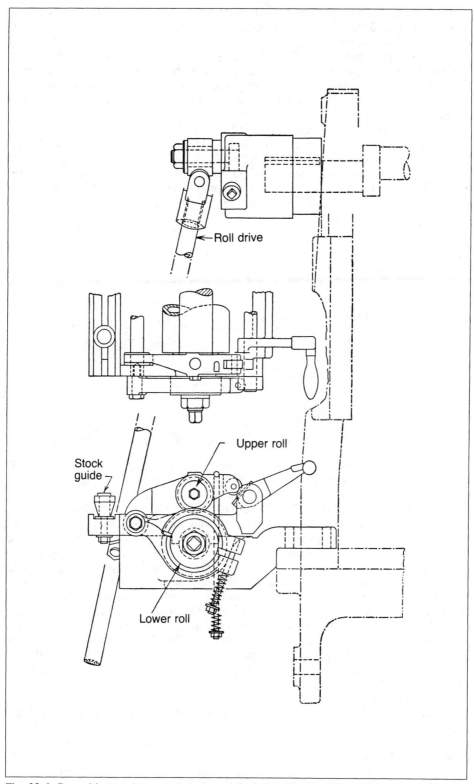

Fig. 23-6. Press-driven and mounted single-tool feed. (*E. W. Bliss Div., Gulf & Western Mfg. Co.*)

Another popular type of hitch feed is air-powered. A large air cylinder powers the back-and-forth feed motion while small cylinders actuate both the stationary and moving grippers. Usually the gripper has a provision for a pilot release that can be actuated by either a press slide mounted limit switch or a programmable rotary limit switch.

Grip Feeds. A grip feed, similar in principle to a hitch feed, incorporates a reciprocating head and grip shoes instead of gripper plates. The pivoted shoes swing down to grip the stock and feed it to the die, but slide over the stock when the head reverses its direction.

A cylinder-type grip feed may be driven from the press crankshaft by an adjustable throw block and lever, or by a cam (Fig. 23-7, D1) actuated by the press slide. On the upstroke, the cylinders (D2) at the left grip and move the stock a feed length determined by adjustment nuts (D3). Movement of the cylinders, carried on a slide (D4), is controlled by a spring (D5). On the downstroke, the stationary cylinders (D6) at the right prevent backward movement of the stock. Both sets of cylinders operate on the principle of the overriding clutch.

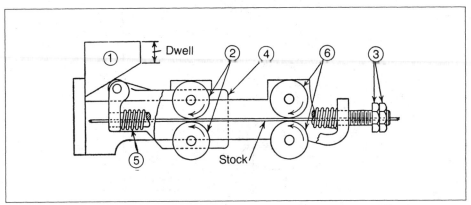

Fig. 23-7. A die or press mounted cylinder-type grip feed. (*H. E. Dickerman Mfg. Co.*)

Slide Feeds. Driven by an adjustable eccentric, this type of feed (Fig. 23-8) incorporates a reciprocating feed block and a hardened blade which engages and moves the stock. Positive stops ensure extreme accuracy of feed. A stock drag or check unit prevents stock movement as the feed blade returns to re-engage the stock.

Slide mechanisms are used to feed precut blanks into a second-operation die. The blanks may be oriented in a magazine chute before being pushed into the die by the feeding slide. The slide may be actuated by cams or linkage connected to the press slide or air cylinders. By changing the shape of the magazine chute and the nest in the feeding slide, this device can be adapted to various shapes of parts.

Slide Feed for Inserting Balls. A slide feed (Fig. 23-9) built in a progressive die to insert balls into a radio-socket contact operates with the air and gravity feed shown in Fig. 23-5. The latter feed directs two balls through a tube (Fig. 23-9, D2) to the proper position in the die. On the downstroke, a cam (D3) contacts a roller (D4) fastened to a pivoted arm (D5). The arm pushes the ball slide (D6) to insert the balls; on the upstroke the ball slide is retracted by a spring (D1) so that the strip may advance for crimping and to allow two more balls to fall into the die.

Pawl Feed. This type of feed uses pawls (Fig. 23-10, D1 and D2) to engage and push the workpiece (a notched strip) to the die. Toothed latches or hooks (instead of the pawls) to engage notches or blanked-out openings would classify the design as a hook feed. The drive mechanism actuates the slide (D5) by means of actuating bars (D4). Adjustment is provided by setscrews (D3).

Dial Feeds. Dial feeds, or circular conveyors carrying workpieces to dies, are used for primary or secondary operations. When many operations are to be performed at one time,

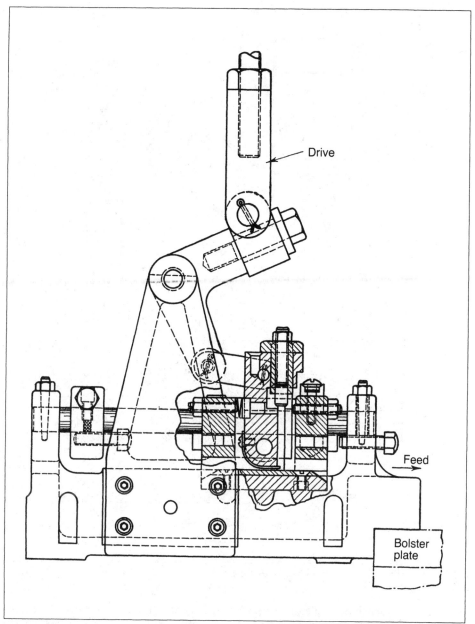

Fig. 23-8. Example of a press-driven slide feed. (*U. S. Tool Co.*)

it is necessary to have a well-supported and guided press slide, since the operations will be located at a considerable distance from the slide's center line. Multiple operations may require positive punch stripping so that the parts are correctly positioned in the dial plate to be conveyed to the next station. Two or more operators may load parts to a dial feed for assembly operations. The use of auxiliary feeds (such as chute, magazine, and hopper feeds) with dial feeds, often in conjunction with sorting devices, results in high production rates.

The revolving member of a dial feeding mechanism generally rotates around a vertical axis actuated by the press ram or an air cylinder. In a typical design, an indexing and locking pawl mechanism, driven by an air cylinder (*D*4, Fig. 23-11), rotates a dial (*D*3),

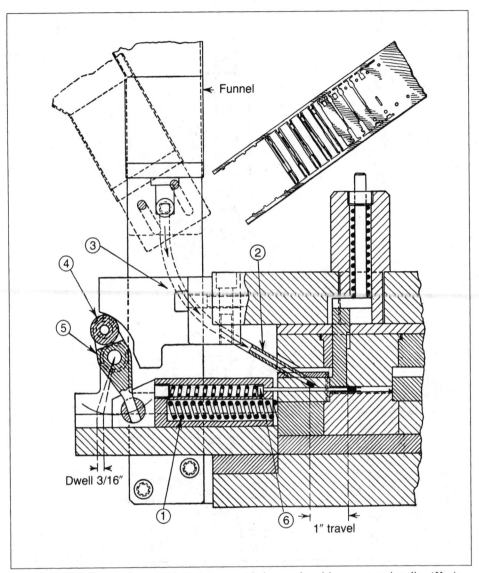

Fig. 23-9. Slide feed for inserting balls into a part being produced in a progressive die. (*Harig Mfg. Co.*)

which circularly positions a die (not shown) under a punch. The control valves for the air cylinder are actuated by solenoids which are controlled by limit switches synchronized with slide travel. Other limit switches (*D*1 and *D*2) prevent the press from being tripped if registration and indexing are incorrect.

A pneumatically operated dial feed may be controlled through a camshaft rotating at a speed determined by the work cycle of the dial. The duration of the work cycle is equal to one revolution of the camshaft. The dial indexing frequency is the sum total of the time (in seconds) for the slowest operation, plus total operational time not acting during this period, plus indexing time, and plus dwell time.

A microprocessor-based programmable limit switch operated in conjunction with a press crankshaft-driven digital resolver is used on most new and rebuilt press installations.

The air-control system should provide a synchronized air blast for injection, as well as controlled air power for auxiliary operations, all of which must be considered in

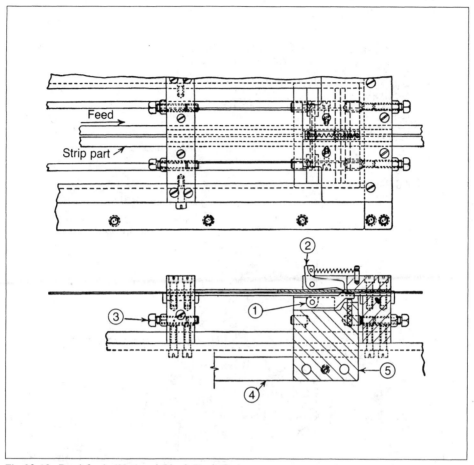

Fig 23-10. Pawl feed. (*National Blank Book Co.*)

calculating time requirements and in designing cams or programming the electronic cam switch.

Rack-driven Dial Feeds. A rack (Fig. 23-12, $D4$) on the press rotates two top gears ($D2$) of a gear train to move the lower rack ($D1$) back and forth. The pawl ($D5$) engages notches in the dial ($D3$) to move it through an arc of 60° on each press stroke. This feed is used with a cam die to pierce 1.080 holes, shown in Fig. 5-11.

Cam-driven Dial Feed. A chain-driven edge cam (Fig. 23-13, $D4$) reciprocates a pawl slide ($D1$) to revolve the dial, which carries 12 stations. A face cam with a slot enclosing the roller could be used instead of the edge cam. The pawl slide is provided with a tension spring ($D2$). The station ($D3$) is for swaging.[6]

Underdriven Dial Feed. Another driving method, using an eccentric arm from the press crankshaft, is shown in Fig. 23-14. The indexing dial is underneath the work dial, which can be easily replaced, allowing other stamping operations with the same feed.

Sheet-feeding Devices. Where large-sized sheets or blanks on pallets are fed to draw dies, an automatic destacker is often used. In one common form of destacker, the sheets are separated by several sheet-fanners using powerful permanent magnets. The individual sheet is lifted by suction cups onto the underside of a magnetic conveyor that transports the blank to the loading station where the blank is pushed into the draw die.

In some systems the sheet is placed directly in the draw die by suction cups. Most systems have an electronic double-blank detector to prevent double blanks being fed into the draw die (Section 26, ref 1 and 3).

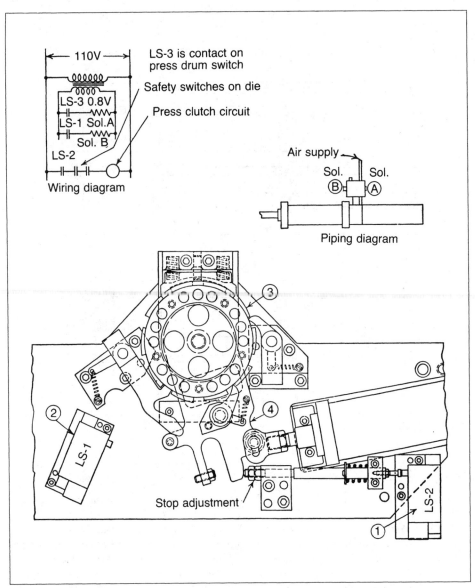

110V

LS-3 is contact on press drum switch

Safety switches on die

Press clutch circuit

LS-3 0.8V
LS-1 Sol.A
Sol. B
LS-2

Wiring diagram

Air supply

Sol. B   Sol. A

Piping diagram

LS-1

LS-2

Stop adjustment

Fig 23-11. Pneumatic-driven indexing dial with electrical interlock. (*Oldsmobile Division, General Motors Corp.*)

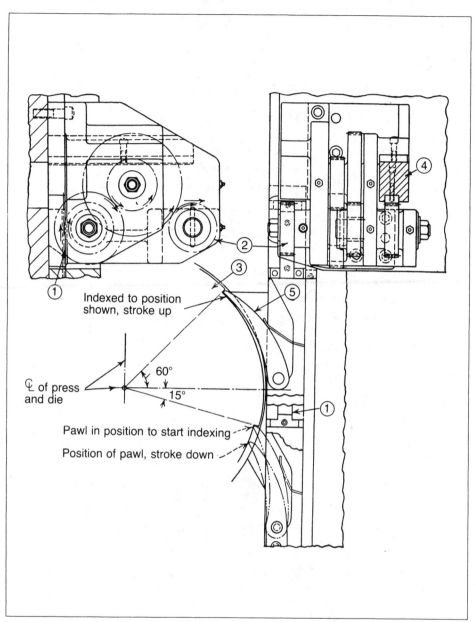

Fig. 23-12. Rack driven dial feed. (*Maytag Co.*)

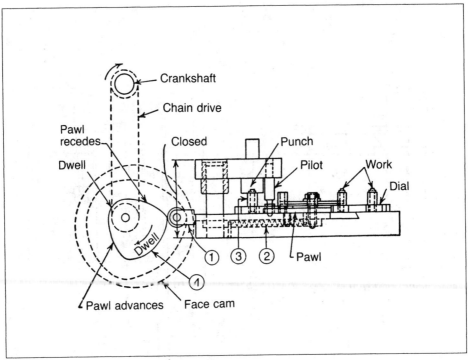

Fig. 23-13. Cam driven dial feed.[6]

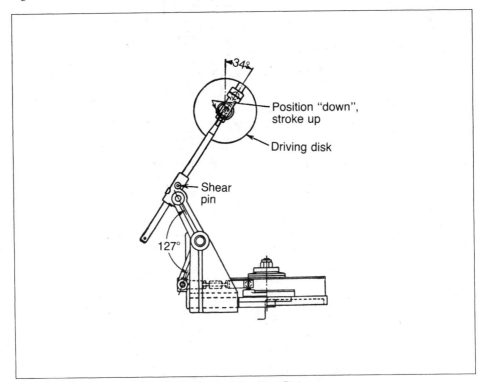

Fig. 23-14. Underdrive dial feed. (*F. J. Littel Machine Co.*)

TRANSFER FEEDS

Some parts are too large or complex to lend themselves to production in a progressive die. The automatic transfer of parts from one die to another by means of a transfer bar, or transfer rails equipped with part-handling fingers, provides the advantages of a progressive die without the need to have a carrier strip transport the parts.

Single-bar Feeds. A single-bar feed, commonly used with inclinable presses, incorporates a cam-actuated bar which reciprocates over the die to receive a part which it then transfers to the next die station. Various ejectors and knockout pins function during the press cycle so that the part can be stripped from a punch and ejected from a die and into the path of the moving bar.

Double-bar Feed. A double-bar feed has a double bar which moves from right to left or vice versa, and also moves front to back or vice versa. Fingers are attached to the bars for gripping and moving the part progressively. This feed type is generally used with straight-side presses.

Often, this feed mechanism is captive to the die and driven by slide motion or an electrical servo drive.

Two-axis Transfer Presses. There is no up-and-down motion available to the feed rail of a two-axis transfer press. As a consequence the parts are dragged along over the lower die surfaces. This generally limits the parts to shallow forms and draws. To detect misfeeds, the die stations are often equipped with proximity switches. The fingers can also be spring loaded and provided with proximity switches to detect spring deflection.

Three-axis Transfer Presses. Presses having built-in transfer feed rails driven by cams powered by the press crankshaft are widely employed in automotive stamping plants. Many assembly plants are integrated with a stamping plant having several such presses for production of the large drawn body panels. The advantages of an integrated stamping and assembly facility include eliminating the shipping costs and possible shipping damage to the panels and inventory reduction since the panels can be made in small lots.

The presses are generally equipped with an extra bolster and set of feed rails to permit the dies and transfer fingers for the next job to be run to be set up, or pre-staged.

The die change is automatic or semi-automatic. The bolsters are self-moving type driven by compressed air motors. Die clamping is automatic.

Die cushion and counterbalance pressure can be automatically set for the job going in from a data base stored in an on-board computer that is a part of the press electrical control system. The slide shut height can also be adjusted automatically, but maintaining all dies at a common shut height is preferred to save time and reduce the chance for damage (Sections 13, 24, and 27 ref 5).

Typical changeover times are under 5 minutes with 10 to 15 changes per 24 hour period.

Designing Dies for Transfer Presses. Some important considerations when designing transfer press dies are:

1. Common pass height.
2. Common shut height.
3. Clearance for the transfer fingers and parts.
4. Designing the dies to produce the part in the available number of stations.

When establishing a common pass height, it is best to establish a common figure for all dies to be run in the press. This will save setup time where common blank feeders and unloaders are used. To increase manufacturing flexibility, this figure should be used on all presses of the same type where possible.

The common shut height should allow for bed and slide deflection. Additional die shut height may be required in the middle dies to compensate for press deflection.[7]

Clearance must be provided for the transfer fingers and part movement. Often the guide pins are placed in the upper die shoe to provide clearance. Computer-aided design (CAD) equipment that can simulate part and finger motion based upon the slide and transfer finger motion curves is the best way to accomplish this (Sections 3, 13, and 24).

If the exact feedrail to bolster and feedrail to slide motion curve is not known, it can be plotted by fastening a felt tip marker to a transfer finger and tracing the motion on a piece of paper fastened to a sheet of plywood supported in the press diespace. Three tracings are needed depicting the motion of the finger relative to the bolster and three relative to the slide.

The product designer must work with the die designer to ensure that the part can be manufactured as intended within the limits of the number of dies that can be accommodated in the transfer press.

Die Lifters for Use in Transfer Presses. Transfer press dies often receive the part in the lifter-up position and drop by pneumatic action as the slide descends. Two or more proximity sensors are mounted on the lifter to detect proper part nesting. The disadvantages of this system are:

1. An emergency stop function must be initiated if the panel does not nest properly on the lifter—this subjects the press and feed rails to severe shock.

2. The delay inherent in pneumatic systems results in variation in lifter timing as a function of press speed, unless an automatic means for correction is provided.

3. If 60 Hz a-c power is used to actuate the pneumatic valves, there is a built-in ambiguity of timing approaching one crankshaft degree that can cause misfeeds in a complex transfer operation.

Some solutions to these problems are:

1. Place the proximity switches in the transfer fingers to permit misfeed sensing in time to initiate a normal top-of-stroke cycle stop.

2. Provide mechanical or gas spring lifters that are driven down by the upper die upon die closure—this can be done by drivers that contact outriggers on the lifter to avoid marking the panel.

3. Power the lifters by means of a servomechanism built into the press bed.

Transfer Finger Construction. Aluminum plate and lightweight metal tubing are common transfer finger construction materials.

Round tubing is a common finger construction material for use on large three-axis machines. It has the advantage of being adjustable in length and position. An inexpensive lightweight construction method makes use of 0.120-in. (3.05-mm) wall steel tubing having an outside diameter of 1.875 in. (47.62 mm). This tubing is used to make feed rail attachments (Fig. 23-15A) and finger attachments (B). The center extension of the finger (C) is made of 2.0-in. (51-mm) outside diameter by 0.058-in. (1.47-mm) wall 6061-T6 aluminum tubing. This lightweight tubing also serves as an easily replaceable fracture link in the event of a misfeed. Additional misfeed protection is provided by the optional spring-loaded base assembly (D). The part is transferred by the pneumatic gripper head (E). Clamping is provided by compressing the outer tubing onto the inner tubing with common automotive muffler clamps (F).

An important consideration in any transfer finger design is to minimize die damage in the event of a misfeed. Practical experience has shown that the gripper jaws (G) or shovel (H) should be of lightweight construction to minimize damage in the event of a mis-hit in the die.

Section 26 has detailed information on a commercially available transfer finger having ruggedized proximity sensors as a built in feature. Universal adjustment is provided by ball-knuckle adjustment joints.

Retrofitting Older Presses With Transfer Rails. Transfer feed rails can be fitted onto existing straight side presses. The most common type of drive uses brushless d-c motors in conjunction with position resolvers on both the feed rail drive and press crankshaft. A digital computer is used to continuously compare and correct the feed rail motion in order to transfer the parts correctly and avoid interference.

Hydraulic cylinder, transfer rail drives are also used in conjunction with position resolvers and modulating valves controlled by a digital computer.

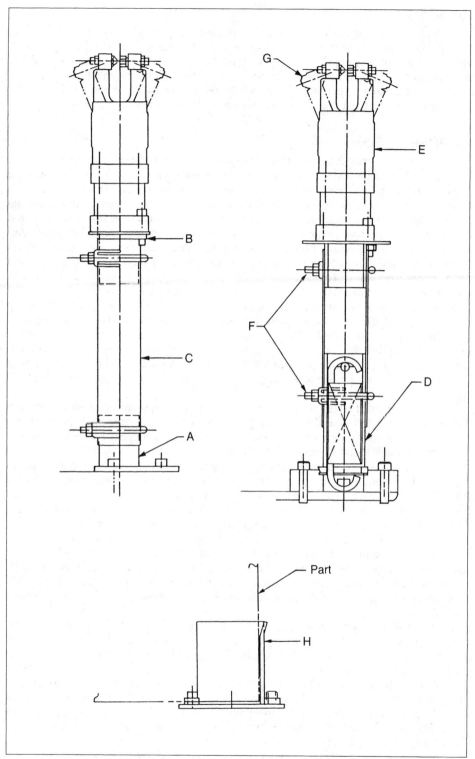

Fig. 23-15. Spring-loaded transfer finger with pneumatic gripper jaw: (A) steel tubing weldment; (B) finger attachment adaptor; (C) aluminum center extension; (D) spring-loaded base assembly; (E) pneumatic gripper head; (F) automotive muffler clamps; (G) gripper jaws; (H) shovel.

ROBOTS

Generally robots can perform any loading, positioning, and unloading function that can be done by a human operator. The main consideration in the design of a die intended for operation by a robot is to provide sensors in the die to insure that the press is loaded and unloaded correctly, if this sensing function is not performed by the robot.[1, 2, 3]

It is important that proper safety barriers be provided, with suitable electrical interlocks to prevent injury to personnel from unexpected robot movement.

HOPPER FEEDS

In general, a hopper is a mechanism into which parts are dumped without regard to orientation. It may be a funnel, a single bin, or a device which, in addition to holding a quantity of parts, performs one or more orientations or selections of position and discharges the parts to tracks for use in machines of all types. Many factors determine the proper hopper design required for a particular job: required rate of feed, required orientation, size of parts, shape of parts, weight of parts, center of gravity, fragility, etc.

The most common types of hoppers and orienting devices depend upon gravity. This limits the rate at which parts may be selected to the rate of fall and amount of fall required for a given orientation. To increase the output of a hopper beyond these limitations requires either additional hoppers or multiple capacity within a hopper

Parts such as bolts and other flanged or headed parts can best be oriented by suspending them by the head. This method is relatively simple and can be incorporated in any hopper. The discharge track usually consists of inclined rails on which the parts slide. Another orienting method considers the center of gravity of the part. If its center of gravity is other than its geometric center, the part can often be allowed to slide or roll down a track with a knife-edge inclined at a compound angle. If the part is not properly oriented, it will fall off the track.

Chamfers, rounded edges, flanges, or grooves are also used for orientation. The part may be supported by a narrow edge or guided by a flange or groove. If not properly oriented, it falls or is scraped off the track and back into the hopper for recirculation.

Oscillating and Rotary Centerboard Hoppers. A blade having a profile which holds parts that fall on it in proper orientation is reciprocated or rotated through a mass of parts. As the blade reaches an appropriate position on its stroke, the parts slide down the blade, through a selective discharge gate, and into a delivery tube or track. These types of hoppers are commonly used for feeding cylinders and tubes having a length greater than their diameter and for balls, slugs, channel shapes, and some irregular forms.

Figure 23-16A illustrates a hopper (D4) having a grooved centerboard blade (D2) which oscillates vertically to pick up parts on its edge. A cam or crank-actuated arm (D1) moves the centerboard up and down. Parts slide down the grooved blade onto the track (D3) and on to the die. When the blade goes down for more parts, the ejector (D5) advances to the end of the track and gently pushes the incorrectly positioned pieces back into the hopper.

A hopper with a rotary centerboard is shown in Fig. 23-16B. The blade (D1) is of the correct thickness and form to suit the part and is indexed clockwise by a geneva cam or a similar intermittent motion mechanism. As the blade indexes to a position opposite the track (D2), the parts slide to the machine.

Barrel Hoppers. Parts are loaded into the barrel, and when the barrel is rotated, vanes or blades on the inner surface of the barrel lift up the parts and drop them as the barrel turns. Some of the parts will drop in proper position on a track, down which they can slide to the die. The rotation of the barrel agitates the parts, thus eliminating a tendency to interlock and become a mass.

Barrel hoppers are suitable for many irregular shapes, stampings, light forgings, and castings.

Figure 23-16C shows a barrel-type hopper. The barrel (D1) has vanes (D2) which pick

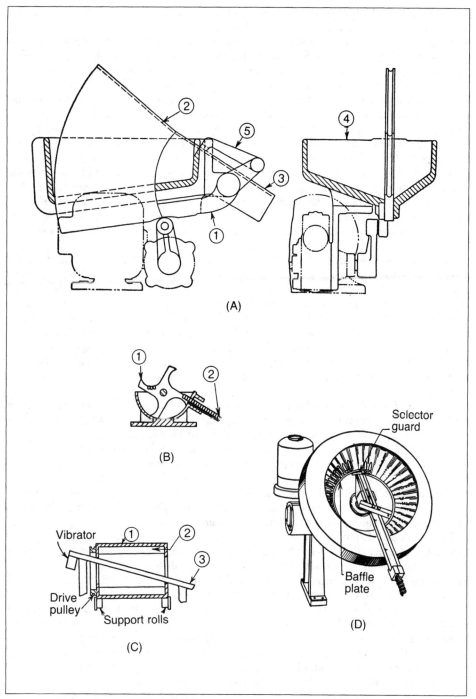

Fig. 23-16. Hopper designs: (A) oscillating centerboard (*J. R. Paquin*); (B) rotary centerboard; (C) barrel hopper;[8] (D) inclined rotary hopper. (*Detroit Power Screwdriver Co.*)

up and, to a certain extent, reorient the parts, and carry them upward until they drop into a chute (D3) which leads them to the press. A continuous belt may be incorporated in the drum in place of a track or chute to carry the parts to the point of use.

Inclined Hoppers. These devices have bowl-type hoppers which are inclined, to help orient the parts. The bowl can either rotate or be stationary while a disk or ring in the bottom of the bowl rotates.

In the rotating ring, a ring revolves, passing specially formed cavities through a mass of parts to be fed. Parts fall into these cavities as the ring revolves, and those properly oriented fall into a discharge chute. Headed parts, spheres, and disks can be selected for feeding by this device. Output depends upon the number of cavities, rate of ring rotation, and part characteristics. The rotary-bowl selector has vanes or cavities by which parts are oriented and fed into a chute or track. There is a minimum of tumbling of parts since they seldom travel more than halfway up the bowl before being accepted or rejected by the selector. Cylinders, tubes, disks, and short-headed parts are all adaptable to this feeder.

Figure 23-16D shows a rotary-ring hopper designed to feed 50-caliber bullet cores point first into the track. Cavities in the ring permit the bullet cores to be picked up as the ring passes through the supply stored in the stationary bowl. The stationary baffle plate inside the rotary ring keeps the parts from falling inward until they are carried up to the end of the track. The selector guard attached to the baffle plate allows those bullet cores which are oriented point first to pass into the track but rejects those having their blunt end first.

Vibratory-bowl Feeder. This type of feeder uses electromagnetic vibration to move parts up an inside spiral track from the convex or flat bottom to the rim of the bowl. The vibratory unit (Fig. 23-17) is contained in the base of the feeder. A potentiometer is used to control the rate of feed; the controller may be built-in or separate from the base unit. Vibratory feeders can be constructed to feed either clockwise or counterclockwise and with single or multiple discharge points. Bowls are available in a wide range of diameters, and each feeder is engineered to handle a specific part of a certain size, weight, and shape. Feeders of this type can handle workpieces of almost any material, including the most fragile parts. Very thin, light parts are not suitable since these will bounce without moving up the track.

As a rule of thumb, the bowl diameter should be approximately ten times the length of the part. This will permit the part to contact the track only and stay away from the bowl wall. The larger bowl also provides storage space for an adequate supply of parts, thus eliminating the need for frequent replenishment. The track in a bowl is tooled spirally around the inner diameter of the bowl in such a manner that the slope is never greater than 5°.

Feed rates, in terms of pieces per minute, depend on size, shape, weight, center of gravity, and other characteristics of the part. The output of the bowl can often be increased by using mechanical aids and vacuum or pneumatic devices.

Parts such as bearing balls will not feed up a bowl track without assistance. To feed them independently, the balls can be loaded into the feeder bowl with a number of pusher parts, such as metal or wooden blocks. The pushers will literally walk the balls up the in-bowl track to the discharge points, where the balls eject into the discharge track, and the pushers drop back into the bowl to assist other balls.

The vibratory-bowl type of feeder controls the number and attitude of parts as they pass a series of engineered obstructions or guides in the bowl track, the ultimate objective being that a sufficient number of correctly oriented parts will arrive at the discharge point of the bowl to satisfy the desired part-feeding rate. Although every feeder bowl is specially developed, a bowl may be converted to handle a slightly different part having the same feed characteristics by changing the hardware in the bowl.

Typical parts that can be fed by vibratory-bowl feeders have been classified into the following categories: (1) cylindrical parts with length greater than 1.2 times the diameter, (2) cylindrical parts with length less than 0.8 times the diameter, (3) cylindrical parts with length ±20% of the diameter, (4) rounded cup parts, (5) disk parts and flat washers, (6) flat rectangular parts, (7) rectangular cup parts, (8) headed parts, and (9) slotted parts.

Figures 23-18 and 23-19 show examples of how a bowl may be selected and tooled to achieve the desired orientation of parts.

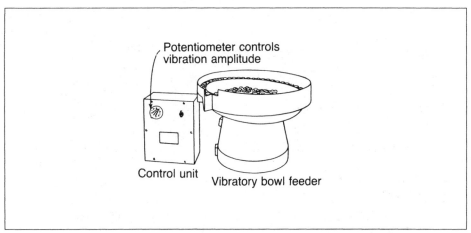

Fig. 23-17. Vibratory-bowl feeder and control unit. (*Automation Devices Inc.*)

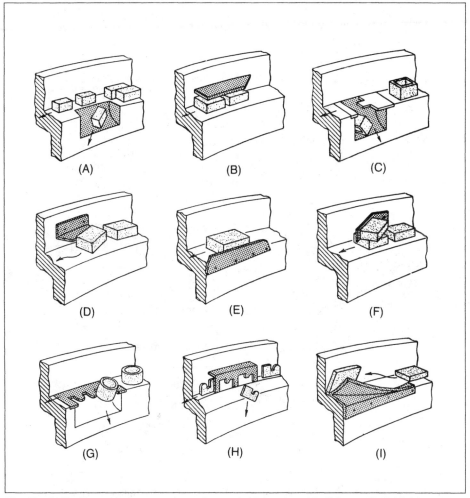

Fig. 23-18. Tooling techniques used to select and orient parts: (*A*) dishout; (*B*) hold-down; (*C*) pocket; (*D*) pressure break; (*E*) retaining rail; (*F*) wiper; (*G*) scallops; (*H*) silhouette; (*I*) roll-up.[9]

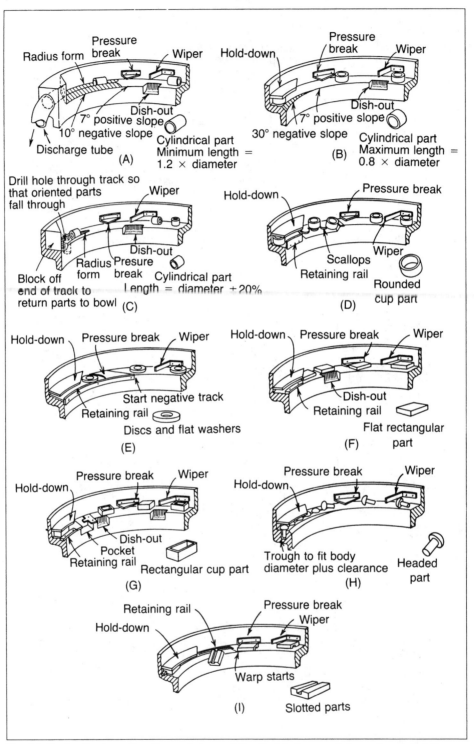

Fig. 23-19. Applications of tooling techniques: (A) cylindrical part, minimum length is 1.2 diameter; (B) cylindrical part, maximum length is 0.8 diameter; (C) cylindrical part, length is diameter ±20%; (D) rounded cup parts; (E) disk and flat washers; (F) flat rectangular parts; (G) rectangular cup parts; (H) headed parts; (I) slotted parts. [9]

UNLOADING DEVICES

The unloading of a finished part from a die can often be achieved by gravity—by allowing the part to fall through the die or slide off an inclined surface in the die, or by inclining the press and die. Larger parts, which may be damaged or those which may require placement in a specific location for subsequent operations, must be removed by hand or mechanical devices.

Air-blast Ejectors. A short blast of air is effective in the removal of small light parts. When the press is near the top of its stroke, a blast of air blows the part out of the die. The air-blast control valve can be actuated by a cam on the crankshaft. This cam is designed to emit a short blast of air just before the top of the stroke or just before the press stops in the case of intermittent operation.

Air-operated Kickers. Parts that are too heavy for air-blast removal may be ejected from the die with some form of air-operated kicker, particularly if the press cannot be inclined. An elementary type is a fast-acting air cylinder which contacts the part through a suitable head. Figure 23-20 illustrates a simple air kicker used to eject a part onto a conveyor after it has been lifted out of the cavity.

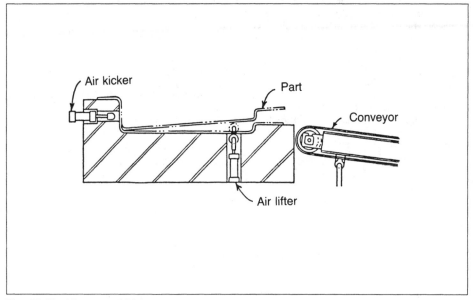

Fig. 23-20. Air-operated kicker.

Mechanical Press-unloading Devices. Many mechanical devices, such as linkages, shedding arms, oscillating arms or trays, pusher slides, and ram-operated devices, have been designed and incorporated into dies to unload the finished parts automatically. Although these devices usually functioned successfully, they were designed for use with a specific die when in a specific press and are difficult to adapt to another die. Therefore, these devices are being replaced with standard unloaders having interchangeable gripping units. The commercial unloaders are actuated by air cylinders, motor driven gear boxes with electrical clutches and brakes, or brushless d-c servo drives. These unloaders can be synchronized with the press stroke. These devices have a swinging or a horizontal reciprocating motion to lift or carry the parts from the dies. The finished parts are raised up out of the die cavities by lifter units built into the die. From this position, jaws grip the parts and carry them from the press onto a conveyor or transfer fixture for feeding a press for a following operation. The normal practice is to synchronize the unloading cycle with the press cycle. Often a common microprocessor-based, programmable logic controller on the press also controls the unloader.[1, 2, 3]

TRANSFER EQUIPMENT BETWEEN PRESSES

Parts may be transferred between presses by conveyors or shuttle-type machines. Conveyors are the simplest means of transferring parts from one press to another, regardless of their orientation to each other. There are many types of commercial conveyors available, such as metal-slat, wooden-slat, rubber- or neoprene-belt, V-belt, and magnetic-belt types.[1, 3]

Turnover Devices. These devices are used to turn a panel upside down as it passes from one press to another. While there are times that the use of a turnover between presses is unavoidable, it must be noted that much of the panel damage and mislocation is chargeable to turnovers.[1]

Transfer Devices. There are many different transfers available for moving a stamping between presses. Each has its own advantages for specific applications. It is of prime importance that complete control of the stamping being handled be maintained at all times.[1]

References

1. Charles Wick, John W. Benedict, and Raymond F. Veilleux, eds., *Tool and Manufacturing Engineers Handbook*, 4th ed., Vol. 2 (Dearborn, MI: Society of Manufacturing Engineers, 1984).
2. *Industrial Robots*, 2nd ed. (Dearborn, MI: Society of Manufacturing Engineers, 1985).
3. Glenn A. Graham, *Automation Encyclopedia: A to Z in Advanced Manufacturing* (Dearborn, MI: Society of Manufacturing Engineers, 1988).
4. Published data of the Liberty Mutual Insurance Co.
5. W.C. Mills, "How to Reduce Costs of Thin Metal Stampings," *American Machinist* (June 3, 1948).
6. C.W. Hinman, *Die Engineering Layout and Formulas* (New York: McGraw-Hill Book Company, 1943).
7. D.A. Smith, "Procedures for Adjusting Dies to a Common Shut Height", SME/FMA Technical Paper TE89-565 (Dearborn, MI: Society of Manufacturing Engineers, 1989).
8. C.C. Kraus, "The Mechanization of Parts Handling," *The Tool Engineer* (May 1950).
9. Floyd E. Smith, "Applying Vibratory Bowl-type Feeders, Part 2. In-bowl Tooling Considerations," *Automation* (December 1962).

SECTION 24
DIE MAINTENANCE— SETTING AND TRYOUT[*]

PLANNING DIE MAINTENANCE

To maintain quality production and to forecast die repair requirements, there must be some method of planning die maintenance. The planning method can be as simple as a single individual in a small shop, such as the press room manager scheduling dies for repair based upon his personal knowledge of problems, or as sophisticated as a computerized data base used to prioritize and schedule needed repairs.

Any good system must take into account key planning factors:
1. Problems and their root causes must be identified.
2. Scheduled inspections and preventive maintenance are important.
3. The capacity to do the repair work must be known.
4. An accurate estimate of the time required to complete each repair is needed.
5. Production scheduling must be considered.
6. There must be a means to prioritize backlogged work.

TEAM METHOD OF STAMPING DIE MAINTENANCE

A Japanese automaker's American integrated stamping and assembly plant depends upon a team approach to solving all die problems relating to quality and productivity. Team interaction is the key to the success of the system. The stamping shop has a total of five transfer press lines, an OBI line, and a blanker. Every die is assigned a home line and seldom runs in other than that line.

Making a Maintenance Request. A three-part Die Maintenance Request form shown in Fig. 24-1 is usually initiated by the transfer press or OBI line production team leader who sketches a description of the problem in space (1) on the form. After signing the form in space (2) he then takes the form to the tool and die team leader for approval.

The team leader reviews the problem with the production team leader, and other persons having knowledge of the problem, as needed. It is his responsibility to plan the required maintenance.

Planning Maintenance. The tool and die team leader must determine the root cause(s) of the problem and identify all dies that will require work to make the correction. The amount of time required to make the correction must also be determined.

Only a two day supply of any given part is usually produced at a time. Production of more than the usual run of parts may be needed to provide additional release time. The tool and die team leader coordinates the required release with the production planning department. When the plan is complete, the tool and die team leader co-signs the form in

*By Jeffrey Gordish, Management Technologies Inc.; Ray Hedding and Eugene J. Narbut, Mazda Motor Manufacturing (USA) Corp.; Roman J. Krygier and Robert W. Prucka, Ford Motor Co.; John McCurdy, W. C. McCurdy Co.; Donald Wilhelm, Helm Instrument Co.

DIE MAINTENANCE REQUEST   *INSPECTION / REPAIR / IMPROVEMENT*
PART 1

		Description of Problem
Part Number		
Part Name		
Press Line		
Die Number		①
Defective Process		
Today's Date	1 Shift ___ 2 Shift___	
Number of Trial Parts/Blanks		Originator ②
Unit Leader		Deadline for Modification ▲

Unit Leader Must Contact Planning Dept. for Dates

DIE MAINTENANCE REQUEST   PART 2

Die Correction Date	D N	D N	Corrective Action	CHECK THE FOLLOWING INSPECTION ITEMS	
Die Correction Time Required				Has Die Been Properly Cleaned?	
Trial Date	D N	D N		Are All Die Parts Secure?	
				No Foreign Items On or In Die	
Trial Time Required				Proper Clearance	
				Are Wear Plates Lubricated?	
Die Maintenance Member	③			Fingers	
				Keeper Pins	
Unit Leader				Lifter Tie Downs	
				Has All Work Been Completed?	
Tryout Area				Has Request Form Been Completed?	
Bench				④	
Tryout Press				▲	
Production Press					

After Correction, Die Maintenance Member Must Check These Items

PRODUCTION   PART 3

		Results of Correction
Production Date	D N	
Unit Leader		⑤
Operation		

White: INFORMATION TO LINE AFTER CORRECTION
(Line → Die Maintenance→ Crane Operator/Die Setter By Mail Post → Line Must Check Results → Die Maintenance)

Fig. 24-1. Three-part form used in a manual die maintenance planning and tracking system. (*Mazda Motor Manufacturing (USA) Corp.*)

blank (3).

One of the three copies is posted on a cork board at the home line and the other two copies are inserted into a pipe attached to the die for that purpose. The required number of sample parts needed for evaluation and tryout are shipped to the dieroom, together with the die(s), by the production team leader. The parts are identified and stored in a designated area.

Tracking and Prioritizing Requests. At the start and near the end of each shift, both the production team leader and the tool and die team leader meet to review problems, including a review of all outstanding maintenance requests posted on the line's cork board. The outstanding requests are prioritized based upon the extent to which quality and production are affected. There are times when the entire correction needed cannot be made due to the size of the job and urgency of other requests. In such a case, the plan may be reduced to a partial correction, with a complete correction scheduled for a later date.

Correction and Follow Up. The diemaker making the repairs records the corrective action on the middle part of the form (4) and fills out the list of inspection items. After the correction is completed, the second copy of the die maintenance request is filed in the die room records for that die and the first copy placed in the pipe attached to the die. When the die is returned to the press line, the first copy is placed in a plastic sleeve attached to the cork board.

The lower space on the form (5) is filled out by the production team leader and line team leader after the success of the repair has been evaluated during a production run. The white copy is then retained in the die repair area as a part of the die's maintenance history.

Cooperation Required. The Japanese system of manufacturing places great emphasis upon each individual being part of a team responsible for quality and productivity. This maintenance system depends upon the cooperation of all employees.

COMPUTERIZED DATABASE DIE MAINTENANCE

A large automaker has improved quality and reduced costs by implementing a computerized die maintenance program in its stamping plants. The basic program strategy was to reduce the amount of breakdown repairs and increase the amount of planned maintenance in order to increase throughput and quality.

The approach revolves around six key steps in the die maintenance workcycle (Fig. 24-2). The work needed is identified and priorities set. The work is planned, followed by scheduling and assigning. Progress is reported by each shift. At the completion of the workcycle, the work performed is evaluated and entered into the history file, by die number.

Daily Operating System. With this maintenance work cycle applied to a typical large tool and die department, a daily operating system has been developed (Fig. 24-3).

First, the work is identified by either preventive maintenance or operational symptoms. A work order is then initiated and entered into the maintenance management system (MMS) die backlog.

Next, a daily planning meeting with personnel from maintenance, operations, and production scheduling is held so that the work order can be planned, prioritized, and scheduled. This planning meeting is a key link in the communications process.

Finally, the work is performed and entered by the planner into history. This provides a basis for analysis of maintenance costs, engineering capabilities and SPC data.

How the Data Is Used. First, a daily work plan is printed out that the supervisor uses to manage his area. It replaces hand written line ups that were formerly passed from shift-to-shift, and clearly avoids misinterpretation. It provides such information as part number, die number, work order number, brief description of work to be done, date required, and a space for hours worked.

Next, there is a die history sheet which provides not only maintenance labor and material costs, but also signals repetitive problems.

Finally, reports can be generated from the database by cause code, type of die, a

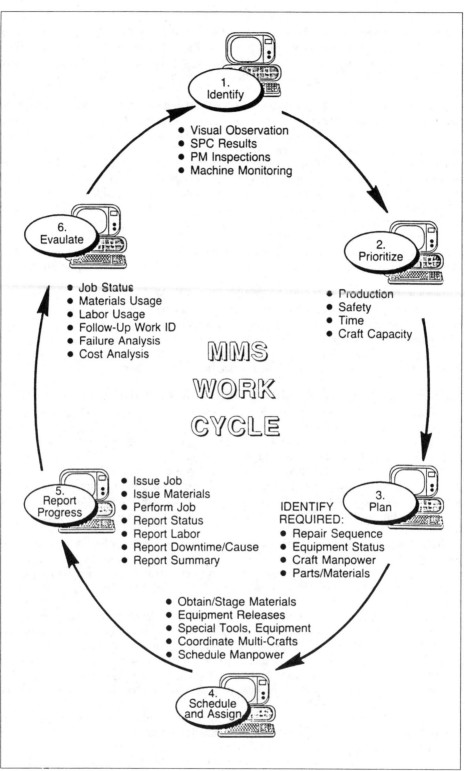

Fig. 24-2. Six key steps in a computerized die maintenance management system work-cycle. (*Management Technologies Inc.*)

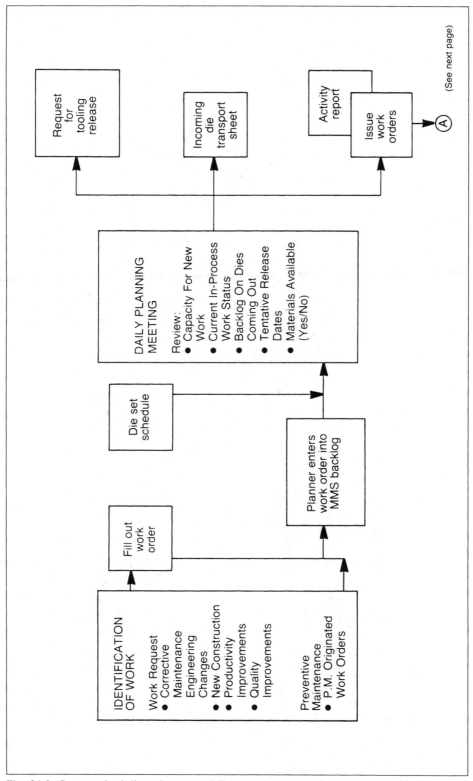

Fig. 24-3. Computerized die maintenance daily operating system flow-chart. (*Ford Motor Co.*)

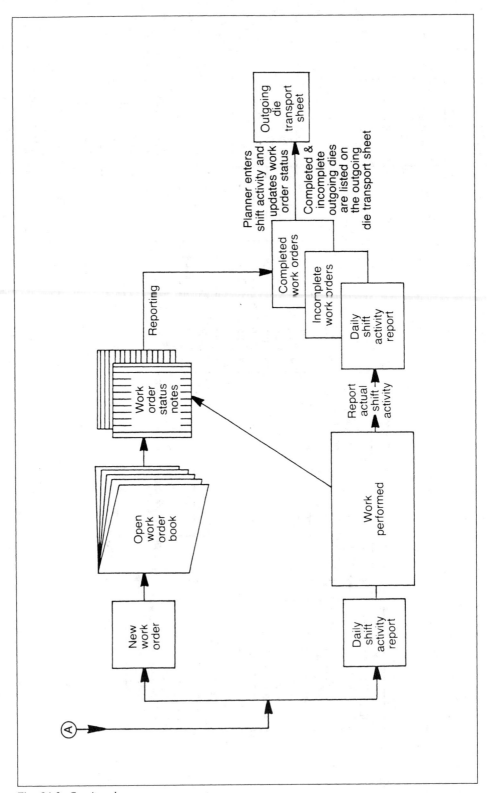

Fig. 24-3. *Continued*.

specific die, cause code as a percentage of total work, and work location.

Additional Features. The MMS ties in with the plant's computer integrated manufacturing (CIM) system. Dieset schedules are prepared, based upon production requirements. The dieset scheduling system (DSS) provides reports on open and completed work orders on tooling scheduled for dieset. The DSS also can forecast future dieset schedules, an important planning tool. Reports on die and automation changeover time performance are also generated to aid in evaluating continuous improvement of manufacturing methods.

To further aid in reducing set up times, a line changeover performance system (LPC) has been developed. Computer-generated data reports on actual set up times, and documents problems that occur during diesetting or upon start up of production.[1]

Choice of Systems. Manual maintenance tracking systems have a limited capacity for information storage and retrieval. Reports on such factors as problem causality and man hours expended are often needed for management decision making. These can be very time consuming to generate manually.

Manual systems generally work very well in shops having up to 15 employees in a repair activity. The actual number depends upon the degree of employee-management cooperative interaction. Manual systems that work well in very large shops often depend on breaking down the maintenance record keeping activity into sub-activities.

The manual system described in this section works well because of strong teamwork, discipline, and commitment to a high and ever-improving quality standard.

The information accessibility needed for wise resource management and efficient plant administration has resulted in CIM systems finding widespread application. Employee acceptance generally is excellent, provided they are involved in the early phases of the system implementation.

The use of a maintenance management data base as part of a CIM system can provide a means to improve quality, reduce cost, and even bring about a needed cultural change.

NEW DIE TRYOUT

The process of adapting new automotive stamping dies to production presses and automation was once very time-onsuming and expensive. The major portion of the actual finishing work was done as part of the die tryout program. For example, metal clearance was hand ground out of draw die reverses and flange die pads. This required days of re-stoning and polishing. Extensive milling of die shoes to provide clearance for automation arms and scrap shed was normal. It was not uncommon to need to relocate a cam to a different operation, or even add an operation because a part could not be extracted from the die. To avoid delaying a program, conservative design practices often resulted in more operations than necessary.

Careful planning and teamwork in the product design, die design, and die construction phases of the program has helped reduce these costs. The use of the high-speed digital computer as a design, manufacturing, and information management tool has been an important factor.

Product Design Factors. This is where the greatest potential savings over the model run is possible. Reduction in component count, complexity, and draw severity all aid manufacturability. Part designs that require deep draws with high strains can add greatly to die tryout time. There are several good sheet metal formability analysis programs available for high-speed digital computers that allow accurate predictions of maximum strains (Sections 3 and 11).

Die Design Factors. Computer-aided design (CAD) correctly applied can avoid many of the design errors that often result in extensive die reworking during a die tryout program. CAD equipment can simulate die, automation, and part motion and assure that proper clearances are provided in the design stage.

Three-axis transfer presses are increasingly used for production of large, automotive body panels. The paths of the transfer rails and fingers in relationship to the die surfaces, are difficult for the die designer to visualize. CAD equipment is used to provide an animated simulation of the three-axis motion of the transfer automation in relationship to die surfaces at any degree of press crankshaft rotation. Any interference that exceeds a specified safety factor can be highlighted (Section 3).

Coordination with the Construction Source. If the dies are correctly designed and constructed, the time required for new die tryout should be minimal. Currently, a common requirement is for the construction source to furnish completed sample panels that agree with the part print.

Traditional Die Tryout Procedure. At one time, the purpose of new die tryout was to take the finished dies from the die construction source that had never been used to form metal and "try out" the dies in production presses. It was considered normal for the production plant to find that the first panels produced had many serious imperfections such as fractures, wrinkles, tight spots, and buckles.

Usually, all dies were routinely spotted by using prussian blue and hand ground using pneumatic grinders. One half of the die, usually the post, or draw punch and upper blankholder in the case of a draw die, was considered the master surface. The prussian blue is applied to the master surface and the die closed.

The prussian blue from the master surface leaves spots on the mating half wherever metal-to-metal contact occurs. After a satisfactory amount of even bearing has been obtained, metal clearance is spotted out.

To grind the needed clearance by hand, a panel is coated on both sides with prussian blue and hit in the die. Spots of blue are transferred to both die surfaces. The panel is examined to determine which spots are true hard bearing, as evidenced by a spot on both sides of the panel in the same place. Only the corresponding true bearing spots on the die half being worked are ground. This procedure is repeated until uniform bearing is obtained over the entire working surface.

Commonality of Shut Height. Checks to verify common die shut height dimensions should be made during both die construction and die tryout. A good way to do this is with a calibrated end measuring rod. This measuring rod can serve as a standard, and if equipped with a dial indicator, can be used to verify that the press is level.

Detailed information of the strategy for obtaining commonality of shut heights is covered later in this section.

Press Alignment. A major source of difficulty when trying out new dies can be traced to press out-of-level problems. The presses that are available for die tryout are often older presses in poor repair.

Experience has shown that even badly worn presses can be used for high-quality die tryout work provided that they are adjusted to be level under load with load cells.[2]

A nominal target figure for press leveling is a maximum error of 0.001 in. (0.02 mm) per foot (305 mm). Provided that the press bed is properly leveled, and in good condition, accuracies as close as 0.008 in (0.20 mm) from corner to corner on a 180 in. (4.6 m) press have been routinely maintained.

The construction source should also have a program to maintain their spotting and tryout presses in good level condition. Verifying this can be an important part of the cooperative follow up work between the production plant and the construction source.

Press Speed Factors. Many presses have a provision to be inched at very low speeds. One term for this feature is micro-inch. Producing test stampings in this manner during die tryout can produce results that cannot be duplicated under full production speeds.

Not only do many drawn stampings show a difference in work hardening rates, but the dynamic mass of the press slide causes a speed-dependent tonnage variation. Very large transfer presses having slide tonnage ratings over 2,000 tons (17.8 MN) have shown a difference of 0.040-in. (1.02 mm) in actual slide position between micro-inch and 16 strokes per minute (spm). Because of this factor, to properly clinch pierce nuts on micro-inch, the slide may need to be lowered.

The root cause is the difference in kinetic energy of the slide under the different speeds.

This effect cannot be compensated for by press counterbalance adjustments. The actual amount of difference can be estimated if the reverse load generated when the slide stops at bottom of stroke and incremental deflection factor of the press are known.

Project Management and Integration. Critical path analysis is a useful tool for determining the required time for a project such as die construction and die tryout (Section 3). Several types of software are available for use in personal computers that are powerful enough to keep track of the critical factors and also report progress and costs.

STEPS TO SET A PROGRESSIVE DIE

Steps Developed by Expert Diesetters. An excellent management procedure for accomplishing cultural change through cooperative interaction with the workforce is to let the employees write their own work-rules. The following rules were developed during a training class at a modern contract stamping shop that is a first-tier supplier to major automotive manufacturers.[3]

1. ACTION: Get ready. Clean up bolster, feeder, and around press. Clean top and bottom of die.
   REASON: A clean area is a safer area. Keep the area clean so no one slips on slugs and gets hurt. Make sure that there are no slugs on die or bolster to spoil parallelism.

2. ACTION: When removing a die, raise slide slightly before inching press on bottom.
   REASON: To avoid sticking press on bottom.

3. ACTION: Inch press on bottom. Adjust for correct shut height plus a small safety factor. Place new die in correct location square with bolster and feeder. If parallels are used, make sure that they are correct for the job. Tie down with secure approved bolts or clamps.
   REASON: To insure a safe repeatable setup.

4. ACTION: If air pins are used, make sure that they are all the correct length. Make sure that press cushion is properly equalized.
   REASON: Unequal or improper pins will cock the die pad or draw ring. The load height will not be correct. The press cushion will be damaged.

5. ACTION: Adjust feed height and lineup to die. Adjust feed pitch to specification stamped on the die.
   REASON: To get the set-up to run without trial and error.

6. ACTION: Place chutes, scrap hopper, and parts container in proper location.
   REASON: To avoid running parts and scrap on the floor.

7. ACTION: Start coil and adjust stock straightener. Feed stock into die to correct location. Make first hit. Advance stock with feeder. Make sure location is correct.
   REASON: To insure that set-up is correct and to be sure not to damage die.

8. ACTION: Run stock through the die and recheck stop blocks. If available, check tonnage meter against specifications stamped on the die. If tonnage reading is not correct, find out why before running production.
   REASON: The second adjustment is to compensate for press deflection under load. Correct tonnage means that the target set-up was repeated.

9. ACTION: Get gages and check first parts. Submit parts for quality control approval only if they are OK. If not, correct reason for parts not being OK.
   REASON: It is a waste of time to submit off-standard parts for quality control approval.

10. ACTION: Double-check set-up. Install part blow-off if needed. Make sure everything runs OK.
    REASON: To insure that set-up is production ready.

COUNTERBALANCE AIR ADJUSTMENT

The correct adjustment of the press counterbalance air pressure is required for the safe operation of a press.

The purpose of the press counterbalance system is to offset or counterbalance the weight of the press slide and upper die.

A pressure gage and adjustable air regulator is provided on the press to permit accurate adjustment to the correct setting. If the setting is too high, excessive clutch wear results. Low settings cause excessive holding brake wear and can cause the holding brake to overheat dangerously.

Many presses have a chart that gives the correct pressure setting for various upper die weights. Dies should have the upper, lower, and total die weight clearly identified to facilitate both safe die handling and correct counterbalance setting.

Common Errors in Counterbalance Adjustment. Large presses have big surge tanks that take a long time to fill. A common mistake is to make a big change in the regulator setting when only a small change, followed by a wait of several minutes to allow the system to stabilize, is needed.

A common source of incorrect pressure settings is caused by inaccurate or missing gages. Press vibration and pressure pulsations can ruin the accuracy of a low-grade gage in a short time. To avoid this problem, only a high-quality, liquid-filled gage with a built-in pulsation snubber should be used.

Some newer presses designed for quick die change feature automatic counterbalance adjustment based upon a computerized database of die numbers. In most cases, the correct pressure must still be determined and entered into the database. Failing to update the database and relying on manual adjustment after problems develop is a common error.

Setting Counterbalance Pressure with an Ammeter. Checking the drive motor current while the press is cycled is an accurate way to find the right air pressure setting. A reading that increases as the slide descends and drops sharply on the upstroke indicates that the pressure is too high.

Amperage readings that are high on the upstroke and increase as the top-of-stroke is approached mean that the setting is too low. When ammeter readings are checked, remember that the press motor must supply the energy lost by the flywheel as the press rolls through the bottom of the stroke. A current surge is normal when that occurs.

Press stroke per minute indicators that operate by measuring the rpm of the drive motor, flywheel, or crankshaft can be used in place of an ammeter. If the press speeds up on the downstroke, the setting is too low. A loss of speed means the air pressure is too high.

Counterbalance Adjustment with a Dial Indicator. The dial indicator method can be used to establish correct counterbalance pressure if the press counterbalance adjustment information is not available. To use this method:

1. Stop the press at 90° on the downstroke.
2. Follow proper lockout procedure.
3. Exhaust the air from the counterbalance but not the holding brake.
4. Place a dial indicator with a magnetic base on the slide so the tip of the indicator touches the slide.
5. Slowly raise the counterbalance air pressure until the dial indicator shows that the counterbalance has lifted the slide.
6. Make a record of this setting so this procedure will not need to be repeated.

This procedure usually establishes the setting very close to that of the ammeter method.[3]

DIE DAMAGE AVOIDANCE

Shorter than normal production runs between die resharpening and refurbishing are often traceable to damage resulting from misfeeds, poor diesetting practices, or bad press conditions. A single mis-hit that results in a sheared cutting edge on a complex die can shorten the expected amount of production from over 100,000 pieces between resharpening to well under 10,000.

As a rule, any lateral force generated in the die should be controlled in the die. This is the function of guide pins and heel blocks. Bad press conditions such as excessive bearing and gibbing clearance can result in lateral forces being transmitted to the die, displacing its precision parts.

Out-of-level conditions can be caused by many factors with the result always being reduced runs between die rework.

Out-of-parallel Conditions. Figure 24-4 illustrates a press slide that is not square with the column or parallel with the press bed. This condition can result from mis-adjustment of the screws or a failed bearing.

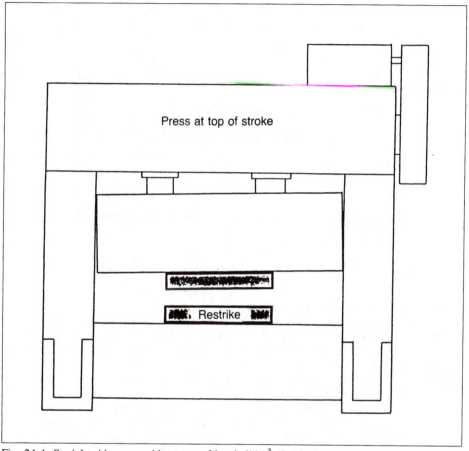

Fig. 24-4. Straight side press with an out-of-level slide.[3] (*Smith & Associates*)

Upon die closure (Fig. 24-5) only the left side of the die closes completely. Reducing press shut height to get the right side to close may result in the condition shown in Fig. 24-6. The left side of the press is badly overloaded resulting in the crown lifting from the column. Continued operation in this manner will damage the die and greatly exacerbate the press problem.

Die Damage. Figure 24-7 shows how a bottoming block can embed in a die shoe as a result of operating a die in an out-of-level press. If the die shoes are not supported firmly by the bolster and slide adaptor plate or parallels, the die shoes can be bent. Also, the upset metal in the shoes will displace and distort the pin and bushing holes resulting in

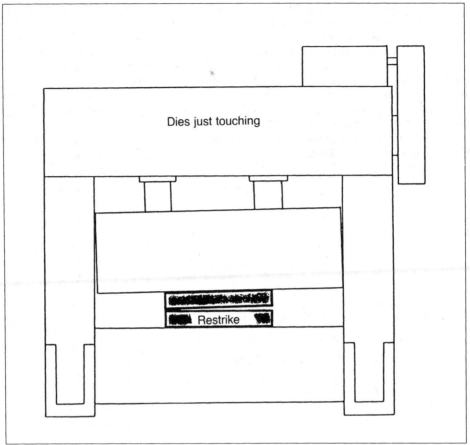

Fig. 24-5. Due to a slide out-of-level condition, only one side of the die makes contact.[3] (*Smith & Associates*)

rapid bushing wear and loss of alignment.

Figure 24-8 illustrates the die damage that can result from attempting to run a pierce die in an out-of-level press. The quality of the part produced may not be acceptable because of a loose burr that will be produced on the left hand side of the pierced opening and a distorted edge on the right hand side.

Stop or equalizing blocks on the corners of the die shoes to limit die closure will correct for minor slide misalignments in some classes of dies. However, damage in cutting dies can still occur, since the cutting is done before the equalizing blocks are contacted.

A figure of permissible slide-to-bed misalignment that is often cited is 0.001 in. per foot or 0.1 mm per meter. A typical straight side press has an incremental deflection factor of 0.001 in. per ton (0.02 mm per 8.9 kN) per corner, or four tons (35.6 kN) total increase in tonnage.

For example, a misalignment of this magnitude could be compensated for in a 48 in. (1.2 m) wide press by using 2 in. (51 mm) diameter stop blocks loaded to a maximum of four tons (35.6 kN) each. Greater misalignments will require proportionally higher tonnages to be wasted on equalizing blocks. A limiting factor in how much misalignment can be tolerated, is how much tonnage can be wasted and still have enough left to make the stamping.

Figure 6-16 illustrates how a press out-of-level condition can result in off-angle flanges.[2]

Foreign Object Under Die. A slug or other foreign object under the die has the same

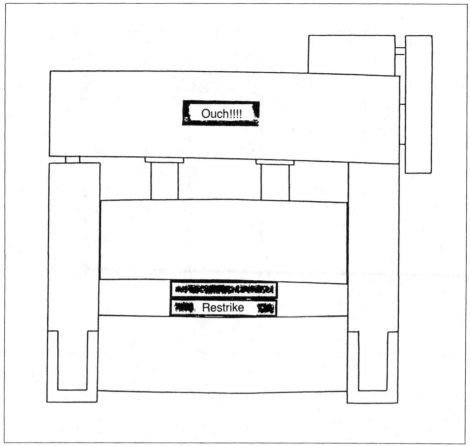

Fig. 24-6. Attempts to form a part in an out-of-level press can damage both the press and die.[3] (*Smith & Associates*)

effect on tooling reliability as an out-of-level slide. Figure 24-9A shows a view of a slug under a die shoe. After the production run, the slug has become imbedded in both the lower die shoe and bolster (Fig. 24-9B). Even though the slug is removed, upset metal remains on the die shoe and bolster where the slug was embedded (Fig. 24-9C).

The area of upset metal should be repaired by peening followed by filing or grinding the spots smooth.

Die Out-of-level. Die out of level conditions are another source of difficulty in stamping shops. There are a number of causes of this condition. Dies that have been spotted or fitted in an out-of-parallel press will reflect the exact out-of-parallel condition of the press. The problem may not be noticed until the die is set in a level press, or worse yet, a press that is out-of-parallel the opposite way. A rigorous program to keep all presses in good alignment is an aid to maintaining stamping flexibility.

Dies may be mis-machined or the die details improperly fitted. It is desirable to check all dies leaving the toolroom after any type of rework that could affect the parallelism of the shoes.

Offset Loading. The ideal condition is to always keep the die load exactly centered under the slide. This is not always possible. Operator loading requirements, automation limitations and scrap chute placement are common reasons why dies are not always centered in a press. In the case of many progressive dies, the heavy loading occurs on the end of the die where the parts exit. In such a case, the die should be offset in order to balance the press loading.

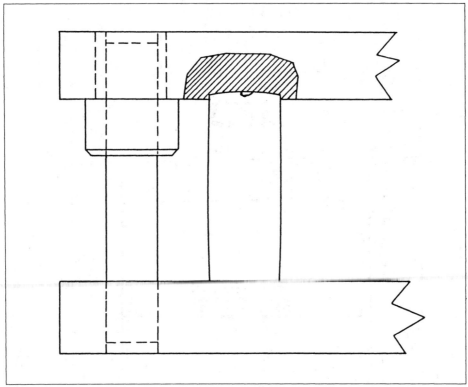

Fig. 24-7. Set-up blocks can become embedded into the die shoe due to press or die out-of-level conditions.[3] (*Smith & Associates*)

Presses are designed to withstand some offset loading. Single point presses are much less able to do so than are presses with multiple connections. Offset loading always causes some degree of out-of-parallel condition. The best plan by far is to balance the loading, particularly in the case of cutting dies.

Figure 24-10 is an exaggerated view of how an offset die load forces the slide out-of-parallel. The die damage that often results is of the type shown in Fig. 24-8.

A possible solution is providing a stock guide to permit the die to be centered (Fig. 24-11).

Taguchi's Parabolic Loss Theory. The concept of striving for perfection in symmetry of press loading should be a part of every continuous improvement program. Figure 24-12 illustrates Dr. Genichi Taguchi's classic parabolic loss curve with upper and lower tolerance limits superimposed. The essence of the theory is that the cost of deviation from the target value increases as the square of the distance from the target value.

In applying this theory to press alignment and symmetry of loading, one should expect the cost of press and die maintenance attributable to offset loading and out-of-parallel conditions to increase in geometric proportion to the severity of the condition. This concept agrees with experience. The Section 3 die engineering formulas show that the die has a very limited ability to align a press slide, and changes in working clearances in the die will result. The latter effect will result in product variability.[2, 4]

Misfeeds. There are many causes of misfeed damage. Failure to eject the part from the die can cause multiple hits. More than one blank can be fed into a die due to blanks sticking together. Slugs can cock in the die opening and cause the next blank to be mislocated.

The best solution where misfeeds in automatic operations can cause damage is to use the most foolproof feeding and ejection systems available and back them up with sensors

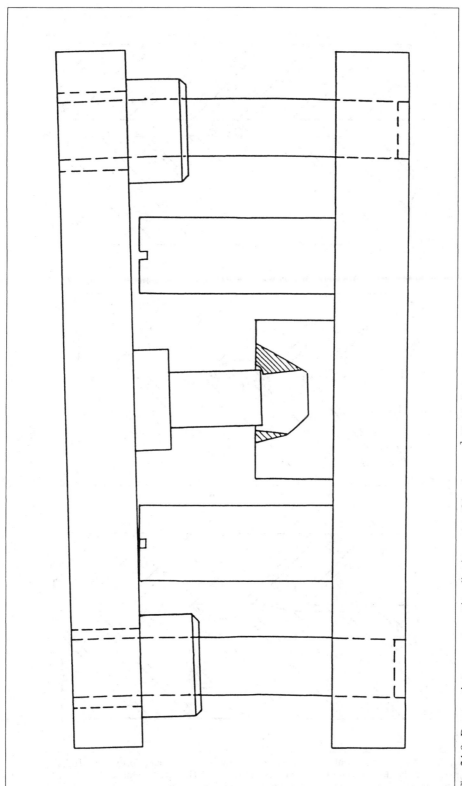

Fig. 24-8. Damage that occurs to cutting dies in an out-of-level press.[3] (*Smith & Associates*)

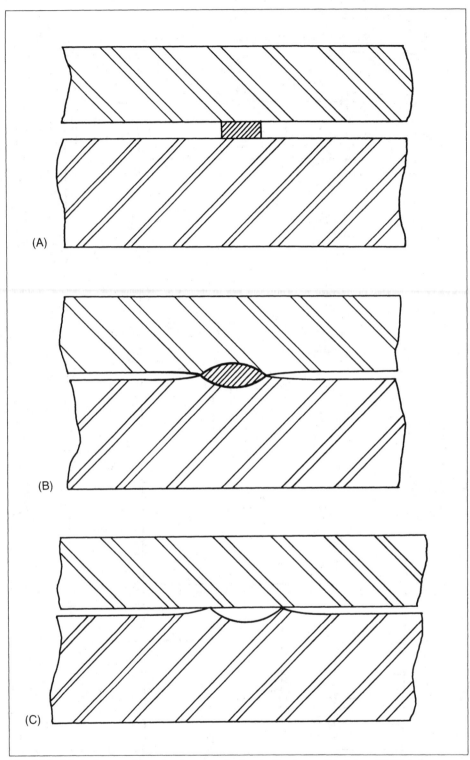

Fig. 24-9. (A) A die set with a slug between the lower die shoe and bolster. (B) The slug imbeds into the die shoe and bolster during the run. (C) The upset metal remains to cause out-of-parallel problems after the next dieset.[3] (*Smith & Associates*)

24-16

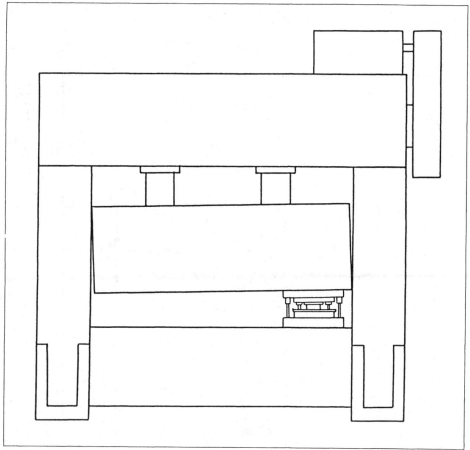

Fig. 24-10. A die set off center causes a slide out of parallel condition.[3] (*Smith & Associates*)

interlocked with the press cycle circuitry. Sections 23 and 26 cover misfeed protection systems in detail.

Damage In Shearing Dies. Figure 24-13A shows a mislocated piece of stock laying in a pierce die. Many press operators do not understand the serious die damage such a mis-hit condition can cause.

As the press cycles, the center punch is deflected to the left (Fig.24-14B) and the stock enters the die opening on the right side of the center die block.

The resulting damage is shown in Fig. 24-15C. Damage of this type is often innocently reported as a broken punch. Operator training is an often overlooked but worth-while investment.

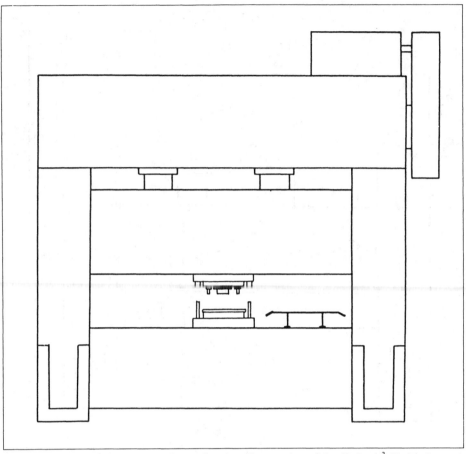

Fig. 24-11. The use of a stock guiding rack to permit centering of the die load.[3] (*Smith & Associates*)

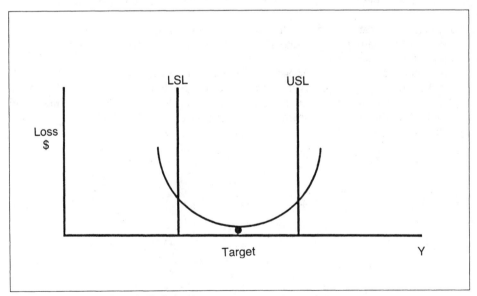

Fig. 24-12. Taguchi's parabolic loss theory.

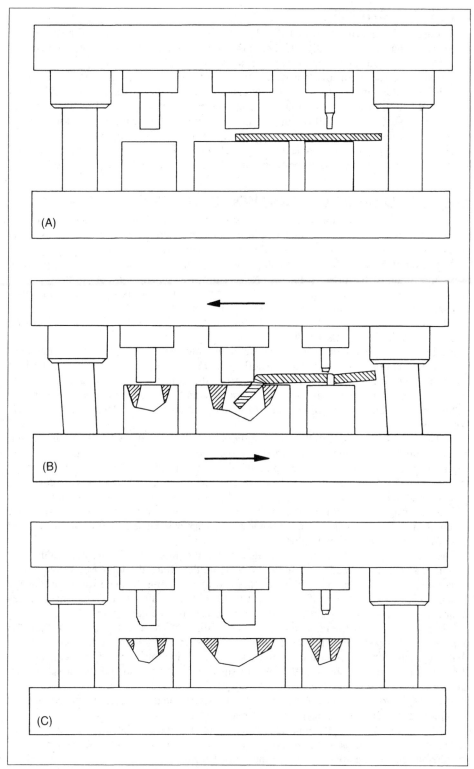

Fig. 24-13. (*A*) A piece of stock out of location in a die. (*B*) The die deflects laterally due to the unbalanced cutting force. (*C*) The resultant damage.[3] (*Smith & Associates*)

QUICK DIE CHANGE

A conservative approach to quick die change is the correct one. Claims of the economic necessity for hardware related QDC methods in order to remain competitive are just idle talk unless it can be shown that a payback actually exists.

For example, adoption of improved clamping methods such as elimination of straps and adoption of a common tie down height can reduce set up time for a straight side press from 45 minutes to under 20 minutes. Adoption of common load and shut height can reduce this figure to under 10 minutes, with the major factor being the threading of the stock. All of these items are low in cost and can be a part of an ongoing standardization program. If you go beyond this level, a careful study of economic order quantities and actual cost of press time should be made in order to justify further expenditures.

ADJUSTING DIES TO A COMMON SHUT HEIGHT

An important way to improve stamping flexibility that has widespread acceptance is to maintain all dies at a common shut height or standard increments of shut heights. This avoids the need to make shut height adjustments when exchanging dies.

Adjustment Procedure. The first step is to decide upon the common shut height to be used, and to adjust the press to this figure with the slide exactly on bottom dead center. An accurate inside micrometer or calibrated end measuring rod used with a toolmaker's adjustable parallel is a good way to make this measurement.

All dies are then designed and built to this common shut height dimension. Existing dies can be adjusted to this dimension by shimming or machining.

Static die and press dimensional measurements alone may not achieve accurate shut heights. Corrections for die and press deflection may be required. In general, the best procedure is to increase the die shut height from the nominal common value enough to obtain the required tonnage. If the required tonnage and the incremental deflection factor of the the press are known, the amount of compensation can be calculated (Section 27, Press Data).

A Tonnage Meter is Helpful. Set the die and bottom carefully to the common shut height while observing the tonnage readings to avoid overload damage. Note the tonnage required to produce a part at the common shut height setting. Compare this tonnage reading with that which is known to be correct for the process. Low readings are corrected by adding shims and high readings corrected by removing shims or re-machining.[4, 5]

Maintaining a Common Shut Height. Once common shut heights are achieved for a group of dies, factors such as die wear and die rework can cause a drift or abrupt change from the established nominal value.

Die wear may result in the shut height becoming less than the normal value. For example, as the draw beads on a draw die blankholder wear, the increased pressure that is required to properly hold the blank is obtained by adjusting the slide for more tonnage which decreases shut height. This is also true of a restrike die. As the working surfaces wear away, increased pressure is needed to produce an acceptable part.

After rework, adjustment of die shims or milling of the die shoe(s) may again be required. In this case, checking the die in a tryout press having an exact means to measure shut height and tonnage prior to returning the die to production, will help to avoid delays. Reworked dies must be bottomed carefully since it must be assumed that a substantial change in shut height may have resulted.

Fine Adjustments. There are several ways to determine the amount of fine adjustment to shut height needed to achieve correct tonnage values.

If the press is equipped with an accurate shut height indicator, the amount of adjustment required to achieve the correct tonnage is exactly equal to the amount of die shut height change needed to achieve the correct figure.

An end measuring rod or inside micrometer can be used to determine the difference between the common shut height figure and the one at which the process tonnage is correct. This is exactly the amount of adjustment needed.

If the incremental deflection factor of the press is known, it can be used directly to determine the amount of shut height change required for the desired tonnage change.

Procedure for Transfer Presses. Transfer presses that have a number of separate dies under a common slide require special consideration to adjust the dies to a common shut height.

The reason for this is that the bed of the transfer press deflects downward much like an archer's bow when stressed. The ram does much the same to a lesser extent depending upon where the loading is in relationship to the pitmans or plungers.

As a result of this deflection under loading, additional tonnage may be required to get the middle dies on bottom. This excessive tonnage results in extreme compression of the other dies. Reduced part quality often results due to coining marks. Extra tonnage can be self-defeating in that it just results in even more press deflection.

One solution is to design the process so that high tonnage operations such as forming, embossing , and coining are at the ends of the press bed. Dies that complete their work before full press closure, such as trim, pierce, and flange dies, are then located in the center of the press. The bed deflection condition is not a problem, since the flanging and trimming operations are completed before the press is completely closed.

Another solution is to shim the dies in the middle of the press to compensate for press deflection.The tonnage for each die when operated alone in a conventional press at a common shut height should equal the total required to produce a good part in a transfer press. However, the required tonnage will very likely be higher, and the quality lower.

Once the dies can be operated in the transfer press at the total tonnage that is known from the database(s) as being correct for the dies operated conventionally, the correctness of the individual shut heights may be assumed.

If fine tuning of the die shut heights is needed to achieve the correct total tonnage, it can be done by carefully examining the part(s) hit in each individual die for hard marks or other signs of over-bottoming. If equalizer blocks are used, check to see if they may be imprinting the striking surfaces. Such a visual examination can provide the basis for a cut and try approach to establishing correct individual shut heights.

If an individual die in the transfer press is suspected of having too much shut height, remove that die and hit the other dies with parts in them while noting how much the total tonnage drops. The amount by which the drop exceeds the measured process tonnage for that die is the excess that must be reduced by lowering that die's shut height.

In-die Force Monitoring in Transfer Presses. Guesswork as to the fine tuning of transfer press shut heights can be greatly lessened if the compression of a die member is measured with a bolt on strain transducer or removable bore hole probe when the die is in tryout in a conventional press.

A good location is in or on the equalizer block(s) or some other member that is compressed in a known way during normal die functioning.

The object is to adjust shut heights of the individual dies to duplicate the strain measurements that were acceptable when the dies were operated in a tryout press. If the readings are given in terms of microstrain, the actual loading of the member can be estimated from the cross-sectional area and type of material, by using simple engineering formulas.

Important Points to Remember. Maintaining dies at a common shut height, or groups of shut heights, will provide benefits that go beyond improvements in stamping flexibility and set-up time reduction. Die change damage avoidance and part quality improvement can also result from adoption of this method.

Adjusting dies to a common shut height may require more work than simply milling or shimming them to a common figure. Even so, the procedure is simple. Never view the procedure as a black art requiring laborious die spotting and shimming in the press.

Several important points are:

1. Always standardize the press shut height with the press empty and on bottom dead

center.

2. As a general rule, flange, pierce, and trim dies can have the same shut height as the press.

3. Presses must deflect to develop tonnage.

4. Dies that require high tonnages may require a slight increase in shut height to compensate for press deflection.

5. If both the press incremental deflection factor, and tonnage required are known, the exact die shut height can be calculated.

6. Multi-die transfer presses require special consideration, but the procedure is still straightforward.

SAFETY WHEN WORKING ON DIES IN PRESSES

Safety of personnel is beyond the scope of this handbook. There are extensive government regulations on the subject that are changed much more frequently than this handbook is revised. A few examples of how manufacturers strive to avoid injury to the personnel who must work on dies that are set in presses will be presented. Pressroom accidents are totally avoidable, and that must be everyone's goal.

Safety Blocks. Extruded magnesium is an excellent material for blocking press slides in the open position. The material is very predictable in compressive loading, and the weight to strength ratio is exceeded only by a few exotic metal composites that are as costly as precious metals.

A few comments for their application:

1. Two blocks should be used and placed across diagonal corners of the press.

2. Safety block pads should be provided in the die to simplify correct block placement.

3. An informal automotive stamping industry standard is to provide a minimum of two safety block pads having a closed height of 6 in. (152 mm).

4. The safety blocks are sized 4.5 in. (114 mm) longer than the press stroke.

5. Hardwood wedges are used to fill in the remaining 1.5 in. (38 mm) space to prevent the slide gaining inertia before contacting the blocks.

6. The compressive strength of the blocks must be sufficient to support the weight of the press slide, attached linkage, and upper die with a safety factor.

7. Since the blocks are not designed to withstand full press tonnage, a safety plug which interrupts the main run circuit is attached by a short chain—the block must not be placed in the die-space with the press running.

8. The compressive strength of the die safety block pads must be sufficient to equal or exceed that of the safety blocks in any press into which the die may be set.

9. Since the length of the blocks is determined by the press stroke, the blocks should be conspicuously identified to prevent use in a press other than the one intended.

10. The blocks should be viewed as the lifeboats on a ship—they should be used for no other purpose.

Figure 24-14 illustrates safety blocks made from commercially available extruded magnesium stock. The aluminum end plates are made from 6061-T6 aluminum plate. These plates become worn and mis-shaped with use and should be replaced. The screws holding the end plates should be secured with medium-strength, thread locking compound to avoid thread damage.

Safety Locks. When working on a die in a press, a means must be provided to lock out the power source. Each employee must have an individual safety lock for this purpose. Multiple safety locks may be required to lock out the energy sources of automation equipment, if the unexpected movement of such equipment could endanger an employee.

Multiple lockout devices to permit each employee assigned to the job to lock out the equipment must be provided.

Die Clamping. To prevent the possible shifting of the die shoe during the operation of the press, the die shoe should be securely clamped to the bolster plate.

If it is necessary to use strap clamps, the hold-down bolt should be as close to the die shoe as possible (Fig. 24-15).

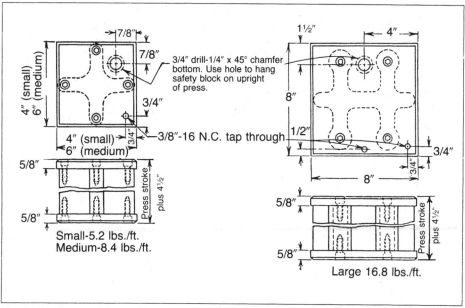

Fig. 24-14. Extruded magnesium safety blocks fitted with aluminum end plates. (*Ford Motor Co.*)

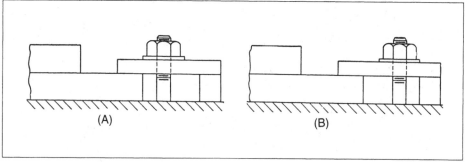

Fig. 24-15. Clamping of die shoe to bolster plate: (*A*) Correct practice with bolt as close to lower shoe as possible; (*B*) Incorrect practice with bolt too far away from lower shoe.

References

1. R.J. Krygier, "Maintenance Management, a Review of the Program Installed at Body Operations Plants, Ford Motor Company", Presented at Society of Manufacturing Engineers, Dearborn, Michigan (August 25, 1989).
2. D.A. Smith, "Why Press Slide Out of Parallel Problems Affect Part Quality and Available Tonnage," SME Technical Paper TE88-442 (Dearborn, MI: Society of Manufacturing Engineers, 1988).
3. D.A. Smith, Training material used through Oakland County (Michigan) Community College and W. S. McCurdy Company, 1989.
4. D.F. Wilhelm, "Measuring Loads On Mechanical Presses," SME Conference Proceeding, Society of Manufacturing Engineers, Dearborn, Michigan, 1988.
5. D.A. Smith, *Adjusting Dies to a Common Shut Height*, SME/FMA Technical Paper TE89-565 (Dearborn, MI: Society of Manufacturing Engineers, 1989).

SECTION 25

LUBRICANTS FOR PRESSWORKING OPERATIONS*

The selection of the proper pressworking lubricant for a specific application hinges on several major factors such as the condition of purchased material, type of tooling, application techniques, environmental restrictions, cleaning and finishing, permitted disposal method, and worker acceptance. All of these factors should be considered when planning the choice of a pressworking lubricant.

One of the newer considerations in determining the choice of a pressworking lubricant is of any possible biochemical effects during the metalforming process. United States federal law requires the lubricant supplier to furnish a completed safety data sheet to the user. This data can provide an important source of biochemical information during the planning process.

GUIDELINES FOR CHOOSING PRESSWORKING LUBRICANTS

Environmental Impact. Current environmental regulations and guidelines for pressworking lubricants are expected to become more stringent and restrictive. The choice of lubricant in many instances may have to be a compromise because of the particular rules and regulations of the federal, state, and local agencies that oversee a particular plant facility.

Tooling Interaction. Modern tooling for high performance presswork may incorporate materials that can be damaged by some pressworking lubricants. Certain formulations of lubricants attack the binding material of tungsten carbide. Electromotive forces (EMF) can be generated between the carbide surface and the machine tool with resulting erosion of the cobalt binder. This same condition can be caused by electrical current being present around the die area in the operation of safety devices, sensing probes, and other electrical hardware. In such cases, the coolant must not act as an electrolyte. Water quality can be a factor. Plant water used for mixing water soluble lubricants may not be suitable for use on carbide tooling.

Unstable water can create a harmful acid condition. Process water is generally unsuitable for this reason. Generally, clean potable water with low mineral content is best.

Presswork operations are becoming more integrated. Some dies, such as those used in multislide work, incorporate an assembly station. Secondary operations, such as tapping, are also being incorporated in tooling. Here again, the lubricant must serve multiple needs.

Transfer press operations require that the lubricant provide correct lubrication for every die and be compatible with the materials used to build them. Some lubricants may attack the self-lubricating graphite composition plugs pressed into guide bushings and

---

* Material in this section was revised by Joseph Ivaska, Vice President of Engineering, Tower Oil and Technology Company, Chicago, Ill.

wear plates. Possible damage to pads and seals made of rubber and elastomer products is a consideration.

In using transfer presses and transfer die systems, careful attention should be given to provide sufficient volumes of cooling and/or lubrication through the use of recirculating systems. This method of applying lubricant is very effective in carrying away the heat of the forming process. This is especially true on high-energy operations such as extruding, coining, punching, ironing, and embossing.

Recirculating systems should be kept clean of metal particles, swarf, oil absorbent, and other harmful contaminants. It may be necessary to install baffles, filters, magnets, and centrifuge equipment to clean the coolant properly. In many instances, the size of the recirculating tank or coolant reservoir must be increased in capacity to provide for sufficient cleansing, cooling, and filtration.

In some cases, it is necessary to install auxiliary lubricating systems to give positive application to selected areas of tooling, especially in progressive dies. The necessary hardware needed for this type of secondary lubrication should be planned for in the design of the die.

Often tooling is treated with special surface coatings such as chrome flashing, ion nitriding, titanium nitride, and titanium carbide. There are specially formulated, high-wetting metalforming lubricants now available in both petroleum-based and water-soluble types for such applications. Many water-based solutions provide excellent wetting and penetration properties.

Consideration should be given to the benefits of washing or steam cleaning dies, followed by re-lubrication, immediately after a production run when water-based metalforming lubricants are used. This family of lubricants is generally alkaline in nature. While they provide excellent cleaning and cooling properties for the die and tooling areas when in production, these same properties can remove machine, guide pin, and die spring lubricants. Water and residual lubricant from synthetics can be trapped in die pockets and if not cleaned, these die areas can be attacked by rust. Rust preventatives can protect the die springs and other components.

Just-in-time production inventory methods require short frequent runs which may preclude frequent washing of dies.

Stamping Material Properties. The four most common categories of material surfaces are:

1. Normal surfaces
2. Active surfaces
3. Inactive surfaces
4. Coated surfaces

The pressworking lubricant must be compatible with the material to be formed (Table 25-1).

In high-performance presswork where transfer die systems, transfer presses, and progressive dies are used, large numbers of piece parts are produced in a relatively short period of time. For this reason, it is extremely important to be sure that the lubricant and material properties are compatible. Extra scrap and re-work can result when piece parts are stained and coated stocks blister and peel.

Normal Surfaces. Most normal surfaces are material surfaces which have a natural affinity to retain lubricant readily and generally do not require special wetting and polarity agents to provide sufficient lubrication.

The most common normal surfaces include all bare mild steel stamping materials, such as cold and hot rolled steel, as well as aluminum-killed and vacuum-degassed steel. These normal surfaces naturally hold lubricant.

Problems arise even on normal surfaces, however, when excessive amounts of contaminants are present. These may include moisture, oils, rust, scale, rust preventatives, and residual pickling acid. These and other surface contaminants can also cause problems in secondary operations such as welding, brazing, painting, plating, or cleaning.

TABLE 25-1
Pressworking Lubricants for Various Materials

Material and its Alloy	Cold	Hot
Aluminum	Water-based solutions Oil solutions Fatty oil emulsions Soap solutions Wax suspension	Graphite suspensions
Copper	Emulsions of fatty oils Solutions of fatty oils Wax suspensions Tallow suspensions	Pigmented pastes Graphite suspensions
Magnesium	Fatty compounds + solvent Oil-based solutions of fatty compounds	Graphite + MoS_2 Soap with wax Tallow with graphite
Nickel	Emulsions Oil-based solutions with EP chlorinated additives Wax conversion coatings with soap	Graphite suspensions MoS_2 suspensions Resin coatings with salts
Low Alloy Steels	Emulsions Soap pastes Waxes Oil-based solution of fatty oils Polymers Conversion coatings with soap MoS_2 + graphite in grease Water-based solutions	Graphite suspensions
Stainless steels	Oil-based solutions of fatty oils Waxes Polymers Oil-based solutions with EP additives Pigmented soaps Conversion coatings with soap	Graphite suspensions
Titanium	Waxes Polymers Pigmented soaps Conversion coatings with soap	Graphite and MoS_2 suspension

(Tower Oil and Technology Company)

Freedom from contamination is especially important if water solubles, synthetic solubles, or chemical solutions are to be used as lubricants.

Active Surfaces. An active material surface is one in which the strength of the bond between the lubricant additive and metal atomic structure is great. Because the attractive energy of the metallic surface is high, surface chemical reactions are encouraged. This may explain why chemically active additives and wetting agents, such as oleic acid, lard oil, emulsifiers, and extreme pressure agents, are so effective in lubricating materials like brass, copper, and terneplate.

However, in many instances, the use of chemically active lubricants on active material surfaces can be troublesome. Often reactive lubricants cause staining, etching, and blushing of copper and brass piece parts, especially when they are nested and stored under humid conditions. This problem is worsened if alkaline vapor or degreaser fumes are present.

Coatings can be classified as belonging to either the active or inactive family, even though the supporting base metal for these coatings is a normal cold rolled material.

Inactive Surfaces. An inactive surface is one in which the strength of the bond between the lubricant additive and the metal atomic structure is low. The attractive energy of the metallic surface is also low. For these reasons, there is much less likelihood of chemical reactions with a lubricant or its chemical additives taking place on an inactive surface.

Typical inactive surfaces are those of stainless steel, aluminum, and nickel. Commonly, the lubricant used on these materials has a high film strength which actually places a mechanical barrier between the tool and piece part to prevent a metal-to-metal contact. High film strength can be obtained by using lubricants containing hydrocarbons, polymers, polar and wetting agents, as well as extreme pressure agents.

Quite often, lubricants containing extreme pressure agents are used on inactive surfaces to provide a mechanical barrier more so than an actual barrier. When working with inactive surfaces at low temperatures, there is very little chemical reactivity taking place with the lubricant. If the lubricant is working properly, it may simply be that it is providing a sufficient mechanical barrier for the operation.

Any lubricant formulation being used on an inactive surface must be compatible with the remainder of the manufacturing process, which may include cleaning, welding, and long term storage.

Hard-to-lubricate Surfaces. Sometimes the surface finish on the material to be formed will not retain lubricant properly. This can apply to brightly polished materials, stainless, tinplate, and bi-metals. Another culprit can be the mill oil itself. Certain coatings may keep water solubles or evaporating compounds from working properly.

Forming some steels can be tricky, especially bright or high-finished stainless steel. The surface of most stainless does not retain lubricant as readily as other metals. For this reason, emulsion-type lubricants with superwetting characteristics have been formulated for forming stainless steels. Heavy duty evaporating compounds containing extreme pressure agents have good anti-wipe properties.

On appliance and automotive trim, felt wipers may have to be used to remove any spotting from either water-based solubles or evaporative compounds. These bright surfaces will show any residual deposits from lubricants used. These very light films may have to be removed by wipers before the parts are wrapped and packed.

Coated Surfaces. The most common coated stock being formed is galvanized steel, and the remainder consists of pre-coated materials. The zinc coating may be applied by hot dip galvanizing or electro-deposition and is primarily intended to serve as a galvanic protective layer for the main metal substrate. Zinc coated steel is widely used where rust protection is required.

Material pick-up can occur when forming galvanized material. Special water soluble lubricants are available which can keep the galvanized metal particles from building up on the tooling. Dry film and water-based solutions can also be successfully used on galvanized stock.

CORROSION OF ZINC COATINGS

Any metal corrosion may be aptly described as the chemical or electrochemical attack by atmospheric moisture or other substances hostile to the surface. "White rust" is the destruction of the surface of galvanized steel or zinc by oxygen or other chemical elements. The reaction is accelerated by the presence of moisture.

The rate of attack is related to the temperature, pH, and composition of moisture and the concentration of dissolved gases within the moisture. The higher the oxygen and carbon dioxide content, the softer the water, and the greater the rate of surface breakdown. The corrosion cycle starts with the formation of zinc oxide, which in turn is converted to zinc hydroxide and then to basic zinc carbonate in the presence of carbon dioxide. The final product prevents the onset of further corrosion at that particular area of attack, provided no highly acidic or alkaline contaminant becomes involved later.

Under unventilated enclosed conditions such as blind angles and box sections, rapid attack by the available oxygen gives rise to a pitting condition, as the exhausted gas is not replaced as fast as it is used. The corrosion pattern, therefore, is non-uniform and typical of "white rust" conditions.

It is essential that piece parts with pockets be drained of as much coolant as possible. Parts in process should be used as rapidly as possible in order to avoid stagnation of any residual fluid. Conditions of high relative humidity and temperature aggravate the condition.

When the metal surface temperature drops in the presence of high humidity, the condensed moisture on the surface is rich in oxygen because the thin water film is exposed to a large volume of air. Conditions are then ideal for extensive uniform corrosion.

Zinc is rapidly attacked in acidic and alkaline aqueous conditions. Storage and manufacture within the vicinity of fumes from plating or pickling shops will accelerate metal surface breakdown.

APPLICATION TECHNIQUES

There are many ways to apply lubricants for pressworking transfer presses, transfer die systems, and progressive die operations. Among the best methods are recirculating systems, roller coaters, and airless sprays.

The use of drip application methods is not as productive and can be a wasteful method of application, unless it is properly installed and controlled. A typical drip lubricator is generally mounted after the stock or roll feed. The drip system is regulated by a petcock or regulating valve which can be adjusted for the flow desired. This method does not provide for automatic shut-off when the equipment stops. This results in wasted lubricant, messy parts, and housekeeping problems. If a large stock area is to be lubricated, a drip lubricator is a very unwise choice.

The lower consumption of lubricant and increased productivity that can be achieved by eliminating costly drip or manual application can often pay back the cost of automatic application equipment in a short time. In laying out a new press line (space permitting), it may be advantageous to install a fully automatic coil feeder, stock straightener, and stock lubricator to obtain the maximum productivity from the press and tooling.

In most instances, transfer presses and transfer die systems work well with recirculating lubrication systems. There are four advantages to be gained: (1) Tooling is kept cool and clean; (2) foreign particles and metal chips are washed away and sent back to the coolant reservoir for coolant conditioning and treatment; (3) savings on consumption of lubricant; and (4) if water-soluble lubricant is used, cleaning and other subsequent operations are much easier to perform.

Preliminary planning of a recirculating lubricating system for a transfer press or transfer die system should take into account many factors. The coolant system capacity should provide a five-minute lubricant cycle time if possible, with a three-minute cycle as the minimum, in order to allow for cleaning and conditioning of the coolant or lubricant. Sizing the system's storage tank at five to ten times the rated capacity of the coolant delivery pump is a good rule.

Roller Coating. This method of application is one of the most popular methods of applying lubricant. No matter what roller coater system is chosen, there are certain design features to consider.

First, the location of the roller coater is important. The preferred position is between the fabricating equipment and the feeding mechanism. Placing the coater before the coil feed can cause the lubricant to be mechanically worked off the metal surface, and this often results in slippage in the coil feeding mechanism.

There should be positive control of the lubricant being applied. The lubricant flow should be stopped when the press is not operating.

The unpowered roller coater is the most widely used. Here the coil feeding mechanism mechanically pushes the stock through the roller coaters. With this type of roller coater, either one or both sides of the material can be lubricated. The stock is coated with lubricant as it passes through the applicator rollers. Any excess lubricant is then squeezed off by wiper rollers and returned to the recirculating reservoir where it is filtered and available for re-application. A typical roller coater unit of this type is shown in Fig. 25-1.

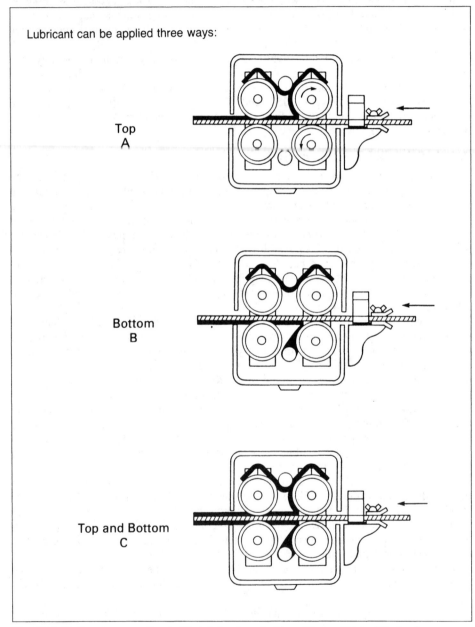

Fig. 25-1. Roll coaters permit versatile lubricant application. (*Tower Oil and Technology Company*)

An important feature of a roller coater is the ability to coat the top, bottom, or both sides of the stock.

The weight of the lubricant being applied by roller coater has a bearing on the pump components involved. When light oils or water-base fluids are involved, a centrifugal pump is adequate for recirculation of the lubricant. When heavy oils are used, a gear type pump may be required.

Common roll materials are steel, neoprene, felt and polyurethane. The surface of the roller can be textured to retain lubricant.

When working nonferrous material or specially coated stock, a soft roller made of either felt, neoprene, or polyurethane should be used so that the material surface will not be scratched or marred.

Airless Sprayers. An airless spray is a mechanical method of producing a finely divided spray of lubricant. Pressure is applied by means of an intensifier and then carried via a high-pressure hose to a tiny orifice in the nozzle, where the lubricant is expelled as a fine spray.

A typical airless set-up consists of an intensifier assembly (piston and barrel), a check valve, a lubricant reservoir of the required size, the various nozzles needed to cover a particular portion of the operation, and an air valve to activate the unit on each stroke of the press.

The spray pattern obtained with the use of airless spray can be either round or fan shaped. With the various spray patterns available, it is an excellent method for spot lubrication, either lubricating the stock before it enters the die or in the die itself.

A modern, airless spray system does not produce mist or fog that results in overspray problems. It can be precisely directed at a target area in the die, and timed to operate in conjunction with the equipment cycle.

CLEANING AND FINISHING REQUIREMENTS

The cleaning and finishing processes being used determine the choice of a pressworking lubricant. For example, low temperature cleaning lines generally are not capable of removing animal fats, residual oils, and certain extreme pressure agents. For low temperature cleaning, specially formulated water solubles are generally recommended.

Hot alkaline wash systems can clean heavy residuals, oils, and certain drawing compounds, but require careful maintenance. The waste disposal costs for alkaline systems can be quite high if skimmers and clarification equipment are not used to lengthen the life of the cleaner.

Solvent vapor degreasing is being phased out by many facilities because of environmental and air quality regulations. If vapor degreasing is used, water-based lubricants, such as water soluble oils, soaps, pastes, and dry films, should not be used to form the part; a lubricant compatible with a vapor system should be selected.

Special Considerations. There are certain restrictive or governing considerations that must be met by the pressworking lubricant for compatibility with secondary operations. Suitability for use with finishing operations, such as painting, powder coating, assembly, or joining, is of prime importance.

The use of some types of water-extendable pressworking lubricants (emulsions and chemical solutions) may allow welding without smoke and weld integrity problems. Painting can be performed without prior cleaning in some cases.

The need for long-term storage and/or long-term rust prevention may be another controlling factor.

Many major manufacturers have established manufacturing standards for the type of pressworking lubricants to be used on their respective components. The ultimate goal is to improve part quality, secondary operations, and ease of finishing.

MICROBIOLOGY OF PRESSWORKING LUBRICANTS

Pressworking lubricants containing water are apt to be affected by any number of the microbes which occur in the environment. Lubricants containing water are affected by contamination, allowing microbes to grow in these fluids. In the case of emulsions containing mineral oil, bacteria are the principal microbe that grows. In water-based solutions, mold and yeast are the principal growing microbiotic organisms.

Lubricant constituents are the food that feeds these microbes. When growth occurs by attacking a constituent, in a short period of time (often as rapidly as one day) a major constituent within the lubricant can be changed chemically by this microbe's feeding, no longer allowing the lubricant to operate as intended. An emulsifier may be unable to promote stable emulsification, rust preventive additives no longer prevent rust formation, or a lubricant additive no longer provides an effective lubricating film.

Anaerobic bacteria are active in the absence of oxygen, often causing the breakdown of sulfur compounds within the fluid. This can result in the strong rotten-egg odor called the "Monday morning smell".

Aerobic bacteria that grow in the presence of oxygen are more aggressive microbes. They can be responsible for the overall chemical breakdown of the lubricant.

Mold and yeast can cause a different type of problem by blocking filters as they continue to grow. They also can cause a slime build-up on tools and dies.

The problem of bacteria, mold, and yeast within the lubricant can be controlled by a balanced formulation of the lubricant. The use of effective biocidal agents together with proper clarification and filtration are good procedures to follow. Other good practices should include clean work habits, which include the exclusion of contaminants such as plant dirt, food, shop rags, and other forms of contamination from the system.

Biological Particulate. The growth of biological organisms, if unchecked, degrades the chemistry of the lubricant to a point where precipitation of unwanted by-products occurs, and particularly in the case of fungi such as mold, may significantly block filters. A completely different effect of biological degradation may be reduced corrosion protection, possibly resulting in the corrosion of piping and other components of the lubrication system, which can generate solid corrosion particulate byproducts. Furthermore, if the lubricant reservoir or flume is constructed of concrete, microbes can attack the concrete, providing a source of lime to further contaminate the lubricant.

LUBRICANT CLARIFICATION RECYCLING AND DISPOSAL

The cost of oil and chemicals and the disposal of pressworking lubricants has increased rapidly. For this reason, it is desirable to conserve and recycle these lubricants. New rules and regulations to protect the environment have also added to the cost of lubricant disposal.

Extending the useful life of pressworking lubricant depends largely on understanding the source of contaminants that can affect its useful life.

The type and choice of clarification, recycling and disposal techniques depends upon the type of lubricant and its contaminants. This can be determined by analysis. If it is oil-based, some of its chemical components may, with time, precipitate and circulate within the fluid, or they may be deposited on surfaces wetted by the fluid. If solid lubricants are used in suspension, these too may be deposited. If emulsions and soluble oils are used, the stability of the emulsion can be reduced and the particle size of the discontinuous phase can increase. These changes influence the choice of the clarification, recycling or disposal process, and equipment. If water-based solutions are used, changes in pH may cause precipitation. In any pressworking operation there is a possibility of biological degradation and resulting particulate contamination which then must be removed to maximize the effectiveness of the lubricant.

The particulates that can contaminate lubricants can include particles from the material that break loose during the pressworking operation. Some typical examples are mill scale, aluminum oxides, and galvanized particulate. The clarification units can remove these particles from the lubricant reservoir by using magnets, filters, etc.

Pressworking lubricants to be recycled may contain a wide variety of solid and liquid contaminants other than those directly resulting from the process. The contaminants present in the fluid may come from varied sources; cleaning fluids, oil absorbent from the floors, food, and various other random contamination sources may end up in recirculating systems. Obviously, these random occurrences are difficult to take into account on a systematic basis, but their occurrence should be considered when the process and equipment selections are made.

Water. The presence of calcium and magnesium promotes the formation of insoluble precipitates, particularly when soaps are present in the emulsion or water-based solution. During recirculation of the pressworking lubricant or the process of clarification on recycling, crystalline precipitates may be produced since their solubility in water is very low.

Understanding these in-process changes of the pressworking lubricant can improve any decision in selecting the proper clarification equipment and the design of the recycling system.

Clarification. As suggested, many factors encourage extending the effective life of a given metalworking lubricant. A significant contribution to extending the life of a lubricating system is the installation of appropriate clarification equipment.

Settling. Particulates in many metalworking lubricant systems are often adequately removed by installation of a simple gravity-settling tank. To facilitate removal of the settled debris, fines, and mold, a drag conveyor is often installed that serves to remove the buildup from the bottom of the tank. Separation of the tank into separate sections by partitioning walls over which the fluid flows increases the effectiveness of the settling system by progressively removing the contaminant that may flow from one section to the other. Several baffles can be installed at the top of the tank. The first baffle can prevent tramp oil from flowing into the tank, while the second drops heavy residuals and metal fines.

Gravity settling systems, because of their low cost and simplicity, are often used as the first stage in a total system which might include cyclone separators, filtration devices and/or a centrifuge. This approach reduces the operating cost of the more sophisticated equipment, and the productivity of the system is frequently improved.

Skimming. Removal of tramp oil from the surface of a gravity-settling tank is simple. It involves the use of a rotating disk or belt wetted by the oil. The belts may be made from such materials as stainless steel, neoprene, or polypropylene. The tramp oil preferentially wets the belt and is continuously removed from the surface of the lubricant.

To facilitate the separation of oil that may have become partially emulsified in the metalworking fluid, aeration units are frequently used. The aeration devices used to facilitate tramp oil removal generate air bubbles to which the tramp oil adheres as they rise to the surface, bringing up not only tramp oil but also fine, precipitated particles of metal oxides and other impurities.

A combination of settling tank, skimmers, and aeration devices will remove coarse particles by settling and fine particles and tramp oil by skimming. This relatively inexpensive combination often provides a sufficiently clean lubricant acceptable for repeated cycling. Magnetic separators can also be incorporated into lubricant clarification systems.

Cyclone Separators. Cyclone separators magnify the effect of gravity in achieving particulate separation through the use of centrifugal force. The contaminated fluid is centrifugally accelerated by a pump through a cone with decreasing diameter. The centrifugal force exerted on the fluid drives the suspended particles to the inner wall of the cone. These particles then build up due to loss of momentum, causing them to fall and be discharged from the bottom. The back pressure resulting from the geometric configuration of the cone and the discharge nozzle causes the cleaned fluid to discharge from the top.

The hydrocyclone may be used most effectively subsequent to the use of settling tanks to remove residual contaminants. The equipment is effective only on relatively light-viscosity fluids of approximately 60-100 SUS at 100° F (38° C).

If tramp oil is present, the hydrocyclone tends to increase its emulsification. This emulsification can lead to problems of fluid breakdown, bacterial growth, corrosion, and staining. Furthermore, unless care is taken, air may be sucked into the hydrocyclone through the bottom exhaust port, creating foaming problems and stable air-in-water mixtures, which may seriously interfere with fluid clarification.

When contamination by tramp fluids is minimal, hydrocyclones provide a relatively simple maintenance-free device for removal of solid contaminants from a low-viscosity metalworking lubricant.

Centrifugation. A centrifuge utilizes centrifugal force to multiply the effect of the difference in specific gravities. It is effective in separating solid particles as well as tramp oil from a lubricant. Furthermore, the centrifuge is effective over a wide viscosity range, and with a self-cleaning design, solid particles can be discharged on a continuous basis. Care must be taken in designing the centrifuge cycle so that only tramp oil is removed, not desirable emulsions. Low-speed centrifuges are designed specifically to remove suspended solids from any type of liquid, except tramp oil. High-speed units, on the other hand, may be used to remove tramp oil as well as solids. Bacteria and mold, if agglomerated, may also be removed with the tramp oil.

Filtration. In positive filtration the metalworking fluid passes through a filter medium driven by gravity, pressure, or vacuum. Solid particles larger than the openings in the filter medium deposit on the filter, forming a filter cake. The most common types of filter media used in the metalworking industry are composed of cloth, paper, organic polymers, or wire screens.

Positive filters are most frequently used, and operate on the same principal as the gravel and sand filter media used for clarifying water. The large swarf collects on the filter media and forms a filter cake, which in turn filters the smaller swarf and grit from the fluid. The thicker the cake, the finer the filtration. The filter medium acts as support for the filter cake.

The efficiency of the unit increases as the cake builds up. Eventually, the density of the cake reaches the point at which the resistance to flow exceeds the vacuum, or pressure, required to maintain the flow. At a predetermined point, the pressure or vacuum is stopped or a level is reached at which the filter medium indexes, exposing fresh medium, and the cycle is repeated. There are various filter systems; some of the more frequently used systems are described below.

Gravity Filters. Because of gravity, the weight of the metalworking fluid is sufficient to provide the force required to penetrate the filter medium. The fluid flows from the machine reservoir to the filter unit, where it passes through the disposable filter medium, supported on a steel belt or wire screen, leaving the swarf on the medium after filtration. The clean fluid is recycled for further use. As the cake builds up, the resistance to flow increases and causes the liquid level in the reservoir to rise. A liquid level switch activates the belt conveyor holding the filter medium so that it indexes, exposing a new filter surface.

The advantages of such a system are principally those of low capital cost and ease of operation. The disadvantages are associated with sizable floor space requirements, the need to dispose of relatively more media and accumulated cake, and the necessity to use a metalworking fluid that has low foaming characteristics to reduce filter medium usage.

Pressure Filters. A typical pressure filter contains two horizontal compartments: a movable top compartment and a stationary bottom compartment. During operation, air pressure seals the two compartments together. The filter medium may be a continuous nylon belt used as is, or one coated with a disposable medium.

Typically, the system consists of a filter unit, a settling tank or sump, and a drag-out chain. The metalworking fluid is pumped to the top shell of the filter under pressure, forcing the fluid through the filter medium. The particulates deposit on the filter medium to form a cake, permitting the cleaned fluid to accumulate in the bottom shell. The fluid is then pumped to a clean supply tank for reuse.

At a predetermined setting, usually 6 to 9 psi (40 to 60 kPa), the index cycle begins. The pump that supplies the metalworking fluid stops, and compressed air blows the fluid from the top shell through the filter medium to the bottom shell. At a predetermined pressure, the upper shell rises; the filter medium advances, exposing a new surface. The shell then closes, and filtration continues. With disposable media, the accumulated cake is fed to a waste tank. If a continuous nylon belt is used alone, compressed air may be used to blow the swarf into the waste tank.

The principal advantages of such a system consist of its ability to remove relatively small particles efficiently. Furthermore, large fluid volumes can be handled with minimal floor space requirements. On the other hand, tramp oil, if not removed, may prematurely plug the filter medium. Hard-water soaps, if formed, may also plug the medium. For efficient operation, chips as well as the smaller fines should be agitated and pumped to the filter unit. Disposable media usage is high because it is advanced along the entire length of the shell. Maintenance costs may also be quite high.

Vacuum Filters. The most common positive filter is powered by vacuum and is often referred to as a suction filter. A typical system may consist of two tanks with a filtering chamber at the bottom, or one covered with either disposable or permanent filter media.

The unit with the disposable medium is designed so that the lubricating fluid is piped to the filter area, where it passes through the medium, leaving the swarf behind. As the cake thickens, resistance to flow increases and causes the vacuum to increase. At a predetermined vacuum level, a vacuum switch activates a conveyor, which moves the dirty medium toward the discharge end, exposing a clean section. During this index cycle, the filtered fluid is piped from the clean tank for further use.

The principal advantages of this system include relatively low capital cost and efficient filtration of particulates from 400 to 980 μin (10 to 25 μm) in size. The disadvantages are related to the need for relatively large floor space, the cost of filter medium, and the need to minimize formation of hard-water soaps that can plug the filter medium.

The positive filter with a permanent medium operates on the same principle. The fluid is pulled through the medium by a vacuum, and a cake is formed on a wire screen. A scraper bar, attached to a chain drag-out, stretches across the wire screen medium. When the resistance to flow reaches a predetermined level, the vacuum is turned off, and the conveyor bar is activated and moves about 1.5 to 2 in. (38 to 50 mm), scraping the swarf off the screen in that area. The vacuum is again activated, and the cycle is completed. The swarf is carried away on the drag bars.

The principal advantages are low capital costs and efficient filtration of the particulates. Furthermore, there is no disposable medium cost. The disadvantages are the relatively large space requirements, and that the fluid selection may be critical to avoid medium plugging.

Coagulation. Lubricants used in the processing of aluminum and aluminum alloys build up fines and metallic soaps over time. A method for coagulating the soaps and fines promotes their separation from the metalworking fluid by centrifugation. Coagulation is achieved by heating the lubricant or other metalworking fluid in contact with very small quantities of a concentrated sodium carbonate solution. The sodium carbonate in contact with the metal fines neutralizes the charge so that coagulation can take place and the suspension is disrupted.

Plant Air. If air surrounding the work site is not controlled, there are several possibilities for introducing operating problems into the manufacturing cycle, which are often erroneously attributed to the lubricant. For example, acid fumes from nearby pickling, plating, or coating facilities may cause rusting on an intermittent basis. High humidity can also cause rust.

Mechanical Design. Lubricant reservoirs should be constructed of materials that are not subject to chemical attack. Concrete tanks are particularly poor choices, since lime in the concrete can be eroded, particulate contamination is increased and lubricant chemistry can be altered.

Flexible piping should be used only if there is no possibility of buildup of microbial contamination on the walls of the pipe that cannot be readily removed by flushing. All pipes and valves should contain as few bends as possible to facilitate cleaning.

Reclamation of metalworking lubricants should be carried out on an ongoing basis so that dead storage of used lubricants does not occur, again to eliminate excessive microbial growth and deterioration of the fluid.

STRUCTURE OF PRESSWORKING LUBRICANTS

Pressworking lubricants fall into three categories: fluids, paste-soaps, and dry films. Each of these families have varying combinations of chemical and physical constituents (Table 25-2).

TABLE 25-2
Comparison of Pressworking Lubricants.

Function	Oil-based Solution	Water-based Solution	Synthetic Solution	Emulsion
Reduce friction between tool and workpiece	1	1	1	2
Reduce heat caused by plastic deformation transferring to the tool	5	1	5	2
Reduce wear and gulling between tool and workpiece due to chemical surface activity	4	1	3	3
Flushing action to prevent buildup of dirt on tooling	5	2	5	3
Minimize subsequent processing costs such as welding and painting	5	1	5	2
Provide lubrication at high pressure boundary conditions	1	4	1	2
Provide a cushion between the workpiece and tool to reduce adhesion and pickup	1	4	1	2
Nonstaining characteristics to protect surface finish	5	1	4	2
Minimize environmental problems with air contamination and disposal problems	5	1	5	2

NOTE: 1 = most effective   5 = least effective. *(Tower Oil and Technology Company)*

Fluids are the most widely used pressworking lubricants. The two major classes of fluids used are solutions and emulsions.

Solutions. There are three types of solutions: oil-based solutions, water-based solutions, and synthetic solutions.

A solution is a fluid in which all of the ingredients are mutually soluble. A typical oil-based solution may consist of a mineral oil base, a wetting agent, a rust inhibitor, and an extreme pressure agent. These are all oil-soluble constituents that can be properly balanced and compounded into a solution.

Oil-based Solutions. These solutions use mineral oil as their base. Mineral oil solutions provide good fluid integrity and are generally safe from biological attack. Oil-based solutions can be recycled and clarified. They provide their own natural lubricant and have good wetting properties. Oil-based solutions have some inherent degree of corrosion protection and can be provided with a wide range of viscosities.

Water-based Solutions. Water-based solutions are true solutions consisting of water surfactants, soluble esters, soluble rust preventatives, and in some cases extreme pressure agents.

These solutions differ greatly from oil-based solutions or emulsions. They need special handling and some additional operating procedures especially during changeover and start-up from conventional fluids.

Synthetic Solutions. These are man-made, compounded pressworking lubricants combining excellent high temperature properties and good boundary lubrication. The main ingredients for these types of synthetics are: synthesized hydrocarbons (polyalphaolefins) and polybutane derivatives. The polyalphaolefins are much like oil-based solutions in their physical characteristics, but have a higher degree of resistance to oxidation. The polyglycols, polyesters, and dibasic acid esters have superior high temperature stability. Synthetic solutions are higher in cost than oil-based solutions.

Emulsions. An emulsion is a fluid system where one immiscible fluid is suspended in another. The mixture with formed droplets is an emulsion. In pressworking lubricants the continuous phase is water, and the suspended phase is oil or a synthetic solution. The continuous phase can contain several additives: extreme pressure agents, cutting agents, fats, or polymers. The balancing of the continuous phase is critical. Stable emulsions require surface-active ingredients as well as special mixing devices and techniques.

There are two families of conventional emulsions: multi-purpose and heavy duty. Multi-purpose emulsions generally consist of an emulsifier, oil, wetting agent, and a rust inhibitor. Heavy-duty emulsions contain emulsifiers, oils, wetting agents, extreme pressure agents, fats or polymers, and non-misting additives.

Invert Emulsions. In this emulsion formulation, the oil is the continuous phase and the water is the discontinuous phase. This type of emulsion is not ideal for drawing, where a good physical barrier is required.

Pastes, Coatings, and Suspensions. When severe pressworking operations are encountered, such as drawing or cold forming, high-film-strength lubricants are necessary for successful operations. Under these conditions, pastes, suspensions, and conversion coatings are often used.

Pastes. Pastes can be made several ways. They can be formulated with oil or be water-based. Pastes can also be pigmented or non-pigmented.

The pigments used in pastes are much like the pigments in paints. They are fine particles of solids which are insoluble in water, fat, or oil. Some of the common pigments are talc, china clay, and lithopone.

Most pigmented pastes come in two forms:

1. Emulsion compounds—These are composed of fats and fatty oils and sometimes mineral oil-pigment, emulsifier, soap, and water. They can be used as supplied, or diluted with water or oil for easier handling and application.
2. Oil compounds—These are pastes consisting of pigments dispersed in fatty oils and/or mineral oils which may be treated with extreme pressure agents (chlorine,

sulfur or phosphate). These compounds are generally used straight or diluted with mineral oil. Dried-on lubricants and pigmented coatings can be used in sheet metal, especially in heavy forming and drawing.

Non-pigmented pastes come in four different categories:

1. Emulsion drawing compounds which are pastes composed of fats and fatty oils and their fatty acids, sometimes containing free mineral oil, various emulsifiers, and water. These products are diluted with water before use.
2. Fats. Fatty oils and fatty acids which are sometimes used straight, but usually are mixed with mineral oil for use.
3. Mineral oil and greases can be used straight when necessary.
4. Emulsions which are usually diluted with water.

Dry-processed Coatings. These coatings are coming rapidly into use because of their easy and economic application, freedom from messy conditions at the press, and ease of handling and cleaning.

For dry soap films, stock or parts should be cleaned in degreasing solution at about 190° F (88° C), rinsed at about 160° to 180° F (71° to 82° C), and water-soluble soap solution applied at temperatures on the order of 180° to 200° F (82° to 93° C). For low production requirements, dip coating may be most economical. For high production, roller coating is preferred for sheet and coil stock.

Wax or wax-fatty coatings for light to medium drawing, especially of nonferrous stock, may be applied by hot dipping at 120° to 150° F (49° to 66° C) by spraying the material hot, or by cold application in a solvent vehicle. In the latter method, the vehicle evaporates, leaving a dry coating.

Phosphate coatings are proper chemical immersion coatings that provide a measure of lubricity.

Graphite coatings are useful under high-temperature and heavy-unit-load conditions where it is unfeasible to use water-base, oil-base, or other solid lubricants. Graphite has the disadvantage of difficult removal and is consequently used for drawing only when absolutely necessary. In fact, the unauthorized use of graphite in automotive body panel drawing dies has caused subsequent paint adhesion problems resulting in customer dissatisfaction.

Suspensions. Suspensions in pressworking lubricants consist of fine particles of various solid lubricants dispersed throughout a carrier fluid. The particles generally vary in size. Quite often the particles being used are insoluble within the suspension–the carrier or vehicle fluid carries the particle in question.

A dispersion may be considered stable when three successful processes are accomplished:

1. Wetting of the particle by the carrier fluid
2. Reduction in the size of particle clusters
3. Stabilization of particle suspension coagulation

Coagulation must be prevented if the suspension is to remain stable for an extended period of time. A repulsive force must be achieved by adding chemically absorbed layers to the dispersed systems. Surface active agents are important in these systems. Each surface active agent is usually designed for a specific system.

Surfactants may be used to promote the wetting of solid by a liquid. These also act as dispersing agents that must have the ability to form energy barriers to prevent aggregation. Some of the typical dispersing agents used in water-based solutions are lignin, sulfonates, acrylic acid-based copolymers, and copolymers of propylene oxide and ethylene oxide. Suspension of solid particles in mineral oil has been achieved through the use of acetyl benzenes and long chain amines.

Warm and hot forging of steel and other metals relies on the use of graphite suspended in mineral oil, and more often water. The structure of graphite, its stability at temperatures over 1,000° F (540° C), and positive side effects of air and water vapor on its lubricity favor the application of graphite with a carrier as a very cost-effective, high-temperature lubricant.

Solid Lubricants. General solid lubricant systems are used in pressworking operations particularly where operations encounter high unit pressures on high temperatures brought on by deformation of metal.

The two major types of solid lubrication are: (1) soaps and soap combinations, (2) graphite and molybdenum disulfide alone or in combination for hot metal working.

Soap Type Pressworking Lubricants. There are four different types of soap lubricants.

1. Dry powders which usually are sodium or other metallic type soaps. They are generally furnished in powder form for use in tube bending and wire drawing.

2. Dried-film compounds which are usually soluble soaps (often mixed with soluble solids, for example borax), containing waxes, wetting agents, and other chemical ingredients. These dry films are used for drawing and can be applied by spraying or dipping with a 10% to 20% hot solution followed by drying prior to the forming or drawing operation.

3. Sodium or potassium soluble soaps; these are diluted up to 10% with water.

4. Metallic soaps such as aluminum or calcium stearate can be used alone or in combination with molybdenum disulfide (MoS_2) and/or graphite. This combination is used in wire drawing and warm forming.

Hot Metalworking Lubricants. Graphite is used in a wide range of temperatures anywhere from 100° to 1,000° F (38° to 538° C) for the hot extrusion of tubing as well as drawing wire. It can also be used at temperatures from 900° F to 1,300° F (482° to 704° C)—the heat range where tungsten and molybdenum can be worked.

Graphite, when mixed with some metal oxides (such as tin oxide and copper oxide, which act as the liquid carrier at elevated temperatures), has been used to extrude steel at temperatures up to 2,100° F (1,150° C). The easy cleavage of the structure of graphite is responsible for its excellent working properties. The presence of oil or water seems to improve its lubricating properties even further. In order to use the advantages of graphite as a pressworking lubricant, it is often blended into a paste or suspension form.

Molybdenum disulfide is another effective solid lubricant up to temperatures of 900° F (482° C), the temperature where it oxidizes. When combined with other solids, molybdenum disulfide is effective up to 1,800° F (982° C).

Glasses and copper coatings are other solid lubricants that are used for pressworking at high temperatures.

Additives. Most of the more successful pressworking lubricants employ some form of additive to achieve better wetting and improved rust protection. In severe forming and drawing operations, extreme pressure additives can be very helpful.

Emulsifiers. When formulated into pressworking lubricants, emulsifiers perform a dual purpose. They promote a stable emulsion while enhancing the cleaning ability of oil-based solutions and emulsions. Emulsifiers can be nonionic or anionic surfactants. The typical nonionic emulsifier can consist of complex esters, fatty acids plus alcohols, monoglycerides, and ethoxycated alcohols. Most nonionic surfactants contain soaps with long chain fatty acids, alkyl aromatic emulsifiers and phosphate esters. Soaps generally are the most widely used anionic emulsifiers. Sometimes they can form hard water soaps in the presence of hard water. In this situation, de-ionized water is recommended.

Extreme Pressure Agents. An extreme pressure additive provides a second line of defense to prevent scoring, galling, and rubbing between the piece part and the tooling, especially in severe drawing and cold forming operations. The chemical reaction produced by the activity of the extreme pressure additive results in a protective film between the tooling and the part being formed. Chlorine, phosphorus, and sulphur containing pressworking lubricants react upon contact with the metal surfaces to form chlorides, phosphides, and sulfides.

When using an extreme pressure additive containing pressworking lubricant, surface staining and corrosion may occur. Cleaning of stampings can be difficult when extreme pressure agents have been used in the pressworking lubricant.

Thickeners. Thickeners can be of great value in heavy forming and drawing operations. They resist lubricant thinning where high heats of deformation are present. They greatly improve pressworking lubricant performance. Polymers, organo-clays, and natural gum, along with metal hydroxides, are used to improve the boundary and viscosity properties of lubricants.

Cleaning of these special thickeners can be a problem, especially where low-temperature cleaning or solvent vapor degreasing is being used.

Anti-foaming Agents. These agents can be very useful where pressworking lubricants are used in recirculating systems. Foam prevents proper bonding of the lubricant to the tooling and the component being worked.

Additives to reduce or prevent foam tend to lower the surface energy of the lubricating film. Some of the additives used are silicones, amides, glycols, and fine particles of solids—usually silica.

Lowering the pumping rate of the recirculating pump and minimizing the free fall of the lubricant into the tank are also helpful.

Corrosion Inhibitors. These additives are used widely to insure quality surface protection during storage and shipping.

They are used in water and oil solutions. In warm-based systems, corrosion protection can become difficult.

Anti-misting Agents. These agents are used to reduce the possibility of introducing airborne particles from pressworking lubricants into the plant environment. These agents are needed when solvents or low viscosity oils are being used. The resulting air mist can be unhealthy and cause housekeeping problems. Small quantities of polymers, such as polybutanes or acrylates, are added to reduce mist. When treated with polymer, the oil or solvent lubricant vehicle forms a larger particle, sharply reducing the formation of lubricant mist.

Passivators. These are useful additives when pressworking coated and nonferrous surfaces. These surfaces must be protected by passivators to reduce the possibility of activity on the metal surface. Organic amines, and sulfur- or nitrogen-containing compounds are the usual passivators. Foreign material suppliers use chromate coatings on coated surfaces such as galvanized steel to protect against attack by salt spray. Passivators can be used in tandem with corrosion preventive additives like complex borates to improve performance of each component in the lubricant.

WATER-BASED PRESSWORKING LUBRICANTS

There are many benefits to be derived from the use of water-based lubricant solutions where possible. When compared to mineral oil-based lubricants the advantages include lower initial cost, more efficient secondary operations, and elimination or reduction of cleaning.

In some instances portions of the production process require modification to permit the use of water-based lubricants. For example, additional clarification and contamination control equipment may have to be installed in order to maintain lubricant stability and product quality.

Tooling Problems. The maintenance and operation of tooling when working with water-based solutions requires several changes in operating procedure. Most synthetics are alkaline in nature and act as detergent soaps. Several changes may be required in tooling maintenance procedures. These solutions remove most conventional greases and machine oils. It may become necessary to protect the tooling and related die components with rust preventatives, especially when the tooling is not in operation during extended periods of time.

Material Considerations. When specifying material for use with water-based lubricant solutions, the material should be ordered clean, dry, and as free as possible from mill oil and rust preventatives.

Paper clad and plastic film protected material finishes can be formed with water-based solutions without damage to the protective coverings, by simply applying the lubricant with roller coaters or spray units.

One area of importance when considering a water-based lubricant is the interaction between the solution and the material surface. Tests should be conducted to determine that the water-based solution will not react in an adverse manner with materials such as galvanized steel and aluminum, resulting in corrosion or staining. The proper dilution strength should be carefully noted. Operating with the proper concentration can be the difference between success and failure. The same surface tests should also be performed on coated stocks.

Contamination Control. While system cleanliness is extremely important, it only goes halfway in assuring success. The overall forming system is also of great importance and should be mechanically designed, modified, or altered in order to operate optimally with the water-based lubricant solution.

In the majority of cases, the contamination control devices required to keep the lubricants clean and effective are quite simple. All that may be needed is a simple floor tank with a few baffles added to drop out metal fines, and a skimmer to drag off the accumulations of tramp oil and greases. These simple measures will lengthen lubricant life, provide better finish to formed parts, and reduce the cost of secondary operations.

There are, of course, exceptions. Some systems may generate large amounts of metal fines. If this is the case, magnetic separators or cyclonic units may be installed to remove the large amount of metallic particles. Media filters are also an effective means of removing foreign particles.

System Start-up. It is possible for contamination to occur during the initial start-up of operations. It is therefore important to make the required adjustments for contamination control. There are three considerations pertaining to system performance that need to receive careful attention when starting up a pressworking operation using water-based solutions:

1. Overall system cleanliness.
2. Machine-lubricant compatibility.
3. Water source analysis.

The importance of starting up the pressworking system in a clean condition cannot be overemphasized. Clean formed parts cannot be obtained unless the system itself is clean, including machinery, tooling, lubricant reservoirs, and any piping that carries lubricant to the forming areas. Generally, the lubricant supplier will specify what type of cleaner is recommended for use in this preparation period, as well as any special cleaning instructions.

As mentioned earlier, most synthetic lubricants are alkaline and will remove most conventional greases and machine lubricants with their "detergency action." They will also flush out metal particles and remove mill oil, tramp oil, and other residue in the system.

The water condition in the plant is another important consideration when starting up. The lubricant supplier should check the plant's water for iron content and hardness. Particular attention should be given to the stability of the water-based lubricant solution in the water intended for use.

Initial Run Observations. It is very important to observe the operating results that are obtained during the initial run of these solutions. It can be readily observed whether the use of the lubricants has been successful or if some changes are needed.

Some problems and their solutions are:

1. Excessive amounts of tramp oil may indicate the need for skimming equipment or even a closed loop clarification system.
2. Solids build-up may show that water treatment or solid removal by centrifuge, media filter, or magnetic separator is needed.

During this period, it can be seen if the pressworking lubricants are compatible with the paint on the presses. Paint can be removed by the fluid if the proper paint was not applied by the machinery builder or the maintenance department.

At this time, it is important to determine how often the concentration level of the water-based solution should be checked. The drag-off of lubricant due to piece part size or configuration may be one factor affecting the level of concentration. Operating at elevated temperatures, and reactions taking place with the microbiological environment, are other considerations that can effect or change the chemical balance of any water-based solution.

It is advisable to check the initial concentration of the solution more frequently at first until a definite pattern or profile is established. This will determine how often the solution should be checked.

Most water-based solutions can be tested and controlled with the aid of a chemical titration kit developed by the manufacturer of the particular solution.

Solids Build-up. This build-up results from the contributions of two different sources. The initial build-up generally occurs from external contaminants such as metal fines, oxides, tramp oil, and grease. The secondary source comes from evaporation of the water-based lubricant solution, thereby resulting in precipitation of the solids originally dissolved in the solution, and through reactions with the components in the water such as iron, calcium, and organic matter. These reactive products then may precipitate out as a solid layer on machine components or on the formed parts.

Skin Irritation. Dermatitis is a skin condition which can occur in manufacturing operations, particularly where operators have poor personal hygiene practices and where plant conditions are dirty. In some instances, skin may be sensitive to specific chemical components, causing inflammation and cracking of the skin. Care in selection of fluid, good housekeeping practices, and an awareness of how different individuals respond to conditions provide a basis for controlling dermatitis.

Corrosion. Water-based solutions, unless properly controlled, may permit corrosion of equipment and tooling, as well as formed parts. Solution concentration must be carefully controlled, as well as the hydrogen ion concentration (pH). Shop personnel should be trained in the proper use of titration testing equipment. Regular checks can usually be conducted by the lubricant supplier's laboratory facilities.

Lubricant Stability. The stability of the fluid is a function of choosing the proper chemical system, keeping the lubricant clean, checking the microbiological balance and the water source, and maintaining the concentration level recommended. Good operating practices are required to maintain lubricant stability.

Foaming. All fluids used in plants are subjected to some degree of stirring. Excessive agitation can result in foaming. Air entrainment and contamination of the fluid is another cause for foaming. As a fluid is being used, the dirt level generally increases, which normally will reduce the foaming problem. If the foaming increases even with a greater dirt load, then an investigation should be carried out to determine the cause for the contamination.

Solution pH. Checking the hydrogen ion concentration (pH) of the synthetic solution is very important. If the solution becomes excessively alkaline (high hydroxyl ion concentration) due to contamination, excessive drying of operator's skin may occur, facilitating dermatitis. If the solution is normally basic and somehow becomes acidic (increasing hydrogen ion concentration), corrosion of metal surfaces may be accelerated. Reduction in basicity is frequently caused by bacterial degradation of the solution chemicals, or may result from contamination. For these reasons, period checks of pH and adjustment of hydrogen ion concentration to the proper levels is required. If gross changes of pH occur, the solution should be discarded and the cause of the contamination should be determined and eliminated.

Subsequent Operations. Some of the greatest benefits of using water-based solutions properly in pressworking applications are the substantial savings that can be realized in the operations that must be performed after the metal forming operation has been completed.

Some typical examples are: possible elimination of degreasing altogether, painting over the lubricant without prior cleaning, easier cleaning and elimination of smoke in the heat treating operation, as well as welding. Not only may smoke be reduced or eliminated, the actual quality of the welding operation and/or the heat treatment may be improved.

Other advantages of water-based solutions are reduced maintenance of equipment and machines because of cleaner operating conditions, and cleaner and safer work areas.

TROUBLESHOOTING DIE AND LUBRICANT PROBLEMS

Scoring. Often, scoring occurs due to lack of lubrication to the tooling area. There can be a number of reasons for a lack of lubrication. The heat of deformation may be too great for the physical and chemical properties of the lubricant. The lubricant may thin excessively on hot die surfaces. Matching the requirements for lubricity with lubricant characteristics at the actual operating temperature can be helpful. Water cooling of the die is another useful technique. The use of water-based lubricants is a good way to reduce die temperatures by evaporative cooling.

Another cause of scoring is poor lubricant application technique. The lubricant must reach the tooling area. Poor penetration can be still another cause for scoring—the lubricant may be too heavy.

In analyzing scoring problems, a good place to start is with the stock condition. Make sure that the stock is free from burrs, protective coatings, excess mill oil, and foreign particles. Another reason for scoring is that the stock thickness may be too great for the clearances provided in the tooling.

In forming hard materials, more force and energy is required to form the part than are needed for soft materials. The lubricant must be able to withstand these extra forces and not wipe or squeeze out, especially in the areas surrounding a tight form radius.

Softer material surfaces, such as clad aluminum alloys and deep drawing grades of electro-galvanized steel, can contribute to scoring. Metal pick-up on the tooling surfaces can generate fine metal particles during drawing and forming operations. These particles must be flushed out by the lubricant and not be allowed to accumulate. Material particles, such as those derived from beryllium copper, can be quite abrasive, resulting in accelerated tool wear.

Tool Wear. Rapid tool wear can be caused by several conditions, including the lack of lubricant. The first and most common occurrence is poor application technique. On high speed applications, lubricant must be present at all forming stations in order to provide good tool life. On very severe operations such as coining, extruding, or tapping, a secondary lubricant may have to be used and applied by some auxiliary means such as an airless sprayer.

The characteristics of the lubricant used can also contribute to tool wear. Large amounts of uninhibited extreme pressure agents can, in some instances, promote tool wear. When using carbide tooling, care should be taken to insure the lubricant is indeed compatible with the tooling surface and the chemical properties are safe for use on carbide tooling.

Metal particles generated during the forming process can bring on rapid wear if they are not properly flushed out by the forming lubricant. This problem can become quite acute when working materials, such as aluminum, hot rolled steel, aluminum-killed stock, and even stainless steel, that generate metal fines on a large scale.

Slug Pulling. Slug pulling is due to the vacuum created by a lubricant seal between the slug and the face of the punch. The solutions to this problem are the use of vented or spring ejector punches, tighter clearances and even air or vacuum ejection. Lubrication can also influence the ease of slug shedding.

In some instances, the physical properties of the lubricant are too adhesive in nature, and therefore, are not satisfactory for high speed perforating. Often just providing correct die clearances will eliminate slug pulling.

Tooling Problems in Drawing. The two most common problems in drawing are wrinkling and fracturing (or tearing).

Wrinkling can have a number of causes. If the draw die is known to produce good parts and the onset of wrinkling occurs at the start of a production run, the binder adjustment is the first thing to check. Generally, wrinkling is a sign that the blankholder is too loose. In severe cases, wrinkling and fracturing may occur simultaneously. This is because the wrinkled stock cannot pass over the draw radius and the metal flow is locked out.

Worn draw beads are a frequent cause of wrinkling in localized areas.

If wrinkling occurs only on one side of the die, check to see if there is a slug or other foreign object under the die. Press out-of-level conditions can also cause wrinkling on one side of the die and splits or fractures on the other side.

Incorrect material specifications and excessive amounts of lubricant, as well as excessive lubricant viscosity, can be a factor in wrinkling. Lubricant must always be applied evenly.

The most frequent cause of splitting or fracturing is excessive blankholder pressure. Splits on one side of the die can be due to a foreign object under the die or a press out-of-level condition. Insufficient clearance between the punch and die can be the cause for fracturing or tearing. The percentage of reduction or depth of draw may be excessive for the type of stock being drawn. The material may be too thin or may be age hardened. Circle grid analysis of the area of the fracture and checking tensile samples against stock specifications are good troubleshooting techniques (Section 11).

The lubricant must be correct for the working temperature of the die. Not enough lubricant due to improper application or poor wetting, or penetration of lubricant can cause fractures. The lubricant properties must provide sufficient barrier or boundary lubrication.

LUBRICANT COMPATIBILITY WITH SUBSEQUENT OPERATIONS

A stamped part is not really completed until other subsequent operations have been performed. If there are problems with cleaning, painting, brazing, or welding, the lubricant could very easily be at fault.

Substantial savings can be realized by understanding what part lubricant properties actually play in finishing operations. The correct lubricant can reduce cost. This is especially true when annealing, stress relieving, or deburring is required.

Annealing. In many instances, annealing operations are performed on a metal formed part before or during the finishing process. The compatibility of the forming lubricant with the annealing operation must not be overlooked.

Annealing is generally performed at temperatures of 1,400° to 1,700° F (760° to 927° C) in a controlled atmosphere. The forming lubricant must burn clean off to obtain a bright anneal. The ingredients in the forming lube must not contaminate the part or the furnace atmosphere. Forming lubricants that contain sulfur, chlorine, animal fats, and pigments can cause adverse surface reactions during annealing, and also contaminate the furnace atmosphere. Before performing any annealing operation, it is good practice to first check the forming lubricant for overall compatibility in both the furnace atmosphere and on the part's surface.

Sometimes, it may be necessary to first clean the formed part of any undesirable substances before the annealing can be done successfully.

Stress Relieving. Stress relieving is another finishing process that requires a careful choice of the forming lubricant properties. Here again, the ultimate goal may be to achieve a clean part before relieving.

The temperatures required for stress relieving are usually in the range of 600° to 800° F (316° to 427° C). This is substantially lower that the temperature required for annealing. The lower temperature means that lubricant residues do not completely burn off, and this can cause contamination problems. These can include staining of parts, furnace atmosphere problems, and eventual damage to the furnace. Among the problem lubricants are sulfurized and chlorinated oils, heavy residual oils, animal fats, graphite, pigments, and soaps.

Deburring Formed Parts. The right choice of a pressworking lubricant can reduce the cost of deburring stamped parts. The proper lubricant can reduce burrs by reducing tooling wear. Deburring costs include the cost of any required cleaning prior to deburring. A lubricant that does not interfere with the deburring operation can reduce costs.

Often, there is a lack of communication between the pressroom and the deburring operation, either in-house or by an outside vendor. Excess lubricant on a piece part may reduce the efficiency of the deburring media. The amount of lubricant on the part should be carefully controlled. Using the minimum amount of lubricant required to achieve the desired results not only saves lubricant, but reduces the cost of subsequent operations.

The economy of being able to use low-temperature cleaning methods prior to deburring should be considered when choosing a lubricant. The use of synthetic forming lubricants or specially formulated lubricants to permit cold or low-temperature cleaning can reduce overall costs.

After deburring, there may be a light water film or moisture deposit on the piece part. A water-displacing rust preventive may be needed to protect the part before final assembly.

Cleaning Stamped Parts. Cleaning is one of the most important operations in preparing a metal formed part for finishing operations such as painting or plating. The need for the forming lubricant to be compatible with the cleaning system cannot be overemphasized.

Low-temperature Cleaning. These systems generally operate at temperatures of approximately 100° F (38° C) and can be quite effective in removing light oils, soluble oils, synthetic lubes, and some specially formulated oils designed for low temperature cleaning. This type of system will not completely remove pigments, pastes, soaps, animal fats, heavy residual oils, and heavily compounded forming lubricants. Low-temperature cleaning systems do conserve valuable energy, but they also have their limitations.

The performance of a low-temperature cleaning system can often be improved either by using a different type of forming lubricant which is more compatible with the system, or by modifying the cleaner to improve its cleaning ability.

Vapor Degreasing This cleaning method uses hot vapors of chlorinated hydrocarbon solvents (generally trichlorethylene or trichlorethane) to remove many types of forming lubricants, generally petroleum-based.

Vapor degreasing can also remove mill oils and rust preventatives, along with some cleaning residues. Vapor degreasing can be a costly operation when too much oil is left on fabricated components before cleaning. This excessive amount of lubricant can result in a build-up of residual oils and other compounds on the boiling or vapor chambers, which can cause foaming or a reduction in the evaporation efficiency. It is not uncommon for a manufacturing plant to increase the life of its chlorinated solvent simply by changing the type of application technique for its forming lubricant, or using a different type of lubricant for the piece part or component in question.

There are some types of lubricants that should not be removed by vapor degreasing. Metalforming lubricants which contain free fatty acids, chlorine, or sulfur can upset the inhibitors in the chlorinated solvent, resulting in hydrochloric acid formation in the degreaser. This can result in corrosion of the heating elements and other working surfaces.

Another family of forming lubricants which can create problems are emulsions. Many of these lubricants are compounded with fatty acids, and also leave some water on piece parts to be cleaned. Even though the chlorinated degreasing solvents are especially inhibited against the effects of hydrochloric acid formation in the presence of water, excess moisture should not enter the degreaser. Another side effect of attempting to remove water-soluble fabricating compounds by vapor degreasing is the white powder deposits left on the piece part after vapor degreasing. These deposits may be due to components of the soap, or may be present in the original fabricating compound.

The operator must be concerned with the following problems when operating a vapor degreaser:

1. Depletion of the stabilizer in the chlorinated solvent
2. Formation of acids
3. Deposits on piece parts
4. Build-up of oily residues in the solvent

Alkaline Cleaning. One common goal in cleaning is to maintain the cleaner properly to extend its life as long as possible. This fundamental rule certainly applies to alkaline cleaning.

Fabricated parts should be drained of excess forming or stamping lubricant whenever possible. Parts can also be stacked in such a way as to maximize draining of the lubricant. Another technique is to blow off the lube before cleaning. Vibratory conveyors can also aid in keeping the amount of lube entering the cleaning tank to a minimum. The recovered lubricant can often be recycled, increasing the savings.

SECTION 26
DIE PROTECTION SYSTEMS[*]

Mechanical switches, electrical contact probes, and electronic sensors have widespread application in metal stamping. Properly applied, die protection systems avoid or limit the severity of die damage in the event of a misfeed or foreign object in the die.

Correctly applied, die protection systems permit one operator to service a number of press operations having fully automatic feeding and part ejection systems. Generally, press and die sensing equipment has an excellent economic payback in increased productivity and die damage cost avoidance.

Any time that an automated pressworking operation requires a human operator to stop the press in the event of a misfeed, a rapid payback on an investment in proper sensing equipment very likely exists and should be investigated.

MECHANICAL LIMIT SWITCHES

Limit switches are used extensively to safeguard the die in case of misfeed or buckling of stock or failure to eject scrap or blanks, and to check the position of parts in assembling dies.

In the progressive die in Fig. 26-1, strips of gilding metal and silver are fed crosswise into the die with automatic feeders. Assembly of a small silver slug into a dovetailed cavity in the gilding metal strip must be performed accurately at each stroke of the press. To prevent inaccurate feeding of the strips and misplacement of the silver slugs, probes on the ends of levers operating the limit switches inspect operations and either allow the press to continue, or stop it if the operations have not been completed. The hinge pads, ($D1$, $D2$, $D3$, and $D4$) are attached to the punch assembly (not shown) and support the actuating levers. There are two probes on the lever ($D5$) which enter the silver strip to assure accurate advance and to check whether or not the slug has been pushed out of the silver strip. The probe ($D6$) checks whether or not the silver slug is in the cavity in the gilding metal strip. The two probes on lever ($D7$) check the advance of the gilding metal strip and whether the blank has dropped out of the strip. The lever to which the probe ($D6$) is attached is jointed and flexes at the center so that the limit switch will remain closed when the probe is resting on the silver slug. Should the silver slug be missing, the end of the probe would assume a lower position, thereby opening the limit switch.

An example of limit switches used to detect misfeed is shown in Fig. 26-2. The lever ($D1$) is held by the tension spring ($D2$) against the limit switch ($D3$) until the stock is fed far enough to open the switch. As long as this switch is closed, the press cannot make another cycle. In case the stock is overfed, the arm opens a limit switch ($D4$), which also prevents the press from operating until the overfeed is corrected.

Figure 26-3 shows another application of a limit switch to detect misfeed. In ordinary operation the pin ($D1$) acts as a pilot for the strip. In case of a misfeed or a broken punch,

* Much of the electronic sensing material was contributed by George Keremedjiev, President, Tecknow Education Services, Inc., Bozeman, Montana.

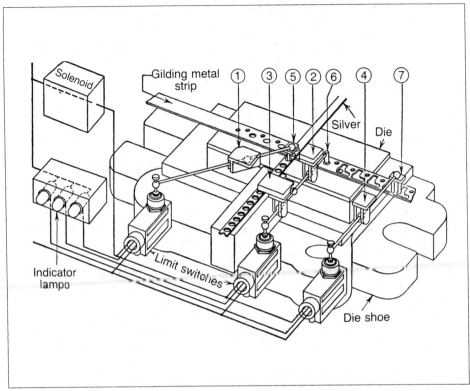

Fig. 26-1. Limit switches guarding a progressive die.

the pilot will not be able to enter the hole in the strip; therefore it will ascend and the tapered head will force the latch (*D2*) to slide outward, depressing the limit switch. The limit switch is wired into the control circuit of the machine and the opening of the switch stops the machine.

Figures 16-14, 16-26, and 16-27 illustrate the use of mechanical limit switches to detect irregular feeding in progressive dies.

METALLIC CONTACT SENSORS

The main advantage of sensing by direct contact with metallic parts is simplicity and low cost.

Transfer press operations have been successfully protected from damage due to mis-hits by sensing the part by means of electrical conduction. It is necessary to insulate the transfer fingers from the feed rail with this system.

A wide variety of contact probes are available including spring-loaded wire types which are capable of withstanding rough handling. Fabrication of custom sensors is quite easily done. Detection of proper feed advance in progressive dies has been successfully accomplished with contact probes fabricated from steel band-strap affixed to a phenol fiber insulating block.

With all types of direct contact sensing, signal conditioning is usually needed to deal with intermittent contact conditions. This can often be accomplished by the inherent delay time of the electromechanical relay used to interrupt the press run circuit in the event of a misfeed.

Junction boxes are available that have electronic signal debouncing circuits. These junction boxes can also provide inputs and power for a number of electronic sensors as well as metallic contact probes.

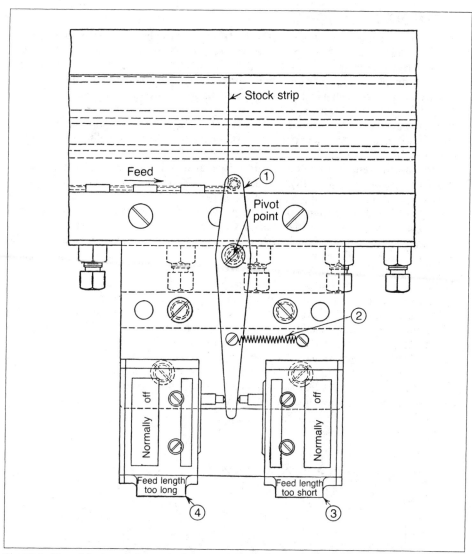

Fig 26-2. Use of limit switches to indicate a misfeed of stock.

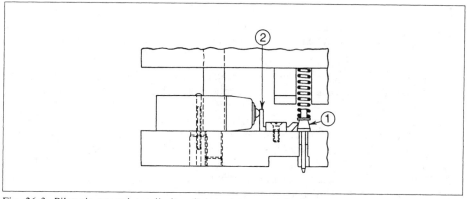

Fig. 26-3. Pilot pin operating a limit switch to detect a misfeed.

ELECTRONIC SENSORS

Advantages of Electronic Sensors Over Mechanical Probes. The primary advantage is that electronic sensors, when properly installed, are not subject to mechanical wear. Once correctly installed in a die, a sensor such as an inductive proximity type will perform its detection duties with a lifetime generally limited only by human mishandling. Since the sensor need make no mechanical contact with its target, its useful lifetime can be viewed as the failure rate of its electronics.

The response speed of electronic sensors is usually faster than a mechanical switch. A limiting factor in high-speed stamping operations usually is the speed with which the press can be stopped, rather than the reaction time of the sensor.

Standardized Sensor Installation. In order to avoid disorganized and randomly conceived wiring schemes, readily available standard connectors and junction boxes can aid a neat and efficient routing of the sensor cables.

The use of standardized interconnection plugs and prewired junction boxes together with a standardized bench testing method will permit diemakers and setup people to troubleshoot a system on the bench. Where possible, the entire sensing system in the die should hook to the press with a single, standardized quick disconnect to facilitate quick die change.

Inductive Proximity Sensors. Of the many types of electronic sensors available for die protection, inductive proximity sensors are by far the most popular.

Self-contained, inductive proximity sensors are available in several types of packaging configurations. For die protection, the most common type is the cylindrical variety as shown in Fig. 26-4. Three basic electronic circuits and a miniature coil serve to generate an inductive field which is sensitive to both ferrous and nonferrous metals. This field emanates from the coil which, along with the three circuits, is potted in epoxy resin in the sensor housing. Because the electronic components are immobilized within the epoxy, the sensor can withstand severe shock and vibration.

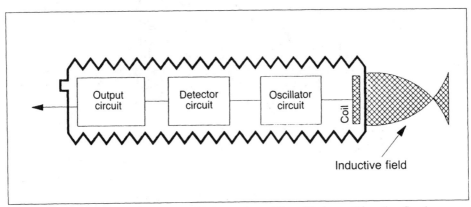

Fig. 26-4. Inductive proximity sensor with internal amplifier. (*Tecknow Education Services, Inc.*)

The housing is made of either plastic or metal. Metal housings are more rugged than plastic and should be used for demanding applications where shock is a problem.

Sensitivity. Generally, the rated target sensitivity for an inductive proximity sensor is based on the response of that sensor to an iron target of a particular dimension. The actual sensitivity varies from this figure when detecting nonferrous and plated ferrous materials.

The inductive field has a three-dimensional shape resembling a fish tail, shown as a cross-hatched area in Fig. 26-4. The target to be sensed must enter the sensing field with enough intrusion to switch the sensor's output status. A target traveling in a lateral direction, such as a strip of material that is being fed into a die, must enter a certain distance into the field in order for detection to occur. The same holds true for target travel that will enter the sensitivity field head on.

The hysteresis of the sensor is the amount of reverse motion of the target which will reverse the sensor status. This on/off detection can be utilized to detect small target dimensional variations as long as they are in excess of the hysteresis number for that sensor.

Since this type of sensor has a fixed size field whose magnitude is determined by the target's geometry and distance, the only way to vary its sensitivity is to physically position it closer or further away from the target. A light which indicates the detection of the target is available on many models of inductive sensors. The exact alignment of the inductive field and the target should be established by trial-and-error on the bench before any permanent modifications are done to the tooling.

Manufacturer's catalogs are an excellent source of information on sensors.

Adjustable Proximity Sensors. Figure 26-5 illustrates an externally amplified proximity sensor. This type of sensor offers the advantage of an electronically variable inductive field size. This feature permits more flexibility in applications where fine adjustments of the field may be necessary due to factors such as tooling wear and material variations. Since the electronics are housed in an external package with only the coil in the sensor housing, this type of sensor is available in miniature sizes as small as 0.120 in (3.05 mm) in diameter. A light indicating object presence is on the amplifier housing. This provides for ease of tooling setup where the indicating light on a self-contained sensor within a die could not be seen.

Specialized types of externally amplified proximity sensors are used for one type of double blank detector on blank feeders. A double blank results in a change in the strength of the inductive field which activates the sensor output. The sensitivity control should be readjusted whenever blank thickness is changed.

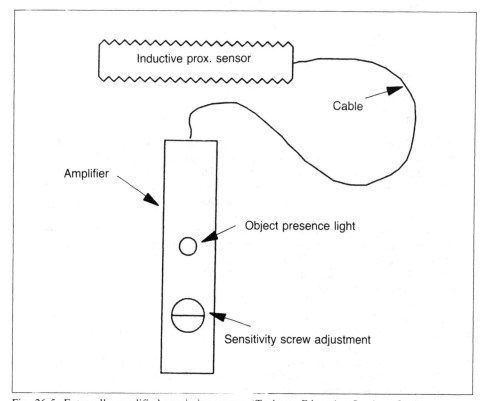

Fig. 26-5. Externally amplified proximity sensor. (*Tecknow Education Services, Inc.*)

Inductive Ring Sensors. Figure 26-6 illustrates an inductive ring sensor. This type of sensor is very useful for the detection of parts being ejected from a die. The sensing coil is large enough for the parts to pass through an opening in the center. The detection of a change in the inductive field is used to verify part ejection.

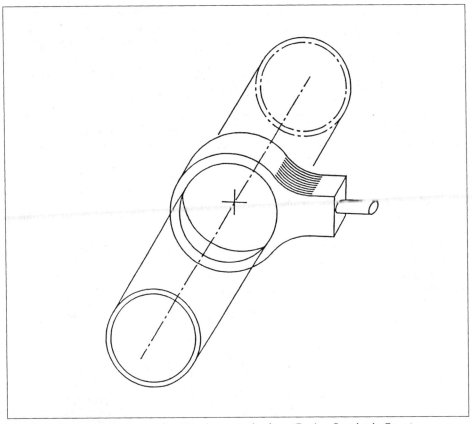

Fig. 26-6. Inductive ring sensor for detecting part ejection. (*Design Standards Corp.*)

Analog Distance Measurement. Inductive proximity sensors generally provide an on/off signal. The hysteresis needed for dependable snap-action operation is provided by a signal conditioning circuit in the detector (Fig. 26-4). No hysteresis is required in an analog output sensor. Instead, the strength of the oscillation is detected and converted to an output, which is a current that varies in proportion to the distance between the sensor and the target.

Specialized Inductive Sensors. Special purpose inductive proximity sensors are available that are built into die and press automation components. Inductive sensors can readily sense the presence of mild steel through a protective housing of nonmagnetic stainless steel. Several new designs, for which patents are pending, integrate sensors with automation components.

Typical applications are draw die blank locating gages that sense the presence of the blank edge, and transfer press fingers that indicate proper part nesting in the part scoop. Figure 26-7 illustrates an inductive proximity sensor having a nonmagnetic stainless steel face integrated into a transfer finger. The sensor (*D*1) fits into a stainless steel part scoop (*D*2). The wiring passes the outside of the hollow tubing that forms the transfer finger body (*D*3) to the transfer rail attachment (*D*4). The attachment device has an electrical connector for the sensor wiring built-in to permit rapid interchanging of automation. A unique feature of this finger design are the ball-and-socket articulated joints that are

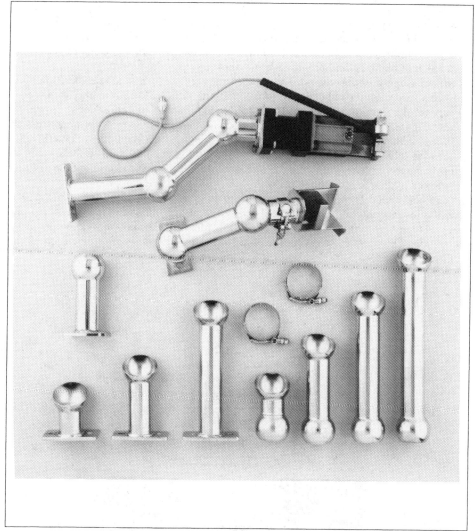

Fig. 26-7. Proximity sensor built into a transfer finger. (*Patent pending, Syron Engineering & Manufacturing Corp.*)

fastened together with soft solder. Adjustment of position is accomplished by warming the joint with a heat gun.

Capacitive Sensors. These sensors are similar in construction to inductive proximity sensors. They are used to detect nonmetallic objects by sensing the increase of electrical capacitance resulting from the object's presence.

Photoelectric Sensors. This is a very versatile family of sensors. Their main advantages over inductive proximity sensors are the ability to sense nonmetallic objects and very long sensing range.

Optical sensors are available in several basic types. The opposed type make use of a separate transmitter (light source), and photoelectric detector (receiver). Retroreflective types either sense objects by detecting the light reflected from the object or by sensing the interruption of a light beam between the sensor and a remote reflector resulting from the presence of an object.

The versatility of optical sensors has been greatly increased by the use of optical fiber light conduits and miniature optical systems to permit very flexible in-die sensing.

Disadvantages include great variations in the sensing range resulting from contamination of optical systems by oil and grime, and higher cost than the inductive proximity type. Color or light reflectance variations of the object to be sensed can also be a problem.

These sensors are widely used to control stock decoilers in order to maintain the correct strip loops in blanking and progressive die applications.

Strain Gage-based Sensors. The measurement of force in a single die station can provide a very sensitive indication of tool condition. Metallic foil strain gages are used as the active elements of custom-made load cells that can be placed under a critical die button. Figure 26-8 illustrates a special load cell for in-die sensing of forming pressure. Such load cells can measure forces from a few pounds (N) to hundreds of tons (kN) full scale, with an accuracy of better than 1%. Major uses of load cells for in-die sensing are determining the force used to score the tops of beverage cans, and measuring the insertion force when assembling cartridge primers.

A noncalibrated type of strain gage sensor is illustrated in Fig. 26-9. This type of sensor is inserted into a drilled hole and secured in place with an expanding tapered plug that is actuated by turning a screw. The output is relative to the pressure developed in the die station. This type of sensor is useful to indicate a change in pressure, such as that resulting from a tool becoming dull or a change in stock condition.[2]

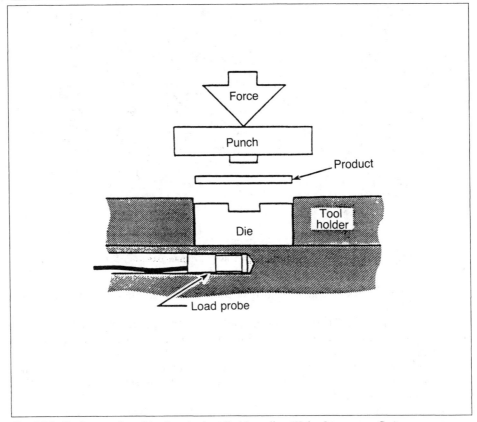

Fig. 26-8. Strain gage-based load probe installed in a die. (*Helm Instrument Co.*)

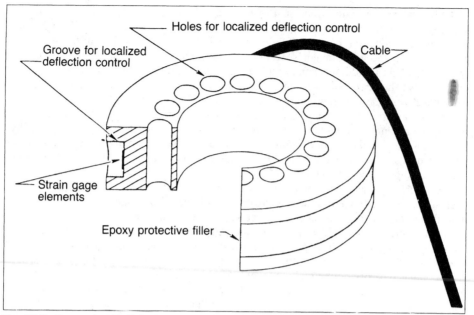

Fig. 26-9. Custom load cell for in-die force measurement. (*Helm Instrument Co.*)

DESIGNING TOOLING TO ACCOMMODATE ELECTRONIC SENSORS AND CABLING

The most effective time to decide where and how to place sensors in a die is at the die design stage. When using computer-aided design (CAD), it is recommended that commonly used sensor types be digitalized into the component libraries for easy retrieval. Thus as the die is being designed, the designer can quickly draw the sensor into its required location.

The routing of sensor cables within the die is best determined in the die design stage. Channels for sensor wiring can be designed easily with a CAD system.

Retrofitting existing dies with electronic sensors can be rather costly. Substantial time is often required to cut pockets for sensors into hardened sections with EDM. Pneumatic hand grinding is often required to cut a channel for sensor wires under hardened die sections. Even so, the cost of retrofitting an existing die is usually more than justified in terms of damage avoidance.

INTERFACING DIE SENSORS TO THE PRESS

A frequently used approach to the routing and interconnecting of sensor cables is the use of individual quick-disconnect connectors on each sensor cable. The lengths of these cables must be sufficient to reach a central junction box which may be mounted on the press or die. The cables external to the die are bundled into a single harness. An alternative to this approach is the use of a single multipin connector to eliminate the need for separate miniature connectors. In this approach, the wiring harness with its individual sensor cables terminates in a central quick-disconnect connector.

Matching Sensors to Monitors. Careful selection of new die protection equipment, or modification of existing models to match the sensor signals, is critical. Where electromechanical feeler probes and switches are simply used to ground a signal, electronic sensors issue outputs which must be compatible with the monitoring controller. The controller can be a commercially available die protection device, a programmable logic controller (PLC), or even a desktop computer. In any case, the signal output from the sensors must be fully compatible with the input requirements of the monitoring device.

Universal Sensor Die Junction Boxes. The sensor interface junction box must be properly mounted on the die or press. Where multiple sensors are used on large dies, mounting the junction box to the die is usually best, provided that the electronic components in the box are not susceptible to damage from shock and vibration. The connecting cable should not remain with the die but be attached to a second junction box on the press. This way, one interconnecting cable can serve many dies and it is not subject to damage from die handling and storage.

The size and particular design characteristics of the junction box, as well as the proper wiring and mounting, are left to the electrical design personnel in each stamping facility. The boxes can either be built in-house, contracted out to an electrical supplier, or purchased as standard units. Certain features are desirable, and should be incorporated in manufacturing engineering standards.

The interface enclosure should be of very robust construction. Often a space for the enclosure can be provided in the die casting or weldment. One damage resistant method is to flange mount the enclosure in an opening cored in the die shoe.

Light emitting diodes (LEDs) can be mounted into the cover plate of the box to show the status of each sensor. This permits easy setup and troubleshooting. Generally the cost of this feature is offset by permitting less costly sensors having no LEDs to be used.

A popular type of commercially available junction box incorporates a low voltage d-c power supply for the sensors. Both normally open and normally closed selector switches allow for quick sensor output selection without the need for rewiring. Many sensors offer both types of outputs and this selector switch can easily assign the particular die protection channel a N.O. (Normally Open) or a N.C. (Normally Closed) function.

EXAMPLES OF SENSOR APPLICATION IN DIES

Strip Thickness and Width Measurements. Both analog inductive and laser sensors are used to monitor the thickness and width of a strip of metal as it is about to enter a die.

Laser sensors make no contact with the strip. By utilizing two sensors, one above and one below the strip, they can actually detect the thickness even if the strip should slightly move up or down within the ranges of the two sensors. The sensors produce a voltage that is proportional to the distance between the sensing surface and the target. By utilizing an electronic comparator unit that sums the two outputs, it is possible to determine the thickness with resolutions of a few millionths of an inch (mm). A disadvantage of laser sensors is that they are rather expensive.

Some users have adapted spring-loaded rollers to make direct contact with the strip (Fig. 26-10). The upper roller moves up and down in proportion to the strip's thickness. A lever provides 4 to 1 mechanical amplification of any movement. The output of the sensor provides a current that varies in proportion to stock thickness. Thickness detection systems of this type are much less expensive than laser systems, but not as accurate.

For width measurements, the sensors are mounted in such a way as to allow them to look at the two edges of the strip. The analog outputs from the sensors will vary in direct proportion to the width of the strip. As in the thickness application, rollers can be used to ride directly on the edges of the strip.

Buckle Detection. Figure 26-11 illustrates the use of an inductive proximity sensor used to detect material buckles. Such an arrangement can be used to detect misfeeds as well as material quality problems.

Form Quality Monitoring. Figure 26-12 illustrates the use of a proximity sensor to detect a poorly formed area or missing form in a progression strip. The sensor is critically positioned in a manner which will allow it to respond to variations on the formed area. By carefully synchronizing the position of the progressive strip with respect to the sensor signal, and choosing a sensor with a resolution that exceeds the part tolerance, it is possible to detect very small variations in the geometry of the part.

In a progressive die, a failure to detect a formed area can be used to detect misfeeds.

Bend Angle Detection. Figure 26-13 illustrates the use of an analog inductive

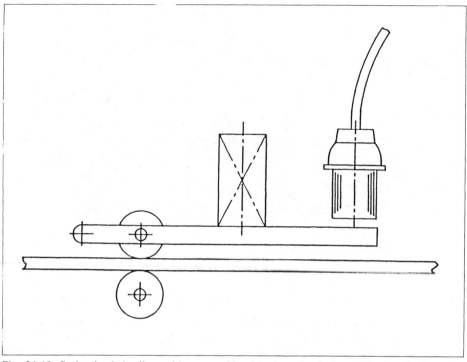

Fig. 26-10. Spring-loaded roller and lever provides 4 to 1 mechanical amplification for detecting stock thickness variation with an analog inductive sensor. (*Design Standards Corp.*)

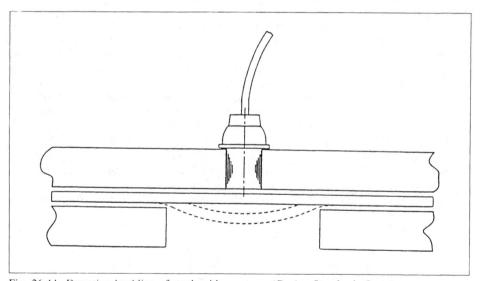

Fig. 26-11. Detecting buckling of stock with a sensor. (*Design Standards Corp.*)

proximity sensor to detect an off-angle flange. Such sensing methods can make possible in-die sensing for SPC data gathering.

Feed Length Sensing. Feed length can be monitored easily with analog sensors as well as the on/off variety of sensors located in the die. These sensors are placed to monitor hole locations or solid strip features for the proper feed length.

Additional sensors can be mounted on the feed mechanism to monitor gripping and sliding motions on air feeds or similar mechanisms on mechanically coupled feeds.

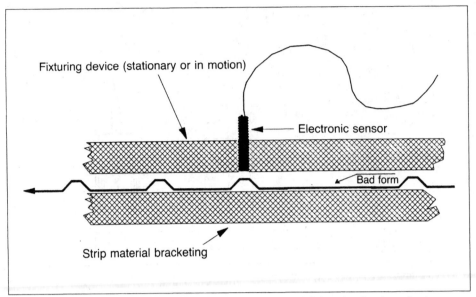

Fixturing device (stationary or in motion)

Electronic sensor

Bad form

Strip material bracketing

Fig. 26-12. Sensing form height and missing forms. (*Tecknow Education Services, Inc.*)

Variations in the material loop can be monitored with ultrasonic sensors as well as analog output photoelectric types. Thus the entire feed cycle can be carefully monitored.

Combining Sensors. The combination of two normally closed sensors with an inexpensive electromechanical relay, to achieve synchronization between two events, is shown in Fig. 26-14. Both sensors must be off in order for the load to be activated. Should any one of the two switch on, the coil in the relay will be activated and the load will be turned off. Thus since these two sensors are of the normally closed variety, the only way in which they will both be off is if both detect the presence of two distinct targets.

Additional Sensing Applications. Sensor utilization in metal stamping includes the detection of punch breakage, part misalignment, burr detection, pad return, and cam return, to name just a few common applications.

Some companies choose to use a minimum of sensors in areas that are very critical to the process. Others have no hesitation about using electronic sensors to monitor all possible problem areas. In both approaches, the operator is given more freedom to pursue areas of responsibility that under a nonautomated system were denied to him or her. SPC charting and material handling duties can often be added to the responsibilities of the operator whose previous function generally consisted largely of sitting for hours in front of the press and acting as nothing more that a human sensor.

Die Setting Applications. The two basic outputs from sensors are on/off for the

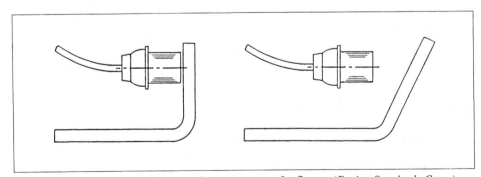

Fig. 26-13. Inductive sensor for measuring squareness of a flange. (*Design Standards Corp.*)

26-12

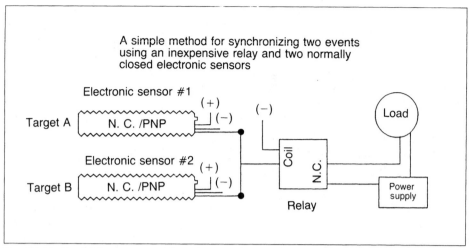

Fig. 26-14. Combining sensors with a relay. (*Tecknow Education Services, Inc.*)

detection of the presence or absence of a target and analog output for the precise measurement of the distance between the target and the sensor. By using a combination of these, it is possible to detect the proper installation of dies within the press as well as the measurement of the shut height. The sensors can be mounted in the press slide, in dies, on bolster plates, and other locations to measure critical setup parameters which normally are done with feeler gages or by visual inspection.

There are solid state miniature memory modules available which can be mounted on each die. These memory devices can store both the identification number for each die and other setup parameters such as feed length, die protection timing, and limit switch angles. The information is read with a special sensor, mounted on the press in such a way that it can read the memory modules mounted on the dies.

Bar coded die identification that is read with a hand-held scanning device is another practical means of causing the appropriate setup parameters to be downloaded from a computerized library of setup information into a microprocessor-based controller.

References

1. P. R. Enzler, "Precision Switches Guard Progressive Die," *Am. Machinist* (April 21, 1949).
2. Donald F. Wilhelm, *Optimized Force Measurement*, SME Conference Proceeding, Society of Manufacturing Engineers, Dearborn, Michigan, 1990.

SECTION 27

PRESS DATA

Presses must be capable of exerting pressure in the amount, location, direction, and period of time needed to accomplish a specific operation. The machines frequently are required to exert not one but several pressures for various intervals of time. Automatic means for feeding blanks and material and ejecting the finished parts are often incorporated in presses (see Section 23).

PRESS CLASSIFICATION

Presses are classified by one or a combination of the following characteristics: source of power, method of actuation of slides, number of slides incorporated, frame type, bed type, and their intended use.

There is extensive technical literature on the subject of presses. Die designers, process engineers, and product designers should become familiar with the many available types of presses and their capabilities.[1, 2, 3]

SOURCE OF POWER

The source of power for presses is either manual, mechanical, or hydraulic.

Manual. These presses are either hand- or foot-powered through levers, screws, or gears. The most common press of this type is the bench-type arbor press.

Mechanical. There are three major types of mechanical press drives. These are the nongeared or flywheel type, single-reduction-gear type, and multiple reduction-gear type which are illustrated schematically in Fig. 27-1. In all three types a flywheel stores energy. The source of power is a motor which returns the flywheel to operating speed between strokes. The permissible slowdown of the flywheel during the work period is about 7% to 10% in nongeared presses and 10% to 20% in geared presses.

The flywheel-type drive (Fig. 27-1A) transmits the energy of the flywheel to the main shaft of the press. The main shaft may be mounted right to left (parallel to the front of the press) with the flywheel on either side, or the shaft may be mounted front to back (at right angles to the front) with the flywheel in the back. Flywheel or nongeared presses are generally applicable to light blanking, operating small, high-speed progressive dies, or other high-speed operations.

The single-reduction-gear drive (Fig. 27-1B and C) transmits the energy of the flywheel to the main shaft through one gear reduction and is recommended for heavier blanking operations or shallow drawing. This drive operates a press at speeds of approximately 30 to 40 strokes per minute and in some high-speed presses up to 150 strokes per minute.

The multiple-gear drive (Fig. 27-1D) transmits the energy of the flywheel to the main shaft through two or more gear reductions. These reductions reduce the strokes per minute of the slide without reducing the flywheel speed. Large presses with heavy slides use a multiple-gear drive since it is impractical to accelerate and decelerate large amounts of

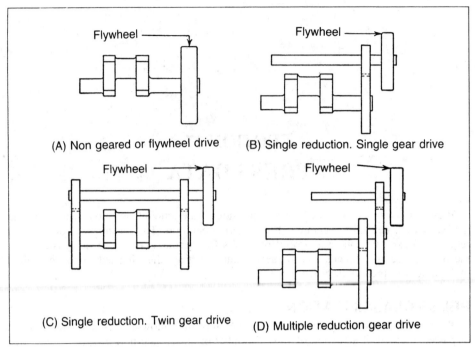

(A) Non geared or flywheel drive (B) Single reduction. Single gear drive

(C) Single reduction. Twin gear drive (D) Multiple reduction gear drive

Fig. 27-1. Types of mechanical press drives.

mass at high speeds. Dependent upon the tonnage capacity of the press and the length of its stroke, speeds normally range from 8 to 20 strokes per minute, although designs are available which have a speed up to 50 strokes per minute.

Whether the press is single- or multiple-geared, energy can be transmitted to the main shaft from one or two gears on each shaft. They are respectively called single or twin drive presses, according to the use of one or two pairs of gears on each shaft to transmit the energy (Fig. 27-1B and C). The twin-drive arrangement is used to reduce the torsional strain on the shaft since it applies power to both sides of the point of resistance. Crank presses of low tonnage capacities, short stroke, and a narrow span of the main shaft from bearing to bearing are generally not twin-driven. The shaft on these presses is not subjected to so great a torsional stress as the shaft on presses with a greater span. It is advisable to twin-drive long-stroke presses, whether or not they have a short or long span between the bearing housings.

Most presses are of the top-drive type in which the driving mechanism is located in the crown and pushes the slide down to perform the operation. The underdrive type has the mechanism under the bed with connecting linkage in the upright to pull the slide downward. The mechanisms of the large underdrive presses are below the floor level, requiring a minimum of headroom above the production floor, but either a basement or trench-type pit is required to service them.

It is desirable to order a hydraulic overload system for presses of this type, since there are no tie rods to heat should the press become stuck on bottom.

Hydraulic. This type of press uses oil pressure in a cylinder with a closed end reacting against a piston to move the slide. A pump supplies the pressure. Hydraulic mechanisms are capable of maintaining constant pressure and speed throughout the entire stroke, or of exerting maximum pressure at any desired point within the limits of slide travel and press capacity.

Modern self-contained hydraulic presses have their own pump, motor, and hydraulic systems. The draw speed of the press is directly proportional to the size of the motor and pump. The tonnage capacity is dependent upon the cross-sectional area of the piston or pistons and the pressure developed by the pump. A few large presses use an accumulator system to supply the energy. This system is capable of supplying a large volume of hydraulic fluid at a relatively constant pressure in a short time. It requires smaller volume capacity pumps than direct-connected systems since the pumps can recharge the accumulator during the time required for loading and unloading the dies.

METHOD OF ACTUATION OF SLIDES

In all power presses the work is performed by the slide or slides through reciprocating motion to and from the press bed or bolster. In mechanical presses, slide motion originates from:

1. Crankshafts incorporating crankpins or eccentrics.
2. Eccentrics cast or welded integrally with or bolted to rotating main gears.
3. A pair of knuckles folded and straightened out by a pitman-crankpin mechanism.
4. Oscillating rocker shafts and toggles.

Main Shafts. Main shafts in single-action presses of the nongeared type are of conventional crank, semi-eccentric, or full-eccentric design. The number of throws or eccentrics on the shaft is determined by the number of points of suspension of the slide. Points of suspension are the number of places where pressure is transmitted by connections to the slide. Press connections, called connecting rods or pitmans, are usually adjustable in length so the shut height of the press can be varied.

The single-point suspension is used in presses with relatively narrow widths between the uprights or with narrow slides. Two-point-suspension presses have connections on each end of the slide. The slides are wide right to left but shallow front to back to accommodate long but relatively narrow dies.

Four-point-suspension presses have connections on each corner of the press slide. These presses can accommodate work in wide (right to left) and deep (front to back) dies, especially when loads are distributed unsymmetrically. The recommended load on any one corner is not more than 30% of the rated tonnage of the press.

Cutting and trimming dies often require much less maintenance when operated in a four-point press as compared to a two-point press. The gibbing wear is also often less.

Eccentric Gear. Eccentric-gear presses incorporate a slide actuated by an eccentric, which is cast or welded integrally with, or keyed and bolted to, the main gears. The gear eccentric is mounted on either a rotating or nonrotating shaft, sometimes used as a power take-off for auxiliary equipment. Torsional stresses are not present in shafts of eccentrically driven presses unlike those of crankshaft-driven presses.

Knuckle Joint. Presses employing knuckle joints develop tremendous pressures near the bottom of the stroke. They are often used for compression operations requiring high pressures during a short portion of the stroke. The force is applied through a crank or eccentric to the joint connecting two levers of equal length. The levers are actuated through a nonadjustable connection from the shaft in the back of the press frame. The slide motion is achieved through the straightening of the two hinged levers which suspends the slide from the crown.

Toggles. A crank or eccentric actuates a series of levers linked in tandem in a sequence of movements through two or more dead-center positions or dissipation points. These are spaced at closely related intervals to accomplish an effective dwell of the holding member. In the toggle mechanism the force is always exerted against one end of a series of levers but should not be confused with the knuckle-joint motion. Figure 27-2 illustrates the actual pressure of a double-action press outer slide measured with load cells while in the dwell portion of the cycle. The three-humped pressure curve is typical of the pressure variation of such presses during dwell. At full tonnage, the variation is under 10%.

Toggle action is widely used in double-action presses to obtain the proper stroking characteristics for the blankholder. Figure 27-3 shows that the reciprocating action of the toggle-driven slide is faster than, and dwells longer than, the crank-driven slide. The curve of the knuckle-joint action shows a shorter degree of dwell than the toggle action.

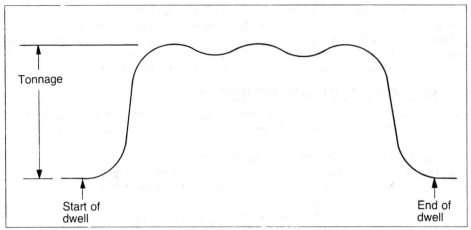

Fig. 27-2. Typical pressure variation of a toggle-actuated slide while on dwell. (*Smith & Associates*)

Cams. Cams are utilized to produce motions at angles and in planes not obtainable for simple crankshaft mechanisms. The motions of cams, cranks, and eccentrics are frequently combined, as on single- and multiple-action presses where one or more dwells or certain hold-down or stripping motions are required. The double-action cam press essentially embodies the same motions as the double-action toggle press but can generally be operated at higher speeds. These presses are usually limited to lower tonnage presses of relatively narrow widths and are primarily used for blanking and cupping operations.

Drag Link. A drag-link mechanism on a press will produce a relatively slow drawing motion with a connection for faster return of the slide. The resulting draw stroke in this type of drive is practically of uniform velocity. It is adaptable for reducing long shells at a maximum and nearly uniform speed and for broaching and burnishing.

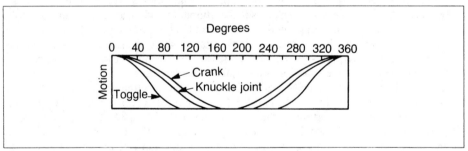

Fig. 27-3. Relative motions and degrees of dwell obtained with crank, knuckle-joint, and toggle actions.[1]

NUMBER OF SLIDES IN ACTION

The classification of a press also includes the number of slides incorporated in the press such as single, double, or triple-action presses.

Single Action. A single-action press is one which has only one slide.

Double Action. A double-action press is one which has two slides. This type of press is usually used for drawing operations during which the outer slide carries the blankholder and the inner slide carries the punch. The outer or blankholder slide, usually having a shorter stroke than the inner or punch-holder slide, dwells to hold the blank while the inner slide continues to descend, carrying the draw punch to perform the drawing operation. The relative motions between the inner and outer slides are illustrated in Fig. 27-4.

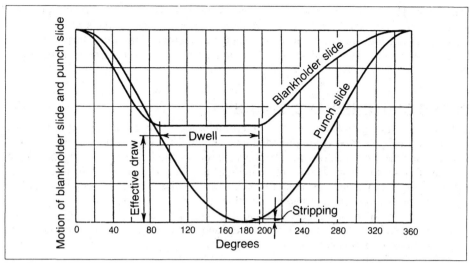

Fig. 27-4. Relationship between the motions of the blankholder or outer slide and the inner slide or plunger of a double-action press. [1]

Triple Action. A triple-action press incorporates three slides having three motions properly synchronized for triple-action drawing, redrawing, and forming. Two slides, the blankholder (outer) and plunger (inner) slides, are located above, and a lower slide is located within the bed. The inner slide develops a small amount of dwell at the bottom of its stroke. The blankholder slide usually has slightly more dwell at the bottom of its stroke, and more dwell than on a comparable double-action press. The lower slide is usually actuated by an eccentric or a crank. The lower slide mechanism as a rule develops a long stroke, and only during the very top portion of the stroke is force transmitted to the lower slide. This is accomplished by limiting the motion of the pressure pin plate and the traveling of the lower slide down from the pin plate proper.

Present-day practice is to synchronize the lower slide motion mechanically with the upper slide motions, although separately driven lower motions, electrically interlocked to the upper motions, are often used.

A disadvantage of a triple action press is that the inner slide always goes into dwell, even though the third action is not being used. At 16 strokes per minute, the inner slide dwell time is approximately 1 second.

Figure 13-15 in Section 13 illustrates a die set-up in a third-action press.

PRESS FRAMES

Power press frames can be broadly classified into two general types, gap frame (or C-frame) and straight side.

Gap Frame. As seen from the side, the housings of a gap-frame press are cut back below the gibs to form the shape of a letter C. This permits feeding of wide strip from the side, or large sheets from the front. Gap-frame presses have a solid back or an open back to permit feeding from front to back or the ejection of finished parts out the back.

Open-back frame presses have frames that are in a fixed vertical position or in a fixed inclined position or a frame that can be inclined. The inclined positions permit finished parts to fall out by gravity, or be blown out by air at the rear of the machine.

Open-back frame presses have the flywheel and gearing, if any, at either the right or left side of the press. The flywheel and gearing on a solid-back press are available on either the side or the back of the press.

Straight Side. This type of press incorporates a slide which travels downward between two straight sides or housings. The frame construction consists of a base, bed, housings, and crown. The frame may be cast or welded in one piece or the individual parts may be joined by steel tie rods. These presses are used for heavy work but the size of the work is limited by the left-to-right distance between the housings and the work must be fed into the die from front to back. Side-to-side feeding may be achieved if there are openings in the housings.

PRESS CLUTCHES AND BRAKES

Clutches. Timing and control of the intermittent reciprocating movement of the slide in a mechanical power press are provided by a clutch. The clutch is inserted between the flywheel and the drive mechanism. The flywheel rotates continuously, and engagement of the clutch causes the drive shaft to rotate and start the slide on its working stroke. As soon as the stroke is completed, the clutch is automatically disengaged and the brake applied unless the press is set for continuous operation.

Types of press clutches are divided into three major groups:
1. Positive clutches, in which the driven and driving members of the clutch are interlocked in engagement.
2. Friction clutches.
3. Eddy-current clutches, in which the driven member follows the driving member by the attraction of two induced magnetic fields.

The slide of a press having a positive clutch is usually stopped at its upper position. Lowering (inching) or backing up (lifting) of the slide during the die setting must be done by turning the flywheel or main gear by hand, chain block, or a hand bar passed through the eye of the crankshaft. Many flywheel-type presses (nongeared) have bar holes provided in the rim of the flywheel into which an inching bar may be placed for turning the press. Presses with friction clutches allow the slides to be stopped, started, or reversed at any point in the stroke. This feature is convenient for setting and adjusting dies, especially in large presses. Modern air-friction clutches are usually considered safer than positive clutches. A well designed air-friction clutch is a "fail-safe" mechanism wherein sudden power or air failure will result in the press stopping immediately. Friction clutches using high-pressure oil or magnetic attraction have been employed to some degree. Presses equipped with eddy-current clutches can be started, stopped, and reversed with variable torsional forces; they can be used in larger-sized presses (500 tons and over). They have no friction surfaces or mechanical connections between driver and driven members, resulting in long, maintenance-free operation. They may be cycled rapidly without serious effects—one of the great limitations of air-friction clutches. With special electronic controls, these clutches can operate under varying degrees of slip resulting in slow draw and fast return motions of the slide.

Brakes. Because of the inertia of the press components which either slide or rotate during the press stroke, a brake is required to stop the slide after the clutch has been disengaged. Brakes in the presses with positive clutches are placed in the outer crankshaft end, opposite to the drive-wheel end. They are, as a rule, a continuous acting type of a yoke, band, or double-shoe construction.

Brakes in presses with friction clutches are single, multiple-disk or drum type, are interlocked with the clutches, and work only when the clutches are disengaged. Often the clutch and brake are a single package operated by a common actuating cylinder.

Eddy-current-type clutches provide dynamic braking within the clutch unit proper. A supplementary holding brake is required to hold the press in position when the rotation has been stopped.

Ineffective brake or clutch action (in hand-fed presses, normally single-stroked) such as an accidental "repeat" can be hazardous to dies, press equipment, and the operator. Current Occupational Safety and Health Act (OSHA) regulations should be closely followed.

DIE CUSHIONS

When a single-action press is used for drawing operations, the manner in which pressure is applied to the blankholder to control the flow of the metal blank is important. The application of pressure to a blankholder is one of the features of a double-action press. Single-action presses lack this feature and require supplementary blankholding equipment.

Dies are sometimes built with a blankholder using compression springs, air cylinders, or high-pressure nitrogen cylinders to supply the holding pressure. This greatly increases the cost of the die. A press cushion can serve every die with this requirement, lowering the cost of tooling.

Pneumatic Die Cushion. This type of die cushion is used with shop air pressure which should not exceed 100 psi (690 kPa). A pneumatic-die-cushion design normally uses either one or two pistons and cylinders. The recommended capacity of a die cushion is about 15% to 20% of the rated press tonnage. The size of the press-bed opening limits the size, type, and capacity of the cushion.

A schematic arrangement of a pneumatic die cushion is shown in Fig. 27-5. This illustration shows an inverted-type cushion in which the downward movement of the blankholder, through pressure pins, forces the cylinder against a cushion of air inside the cylinder, and moves the air back into the surge tank. On the upstroke, the air in the surge tank returns the cylinder. Other designs function without surge tanks.[4]

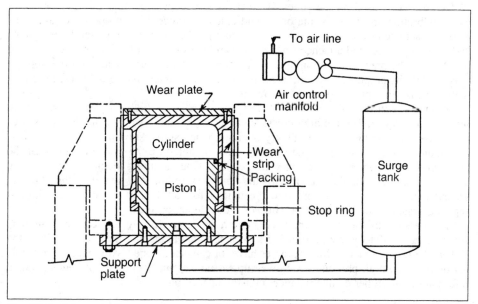

Fig. 27-5. Schematic diagram of a pneumatic die cushion.[4]

It is important to load the cushion evenly in order to avoid premature wear and cushion failure. The die designer should incorporate equalizing pins in the lower shoe if needed to accomplish equalization of loading. Often these can be actuated by the upper heel blocks. In some cases, special pin drivers made of structural tubing may be attached to the slide to actuate equalizing pins. Operator safety must be considered when doing this because additional pinch points are created.

Die cushions are often used in double-action presses to keep the bottom of the stamping flat, or to hold it to shape, or to prevent distortion and slippage while drawing. They are also used to actuate liftout rods or plates to push the finished drawn part out of the die cavity.

A few presses provide for the removal of scrap through the bolster on presses equipped with die cushions by means of a special stool which may be installed on top of the cushion to allow the slugs to fall out the side of the press. Another means of providing for scrap removal is a slug tube through the cushion to allow smaller pieces of scrap to fall straight through the press bed.[5]

Hydropneumatic Die Cushions. These die cushions are recommended where the capacity required is greater than can be obtained with 100 psi (690 kPa) air pressure on a pneumatic cushion. Hydropneumatic cushions are slower acting than the pneumatic cushions, and as a result are usually used on larger, slower presses. They can be adjusted to hold a large light blank for deep drawing or shallow forming, or to grip heavy-gage material as tightly as required for curved-surface or flat-bottom forming.

The use of hydropneumatic cushions is limited to slow operating speeds. Where long cushion strokes are required, slow operating speeds must be used. Conversely, faster operating speeds can be used only with very short strokes.[1]

Locking Devices for Die Cushions. A locking device is a hydraulic unit which controls and delays the stripping action of pneumatic and hydropneumatic die cushions by delaying the upstroke of the cushion with respect to the upstroke of the press slide. The net result prevents the damaging of a drawn shell which otherwise might occur if the cushion were permitted to exert pressure on the shell during the upstroke of the press.[1, 6]

SLIDE COUNTERBALANCE

This device is used on the slides of all presses, except those that are very small, to reduce vibration and to assist the brake and clutch in functioning properly. Counterbalances are actuated by springs or air pressure, and provide a means to offset the weight of the slide, upper die, and attached linkage. This relieves much of the load of the slide and punch from the press connection and shaft, reducing the amount of load on the holding brake. With the proper adjustment of air pressure in a counterbalance cylinder, problems with backlash in a press drive can be minimized.

Generally, adjusting the counterbalance pressure slightly higher is less harmful than setting it too low. Excessive high pressure causes rapid clutch wear, while low pressures will cause the holding brake to quickly overheat when single-stroking.

Counterbalances consist of one or more units mounted on the crown or frame of the press with the piston shafts connected to the slide. One mounting method is to place the units on either side of the crown with the shafts extending downward to the slide. A second method is to support or build the units into each end of the press housing. A separate system is needed for each slide of a double-action press.

Air pressure to the counterbalance system is controlled by a pressure regulator. Approximately 40 to 50 psi (276 to 345 kPa) is required to balance the dead weight of the slide and attached linkage. Many presses have, or can be equipped with, a tag with a table giving the correct air pressure for a given upper die weight. Section 24 provides detailed information on proper counterbalance adjustment methods.

HYDRAULIC OVERLOAD SYSTEMS

Figure 27-6 illustrates a slide from a single-action straight-side machine that has hydraulic overload cylinders built into the slide connection. This system has an advantage over a hydraulic overload-relief bed operating on a similar principle in that an overload cylinder is provided for each connection point. This provides superior overload sensing and protection.

In the event of an overload, oil is dumped very rapidly, through a quick-action valve allowing the overload cylinder to collapse about 0.75 in. (19.0 mm) and relieving the overload condition. It is important that all cylinders be dumped simultaneously, even though only one initiated the dumping process, so as not to create a serious out-of-parallel condition that can damage the die.

It is feasible to dump the system remotely, by an electrical signal from a strain gage-

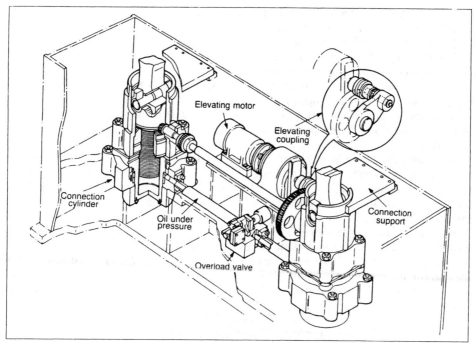

Fig. 27-6. Permanent overload protection unit has pre-stressed hydraulic cylinders. (*Verson Allsteel Press Co.*)[1]

based tonnage metering system. Such an arrangement provides a superior die and press protection system.

JIC STANDARD PRESS DIMENSIONS

The Joint Industry Conference (JIC) of press users and press manufacturers has developed standards on the following classes of mechanical presses:
1. Open-back inclinable presses.
2. Straight-side, single-action, single-point presses.
3. Straight-side, single-action, multiple-point presses.
4. Straight-side, double-action, single- and multiple-point presses.
5. Rail or frame presses.
6. Triple-action, multiple-point presses.

The standards developed are for the working area of the press; namely, die space, stroke, bed and slide areas, and other dimensions essential to obtaining interchangeability of dies between all presses of like tonnage and die-space areas, regardless of make. All of the following JIC standards are given in English units. There are presently over 300,000 such presses in use in North America, nearly all of which are built to this standard.

A visible means of identifying a press according to the action, points of suspension, tonnage, and dimensions of the bed was suggested by the JIC and is in general usage by the press manufacturers.

The following is the recommended method of press marking and is to be placed at a central position on the front of the press.

NAME OF MANUFACTURER
D2-600-96-72

As these symbols appear left to right they represent:
1. Action and points of suspension.
2. Tonnage in standard short tons.
3. Left-to-right dimension in inches.
4. Front-to-back dimension in inches.

In item 1, the letter "S" is used to denote single-action; "D," double-action; "T," triple-action; and "O.B.I.," open-back inclinable.

Specification tags are attached to the press which state tonnage, stroke, shut height, adjustment, size, and strokes per minute for the press slide or slides as the case may be. An additional tag is used to give the die-cushion data.

A standard distance above the bottom of the maximum stroke has been established; at this point the press will exert the full rated tonnage. Single-action, straight-side presses of the single-end-geared single-point type will exert full tonnage 0.25 in. (6.4 mm) up from the bottom of the stroke. The double-point eccentric and twin-drive types will exert full tonnage 0.5 in. (12.7 mm) above the bottom of the stroke. The main or inner slide of double and triple action presses will exert full tonnage 0.5 in. (12.7 mm) above the bottom of stroke, and 0.25 in. (6.4 mm) is the corresponding distance for the blankholder or outer slide for the same press. The lower slide of triple-action presses will exert full tonnage 0.5 in. (12.7 mm) from the top of the stroke. These dimensions relate to presses with JIC standard strokes. The measurements for open-back inclinable presses are shown in Table 27-1.

TABLE 27-1
Measurement of Maximum Rated Tonnage
For Open-back Inclinable Presses

Tonnage	Above Bottom of Stroke, in.	
	Nongeared Type	Geared Type
22	0.031	0.125
32	0.031	0.125
45	0.062	0.250
60	0.062	0.250
75	0.062	0.250
110	0.062	0.250
150	. . .	0.250
200	. . .	0.250

Bolster Plates. Bolster plates are centrally located on the top of the bed by two dowels, one located at the front right-hand side and the other at the rear left-hand side. The dowel-pin centers are measured from the center lines of the bolster at a given fixed dimension, held to a tolerance of ±0.015 in. (±0.38 mm), for each of the various widths and depths. T-slots are machined from front to back to receive 0.75- and 1-in. (19.0- and 25.4-mm) T-bolts. The center slot is located on the center line and additional slots continue at 6-in. (152-mm) spacings on both sides of this slot. The same pattern for T-slots is followed in the slide. Each slot is measured from the center line and the dimension held within a tolerance of ±0.015 in. (±0.38 mm). The bolster plates for open-back inclinable and single-action, straight-side, single-point light-series presses have the T-slots machined a fixed distance toward center from all four sides.

The drilling of holes in the bolster plate for pressure pins follows a similar pattern in the medium- and heavy-tonnage presses. Holes are located 3 in. (76 mm) from the bolster center lines and continue on both sides of these holes at 6-in. (152-mm) centers. The rows of holes are dimensioned from the center lines and are held within a tolerance of ±0.015 in. (0.38 mm). This spacing places rows of holes extending front to back between the

T-slots. Light-tonnage, straight-side, and gap-frame presses have holes drilled at 3 in. (76 mm) centers with the same dimensional tolerances.

Open-back Inclinable Presses. The JIC standard dimensions for the open-back inclinable (gap-frame) presses are shown in Fig. 27-7. The dimensions are given according to the established tonnage capacity of the press.

The layout of the ram or slide face is shown in Fig. 27-8, and corresponding tabular data are given in Table 27-2. Included in these data are the anchorage holes and the T-slots, if any. The size and depth of the punch or upper shoe stem hole are given in Fig. 27-7. The T-slots and anchorage holes are dimensioned from the center line and held within the ±0.015 in. (±0.38 mm) tolerance.

The layout of the bolster plates for the open-back inclinable presses is shown in Fig. 27-9, with pertinent tabular data listed in Table 27-3. The pressure pinholes and T-slots are machined on the bolster center lines and extend on both sides of the center lines as mentioned. The ASA standard 0.75-in. (19.0-mm) T-slots are spaced at 6-in. (152-mm) increments with the spacing held within a tolerance of ±0.015 in. (±0.38 mm) on their distance from respective center lines. The pressure pinholes are drilled at 3-in. (76-mm) centers and are dimensioned from the center lines and held to within ±0.015 in. (±0.38 mm) tolerance. The bolster-plate thicknesses are given in Fig. 27-9.

Straight-side, Single-action, Single-point Presses. This press classification has been divided into 10 bed sizes with 15 tonnage ratings. A schematic layout of the bed and slide for this press classification is shown in Fig. 27-10, with its corresponding tabular data given in Table 27-4. The total number of die-anchorage holes in the slide is given, and they are evenly spaced to the right and left of the center line along the front and back edges of the slide face. T-slots machined in the slide face extending from front to back, and the number of T-slots listed in the tabulation, are evenly spaced on each side of the front-to-back center line. Holes for knockout pins through the slide are provided as shown.

The bolster-plate dimensions, number of T-slots, and number of rows of pressure pinholes are given in Table 27-5 and the tabulated layout is shown in Fig. 27-11 for this press classification.

Straight-side, Single-action, Multiple-point Presses. A schematic layout of the press bed, bolster plate, and press ram or slide for this press classification is shown in Fig. 27-12 and the JIC standardized dimensions are listed in Table 27-6.

Shown on the bolster-plate layout is the method of identifying and marking the rows of pressure pinholes. The even numbers identify the pressure pinholes to the right of the front-to-back center line and odd numbers are used to identify those to the left of the center line. The letter A and alternate letters are used to identify the rows of holes back of the right-to-left center line. The letter B and alternate letters are used to identify the rows in front of the center line. The letters and numerals are to be stamped in the proper areas using 0.5-in. (12.7-mm) high characters.

Table 27-7 is supplementary to Table 27-6 and supplies the bolster thickness and rows of pressure pinholes front to back in relation to the rated tonnage of the press.

The JIC standard lengths of stroke of the press slide for straight-side single-action presses are given in Table 27-8. These strokes have been established in relation to the rated tonnage of the press.

The bolster plates on straight-side presses have horizontally tapped holes for attaching of mounting hooks. The location and size of these holes are shown in Fig. 27-13.

Straight-side, Double-action, Single- and Multiple-point Presses. The JIC standard dimensions for the working area of this press classification are listed in Tables 27-9 and 27-10. The schematic layout of the slide, showing the sizes of the inner slide, T-slot, and anchorage-hole arrangement, is shown in Fig. 27-14. The schematic arrangement of the bolster plate and press bed of double-action presses is similar to that of single-action presses; therefore part of Fig. 27-12 applies to double-action presses as well as single-action presses.

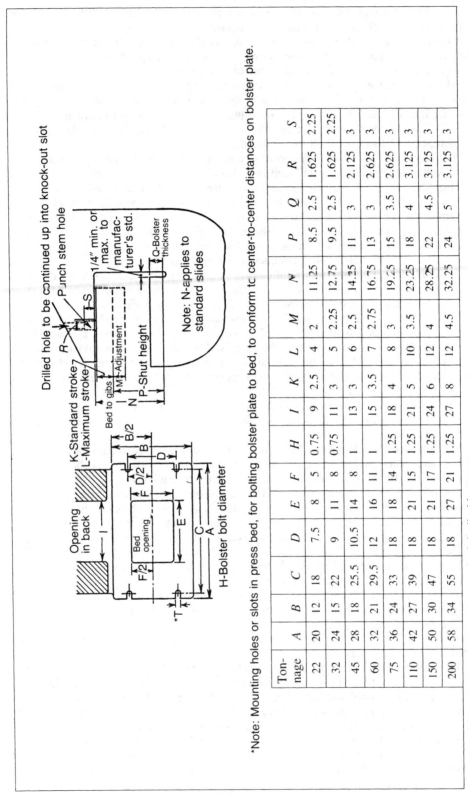

Drilled hole to be continued up into knock-out slot

Punch stem hole

1/4" min. or max. to manufacturer's std.

Q-Bolster thickness

P-Shut height

M-Adjustment

Bed to gibs

Note: N-applies to standard slides

K-Standard stroke
L-Maximum stroke

Opening in back

Bed opening

H-Bolster bolt diameter

Note: Mounting holes or slots in press bed, for bolting bolster plate to bed, to conform to center-to-center distances on bolster plate.

Tonnage	A	B	C	D	E	F	H	I	K	L	M	N	P	Q	R	S
22	20	12	18	7.5	8	5	0.75	9	2.5	4	2	11.25	8.5	2.5	1.625	2.25
32	24	15	22	9	11	8	0.75	11	3	5	2.25	12.75	9.5	2.5	1.625	2.25
45	28	18	25.5	10.5	14	8	1	13	3	6	2.5	14.25	11	3	2.125	3
60	32	21	29.5	12	16	11	1	15	3.5	7	2.75	16.75	13	3	2.625	3
75	36	24	33	18	18	14	1.25	18	4	8	3	19.25	15	3.5	2.625	3
110	42	27	39	18	21	15	1.25	21	5	10	3.5	23.25	18	4	3.125	3
150	50	30	47	18	21	17	1.25	24	6	12	4	28.25	22	4.5	3.125	3
200	58	34	55	18	27	21	1.25	27	8	12	4.5	32.25	24	5	3.125	3

*Note: Mounting holes or slots in press bed, for bolting bolster plate to bed, to conform to center-to-center distances on bolster plate.

Fig. 27-7. JIC standard dimensions for open-back inclinable presses.

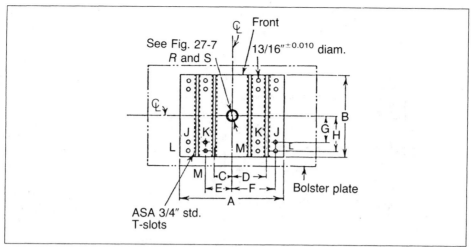

Fig. 27-8. Layout of slide for open-back inclinable presses.

TABLE 27-2
Dimensions of the Slide for an
Open-back Inclinable Press (See Fig. 27-8)

Tonnage	A	B	C	D	E	F	G	H	Hole Pattern	No. T-Slots
22	12	10	. .	. .	. .	4.750	. . .	3.750	L	0
32	15	12	. .	. .	. .	6	. . .	4.500	L	0
45	18	14	. .	. .	. .	7.500	3	6	J–L	0
60	21	16	. .	. .	. .	9	3	6	J–L	0
75	24	18	6	. .	. .	9	6	7.500	J–L	2
110	28	21	6	. .	9	12	7.500	9	J–M	2
150	34	24	6	12	9	15	. . .	9	L–M	4
200	36	28	6	12	9	15	9	12	J–K–L–M	4

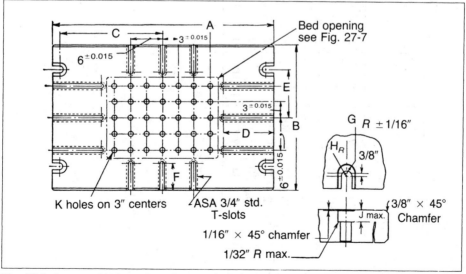

Fig. 27-9. Layout of bolster plates for open-back inclinable presses.

A	B	C	D	E	F	G	H	J	K	Left to Right		Front to Back	
										T-Slots	Pressure Pinholes	T-Slots	Pressure Pinholes
20	12	9	5	3.750	3	0.750	0.406	1.250	. . . . . .	2	0	2	0
24	15	11	6	4.500	2.500	0.750	0.406	1.250	0.812	2	3	2	3
28	18	12.750	6	5.250	4	0.875	0.562	1.250	0.812	2	5	2	3
32	21	14.750	7	6	4	0.875	0.562	1.250	0.812	2	5	2	3
36	24	16.500	8	9	4	1.062	0.688	1.500	0.812	6	5	6	5
42	27	19.500	9.500	9	5	1.062	0.688	1.500	0.812	6	7	6	5
50	30	23.500	14	9	6	1.062	0.688	1.500	1.062	6	7	6	5
58	34	27.500	14.500	9	6	1.062	0.688	1.500	1.062	10	9	6	7

TABLE 27-4
Bed and Slide Dimensions for Straight-side, Single-action, Single-point Presses (See Fig. 27-10)

Left to Right				Anchorage Holes, Slide		T-Slots Slide		Front to Back				Knock-Out Pin-Holes	Tonnage	
I ±0.250	B min	C ±0.015	D ±0.032	No.	Diam.	No.	Size	H ±0.250 Bed	H ±0.250 Slide	J	K*		Min	Max
21	14	3	8.500	8	0.812	3	0.750	24	21	14	1.500	1	...	75
24	18	4	9	8	0.812	3	0.750	30	24	18	1.500	1	100	125
27	21	4	9	8	0.812	3	0.750	33	27	21	1.500	1	...	150
30	21	4	9	8	1.125	5	1	36	30	21	1.500	1	200	250
36	24	9	15	12	1.125	5	1	42	36	24	1.500	1	300	400
42	27	9	15	12	1.125	5	1	48	42	27	3	9	500	600
48	33	15	21	16	1.125	7	1	54	48	33	3	9	800	1,000
54	39	15	21	16	1.125	7	1	60	54	39	3	9	...	1,250
60	43	15	21	20	1.125	9	1	66	60	43	3	9	...	1,600
66	47	21	27	20	1.125	9	1	72	66	47	3	9	...	2,000

* Reference dimension only; holes located from left-to-right center line and held to a tolerance of ±0.015 in.

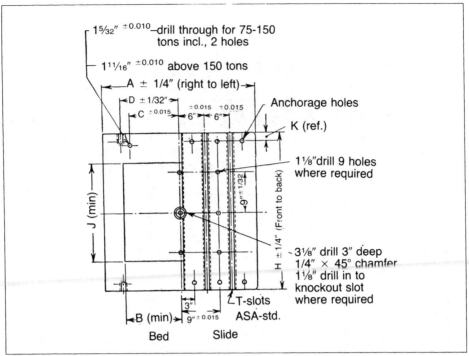

Fig. 27-10. Layout of bed and slide for straight-side, single-action, single-point presses, light and heavy series.

TABLE 27-5
Bolster-Plate Data for Straight-side, Single-action,
Single-point Presses (See Fig. 27-11)

		Left to Right					Front to Back		
A	*C*	*D*	No. T-Slots	Rows Pressure Pinholes	*H*	No. T-Slots	Rows Pressure Pinholes	Thickness*
21	3	8.5	6	5	24	6	5	3.5
24	4	9	6	5	30	6	5	4 (100*T*) 4.5 (125*T*)
27	4	9	6	7	33	10	7	5
30	4	9	5	4	36	. .	4	5 (200*T*) 5.5 (250*T*)
36	9	15	5	4	42	. .	4	6 (300*T*) 6.5 (400*T*)
42	9	15	7	4	48	. .	4	7 (500*T*) 7.5 (600*T*)
48	15	21	7	6	54	. .	6	8.5 (800*T*) 9 (1,000*T*)
54	15	21	9	6	60	. .	6	9.5
60	15	21	9	6	66	. .	6	10
66	21	27	11	8	72	. .	8	11

* Values in parentheses denote press size or tonnage.

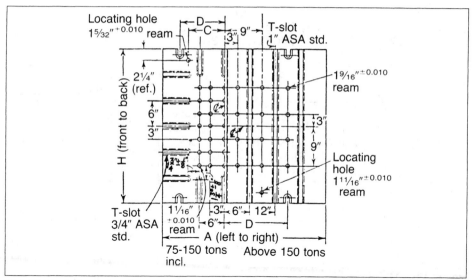

Fig. 27-11. Bolster-plate for straight-side, single-action, single-point presses.

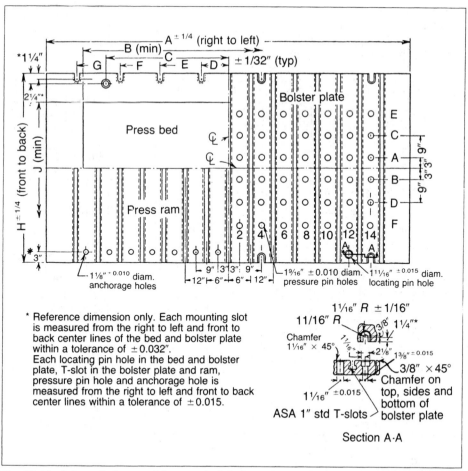

* Reference dimension only. Each mounting slot
is measured from the right to left and front to
back center lines of the bed and bolster plate
within a tolerance of ±0.032".
Each locating pin hole in the bed and bolster
plate, T-slot in the bolster plate and ram,
pressure pin hole and anchorage hole is
measured from the right to left and front to back
center lines within a tolerance of ±0.015.

Fig. 27-12. Schematic layout of the press bed, bolster plate, and press slide of straight-side,
single-action, multiple-point presses.

TABLE 27-6
Tooling Dimensions for Single-action
Multiple-point Presses*

											Front to Back		Range of Tonnages†	
				Left to Right										
A	B	C	D	E	F	G	No. Mounting Slots, Bed	Rows of Pressure Pinholes	Total Anchorage Holes, Slide	T-Slots Ram and Bolster	H	J	Min	Max
48	33	9	15	..	..	...	4	6	12	7	30	21	50	125
											36	27	75	125
											36	21	150	200
											42	33	100	125
											42	27	150	250
											48	33	150	250
60	45	15	21	.		...	4	8	16	9	30	21	50	125
											36	27	75	125
											36	21	150	200
											42	33	100	125
											42	27	150	250
											48	33	150	300
											54	39	200	300
											60	45	250	300
72	57	21	9	27	..	...	8	10	20	11	36	21	75	200
											42	27	100	250
											48	33	150	300
											54	39	200	300
											54	33	400	500
											60	45	250	300
											60	39	400	500
84	69	27	9	33	..	...	8	12	24	13	36	21	75	200
											42	27	100	250
											48	33	150	300
											54	39	200	300
											54	33	400	500
											60	45	250	300
											60	39	400	600
											66	51	250	300
											66	45	400	600
											72	51	400	600
96	81	33	9	39	..	...	8	14	23	15	48	33	150	300
											54	39	200	300
											54	33	400	500
											60	45	250	300
											60	39	400	800
											66	51	250	300
											66	45	400	800
											72	51	400	800

TABLE 27-6
Tooling Dimensions for Single-action
Multiple-point Presses* *(Continued)*

												Front to Back		Range of Tonnages†	
	Left to Right														
A	B	C	D	E	F	G	No. Mounting Slots, Bed	Rows of Pressure Pinholes	Total Anchorage Holes, Slide	T-Slots Ram and Bolster		H	J	Min	Max
108	93	39	9	45	. .	. . .	8	16	32	17		48	33	150	300
												54	39	200	300
												54	33	400	500
												60	45	250	300
												60	39	400	800
												66	51	250	300
												66	45	400	800
												72	51	400	1,250
												78	57	400	1,250
												84	63	500	1,250
120	105	45	9	27	51	. . .	12	18	36	19		60	45	250	300
												60	39	400	800
												66	51	250	300
												66	45	400	800
												72	51	400	1,250
												78	57	400	1,250
												84	63	500	1,250
132	117	51	9	33	57	. . .	12	20	40	21		60	45	250	300
												60	39	400	800
												66	51	250	300
												66	45	400	800
												72	51	400	1,250
												78	57	400	1,250
												84	63	500	1,250
												90	69	500	1,250
												96	75	500	1,250
144	129	57	9	33	63	. . .	12	22	44	23		72	51	400	1,250
												78	57	400	1,250
												84	63	500	1,250
												90	69	500	1,250
												90	63	1,600	2,000
												96	75	500	1,250
												96	69	1,600	2,000
156	141	63	9	39	69	. .	12	24	48	25		72	51	400	1,250
												78	57	400	1,250
												84	63	500	1,250
												90	69	500	1,250
												90	63	1,600	2,000
												96	75	500	1,250
												96	69	1,600	2,000
												102	81	500	1,250
												102	75	1,600	2,000
												108	87	500	1,250
												108	81	1,600	2,000

TABLE 27-6
Tooling Dimensions for Single-action
Multiple-point Presses* *(Continued)*

														Front to Back		Range of Tonnages†	
						Left to Right											
A	B	C	D	E	F	G	No. Mounting Slots, Bed	Rows of Pressure Pinholes	Total Anchorage Holes, Slide	T-Slots Ram and Bolster	H	J	Min	Max
180	165	75	9	33	57	81	16	28	56	29	84	63	500	1,250
											90	69	500	1,250
											90	63	1,600	2,000
											96	75	500	1,250
											96	69	1,600	2,000
											102	81	500	1,250
											102	75	1,600	2,000
											108	87	500	1,250
											108	81	1,600	2,000
204	189	87	9	33	63	93	16	32	64	33	96	75	500	1,250
											96	69	1,600	2,000
											102	81	500	1,250
											102	75	1,600	2,000
											108	87	500	1,250
											108	81	1,600	2,000
228	213	99	9	45	75	105	16	36	72	37	96	75	1,000	1,250
											96	69	1,600	2,000
											102	81	1,000	1,250
											102	75	1,600	2,000
											108	87	1,000	1,250
											108	81	1,600	2,000

* For rows of pressure pinholes, front to back, and thickness of bolster see Table 27-9.

† Increments, tons:

From	50 tons to	150 tons, incl.:	increment, 25 tons
Above	150 tons to	300 tons, incl.:	increment, 50 tons
Above	300 tons to	600 tons, incl.:	increment, 100 tons
Above	600 tons to 1,000 tons, incl.:		increment, 200 tons
Above 1,000 tons to 1,250 tons, incl.:			increment, 250 tons
Above 1,250 tons to 1,600 tons, incl.:			increment, 350 tons
Above 1,600 tons to 2,000 tons, incl.:			increment, 400 tons

TABLE 27-7
Single-action Multiple-point Press Bolster-plate Thickness and Rows of Pressure Pinholes Front to Back in Relation to Press Capacity and Depth

Capacity of Press, tons	Thickness of Bolster Plate	Bolster Plate	
		Total Depth Front to Back	Total Rows of Pressure Pinholes Front to Back
50	3.5	30	4
75	4	30, 36	4
100	4.5	30, 36	4
100	4.5	42	6
100	4.5	42	4*
125	5	30, 36	4
125	5	42	6
125	5	42	4*
150	5.5	36, 42	4
150	5.5	48	6
200	6	36, 42	4
200	6	48, 54	6
250	6.5	42	4
250	6.5	48, 54	6
250	7	60, 66	8
300	7	48, 54	6
300	7	60	8
300	7.5	66	8
300	7.5	72	10
400	7.5	54, 60	6
400	7.5	66	8
400	8	72	8
400	8	78	10
500	8	54, 60	6
500	8	66	8
500	9.5	72	8
500	9.5	78, 84	10
500	9.5	90	12
500	11.5	96	12
500	11.5	102, 108	14
600	8.5	60	6
600	8.5	66	8
600	10	72	8
600	10	78, 84	10
600	10	90	12
600	12	96	12
600	12	102, 108	14

TABLE 27-7
Single-action Multiple-point Press Bolster-plate Thickness and Rows of Pressure Pinholes Front to Back in Relation to Press Capacity and Depth *(Continued)*

Capacity of Press, tons	Thickness of Bolster Plate	Bolster Plate	
		Total Depth Front to Back	Total Rows of Pressure Pinholes Front to Back
800	9	60	6
800	9	66	8
800	10.5	72	8
800	10.5	78, 84	10
800	10.5	90	12
800	12	96	12
800	12	102, 108	14
1,000	11	72	8
1,000	11	78, 84	10
1,000	11	90	12
1,000	12	96	12
1,000	12	102, 108	14
1,250	11	72	8
1,250	11	78, 84	10
1,250	12	90, 96	12
1,250	12	102, 108	14
1,600	12	84, 90	10
1,600	12	96, 102	12
1,600	12	103	14
2,000	12	90	10
2,000	12	96, 102	12
2,000	12	108	14

* Applies to 72 by 42 and 84 by 42 press sizes.

TABLE 27-8
JIC Standard Strokes
For Straight-side Single-action Presses

Capacity of Press, Tons	Length of Stroke, In.
50– 75	4–6–8
100– 125	4–6–8–10
150	4–8–12
200	4–8–12–16
250– 300	8–12–16
400– 500	8–12–16–20
600– 800	12–16–20–24
1,000–1,250	16–20–24–28
1,600–2,000	16–20–24–28–32

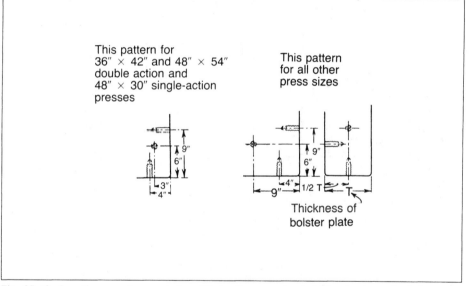

Fig. 27-13. Location, spacing, and size of holes for mounting handling hooks to the bolster plate in straight-side presses.

TABLE 27-9
Tooling Dimensions for Double-action Single- And Multiple-point Presses

	Left to Right												Front to Back			Capacity, tons	
A	B	C	D	E	F	G	K	No. Mounting Slots, Bed	Rows of Pressure Pinholes	Anchorage Holes, Slide Outer	T-Slots Inner Slide and Bolster	H	J	L	Outer Slide	Inner Slide

Single-Point Presses

A	B	C	D	E	F	G	K	No. Mounting Slots, Bed	Rows of Pressure Pinholes	Anchorage Holes, Slide Outer	T-Slots Inner Slide and Bolster	H	J	L	Outer Slide	Inner Slide
36	27	9	14				24	4	4	8	3-slide 5-bolster	42	27	30	125 150	250 300
48	40	9	15				36	4	6	8	5-slide 7-bolster	54	33	42	200 250	400 500
60	52	15	21				42	4	8	12	7-slide 9-bolster	66	45	48	300 400 500	600 800 1,000
72	60	21	9	27			54	8	10	12	9-slide 11-bolster	78	51	60	625 800 1,000	1,250 1,600 2,000

Multiple-Point Presses

A	B	C	D	E	F	G	K	No. Mounting Slots, Bed	Rows of Pressure Pinholes	Anchorage Holes, Slide Outer	T-Slots Inner Slide and Bolster	H	J	L	Outer Slide	Inner Slide
72	57	21	9	27			60	8	10	12	9-slide	48	33	36	125 125 150 175	200 200 250 300
											11-bolster	60	45	48		

Left section

84	69	27	9	33		72	8	12	16	11-slide 13-bolster
96	81	33	9	39		78	8	14	16	13-slide 15-bolster
108	93	39	9	45	51	90	8	15	20	15-slide 17-bolster
120	105	45	9	27		102	12	18	20	17-slide 19-bolster

Right section (gear / capacity)

Group					
11-slide 13-bolster	48	33	36	175	300
	48	27	36	225	400
	60	45	48	175	300
	60	39	48	225	400
13-slide 15-bolster	48	33	36	175	300
	48	27	36	225	400
	60	45	48	175	300
	60	39	48	225	400
	72	57	54	175	300
	72	51	54	225	400
15-slide 17-bolster	60	39	48	225	400
				350	600
	72	51	54	225	400
				350	600
17-slide 19-bolster	60	39	48	225	400
				350	600
				500	800
	72	51	54	225	400
				350	600
				500	800
	84	63	66	225	400
				350	600
				500	800

A	B	C	D	E	F	G	K	No. Mounting Slots, Bed	Rows of Pressure Pinholes	Anchorage Holes, Slide Outer	T-Slots Inner Slide and Bolster	H	J	L	Outer Slide	Inner Slide
132	117	51	9	33	57		114	12	20	24	19-slide 21-bolster	72	51	54	350 500 600	600 800 1,000
												84	63	66	350 500 600	600 800 1,000
144	129	57	9	33	63		126	12	22	24	21-slide 23-bolster	72	51	54	350 500 600 700 350	600 800 1,000 1,250 600
												84	63	66	500 600 700 350	800 1,000 1,250 600
												96	75	78	500 600 700	800 1,000 1,250

Columns A–K: Left to Right. Columns H, J, L: Front to Back. Outer Slide / Inner Slide: Capacity, tons.

600/800/1,000/1,250	350/500/600/700				Slide / Bolster										
600	350	66	63	84	23-slide / 25-bolster	28	24	12	138		69	39	9	63	156 ǀ 141
800	500														
1,000	600														
1,250	700														
600	350	78	75	96											
800	500														
1,000	600														
1,250	700														
600	350	78	75	96	25-slide / 29-bolster	32	28	16	156	81	57	33	9	75	180 ǀ 165
800	500														
1,000	600														
1,250	700														
600	350	84	87	108											
800	500														
1,000	600														
1,250	700														
800	500	84	87	108	29-slide / 33-bolster	36	32	16	180	93	63	33	9	87	204 ǀ 189
1,000	600														
1,250	700	84	81	108											
1,600	900														
1,600	900	84	81	108	33-slide / 37-bolster	40	36	16	204	105	75	45	9	99	228 ǀ 213

TABLE 27-10
Bolster-plate Thickness and Number of Rows
Of Pressure Pinholes Front to Back for Double-action Presses

Capacity, tons		Bolster Plate		
Outer Ram	Inner Ram	Thickness	Total Depth Front to Back	Total Rows of Pressure Pinholes Front to Back
Single Point				
125	250	7	42	4
150	300	7.5	42	4
200	400	8	54	6
250	500	8.5	54	6
300	600	9	66	8
400	800	11	66	8
500	1,000	11	66	8
625	1,250	12	78	8
800	1,600	12	78	8
1,000	2,000	12	78	8
Multiple Point				
125	200	6.5	48	6
125	200	6.5	60	8
150	250	7	60	8
175	300	7.5	48	6
175	300	7.5	60	8
175	300	8	72	10
225	400	8	48	4
225	400	8	60	6
225	400	9.5	72	8
225	400	9.5	84	10
350	600	9	60	6
350	600	11	72	8
350	600	11	84	10
350	600	12	96	12
350	600	12	108	14
500	800	11	60	6
500	800	12	72	8
500	800	12	84	10
500	800	12	96	12
500	800	12	108	14
600	1,000	12	72	8
600	1,000	12	84	10
600	1,000	12	96	12
600	1,000	12	108	14
700	1,250	12	72	8
700	1,250	12	84	10
700	1,250	12	96	12
700	1,250	12	108	14
900	1,600	12	108	14

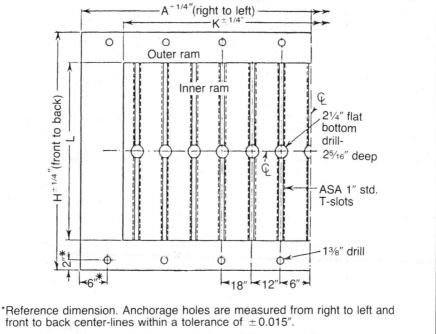

Fig. 27-14. Slide layout for double-action press.

RETROFITTING OLDER PRESSES

Unless it is subjected to extreme abuse, it is almost impossible to wear out a large press of good mechanical design to the extent that it is not economically feasible to rebuild it. Generally, unless a number of major components such as the crown, crankshaft, slide, gears, and bed are damaged beyond repair, it is worth considering.

During the rebuilding process, the press can be fitted with such improvements as solid state controls, a hydraulic overload system, quick die change equipment, and an improved clutch and brake.[2]

PRESS DEFLECTION

Incremental Deflection Factor. Press deflection relates directly to the amount of tonnage that the press is developing. When the press is adjusted so that the dies just make contact, no tonnage is developed. To accomplish this, the slide must be adjusted below the setting where die contact first occurs.

When tonnage is developed, the press members all are distorted or deflected slightly. Just as a coil spring must change shape to develop pressure, press members are deflected when a press develops tonnage.

By definition, incremental deflection factor is the amount of press shut height decrease, or die shut height increase that will result in a press tonnage increase of four tons (35.6 kN) total or one ton (8.9 kN) per corner. The best way to make an exact determination is by using calibration load cells, although an accurate tonnage meter can also provide useful data.

Measuring Press Deflection With Load Cells. The load cells are placed in the die space on strong supports (Fig. 27-15). One load cell per slide connection should be used. A 0.030 in. (0.76 mm) thick shim is placed under each load cell. The press is adjusted to produce full rated press tonnage when striking the load cells (Fig. 27-16). Additional shims are added as needed to obtain equal tonnage on all load cells. The 0.030 in. shims are removed and the total drop in tonnage noted.

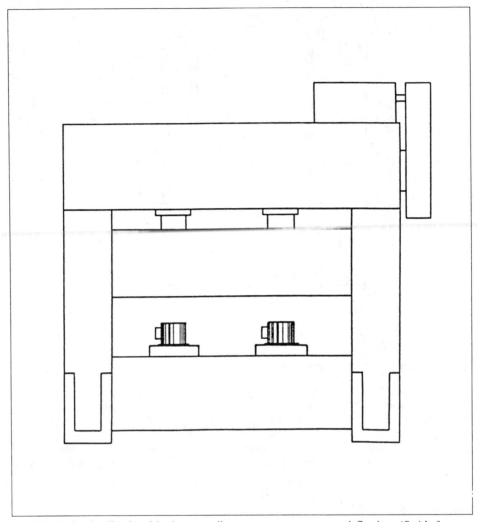

Fig. 27-15. Load cells placed in the press die-space to measure press deflection. (*Smith & Associates*)

The total drop in tonnage is divided into the thickness of the test shim material in thousandths of an inch. The result is the amount of press deflection that occurs for each ton of increased pressure.

It is often useful to think of deflection in terms of tons per end or corner of the slide, depending on the number of slide connections. To obtain incremental deflection factor in these terms, simply multiply the incremental deflection factor by 2 or 4 respectively. This figure is useful when analyzing the effect of slide out-of-level or offset loading conditions.

An Approximate Figure. The results of many tests made with load cells in straight side presses show a tight grouping of incremental deflection factors. Expressed as deflection per ton per corner they range from 0.0008 in. (0.020 mm) for presses with cast steel members to 0.0015 in. (0.038 mm) for light-weight welded construction. An approximate figure is 0.001 in. (0.025 mm) per ton per corner.[7, 8]

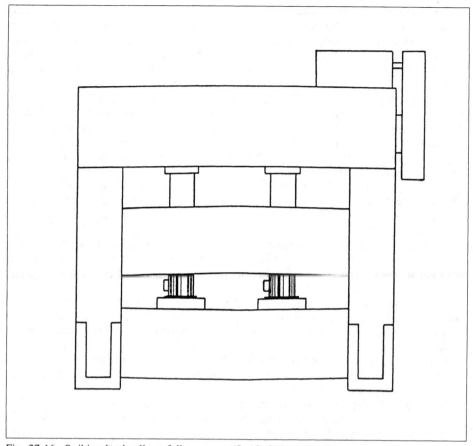

Fig. 27-16. Striking load cells at full tonnage. (*Smith & Associates*)

MEASURING LOADS ON MECHANICAL PRESSES

There are many reasons why it is important to measure loads on stamping presses.

1. Forces required to make parts must not exceed the capacity of the machine.

2. Every effort should be made to assure only the minimum force is used to make the part. This increases tool life and minimizes press component wear as well.

3. If the tonnage requirement of a particular tool is known, it aids in selecting the proper size press.

4. One of the more important advantages of being able to measure loads is the ability to determine the condition of the press and recognize press problems. Press analysis can be an important part of a preventative maintenance program.

5. The ability to accurately measure loads enables the operator to adjust the stamping process to assure proper tool and press operation, resulting in better part quality.

6. Energy Savings. The forming of parts in the metalworking process is the result of using the kinetic energy available from the flywheel. Consuming this energy in the working cycle reduces the flywheel speed, hence its energy.

7. Die set up time can be greatly reduced and jobs can be set to previously established optimum conditions of loading and load distribution. Off-center loading can easily be detected.

8. Early detection of change in stock thickness, die lubrication, or metal characteristic help maintain quality.

9. Accurate parts counting can also be done by allowing the counter to operate only after the press has reached a pre-determined force.

Snap-through. Reverse load forces occur during blanking operations when the punches break through the part. The energy stored as strain in the die and press is suddenly released subjecting the press to a reverse load phenomenon. Section 4 has detailed information on dealing with snap-through problems.

It is important that force measuring instruments be capable of reading these peak reverse forces.

Tonnage Meter Operating Principles. Most tonnage meters operate by measuring slight physical changes in the press structure with a electrical resistance foil strain gage. The strain gage is bonded to the press structure at a location where extension or compression of the structure relates directly to actual operating tonnage. To avoid temperature differential effects and drift, four gages are used in a wheatstone bridge configuration.

When a press force is generated the structure deforms and the resulting strain unbalances the bridge. A minute voltage, proportional to this force, appears at the output terminals. This output is fed via shielded cables (to reduce spurious signals) to the input of the tonnage meter.

Bolt-on Strain Transducers. Because of the delicate nature of installing strain gages directly to press members, bolt-on strain transducers are often used in place of individual strain gages. These sensors are affixed by screws. Pre-tapped weld pads positioned with a weld jig greatly speeds up the work.

The bolt-on sensors are produced by bonding the strain gages under controlled conditions to an intermediate structure which can be bolted to the machine structure under test. This method also allows for mechanical amplification, further increasing the measurement accuracy.

An advantage in the use of bolt-on strain gage transducers is that the individual units are interchangeable. Once a machine is calibrated, transducers can be interchanged without re-calibration.

Transducer Locations. The press pitmans are a popular location for strain transducers. This works particularly well if transducers are mounted on opposite sides of the pitman and wired in parallel to cancel out errors due to bending.

A good means of protecting the signal wires connected to a moving press member from damage is to run the wire through coiled nylon air hose between junction boxes. The transition at the junction box knock-out openings can be made with standard pipe fitting adapters.

Pull rods are a popular location on underdriven presses. To eliminate any error due to bending, it is best to place dual gages or transducers on opposite sides of the rod or link. One manufacturer claims good results by applying Micro-Measurements brand spot-weldable metal foil backed strain gages with a 50 joule capacitive-discharge spot-welder. One-half of the bridge is placed on each side of the pull rod or link. This method takes advantage of engineering calculations based upon pull rod cross sectional area and Young's Modulus for the material involved to determine calibration factors.

A location shown to give similar results to that of the pull rod location is on the top of the slide near the pull-rod attachment point. [8]

Column mounted sensors give results representative of die and press conditions, provided the mounting locations are chosen with care. Many presses have box-like heavy plate shrouds surrounding the tie-rods with much lighter fabricated plates used to make up the rest of the column structure. This forms a housing for mechanical equipment and provides lateral bracing. A good location for the sensor is on the heavy shroud just below the gibbing.

If the sensor is located on the light weight plate on the outside of the columns, the results will be much less representative of actual press and die conditions. This is because the bed and crown bow under loading, relieving the preload on the tie rod shrouds at a much faster rate than occurs on the lightweight outside plate. The bowing of the bed and crown tends to drive the outside of the column into compression.

The press bed is generally not a good location for sensors because the calibration is effected by the size and location of the lower die shoe.

Gap and OBI presses are usually gaged near the throat opening on the side of the press. This location is as good as any on the frame, but any location other than the pitman will be affected by die location. The calibration should be done with a load cell directly under the connection. If the die is set toward the rear of the bolster, the tonnage meter will read erroneously low. Presses have been damaged because diesetters were not aware of this problem.

Sensor locations should be guided by sound mechanical principles, not ease of installation. Otherwise the press room personnel will not be able to correlate the results on the meter with actual conditions, and the tonnage meter will fall into disuse.

Calibration. The process of adjusting the force measurement instrument to agree with the actual load on the press is referred to as calibration.

Two common ways to calibrate a mechanical press are dynamic calibration with load cells and static calibration with hydraulic jacks. With either method, the purpose is to apply a known force to the press structure.

The sensitivity of the measurement instrument is then adjusted so the indicated meter force reading corresponds to the known force applied to the press structure.

Calibration Numbers. The resistance of the wheatstone bridge is nearly always a nominal 350 ohms. To verify or re-establish the correct tonnage meter gain setting, an artificial tonnage is created by shunting a high value resistor across one arm of the bridge. The values range from 50,000 to 1,000,000 ohms. This is usually done immediately after calibration with load cells and the number is recorded inside the instrument cabinet for reference.

If the sensors are carefully located with attention to exact symmetry, the calibration numbers should be within 10% of each other, provided the press is in good condition. A large scatter indicates there is a press problem.

Interpreting Readings. Many types of tonnage meters sum the analog signals from the several channels to provide a total peak reading. Often this reading is much lower than the arithmetical sum of the individual channels. This is normal, and is a result of the peak tonnages on the individual channels not occurring simultaneously.

DETERMINING PRESS CONDITION

The following is a suggested method of dynamically measuring the parallelism of slide to bolster.

Testing Presses With Load Cells. The traditional method of checking parallelism is to sweep the opening from slide to bolster with a dial indicator while the press is not running. This does not necessarily take into account the difference in bearing clearances. In those cases where blocks or jacks are used to support the slide, care must be exercised not to introduce bending and deflections that would mask the true alignment condition. Static checking of parallelism does not take into account those deflections that are a result of mass in motion creating forces found only when the press is cycling.[9]

To determine the condition of parallelism under load, two to four calibration load cells are placed in the diespace on strong supports. Shims are added under each load cell and the slide adjusted downward to obtain the desired loading when striking the load cells. The amount of shims needed in each individual load cell is equal to the the amount of out-of-parallel condition at each load cell.

In the case of OBI or gap frame presses it is important to recognize the angular deflections that occur in this style press. Despite lack of an industry standard for angular deflections, most press manufacturers design around the value of 0.0015-in. per inch (0.038 mm per 25.4 mm) of throat depth at rated capacity. Considering a nominal throat depth of 20 in. (508 mm), this could mean an angular deflection of as much as 0.030 in. (0.76 mm).

The tool may not accept this much misalignment. There is also the misconception that guide pins and bushings can compensate for this misalignment. The size of the pin and bushing large enough to do so, however, could easily consume all of the working area of the press leaving no space to make parts (Section 3).

WAVEFORM SIGNATURE ANALYSIS

Nearly all tonnage meters have an analog output to permit analysis of the waveform signature. The output voltage ranges from 2.5 to 5 volts at 100% capacity, depending upon the manufacturer's specifications. This signal is fed to an oscilloscope, strip chart recorder, or digital recorder. Digital recorders have the advantage of permitting computer-aided analysis such as modal analysis, modem transmission, and archival storage.

Figure 27-2 and Figs. 4-30 and 4-31 are idealized examples of waveform signatures. Actual signatures are often somewhat ragged due to to the nature of the actual forces being measured. This procedure is excellent for analysis of a number of press and die problems such as snap-through energy, loose tie rods, counterbalance problems, gib chatter, and floating draw ring impact problems, to name a few.

Tonnage Curve Analysis. The widespread use of inverted draw and stretch-draw dies in automotive stamping plants has made analysis of forming forces throughout the press stroke a necessity. Attempts to lessen the impact of the blankholder on the draw ring by slowing the press down or by using compression shock-absorbers or rubber have made the problem worse in some, but not all cases.

The kinetic energy of the slide can often be used to advantage in the process. Measured values of up to 10% of slide tonnage have been observed when contacting the work fairly high in the press stroke, without inducing a strain in the press frame. The press is acting much like a drop-hammer in such cases. This phenomenon can be confusing because it doesn't show as a peak when reading data from column mounted sensors. A normal peak will be seen from in-die sensor data.

To analyze the operation for a problem, a linear transducer that provides a voltage output relative to position can be used with a multiple channel recorder. This will juxtapose slide distance from bottom of stroke with actual instantaneous tonnage. This information can be compared to the tonnage curve supplied by the press manufacturer.

Some tonnage meters can be programmed with the press tonnage curve and are capable of continuously comparing tonnage with crankshaft rotation. An alarm signal is generated if there is a tonnage curve violation.

Draw ring bounce or rebound upon initial contact may occur. This can be a problem.

Correction of measured tonnage curve violations may involve stepping cylinder engagement, lowering the ring, changing the press speed, modifying the press, or specifying a larger press.

PRESS BED LEVELING

The necessity of frequent re-leveling of large precision machine tools is well known and accepted. Like machine tools, press beds require periodic re-levelings to compensate for foundation settling.

Figure 27-17 illustrates a straight side with vibration isolating leveling devices installed. These devices allow easy adjustment to be made when leveling the press bed. They are available in sizes for all presses including the largest transfer presses built to date.

The larger sizes are equipped with internal hydraulic cylinders to permit the screw to be adjusted with ease. All sizes can be equipped with strain gages installed as an option. The latter feature permits electronic weighing of each corner of the press separately to spread the weight evenly. A cut-away view is shown in Fig. 27-18.

Figure 27-19 illustrates a press bed mounted on an uneven surface. While the twist or

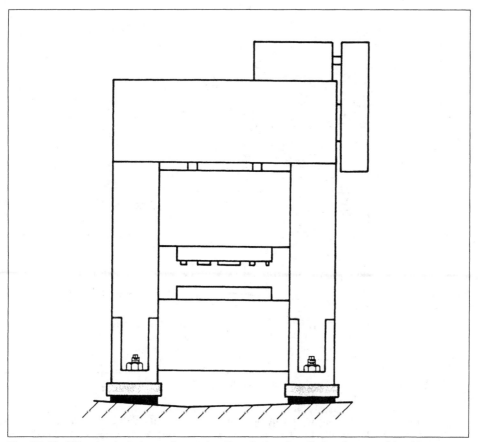

Fig. 27-17. Vibration isolation and press leveling devices installed under a straight-side press on an uneven surface.[10] (*Vibro/Dynamics Corp.*)

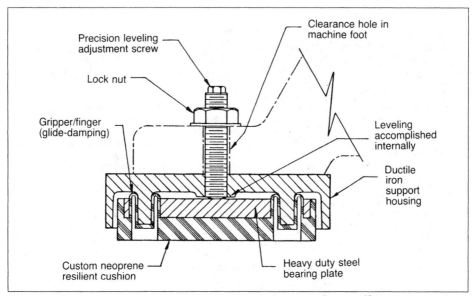

Fig. 27-18. Cutaway view of press leveling and vibration isolating device.[10] (*Vibro/Dynamics Corp.*)

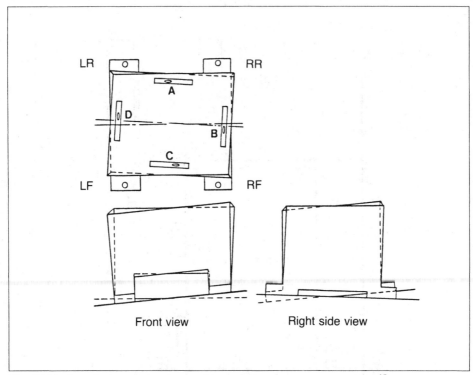

LR   RR

A

D

B

C

LF   RF

Front view   Right side view

Fig. 27-19. Press bed warped and twisted by installing on an uneven surface.[10] *(Vibro/Dynamics Corp.)*

warpage may not be visible to the unaided eye, it is readily detected with a precision machinist's level. When the twist is properly adjusted out of the bed, (Fig. 27-20) the machinist's level will indicate all four sides of the press bed are level within close limits.

References

1. Charles Wick, John W. Benedict, and Raymond F. Veilleux, eds., *Tool and Manufacturing Engineers Handbook*, 4th ed., Volume 2, (Dearborn, MI: Society of Manufacturing Engineers, 1984).
2. D. F. Wilhelm, *Understanding Presses and Press Operations* (Dearborn, MI: Society of Manufacturing Engineers, 1981).
3. *Presses* (Dearborn, MI: Society of Manufacturing Engineers, 1985).
4. S. D. Brootzkoos, "How Modern Mechanical Presses Operate," *Am. Machinist* (Nov. 17, 1949).
5. *Die Cushions for Mechanical Presses*, Danly Machine Specialties, Inc., 1950.
6. *Bliss Power Press Handbook*, E. W. Bliss Co., Canton, OH, 1950.
7. D. A. Smith, *Procedures for Adjusting Dies to a Common Shut Height*, SME/FMA Technical Paper TE89-565, 1989.
8. D. A. Smith, *Why Press Slide Out of Parallel Problems Affect Part Quality and Available Tonnage*, SME Technical Paper TE88-442, 1988.
9. D. F. Wilhelm, "Measuring Loads on Mechanical Presses", SME Conference Proceeding, Society of Manufacturing Engineers, Dearborn, MI, 1988.
10. M. Young, "Press Mounting Techniques for Improved Die Life", SME Conference Proceeding, Society of Manufacturing Engineers, Dearborn, MI, 1988.

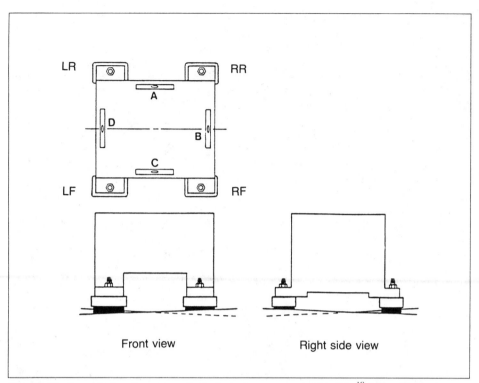

LR

A

D

B

C

LF RF

RR

Front view

Right side view

Fig. 27-20. Properly leveling a press bed eliminates warpage and twists.[10] *(Vibro/Dynamics Corp.)*

SECTION 28
FERROUS DIE MATERIALS*

Ferrous materials used in stamping operations may be classified as (1) wrought tool and die steels, (2) wrought and cast carbon and low-alloy steels, (3) plain and alloyed cast irons, and (4) plain and alloyed ductile (nodular) irons.

Tool steels find wide application in the stamping of the large volumes of small- and medium-sized parts and as inserts in larger dies. These steels are designed especially to develop high hardness levels and abrasion resistance, both through heat treatment and through the existence of hard, stable, and complex chromium, tungsten, molybdenum, and vanadium carbides.

The wrought carbon and low-alloy steels, readily weldable and responsive to a wide variety of heat treatment, are used for machine parts, keys, bolts, retainers, and support tooling. They are often case-hardened or quenched and tempered to increase wear resistance and strength.

Cast-steel dies are used for large drawing and forming dies where maximum impact toughness is required. At carbon levels of 0.35% and higher, cast-alloy-steel dies can be effectively flame-hardened at points of wear, thus increasing their resistance to wear.

Cast irons are used for shoes, plates, dies, adapters, and other large components; they are alloyed and hardened for large sheet-metal drawing and forming dies. The cast irons have the advantages of good machinability, low initial cost, and ability to be cast into difficult and complicated shapes. They have high compressive strength and good resistance to galling, and the alloy cast-iron dies can be hardened locally to increase wear resistance in selected areas.

The ductile (nodular) irons retain the casting advantages of cast iron, but because the free graphite is present in spheroidal shape rather than flake, this material develops toughness, stiffness, and strength levels approaching those of steel. Alloyed ductile irons can be effectively flame hardened to improve their resistance, though hardening is also usefully applied to ductile-iron castings to develop high strength levels and resistance to wear.

Almost all the ferrous die materials can be hardened, softened, strengthened, or toughened by heat treatment. In general, the principles are the same for all, though the specific handling and heat-treatment cycles vary considerably.

AVAILABILITY AND SELECTION OF TOOL AND DIE STEELS

The steels listed in Table 28-1 are used in the great majority of the metal-stamping operations where tool and die steels are required. The list contains 27 steels, which are readily available from almost all tool steel sources. These steels represent slight variations for improved performance in certain instances, and their use is sometimes justified because of special considerations. They may have exceptionally heavy usage for certain

* Revised by Rollin Bondar, MPD Welding Inc.; Rod Denton and Timothy Zemaitis, Sun Steel Treating Inc.

TABLE 28-1
Tool and Die Steels

AISI Steel Type	Nominal Composition, Percent							
	C	Mn	Si	W	Cr	Mo	V	Other
W1	1.05	0.25	0.20	—	0.20	—	0.05	—
W2	1.00	0.25	0.20	—	—	—	0.20	—
O1	0.90	1.25	0.30	0.50	0.50	—	—	—
O2	0.90	1.60	—	—	—	—	—	—
O6	1.45	0.80	1.10	—		0.25	—	—
A2	1.00	0.60	0.30	—	5.25	1.10	0.25	—
A8	0.55	—	1.00	1.25	5.00	1.25	—	—
A9	0.50	—	1.00	—	5.20	1.40	1.00	1.40 Ni
D2	1.55	0.30	0.40	—	11.50	0.80	0.90	—
D3	2.10	0.40	0.90	0.80	11.70	—	—	—
D4	2.25	0.35	0.50	—	11.50	0.80	0.20	—
D7	2.30	0.40	0.40	—	12.50	1.10	4.00	—
S1	0.50	—	0.50	2.25	1.30	—	0.25	—
S5	0.60	0.85	1.90	—	0.20	0.50	0.20	—
S7	0.50	0.75	0.30	—	3.25	1.40	—	—
T1	0.75	—	—	18.00	4.00	—	1.10	—
T15	1.55	—	—	12.25	4.00	—	5.00	5.00 Co
M1	0.80	—	—	1.50	3.75	8.50	1.05	—
M2	0.85	—	—	6.25	4.15	5.00	1.90	—
M4	1.30	—	—	5.50	4.00	4.50	4.00	—
M7	1.00	—	—	1.70	3.75	8.75	2.00	—
M42	1.08	—	—	1.50	3.75	9.50	1.10	8.00 Co
H12	0.35	—	1.00	1.25	5.00	1.35	0.30	—
H13	0.38	—	1.00	—	5.20	1.25	1.00	—
H21	0.30	0.25	0.30	9.00	3.25	—	0.25	—
H26	0.53	—	—	18.00	4.00	—	1.00	—
L6	0.75	0.70	—	—	0.80	0.30	—	1.50 Ni

W — Water Hardening   D — High Carbon, High Chromium Die Steels
O — Oil Hardening   S — Shock Resisting
A — Air Hardening   T — Tungsten-base, High Speed
H — Hot Working   M — Molybdenum-base, High Speed
     L — Special-purpose, Low-alloy

types of metal-stamping or forming operations or are recommended in other instances.

The steels are identified by letter and number symbols. The letter represents the group of the steel involved. The number indicates a separation of one grade or type from another.

All the steels in the list can be heat-treated to a hardness greater than 50 HRC and accordingly they are hard, strong, wear-resistant materials. Some, such as T15, are capable of heat treatment to a hardness as high as Rockwell C68. Frequently hardness is proportional to wear resistance, but this is not always true, for the wear resistance usually also increases as the alloy content—particularly the carbon content—increases. The toughness of the steels, on the other hand, is, with a few exceptions, inversely proportional to the hardness, and increases markedly as the alloy content or the carbon content is lowered.

CHARACTERISTICS OF TOOL AND DIE STEELS

Table 28-2 lists the hardening the tempering treatments, and Table 28-3 the basic characteristics, of the various tool steels listed. A careful study of these characteristics will be of great assistance in making the selection of the proper grade. For each job a different basic characteristic may assume major importance, depending upon the material being formed or worked; the requirements for economy, toughness, and wear resistance, or the design, which may affect the heat-treating characteristics or the need for machinability. Table 28-4 shows typical press operations and recommended tool steels and operating hardness levels. The steels have been arranged in order of increasing wear resistance. For maximum life between resharpenings, the last steel listed under each application should generally be selected.

A brief statement of the merits of the different groups follows.

W, Water-hardening Tool Steels. W1 and W2 are both readily available and of low cost. W2 contains vanadium and is more uniform in response to heat treatment; it is of a finer grain size with a higher toughness. Both are shallow-hardening and, when hardened with a hard case and a softer internal core, have high toughness. They are quenched in water or brine and should be applied where movement in hardening is not important.

O, Oil-hardening Tool Steels. Steels O1 and O2 have, for many years, been the workhorses of the die-steel industry and are known familiarly as manganese oil-hardening tool steel. Readily available and of low cost, these steels, which are hardened in oil, have less movement than the water-hardening steels and are of equal toughness when the water-hardening steels are hardened throughout. For special applications, type O6, which contains free carbon in the form of graphite, has been used successfully.

A, Air-hardening Die Steels. The principal air-hardening die steel employed is steel A2. This steel has a minimum movement in hardening and has higher toughness than the oil-hardening die steels, with equal or greater wear resistance. It has a slightly higher hardening temperature than the manganese types. The availability of the popular A2 steel is excellent. Type A8 is the toughest steel in this group, but its low carbon content makes it less wear-resistant than A2.

D, High-carbon High-chromium Die Steels. The principal steels of wide application for long-run dies are steels in this group. Grade D2 containing 1.50% carbon is of moderate toughness and intermediate wear resistance, whereas steels D3, D4, and D7, containing additional carbon, are of very high wear resistance and somewhat lower toughness. Selection between these is based on the length of run desired and machining and grinding problems. D2 and D4, containing molybdenum, are air-hardening and have minimum movement in hardening.

S, Shock-resisting Tool Steels. These steels contain less carbon and have higher toughness. They are employed where heavy cutting or forming operations are required and where breakage is a serious problem with higher-carbon materials that might have longer life through higher wear resistance alone. Choice among the grades is a matter of experience. All steels are readily available, with steel S5 being widely employed. This

TABLE 28-2
Hardening and Tempering Treatments for Tool and Die Steels

AISI Steel Type	Preheat Temperature, Degrees F	Rate of Heating for Hardening	Hardening Temperature, Degrees F	Minimum Time at Temperature, Minutes	Quenching Medium
W1	1200°/1250°	Slow	1425°–1500°	15 min.	Brine or Water
W2	1200°/1250°	Slow	1425°–1500°	15 min.	Brine or Water
O1	1200°/1250°	Slow	1450°–1500°	15 min.	Warm Oil
O2	1200°/1250°	Slow	1400°–1475°	15 min.	Warm Oil
O6	1200°/1250°	Slow	1450°–1500°	30 min.	Warm Oil
A2	1450°/1500°	Slow	1725°–1800°	30 min.	Air, Oil, Salt
A8	1450°/1500°	Slow	1800°–1850°	30 min.	Air, Oil, Salt
A9	1450°/1500°	Slow	1800°–1850°	30–60 min.	Air, Salt
D2	1450°/1500°	Slow	1800°–1875°	30 min.	Air, Salt
D3	1450°	Slow	1750°	30 min.	Warm Oil
D4	1450°/1500°	Slow	1800°–1850°	30 min.	Air, Salt, Oil
D7	1450°	Slow	1875°–2000°	45 min.	Air, Oil
S1	1400°	Slow	1700°–1750°	15–20 min.	Salt, Warm Oil
S5	1250°	Slow	1550°–1650°	15–20 min.	Salt, Oil, Water
S7	1200°/1300°	Slow	1725°–1750°	20 min.	Oil, Salt
T1	1500°/1600°	Quickly from preheat	2100°–2350°	3–7min.	Salt, Warm Oil
T15	1450°/1550°	Quickly from preheat	2175°–2275°	3–5 min.	Salt, Warm Oil
M1	1500°/1550°	Quickly from preheat	2150°–2225°	2–5 min.	Salt, Warm Oil
M2	1500°/1550°	Quicly from preheat	2175°—2275°	3–7 min.	Salt, Warm Oil
M4	1500°/1550°	Quickly from preheat	2150°–2250°	3–7 min.	Salt, Warm Oil
M7	1500°/1550°	Quickly from preheat	2175°–2225°	3–7 min.	Salt, Warm Oil
M42	1500°/1550°	Quickly from preheat	2150°–2200°	3–7 min.	Salt, Warm Oil
H12	1400°/1500°	Slow	1825°–1875°	30–60 min.	Salt, Air
H13	1400°	Slow	1825°–1875°	30–60 min.	Salt, Air
H21	1500°	Slow	2000°–2250°	10–30 min.	Oil, Air
H26	1500°	Slow	2100°–2250°	10–30 min.	Oil, Air
L6	1200°/1250°	Slow	1450°–1550°	30 min.	Warm Oil

TABLE 28-2
(Continued)

AISI Steel Type	Tempering Temperature, Degrees F	Depth of Hardening	Resistance to Decarburization	As Quenched Hardness HRC	Maximum Tempered Hardness HRC
W1	350°–550°	Shallow	Best	65–67	62–64
W2	350°–550°	Shallow	Best	65–67	62–64
O1	350°–600°	Medium	Good	63–65	62–64
O2	350°–600°	Medium	Good	63–65	62–64
O6	350°–600°	Medium	Good	63–65	61–63
A2	350°–950°	Deep	Fair/Good	63–64	62–63
A8	350°–950°	Deep	Fair	62	60–62
A9	350°–1000°	Deep	Poor/Fair	58	56–58
D2	350°–1000°	Deep	Fair	62–64	61–63
D3	350°–950°	Deep	Fair	63–65	61–63
D4	400°–1000°	Deep	Fair	63–65	61–63
D7	300°–1000°	Deep	Fair	65–67	65
S1	300°–800°	Medium	Fair	57–59	56–58
S5	300°–800°	Medium	Fair	60–62	59–61
S7	400°–1000°	Medium	Poor/Fair	58–60	57–59
T1	1000°–1100°	Deep	Good	64	64–65
T15	1000°–1100°	Deep	Fair/Good	64–66	66–68
M1	1000°–1150°	Deep	Fair	64–66	65–67
M2	1000°–1100°	Deep	Fair	65–66	64–65
M4	1000°–1100°	Deep	Fair	65–66	65–66
M7	1000°–1100°	Deep	Fair/Good	66–67	65–67
M42	975°–1075°	Deep	Fair/Good	64–66	67–69
H12	1000°–1100°	Deep	Fair	51–53	51–53
H13	1000°–1150°	Deep	Fair	54–56	53–56
H21	950°–1250°	Med./Deep	Fair	53–55	54–55.5
H26	1000°–1200°	Deep	Fair/Good	57–59	56–58
L6	350°–600°	Medium	Good	62–63	62

TABLE 28-3
Comparison of Basic Characteristics of Tool and Die Steels

AISI Steel Type	Non Deforming Properties	Safety in Hardening	Toughness	Resistance to Softening Effect of Heat	Wear Resistance	Decarburization Risk During Heat Treatment	Brittleness	Machin-ability
W1	L	F	G	L	G	L	L / M	B
W2	L	F	G	L	G	L	L / M	B
O1	G	G	G	L	F	L	L	G
O2	G	G	G	L	F	L	L	G
O6	G	G	G	L	F	L	L	B
A2	B	B	F / G	F	G	L	L	F / G
A8	G / B	B	B	G	G	M	L	F / G
A9	B	B	B	B	G	M	L	F / G
D2	B	B	L / F	F	G	L	L	L
D3	G	B	L	F	G / B	L	L	L
D4	B	B	L	F	G / B	L	L	L
D7	B	B	L / F	F	B	L	L	L
S1	F	F / G	B	F	F / G	M	L	G
S5	F	G	B	F	G	L	L	G
S7	F	G	B	F	G	M	L	G
T1	G	G	G	B	G / B	L	L	F
T15	G	G	L	B	B	L	M	L / F
M1	G	G	G	B	G / B	L	L	F
M2	G	G	G	B	G / B	L	L	F
M4	G	G	G	B	G / B	L	L	F
M7	G	G	G	G / B	B	L	L	F
M42	G	G	G	G / B	G / B	L	M	F
H12	G	G / B	G / B	G / B	F / G	L / M	L	F / G
H13	G	G / B	G / B	G / B	F / G	L / M	L	F / G
H21	G	G / B	G	G / B	F / G	L / M	L	F / G
H26	G	G	F	B	B	L / M	L	F / G
L6	F / G	G / B	G	L	F	L / M	L	G

B = BEST   F = FAIR   G = GOOD   L = LOW   M = MEDIUM

grade is an oil-hardening type of silicomanganese steel and is more economical than steel S1, which has equivalent toughness properties with greater wear resistance.

T and M, Tungsten and Molybdenum High-speed Steels. Steels T1 and M2 are equivalent in performance and represent standard high-speed steels which have excellent properties for cold-working dies. They have higher toughness than many of the other die

TABLE 28-4
Recommended Tool Steels and Hardness

Application	AISI Steel Type, Group	Hardness R_C
Blanking Dies and Punches (Short Run)	O, A, W	58–60
Blanking Dies and Punches (Long Run)	A, D	60–62
	M	61–63
Bending Dies	O, A, D	58–60
Coining Dies	S	52–54
	W	57–59
	D	58–60
Drawing Dies	H	52–54
	W, O	58–60
	D	60–62
Dies — Cold Extrusion	W	59–61
	D	60–62
	M	62–64
— Embossing	W, L, O, A, D	58–62
— Lamination	L	58–60
	D, A, M	60–62
— Sizing	M, D	60–62
	W	59–61
Punches — Embossing	S	58–60
— Trimming	W, D, O, A	58–60
— Notching	M	60–62

steels, combined with excellent wear resistance. While they are expensive, they are readily available. T15 and M4 are hardened by the standard method rather than carburizing, because they already have a very high carbon content combined with a high vanadium content.

Type M1 may occasionally be used in place of T1 and M2, but it is more susceptible to decarburization. Steel T15 is the most wear-resistant of all steels in the list, and steel M4 is slightly greater in wear resistance than a steel such as D4. These steels are more difficult to machine and grind than the other high-speed steels, but the improved performance obtainable very often justifies the extra machining expense.

L, Low-alloy Tool Steels. Of the many low-alloy steels effective as die materials, steel L6 is a chromium-nickel steel. In large sizes it is water-quenched and has a hard case and a soft core, with an attendant high overall toughness. In small sizes it may be oil-quenched.

H, Hot Working Steels. Die casting dies, extrusion dies, hot forming dies, and hot drawing mandrels are typical hot-work applications.

HEAT TREATMENT OF DIE STEELS

Simplified Theory of Hardening Steel. Iron has two distinct and different atomic arrangements—one existing at room temperature (and again near the melting point) and one above the transformation temperature. Without this phenomenon it would be impossible to harden iron-based alloys by heat treatment.

Briefly, what happens in the heat treatment of die steels is represented graphically by Fig. 28-1. Starting in the annealed machinable condition, A, the steel is soft, consisting

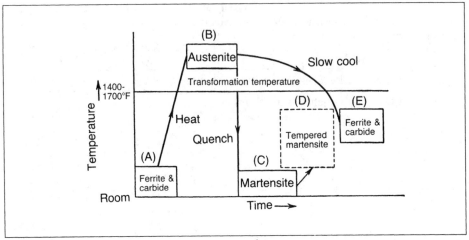

Fig. 28-1. Simplified chart of the hardening of steel.[1]

internally of an aggregate of ferrite and ferrite carbide. Upon heating above the transformation temperature, 1414° to 1700° F (768° to 927° C), the crystal structure of ferrite changes, becomes austenite, B, and dissolves a large portion of the carbide. This new structure, austenite, is always a prerequisite for hardening. By quenching, cooling it rapidly to room temperature, the carbon is retained in solution, and the structure known as martensite, C, results. This is the hard-matrix structure in steels. The rapid cooling results in high internal stresses, because the transformation from austenite to martensite involves some volumetric expansion against the natural stiffness of the steel. To overcome this problem, the steel is reheated to an intermediate temperature, D, to soften it slightly and relieve those residual stresses which otherwise would embrittle the steel.

If quenching is not rapid enough, the austenite reverts to ferrite and carbide, E, and high hardness is not obtained. The rate of quenching required to produce martensite depends primarily on the alloy content. Low alloy steels require rapid cooling in water or oil, while highly alloyed steel usually can be air-quenched at a much slower rate. The alloying elements make the reactions more sluggish.

Throughout all these heat-treating reactions, most die steels retain excess, or undissolved, carbides which take no direct part in the hardening. The high-carbon, high-chromium steels, for example, have large quantities of excess iron-chromium carbide, which give them in large measure the high degree of abrasion resistance possessed by this class of steel.

Steels having an appreciable vanadium content show even greater abrasion resistance because of excess vanadium carbide. Vanadium carbide is exceedingly hard, having higher hardness than either tungsten carbide or the silicon carbide used to make grinding wheels.

Influence of Heat Treatment on Die Life. Each type of die steel must be handled slightly differently from any other for optimum results. Different temperatures, different heating and cooling rates, and variable tempering procedures must be used. The properties of die steels as developed in heat treatment have an important and direct effect on die life. In general, it may be said that the harder a given die, the longer it will wear, while the softer a die is, the tougher it becomes. Thus, assuming the proper die steel is being used and no other factors are operative, dies which are wearing out too quickly should be made harder for improved life, and dies which are breaking or cracking should be made softer.

However, dies frequently spall or break because they first become dull, and extreme pressures build up. This is evidence for increasing, not decreasing, the hardness for added die life.

Within limits, heat treatment can be used to adjust these variables to best advantage.

An oil-hardening steel may work best on one application at Rockwell C62 and on another involving higher stresses and shock at Rockwell C58. Adjustments of the tempering temperature easily produce the hardness desired.

Double tempering and in some instances triple tempering is a requirement for tool service application. This is because these steels retain austenite in the quench. The first temper affects the martensite formed in the quench and conditions the austenite so that it transforms upon air cooling from the temper. Double temper is necessary to affect the martensite which forms after the first temper. Triple tempering relinquishes any retention of austenite and promotes part toughness.

Hardness is not the only measure of effective heat treating, for it is possible to produce equivalent hardness in nearly all die steels by different combinations of hardening and tempering temperatures and times, and usually one of these is superior to the rest. Many steels are very sensitive to slight overheating with their toughness dropping off considerably, yet the hardness may be virtually unaffected. There is no simple test to determine whether a die has been slightly overheated in hardening, though such symptoms as undue size changes, abnormal response to tempering, or a decrease in magnetism are indicators of overheating.

Temperature control is of very great importance in heat treating, and every effort should be made to use the proper temperatures for each grade of steel as recommended. Pyrometers must be properly maintained and frequently checked for accuracy, for at the high temperatures involved in heat-treating dies, it is easy for the thermocouples to become contaminated and lose their calibration.

Control of Surface Chemistry. In addition to the obvious scaling which occurs when dies are heated in furnaces having no protective atmospheres, surface decarburization to a considerable depth can occur because of the oxidizing effects of free oxygen, water vapor, or carbon monoxide. Typical decarburization rates are 0.010 to 0.030 in. (0.25 to 0.76 mm) per hour. The loss of surface carbon can produce poor wear resistance through the loss of excess carbide and/or the inability to develop the normal as-quenched hardness for the grade. It is customary to make tools sufficiently oversize so that they can be ground to remove any surface decarburization which occurred during heat treatment.

If the atmosphere used to heat dies for hardening is strongly reducing, it is possible to carburize the surface even though the carbon content is already quite high. In some instances a carburized case may increase wear resistance. However, if uncontrolled, it can lead to excessive brittleness and consequent tool breakage or to a "soft skin" resulting from austenite retention with consequent poor wear resistance.

Molten salt baths provide decarburization protection for die steels. Careful maintenance involving desludging and rectification is necessary to maintain a neutral condition.

Control of Dimensional Change. Dimensional changes occur during the hardening operation on dies because the hardened steel occupies a greater volume than the annealed steel from which it came. Unfortunately, the dimensional changes which result from the volume of change are usually not the same in all three directions, making it impossible to predict the changes accurately. The size and shape of the individual workpiece are the most important factors influencing this variable.

The dimensional changes resulting from hardening may vary from zero to approximately three parts per thousand. Dimensional changes of this amount are of no concern, provided enough grind stock is provided to allow for scale and decarburization. If high-quality furnaces are used which produce little or no scale or decarburization, grind stock may not be required.

Air-hardening tool steels usually produce dimensional changes which are less than one part per thousand, and therefore are widely used where precision is needed. It is also possible to control the size changes in tools made of type D2 high-carbon, high-chromium steel to so-called "zero size change" by an austenite-martensite balance control obtained by multiple tempering.[2] On a series of identical tools made of air-hardening steel, it is customary to make allowance for expected dimensional changes on the basis of the experience from the first part treated.

Gas Nitriding. The use of nitriding to produce a hard, wear-resisting case on steels is well known. This procedure can be used also on some tool steels to improve wear resistance. Gas nitriding can be used advantageously only on tool steels which do not temper back excessively at the nitriding temperature, 975° F (524° C). This limits gas nitriding largely to the hot-work steels and the high-carbon, high-chromium grades. High-speed steels form an exceptionally brittle nitrided case; therefore, gas nitriding should not be used on these steels.

Gas nitriding of tool steels is usually carried out for a period of from 10 to 72 hours, producing a case depth of from 0.002 to 0.018 in. (0.05 to 0.46 mm). Whenever deep-nitrided cases are used, caution must be exercised to avoid production of a white layer on the surface of the nitrided case, or else provision must be made for carefully grinding the white layer away before the tool is used. When surface brittleness is not a factor, nitrided tools often outwear normally treated tools up to 10 to 1.

Gas nitriding of tool steels has largely been supplanted by ion nitriding, as well as the titanium nitride (TiN) and titanium carbide (TiC) coatings covered later in this section.

Die Design for Successful Heat Treatment. Tools cannot be designed on the basis of dimension alone. The designer must consider the type of steel which will be used, with particular regard as to whether the steel will be hardened by quenching in water, oil, or air. Generally, liquid-hardened tools must be conservatively designed, while air-hardened tools can tolerate some features which would be improper for liquid quenching. The design of a tool must also take into account the equipment available for heat treatment.

The following rules are offered for guidance in die design:

1. Avoid sharp corners wherever feasible, using the largest fillets possible. On some tools it is possible to fillet the corners and then grind the fillet out after hardening.

2. Avoid extreme section changes or heavily unbalanced masses. Use a two-piece assembly if possible.

3. Holes should be designed to provide for effective internal quenching if the internal surface must be hard, or they should be packed before hardening to prevent nonuniformity of cooling during the quench.

4. Blind holes should be avoided and should be drilled through if possible.

5. Locate holes as follows:
   a. Holes near the edges of dies should be located so that the wall between the hole and the die edge is greater than the hole diameter.
   b. Holes near corners of dies should be located so that the wall between the hole and the die edge is equal to 2 times the hole diameter or greater.
   c. Adjacent holes in dies should be located so that the wall between the holes is greater than the thickness of the die.
   d. Stagger holes, rather than placing them in line.

6. Avoid deep scratches, tool marks, and deep stamp marks.

7. Order stock large enough to allow for machining to remove decarburized surfaces.

REPAIRING DIES BY WELDING

Basically three welding methods are used for die repair. The American Welding Society classifies them using the following lettering system.

1. SMAW–Shielded metal arc welding or stick welding.
2. GMAW–Gas metal arc welding, also known as MIG.
3. GTAW–Gas tungsten arc welding, also called TIG or heliarc welding.

The SMAW or stick welding is the most popular method since it is extremely versatile. This method also offers a wide selection of filler metal analysis. The GMAW or MIG process is usually superior in terms of speed and quality, however. The filler metals are available as either solid or tubular wire. The tubular wire filler metal offers the widest range of alloy selection.

The GTAW or TIG process offers the operator optimum control and inclusion-free welds. The selection of bare metal filler rods is somewhat limited.

Die Welding Applications. In die welding, wear resistant alloys are applied to the surface of dies to increase service life, avoid down time, or to rebuild or repair dies which have been damaged in service. Welding is also used to correct machining errors, and to increase the wear resistance of die surfaces. Varying degrees of hardness, toughness, and wear resistance are available in welding alloys depending upon the application.

Die Welding Materials. Tool steel welding materials are normally heat treatable alloys. Alloys for SMAW (stick) welding have a wire core covered with a flux coating which may also contain alloy constituents, including rare earth elements. Alloying elements in the flux coating are passed across the arc and form a homogeneous weld deposit.

GMAW arc welding alloys for fully and semiautomatic MIG welding are available in in tubular and solid wire products. This process requires an external shielding gas. Additional alloying elements can be incorporated in the tubular flux. Solid MIG welding wire does not provide additional alloying.

GTAW (TIG) alloys are made of bare wire cut to 36 in. (914 mm) long. Small diameter sizes are available that permit the operator the utmost control for precise build-ups and contours. All basic tool steel alloys are available.

The tool steel groups most commonly repaired by welding are water, oil, and air hardening steel. The degree of welding difficulty is dictated by the alloy content of the base metal. Usually steels with higher carbon contents require higher preheats before welding, more care during welding, and consideration of tempering temperatures after welding.

Base Metal. When the base metal for an intended application is carbon steel, there are no particular restrictions in the choice of an alloy tool steel welding material. There are, however, some restrictions with other base metals, including stainless steels. Tool steel welding requires taking precautions similar to those necessary with any other type of welding process to prevent cracking of the base metal during heating and cooling.

All base metals with a carbon content higher than 0.35% should be pre-heated and post-heated to decrease brittleness in the base metal near the heat affected zone. High-carbon alloy steels such as air-hardening tool steels are more difficult to weld because of the likelihood of cracking. Tool steel welding is used successfully in most instances for the repair of tools made of air-hardening steel. Hot extrusion tools are a notable example of successful use of this type of repair.

Cast-iron dies can be welded, but the method of application of the weld differs from that for steel because cast iron melts at a lower temperature. Nickel materials with lower melting temperatures are recommended for underlay on all cast irons.

Generally, after careful preparation involving grinding out the area to be repaired and preheating, a deposition of nickel alloy weld is made followed by the desired welding material required to obtain the needed surface properties. Where a very high-strength bond with the cast iron must be obtained, the iron may be studded with mild steel threaded inserts. An easy way to do this is to drill and tap the iron casting for mild steel cap-screws. The screws are tightened and then cut off flush with a cold chisel. It is very important to not use alloy steel screws because a brittle weld may result. This method is excellent when repairing old, oil-soaked cast iron.

Large, broken, iron die castings can be repaired by brazing. Preparation by machining or grinding is required to provide a large area for adhesion of the braze material. Careful pre-heating and post-heating is an absolute necessity. Charcoal is an excellent preheating fuel source for occasional repairs on large castings.

Welded Trim Edges. The dies used to trim large, irregularly-shaped drawn panels are made of grey cast iron. The cutting steels on the punch shoe are either of cast alloy steel or wrought composite tool steel construction. Mounted to the lower die shoe is a grey iron casting called a post, onto which the panel is placed for trimming. The post has a milled

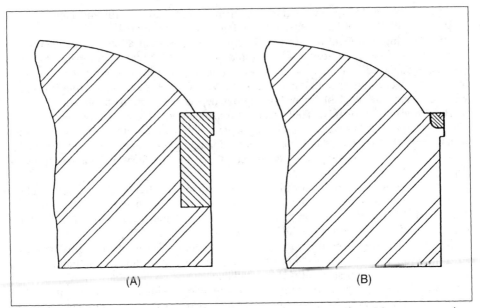

(A)   (B)

Fig. 28-2. Composite tool steel welded trim edge: (A) conventional tool steel insert construction; (B) welded tool steel composite edge. (*MPD Welding, Inc.*)

ledge or pockets into which the tool steel lower trim inserts are fastened (Fig. 28-2A).

An alternative construction method is to eliminate the post inserts, and provide a tool steel cutting edge on the grey iron post by welding (Fig. 28-2B). Success in producing a cutting edge by the welding method depends upon following strict procedures in preheating, welding materials, welding technique, peening, and cooling.

A U-groove is first ground or machined around the post. This is most economically done on the CNC profile mill used to machine the finished contour of the post. A vertical land is provided as a reference of the trim line location for finish grinding after welding.

To avoid producing a layer of brittle, white chilled iron under the weld, the casting must be preheated to 750° F (400° C). A good method is to place the casting on a special table fitted with a number of small gas burners. To conserve heat, the casting is covered with mineral fiber blankets except for the area being welded.

Temperature control can be provided by a thermocouple probe which can also automatically regulate the burners. Many skilled operators control the burners manually and determine the heat with temperature indicating marking sticks.

Two types of weld are deposited. A layer of unique nickel alloy weld is deposited first to act as a buffer for the hard weld material that forms the cutting edge. Materials commonly used include AISI-S.7, H12, and H19.

The buffer layer serves to prevent excess carbon and other elements from the grey iron from mixing with the air hard weld and changing its properties. The nickel alloy layer also provides a good bond with the casting.

The buffer layer is carefully deposited onto one small area at a time. To avoid the production of white chilled iron, several passes are deposited in sequence. Each successive pass serves to anneal the iron under the weld.

To relieve the tensile stress that would occur upon cooling, the weld metal is upset by careful peening with a pneumatic hand tool between each pass. The finished weld must be as stress free as possible, although slight compressive stresses are not considered harmful.

The final cutting edge is built up with air hard weld and after cooling to room temperature is carefully tempered.

If trim changes or repairs are needed, the casting should be ground out to expose sound metal and the initial welding procedure including preheating, and slow cooling followed. Small chipped spots can be repaired by TIG welding and peening as an emergency measure although the quality of the repair is compromised.

CHROMIUM PLATING

A thin chromium plate is often applied to forming and drawing dies in order to increase wear resistance and reduce galling. This chromium-plated surface has a very low coefficient of friction with excellent nongalling characteristics. The usual practice is to apply a layer of chromium 0.0005 to 0.001 in. (0.013 to 0.021 mm) thick to a very finely ground or polished surface. The time taken to polish the surface is well spent, as any defects or irregularities present on the surface prior to chromium plating will show through and actually be aggravated during the plating operation.

Chromium plating is also used to repair worn dies. It is possible to build up a layer of chromium of as much as 0.010 in. (0.25 mm) on a worn surface and thus increase the total production life of the die.

When chromium plating is used to improve the wearing and surface friction characteristics of large sheet metal drawing dies, the amount of drawing lubricant needed is often reduced or eliminated altogether. A disadvantage of chromium plating such dies is that repairs made by welding or oilstoning will destroy the plating. When such drawing dies are repaired by welding, care should be taken that high nickel weld is not used as chromium will not adhere well to such surfaces.

ION NITRIDING[3]

There are a wide range of surface hardening applications that can be addressed by ion nitriding. Uses range from improving the wear resistance of small tool steel die sections to large iron alloy drawing punches weighing 10 or more tons. In the latter application, it is often employed as an alternative to hard chromium plating to reduce surface friction as well as metal pickup when drawing zinc coated steel. Ion nitriding is a surface hardening method, but it is basically different from other types of surface hardening in that it does not perform hardening by changing the available structure near the material surface as do other methods, but performs hardening by changing the material composition through the penetration and diffusion of nitrogen into the material.

Principle of Operation. A *glow discharge* or *ion processing* takes place when a d-c voltage is applied between the furnace as the anode and the workpiece as the cathode, both placed in a low-pressure nitrogen gas atmosphere. The nitrogen gas in the furnace is ionized and emits electrons (with a negative charge) and ions of nitrogen (positively charged) which move toward the cathode (i.e., the workpiece), and are accelerated with high speed by the sharp cathode drop just in front of the cathode surface, and bombard the workpiece.

The high-speed kinetic energy of the ions is converted into thermal energy, which not only heats the workpiece from its surface, but also partly implants ions directly on to the workpiece surface. This bombardment also causes cathode sputtering, thus driving out electrons and elements such as iron, carbon, and oxygen from the workpiece surface. The iron atoms thus driven out combine with the atomic nitrogen in the plasma produced by the glow discharge and form iron nitride (Fe-N) which is absorbed on the outermost surface (Fig. 28-3). The iron nitride which is absorbed on the outermost layer is gradually decomposed into nitrides of lower orders by the temperature increase, and ion bombardment and a portion of the nitrogen penetrates and diffuses into the workpiece. The depth of diffusion and the type of tissue resulting depends upon the temperature, gas composition, and time. A portion of the nitrogen returns into the discharge plasma and combines with increasingly sputtered iron, to repeat the above-mentioned phenomenon.

The process is versatile in that it allows the controlling of layer composition and

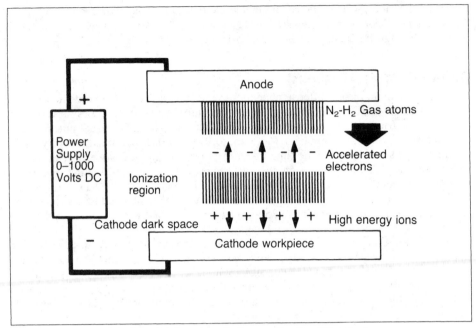

Fig. 28-3. Physics of the ion nitriding glow discharge. (*Sun Steel Treating*)

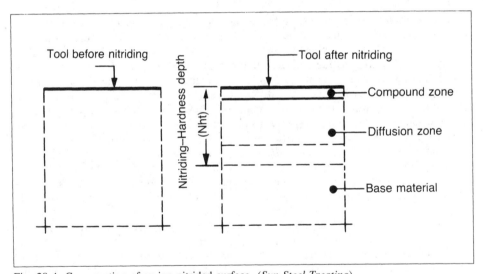

Fig. 28-4. Cross-section of an ion nitrided surface. (*Sun Steel Treating*)

structure through the altering of the ion parameters. It's these parameters that make the process flexible in developing the required properties for a particular part or use. The nitride case consists of two zones, the compound zone and the diffusion zone (Fig. 28-4). A carbon-free atmosphere coupled with low levels of nitrogen will generate gamma prime which is ductile and has excellent wear characteristics for use on higher alloyed steels. A high carbon atmosphere, in the form of methane or propane and a high level of nitrogen, will result in an epsilon layer, excellent for resistance to stress, shock loading, impact and wear, but less ductile than a gamma prime layer and used on lower alloys at shallower case depths. Depending on the type of irons or steels being treated and the desired results, the process can be tailored as to time, temperature, gas mix, and pressure to achieve either

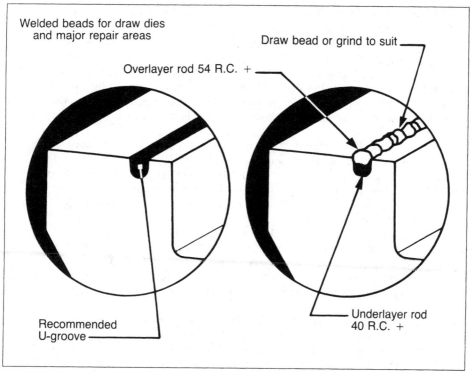

Fig. 28-5. Recommended repair procedure for ion nitrided dies. (*Sun Steel Treating*)

an epsilon or gamma prime zone or a customized process, blending both epsilon and gamma prime to get the desired results.

Large Die Applications. The automotive stamping industry requires a surface treatment which allows normal processing and finishing of dies. The large dies, usually made of a high grade iron alloy, must be capable of being repaired by welding and polishing without destroying the hard nitrided surface. The low-temperature vacuum process produces a clean, distortion-free surface with case depths ranging from 0.010 to 0.030 in. (0.25 to 0.76 mm) at temperatures ranging from 750° to 950° F (400° to 510° C), producing a hardness of 58-65 R_C.

Nitriding equipment for very large dies ranges in size from 18 feet (5.5 m) long and 10 feet (3 m) wide and high, to specialty units of up to 30 feet (9.1 m) high and 6 feet (1.8 m) wide.

An ion-nitrided surface can be welded, for repair purposes, before or after nitriding without removing the die from production if so desired. Opening up the repair area, by grinding down to base metal below the nitride case, then filling the area with a soft filler rod followed by a hard cap rod, will serve to return the die to operation more quickly than any other method. Class A surfaces are better served by the harder weld in that they do not act as pick-up points for electro-galvanized, zinc coated metals or dirt, which could lead to scoring or galling. Fig. 28-5 illustrates the recommended method of adding draw beads or making welded repairs to ion nitrided draw dies.

A frequent cause of damage to the draw dies for large parts such as automotive panels is slug marks resulting from hitting a folded blank or foreign object in the die. When this happens, a low spot on the draw punch or a reverse often results. This low spot can often be repaired in the press by heating the low spot with an oxy-acetylene torch to pop the slug mark, followed by light oilstoning. An advantage of ion nitriding over hard chromium plating is that this repair procedure will not destroy the nitrided surface. In fact, ion nitrided surfaces can be flame hardened.

TITANIUM NITRIDE AND TITANIUM CARBIDE COATINGS

Titanium nitride (TiN) and titanium carbide (TiC) improve the life of tools by acting as a chemical and thermal barrier to diffusion and fusion. The coatings are very thin, typically 0.0001 to 0.0003 in. (0.002 to 0.008 mm) in thickness, and quite hard. Although the thin coatings are very brittle, they tend to assume the ductility and deformation characteristics of the substrate material. The coatings are also quite lubricous, serving to lower the coefficient of friction between the tool and the workpiece. By depositing TiN or TiC onto a steel or carbide tool, the improvement in lubricity causes the tool to resist galling.

Physical Vapor Deposition (PVD) of Titanium Nitride. This coating process is carried out in a high vacuum at temperatures between 400° and 900° F (204° and 482° C). This range of temperatures does not exceed those used to draw hardened high speed tool steel. Because there is very little distortion or size change on the workpiece, this coating process is frequently used to coat finished punches and buttons whenever rapid wear or galling is found to be a problem.

The plasma source coats the workpiece in a straight line of sight process. Special rotating fixtures with water cooling may be required, to ensure that all surfaces are evenly coated and that small sections are not overheated.

Titanium nitride (TiN) coatings may also be applied to nonferrous materials such as bronze. A TiN coating deposited by the PVD process is easily recognized by its gold color.

Chemical Vapor Deposition (CVD) of Titanium Carbide and Nitride. This coating process is done at much higher temperatures — 1,740° to 1,920° F (949° to 1,049° C) — than the PVD process. For this reason, it is a normal practice to follow the coating procedure with a conventional heat treatment of the tool steel substrate.

TiC is limited to tool steel and solid carbide die materials because the substrate surface must act as a catalyst. It is superior to PVD coatings when extreme abrasive wear is a problem. The coating is deposited from a vapor, so uniform coating including blind slots and blind holes is possible.

Because it has less lubricity than TiN, TiC is a better choice for thread roll dies where a reduction in friction is detrimental to the life of the tool. Multi-layer coatings done in conjunction with TiN are available.

A CVD coating is dull grey in color. When a CVD-coated tool is polished, the resultant tool is silver, quite indistinguishable from the base metal.

Surface Preparation. Before either of these coating processes are begun it is important that the tool be cleaned, and sometimes a pre-coating surface polishing is required. Any coating is only as good as the substrate surface on which it is placed. Coating cannot make up for imperfections in the tool. A good uncoated tool results in a better performing coated tool.

Choice of Methods. The CVD coating method can deposit both TiN and TiC. Coatings can be applied to all tool steels as well as solid carbide tooling. When very high wearability qualities are required, and the distortion caused by the post coating heat treatment that is usually needed is not a problem, CVD may be the best choice.

PVD is a low temperature process that can be applied to all tool steels, but is generally used to increase the wearability of finished, high-speed steel parts, solid carbide, and brazed carbide tooling.

CRYOGENIC TREATMENT

Cryogenic treatment will maintain hardness while actually improving the toughness and fatigue strength of hardened tool steel. This improvement occurs due to the continued transformation of retained austenite into the more desirable martensite, at temperatures ranging from −120° to −300° F (−84° to −184° C).

Size change is negligible but can occur if the material is not properly heat treated. Cryogenic treatment will not correct poor heat treatment practices. Multiple draws or tempers should not be eliminated knowing that the cold process will convert austenite to martensite. Cryogenic treatment must begin almost immediately after heat treatment to be effective; it is often performed between the last two draws.

POWDER METALLURGY PROCESS

The powder metallurgy process provides a much more uniform steel than can be produced by casting the steel into ingots. Cast tool steel ingots have all of the problems of segregation of alloying constituents upon cooling that are associated with mild steel ingot production.

The molten tool steel is first atomized in a nonreactive gas atmosphere into very fine steel shot having uniform properties. This shot is placed in large steel canisters which are evacuated and sealed shut. The canisters are placed into a furnace and slowly brought up to the welding temperature of the alloy being produced.

As the canister heats up, it collapses around the hot steel shot, fusing it into a homogeneous mass. The material from the steel canister itself becomes a decarburized layer on the surface of the rolled tool steel which can be removed by machining or grinding.

Material produced by the powder metallurgy process exhibits improved grindability in the heat treated condition. Annealed material also exhibits excellent machinability. The chief advantages of the product is the extreme uniformity and freedom from imperfections die-to-element segregation, and pipe inclusions.

WROUGHT CARBON AND LOW-ALLOY STEELS

Wrought-steel plate, rounds, and shapes are often used in the fabrication of brackets; frames; feeding, ejecting, and transfer mechanisms, and other auxiliaries to the dies where structural strength and weldability, rather than wear resistance, are the primary requirements. Short-run steel dies are sometimes made with carburized, hot-rolled-steel wear surfaces.

Where the properties of AISI-SAE 1018 or similar steel (boiler plate) suffice, they have the advantage of being readily available and economical. Higher strength levels with optimum impact properties are achieved, usually through heat treatment, if alloy steels such as those listed in Table 28-5 are used. These grades are often carried in stock in the heat treated condition for the convenience of users, and resulfurized grades are available for improved machinability in the hardened condition.

CAST CARBON AND LOW-ALLOY STEELS

Cast-steel shoes and other die components are often used for large drawing, forming, or trimming dies where a combination of high toughness and strength is required. These steel castings are usually annealed or normalized to provide a homogeneous structure, free from casting stresses. Heat treating or flame hardening is often employed to obtain the desired strength, wear resistance, and toughness.

Table 28-5 lists several widely used cast steels with their characteristics. The basic principles of heat treatment previously discussed apply to the cast steels as well as the tool steels, except that these steel castings are usually lower in alloy content and often are so large that localized surface hardening is more practical than through-hardening.

Trim steels are also cast of tool steel as an alternative to more expensive, composite steel forgings. The alloys most commonly used are AISI-SAE 4150 and 4340, which have good flame hardening properties (Table 28-5).

Composite forgings are usually specified for trim die steels for heavy drawn stampings over 0.050 in. (1.27 mm) thick because of superior shock and wear resistance.

TABLE 28-5
Typical Wrought and Cast Steels Used for Dies and Machine Parts

AISI-SAE Designation	Average Composition, Percent						Characteristics
	C	Mn	Si	Cr	Ni	Mo	
Wrought Steels							
1018	0.15–0.20	0.60–0.90	0.15–0.17	....	....	....	Inexpensive, readily weldable with good strength and toughness case hardenable by carburizing, low hardenability.
1045	0.43–0.50	0.60–0.90	0.20–0.23	....	....	....	Middle range strength machinery steel. Low hardenability.
3310	0.08–0.13	0.45–0.60	0.20–0.35	1.4–1.75	3.25–3.75	....	Carburizing steel of high hardenability. Uses include heavy-duty bearings and gears.
4140	0.38–0.43	0.75–1.0	0.20–0.35	0.80–1.1	....	0.15–0.25	Middle range hardenability. Shafting is primary use.
4340	0.38–0.43	0.65–0.85	0.20–0.35	0.70–0.90	1.65–2.0	0.20–0.30	Tough machinery steel used for gears, splines, and shafts.
8620	0.17–0.23	0.70–0.90	0.20–0.35	0.40–0.60	0.40–0.70	0.15–0.25	Carburizing steel of intermediate hardenability for gears.
52100	0.98–1.0	0.30–0.40	0.25–0.30	1.35–1.5	....	....	Ball and roller-bearing steel used in wear applications without carburizing.
Cast Steels							
1070	0.68–0.72	0.72–0.78	0.30–0.40	....	....	....	Readily weldable, good strength and toughness, low hardenability.
4150	0.48–0.53	0.75–1.0	0.20–0.35	0.80–1.1	....	0.15–0.25	Flame hardenable. Has intermediate hardenability.
4340	0.38–0.43	0.65–0.85	0.20–0.35	0.70–0.90	1.65–2.0	0.20–0.30	High hardenability, high strength and toughness.
6150	0.48–0.53	0.70–0.90	0.20–0.35	0.80–1.1	....	....	Intermediate hardenability. Also flame hardenable.
8640	0.38–0.43	0.85–0.95	0.30–0.40	0.40–0.60	0.40–0.70	0.15–0.25	High hardenability, high strength and toughness.

Steels requiring a severe quench to achieve hardness are more likely to need straightening after heat treatment.

CAST IRONS

The high compressive strengths, manufacturing economy, and ease of casting of gray cast iron make this material useful, especially in large forming and drawing dies.

Soft, unalloyed gray irons are widely used for plates, jigs, spacers, and other die parts where no wear is involved. Table 28-6 shows properties with some typical applications of gray cast irons.

Cast irons or die components are usually in the composition range listed in Table 28-7. The composition is secondary to the properties, however, and is usually left to the discretion of the foundry. For uniform mechanical properties and machinability, it is important that the structure be free from excessively large flake graphite, large primary carbides (chill), and excessive amounts of steadite (iron phosphide). The foundry achieves this by properly balancing the iron's composition, using foundry-floor chill tests and Brinell hardness measurements for control.

The metallurgical constituents found in the cast irons affect the properties important to die life and economy as shown in Table 28-8.

Fully pearlitic irons with uniform flake-graphite structures are excellent for wear resistance. Wear resistance can be, and often is, significantly improved by flame hardening draw radii or other wear areas. Alloy additions of chromium, molybdenum, and nickel are commonly used to produce uniform pearlite structures and to improve the iron's response to flame hardening.

Elements found in—or added to—cast iron can have an effect on the properties designed into dies, as shown in Table 28-9.

The use of alloys in cast-iron dies is illustrated by the die irons listed in Table 28-7.

In the chromium-molybdenum alloy iron, the chromium is present in an amount which will effectively stabilize pearlite but which will not form massive carbides. Both the chromium and molybdenum contribute to the hardenability for flame hardening.

In the nickel-chromium iron, the nickel makes a higher chromium level possible without the danger of massive carbides, and also refines and toughens the structure. Again, both alloying elements contribute to the iron's hardenability, and the nickel contributes latitude to the flame-hardening procedure, which is especially advantageous to shop hardening.

The acicular structure in cast irons is sometimes developed by heat treating and can be attained as cast with higher alloying. Though acicular irons are far tougher and stronger than any of the pearlitic irons, the effect of flake graphite in cast iron limits the extent to which higher strength or toughness levels can be developed.

The silicon and carbon levels in cast iron are balanced by the foundry, depending on the section size of the casting to be made and the hardness level specified.

TABLE 28-6
Nominal Mechanical Properties of ASTM Classes of Gray Cast Irons

Property	ASTM Class 20	ASTM Class 25	ASTM Class 30	ASTM Class 35	ASTM Class 40	ASTM Class 50	ASTM Class 60
Tensile Strength, psi (MPa)	20,000 to 24,000 (138 to 165)	25,000 to 30,000 (172 to 207)	30,000 to 34,000 (207 to 234)	35,000 to 40,000 (241 to 276)	40,000 to 46,000 (276 to 317)	50,000 to 56,000 (345 to 386)	60,000 to 65,000 (414 to 448)
Compressive Strength, psi (MPa)	To 85,000 (586)	To 100,000 (689)	To 115,000 (793)	To 130,000 (896)	To 150,000 (1,034)	To 170,000 (1,172)	To 190,000 (1,310)
Brinell Hardness	To 180	To 205	To 225	To 230	To 235	To 265	To 285
Modulus of Elasticity, psi (MPa)	10×10^6 (68,940)	12×10^6 (82,728)	13×10^6 (89,622)	15×10^6 (103,410)	16×10^6 (110,304)	18.5×10^6 (127,539)	20.3×10^6 (139,948)
Machinability	Fair	Fair	Fair / Good	Good	Good	Good / Excellent	Excellent
Some Applications	Close Dimensional Thin Section Castings	Close Dimensional Thin Section Castings	Machinery	Automotive	Machine Tools	Machine Tools	Presses Crankshafts Dies

TABLE 28-7
Typical Compositional Ranges of Cast Irons Used For Die Components*

Composition, %	Unalloyed Cast Irons Used For Plates, Jigs, Spacers, Etc.	Alloy Cast Irons Used For Sheet-Metal Dies, Rings, Shoes; Often Flame-Hardened		
		Pearlitic Cr-Mo	Pearlitic Ni-Cr	Acicular Ni-Mo
Total carbon	2.90–3.50	2.80–3.50	2.80–3.25	2.90–3.10
Combined carbon		0.60, min	0.60, min	
Silicon	1.25–2.25	1.50–2.25	1.50–2.20	1.40–1.60
Manganese	0.60–0.90	0.60–0.90	0.60–0.90	0.80–1.00
Sulfur	0.20, max	0.12, max	0.12, max	0.12, max
Phosphorus	0.20, max	0.12, max	0.12, max	0.05, max
Chromium		0.35–0.50	0.50–0.70	
Molybdenum		0.35–0.50		0.75–1.70
Nickel			1.25–1.75	0.75–3.00
Bhn	150–200	205–255	200–250	250–320
Flame-hardenable	Not readily flame-hardenable	50 R_C, min	50 R_C, min	50 R_C, min

* Based on specifications of several major automotive companies and data of International Nickel Company.

DUCTILE IRONS

The ductile (nodular) irons retain the casting advantages of cast iron, but because the free graphite is present in spheroidal shape rather than in flake form, this material develops toughness and strength levels approaching those of steel.

This combination of properties is especially useful in large forming and drawing dies where heavy impact loads or high transverse stresses are encountered or where breakage has occurred with grey iron castings.

Table 28-10 shows three ductile iron types that have been found useful and most satisfactory in die work. Their general applications are as follows:

1. Type I. Punch shoes, retaining slides, die pads, jigs, fixtures, and other parts where maximum shock resistance, machinability, or weldability is required. Explosive forming dies.

2. Type II. Punches, dies, die blocks, blankholders, hold-down rings, guides, redrawing dies, slides, gears, and other parts where higher strength, excellent wear resistance, and response to flame hardening are required. Explosive forming dies.

3. Type III. Applications similar to type II but where even higher, deeper hardening response is required for maximum wear resistance. Can be air-hardened and drawn to 275 to 350 Bhn. Small parts often through-hardened by oil quench and temper.

Flame-hardened ductile-iron dies have performed well in place of iron or steel dies containing tool-steel inserts; considerable cost savings have been realized through the elimination of machining in these applications. For satisfactory response to flame hardening, casting structure and chemical composition are important. As in cast steel and cast iron, a pearlitic structure is necessary if flame hardening to Rockwell C50 or over is desired. Table 28-10 shows hardness levels that can be expected from flame hardening ductile irons. Smaller ductile-iron castings also can be through-hardened to develop high strength levels and resistance to wear.

TABLE 28-8

Effects of Metallurgical Constituents on Properties of Cast-iron Dies

Constituent	Property					
	Impact Resistance	Hardness Level	Wear and Gall Resistance	Machinability	Flame Hardenability	Strength
Ferrite	Improves	Lowers	Lowers	Improves	Lowers	Lowers
Pearlite		Raises	Improves	Lowers	Improves	Improves
Graphite	Lowers	Lowers	Improves	Improves	. . . . . .	Lowers
Other:						
Massive carbides	Drastically lowers	Drastically raises	Lowers	Drastically lowers	. . . . . .	. . . . . .
Fine carbides	Lowers	Raises	Improves	Lowers	. . . . . .	. . . . . .
Steadite	Lowers	Raises	Improves	Lowers	Lowers if excessive	Lowers
Acicular	Improves	Raises	Improves	Lowers	Improves	Improves

28-22

TABLE 28-9
Effects of Alloys on Properties of Cast-iron Dies

Alloying Element	Die Property								
	Casting-section Uniformity	Chill Propensity	Pearlite Stability	Machinability	Wear Resistance	Hardness Level	Hardenability	Ability To Accept Variations In Flame-Hardening Practice	Strength
Silicon	Decreases	Lowers	Lowers	Improves	Lowers	Lowers	Lowers	Lowers	Lowers
Manganese	Improves	. . . . .	Improves	. . . . .	. . . . .	Higher	Higher	. . . . .	Higher
Chromium	Decreases	Much higher	Improves	Lowers	Improves	Higher	Higher	. . . . .	Higher
Nickel	Improves	Lowers	Improves	Improves	Improves	Higher	Higher	Improves	Higher
Molybdenum	Improves	Higher	Improves	Lowers	Improves	Higher	Higher	Improves	Higher
Copper	Improves	. . . . .	Improves	Improves	Improves	Higher	. . . . .	. . . . .	Higher
Vanadium	. . . . .	Higher	. . . . .	. . . . .	. . . . .	Higher	Higher	. . . . .	Higher
Phosphorus	. . . . .	. . . . .	. . . . .	. . . . .	Improves at low levels	. . . . .	. . . . .	Lowers	. . . . .

TABLE 28-10
Properties and Composition of Ductile (Nodular) Irons[4]

	Type I*	Type II†	Type III‡
Mechanical Properties			
Tensile strength, ultimate, psi (MPa)	60,000 (414)	80,000 (552)	95,000 (655)
Yield strength at 0.2% offset, psi (MPa)	45,000 (310)	55,000 (379)	65,000 (448)
Elongation, min, %	10	5	4
Bhn	143–190	190–241	225–285
Ratio of yield strength in compression to yield strength in tension	1.2	1.2	1.2
Modulus of elasticity, psi (MPa)			
Tension	22–24×10^6 (151,668–165,456)	23–25×10^6 (158,562–172,350)	23–25×10^6 (158,562–172,350)
Torsion	9.2–9.5×10^6 (63,425–64,493)	9.2–9.5×10^6 (63,425–64,493)	9.2–9.5×10^6 (63,425–64,493)
Ratio of torsional strength (in shear) to ultimate tensile strength	0.9	0.9	0.9
Proportional limit, psi (MPa)	35,000 (241)	40,000 (276)	45,000 (310)
Poisson's ratio	0.28–0.29	0.28–0.29	0.28–0.29
Ratio of endurance limit to ultimate tensile strength			
Unnotched	0.50	0.45	0.40
Notched	0.30	0.26	0.22
Chemical Composition, %			
C (total)	3.25–4.00	3.25–4.00	3.25–4.00
Si	2.60 max	2.60 max	2.60 max
Mn	0.20–0.70	0.35–0.80	0.35–0.80
P	0.08 max	0.08 max	0.08 max
Ni	0.50–1.00	1.00–2.00	2.50–3.50
Mo	. . . . . .	0–0.50	0.30–0.60
Heat Treatment			
Required	Anneal, 1,300° F (704° C) 5 hr	Stress-relieve at 1,050°F (566° C)	Stress-relieve at 1,050°F (566° C) or normalize and temper at 1,050°F

* Ferritic, toughest and most ductile.
† Pearlitic, surface-hardenable.
‡ High alloy, high hardenability.

References

1. S.G. Fletcher, "The Selection and Treatment of Die Steels," *The Tool Engineer* (April 1, 1952).
2. J.Y. Riedel, "Distortion of Tool Steels in Heat Treatment," *Metal Treating* (July-August and September-October 1951).
3. R. Denton, "Applications of Ion Nitriding," Society of Manufacturing Engineers, Selecting Tooling Materials and Tooling Treatments for Increased Tool Performance Clinic (November 1989).
4. C.R. Isleib, "The Application and Design of Ductile Iron Drawing and Forming Dies," Society of Manufacturing Engineers, Creative Manufacturing Seminar (February 1963).

SECTION 29
NONFERROUS AND NONMETALLIC DIE MATERIALS

Dies made of nonferrous materials are used for a wide variety of reasons. They are economical for limited production runs, they can be used as experimental models, and they have superior functioning, such as preservation of part finish, relatively light weight and portability for extremely large tools, corrosion resistance, ease of fabrication, and low lead time requirements. They also permit fast, easy corrections when design changes are necessary. Nonferrous die materials include aluminum alloys, zinc-based alloys, lead-based alloys, bismuth alloys, cast-beryllium alloys, copper-based alloys, plastics, elastomers, and tungsten carbides.

Dies made with tungsten carbide elements are most economical for large production quantities and for stampings having critical tolerances. The tungsten carbide parts are expensive because of their hardness (which makes them difficult to machine) and the close tolerances to which they are normally held.

CEMENTED CARBIDES

Cemented carbides consist of finely divided hard particles of carbide of a refractory metal sintered with one or more metals such as iron, nickel, or cobalt as a binder, forming a body of high hardness and compressive strength. The hard particles are carbides of tungsten, titanium, and tantalum, with some occasional specialized use of the carbides of columbium, molybdenum, vanadium, chromium, zirconium, and hafnium.

The selection of the best carbide grade for a specific job is sometimes a difficult task. The wide range of carbide grades available and the continuous development of new grades have prevented or arrested the development of an adequate standard classification system. Therefore, knowledge of actual job-specific performance of carbide dies and close liaison with the supplier regarding application may be required. Table 29-1 lists the properties of several groups of cemented carbide materials used in forming tools.

Variation in grain size can affect the properties and performance of an otherwise constant material. As the grain size decreases, the binder metal between the grains becomes thinner, making a more wear-resistant and more rigid carbide material. Conversely, a larger average grain size increases the thickness of the boundary layer and also increases shock resistance, but it decreases the stiffness and makes the edge more vulnerable to thermal deformation. A proper blending of grain size can result in desirable combinations of strength and wear resistance.[1]

Thermal expansion is an important physical characteristic of carbide, and, for most carbide grades, it ranges from one-third to one-half that of steel. This must be considered when carbide is attached to a steel support or body. Attaching the carbide to a steel body by clamps, wedges, or screws minimizes the effects of strain caused by the dissimilar expansion of carbide and steel.

Restrictions exist on the use of lubricants on some carbide die materials, particularly

TABLE 29-1

Compositions and Properties of Cemented Carbides for Forming[2, 3]

Group	A	B	C	D	E	F
Composition,%:						
Co	5–7	8–10	11 – 13	14–17	18–22	23–30
Ta or NbC	0–2.5	0–2.5	0–2.5	0–2.5	0–2.5	0–2.5
TiC, max	1	1	1	1	1	1
WC	Balance	Balance	Balance	Balance	Balance	Balance
Average WC grain size, μ m (μ in.)	1–5 (0.00004–0.0002)	1–5 (0.00004–0.0002)	1–5 (0.00004–0.0002)	1–5 (0.00004–0.0002)	1–5 (0.00004–0.0002)	1–5 (0.00004–0.0002)
Density, g/cm^3 (lb/in.^{3})	14.7–15.1 (0.53–0.55)	14.4–14.7 (0.52–0.53)	14.1–14.4 (0.51–0.52)	13.8–14.2 (0.50–0.51)	13.3–13.8 (0.48–0.50)	12.6–13 (0.46–0.47)
Hardness:						
room temperature:						
Vickers	1,450–1,550	1,300–1,400	1,200–1,250	1,100–1,150	950–1,000	800–850
R_A	90.5–9.10	89.0–90.0	88.0–88.5	87.0–87.5	85.5–86.0	83.5–84.0
at 300° C (572° F):						
Vickers	1200	1050	1000	900	750	600
Transverse rupture strength, N/mm^2 (ksi):						
room temperature	>1,800 (261)	>2,000 (290)	>2,200 (319)	>2,500 (363)	>2,500 (363)	>2,300 (334)
300° C (572° F)	>1,600 (232)	>1,800 (261)	>2,000 (290)	>2,300 (334)	>2,200 (319)	>1,900 (276)
Compressive strength, N/mm^2 (ksi)						
room temperature	>4,100 (595)	>3,800 (551)	>3,600 (522)	>3,300 (479)	>3,100 (450)	>2,700 (392)
300° F (572°)	>3,100 (450)	>2,900 (421)	>2,800 (406)	>2,700 (392)	>2,500 (363)	>2,300 (334)
Young's modulus, N/mm^2 (ksi)	630,000 (91,400)	610,000 (88,500)	570,000 (82,700)	540,000 (78,300)	500,000 (72,500)	450,000 (65,300)
Thermal expansion coefficient, room temperatureto 400° C (752° F):						
per ° C	4.9×10^{-6}	5.0×10^{-6}	5.4×10^{-6}	5.6×10^{-6}	6.3×10^{-6}	7.0×10^{-6}
per ° F	8.8×10^{-6}	9.0×10^{-6}	9.7×10^{-6}	10.1×10^{-6}	11.3×10^{-6}	12.6×10^{-6}

Note: N/mm^2 is equivalent in value to MPa.

those containing sulphur, on some carbide die materials. While tungsten carbide is essentially an inert material, the cobalt binder material is attacked by some lubricants. Section 25 has detailed information on the compatibility of lubricants with carbide tooling. Electrolytic corrosion can also result from stray currents from electrical part sensing. If such a problem is suspected, eddy current, magnetic, or light beam sensors should be employed as an alternative detection means (Section 26).

MACHINABLE CARBIDES

These are machinable ferrous alloys sold under the trade name of Ferro-Tic® and are made in several grades. These range from 20%-70% (by volume) titanium carbide, tungsten carbide, titanium-tungsten double carbides, or other refractory carbides as the hard phase. These are contained in a heat-treatable matrix or binder that is mainly iron.

The grade most used for dies and molds contains neither tungsten or cobalt. The main constituent of this grade is approximately 45% titanium carbide and the balance mainly alloy tool steel.

The heat treatment required to harden the matrix is much the same as for conventional tool steels. The manufacturer's recommendations should be closely followed for proper product application, machining, and heat treatment.

NONFERROUS CAST DIE MATERIALS

Cast Aluminum Bronzes. Proprietary bronze (e.g., Ampco® metal) cast to the die shape is used for forming and drawing stainless steel and other difficult-to-work materials without scratching or galling. These alloys are also used for die wear plates and nitrogen cylinder shaft fore-bushings where high load-carrying capacity is required.

The coefficient of expansion is 9 parts per million (ppm) per degree Fahrenheit and 16.2 ppm per degree Centigrade. Table 29-2 provides data on four common grades of Ampco® metal.

TABLE 29-2
Compositions, Properties, and Applications of Aluminum Bronzes for Dies[2]

Ampco Alloy	Chemical Composition, %				Ultimate Compressive Strength, ksi (MPa)	Brinell Hardness, 3,000 kg Load	Rockwell Hardness	Applications
	Al	Fe	Others	Cu				
20: sand cast centrifugally cast extruded	11.0–12.0	3.2–4.5	0.50	Balance	150 (1,034) 155 (1,069) 155 (1,069)	212 223 207	B96 B97 B95	For short-run production and medium requirements
21: sand cast centrifugally cast extruded	12.5–13.5	3.7–5.2	0.50	175 (1,207) Balance	286 190 (1,310) 190 (1,310)	C29 286 286	C29 C29	
22: sand cast centrifugally cast extruded	13.6–14.6	4.2–5.7	0.50	Balance	200 (1,379) 210 (1,448) 210 (1,448)	332 332 332	C35 C35 C35	For long-run production and severe-duty requirements
25: sand cast centrifugally cast extruded	14.5–16.0	4.0–7.0	3.8–8.0	Balance	220 (1,517) 225 (1,551) 225 (1,551)	364 375 375	C38 C39 C39	

(Ampco Metal Div., Ampco—Pittsburgh Corp.)

Zinc-based Alloys. Zinc-alloy die materials have a higher tensile strength and impact resistance than pure zinc. The alloys can be cast into dies for blanking and drawing a variety of aluminum and steel parts, especially complicated shapes and deeper draws than are possible with plastic or wooden dies. The working surface is dense and smooth and requires only surface machining and polishing. Dies made of this material and mounted on die sets are frequently used for blanking light gages of aluminum.

Frequently one member of the die set is composed of a zinc alloy, and the other member is made of a softer material. This is particularly so in drophammer operations where soft metals are encountered. Harder punches are required for forming steel sheets or where sharp definition is necessary. Because of a tendency to spread, close tolerances cannot be maintained in blanking operations. Worn and obsolete dies made of zinc alloy can be remelted to achieve nearly 100% material salvage.

Various proprietary zinc alloys, such as those called Kirksite alloys, are available. A typical composition is 3.5% to 4.5% aluminum, 2.5% to 3.5% copper, and 0.02% to 0.10% magnesium. The remainder is 99.99% pure zinc. In order to minimize health hazards from grinding dusts generated when working these materials, the impurity levels of lead and cadmium must be kept very low.

Lead-based Alloys. Lead punches, composed of 6% to 7% antimony, 0.04% impurities, and the remainder lead, have been used with zinc alloy dies. Strict precautions must be followed to avoid exposing personnel to the hazards of lead toxicity when melting, casting, and working with such materials.

Cast Beryllium Copper Alloys. Cast alloys of beryllium, cobalt, and copper have characteristics comparable to those of the proprietary aluminum bronze alloys previously discussed. Beryllium is a very toxic substance, and proper precautions regarding ventilation and industrial hygiene must be taken when working with these alloys.

Bismuth Alloys. The alloys of bismuth are used chiefly as matrix material for securing punch and die parts in a die assembly and as-cast punches and dies for short-run forming and drawing operations.

These alloys are classified as low-melting-point alloys. The melting point of half of the alloys listed in Table 29-3 is below the boiling point of water.

TABLE 29-3
Compositions and Properties of some Bismuth Alloys[2, 3]

Composition, %								
Bismuth	44.7	49.0	50.0	42.5	48.0	55.5	58.0	44.0
Lead	22.6	18.0	26.7	37.7	28.5	44.5	—	—
Tin	8.3	12.0	13.3	11.3	14.5	—	42.0	60.0
Cadmium	5.3	—	10.0	8.5	—	—	—	—
Other	19.1	21.0	—	—	9.0	—	—	—
Properties								
Melting temperature, °F	117	136	158	159–194	217–440	255	281	281–388
(°C)	(47)	(58)	(70)	(70–90)	(103–227)	(124)	(138)	(138–198)
Tensile strength, ksi (MPa)	5.4 (37.2)	6.3 (43.4)	5.99 (41.3)	5.4 (37.2)	13 (89.6)	6.4 (44.1)	8 (55.2)	8 (55.2)
Bhn	12	14	9.2	9	19	10.2	22	2

Alloys of this type also find use in bending metal tubes into complex shapes without kinking. The tube is first filled with the alloy material and bent after it solidifies. The alloy can easily be melted out after the bending is completed.

NONMETALLIC DIE MATERIALS

Hardboard Laminates. High density panels composed of compressed wood fiber and lignin are used for jigs, dies, fixtures, templates, patterns, molds, and spinning chucks. This material has performed satisfactorily as dies and molds in deep drawing, metal forming (form blocks, dies for rubber forming, and stretch forms) and plastic forming (vacuum and compression), assembly fixtures in aircraft, welding fixtures, and large spinning chucks. Properties and available thicknesses of typical hardboard laminate die stocks are listed in Table 29-4.

Hardboard can be used in single- or double-member dies with or without metal facings. Steel cutting plates are usually inserted in blanking and piercing dies. While hardboard laminates, by themselves, are excellent for short- and intermediate-length production runs, beads and plastic or metal faces are frequently incorporated for long production runs.

TABLE 29-4
Average Properties of Hardening Die Stock[5]

Property	Thickness, Nominal, in. (mm)									
	0.125 (3.18)	0.250 (6.35)	0.375 (9.52)	0.500 (12.7)	0.625 (15.88)	0.750 (19.05)	1 (25.4)	1.250 (31.75)	1.50 (38.1)	2 (51)
Weight, lb/ft² (kg/m²)	0.83 (4.0)	1.76 (8.6)	2.80 (13.7)	3.71 (18.1)	4.70 (22.9)	5.63 (27.5)	7.53 (36.8)	9.35 (45.6)	11.2 (54.7)	15.0 (73.2)
Specific gravity	1.21	1.38	1.39	1.40	1.41	1.41	1.41	1.41	1.41	1.41
Flexural strength psi (MPa)*	11,700 (81)	12,500 (86)	12,500 (86)	12,500 (86)	12,500 (86)	12,500 (86)	12,500 (86)	12,000 (83)	12,000 (83)	10,000 (69)
Bond strength psi (MPa)†	450 (3.1)	450 (3.1)	480 (3.3)	500 (3.4)	500 (3.4)	500 (3.4)	500 (3.4)	500 (3.4)	500 (3.4)	500 (3.4)
Water absorption, % in 24 hrs.	6.0	1.2	0.9	0.9	0.8	0.5	0.4	0.4	0.4	0.3

* ASTM D-790
† ASTM D-952
Tensile strength, 7,700 psi (53 MPa)
Compressive strength, ult., 26,500 psi (193 MPa)
Shear strength, 7,000 psi (48 MPa)

Modulus of elasticity, 1,250,000 psi (8,618 MPa)
Impact, Izod: ft-lb/in.: edgewise 0.6;
flatwise 1.5
Hardness, Rockwell M: 95

One widely known hardboard, Masonite, is a cellulose semiplastic material available in thicknesses of 0.0125 to 2 in. (0.317 to 50.8 mm). It can be readily laminated with cold-setting adhesives or epoxies. High dimensional stability, light weight (one-sixth that of steel and one half that of aluminum), hard-smooth surface, uniform density throughout the panel, and abrasion resistance make hardboard laminates an excellent die material.

Densified Wood. Various woods are impregnated with a phenolic resin after which the laminated assembly is compressed to about 50% of the original thicknesses of the wood layers.

Densified wood punches and dies are used in forming and drawing dies; in the latter, scoring of the part is infrequent because of the low coefficient of friction of densified wood when properly finished.

Typical compressive strengths are 28,000 psi (193 MPa) and the specific gravity is 1.35.[6]

Hardwood. Hardwood can be used for form blocks, but laminated impregnated wood, hardboard, and plastics have largely replaced it. Hard maple and beech are good woods for die applications if they are carefully selected for close grain structure.

Rubber. Molded rubber female dies and rubber-covered punches are used in difficult forming operations, such as the forming of deeply fluted lighting reflectors. Specifications for rubbers used in the conventional Guerin, Marform, and Hydroform processes, are confined generally to performance needed for proper fabrication of a pressed material.

Forming pads are supplied by manufacturers under their own specifications. The manufacturer's data should be followed.

Other rubbers, including synthetic rubber, can be furnished to give good service for a given pressworking operation including the hot forming of magnesium.

Cork. Soft, medium, and hard cork layers, compressed into sheet form, are sometimes used with, or in place of, rubber pads. Cork deforms only slightly in any direction other than that of the applied load, while rubber flows in all directions.

Plastics. The use of plastic materials for dies is not advisable unless the specifications for the materials are within the physical limits of the application requirements. Selection of plastics is based on economy relative to die life expectancy, production volumes, original and maintenance costs, degree of care required to prevent damage to the product (scratching), and lead time available.[7]

The application of plastic materials has certain limitations. The limitations for one industry, however, need not be necessarily true for another. They must be interpreted and applied to specific problems or applications. Plastic dies in many instances have proven less costly and have been satisfactory.

Draw radii can be expected to be a primary source of concern because maximum loads and abrasion occur in these areas, resulting from the combined forces of compression, shear, and friction. Heat distortion is a prominent factor in the limitation of plastic die life. Spanking or bottoming of a plastic die is of little value for the removal of wrinkles formed in a metal stamping. Ironing of wall sections is not recommended since a sufficient compressive force will not be obtained to preform the ironing or sizing required by the metal. Excessive burrs on metal blanks must also be considered a limitation since the low tensile properties and comparatively soft die surface will not withstand the shearing action of the burrs. Surface pickup and filler material must be observed with caution to assure blank cleanliness.[8]

Polyesters. The chief advantage of this material has over the stronger and more stable epoxy resins is cost. Shrinkage rates vary from 2% to 7%. This shrinkage can cause large internal stresses in large cast sections. Inorganic filler materials such as fiberglass and magnesia help reduce these stresses.

Polyester resins are still widely used as surfacing materials on tooling aids.

Epoxies. Epoxies are used for capped forming tools, metal bonding tools, vacuum forming tools, molding dies, and all plastic impact tools. Draw dies having a metal core of either ferrous or zinc-alloy materials, capped with a working face of epoxy, are used in the aircraft, appliance, and automotive industries. Rubber forming dies are made of combinations of cast and laminated epoxy applied to a heavy steel base.

Epoxies, together with polyurethanes (discussed next), have virtually preempted the field of plastics used in die construction. Advantages of epoxies when compared to other materials such as polyester resins include high dimensional stability, low shrinkage, and excellent chemical resistance.

The excellent wettability of the resins facilitates the use of fiberglass cloth as a laminating and reinforcing material.

Typical tensile strengths range from 9,000 to 12,000 psi (62 to 83 MPa) with compressive strengths from 15,000 to 18,000 psi (103 to 124 MPa).

Care is required when working with the uncured materials to avoid allergic reactions and dermatitis problems. Some individuals are very sensitive to these materials, and many persons become sensitive over a period of time unless strict hygiene practices are followed.

Polyurethanes. Fluid or liquid polyurethane elastomers are reaction products of polyisocyanates and low molecular weight resins containing two or more hydroxyl groups per molecule. They combine many of the good properties of both elastomers and plastics and have demonstrated a unique combination of abrasion resistance, tensile strength, and high load bearing capacity not available in conventional elastomers, plus impact resistance and resilience not available in plastics.

Because of their liquid, uncured form and their excellent cured properties, these polymers are useful in draw dies, drophammer dies, forming and stamping pads, press-brake forming dies, mandrels, expanding punches, and other tool design applications.

A major use of this material is for die automation components such as kickers, lifter heads and rollers, where the excellent wear-resistance of the material together with its non-marking characteristics make it very well suited. Polyurethane is available as a two-part liquid formulation that can be mixed and cast in place to form custom made die pads and transfer automation jaws.

Polyurethanes are resistant to oil, heat, oxidation, and ozone.[9]

References

1. Society of Manufacturing Engineers, *Machining with Carbides and Oxides* (New York: McGraw-Hill Book Company, 1962).
2. Charles Wick, John W. Benedict, and Raymond F. Veilleux, eds., "Die and Mold Materials," *Tool and Manufacturing Engineers Handbook*, 4th ed., Volume 2 (Dearborn, MI: Society of Manufacturing Engineers, 1984).
3. *General Aspects of Tool Design and Tool Materials for Cold and Warm Forging*, Document No. 4/82, International Cold Forging Group (Surrey, England: Portcullis Press Ltd., 1982).
4. O.J. Seeds, "How to Select and Use Low-melting Alloys as Production Aids," *Mater. Methods* (September 1950).
5. Published Data of Masonite Corp., Chicago, Ill.
6. "Densified Wood Applied to Metal-working Operations," *Machinery* (March 1949).
7. Society of Manufacturing Engineers, *Tooling for Aircraft and Missile Manufacturing* (New York: McGraw-Hill Book Company, 1964).
8. William Weaver, *Known Limitations of Plastic Tooling*, Plastic Tooling Symposium, Paper No. 34, Society of Manufacturing Engineers, 1957.
9. R.J. Gallagher, *Polyurethane Elastomers for Tooling*, Fourth Annual Seminar on Plastics for Tooling, Society of the Plastics Industry, June, 1962.

INDEX